Rational (Reciprocal) Function

$$f(x) = \frac{1}{x}$$

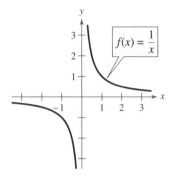

Domain: $(-\infty, 0) \cup (0, \infty)$
Range: $(-\infty, 0) \cup (0, \infty)$
No intercepts
Decreasing on $(-\infty, 0)$ and $(0, \infty)$
Odd function
Origin symmetry
Vertical asymptote: y-axis
Horizontal asymptote: x-axis

Exponential Function

$$f(x) = a^x, \ a > 0, \ a \neq 1$$

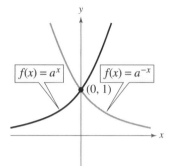

Domain: $(-\infty, \infty)$
Range: $(0, \infty)$
Intercept: $(0, 1)$
Increasing on $(-\infty, \infty)$
   for $f(x) = a^x$
Decreasing on $(-\infty, \infty)$
   for $f(x) = a^{-x}$
Horizontal asymptote: x-axis
Continuous

Logarithmic Function

$$f(x) = \log_a x, \ a > 0, \ a \neq 1$$

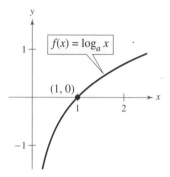

Domain: $(0, \infty)$
Range: $(-\infty, \infty)$
Intercept: $(1, 0)$
Increasing on $(0, \infty)$
Vertical asymptote: y-axis
Continuous
Reflection of graph of $f(x) = a^x$
   in the line $y = x$

Sine Function

$$f(x) = \sin x$$

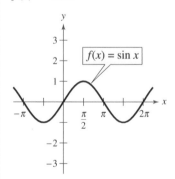

Domain: $(-\infty, \infty)$
Range: $[-1, 1]$
Period: 2π
x-intercepts: $(n\pi, 0)$
y-intercept: $(0, 0)$
Odd function
Origin symmetry

Cosine Function

$$f(x) = \cos x$$

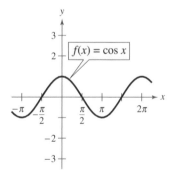

Domain: $(-\infty, \infty)$
Range: $[-1, 1]$
Period: 2π
x-intercepts: $\left(\frac{\pi}{2} + n\pi, 0\right)$
y-intercept: $(0, 1)$
Even function
y-axis symmetry

Tangent Function

$$f(x) = \tan x$$

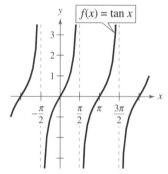

Domain: all $x \neq \dfrac{\pi}{2} + n\pi$

Range: $(-\infty, \infty)$
Period: π
x-intercepts: $(n\pi, 0)$
y-intercept: $(0, 0)$
Vertical asymptotes:

$$x = \frac{\pi}{2} + n\pi$$

Odd function
Origin symmetry

Cosecant Function

$f(x) = \csc x$

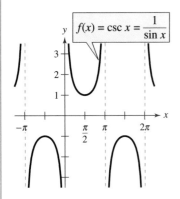

Domain: all $x \neq n\pi$
Range: $(-\infty, -1] \cup [1, \infty)$
Period: 2π
No intercepts
Vertical asymptotes: $x = n\pi$
Odd function
Origin symmetry

Secant Function

$f(x) = \sec x$

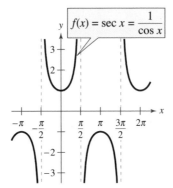

Domain: all $x \neq \dfrac{\pi}{2} + n\pi$

Range: $(-\infty, -1] \cup [1, \infty)$
Period: 2π
y-intercept: $(0, 1)$
Vertical asymptotes:

$$x = \frac{\pi}{2} + n\pi$$

Even function
y-axis symmetry

Cotangent Function

$f(x) = \cot x$

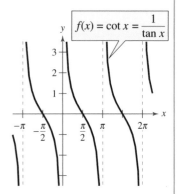

Domain: all $x \neq n\pi$
Range: $(-\infty, \infty)$
Period: π

x-intercepts: $\left(\dfrac{\pi}{2} + n\pi, 0 \right)$

Vertical asymptotes: $x = n\pi$
Odd function
Origin symmetry

Inverse Sine Function

$f(x) = \arcsin x$

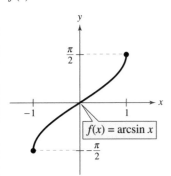

Domain: $[-1, 1]$
Range: $\left[-\dfrac{\pi}{2}, \dfrac{\pi}{2} \right]$
Intercept: $(0, 0)$
Odd function
Origin symmetry

Inverse Cosine Function

$f(x) = \arccos x$

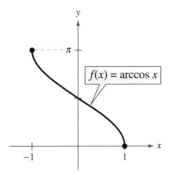

Domain: $[-1, 1]$
Range: $[0, \pi]$

y-intercept: $\left(0, \dfrac{\pi}{2} \right)$

Inverse Tangent Function

$f(x) = \arctan x$

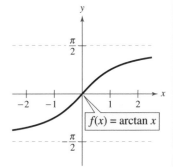

Domain: $(-\infty, \infty)$
Range: $\left(-\dfrac{\pi}{2}, \dfrac{\pi}{2} \right)$
Intercept: $(0, 0)$
Horizontal asymptotes:

$$y = \pm\frac{\pi}{2}$$

Odd function
Origin symmetry

Instructor's Annotated Edition

Trigonometry

Seventh Edition

Ron Larson

The Pennsylvania State University
The Behrend College

Robert Hostetler

The Pennsylvania State University
The Behrend College

With the assistance of David C. Falvo

The Pennsylvania State University
The Behrend College

Houghton Mifflin Company   Boston   New York

Publisher: Richard Stratton
Sponsoring Editor: Cathy Cantin
Development Manager: Maureen Ross
Development Editor: Lisa Collette
Editorial Associate: Elizabeth Kassab
Supervising Editor: Karen Carter
Senior Project Editor: Patty Bergin
Editorial Assistant: Julia Keller
Art and Design Manager: Gary Crespo
Executive Marketing Manager: Brenda Bravener-Greville
Director of Manufacturing: Priscilla Manchester
Cover Design Manager: Tony Saizon

Cover Image: Ryuichi Okano/Amana Japan

Printed in the U.S.A.

Library of Congress Catalog Card Number: 2005929681

Instructor's exam copy;

ISBN 13: 978-0-618-64334-9
ISBN 10: 0-618-64334-6

For orders, use student text ISBNs:

ISBN 13: 978-0-618-64332-5
ISBN 10: 0-618-64332-X

123456789–DOW–10 09 08 07 06

Contents

CONTENTS

A Word from the Authors

Welcome to *Trigonometry*: Seventh Edition. We are pleased to present this new edition of our textbook in which we focus on making the mathematics accessible, supporting student success, and offering instructors flexible teaching options.

Accessible to Students

Over the years we have taken care to write this text with the student in mind. Paying careful attention to the presentation, we use precise mathematical language and a clear writing style to develop an effective learning tool. We believe that every student can learn mathematics, and we are committed to providing a text that makes the mathematics of the trigonometry course accessible to all students. For the Seventh Edition, we have revised and improved many text features designed for this purpose.

Throughout the text, we now present solutions to many examples from multiple perspectives—algebraically, graphically, and numerically. The side-by-side format of this pedagogical feature helps students to see that a problem can be solved in more than one way and to see that different methods yield the same result. The side-by-side format also addresses many different learning styles.

We have found that many trigonometry students grasp mathematical concepts more easily when they work with them in the context of real-life situations. Students have numerous opportunities to do this throughout the Seventh Edition. The new *Make a Decision* feature has been added to the text in order to further connect real-life data and applications and motivate students. They also offer students the opportunity to generate and analyze mathematical models from large data sets. To reinforce the concept of functions, each function is introduced at the first point of use in the text with a definition and description of basic characteristics. Also, all elementary functions are presented in a summary on the endpapers of the text for convenient reference.

We have carefully written and designed each page to make the book more readable and accessible to students. For example, to avoid unnecessary page turning and disruptions to students' thought processes, each example and corresponding solution begins and ends on the same page.

Supports Student Success

During more than 30 years of teaching and writing, we have learned many things about the teaching and learning of mathematics. We have found that students are most successful when they know what they are expected to learn and why it is important to learn the concepts. With that in mind, we have enhanced the thematic study thread throughout the Seventh Edition.

Each chapter begins with a list of applications that are covered in the chapter and serve as a motivational tool by connecting section content to real-life situations. Using the same pedagogical theme, each section begins with a set of

PREFACE

section learning objectives—*What You Should Learn*. These are followed by an engaging real-life application—*Why You Should Learn It*—that motivates students and illustrates an area where the mathematical concepts will be applied in an example or exercise in the section. The *Chapter Summary—What Did You Learn?*—at the end of each chapter is a section-by-section overview that ties the learning objectives from the chapter to sets of *Review Exercises* at the end of each chapter.

Throughout the text, other features further improve accessibility. *Study Tips* are provided throughout the text at point-of-use to reinforce concepts and to help students learn how to study mathematics. *Technology, Writing About Mathematics, Historical Notes,* and *Explorations* have been expanded in order to reinforce mathematical concepts. Each example with worked-out solution is now followed by a *Checkpoint*, which directs the student to work a similar exercise from the exercise set. The *Section Exercises* now begin with a *Vocabulary Check*, which gives the students an opportunity to test their understanding of the important terms in the section. A new *Prerequisite Skills Review* is offered at the beginning of each exercise set. *Synthesis Exercises* check students' conceptual understanding of the topics in each section. The new *Make a Decision* exercises further connect real-life data and applications and motivate students. *Skills Review Exercises* provide additional practice with the concepts in the chapter or previous chapters. *Chapter Tests*, at the end of each chapter, and periodic *Cumulative Tests* offer students frequent opportunities for self-assessment and to develop strong study- and test-taking skills.

The use of technology also supports students with different learning styles. *Technology* notes are provided throughout the text at point-of-use. These notes call attention to the strengths and weaknesses of graphing technology, as well as offer alternative methods for solving or checking a problem using technology. These notes also direct students to the *Graphing Technology Guide*, on the textbook website, for keystroke support that is available for numerous calculator models. The use of technology is optional. This feature and related exercises can be omitted without the loss of continuity in coverage of topics.

Numerous additional text-specific resources are available to help students succeed in the trigonometry course. These include "live" online tutoring, instructional DVDs, and a variety of other resources, such as tutorial support and self-assessment, which are available on the HM mathSpace® CD-ROM, the Web, and in Eduspace®. In addition, the *Online Notetaking Guide* is a notetaking guide that helps students organize their class notes and create an effective study and review tool.

Flexible Options for Instructors

From the time we first began writing textbooks in the early 1970s, we have always considered it a critical part of our role as authors to provide instructors with flexible programs. In addition to addressing a variety of learning styles, the optional features within the text allow instructors to design their courses to meet their instructional needs and the needs of their students. For example, the

Explorations throughout the text can be used as a quick introduction to concepts or as a way to reinforce student understanding.

Our goal when developing the exercise sets was to address a wide variety of learning styles and teaching preferences. New to this edition are the *Vocabulary Check* questions, which are provided at the beginning of every exercise set to help students learn proper mathematical terminology. In each exercise set we have included a variety of exercise types, including questions requiring writing and critical thinking, as well as real-data applications. The problems are carefully graded in difficulty from mastery of basic skills to more challenging exercises. Some of the more challenging exercises include the *Synthesis Exercises* that combine skills and are used to check for conceptual understanding and the new *Make a Decision* exercises that further connect real-life data and applications and motivate students. *Skills Review Exercises*, placed at the end of each exercise set, reinforce previously learned skills. In addition, Houghton Mifflin's Eduspace® website offers instructors the option to assign homework and tests online—and also includes the ability to grade these assignments automatically.

Several other print and media resources are also available to support instructors. The *Online Instructor Success Organizer* includes suggested lesson plans and is an especially useful tool for larger departments that want all sections of a course to follow the same outline. The *Instructor's Edition* of the *Student Notetaking Guide* can be used as a lecture outline for every section of the text and includes additional examples for classroom discussion and important definitions. This is another valuable resource for schools trying to have consistent instruction and it can be used as a resource to support less experienced instructors. When used in conjunction with the *Student Notetaking Guide* these resources can save instructors preparation time and help students concentrate on important concepts.

Instructors who stress applications and problem solving, or exploration and technology, coupled with more traditional methods will be able to use this text successfully.

We hope you enjoy the Seventh Edition.

Ron Larson
Robert Hostetler

Acknowledgments

We would like to thank the many people who have helped us prepare the text and the supplements package. Their encouragement, criticisms, and suggestions have been invaluable to us.

Reviewers

Yvonne Aucoin, *Tidewater Community College*; Karen Emerson, *St. Petersburg College*; Roger Goldwyn, *Florida Atlantic University*; John Gordon, *Southern Polytechnic State University*; Sheyleah Harris, *South Plains College*; Peggy Hart, *Doane College*; Constance Meade, *College of Southern Idaho*; Peggy Miller, *University of Nebraska at Kearney*; Moe Najafi, *Kent State University*; Michael Sakowski, *Lake Superior College*

We would like to thank the staff of Larson Texts, Inc. who assisted in preparing the manuscript, rendering the art package, and typesetting and proofreading the pages and supplements.

On a personal level, we are grateful to our wives, Deanna Gilbert Larson and Eloise Hostetler for their love, patience, and support. Also, a special thanks goes to R. Scott O'Neil.

If you have suggestions for improving this text, please feel free to write us. Over the past three decades we have received many useful comments from both instructors and students, and we value these very much.

Ron Larson
Robert Hostetler

Textbook Features and Highlights

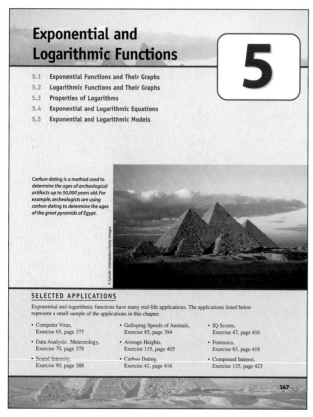

Exponential and Logarithmic Functions — 5

5.1 Exponential Functions and Their Graphs
5.2 Logarithmic Functions and Their Graphs
5.3 Properties of Logarithms
5.4 Exponential and Logarithmic Equations
5.5 Exponential and Logarithmic Models

Carbon dating is a method used to determine the ages of archeological artifacts up to 50,000 years old. For example, archeologists are using carbon dating to determine the ages of the great pyramids of Egypt.

© Sylvain Grandadam/Getty Images

SELECTED APPLICATIONS
Exponential and logarithmic functions have many real-life applications. The applications listed below represent a small sample of the applications in this chapter.

• Computer Virus,
  Exercise 65, page 377
• Data Analysis: Meteorology,
  Exercise 70, page 378
• Sound Intensity,
  Exercise 90, page 388

• Galloping Speeds of Animals,
  Exercise 85, page 394
• Average Heights,
  Exercise 115, page 405
• Carbon Dating,
  Exercise 41, page 416

• IQ Scores,
  Exercise 47, page 416
• Forensics,
  Exercise 63, page 418
• Compound Interest,
  Exercise 135, page 423

367

• Chapter Opener

Each chapter begins with a comprehensive overview of the chapter concepts. The photograph and caption illustrate a real-life application of a key concept. Section references help students prepare for the chapter.

• Applications List

An abridged list of applications, covered in the chapter, serve as a motivational tool by connecting section content to real-life situations.

• "What You Should Learn" and "Why You Should Learn It"

Sections begin with *What You Should Learn*, an outline of the main concepts covered in the section, and *Why You Should Learn It*, a real-life application or mathematical reference that illustrates the relevance of the section content.

Section 5.3   Properties of Logarithms   **389**

5.3   Properties of Logarithms

What you should learn
• Use the change-of-base formula to rewrite and evaluate logarithmic expressions.
• Use properties of logarithms to evaluate or rewrite logarithmic expressions.
• Use properties of logarithms to expand or condense logarithmic expressions.
• Use logarithmic functions to model and solve real-life problems.

Why you should learn it
Logarithmic functions can be used to model and solve real-life problems. For instance, in Exercises 81–83 on page 394, a logarithmic function is used to model the relationship between the number of decibels and the intensity of a sound.

AP Photo/Stephen Chernin

Change of Base

Most calculators have only two types of log keys, one for common logarithms (base 10) and one for natural logarithms (base e). Although common logs and natural logs are the most frequently used, you may occasionally need to evaluate logarithms to other bases. To do this, you can use the following **change-of-base formula**.

Change-of-Base Formula

Let a, b, and x be positive real numbers such that $a \neq 1$ and $b \neq 1$. Then $\log_a x$ can be converted to a different base as follows.

Base b	Base 10	Base e
$\log_a x = \dfrac{\log_b x}{\log_b a}$	$\log_a x = \dfrac{\log x}{\log a}$	$\log_a x = \dfrac{\ln x}{\ln a}$

One way to look at the change-of-base formula is that logarithms to base a are simply *constant multiples* of logarithms to base b. The constant multiplier is $1/(\log_b a)$.

Example 1   Changing Bases Using Common Logarithms

a. $\log_4 25 = \dfrac{\log 25}{\log 4}$ $\log_a x = \frac{\log x}{\log a}$

  $\approx \dfrac{1.39794}{0.60206}$ Use a calculator.

  ≈ 2.3219 Simplify.

b. $\log_2 12 = \dfrac{\log 12}{\log 2} \approx \dfrac{1.07918}{0.30103} \approx 3.5850$

✓CHECKPOINT   Now try Exercise 1(a).

Example 2   Changing Bases Using Natural Logarithms

a. $\log_4 25 = \dfrac{\ln 25}{\ln 4}$ $\log_a x = \frac{\ln x}{\ln a}$

  $\approx \dfrac{3.21888}{1.38629}$ Use a calculator.

  ≈ 2.3219 Simplify.

b. $\log_2 12 = \dfrac{\ln 12}{\ln 2} \approx \dfrac{2.48491}{0.69315} \approx 3.5850$

✓CHECKPOINT   Now try Exercise 1(b).

• Examples

Many examples present side-by-side solutions with multiple approaches—algebraic, graphical, and numerical. This format addresses a variety of learning styles and shows students that different solution methods yield the same result.

• Checkpoint

The *Checkpoint* directs students to work a similar problem in the exercise set for extra practice.

• Explorations

The *Exploration* engages students in active discovery of mathematical concepts, strengthens critical thinking skills, and helps them to develop an intuitive understanding of theoretical concepts.

• Study Tips

Study Tips reinforce concepts and help students learn how to study mathematics.

• Technology

The *Technology* feature gives instructions for graphing utilities at point of use.

• Additional Features

Additional carefully crafted learning tools, designed to connect concepts, are placed throughout the text. These learning tools include *Writing About Mathematics*, *Historical Notes*, and an extensive art program.

• Real-Life Applications

A wide variety of real-life applications, many using current real data, are integrated throughout the examples and exercises. The 🌐 indicates an example that involves a real-life application.

• Algebra of Calculus

Throughout the text, special emphasis is given to the algebraic techniques used in calculus. Algebra of Calculus examples and exercises are integrated throughout the text and are identified by the symbol ∯.

• Section Exercises

The section exercise sets consist of a variety of computational, conceptual, and applied problems.

• Vocabulary Check

Section exercises begin with a *Vocabulary Check* that serves as a review of the important mathematical terms in each section.

• Prerequisite Skills Review

Extra practice and a review of algebra skills, needed to complete the section exercise sets, are offered to the students and available in Eduspace®.

• Model It

These multi-part applications that involve real data offer students the opportunity to generate and analyze mathematical models.

• Synthesis and Skills Review Exercises

Each exercise set concludes with the two types of exercises.

Synthesis exercises promote further exploration of mathematical concepts, critical thinking skills, and writing about mathematics. The exercises require students to show their understanding of the relationships between many concepts in the section.

Skills Review Exercises reinforce previously learned skills and concepts.

Make a Decision exercises, found in selected sections, further connect real-life data and applications and motivate students. They also offer students the opportunity to generate and analyze mathematical models from large data sets.

FEATURES

• Chapter Summary

The *Chapter Summary* "*What Did You Learn?*" is a section-by-section overview that ties the learning objectives from the chapter to sets of Review Exercises for extra practice.

• Review Exercises

The chapter *Review Exercises* provide additional practice with the concepts covered in the chapter.

• Chapter Tests and Cumulative Tests

Chapter Tests, at the end of each chapter, and periodic *Cumulative Tests* offer students frequent opportunities for self-assessment and to develop strong study and test-taking skills.

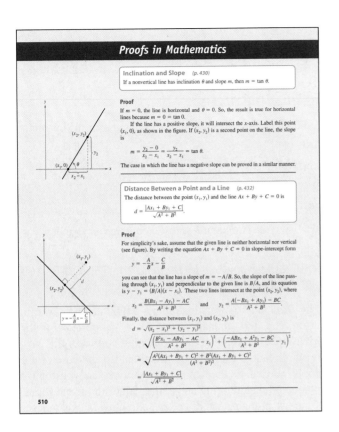

Proofs in Mathematics

Inclination and Slope (p. 430)

If a nonvertical line has inclination θ and slope m, then $m = \tan \theta$.

Proof

If $m = 0$, the line is horizontal and $\theta = 0$. So, the result is true for horizontal lines because $m = 0 = \tan 0$.

If the line has a positive slope, it will intersect the x-axis. Label this point $(x_1, 0)$, as shown in the figure. If (x_2, y_2) is a second point on the line, the slope is

$$m = \frac{y_2 - 0}{x_2 - x_1} = \frac{y_2}{x_2 - x_1} = \tan \theta.$$

The case in which the line has a negative slope can be proved in a similar manner.

Distance Between a Point and a Line (p. 432)

The distance between the point (x_1, y_1) and the line $Ax + By + C = 0$ is

$$d = \frac{|Ax_1 + By_1 + C|}{\sqrt{A^2 + B^2}}.$$

Proof

For simplicity's sake, assume that the given line is neither horizontal nor vertical (see figure). By writing the equation $Ax + By + C = 0$ in slope-intercept form

$$y = -\frac{A}{B}x - \frac{C}{B}$$

you can see that the line has a slope of $m = -A/B$. So, the slope of the line passing through (x_1, y_1) and perpendicular to the given line is B/A, and its equation is $y - y_1 = (B/A)(x - x_1)$. These two lines intersect at the point (x_2, y_2), where

$$x_2 = \frac{B(Bx_1 - Ay_1) - AC}{A^2 + B^2} \quad \text{and} \quad y_2 = \frac{A(-Bx_1 + Ay_1) - BC}{A^2 + B^2}.$$

Finally, the distance between (x_1, y_1) and (x_2, y_2) is

$$d = \sqrt{(x_2 - x_1)^2 + (y_2 - y_1)^2}$$

$$= \sqrt{\left(\frac{B^2x_1 - ABy_1 - AC}{A^2 + B^2} - x_1\right)^2 + \left(\frac{-ABx_1 + A^2y_1 - BC}{A^2 + B^2} - y_1\right)^2}$$

$$= \sqrt{\frac{A^2(Ax_1 + By_1 + C)^2 + B^2(Ax_1 + By_1 + C)^2}{(A^2 + B^2)^2}}$$

$$= \frac{|Ax_1 + By_1 + C|}{\sqrt{A^2 + B^2}}.$$

510

• **Proofs in Mathematics**

At the end of every chapter, proofs of important mathematical properties and theorems are presented as well as discussions of various proof techniques.

• **P.S. Problem Solving**

Each chapter concludes with a collection of thought-provoking and challenging exercises that further explore and expand upon the chapter concepts. These exercises have unusual characteristics that set them apart from traditional text exercises.

P.S. Problem Solving

This collection of thought-provoking and challenging exercises further explores and expands upon concepts learned in this chapter.

1. As a salesperson, you receive a monthly salary of $2000, plus a commission of 7% of sales. You are offered a new job at $2300 per month, plus a commission of 5% of sales.
   (a) Write a linear equation for your current monthly wage W_1 in terms of your monthly sales S.
   (b) Write a linear equation for the monthly wage W_2 of your new job offer in terms of the monthly sales S.
   (c) Use a graphing utility to graph both equations in the same viewing window. Find the point of intersection. What does it signify?
   (d) You think you can sell $20,000 per month. Should you change jobs? Explain.

2. For the numbers 2 through 9 on a telephone keypad (see figure), create two relations: one mapping numbers onto letters, and the other mapping letters onto numbers. Are both relations functions? Explain.

3. What can be said about the sum and difference of each of the following?
   (a) Two even functions   (b) Two odd functions
   (c) An odd function and an even function

4. The two functions given by
   $$f(x) = x \quad \text{and} \quad g(x) = -x$$
   are their own inverse functions. Graph each function and explain why this is true. Graph other linear functions that are their own inverse functions. Find a general formula for a family of linear functions that are their own inverse functions.

5. Prove that a function of the following form is even.
   $$y = a_{2n}x^{2n} + a_{2n-2}x^{2n-2} + \cdots + a_2x^2 + a_0$$

6. A miniature golf professional is trying to make a hole-in-one on the miniature golf green shown. A coordinate plane is placed over the golf green. The golf ball is at the point $(2.5, 2)$ and the hole is at the point $(9.5, 2)$. The professional wants to bank the ball off the side wall of the green at the point (x, y). Find the coordinates of the point (x, y). Then write an equation for the path of the ball.

FIGURE FOR 6

7. At 2:00 P.M. on April 11, 1912, the *Titanic* left Cobh, Ireland, on her voyage to New York City. At 11:40 P.M. on April 14, the *Titanic* struck an iceberg and sank, having covered only about 2100 miles of the approximately 3400-mile trip.
   (a) What was the total length of the *Titanic*'s voyage in hours?
   (b) What was the *Titanic*'s average speed in miles per hour?
   (c) Write a function relating the *Titanic*'s distance from New York City and the number of hours traveled. Find the domain and range of the function.
   (d) Graph the function from part (c).

8. Consider the function given by $f(x) = -x^2 + 4x - 3$. Find the average rate of change of the function from x_1 to x_2.
   (a) $x_1 = 1, x_2 = 2$ (b) $x_1 = 1, x_2 = 1.5$
   (c) $x_1 = 1, x_2 = 1.25$
   (d) $x_1 = 1, x_2 = 1.125$
   (e) $x_1 = 1, x_2 = 1.0625$
   (f) Does the average rate of change seem to be approaching one value? If so, what value?
   (g) Find the equation of the secant line through the points $(x_1, f(x_1))$ and $(x_2, f(x_2))$ for parts (a)–(e).
   (h) Find the equation of the line though the point $(1, f(1))$ using your answer from part (g) as the slope of the line.

9. Consider the functions given by $f(x) = 4x$ and $g(x) = x + 6$.
   (a) Find $(f \circ g)(x)$.
   (b) Find $(f \circ g)^{-1}(x)$.
   (c) Find $f^{-1}(x)$ and $g^{-1}(x)$.
   (d) Find $(g^{-1} \circ f^{-1})(x)$ and compare the result with that of part (b).
   (e) Repeat parts (a) through (d) for $f(x) = x^3 + 1$ and $g(x) = 2x$.
   (f) Write two one-to-one functions f and g, and repeat parts (a) through (d) for these functions.
   (g) Make a conjecture about $(f \circ g)^{-1}(x)$ and $(g^{-1} \circ f^{-1})(x)$.

127

10. You are in a boat 2 miles from the nearest point on the coast. You are to travel to a point Q, 3 miles down the coast and 1 mile inland (see figure). You can row at 2 miles per hour and walk at 4 miles per hour.

    (a) Write the total time T of the trip as a function of x.
    (b) Determine the domain of the function.
    (c) Use a graphing utility to graph the function. Be sure to choose an appropriate viewing window.
    (d) Use the *zoom* and *trace* features to find the value of x that minimizes T.
    (e) Write a brief paragraph interpreting these values.

11. The **Heaviside function** $H(x)$ is widely used in engineering applications. (See figure.) To print an enlarged copy of the graph, go to the website www.mathgraphs.com.
    $$H(x) = \begin{cases} 1, & x \geq 0 \\ 0, & x < 0 \end{cases}$$
    Sketch the graph of each function by hand.
    (a) $H(x) - 2$ (b) $H(x - 2)$ (c) $-H(x)$
    (d) $H(-x)$ (e) $\frac{1}{2}H(x)$ (f) $-H(x - 2) + 2$

12. Let $f(x) = \dfrac{1}{1 - x}$.
    (a) What are the domain and range of f?
    (b) Find $f(f(x))$. What is the domain of this function?
    (c) Find $f(f(f(x)))$. Is the graph a line? Why or why not?

13. Show that the Associative Property holds for compositions of functions—that is,
    $$(f \circ (g \circ h))(x) = ((f \circ g) \circ h)(x).$$

14. Consider the graph of the function f shown in the figure. Use this graph to sketch the graph of each function. To print an enlarged copy of the graph, go to the website www.mathgraphs.com.
    (a) $f(x + 1)$ (b) $f(x) + 1$ (c) $2f(x)$ (d) $f(-x)$
    (e) $-f(x)$ (f) $|f(x)|$ (g) $f(|x|)$

15. Use the graphs of f and f^{-1} to complete each table of function values.

(a)

x	−4	−2	0	4
$(f(f^{-1}(x))$				

(b)

x	−3	−2	0	1
$(f + f^{-1})(x)$				

(c)

x	−3	−2	0	1
$(f \cdot f^{-1})(x)$				

(d)

x	−4	−3	0	4
$|f^{-1}(x)|$				

128

FEATURES

Supplements

Supplements for the Instructor

Trigonometry, Seventh Edition, has an extensive support package for the instructor that includes:

Instructor's Annotated Edition (IAE)

Online Complete Solutions Guide

Online Instructor Success Organizer

Online Teaching Center: This free companion website contains an abundance of instructor resources.

HM ClassPrep™ with HM Testing (powered by Diploma™): This CD-ROM is a combination of two course management tools.

- *HM Testing* (powered by *Diploma™*) offers instructors a flexible and powerful tool for test generation and test management. Now supported by the Brownstone Research Group's market-leading *Diploma™* software, this new version of *HM Testing* significantly improves on functionality and ease of use by offering all the tools needed to create, author, deliver, and customize multiple types of tests—including authoring and editing algorithmic questions. *Diploma™* is currently in use at thousands of college and university campuses throughout the United States and Canada.

- HM ClassPrep™ also features supplements and text-specific resources for the instructor.

Eduspace®: Eduspace®, powered by Blackboard®, is Houghton Mifflin's customizable and interactive online learning tool. Eduspace® provides instructors with online courses and content. By pairing the widely recognized tools of Blackboard® with quality, text-specific content from Houghton Mifflin Company, Eduspace® makes it easy for instructors to create all or part of a course online. This online learning tool also contains ready-to-use homework exercises, quizzes, tests, tutorials, and supplemental study materials.

Visit *www.eduspace.com* for more information.

Eduspace® with eSolutions: Eduspace® with eSolutions combines all the features of Eduspace® with an electronic version of the textbook exercises and the complete solutions to the odd-numbered exercises, providing students with a convenient and comprehensive way to do homework and view course materials.

Supplements for the Student

Trigonometry, Seventh Edition, has an extensive support package for the student that includes:

Study and Solutions Guide

Online Student Notetaking Guide

Instructional DVDs

Online Study Center: This free companion website contains an abundance of student resources.

HM mathSpace® CD-ROM: This tutorial CD-ROM provides opportunities for self-paced review and practice with algorithmically generated exercises and step-by-step solutions.

Eduspace®: Eduspace®, powered by Blackboard®, is Houghton Mifflin's customizable and interactive online learning tool for instructors and students. Eduspace® is a text-specific, web-based learning environment that your instructor can use to offer students a combination of practice exercises, multimedia tutorials, video explanations, online algorithmic homework and more. Specific content is available 24 hours a day to help you succeed in your course.

Eduspace® with eSolutions: Eduspace® with eSolutions combines all the features of Eduspace® with an electronic version of the textbook exercises and the complete solutions to the odd-numbered exercises. The result is a convenient and comprehensive way to do homework and view your course materials.

Smarthinking®: Houghton Mifflin has partnered with Smarthinking® to provide an easy-to-use, effective, online tutorial service. Through state-of-the-art tools and whiteboard technology, students communicate in real-time with qualified e-instructors who can help the students understand difficult concepts and guide them through the problem-solving process while studying or completing homework.

Three levels of service are offered to the students.

Live Tutorial Help provides real-time, one-on-one instruction.

Question Submission allows students to submit questions to the tutor outside the scheduled hours and receive a reply usually within 24 hours.

Independent Study Resources connects students around-the-clock to additional educational resources, ranging from interactive websites to Frequently Asked Questions.

Visit *smarthinking.com* for more information.

Limits apply; terms and hours of SMARTHINKING® service are subject to change.

Prerequisites

P

Functions play a primary role in modeling real-life situations. The estimated growth in the number of digital music scales in the United States can be modeled by a cubic function.

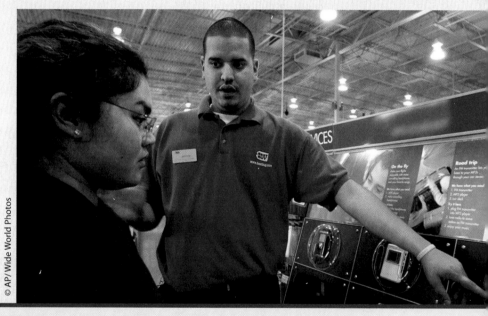

© AP/Wide World Photos

SELECTED APPLICATIONS

Basic algebra concepts have many real-life applications. The applications listed below represent a small sample of the applications in this chapter.

P.1   Review of Real Numbers and Their Properties

What you should learn

- Represent and classify real numbers.
- Order real numbers and use inequalities.
- Find the absolute values of real numbers and find the distance between two real numbers.
- Evaluate algebraic expressions.
- Use the basic rules and properties of algebra.

Why you should learn it

Real numbers are used to represent many real-life quantities. For example, in Exercise 65 on page 10, you will use real numbers to represent the federal deficit.

The *HM mathSpace®* CD-ROM and *Eduspace®* for this text contain additional resources related to the concepts discussed in this chapter.

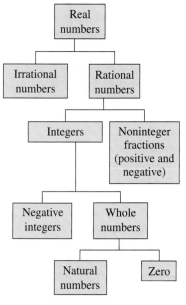

FIGURE P.1   *Subsets of real numbers*

Real Numbers

Real numbers are used in everyday life to describe quantities such as age, miles per gallon, and population. Real numbers are represented by symbols such as

$$-5, 9, 0, \frac{4}{3}, 0.666\ldots, 28.21, \sqrt{2}, \pi, \text{ and } \sqrt[3]{-32}.$$

Here are some important **subsets** (each member of subset B is also a member of set A) of the real numbers. The three dots, called *ellipsis points*, indicate that the pattern continues indefinitely.

$$\{1, 2, 3, 4, \ldots\} \qquad \text{Set of natural numbers}$$
$$\{0, 1, 2, 3, 4, \ldots\} \qquad \text{Set of whole numbers}$$
$$\{\ldots, -3, -2, -1, 0, 1, 2, 3, \ldots\} \qquad \text{Set of integers}$$

A real number is **rational** if it can be written as the ratio p/q of two integers, where $q \neq 0$. For instance, the numbers

$$\frac{1}{3} = 0.3333\ldots = 0.\overline{3}, \frac{1}{8} = 0.125, \text{ and } \frac{125}{111} = 1.126126\ldots = 1.\overline{126}$$

are rational. The decimal representation of a rational number either repeats $\left(\text{as in } \frac{173}{55} = 3.1\overline{45}\right)$ or terminates $\left(\text{as in } \frac{1}{2} = 0.5\right)$. A real number that cannot be written as the ratio of two integers is called **irrational.** Irrational numbers have infinite nonrepeating decimal representations. For instance, the numbers

$$\sqrt{2} = 1.4142135\ldots \approx 1.41 \quad \text{and} \quad \pi = 3.1415926\ldots \approx 3.14$$

are irrational. (The symbol $\approx$ means "is approximately equal to.") Figure P.1 shows subsets of real numbers and their relationships to each other.

Real numbers are represented graphically by a **real number line.** The point 0 on the real number line is the **origin.** Numbers to the right of 0 are positive, and numbers to the left of 0 are negative, as shown in Figure P.2. The term **nonnegative** describes a number that is either positive or zero.

Origin

Negative direction ← −4 −3 −2 −1 0 1 2 3 4 → Positive direction

FIGURE P.2   *The real number line*

As illustrated in Figure P.3, there is a *one-to-one correspondence* between real numbers and points on the real number line.

$-\frac{5}{3}$ 0.75 π
−3 −2 −1 0 1 2 3

Every real number corresponds to exactly one point on the real number line.

FIGURE P.3   *One-to-one*

-2.4 $\sqrt{2}$
−3 −2 −1 0 1 2 3

Every point on the real number line corresponds to exactly one real number.

Ordering Real Numbers

One important property of real numbers is that they are *ordered*.

> ### Definition of Order on the Real Number Line
>
> If a and b are real numbers, a is less than b if $b - a$ is positive. The **order** of a and b is denoted by the **inequality** $a < b$. This relationship can also be described by saying that b is *greater than a* and writing $b > a$. The inequality $a \leq b$ means that a is *less than or equal to b*, and the inequality $b \geq a$ means that b is *greater than or equal to a*. The symbols $<$, $>$, $\leq$, and $\geq$ are *inequality symbols*.

FIGURE P.4 $a < b$ if and only if a lies to the left of b.

Geometrically, this definition implies that $a < b$ if and only if a lies to the *left* of b on the real number line, as shown in Figure P.4.

Example 1   Interpreting Inequalities

Describe the subset of real numbers represented by each inequality.

a. $x \leq 2$ **b.** $-2 \leq x < 3$

FIGURE P.5

Solution

a. The inequality $x \leq 2$ denotes all real numbers less than or equal to 2, as shown in Figure P.5.

b. The inequality $-2 \leq x < 3$ means that $x \geq -2$ *and* $x < 3$. This "double inequality" denotes all real numbers between -2 and 3, including -2 but not including 3, as shown in Figure P.6.

FIGURE P.6

✓**CHECKPOINT**   Now try Exercise 19.

Inequalities can be used to describe subsets of real numbers called **intervals**. In the bounded intervals below, the real numbers a and b are the **endpoints** of each interval. The endpoints of a closed interval are included in the interval, whereas the endpoints of an open interval are not included in the interval.

> ### Bounded Intervals on the Real Number Line
>
> | Notation | Interval Type | Inequality | Graph |
> |---|---|---|---|
> | $[a, b]$ | Closed | $a \leq x \leq b$ | |
> | (a, b) | Open | $a < x < b$ | |
> | $[a, b)$ | | $a \leq x < b$ | |
> | $(a, b]$ | | $a < x \leq b$ | |

STUDY TIP

The reason that the four types of intervals at the right are called *bounded* is that each has a finite length. An interval that does not have a finite length is *unbounded* (see page 4).

The symbols ∞, **positive infinity,** and $-\infty$, **negative infinity,** do not represent real numbers. They are simply convenient symbols used to describe the unboundedness of an interval such as $(1, \infty)$ or $(-\infty, 3]$.

> ### Unbounded Intervals on the Real Number Line
>
> | Notation | Interval Type | Inequality | Graph |
> |----------|---------------|------------|-------|
> | $[a, \infty)$ | | $x \geq a$ | |
> | (a, ∞) | Open | $x > a$ | |
> | $(-\infty, b]$ | | $x \leq b$ | |
> | $(-\infty, b)$ | Open | $x < b$ | |
> | $(-\infty, \infty)$ | Entire real line | $-\infty < x < \infty$ | |

Example 2   **Using Inequalities to Represent Intervals**

Use inequality notation to describe each of the following.

a. c is at most 2.   **b.** m is at least -3.

c. All x in the interval $(-3, 5]$

Solution

a. The statement "c is at most 2" can be represented by $c \leq 2$.

b. The statement "m is at least -3" can be represented by $m \geq -3$.

c. "All x in the interval $(-3, 5]$" can be represented by $-3 < x \leq 5$.

✓**CHECKPOINT**   Now try Exercise 31.

Example 3   **Interpreting Intervals**

Give a verbal description of each interval.

a. $(-1, 0)$ **b.** $[2, \infty)$ **c.** $(-\infty, 0)$

Solution

a. This interval consists of all real numbers that are greater than -1 and less than 0.

b. This interval consists of all real numbers that are greater than or equal to 2.

c. This interval consists of all negative real numbers.

✓**CHECKPOINT**   Now try Exercise 29.

The **Law of Trichotomy** states that for any two real numbers a and b, *precisely* one of three relationships is possible:

$$a = b, \quad a < b, \quad \text{or} \quad a > b. \qquad \text{Law of Trichotomy}$$

Absolute Value and Distance

The **absolute value** of a real number is its *magnitude*, or the distance between the origin and the point representing the real number on the real number line.

Exploration

Absolute value expressions can be evaluated on a graphing utility. When an expression such as $|3 - 8|$ is evaluated, parentheses should surround the expression, as shown below.

```
abs(3-8)
                    5
```

Evaluate each expression. What can you conclude?

a. $|6|$ **b.** $|-1|$
c. $|5 - 2|$ **d.** $|2 - 5|$

Definition of Absolute Value

If a is a real number, then the absolute value of a is

$$|a| = \begin{cases} a, & \text{if } a \geq 0 \\ -a, & \text{if } a < 0 \end{cases}.$$

Notice in this definition that the absolute value of a real number is never negative. For instance, if $a = -5$, then $|-5| = -(-5) = 5$. The absolute value of a real number is either positive or zero. Moreover, 0 is the only real number whose absolute value is 0. So, $|0| = 0$.

Example 4   Evaluating the Absolute Value of a Number

Evaluate $\dfrac{|x|}{x}$ for (a) $x > 0$ and (b) $x < 0$.

Solution

a. If $x > 0$, then $|x| = x$ and $\dfrac{|x|}{x} = \dfrac{x}{x} = 1$.

b. If $x < 0$, then $|x| = -x$ and $\dfrac{|x|}{x} = \dfrac{-x}{x} = -1$.

✓CHECKPOINT   Now try Exercise 47.

Properties of Absolute Values

1. $|a| \geq 0$ **2.** $|-a| = |a|$

3. $|ab| = |a||b|$ **4.** $\left|\dfrac{a}{b}\right| = \dfrac{|a|}{|b|}, \quad b \neq 0$

Absolute value can be used to define the distance between two points on the real number line. For instance, the distance between -3 and 4 is

$$|-3 - 4| = |-7|$$
$$= 7$$

as shown in Figure P.7.

```
        7
  ┌──────────────┐
  ●──┼──┼──┼──┼──┼──┼──●
 -3  -2  -1  0   1   2   3   4
```

FIGURE P.7   *The distance between* -3 *and 4 is 7.*

Distance Between Two Points on the Real Number Line

Let a and b be real numbers. The **distance between a and b** is

$$d(a, b) = |b - a| = |a - b|.$$

Algebraic Expressions

One characteristic of algebra is the use of letters to represent numbers. The letters are **variables,** and combinations of letters and numbers are **algebraic expressions.** Here are a few examples of algebraic expressions.

$$5x, \qquad 2x - 3, \qquad \frac{4}{x^2 + 2}, \qquad 7x + y$$

Definition of an Algebraic Expression

An **algebraic expression** is a collection of letters (**variables**) and real numbers (**constants**) combined using the operations of addition, subtraction, multiplication, division, and exponentiation.

The **terms** of an algebraic expression are those parts that are separated by *addition.* For example,

$$x^2 - 5x + 8 = x^2 + (-5x) + 8$$

has three terms: x^2 and $-5x$ are the **variable terms** and 8 is the **constant term.** The numerical factor of a variable term is the **coefficient** of the variable term. For instance, the coefficient of $-5x$ is -5, and the coefficient of x^2 is 1.

To **evaluate** an algebraic expression, substitute numerical values for each of the variables in the expression. Here are two examples.

Expression	Value of Variable	Substitute	Value of Expression
$-3x + 5$	$x = 3$	$-3(3) + 5$	$-9 + 5 = -4$
$3x^2 + 2x - 1$	$x = -1$	$3(-1)^2 + 2(-1) - 1$	$3 - 2 - 1 = 0$

When an algebraic expression is evaluated, the **Substitution Principle** is used. It states that "If $a = b$, then a can be replaced by b in any expression involving a." In the first evaluation shown above, for instance, 3 is *substituted* for x in the expression $-3x + 5$.

Basic Rules of Algebra

There are four arithmetic operations with real numbers: *addition, multiplication, subtraction,* and *division,* denoted by the symbols $+$, $\times$ or $\cdot$, $-$, and $\div$ or $/$. Of these, addition and multiplication are the two primary operations. Subtraction and division are the inverse operations of addition and multiplication, respectively.

Definitions of Subtraction and Division

Subtraction: Add the opposite.   **Division:** Multiply by the reciprocal.

$$a - b = a + (-b) \qquad\qquad \text{If } b \neq 0, \text{ then } a/b = a\left(\frac{1}{b}\right) = \frac{a}{b}.$$

In these definitions, $-b$ is the **additive inverse** (or opposite) of b, and $1/b$ is the **multiplicative inverse** (or reciprocal) of b. In the fractional form a/b, a is the **numerator** of the fraction and b is the **denominator.**

Because the properties of real numbers below are true for variables and algebraic expressions as well as for real numbers, they are often called the **Basic Rules of Algebra.** Try to formulate a verbal description of each property. For instance, the first property states that *the order in which two real numbers are added does not affect their sum.*

Basic Rules of Algebra

Let a, b, and c be real numbers, variables, or algebraic expressions.

Property		*Example*
Commutative Property of Addition:	$a + b = b + a$	$4x + x^2 = x^2 + 4x$
Commutative Property of Multiplication:	$ab = ba$	$(4 - x)x^2 = x^2(4 - x)$
Associative Property of Addition:	$(a + b) + c = a + (b + c)$	$(x + 5) + x^2 = x + (5 + x^2)$
Associative Property of Multiplication:	$(ab)c = a(bc)$	$(2x \cdot 3y)(8) = (2x)(3y \cdot 8)$
Distributive Properties:	$a(b + c) = ab + ac$	$3x(5 + 2x) = 3x \cdot 5 + 3x \cdot 2x$
	$(a + b)c = ac + bc$	$(y + 8)y = y \cdot y + 8 \cdot y$
Additive Identity Property:	$a + 0 = a$	$5y^2 + 0 = 5y^2$
Multiplicative Identity Property:	$a \cdot 1 = a$	$(4x^2)(1) = 4x^2$
Additive Inverse Property:	$a + (-a) = 0$	$5x^3 + (-5x^3) = 0$
Multiplicative Inverse Property:	$a \cdot \dfrac{1}{a} = 1, \quad a \neq 0$	$(x^2 + 4)\left(\dfrac{1}{x^2 + 4}\right) = 1$

Because subtraction is defined as "adding the opposite," the Distributive Properties are also true for subtraction. For instance, the "subtraction form" of $a(b + c) = ab + ac$ is $a(b - c) = ab - ac$.

Properties of Negation and Equality

Let a and b be real numbers, variables, or algebraic expressions.

Property	*Example*
1. $(-1)a = -a$	$(-1)7 = -7$
2. $-(-a) = a$	$-(-6) = 6$
3. $(-a)b = -(ab) = a(-b)$	$(-5)3 = -(5 \cdot 3) = 5(-3)$
4. $(-a)(-b) = ab$	$(-2)(-x) = 2x$
5. $-(a + b) = (-a) + (-b)$	$-(x + 8) = (-x) + (-8)$
	$\qquad = -x - 8$
6. If $a = b$, then $a \pm c = b \pm c$.	$\frac{1}{2} + 3 = 0.5 + 3$
7. If $a = b$, then $ac = bc$.	$4^2 \cdot 2 = 16 \cdot 2$
8. If $a \pm c = b \pm c$, then $a = b$.	$1.4 - 1 = \frac{7}{5} - 1 \Rightarrow 1.4 = \frac{7}{5}$
9. If $ac = bc$ and $c \neq 0$, then $a = b$.	$3x = 3 \cdot 4 \Rightarrow x = 4$

STUDY TIP

Notice the difference between the *opposite of a number* and a *negative number*. If a is already negative, then its opposite, $-a$, is positive. For instance, if $a = -5$, then

$$-a = -(-5) = 5.$$

Activities

1. Evaluate: $4 - |-3|$.
   Answer: 1
2. Find the distance between -41 and 16.
   Answer: 57
3. Use inequality notation to describe the set of nonnegative numbers.
   Answer: $x \geq 0$
4. Use interval notation to describe the inequality $-6 < x \leq 13$.
   Answer: $(-6, 13]$

Point out to students that to add or subtract fractions with unlike denominators, they can use Property 5 of fractions, as in Example 5(c), or they can rewrite the fractions with like denominators using the least common denominator (LCD) of the fractions.

Properties of Zero

Let a and b be real numbers, variables, or algebraic expressions.

1. $a + 0 = a$ and $a - 0 = a$ **2.** $a \cdot 0 = 0$

3. $\dfrac{0}{a} = 0, \qquad a \neq 0$ **4.** $\dfrac{a}{0}$ is undefined.

5. Zero-Factor Property: If $ab = 0$, then $a = 0$ or $b = 0$.

Properties and Operations of Fractions

Let a, b, c, and d be real numbers, variables, or algebraic expressions such that $b \neq 0$ and $d \neq 0$.

1. Equivalent Fractions: $\dfrac{a}{b} = \dfrac{c}{d}$ if and only if $ad = bc$.

2. Rules of Signs: $-\dfrac{a}{b} = \dfrac{-a}{b} = \dfrac{a}{-b}$ and $\dfrac{-a}{-b} = \dfrac{a}{b}$

3. Generate Equivalent Fractions: $\dfrac{a}{b} = \dfrac{ac}{bc}, \qquad c \neq 0$

4. Add or Subtract with Like Denominators: $\dfrac{a}{b} \pm \dfrac{c}{b} = \dfrac{a \pm c}{b}$

5. Add or Subtract with Unlike Denominators: $\dfrac{a}{b} \pm \dfrac{c}{d} = \dfrac{ad \pm bc}{bd}$

6. Multiply Fractions: $\dfrac{a}{b} \cdot \dfrac{c}{d} = \dfrac{ac}{bd}$

7. Divide Fractions: $\dfrac{a}{b} \div \dfrac{c}{d} = \dfrac{a}{b} \cdot \dfrac{d}{c} = \dfrac{ad}{bc}, \qquad c \neq 0$

Example 5   Properties and Operations of Fractions

a. Equivalent fractions: $\dfrac{x}{5} = \dfrac{3 \cdot x}{3 \cdot 5} = \dfrac{3x}{15}$ **b.** Divide fractions: $\dfrac{7}{x} \div \dfrac{3}{2} = \dfrac{7}{x} \cdot \dfrac{2}{3} = \dfrac{14}{3x}$

c. Add fractions with unlike denominators: $\dfrac{x}{3} + \dfrac{2x}{5} = \dfrac{5 \cdot x + 3 \cdot 2x}{3 \cdot 5} = \dfrac{11x}{15}$

✓**CHECKPOINT**   Now try Exercise 103.

If a, b, and c are integers such that $ab = c$, then a and b are **factors** or **divisors** of c. A **prime number** is an integer that has exactly two positive factors—itself and 1—such as 2, 3, 5, 7, and 11. The numbers 4, 6, 8, 9, and 10 are **composite** because each can be written as the product of two or more prime numbers. The number 1 is neither prime nor composite. The **Fundamental Theorem of Arithmetic** states that every positive integer greater than 1 can be written as the product of prime numbers in precisely one way (disregarding order). For instance, the *prime factorization* of 24 is $24 = 2 \cdot 2 \cdot 2 \cdot 3$.

| P.1 | Exercises | The *HM mathSpace*® CD-ROM and *Eduspace*® for this text contain step-by-step solutions to all odd-numbered exercises. They also provide Tutorial Exercises for additional help. |

VOCABULARY CHECK: Fill in the blanks.

1. A real number is _____ if it can be written as the ratio $\frac{p}{q}$ of two integers, where $q \neq 0$.

2. _____ numbers have infinite nonrepeating decimal representations.

3. The distance between a point on the real number line and the origin is the _____ _____ of the real number.

4. A number that can be written as the product of two or more prime numbers is called a _____ number.

5. An integer that has exactly two positive factors, the integer itself and 1, is called a _____ number.

6. An algebraic expression is a collection of letters called _____ and real numbers called _____.

7. The _____ of an algebraic expression are those parts separated by addition.

8. The numerical factor of a variable term is the _____ of the variable term.

9. The _____ _____ states that if $ab = 0$, then $a = 0$ or $b = 0$.

In Exercises 1–6, determine which numbers in the set are (a) natural numbers, (b) whole numbers, (c) integers, (d) rational numbers, and (e) irrational numbers.

1. $-9, -\frac{7}{2}, 5, \frac{2}{3}, \sqrt{2}, 0, 1, -4, 2, -11$

2. $\sqrt{5}, -7, -\frac{7}{3}, 0, 3.12, \frac{5}{4}, -3, 12, 5$

3. $2.01, 0.666 \ldots, -13, 0.010110111 \ldots, 1, -6$

4. $2.3030030003 \ldots, 0.7575, -4.63, \sqrt{10}, -75, 4$

5. $-\pi, -\frac{1}{3}, \frac{6}{3}, \frac{1}{2}\sqrt{2}, -7.5, -1, 8, -22$

6. $25, -17, -\frac{12}{5}, \sqrt{9}, 3.12, \frac{1}{2}\pi, 7, -11.1, 13$

In Exercises 7–10, use a calculator to find the decimal form of the rational number. If it is a nonterminating decimal, write the repeating pattern.

7. $\frac{5}{8}$ 8. $\frac{1}{3}$

9. $\frac{41}{333}$ 10. $\frac{6}{11}$

In Exercises 11 and 12, approximate the numbers and place the correct symbol (< or >) between them.

11.

12.

In Exercises 13–18, plot the two real numbers on the real number line. Then place the appropriate inequality symbol (< or >) between them.

13. $-4, -8$ 14. $-3.5, 1$

15. $\frac{3}{2}, 7$ 16. $1, \frac{16}{3}$

17. $\frac{5}{6}, \frac{2}{3}$ 18. $-\frac{8}{7}, -\frac{3}{7}$

In Exercises 19–30, (a) give a verbal description of the subset of real numbers represented by the inequality or the interval, (b) sketch the subset on the real number line, and (c) state whether the interval is bounded or unbounded.

19. $x \leq 5$ 20. $x \geq -2$

21. $x < 0$ 22. $x > 3$

23. $[4, \infty)$ 24. $(-\infty, 2)$

25. $-2 < x < 2$ 26. $0 \leq x \leq 5$

27. $-1 \leq x < 0$ 28. $0 < x \leq 6$

29. $[-2, 5)$ 30. $(-1, 2]$

In Exercises 31–38, use inequality notation to describe the set.

31. All x in the interval $(-2, 4]$

32. All y in the interval $[-6, 0)$

33. y is nonnegative.   34. y is no more than 25.

35. t is at least 10 and at most 22.

36. k is less than 5 but no less than -3.

37. The dog's weight W is more than 65 pounds.

38. The annual rate of inflation r is expected to be at least 2.5% but no more than 5%.

In Exercises 39–48, evaluate the expression.

39. $|-10|$ 40. $|0|$

41. $|3 - 8|$ 42. $|4 - 1|$

43. $|-1| - |-2|$ 44. $-3 - |-3|$

45. $\dfrac{-5}{|-5|}$ 46. $-3|-3|$

47. $\dfrac{|x + 2|}{x + 2}$, $x < -2$ 48. $\dfrac{|x - 1|}{x - 1}$, $x > 1$

In Exercises 49–54, place the correct symbol (<, >, or =) between the pair of real numbers.

49. $|-3| \quad -|-3|$

50. $|-4| \quad |4|$

51. $-5 \quad -|5|$

52. $-|-6| \quad |-6|$

53. $-|-2| \quad -|2|$

54. $-(-2) \quad -2$

In Exercises 55–60, find the distance between a and b.

55. $a = 126, b = 75$

56. $a = -126, b = -75$

57. $a = -\frac{5}{2}, b = 0$

58. $a = \frac{1}{4}, b = \frac{11}{4}$

59. $a = \frac{16}{5}, b = \frac{112}{75}$

60. $a = 9.34, b = -5.65$

Budget Variance In Exercises 61–64, the accounting department of a sports drink bottling company is checking to see whether the actual expenses of a department differ from the budgeted expenses by more than $500 or by more than 5%. Fill in the missing parts of the table, and determine whether each actual expense passes the "budget variance test."

		Budgeted Expense, b	Actual Expense, a	$\lvert a - b \rvert$	$0.05b$
61.	Wages	$112,700	$113,356		
62.	Utilities	$9,400	$9,772		
63.	Taxes	$37,640	$37,335		
64.	Insurance	$2,575	$2,613		

Model It

65. ***Federal Deficit*** The bar graph shows the federal government receipts (in billions of dollars) for selected years from 1960 through 2000. (Source: U.S. Office of Management and Budget)

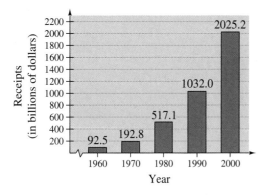

(a) Complete the table. (*Hint:* Find $\lvert$Receipts $-$ Expenditures$\rvert$.)

Model It (continued)

Year	Expenditures (in billions)	Surplus or deficit (in billions)
1960	$92.2	
1970	$195.6	
1980	$590.9	
1990	$1253.2	
2000	$1788.8	

(b) Use the table in part (a) to construct a bar graph showing the magnitude of the surplus or deficit for each year.

66. ***Veterans*** The table shows the number of living veterans (in thousands) in the United States in 2002 by age group. Construct a circle graph showing the percent of living veterans by age group as a fraction of the total number of living veterans. (Source: Department of Veteran Affairs)

Age group	Number of veterans
Under 35	2213
35–44	3290
45–54	4666
55–64	5665
65 and older	9784

In Exercises 67–72, use absolute value notation to describe the situation.

67. The distance between x and 5 is no more than 3.

68. The distance between x and -10 is at least 6.

69. y is at least six units from 0.

70. y is at most two units from a.

71. While traveling on the Pennsylvania Turnpike, you pass milepost 326 near Valley Forge, then milepost 351 near Philadelphia. How many miles do you travel during that time period?

72. The temperature in Chicago, Illinois was 48° last night at midnight, then 82° at noon today. What was the change in temperature over the 12-hour period?

In Exercises 73–78, identify the terms. Then identify the coefficients of the variable terms of the expression.

73. $7x + 4$

74. $6x^3 - 5x$

75. $\sqrt{3}x^2 - 8x - 11$

76. $3\sqrt{3}x^2 + 1$

77. $4x^3 + \dfrac{x}{2} - 5$

78. $3x^4 - \dfrac{x^2}{4}$

In Exercises 79–84, evaluate the expression for each value of x. (If not possible, state the reason.)

Expression	Values
79. $4x - 6$	(a) $x = -1$ (b) $x = 0$
80. $9 - 7x$	(a) $x = -3$ (b) $x = 3$
81. $x^2 - 3x + 4$	(a) $x = -2$ (b) $x = 2$
82. $-x^2 + 5x - 4$	(a) $x = -1$ (b) $x = 1$
83. $\dfrac{x + 1}{x - 1}$	(a) $x = 1$ (b) $x = -1$
84. $\dfrac{x}{x + 2}$	(a) $x = 2$ (b) $x = -2$

In Exercises 85–96, identify the rule(s) of algebra illustrated by the statement.

85. $x + 9 = 9 + x$

86. $2\left(\tfrac{1}{2}\right) = 1$

87. $\dfrac{1}{h + 6}(h + 6) = 1, \quad h \neq -6$

88. $(x + 3) - (x + 3) = 0$

89. $2(x + 3) = 2 \cdot x + 2 \cdot 3$

90. $(z - 2) + 0 = z - 2$

91. $1 \cdot (1 + x) = 1 + x$

92. $(z + 5)x = z \cdot x + 5 \cdot x$

93. $x + (y + 10) = (x + y) + 10$

94. $x(3y) = (x \cdot 3)y = (3x)y$

95. $3(t - 4) = 3 \cdot t - 3 \cdot 4$

96. $\tfrac{1}{7}(7 \cdot 12) = \left(\tfrac{1}{7} \cdot 7\right)12 = 1 \cdot 12 = 12$

In Exercises 97–104, perform the operation(s). (Write fractional answers in simplest form.)

97. $\dfrac{3}{16} + \dfrac{5}{16}$

98. $\dfrac{6}{7} - \dfrac{4}{7}$

99. $\dfrac{5}{8} - \dfrac{5}{12} + \dfrac{1}{6}$

100. $\dfrac{10}{11} + \dfrac{6}{33} - \dfrac{13}{66}$

101. $12 \div \dfrac{1}{4}$

102. $-\left(6 \cdot \dfrac{4}{8}\right)$

103. $\dfrac{2x}{3} - \dfrac{x}{4}$

104. $\dfrac{5x}{6} \cdot \dfrac{2}{9}$

105. (a) Use a calculator to complete the table.

n	1	0.5	0.01	0.0001	0.000001
$5/n$					

(b) Use the result from part (a) to make a conjecture about the value of $5/n$ as n approaches 0.

106. (a) Use a calculator to complete the table.

n	1	10	100	10,000	100,000
$5/n$					

(b) Use the result from part (a) to make a conjecture about the value of $5/n$ as n increases without bound.

Synthesis

True or False? **In Exercises 107 and 108, determine whether the statement is true or false. Justify your answer.**

107. If $a < b$, then $\dfrac{1}{a} < \dfrac{1}{b}$, where $a \neq b \neq 0$.

108. Because $\dfrac{a + b}{c} = \dfrac{a}{c} + \dfrac{b}{c}$, then $\dfrac{c}{a + b} = \dfrac{c}{a} + \dfrac{c}{b}$.

109. ***Exploration*** Consider $|u + v|$ and $|u| + |v|$, where $u \neq v \neq 0$.

(a) Are the values of the expressions always equal? If not, under what conditions are they unequal?

(b) If the two expressions are not equal for certain values of u and v, is one of the expressions always greater than the other? Explain.

110. ***Think About It*** Is there a difference between saying that a real number is positive and saying that a real number is nonnegative? Explain.

111. ***Think About It*** Because every even number is divisible by 2, is it possible that there exist any even prime numbers? Explain.

112. ***Writing*** Describe the differences among the sets of natural numbers, whole numbers, integers, rational numbers, and irrational numbers.

In Exercises 113 and 114, use the real numbers A, B, and C shown on the number line. Determine the sign of each expression.

113. (a) $-A$

(b) $B - A$

114. (a) $-C$

(b) $A - C$

115. ***Writing*** Can it ever be true that $|a| = -a$ for a real number a? Explain.

P.2 Solving Equations

What you should learn

- Identify different types of equations.
- Solve linear equations in one variable and equations that lead to linear equations.
- Solve quadratic equations by factoring, extracting square roots, completing the square, and using the Quadratic Formula.
- Solve polynomial equations of degree three or greater.
- Solve equations involving radicals.
- Solve equations with absolute values.

Why you should learn it

Linear equations are used in many real-life applications. For example, in Exercise 185 on page 24, linear equations can be used to model the relationship between the length of a thigh-bone and the height of a person, helping researchers learn about ancient cultures.

Equations and Solutions of Equations

An **equation** in x is a statement that two algebraic expressions are equal. For example

$$3x - 5 = 7, \quad x^2 - x - 6 = 0, \quad \text{and} \quad \sqrt{2x} = 4$$

are equations. To **solve** an equation in x means to find all values of x for which the equation is true. Such values are **solutions.** For instance, $x = 4$ is a solution of the equation

$$3x - 5 = 7$$

because $3(4) - 5 = 7$ is a true statement.

The solutions of an equation depend on the kinds of numbers being considered. For instance, in the set of rational numbers, $x^2 = 10$ has no solution because there is no rational number whose square is 10. However, in the set of real numbers, the equation has the two solutions $x = \sqrt{10}$ and $x = -\sqrt{10}$.

An equation that is true for *every* real number in the *domain* of the variable is called an **identity.** The domain is the set of all real numbers for which the equation is defined. For example

$$x^2 - 9 = (x + 3)(x - 3) \qquad \text{Identity}$$

is an identity because it is a true statement for any real value of x. The equation

$$\frac{x}{3x^2} = \frac{1}{3x} \qquad \text{Identity}$$

where $x \neq 0$, is an identity because it is true for any nonzero real value of x.

An equation that is true for just *some* (or even none) of the real numbers in the domain of the variable is called a **conditional equation.** For example, the equation

$$x^2 - 9 = 0 \qquad \text{Conditional equation}$$

is conditional because $x = 3$ and $x = -3$ are the only values in the domain that satisfy the equation. The equation $2x - 4 = 2x + 1$ is conditional because there are no real values of x for which the equation is true. Learning to solve conditional equations is the primary focus of this section.

Linear Equations in One Variable

> **Definition of a Linear Equation**
>
> A **linear equation in one variable** x is an equation that can be written in the standard form
>
> $$ax + b = 0$$
>
> where a and b are real numbers with $a \neq 0$.

A linear equation has exactly one solution. To see this, consider the following steps. (Remember that $a \neq 0$.)

$$ax + b = 0 \qquad \text{Write original equation.}$$

$$ax = -b \qquad \text{Subtract } b \text{ from each side.}$$

$$x = -\frac{b}{a} \qquad \text{Divide each side by } a.$$

To solve a conditional equation in x, isolate x on one side of the equation by a sequence of **equivalent** (and usually simpler) **equations,** each having the same solution(s) as the original equation. The operations that yield equivalent equations come from the Substitution Principle and the Properties of Equality studied in Section P.1.

British Museum

Historical Note

This ancient Egyptian papyrus, discovered in 1858, contains one of the earliest examples of mathematical writing in existence. The papyrus itself dates back to around 1650 B.C., but it is actually a copy of writings from two centuries earlier. The algebraic equations on the papyrus were written in words. Diophantus, a Greek who lived around A.D. 250, is often called the Father of Algebra. He was the first to use abbreviated word forms in equations.

Generating Equivalent Equations

An equation can be transformed into an *equivalent equation* by one or more of the following steps.

	Given Equation	*Equivalent Equation*
1. Remove symbols of grouping, combine like terms, or simplify fractions on one or both sides of the equation.	$2x - x = 4$	$x = 4$
2. Add (or subtract) the same quantity to (from) *each* side of the equation.	$x + 1 = 6$	$x = 5$
3. Multiply (or divide) *each* side of the equation by the same *nonzero* quantity.	$2x = 6$	$x = 3$
4. Interchange the two sides of the equation.	$2 = x$	$x = 2$

STUDY TIP

After solving an equation, you should check each solution in the original equation. For instance, you can check the solution to Example 1(a) as follows.

$$3x - 6 = 0 \qquad \text{Write original equation.}$$

$$3(2) - 6 \stackrel{?}{=} 0 \qquad \text{Substitute 2 for } x.$$

$$0 = 0 \qquad \text{Solution checks. } \checkmark$$

Try checking the solution to Example 1(b).

Example 1   Solving a Linear Equation

a. $\quad 3x - 6 = 0 \qquad$ Original equation

$$3x = 6 \qquad \text{Add 6 to each side.}$$

$$x = 2 \qquad \text{Divide each side by 3.}$$

b. $\quad 5x + 4 = 3x - 8 \qquad$ Original equation

$$2x + 4 = -8 \qquad \text{Subtract } 3x \text{ from each side.}$$

$$2x = -12 \qquad \text{Subtract 4 from each side.}$$

$$x = -6 \qquad \text{Divide each side by 2.}$$

✓CHECKPOINT   Now try Exercise 13.

To solve an equation involving fractional expressions, find the least common denominator (LCD) of all terms and multiply every term by the LCD. This process will clear the original equation of fractions and produce a simpler equation to work with.

STUDY TIP

An equation with a *single fraction* on each side can be cleared of denominators by **cross multiplying,** which is equivalent to multiplying by the LCD and then dividing out. To do this, multiply the left numerator by the right denominator and the right numerator by the left denominator as follows.

$$\frac{a}{b} = \frac{c}{d} \qquad \text{LCD is } bd.$$

$$\frac{a}{b} \cdot bd = \frac{c}{d} \cdot bd \quad \text{Multiply by LCD.}$$

$$ad = cb \qquad \begin{array}{l}\text{Divide out} \\ \text{common factors.}\end{array}$$

Example 2 An Equation Involving Fractional Expressions

Solve $\dfrac{x}{3} + \dfrac{3x}{4} = 2$.

Solution

$$\frac{x}{3} + \frac{3x}{4} = 2 \qquad\qquad \text{Write original equation.}$$

$$(12)\frac{x}{3} + (12)\frac{3x}{4} = (12)2 \qquad \text{Multiply each term by the LCD of 12.}$$

$$4x + 9x = 24 \qquad\qquad \text{Divide out and multiply.}$$

$$13x = 24 \qquad\qquad \text{Combine like terms.}$$

$$x = \frac{24}{13} \qquad\qquad \text{Divide each side by 13.}$$

The solution is $x = \frac{24}{13}$. Check this in the original equation.

✓CHECKPOINT   Now try Exercise 21.

When multiplying or dividing an equation by a *variable* quantity, it is possible to introduce an extraneous solution. An **extraneous solution** is one that does not satisfy the original equation. Therefore, it is essential that you check your solutions.

Example 3 An Equation with an Extraneous Solution

Solve $\dfrac{1}{x-2} = \dfrac{3}{x+2} - \dfrac{6x}{x^2-4}$.

Solution

The LCD is $x^2 - 4$, or $(x+2)(x-2)$. Multiply each term by this LCD.

$$\frac{1}{x-2}(x+2)(x-2) = \frac{3}{x+2}(x+2)(x-2) - \frac{6x}{x^2-4}(x+2)(x-2)$$

$$x + 2 = 3(x - 2) - 6x, \qquad x \neq \pm 2$$

$$x + 2 = 3x - 6 - 6x$$

$$x + 2 = -3x - 6$$

$$4x = -8 \quad \Longrightarrow \quad x = -2 \qquad \text{Extraneous solution}$$

In the original equation, $x = -2$ yields a denominator of zero. So, $x = -2$ is an extraneous solution, and the original equation has *no solution*.

✓CHECKPOINT   Now try Exercise 37.

STUDY TIP

Recall that the least common denominator of two or more fractions consists of the product of all prime factors in the denominators, with each factor given the highest power of its occurrence in any denominator. For instance, in Example 3, by factoring each denominator you can determine that the LCD is $(x + 2)(x - 2)$.

Quadratic Equations

A **quadratic equation** in x is an equation that can be written in the general form

$$ax^2 + bx + c = 0$$

where a, b, and c are real numbers, with $a \neq 0$. A quadratic equation in x is also known as a **second-degree polynomial equation** in x.

You should be familiar with the following four methods of solving quadratic equations.

Solving a Quadratic Equation

Factoring: If $ab = 0$, then $a = 0$ or $b = 0$.

Example:

$$x^2 - x - 6 = 0$$
$$(x - 3)(x + 2) = 0$$
$$x - 3 = 0 \quad \Longrightarrow \quad x = 3$$
$$x + 2 = 0 \quad \Longrightarrow \quad x = -2$$

Square Root Principle: If $u^2 = c$, where $c > 0$, then $u = \pm\sqrt{c}$.

Example:

$$(x + 3)^2 = 16$$
$$x + 3 = \pm 4$$
$$x = -3 \pm 4$$
$$x = 1 \quad \text{or} \quad x = -7$$

Completing the Square: If $x^2 + bx = c$, then

$$x^2 + bx + \left(\frac{b}{2}\right)^2 = c + \left(\frac{b}{2}\right)^2 \qquad \text{Add } \left(\frac{b}{2}\right)^2 \text{ to each side.}$$

$$\left(x + \frac{b}{2}\right)^2 = c + \frac{b^2}{4}.$$

Example:

$$x^2 + 6x = 5$$
$$x^2 + 6x + 3^2 = 5 + 3^2 \qquad \text{Add } \left(\frac{6}{2}\right)^2 \text{ to each side.}$$
$$(x + 3)^2 = 14$$
$$x + 3 = \pm\sqrt{14}$$
$$x = -3 \pm \sqrt{14}$$

Quadratic Formula: If $ax^2 + bx + c = 0$, then $x = \dfrac{-b \pm \sqrt{b^2 - 4ac}}{2a}$.

Example:

$$2x^2 + 3x - 1 = 0$$
$$x = \frac{-3 \pm \sqrt{3^2 - 4(2)(-1)}}{2(2)}$$
$$= \frac{-3 \pm \sqrt{17}}{4}$$

STUDY TIP

The Square Root Principle is also referred to as *extracting square roots.*

STUDY TIP

You can solve every quadratic equation by completing the square or using the Quadratic Formula.

Example 4   **Solving a Quadratic Equation by Factoring**

a.
$$2x^2 + 9x + 7 = 3$$ Original equation

$$2x^2 + 9x + 4 = 0$$ Write in general form.

$$(2x + 1)(x + 4) = 0$$ Factor.

$$2x + 1 = 0 \implies x = -\frac{1}{2}$$ Set 1st factor equal to 0.

$$x + 4 = 0 \implies x = -4$$ Set 2nd factor equal to 0.

The solutions are $x = -\frac{1}{2}$ and $x = -4$. Check these in the original equation.

b.
$$6x^2 - 3x = 0$$ Original equation

$$3x(2x - 1) = 0$$ Factor.

$$3x = 0 \implies x = 0$$ Set 1st factor equal to 0.

$$2x - 1 = 0 \implies x = \frac{1}{2}$$ Set 2nd factor equal to 0.

The solutions are $x = 0$ and $x = \frac{1}{2}$. Check these in the original equation.

✓CHECKPOINT   Now try Exercise 57.

A common error occurs when there are only two terms in an equation, as in Example 1(b).

Sample Error:

$6x^2 - 3x = 0$

$6x^2 = 3x$

$2x = 1$

$x = \dfrac{1}{2}$

The solution $x = 0$ is lost when $6x^2 = 3x$ is divided by $3x$.

Note that the method of solution in Example 4 is based on the Zero-Factor Property from Section P.1. Be sure you see that this property works *only* for equations written in general form (in which the right side of the equation is zero). So, all terms must be collected on one side *before* factoring. For instance, in the equation $(x - 5)(x + 2) = 8$, it is *incorrect* to set each factor equal to 8. Try to solve this equation correctly.

Example 5   **Extracting Square Roots**

Solve each equation by extracting square roots.

a. $4x^2 = 12$ **b.** $(x - 3)^2 = 7$

Solution

a. $4x^2 = 12$ Write original equation.

$\quad\quad x^2 = 3$ Divide each side by 4.

$\quad\quad x = \pm\sqrt{3}$ Extract square roots.

When you take the square root of a variable expression, you must account for both positive and negative solutions. So, the solutions are $x = \sqrt{3}$ and $x = -\sqrt{3}$. Check these in the original equation.

b. $(x - 3)^2 = 7$ Write original equation.

$\quad\quad x - 3 = \pm\sqrt{7}$ Extract square roots.

$\quad\quad x = 3 \pm \sqrt{7}$ Add 3 to each side.

The solutions are $x = 3 \pm \sqrt{7}$. Check these in the original equation.

✓CHECKPOINT   Now try Exercise 77.

When solving quadratic equations by completing the square, you must add $(b/2)^2$ to *each side* in order to maintain equality. If the leading coefficient is *not* 1, you must divide each side of the equation by the leading coefficient *before* completing the square, as shown in Example 7.

Example 6   Completing the Square: Leading Coefficient Is 1

Additional Example: Completing the Square Within an Algebraic Expression

Rewrite the denominator as the sum or difference of two squares.

$$\frac{1}{x^2 - 2x - 3}$$

Solution

Complete the square of the denominator by adding and subtracting $(-1)^2$ to the expression.

$$x^2 - 2x - 3 = x^2 - 2x + (-1)^2$$
$$- 3 - (-1)^2$$
$$= (x^2 - 2x + 1) - 4$$
$$= (x - 1)^2 - 2^2$$

The original expression can be written as

$$\frac{1}{(x - 1)^2 - 2^2}.$$

Solve $x^2 + 2x - 6 = 0$ by completing the square.

Solution

$$x^2 + 2x - 6 = 0 \qquad \text{Write original equation.}$$
$$x^2 + 2x = 6 \qquad \text{Add 6 to each side.}$$
$$x^2 + 2x + 1^2 = 6 + 1^2 \qquad \text{Add } 1^2 \text{ to each side.}$$

(half of 2)^2

$$(x + 1)^2 = 7 \qquad \text{Simplify.}$$
$$x + 1 = \pm\sqrt{7} \qquad \text{Take square root of each side.}$$
$$x = -1 \pm \sqrt{7} \qquad \text{Subtract 1 from each side.}$$

The solutions are $x = -1 \pm \sqrt{7}$. Check these in the original equation.

✓CHECKPOINT   Now try Exercise 85.

Example 7   Completing the Square: Leading Coefficient Is Not 1

$$3x^2 - 4x - 5 = 0 \qquad \text{Original equation}$$
$$3x^2 - 4x = 5 \qquad \text{Add 5 to each side.}$$
$$x^2 - \frac{4}{3}x = \frac{5}{3} \qquad \text{Divide each side by 3.}$$
$$x^2 - \frac{4}{3}x + \left(-\frac{2}{3}\right)^2 = \frac{5}{3} + \left(-\frac{2}{3}\right)^2 \qquad \text{Add } \left(-\frac{2}{3}\right)^2 \text{ to each side.}$$

$\left(\text{half of } -\frac{4}{3}\right)^2$

$$x^2 - \frac{4}{3}x + \frac{4}{9} = \frac{19}{9} \qquad \text{Simplify.}$$
$$\left(x - \frac{2}{3}\right)^2 = \frac{19}{9} \qquad \text{Perfect square trinomial.}$$
$$x - \frac{2}{3} = \pm\frac{\sqrt{19}}{3} \qquad \text{Extract square roots.}$$
$$x = \frac{2}{3} \pm \frac{\sqrt{19}}{3} \qquad \text{Solutions}$$

✓CHECKPOINT   Now try Exercise 91.

Example 8   The Quadratic Formula: Two Distinct Solutions

Use the Quadratic Formula to solve $x^2 + 3x = 9$.

Solution

$$x^2 + 3x = 9$$ Write original equation.

$$x^2 + 3x - 9 = 0$$ Write in general form.

$$x = \frac{-b \pm \sqrt{b^2 - 4ac}}{2a}$$ Quadratic Formula

$$x = \frac{-3 \pm \sqrt{(3)^2 - 4(1)(-9)}}{2(1)}$$ Substitute $a = 1$, $b = 3$, and $c = -9$.

$$x = \frac{-3 \pm \sqrt{45}}{2}$$ Simplify.

$$x = \frac{-3 \pm 3\sqrt{5}}{2}$$ Simplify.

The equation has two solutions:

$$x = \frac{-3 + 3\sqrt{5}}{2} \quad \text{and} \quad x = \frac{-3 - 3\sqrt{5}}{2}.$$

Check these in the original equation.

✓CHECKPOINT   Now try Exercise 101.

Example 9   The Quadratic Formula: One Solution

Use the Quadratic Formula to solve $8x^2 - 24x + 18 = 0$.

Solution

$$8x^2 - 24x + 18 = 0$$ Write original equation.

$$4x^2 - 12x + 9 = 0$$ Divide out common factor of 2.

$$x = \frac{-b \pm \sqrt{b^2 - 4ac}}{2a}$$ Quadratic Formula

$$x = \frac{-(-12) \pm \sqrt{(-12)^2 - 4(4)(9)}}{2(4)}$$ Substitute $a = 4$, $b = -12$, and $c = 9$.

$$x = \frac{12 \pm \sqrt{0}}{8} = \frac{3}{2}$$ Simplify.

This quadratic equation has only one solution: $x = \frac{3}{2}$. Check this in the original equation.

✓CHECKPOINT   Now try Exercise 105.

Note that Example 9 could have been solved without first dividing out a common factor of 2. Substituting $a = 8$, $b = -24$, and $c = 18$ into the Quadratic Formula produces the same result.

Activities

1. Solve $3x^2 = 5x$.

   Answer: $x = 0, \frac{5}{3}$

2. Solve $(x + 3)^2 = 12$.

   Answer: $x = -3 \pm 2\sqrt{3}$

3. Complete the square for the quadratic portion of $\dfrac{1}{\sqrt{x^2 - 6x + 13}}$.

   Answer: $\dfrac{1}{\sqrt{(x - 3)^2 + 4}}$

4. Use the Quadratic Formula to solve $3x^2 = 3x + 1$.

   Answer: $x = \dfrac{3 \pm \sqrt{21}}{6}$

5. If the perimeter of a right triangle is 42 units and the length of the hypotenuse is 20 units, find the lengths of the other two sides of the triangle.

   Answer: 19.888; 2.112

Polynomial Equations of Higher Degree

The methods used to solve quadratic equations can sometimes be extended to solve polynomial equations of higher degree.

STUDY TIP

A common mistake that is made in solving an equation such as that in Example 10 is to divide each side of the equation by the variable factor x^2. This loses the solution $x = 0$. When solving an equation, always write the equation in general form, then factor the equation and set each factor equal to zero. Do not divide each side of an equation by a variable factor in an attempt to simplify the equation.

Example 10   **Solving a Polynomial Equation by Factoring**

Solve $3x^4 = 48x^2$.

Solution

First write the polynomial equation in general form with zero on one side, factor the other side, and then set each factor equal to zero and solve.

$$3x^4 = 48x^2$$ Write original equation.

$$3x^4 - 48x^2 = 0$$ Write in general form.

$$3x^2(x^2 - 16) = 0$$ Factor out common factor.

$$3x^2(x + 4)(x - 4) = 0$$ Write in factored form.

$$3x^2 = 0 \implies x = 0$$ Set 1st factor equal to 0.

$$x + 4 = 0 \implies x = -4$$ Set 2nd factor equal to 0.

$$x - 4 = 0 \implies x = 4$$ Set 3rd factor equal to 0.

You can check these solutions by substituting in the original equation, as follows.

Check

$$3(0)^4 = 48(0)^2$$ 0 checks. ✓

$$3(-4)^4 = 48(-4)^2$$ -4 checks. ✓

$$3(4)^4 = 48(4)^2$$ 4 checks. ✓

So, you can conclude that the solutions are $x = 0$, $x = -4$, and $x = 4$.

✓**CHECKPOINT**   Now try Exercise 135.

Example 11   **Solving a Polynomial Equation by Factoring**

Solve $x^3 - 3x^2 - 3x + 9 = 0$.

Solution

$$x^3 - 3x^2 - 3x + 9 = 0$$ Write original equation.

$$x^2(x - 3) - 3(x - 3) = 0$$ Factor by grouping.

$$(x - 3)(x^2 - 3) = 0$$ Distributive Property

$$x - 3 = 0 \implies x = 3$$ Set 1st factor equal to 0.

$$x^2 - 3 = 0 \implies x = \pm\sqrt{3}$$ Set 2nd factor equal to 0.

The solutions are $x = 3$, $x = \sqrt{3}$, and $x = -\sqrt{3}$. Check these in the original equation.

✓**CHECKPOINT**   Now try Exercise 143.

Equations Involving Radicals

Operations such as squaring each side of an equation, raising each side of an equation to a rational power, and multiplying each side of an equation by a variable quantity all can introduce extraneous solutions. So, when you use any of these operations, checking your solutions is crucial.

Example 12 Solving Equations Involving Radicals

a.

$\sqrt{2x+7} - x = 2$	Original equation
$\sqrt{2x+7} = x + 2$	Isolate radical.
$2x + 7 = x^2 + 4x + 4$	Square each side.
$0 = x^2 + 2x - 3$	Write in general form.
$0 = (x+3)(x-1)$	Factor.
$x + 3 = 0 \quad\Longrightarrow\quad x = -3$	Set 1st factor equal to 0.
$x - 1 = 0 \quad\Longrightarrow\quad x = 1$	Set 2nd factor equal to 0.

To show why the radical should be isolated, have students square each side of the equation before isolating the radical. Compare problem-solving strategies to help convince students of the need to isolate the radical.

By checking these values, you can determine that the only solution is $x = 1$.

b.

$\sqrt{2x-5} - \sqrt{x-3} = 1$	Original equation
$\sqrt{2x-5} = \sqrt{x-3} + 1$	Isolate $\sqrt{2x-5}$.
$2x - 5 = x - 3 + 2\sqrt{x-3} + 1$	Square each side.
$2x - 5 = x - 2 + 2\sqrt{x-3}$	Combine like terms.
$x - 3 = 2\sqrt{x-3}$	Isolate $2\sqrt{x-3}$.
$x^2 - 6x + 9 = 4(x-3)$	Square each side.
$x^2 - 10x + 21 = 0$	Write in general form.
$(x-3)(x-7) = 0$	Factor.
$x - 3 = 0 \quad\Longrightarrow\quad x = 3$	Set 1st factor equal to 0.
$x - 7 = 0 \quad\Longrightarrow\quad x = 7$	Set 2nd factor equal to 0.

STUDY TIP

When an equation contains two radicals, it may not be possible to isolate both. In such cases, you may have to raise each side of the equation to a power at *two* different stages in the solution, as shown in Example 12(b).

The solutions are $x = 3$ and $x = 7$. Check these in the original equation.

✓CHECKPOINT Now try Exercise 155.

Example 13 Solving an Equation Involving a Rational Exponent

$(x-4)^{2/3} = 25$	Original equation
$\sqrt[3]{(x-4)^2} = 25$	Rewrite in radical form.
$(x-4)^2 = 15{,}625$	Cube each side.
$x - 4 = \pm 125$	Take square root of each side.
$x = 129, \quad x = -121$	Add 4 to each side.

✓CHECKPOINT Now try Exercise 163.

Additional Examples

a. $\dfrac{7}{x} - 2 = \dfrac{10}{3x - 1}$

$x(3x - 1)\dfrac{7}{x} - x(3x - 1)2$

$\quad = x(3x - 1)\left(\dfrac{10}{3x - 1}\right)$

$7(3x - 1) - 2x(3x - 1) = 10x$

$21x - 7 - 6x^2 + 2x = 10x$

$-6x^2 + 13x - 7 = 0$

$6x^2 - 13x + 7 = 0$

$(6x - 7)(x - 1) = 0$

$6x - 7 = 0 \implies x = \dfrac{7}{6}$

$x - 1 = 0 \implies x = 1$

b. $\dfrac{3x}{x + 1} = \dfrac{12}{x^2 - 1} + 2$

$(x^2 - 1)\dfrac{3x}{x + 1} = (x^2 - 1)\dfrac{12}{x^2 - 1}$

$\quad\quad\quad\quad\quad + (x^2 - 1)2$

$3x(x - 1) = 12 + (x^2 - 1)2$

$3x^2 - 3x = 12 + 2x^2 - 2$

$x^2 - 3x - 10 = 0$

$(x + 2)(x - 5) = 0$

$x + 2 = 0 \implies x = -2$

$x - 5 = 0 \implies x = 5$

Activities

1. Solve for x: $9x^4 - 9x^2 + 2 = 0$.

   Answer: $x = \pm\dfrac{\sqrt{3}}{3}, \pm\dfrac{\sqrt{6}}{3}$

2. Solve for x: $\sqrt{3x + 10} - x = 4$.

   Answer: $x = -2, -3$

3. Solve for x: $|5x + 2| = 22$.

   Answer: $x = -\dfrac{24}{5}, 4$

Equations with Absolute Values

To solve an equation involving an absolute value, remember that the expression inside the absolute value signs can be positive or negative. This results in *two* separate equations, each of which must be solved. For instance, the equation

$$|x - 2| = 3$$

results in the two equations $x - 2 = 3$ and $-(x - 2) = 3$, which implies that the equation has two solutions: $x = 5$ and $x = -1$.

Example 14   **Solving an Equation Involving Absolute Value**

Solve $|x^2 - 3x| = -4x + 6$.

Solution

Because the variable expression inside the absolute value signs can be positive or negative, you must solve the following two equations.

First Equation

$x^2 - 3x = -4x + 6$	Use positive expression.
$x^2 + x - 6 = 0$	Write in general form.
$(x + 3)(x - 2) = 0$	Factor.
$x + 3 = 0 \implies x = -3$	Set 1st factor equal to 0.
$x - 2 = 0 \implies x = 2$	Set 2nd factor equal to 0.

Second Equation

$-(x^2 - 3x) = -4x + 6$	Use negative expression.
$x^2 - 7x + 6 = 0$	Write in general form.
$(x - 1)(x - 6) = 0$	Factor.
$x - 1 = 0 \implies x = 1$	Set 1st factor equal to 0.
$x - 6 = 0 \implies x = 6$	Set 2nd factor equal to 0.

Check

$|(-3)^2 - 3(-3)| \overset{?}{=} -4(-3) + 6$	Substitute -3 for x.
$18 = 18$	-3 checks. ✓
$|(2)^2 - 3(2)| \overset{?}{=} -4(2) + 6$	Substitute 2 for x.
$2 \neq -2$	2 does not check.
$|(1)^2 - 3(1)| \overset{?}{=} -4(1) + 6$	Substitute 1 for x.
$2 = 2$	1 checks. ✓
$|(6)^2 - 3(6)| \overset{?}{=} -4(6) + 6$	Substitute 6 for x.
$18 \neq -18$	6 does not check.

The solutions are $x = -3$ and $x = 1$.

✓CHECKPOINT   Now try Exercise 181.

P.2 Exercises

VOCABULARY CHECK: Fill in the blanks.

1. An _____ is a statement that equates two algebraic expressions.

2. To find all values that satisfy an equation is to _____ the equation.

3. There are two types of equations, _____ and _____ equations.

4. A linear equation in one variable is an equation that can be written in the standard from _____.

5. When solving an equation, it is possible to introduce an _____ solution, which is a value that does not satisfy the original equation.

6. An equation of the form $ax^2 + bx + c = 0$, $a \neq 0$ is a _____ _____, or a second-degree polynomial equation in x.

7. The four methods that can be used to solve a quadratic equation are _____, _____, _____, and the _____.

In Exercises 1–10, determine whether the equation is an identity or a conditional equation.

1. $2(x - 1) = 2x - 2$

2. $3(x + 2) = 5x + 4$

3. $-6(x - 3) + 5 = -2x + 10$

4. $3(x + 2) - 5 = 3x + 1$

5. $4(x + 1) - 2x = 2(x + 2)$

6. $-7(x - 3) + 4x = 3(7 - x)$

7. $x^2 - 8x + 5 = (x - 4)^2 - 11$

8. $x^2 + 2(3x - 2) = x^2 + 6x - 4$

9. $3 + \dfrac{1}{x + 1} = \dfrac{4x}{x + 1}$

10. $\dfrac{5}{x} + \dfrac{3}{x} = 24$

In Exercises 11–26, solve the equation and check your solution.

11. $x + 11 = 15$

12. $7 - x = 19$

13. $7 - 2x = 25$

14. $7x + 2 = 23$

15. $8x - 5 = 3x + 20$

16. $7x + 3 = 3x - 17$

17. $2(x + 5) - 7 = 3(x - 2)$

18. $3(x + 3) = 5(1 - x) - 1$

19. $x - 3(2x + 3) = 8 - 5x$

20. $9x - 10 = 5x + 2(2x - 5)$

21. $\dfrac{5x}{4} + \dfrac{1}{2} = x - \dfrac{1}{2}$

22. $\dfrac{x}{5} - \dfrac{x}{2} = 3 + \dfrac{3x}{10}$

23. $\frac{3}{2}(z + 5) - \frac{1}{4}(z + 24) = 0$

24. $\dfrac{3x}{2} + \dfrac{1}{4}(x - 2) = 10$

25. $0.25x + 0.75(10 - x) = 3$

26. $0.60x + 0.40(100 - x) = 50$

In Exercises 27–48, solve the equation and check your solution. (If not possible, explain why.)

27. $x + 8 = 2(x - 2) - x$

28. $8(x + 2) - 3(2x + 1) = 2(x + 5)$

29. $\dfrac{100 - 4x}{3} = \dfrac{5x + 6}{4} + 6$

30. $\dfrac{17 + y}{y} + \dfrac{32 + y}{y} = 100$

31. $\dfrac{5x - 4}{5x + 4} = \dfrac{2}{3}$

32. $\dfrac{10x + 3}{5x + 6} = \dfrac{1}{2}$

33. $10 - \dfrac{13}{x} = 4 + \dfrac{5}{x}$

34. $\dfrac{15}{x} - 4 = \dfrac{6}{x} + 3$

35. $3 = 2 + \dfrac{2}{z + 2}$

36. $\dfrac{1}{x} + \dfrac{2}{x - 5} = 0$

37. $\dfrac{x}{x + 4} + \dfrac{4}{x + 4} + 2 = 0$

38. $\dfrac{7}{2x + 1} - \dfrac{8x}{2x - 1} = -4$

39. $\dfrac{2}{(x - 4)(x - 2)} = \dfrac{1}{x - 4} + \dfrac{2}{x - 2}$

40. $\dfrac{4}{x - 1} + \dfrac{6}{3x + 1} = \dfrac{15}{3x + 1}$

41. $\dfrac{1}{x - 3} + \dfrac{1}{x + 3} = \dfrac{10}{x^2 - 9}$

42. $\dfrac{1}{x - 2} + \dfrac{3}{x + 3} = \dfrac{4}{x^2 + x - 6}$

43. $\dfrac{3}{x^2 - 3x} + \dfrac{4}{x} = \dfrac{1}{x - 3}$

44. $\dfrac{6}{x} - \dfrac{2}{x + 3} = \dfrac{3(x + 5)}{x^2 + 3x}$

45. $(x + 2)^2 + 5 = (x + 3)^2$

46. $(x + 1)^2 + 2(x - 2) = (x + 1)(x - 2)$

47. $(x + 2)^2 - x^2 = 4(x + 1)$

48. $(2x + 1)^2 = 4(x^2 + x + 1)$

In Exercises 49–54, write the quadratic equation in general form.

49. $2x^2 = 3 - 8x$

50. $x^2 = 16x$

51. $(x - 3)^2 = 3$

52. $13 - 3(x + 7)^2 = 0$

53. $\frac{1}{5}(3x^2 - 10) = 18x$

54. $x(x + 2) = 5x^2 + 1$

In Exercises 55–68, solve the quadratic equation by factoring.

55. $6x^2 + 3x = 0$

56. $9x^2 - 1 = 0$

57. $x^2 - 2x - 8 = 0$

58. $x^2 - 10x + 9 = 0$

59. $x^2 + 10x + 25 = 0$

60. $4x^2 + 12x + 9 = 0$

61. $3 + 5x - 2x^2 = 0$

62. $2x^2 = 19x + 33$

63. $x^2 + 4x = 12$

64. $-x^2 + 8x = 12$

65. $\frac{3}{4}x^2 + 8x + 20 = 0$

66. $\frac{1}{8}x^2 - x - 16 = 0$

67. $x^2 + 2ax + a^2 = 0$, a is a real number

68. $(x + a)^2 - b^2 = 0$, a and b are real numbers

In Exercises 69–82, solve the equation by extracting square roots.

69. $x^2 = 49$

70. $x^2 = 169$

71. $x^2 = 11$

72. $x^2 = 32$

73. $3x^2 = 81$

74. $9x^2 = 36$

75. $(x - 12)^2 = 16$

76. $(x + 13)^2 = 25$

77. $(x + 2)^2 = 14$

78. $(x - 5)^2 = 30$

79. $(2x - 1)^2 = 18$

80. $(4x + 7)^2 = 44$

81. $(x - 7)^2 = (x + 3)^2$

82. $(x + 5)^2 = (x + 4)^2$

In Exercises 83–92, solve the quadratic equation by completing the square.

83. $x^2 + 4x - 32 = 0$

84. $x^2 - 2x - 3 = 0$

85. $x^2 + 12x + 25 = 0$

86. $x^2 + 8x + 14 = 0$

87. $9x^2 - 18x = -3$

88. $9x^2 - 12x = 14$

89. $8 + 4x - x^2 = 0$

90. $-x^2 + x - 1 = 0$

91. $2x^2 + 5x - 8 = 0$

92. $4x^2 - 4x - 99 = 0$

In Exercises 93–116, use the Quadratic Formula to solve the equation.

93. $2x^2 + x - 1 = 0$

94. $2x^2 - x - 1 = 0$

95. $16x^2 + 8x - 3 = 0$

96. $25x^2 - 20x + 3 = 0$

97. $2 + 2x - x^2 = 0$

98. $x^2 - 10x + 22 = 0$

99. $x^2 + 14x + 44 = 0$

100. $6x = 4 - x^2$

101. $x^2 + 8x - 4 = 0$

102. $4x^2 - 4x - 4 = 0$

103. $12x - 9x^2 = -3$

104. $16x^2 + 22 = 40x$

105. $9x^2 + 24x + 16 = 0$

106. $36x^2 + 24x - 7 = 0$

107. $4x^2 + 4x = 7$

108. $16x^2 - 40x + 5 = 0$

109. $28x - 49x^2 = 4$

110. $3x + x^2 - 1 = 0$

111. $8t = 5 + 2t^2$

112. $25h^2 + 80h + 61 = 0$

113. $(y - 5)^2 = 2y$

114. $(z + 6)^2 = -2z$

115. $\frac{1}{2}x^2 + \frac{3}{8}x = 2$

116. $\left(\frac{5}{7}x - 14\right)^2 = 8x$

In Exercises 117–124, use the Quadratic Formula to solve the equation. (Round your answer to three decimal places.)

117. $5.1x^2 - 1.7x - 3.2 = 0$

118. $2x^2 - 2.50x - 0.42 = 0$

119. $-0.067x^2 - 0.852x + 1.277 = 0$

120. $-0.005x^2 + 0.101x - 0.193 = 0$

121. $422x^2 - 506x - 347 = 0$

122. $1100x^2 + 326x - 715 = 0$

123. $12.67x^2 + 31.55x + 8.09 = 0$

124. $-3.22x^2 - 0.08x + 28.651 = 0$

In Exercises 125–134, solve the equation using any convenient method.

125. $x^2 - 2x - 1 = 0$

126. $11x^2 + 33x = 0$

127. $(x + 3)^2 = 81$

128. $x^2 - 14x + 49 = 0$

129. $x^2 - x - \frac{11}{4} = 0$

130. $x^2 + 3x - \frac{3}{4} = 0$

131. $(x + 1)^2 = x^2$

132. $a^2x^2 - b^2 = 0$, a and b are real numbers

133. $3x + 4 = 2x^2 - 7$

134. $4x^2 + 2x + 4 = 2x + 8$

In Exercises 135–152, find all solutions of the equation. Check your solutions in the original equation.

135. $4x^4 - 18x^2 = 0$

136. $20x^3 - 125x = 0$

137. $x^4 - 81 = 0$

138. $x^6 - 64 = 0$

139. $x^3 + 216 = 0$

140. $27x^3 - 512 = 0$

141. $5x^3 + 30x^2 + 45x = 0$

142. $9x^4 - 24x^3 + 16x^2 = 0$

143. $x^3 - 3x^2 - x + 3 = 0$

144. $x^3 + 2x^2 + 3x + 6 = 0$

145. $x^4 - x^3 + x - 1 = 0$

146. $x^4 + 2x^3 - 8x - 16 = 0$

147. $x^4 - 4x^2 + 3 = 0$

148. $x^4 + 5x^2 - 36 = 0$

149. $4x^4 - 65x^2 + 16 = 0$

150. $36t^4 + 29t^2 - 7 = 0$

151. $x^6 + 7x^3 - 8 = 0$

152. $x^6 + 3x^3 + 2 = 0$

In Exercises 153–184, find all solutions of the equation. Check your solutions in the original equation.

153. $\sqrt{2x} - 10 = 0$

154. $4\sqrt{x} - 3 = 0$

155. $\sqrt{x - 10} - 4 = 0$

156. $\sqrt{5 - x} - 3 = 0$

157. $\sqrt[3]{2x + 5} + 3 = 0$

158. $\sqrt[3]{3x + 1} - 5 = 0$

159. $-\sqrt{26 - 11x} + 4 = x$

160. $x + \sqrt{31 - 9x} = 5$

161. $\sqrt{x+1} = \sqrt{3x+1}$ **162.** $\sqrt{x+5} = \sqrt{x-5}$

163. $(x-5)^{3/2} = 8$ **164.** $(x+3)^{3/2} = 8$

165. $(x+3)^{2/3} = 8$ **166.** $(x+2)^{2/3} = 9$

167. $(x^2-5)^{3/2} = 27$ **168.** $(x^2-x-22)^{3/2} = 27$

169. $3x(x-1)^{1/2} + 2(x-1)^{3/2} = 0$

170. $4x^2(x-1)^{1/3} + 6x(x-1)^{4/3} = 0$

171. $x = \dfrac{3}{x} + \dfrac{1}{2}$ **172.** $\dfrac{4}{x} - \dfrac{5}{3} = \dfrac{x}{6}$

173. $\dfrac{1}{x} - \dfrac{1}{x+1} = 3$ **174.** $\dfrac{4}{x+1} - \dfrac{3}{x+2} = 1$

175. $\dfrac{20-x}{x} = x$ **176.** $4x+1 = \dfrac{3}{x}$

177. $\dfrac{x}{x^2-4} + \dfrac{1}{x+2} = 3$ **178.** $\dfrac{x+1}{3} - \dfrac{x+1}{x+2} = 0$

179. $|2x-1| = 5$ **180.** $|3x+2| = 7$

181. $|x| = x^2 + x - 3$ **182.** $|x^2 + 6x| = 3x + 18$

183. $|x+1| = x^2 - 5$ **184.** $|x-10| = x^2 - 10x$

Model It

185. *Anthropology*   The relationship between the length of an adult's femur (thigh bone) and the height of the adult can be approximated by the linear equations

$y = 0.432x - 10.44$ Female

$y = 0.449x - 12.15$ Male

where y is the length of the femur in inches and x is the height of the adult in inches (see figure).

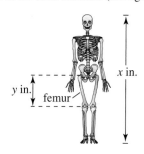

(a) An anthropologist discovers a femur belonging to an adult human female. The bone is 16 inches long. Estimate the height of the female.

(b) From the foot bones of an adult human male, an anthropologist estimates that the person's height was 69 inches. A few feet away from the site where the foot bones were discovered, the anthropologist discovers a male adult femur that is 19 inches long. Is it likely that both the foot bones and the thigh bone came from the same person?

Model It (continued)

(c) Complete the table to determine if there is a height of an adult for which an anthropologist would not be able to determine whether the femur belonged to a male or a female.

Height, x	Female femur length, y	Male femur length, y
60		
70		
80		
90		
100		
110		

---

186. *Operating Cost*   A delivery company has a fleet of vans. The annual operating cost C per van is $C = 0.32m + 2500$ where m is the number of miles traveled by a van in a year. What number of miles will yield an annual operating cost of $10,000?

187. *Flood Control*   A river has risen 8 feet above its flood stage. The water begins to recede at a rate of 3 inches per hour. Write a mathematical model that shows the number of feet above flood stage after t hours. If the water continually recedes at this rate, when will the river be 1 foot above its flood stage?

188. *Floor Space*   The floor of a one-story building is 14 feet longer than it is wide. The building has 1632 square feet of floor space.

(a) Draw a diagram that gives a visual representation of the floor space. Represent the width as w and show the length in terms of w.

(b) Write a quadratic equation in terms of w.

(c) Find the length and width of the floor of the building.

189. *Packaging*   An open box with a square base (see figure) is to be constructed from 84 square inches of material. The height of the box is 2 inches. What are the dimensions of the box? (*Hint:* The surface area is $S = x^2 + 4xh$.)

190. *Geometry*   The hypotenuse of an isosceles right triangle is 5 centimeters long. How long are its sides?

191. *Geometry*   An equilateral triangle has a height of 10 inches. How long is one of its sides? (*Hint:* Use the height of the triangle to partition the triangle into two congruent right triangles.)

192. *Flying Speed*   Two planes leave simultaneously from Chicago's O'Hare Airport, one flying due north and the other due east (see figure). The northbound plane is flying 50 miles per hour faster than the eastbound plane. After 3 hours, the planes are 2440 miles apart. Find the speed of each plane.

193. *Voting Population*   The total voting-age population P (in millions) in the United States from 1990 to 2002 can be modeled by

$$P = \frac{182.45 - 3.189t}{1.00 - 0.026t}, \quad 0 \le t \le 12$$

where t represents the year, with $t = 0$ corresponding to 1990.   (Source: U.S. Census Bureau)

(a) In which year did the total voting-age population reach 200 million?

(b) Use the model to predict when the total voting-age population will reach 230 million. Is this prediction reasonable? Explain.

194. *Airline Passengers*   An airline offers daily flights between Chicago and Denver. The total monthly cost C (in millions of dollars) of these flights is $C = \sqrt{0.2x + 1}$ where x is the number of passengers (in thousands). The total cost of the flights for June is 2.5 million dollars. How many passengers flew in June?

195. *Demand*   The demand equation for a video game is modeled by $p = 40 - \sqrt{0.01x + 1}$ where x is the number of units demanded per day and p is the price per unit. Approximate the demand when the price is $37.55.

196. *Demand*   The demand equation for a high definition television set is modeled by

$$p = 800 - \sqrt{0.01x + 1}$$

where x is the number of units demanded per month and p is the price per unit. Approximate the demand when the price is $750.

Synthesis

True or False?   **In Exercises 197–200, determine whether the statement is true or false. Justify your answer.**

197. The equation $x(3 - x) = 10$ is a linear equation.

198. If $(2x - 3)(x + 5) = 8$, then either $2x - 3 = 8$ or $x + 5 = 8$.

199. An equation can never have more than one extraneous solution.

200. When solving an absolute value equation, you will always have to check more than one solution.

201. *Think About It*   What is meant by *equivalent equations*? Give an example of two equivalent equations.

202. *Writing*   Describe the steps used to transform an equation into an equivalent equation.

203. To solve the equation $2x^2 + 3x = 15x$, a student divides each side by x and solves the equation $2x + 3 = 15$. The resulting solution ($x = 6$) satisfies the original equation. Is there an error? Explain.

204. Solve $3(x + 4)^2 + (x + 4) - 2 = 0$ in two ways.

(a) Let $u = x + 4$, and solve the resulting equation for u. Then solve the u-solution for x.

(b) Expand and collect like terms in the equation, and solve the resulting equation for x.

(c) Which method is easier? Explain.

Think About It   **In Exercises 205–210, write a quadratic equation that has the given solutions. (There are many correct answers.)**

205. -3 and 6

206. -4 and -11

207. 8 and 14

208. $\frac{1}{6}$ and $-\frac{2}{5}$

209. $1 + \sqrt{2}$ and $1 - \sqrt{2}$

210. $-3 + \sqrt{5}$ and $-3 - \sqrt{5}$

In Exercises 211 and 212, consider an equation of the form $x + |x - a| = b$, where a and b are constants.

211. Find a and b when the solution of the equation is $x = 9$. (There are many correct answers.)

212. *Writing*   Write a short paragraph listing the steps required to solve this equation involving absolute values and explain why it is important to check your solutions.

213. Solve each equation, given that a and b are not zero.

(a) $ax^2 + bx = 0$

(b) $ax^2 - ax = 0$

P.3   **The Cartesian Plane and Graphs of Equations**

What you should learn

- Plot points in the Cartesian plane.
- Use the Distance Formula to find the distance between two points.
- Use the Midpoint Formula to find the midpoint of a line segment.
- Use a coordinate plane to model and solve real-life problems.
- Sketch graphs of equations.
- Find x- and y-intercepts of graphs of equations.
- Use symmetry to sketch graphs of equations.
- Find equations of and sketch graphs of circles.

Why you should learn it

The graph of an equation can help you see relationships between real-life quantities. For example, in Exercise 100 on page 39, a graph can be used to estimate the life expectancies of children who are born in the years 2005 and 2010.

The Cartesian Plane

Just as you can represent real numbers by points on a real number line, you can represent ordered pairs of real numbers by points in a plane called the **rectangular coordinate system,** or the **Cartesian plane,** named after the French mathematician René Descartes (1596–1650).

The Cartesian plane is formed by using two real number lines intersecting at right angles, as shown in Figure P.8. The horizontal real number line is usually called the **x-axis,** and the vertical real number line is usually called the **y-axis.** The point of intersection of these two axes is the **origin,** and the two axes divide the plane into four parts called **quadrants.**

FIGURE P.8   FIGURE P.9

Each point in the plane corresponds to an **ordered pair** (x, y) of real numbers x and y, called **coordinates** of the point. The **x-coordinate** represents the directed distance from the y-axis to the point, and the **y-coordinate** represents the directed distance from the x-axis to the point, as shown in Figure P.9.

The notation (x, y) denotes both a point in the plane and an open interval on the real number line. The context will tell you which meaning is intended.

Example 1   Plotting Points in the Cartesian Plane

Plot the points $(-1, 2)$, $(3, 4)$, $(0, 0)$, $(3, 0)$, and $(-2, -3)$.

Solution

To plot the point $(-1, 2)$, imagine a vertical line through -1 on the x-axis and a horizontal line through 2 on the y-axis. The intersection of these two lines is the point $(-1, 2)$. The other four points can be plotted in a similar way, as shown in Figure P.10.

✓CHECKPOINT   Now try Exercise 3.

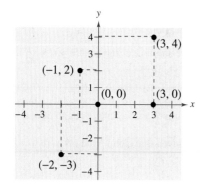

FIGURE P.10

The beauty of a rectangular coordinate system is that it allows you to *see* relationships between two variables. It would be difficult to overestimate the importance of Descartes's introduction of coordinates in the plane. Today, his ideas are in common use in virtually every scientific and business-related field.

Example 2   Sketching a Scatter Plot

From 1990 through 2003, the amounts A (in millions of dollars) spent on skiing equipment in the United States are shown in the table, where t represents the year. Sketch a scatter plot of the data. (Source: National Sporting Goods Association)

Solution

To sketch a *scatter plot* of the data shown in the table, you simply represent each pair of values by an ordered pair (t, A) and plot the resulting points, as shown in Figure P.11. For instance, the first pair of values is represented by the ordered pair $(1990, 475)$. Note that the break in the t-axis indicates that the numbers between 0 and 1990 have been omitted.

Year, t	Amount, A
1990	475
1991	577
1992	521
1993	569
1994	609
1995	562
1996	707
1997	723
1998	718
1999	648
2000	495
2001	476
2002	527
2003	464

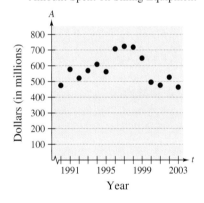

FIGURE P.11

✓CHECKPOINT   Now try Exercise 17.

In Example 2, you could have let $t = 1$ represent the year 1990. In that case, the horizontal axis would not have been broken, and the tick marks would have been labeled 1 through 14 (instead of 1990 through 2003).

Technology

The scatter plot in Example 2 is only one way to represent the data graphically. You could also represent the data using a bar graph and a line graph. If you have access to a graphing utility, try using it to represent graphically the data given in Example 2.

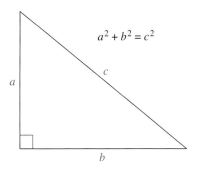

$$a^2 + b^2 = c^2$$

FIGURE P.12

The Distance Formula

Recall from the Pythagorean Theorem that, for a right triangle with hypotenuse of length c and sides of lengths a and b, you have

$$a^2 + b^2 = c^2 \qquad \text{Pythagorean Theorem}$$

as shown in Figure P.12. (The converse is also true. That is, if $a^2 + b^2 = c^2$, then the triangle is a right triangle.)

Suppose you want to determine the distance d between two points (x_1, y_1) and (x_2, y_2) in the plane. With these two points, a right triangle can be formed, as shown in Figure P.13. The length of the vertical side of the triangle is $|y_2 - y_1|$, and the length of the horizontal side is $|x_2 - x_1|$. By the Pythagorean Theorem, you can write

$$d^2 = |x_2 - x_1|^2 + |y_2 - y_1|^2$$
$$d = \sqrt{|x_2 - x_1|^2 + |y_2 - y_1|^2} = \sqrt{(x_2 - x_1)^2 + (y_2 - y_1)^2}.$$

This result is the **Distance Formula.**

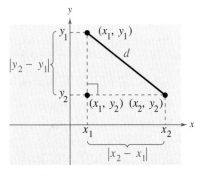

FIGURE P.13

> ### The Distance Formula
> The distance d between the points (x_1, y_1) and (x_2, y_2) in the plane is
> $$d = \sqrt{(x_2 - x_1)^2 + (y_2 - y_1)^2}.$$

Example 3   Finding a Distance

Find the distance between the points $(-2, 1)$ and $(3, 4)$.

Algebraic Solution

Let $(x_1, y_1) = (-2, 1)$ and $(x_2, y_2) = (3, 4)$. Then apply the Distance Formula.

$$d = \sqrt{(x_2 - x_1)^2 + (y_2 - y_1)^2} \qquad \text{Distance Formula}$$
$$= \sqrt{[3 - (-2)]^2 + (4 - 1)^2} \qquad \begin{array}{l}\text{Substitute for}\\ x_1, y_1, x_2, \text{and } y_2.\end{array}$$
$$= \sqrt{(5)^2 + (3)^2} \qquad \text{Simplify.}$$
$$= \sqrt{34} \qquad \text{Simplify.}$$
$$\approx 5.83 \qquad \text{Use a calculator.}$$

So, the distance between the points is about 5.83 units. You can use the Pythagorean Theorem to check that the distance is correct.

$$d^2 \overset{?}{=} 3^2 + 5^2 \qquad \text{Pythagorean Theorem}$$
$$\left(\sqrt{34}\right)^2 \overset{?}{=} 3^2 + 5^2 \qquad \text{Substitute for } d.$$
$$34 = 34 \qquad \text{Distance checks. } \checkmark$$

Graphical Solution

Use centimeter graph paper to plot the points $A(-2, 1)$ and $B(3, 4)$. Carefully sketch the line segment from A to B. Then use a centimeter ruler to measure the length of the segment.

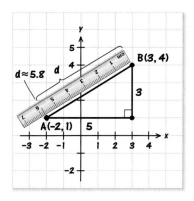

FIGURE P.14

The line segment measures about 5.8 centimeters, as shown in Figure P.14. So, the distance between the points is about 5.8 units.

✓CHECKPOINT   Now try Exercises 25(a) and (b).

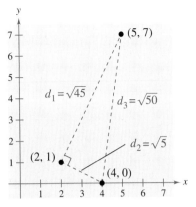

FIGURE P.15

An overhead projector is useful for showing how to plot points and equations. Try projecting a grid onto the chalkboard and then plotting points on the chalkboard, or try using overhead markers and graph directly on the transparency. A viewscreen, a device used with an overhead projector to project a graphing calculator's screen image, is also useful.

Example 4   **Verifying a Right Triangle**

Show that the points $(2, 1)$, $(4, 0)$, and $(5, 7)$ are vertices of a right triangle.

Solution

The three points are plotted in Figure P.15. Using the Distance Formula, you can find the lengths of the three sides as follows.

$$d_1 = \sqrt{(5 - 2)^2 + (7 - 1)^2} = \sqrt{9 + 36} = \sqrt{45}$$

$$d_2 = \sqrt{(4 - 2)^2 + (0 - 1)^2} = \sqrt{4 + 1} = \sqrt{5}$$

$$d_3 = \sqrt{(5 - 4)^2 + (7 - 0)^2} = \sqrt{1 + 49} = \sqrt{50}$$

Because

$$(d_1)^2 + (d_2)^2 = 45 + 5 = 50 = (d_3)^2$$

you can conclude by the Pythagorean Theorem that the triangle must be a right triangle.

✓*CHECKPOINT*   Now try Exercise 35.

The Midpoint Formula

To find the **midpoint** of the line segment that joins two points in a coordinate plane, you can simply find the average values of the respective coordinates of the two endpoints using the **Midpoint Formula.**

The Midpoint Formula

The midpoint of the line segment joining the points (x_1, y_1) and (x_2, y_2) is given by the Midpoint Formula

$$\text{Midpoint} = \left(\frac{x_1 + x_2}{2}, \frac{y_1 + y_2}{2} \right).$$

For a proof of the Midpoint Formula, see Proofs in Mathematics on page 126.

Example 5   **Finding a Line Segment's Midpoint**

Find the midpoint of the line segment joining the points $(-5, -3)$ and $(9, 3)$.

Solution

Let $(x_1, y_1) = (-5, -3)$ and $(x_2, y_2) = (9, 3)$.

$$\text{Midpoint} = \left(\frac{x_1 + x_2}{2}, \frac{y_1 + y_2}{2} \right) \qquad \text{Midpoint Formula}$$

$$= \left(\frac{-5 + 9}{2}, \frac{-3 + 3}{2} \right) \qquad \text{Substitute for } x_1, y_1, x_2, \text{ and } y_2.$$

$$= (2, 0) \qquad \text{Simplify.}$$

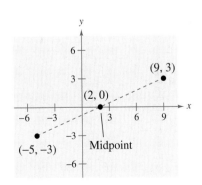

FIGURE P.16

The midpoint of the line segment is $(2, 0)$, as shown in Figure P.16.

✓*CHECKPOINT*   Now try Exercise 25(c).

Applications

Example 6 Finding the Length of a Pass

During the third quarter of the 2004 Sugar Bowl, the quarterback for Louisiana State University threw a pass from the 28-yard line, 40 yards from the sideline. The pass was caught by the wide receiver on the 5-yard line, 20 yards from the same sideline, as shown in Figure P.17. How long was the pass?

Solution

You can find the length of the pass by finding the distance between the points (40, 28) and (20, 5).

$$d = \sqrt{(x_2 - x_1)^2 + (y_2 - y_1)^2} \qquad \text{Distance Formula}$$

$$= \sqrt{(40 - 20)^2 + (28 - 5)^2} \qquad \text{Substitute for } x_1, y_1, x_2, \text{ and } y_2.$$

$$= \sqrt{400 + 529} \qquad \text{Simplify.}$$

$$= \sqrt{929} \qquad \text{Simplify.}$$

$$\approx 30 \qquad \text{Use a calculator.}$$

So, the pass was about 30 yards long.

✓CHECKPOINT Now try Exercise 39.

In Example 6, the scale along the goal line does not normally appear on a football field. However, when you use coordinate geometry to solve real-life problems, you are free to place the coordinate system in any way that is convenient for the solution of the problem.

Example 7 Estimating Annual Revenue

FedEx Corporation had annual revenues of $20.6 billion in 2002 and $24.7 billion in 2004. Without knowing any additional information, what would you estimate the 2003 revenue to have been? (Source: FedEx Corp.)

Solution

One solution to the problem is to assume that revenue followed a linear pattern. With this assumption, you can estimate the 2003 revenue by finding the midpoint of the line segment connecting the points (2002, 20.6) and (2004, 24.7).

$$\text{Midpoint} = \left(\frac{x_1 + x_2}{2}, \frac{y_1 + y_2}{2} \right) \qquad \text{Midpoint Formula}$$

$$= \left(\frac{2002 + 2004}{2}, \frac{20.6 + 24.7}{2} \right) \qquad \text{Substitute for } x_1, y_1, x_2, \text{ and } y_2.$$

$$= (2003, 22.65) \qquad \text{Simplify.}$$

So, you would estimate the 2003 revenue to have been about $22.65 billion, as shown in Figure P.18. (The actual 2003 revenue was $22.5 billion.)

✓CHECKPOINT Now try Exercise 41.

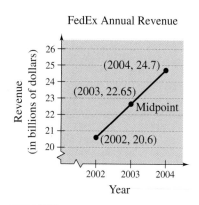

Football Pass

FIGURE P.17

FedEx Annual Revenue

FIGURE P.18

The Graph of an Equation

Earlier in this section, you used a coordinate system to represent graphically the relationship between two quantities. There, the graphical picture consisted of a collection of points in a coordinate plane (see Example 2).

Frequently, a relationship between two quantities is expressed as an **equation in two variables.** For instance, $y = 7 - 3x$ is an equation in x and y. An ordered pair (a, b) is a **solution** or **solution point** of an equation in x and y if the equation is true when a is substituted for x and b is substituted for y. For instance, $(1, 4)$ is a solution of $y = 7 - 3x$ because $4 = 7 - 3(1)$ is a true statement.

In the remainder of this section you will review some basic procedures for sketching the graph of an equation in two variables. The **graph of an equation** is the set of all points that are solutions of the equation. The basic technique used for sketching the graph of an equation is the **point-plotting method.** To sketch a graph using the point-plotting method, first, if possible, rewrite the equation so that one of the variables is isolated on one side of the equation. Next, make a table of values showing several solution points. Then plot the points from your table on a rectangular coordinate system. Finally, connect the points with a smooth curve or line.

Example 8   Sketching the Graph of an Equation

Sketch the graph of $y = x^2 - 2$.
Solution

Because the equation is already solved for y, begin by constructing a table of values.

x	-2	-1	0	1	2	3
$y = x^2 - 2$	2	-1	-2	-1	2	7
(x, y)	$(-2, 2)$	$(-1, -1)$	$(0, -2)$	$(1, -1)$	$(2, 2)$	$(3, 7)$

Next, plot the points given in the table, as shown in Figure P.19. Finally, connect the points with a smooth curve, as shown in Figure P.20.

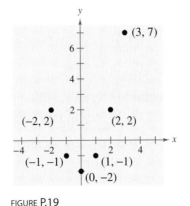

FIGURE P.19   FIGURE P.20

✓CHECKPOINT   Now try Exercise 47.

No x-intercepts; one y-intercept

Three x-intercepts; one y-intercept

One x-intercept; two y-intercepts

No intercepts

FIGURE P.21

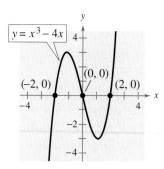

FIGURE P.22

Intercepts of a Graph

It is often easy to determine the solution points that have zero as either the *x*-coordinate or the *y*-coordinate. These points are called **intercepts** because they are the points at which the graph intersects or touches the *x*- or *y*-axis. It is possible for a graph to have no intercepts, one intercept, or several intercepts, as shown in Figure P.21.

Note that an *x*-intercept can be written as the ordered pair $(x, 0)$ and a *y*-intercept can be written as the ordered pair $(0, y)$. Some texts denote the *x*-intercept as the *x*-coordinate of the point $(a, 0)$ [and the *y*-intercept as the *y*-coordinate of the point $(0, b)$] rather than the point itself. Unless it is necessary to make a distinction, we will use the term *intercept* to mean either the point or the coordinate.

Finding Intercepts

1. To find *x*-intercepts, let *y* be zero and solve the equation for *x*.

2. To find *y*-intercepts, let *x* be zero and solve the equation for *y*.

Example 9 **Finding x- and y-Intercepts**

Find the *x*- and *y*-intercepts of the graph of $y = x^3 - 4x$.

Solution

Let $y = 0$. Then

$$0 = x^3 - 4x = x(x^2 - 4)$$

has solutions $x = 0$ and $x = \pm 2$.

x-intercepts: $(0, 0), (2, 0), (-2, 0)$

Let $x = 0$. Then

$$y = (0)^3 - 4(0)$$

has one solution, $y = 0$.

y-intercept: $(0, 0)$ See Figure P.22.

✓CHECKPOINT   Now try Exercise 51.

Symmetry

Graphs of equations can have **symmetry** with respect to one of the coordinate axes or with respect to the origin. Symmetry with respect to the x-axis means that if the Cartesian plane were folded along the x-axis, the portion of the graph above the x-axis would coincide with the portion below the x-axis. Symmetry with respect to the y-axis or the origin can be described in a similar manner, as shown in Figure P.23.

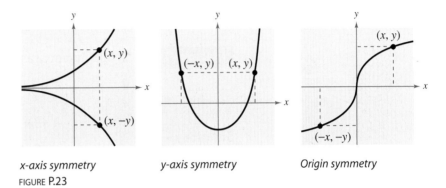

x-axis symmetry   *y-axis symmetry*   *Origin symmetry*

FIGURE P.23

Knowing the symmetry of a graph *before* attempting to sketch it is helpful, because then you need only half as many solution points to sketch the graph. There are three basic types of symmetry, described as follows.

Graphical Tests for Symmetry

1. A graph is **symmetric with respect to the x-axis** if, whenever (x, y) is on the graph, $(x, -y)$ is also on the graph.

2. A graph is **symmetric with respect to the y-axis** if, whenever (x, y) is on the graph, $(-x, y)$ is also on the graph.

3. A graph is **symmetric with respect to the origin** if, whenever (x, y) is on the graph, $(-x, -y)$ is also on the graph.

Example 10   Testing for Symmetry

The graph of $y = x^2 - 2$ is symmetric with respect to the y-axis because the point $(-x, y)$ is also on the graph of $y = x^2 - 2$. (See Figure P.24.) The table below confirms that the graph is symmetric with respect to the y-axis.

x	-3	-2	-1	1	2	3
y	7	2	-1	-1	2	7
(x, y)	$(-3, 7)$	$(-2, 2)$	$(-1, -1)$	$(1, -1)$	$(2, 2)$	$(3, 7)$

✓CHECKPOINT   Now try Exercise 61.

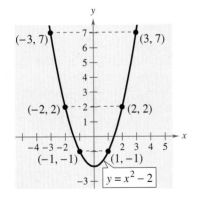

FIGURE P.24   *y-axis symmetry*

Algebraic Tests for Symmetry

1. The graph of an equation is symmetric with respect to the *x*-axis if replacing *y* with $-y$ yields an equivalent equation.

2. The graph of an equation is symmetric with respect to the *y*-axis if replacing *x* with $-x$ yields an equivalent equation.

3. The graph of an equation is symmetric with respect to the origin if replacing *x* with $-x$ and *y* with $-y$ yields an equivalent equation.

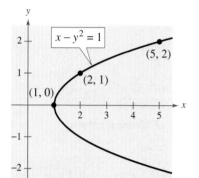

FIGURE P.25

Example 11   Using Symmetry as a Sketching Aid

Use symmetry to sketch the graph of

$$x - y^2 = 1.$$

Solution

Of the three tests for symmetry, the only one that is satisfied is the test for *x*-axis symmetry because $x - (-y)^2 = 1$ is equivalent to $x - y^2 = 1$. So, the graph is symmetric with respect to the *x*-axis. Using symmetry, you only need to find the solution points above the *x*-axis and then reflect them to obtain the graph, as shown in Figure P.25.

y	$x = y^2 + 1$	(x, y)
0	1	$(1, 0)$
1	2	$(2, 1)$
2	5	$(5, 2)$

✓CHECKPOINT   Now try Exercise 77.

STUDY TIP

Notice that when creating the table in Example 11, it is easier to choose *y*-values and then find the corresponding *x*-values of the ordered pairs.

Example 12   Sketching the Graph of an Equation

Sketch the graph of

$$y = |x - 1|.$$

Solution

This equation fails all three tests for symmetry and consequently its graph is not symmetric with respect to either axis or to the origin. The absolute value sign indicates that *y* is always nonnegative. Create a table of values and plot the points as shown in Figure P.26. From the table, you can see that $x = 0$ when $y = 1$. So, the *y*-intercept is $(0, 1)$. Similarly, $y = 0$ when $x = 1$. So, the *x*-intercept is $(1, 0)$.

x	-2	-1	0	1	2	3	4
$y = |x - 1|$	3	2	1	0	1	2	3
(x, y)	$(-2, 3)$	$(-1, 2)$	$(0, 1)$	$(1, 0)$	$(2, 1)$	$(3, 2)$	$(4, 3)$

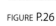

FIGURE P.26

✓CHECKPOINT   Now try Exercise 81.

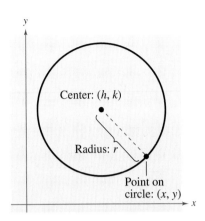

FIGURE P.27

Throughout this course, you will learn to recognize several types of graphs from their equations. For instance, you will learn to recognize that the graph of a second-degree equation of the form

$$y = ax^2 + bx + c$$

is a parabola (see Example 8). The graph of a **circle** is also easy to recognize.

Circles

Consider the circle shown in Figure P.27. A point (x, y) is on the circle if and only if its distance from the center (h, k) is r. By the Distance Formula,

$$\sqrt{(x - h)^2 + (y - k)^2} = r.$$

By squaring each side of this equation, you obtain the **standard form of the equation of a circle.**

> ### Standard Form of the Equation of a Circle
> The point (x, y) lies on the circle of **radius** r and **center** (h, k) if and only if
> $$(x - h)^2 + (y - k)^2 = r^2.$$

From this result, you can see that the standard form of the equation of a circle *with its center at the origin,* $(h, k) = (0, 0)$, is simply

$$x^2 + y^2 = r^2. \qquad \text{Circle with center at origin}$$

STUDY TIP

To find the correct h and k, from the equation of the circle in Example 13, it may be helpful to rewrite the quantities $(x + 1)^2$ and $(y - 2)^2$, using subtraction.

$$(x + 1)^2 = [x - (-1)]^2,$$
$$(y - 2)^2 = [y - (2)]^2$$

So, $h = -1$ and $k = 2$.

Example 13   **Finding the Equation of a Circle**

The point $(3, 4)$ lies on a circle whose center is at $(-1, 2)$, as shown in Figure P.28. Write the standard form of the equation of this circle.

Solution

The radius of the circle is the distance between $(-1, 2)$ and $(3, 4)$.

$$
\begin{aligned}
r &= \sqrt{(x - h)^2 + (y - k)^2} & \text{Distance Formula} \\
&= \sqrt{[3 - (-1)]^2 + (4 - 2)^2} & \text{Substitute for } x, y, h, \text{ and } k. \\
&= \sqrt{4^2 + 2^2} & \text{Simplify.} \\
&= \sqrt{16 + 4} & \text{Simplify.} \\
&= \sqrt{20} & \text{Radius}
\end{aligned}
$$

Using $(h, k) = (-1, 2)$ and $r = \sqrt{20}$, the equation of the circle is

$$
\begin{aligned}
(x - h)^2 + (y - k)^2 &= r^2 & \text{Equation of circle} \\
[x - (-1)]^2 + (y - 2)^2 &= \left(\sqrt{20}\right)^2 & \text{Substitute for } h, k, \text{ and } r. \\
(x + 1)^2 + (y - 2)^2 &= 20. & \text{Standard form}
\end{aligned}
$$

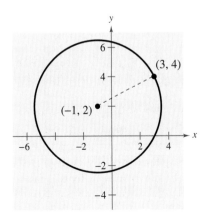

FIGURE P.28

✓CHECKPOINT   Now try Exercise 87.

P.3 | Exercises

VOCABULARY CHECK: Fill in the blanks.

1. An ordered pair of real numbers can be represented in a plane called the rectangular coordinate system or the _____ plane.

2. The _____ _____ is a result derived from the Pythagorean Theorem.

3. Finding the average values of the respective coordinates of the two endpoints of a line segment in a coordinate plane is also known as using the _____ _____.

4. An ordered pair (a, b) is a _____ of an equation in x and y if the equation is true when a is substituted for x and b is substituted for y.

5. The set of all solution points of an equation is the _____ of the equation.

6. The points at which a graph intersects or touches an axis are called the _____ of the graph.

7. A graph is symmetric with respect to the _____ if, whenever (x, y) is on the graph, $(-x, y)$ is also on the graph.

8. The equation $(x - h)^2 + (y - k)^2 = r^2$ is the standard form of the equation of a _____ with center _____ and radius _____.

In Exercises 1 and 2, approximate the coordinates of the points.

1.

2.
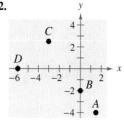

In Exercises 3–6, plot the points in the Cartesian plane.

3. $(-4, 2)$, $(-3, -6)$, $(0, 5)$, $(1, -4)$

4. $(0, 0)$, $(3, 1)$, $(-2, 4)$, $(1, -1)$

5. $(3, 8)$, $(0.5, -1)$, $(5, -6)$, $(-2, 2.5)$

6. $\left(1, -\frac{1}{3}\right)$, $\left(\frac{3}{4}, 3\right)$, $(-3, 4)$, $\left(-\frac{4}{3}, -\frac{3}{2}\right)$

In Exercises 7 and 8, find the coordinates of the point.

7. The point is located three units to the left of the y-axis and four units above the x-axis.

8. The point is located eight units below the x-axis and four units to the right of the y-axis.

In Exercises 9–16, determine the quadrant(s) in which (x, y) is located so that the condition(s) is (are) satisfied.

9. $x > 0$ and $y < 0$

10. $x < 0$ and $y < 0$

11. $x = -4$ and $y > 0$

12. $x > 2$ and $y = 3$

13. $x < 0$ and $-y > 0$

14. $-x > 0$ and $y < 0$

15. $xy > 0$

16. $xy < 0$

In Exercises 17 and 18, sketch a scatter plot of the data.

17. **Number of Stores** The table shows the number y of Wal-Mart stores for each year x from 1996 through 2003. (Source: Wal-Mart Stores, Inc.)

Year, x	Number of stores, y
1996	3054
1997	3406
1998	3599
1999	3985
2000	4189
2001	4414
2002	4688
2003	4906

18. **Meteorology** The lowest temperature on record y (in degrees Fahrenheit) in Duluth, Minnesota, for each month x, where $x = 1$ represents January, are shown as data points (x, y). (Source: NOAA)

$(1, -39)$, $(2, -39)$, $(3, -29)$, $(4, -5)$, $(5, 17)$, $(6, 27)$, $(7, 35)$, $(8, 32)$, $(9, 22)$, $(10, 8)$, $(11, -23)$, $(12, -34)$

In Exercises 19–22, find the distance between the points. (*Note:* In each case, the two points lie on the same horizontal or vertical line.)

19. $(6, -3), (6, 5)$ **20.** $(1, 4), (8, 4)$

21. $(-3, -1), (2, -1)$ **22.** $(-3, -4), (-3, 6)$

In Exercises 23 and 24, (a) find the length of each side of the right triangle, and (b) show that these lengths satisfy the Pythagorean Theorem.

23.

24.

In Exercises 25–34, (a) plot the points, (b) find the distance between the points, and (c) find the midpoint of the line segment joining the points.

25. $(1, 1), (9, 7)$ **26.** $(1, 12), (6, 0)$

27. $(-4, 10), (4, -5)$ **28.** $(-7, -4), (2, 8)$

29. $(-1, 2), (5, 4)$ **30.** $(2, 10), (10, 2)$

31. $\left(\frac{1}{2}, 1\right), \left(-\frac{5}{2}, \frac{4}{3}\right)$ **32.** $\left(-\frac{1}{3}, -\frac{1}{3}\right), \left(-\frac{1}{6}, -\frac{1}{2}\right)$

33. $(6.2, 5.4), (-3.7, 1.8)$

34. $(-16.8, 12.3), (5.6, 4.9)$

In Exercises 35 and 36, show that the points form the vertices of the indicated polygon.

35. Right triangle: $(4, 0), (2, 1), (-1, -5)$

36. Isosceles triangle: $(1, -3), (3, 2), (-2, 4)$

Advertising   In Exercises 37 and 38, use the graph below, which shows the costs of a 30-second television spot (in thousands of dollars) during the Super Bowl from 1989 to 2003. (Source: USA Today Research and CNN)

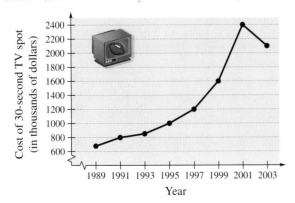

37. Approximate the percent increase in the cost of a 30-second spot from Super Bowl XXIII in 1989 to Super Bowl XXXV in 2001.

38. Estimate the percent increase in the cost of a 30-second spot (a) from Super Bowl XXIII in 1989 to Super Bowl XXVII in 1993 and (b) from Super Bowl XXVII in 1993 to Super Bowl XXXVII in 2003.

39. *Sports*   A soccer player passes the ball from a point that is 18 yards from the endline and 12 yards from the sideline. The pass is received by a teammate who is 42 yards from the same endline and 50 yards from the same sideline, as shown in the figure. How long is the pass?

40. *Flying Distance*   An airplane flies from Naples, Italy in a straight line to Rome, Italy, which is 120 kilometers north and 150 kilometers west of Naples. How far does the plane fly?

41. *Sales*   Pepsi Bottling Group, Inc. had sales of $6603 million in 1996 and $10,800 million in 2004. Use the Midpoint Formula to estimate the sales in 1998, 2000, and 2002. Assume that the sales followed a linear pattern. (Source: Pepsi Bottling Group, Inc.)

42. *Sales*   The Coca-Cola Company had sales of $18,546 million in 1996 and $21,900 million in 2004. Use the Midpoint Formula to estimate the sales in 1998, 2000, and 2002. Assume that the sales followed a linear pattern. (Source: The Coca-Cola Company)

In Exercises 43–46, determine whether each point lies on the graph of the equation.

	Equation		*Points*	
43.	$y = \sqrt{x + 4}$	(a) $(0, 2)$	(b) $(5, 3)$
44.	$y = x^2 - 3x + 2$	(a) $(2, 0)$	(b) $(-2, 8)$
45.	$y = 4 - |x - 2|$	(a) $(1, 5)$	(b) $(6, 0)$
46.	$y = \frac{1}{3}x^3 - 2x^2$	(a) $\left(2, -\frac{16}{3}\right)$	(b) $(-3, 9)$

In Exercises 47–48, complete the table. Use the resulting solution points to sketch the graph of the equation.

47. $y = -2x + 5$

x	-1	0	1	2	$\frac{5}{2}$
y					
(x, y)					

48. $y = x^2 - 3x$

x	-1	0	1	2	3
y					
(x, y)					

In Exercises 49–60, find the x- and y-intercepts of the graph of the equation.

49. $y = 16 - 4x^2$　　　**50.** $y = (x + 3)^2$

51. $y = 5x - 6$　　　**52.** $y = 8 - 3x$

53. $y = \sqrt{x + 4}$　　　**54.** $y = \sqrt{2x - 1}$

55. $y = |3x - 7|$　　　**56.** $y = -|x + 10|$

57. $y = 2x^3 - 4x^2$　　　**58.** $y = x^4 - 25$

59. $y^2 = 6 - x$　　　**60.** $y^2 = x + 1$

In Exercises 61–64, assume that the graph has the indicated type of symmetry. Sketch the complete graph of the equation. To print an enlarged copy of the graph, go to the website www.mathgraphs.com.

61.

y-Axis symmetry

62.

x-Axis symmetry

63.

Origin symmetry

64.

y-Axis symmetry

In Exercises 65–72, use the algebraic tests to check for symmetry with respect to both axes and the origin.

65. $x^2 - y = 0$　　　**66.** $x - y^2 = 0$

67. $y = x^3$　　　**68.** $y = x^4 - x^2 + 3$

69. $y = \dfrac{x}{x^2 + 1}$　　　**70.** $y = \dfrac{1}{x^2 + 1}$

71. $xy^2 + 10 = 0$　　　**72.** $xy = 4$

In Exercises 73–84, use symmetry to sketch the graph of the equation.

73. $y = -3x + 1$　　　**74.** $y = 2x - 3$

75. $y = x^2 - 2x$　　　**76.** $y = -x^2 - 2x$

77. $y = x^3 + 3$　　　**78.** $y = x^3 - 1$

79. $y = \sqrt{x - 3}$　　　**80.** $y = \sqrt{1 - x}$

81. $y = |x - 6|$　　　**82.** $y = 1 - |x|$

83. $x = y^2 - 1$　　　**84.** $x = y^2 - 5$

In Exercises 85–90, write the standard form of the equation of the circle with the given characteristics.

85. Center: $(2, -1)$; radius: 4

86. Center: $(-7, -4)$; radius: 7

87. Center: $(-1, 2)$; solution point: $(0, 0)$

88. Center: $(3, -2)$; solution point: $(-1, 1)$

89. Endpoints of a diameter: $(0, 0)$, $(6, 8)$

90. Endpoints of a diameter: $(-4, -1)$, $(4, 1)$

In Exercises 91–96, find the center and radius of the circle, and sketch its graph.

91. $x^2 + y^2 = 25$　　　**92.** $x^2 + y^2 = 16$

93. $(x - 1)^2 + (y + 3)^2 = 9$　**94.** $x^2 + (y - 1)^2 = 1$

95. $\left(x - \frac{1}{2}\right)^2 + \left(y - \frac{1}{2}\right)^2 = \frac{9}{4}$　**96.** $(x - 2)^2 + (y + 3)^2 = \frac{16}{9}$

97. *Depreciation* A manufacturing plant purchases a new molding machine for $225,000. The depreciated value y (drop in value) after t years is given by $y = 225{,}000 - 20{,}000t$, $0 \le t \le 8$. Sketch the graph of the equation.

98. *Consumerism* You purchase a jet ski for $8100. The depreciated value y after t years is given by $y = 8100 - 929t$, $0 \le t \le 6$. Sketch the graph of the equation.

99. *Electronics* The resistance y (in ohms) of 1000 feet of solid copper wire at 68 degrees Fahrenheit can be approximated by the model $y = \dfrac{10{,}770}{x^2} - 0.37$, $5 \le x \le 100$ where x is the diameter of the wire in mils (0.001 inch). (Source: American Wire Gage)

(a) Complete the table.

x	5	10	20	30	40	50
y						

x	60	70	80	90	100
y					

(b) Use the table of values in part (a) to sketch a graph of the model. Then use your graph to estimate the resistance when $x = 85.5$.

(c) Use the model to confirm algebraically the estimate you found in part (b).

(d) What can you conclude in general about the relationship between the diameter of the copper wire and the resistance?

Model It

100. Population Statistics   The table shows the life expectancies of a child (at birth) in the United States for selected years from 1920 to 2000.   (Source: U.S. National Center for Health Statistics)

Year	Life expectancy, y
1920	54.1
1930	59.7
1940	62.9
1950	68.2
1960	69.7
1970	70.8
1980	73.7
1990	75.4
2000	77.0

A model for the life expectancy during this period is
$$y = -0.0025t^2 + 0.574t + 44.25, \quad 20 \le t \le 100,$$
where y represents the life expectancy and t is the time in years, with $t = 20$ corresponding to 1920.

(a) Sketch a scatter plot of the data.

(b) Graph the model for the data and compare the scatter plot and the graph.

(c) Determine the life expectancy in 1948 both graphically and algebraically.

(d) Use the graph of the model to estimate the life expectancies of a child for the years 2005 and 2010.

(e) Do you think this model can be used to predict the life expectancy of a child 50 years from now? Explain.

Synthesis

True or False?   In Exercises 101–104, determine whether the statement is true or false. Justify your answer.

101. In order to divide a line segment into 16 equal parts, you would have to use the Midpoint Formula 16 times.

102. The points $(-8, 4)$, $(2, 11)$, and $(-5, 1)$ represent the vertices of an isosceles triangle.

103. A graph is symmetric with respect to the x-axis if, whenever (x, y) is on the graph, $(-x, y)$ is also on the graph.

104. A graph of an equation can have more than one y-intercept.

105. Think About It   What is the y-coordinate of any point on the x-axis? What is the x-coordinate of any point on the y-axis?

106. Think About It   When plotting points on the rectangular coordinate system, is it true that the scales on the x- and y-axes must be the same? Explain.

107. Proof   Prove that the diagonals of the parallelogram in the figure intersect at their midpoints.

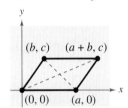

108. Think About It   Suppose you correctly enter an expression for the variable y on a graphing utility. However, no graph appears on the display when you graph the equation. Give a possible explanation and the steps you could take to remedy the problem. Illustrate your explanation with an example.

109. Think About It   Find a and b if the graph of $y = ax^2 + bx^3$ is symmetric with respect to (a) the y-axis and (b) the origin. (There are many correct answers.)

110. Make a Conjecture   Plot the points $(2, 1)$, $(-3, 5)$, and $(7, -3)$ on a rectangular coordinate system. Then change the sign of the x-coordinate of each point and plot the three new points on the same rectangular coordinate system. Make a conjecture about the location of a point when each of the following occurs.

(a) The sign of the x-coordinate is changed.

(b) The sign of the y-coordinate is changed.

(c) The signs of both the x- and y-coordinates are changed.

P.4   Linear Equations in Two Variables

What you should learn

- Use slope to graph linear equations in two variables.
- Find slopes of lines.
- Write linear equations in two variables.
- Use slope to identify parallel and perpendicular lines.
- Use slope and linear equations in two variables to model and solve real-life problems.

Why you should learn it

Linear equations in two variables can be used to model and solve real-life problems. For instance, in Exercise 109 on page 52, you will use a linear equation to model student enrollment at the Pennsylvania State University.

Courtesy of Pennsylvania State University

The *HM mathSpace*® CD-ROM and *Eduspace*® contain additional resources related to the concepts discussed in this chapter.

Using Slope

The simplest mathematical model for relating two variables is the **linear equation in two variables** $y = mx + b$. The equation is called *linear* because its graph is a line. (In mathematics, the term *line* means *straight line*.) By letting $x = 0$, you can see that the line crosses the y-axis at $y = b$, as shown in Figure P.29. In other words, the y-intercept is $(0, b)$. The steepness or slope of the line is m.

$$y = mx + b$$

Slope ⎯⎯   ⎯⎯ y-Intercept

The **slope** of a nonvertical line is the number of units the line rises (or falls) vertically for each unit of horizontal change from left to right, as shown in Figure P.29 and Figure P.30.

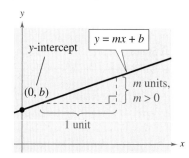

Positive slope, line rises.
FIGURE P.29

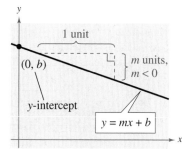

Negative slope, line falls.
FIGURE P.30

A linear equation that is written in the form $y = mx + b$ is said to be written in **slope-intercept form.**

The Slope-Intercept Form of the Equation of a Line

The graph of the equation

$$y = mx + b$$

is a line whose slope is m and whose y-intercept is $(0, b)$.

Exploration

Use a graphing utility to compare the slopes of the lines $y = mx$, where $m = 0.5, 1, 2,$ and 4. Which line rises most quickly? Now, let $m = -0.5, -1, -2,$ and -4. Which line falls most quickly? Use a square setting to obtain a true geometric perspective. What can you conclude about the slope and the "rate" at which the line rises or falls?

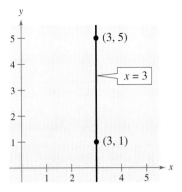

FIGURE P.31   *Slope is undefined.*

Common Error

Many students confuse the line $x = a$ with the point $x = a$ on the real number line, or the line $y = b$ with the point $y = b$. Point out to students that they need to be aware of the context in which $x = a$ or $y = b$ is presented to know whether it refers to the line in the plane or the point on the real number line.

Once you have determined the slope and the y-intercept of a line, it is a relatively simple matter to sketch its graph. In the next example, note that none of the lines is vertical. A vertical line has an equation of the form

$$x = a. \qquad \text{Vertical line}$$

The equation of a vertical line cannot be written in the form $y = mx + b$ because the slope of a vertical line is undefined, as indicated in Figure P.31.

Example 1   **Graphing a Linear Equation**

Sketch the graph of each linear equation.

a. $y = 2x + 1$

b. $y = 2$

c. $x + y = 2$

Solution

a. Because $b = 1$, the y-intercept is $(0, 1)$. Moreover, because the slope is $m = 2$, the line *rises* two units for each unit the line moves to the right, as shown in Figure P.32.

b. By writing this equation in the form $y = (0)x + 2$, you can see that the y-intercept is $(0, 2)$ and the slope is zero. A zero slope implies that the line is horizontal—that is, it doesn't rise *or* fall, as shown in Figure P.33.

c. By writing this equation in slope-intercept form

$$x + y = 2 \qquad \text{Write original equation.}$$

$$y = -x + 2 \qquad \text{Subtract } x \text{ from each side.}$$

$$y = (-1)x + 2 \qquad \text{Write in slope-intercept form.}$$

you can see that the y-intercept is $(0, 2)$. Moreover, because the slope is $m = -1$, the line *falls* one unit for each unit the line moves to the right, as shown in Figure P.34.

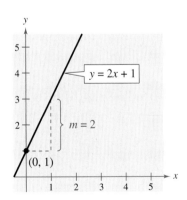

When m is positive, the line rises.
FIGURE P.32

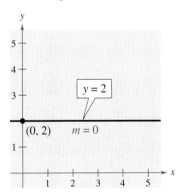

When m is 0, the line is horizontal.
FIGURE P.33

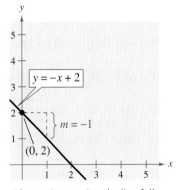

When m is negative, the line falls.
FIGURE P.34

✓CHECKPOINT   Now try Exercise 9.

Finding the Slope of a Line

Given an equation of a line, you can find its slope by writing the equation in slope-intercept form. If you are not given an equation, you can still find the slope of a line. For instance, suppose you want to find the slope of the line passing through the points (x_1, y_1) and (x_2, y_2), as shown in Figure P.35. As you move from left to right along this line, a change of $(y_2 - y_1)$ units in the vertical direction corresponds to a change of $(x_2 - x_1)$ units in the horizontal direction.

$$y_2 - y_1 = \text{the change in } y = \text{rise}$$

and

$$x_2 - x_1 = \text{the change in } x = \text{run}$$

The ratio of $(y_2 - y_1)$ to $(x_2 - x_1)$ represents the slope of the line that passes through the points (x_1, y_1) and (x_2, y_2).

$$\text{Slope} = \frac{\text{change in } y}{\text{change in } x}$$

$$= \frac{\text{rise}}{\text{run}}$$

$$= \frac{y_2 - y_1}{x_2 - x_1}$$

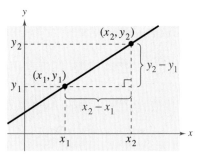

FIGURE P.35

The Slope of a Line Passing Through Two Points

The **slope** m of the nonvertical line through (x_1, y_1) and (x_2, y_2) is

$$m = \frac{y_2 - y_1}{x_2 - x_1}$$

where $x_1 \neq x_2$.

When this formula is used for slope, the *order of subtraction* is important. Given two points on a line, you are free to label either one of them as (x_1, y_1) and the other as (x_2, y_2). However, once you have done this, you must form the numerator and denominator using the same order of subtraction.

$$m = \frac{y_2 - y_1}{x_2 - x_1} \qquad m = \frac{y_1 - y_2}{x_1 - x_2} \qquad m = \frac{y_2 - y_1}{x_1 - x_2}$$

Correct   Correct   Incorrect

For instance, the slope of the line passing through the points $(3, 4)$ and $(5, 7)$ can be calculated as

$$m = \frac{7 - 4}{5 - 3} = \frac{3}{2}$$

or, reversing the subtraction order in both the numerator and denominator, as

$$m = \frac{4 - 7}{3 - 5} = \frac{-3}{-2} = \frac{3}{2}.$$

Example 2   Finding the Slope of a Line Through Two Points

Find the slope of the line passing through each pair of points.

a. $(-2,\ 0)$ and $(3, 1)$ **b.** $(-1,\ 2)$ and $(2, 2)$

c. $(0, 4)$ and $(1, -1)$ **d.** $(3, 4)$ and $(3, 1)$

Solution

a. Letting $(x_1, y_1) = (-2,\ 0)$ and $(x_2, y_2) = (3, 1)$, you obtain a slope of

$$m = \frac{y_2 - y_1}{x_2 - x_1} = \frac{1 - 0}{3 - (-2)} = \frac{1}{5}.$$ See Figure P.36.

b. The slope of the line passing through $(-1, 2)$ and $(2, 2)$ is

$$m = \frac{2 - 2}{2 - (-1)} = \frac{0}{3} = 0.$$ See Figure P.37.

c. The slope of the line passing through $(0, 4)$ and $(1, -1)$ is

$$m = \frac{-1 - 4}{1 - 0} = \frac{-5}{1} = -5.$$ See Figure P.38.

d. The slope of the line passing through $(3, 4)$ and $(3, 1)$ is

$$m = \frac{1 - 4}{3 - 3} = \frac{-3}{0}.$$ See Figure P.39.

Because division by 0 is undefined, the slope is undefined and the line is vertical.

Common Error

A common error when finding the slope of a line is combining x- and y-coordinates in either the numerator or denominator, or both, as in

$$m = \frac{y_2 - x_1}{x_2 - y_1}.$$

STUDY TIP

In Figures P.36 to P.39, note the relationships between slope and the orientation of the line.

a. Positive slope: line rises from left to right

b. Zero slope: line is horizontal

c. Negative slope: line falls from left to right

d. Undefined slope: line is vertical

FIGURE **P.36**

FIGURE **P.37**

FIGURE **P.38**

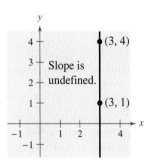

FIGURE **P.39**

✓CHECKPOINT   Now try Exercise 21.

Writing Linear Equations in Two Variables

If (x_1, y_1) is a point on a line of slope m and (x, y) is *any other* point on the line, then

$$\frac{y - y_1}{x - x_1} = m.$$

This equation, involving the variables x and y, can be rewritten in the form

$$y - y_1 = m(x - x_1)$$

which is the **point-slope form** of the equation of a line.

> **Point-Slope Form of the Equation of a Line**
>
> The equation of the line with slope m passing through the point (x_1, y_1) is
>
> $$y - y_1 = m(x - x_1).$$

The point-slope form is most useful for *finding* the equation of a line. You should remember this form.

Example 3   Using the Point-Slope Form

Find the slope-intercept form of the equation of the line that has a slope of 3 and passes through the point $(1, -2)$.

Solution

Use the point-slope form with $m = 3$ and $(x_1, y_1) = (1, -2)$.

$$y - y_1 = m(x - x_1) \qquad \text{Point-slope form}$$
$$y - (-2) = 3(x - 1) \qquad \text{Substitute for } m, x_1, \text{ and } y_1.$$
$$y + 2 = 3x - 3 \qquad \text{Simplify.}$$
$$y = 3x - 5 \qquad \text{Write in slope-intercept form.}$$

The slope-intercept form of the equation of the line is $y = 3x - 5$. The graph of this line is shown in Figure P.40.

✓*CHECKPOINT*   Now try Exercise 39.

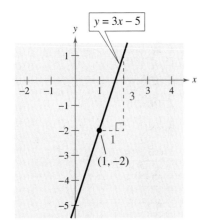

FIGURE P.40

The point-slope form can be used to find an equation of the line passing through two points (x_1, y_1) and (x_2, y_2). To do this, first find the slope of the line

$$m = \frac{y_2 - y_1}{x_2 - x_1}, \qquad x_1 \neq x_2$$

and then use the point-slope form to obtain the equation

$$y - y_1 = \frac{y_2 - y_1}{x_2 - x_1}(x - x_1). \qquad \text{Two-point form}$$

This is sometimes called the **two-point form** of the equation of a line.

STUDY TIP

When you find an equation of the line that passes through two given points, you only need to substitute the coordinates of one of the points into the point-slope form. It does not matter which point you choose because both points will yield the same result.

Find d_1 and d_2 in terms of m_1 and m_2, respectively (see figure). Then use the Pythagorean Theorem to find a relationship between m_1 and m_2.

Parallel and Perpendicular Lines

Slope can be used to decide whether two nonvertical lines in a plane are parallel, perpendicular, or neither.

Parallel and Perpendicular Lines

1. Two distinct nonvertical lines are **parallel** if and only if their slopes are equal. That is, $m_1 = m_2$.

2. Two nonvertical lines are **perpendicular** if and only if their slopes are negative reciprocals of each other. That is, $m_1 = -1/m_2$.

Example 4   **Finding Parallel and Perpendicular Lines**

Find the slope-intercept forms of the equations of the lines that pass through the point $(2, -1)$ and are (a) parallel to and (b) perpendicular to the line $2x - 3y = 5$.

Solution

By writing the equation of the given line in slope-intercept form

$2x - 3y = 5$	Write original equation.
$-3y = -2x + 5$	Subtract $2x$ from each side.
$y = \frac{2}{3}x - \frac{5}{3}$	Write in slope-intercept form.

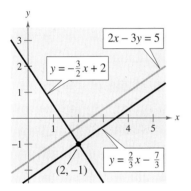

FIGURE **P.41**

you can see that it has a slope of $m = \frac{2}{3}$, as shown in Figure P.41.

a. Any line parallel to the given line must also have a slope of $\frac{2}{3}$. So, the line through $(2, -1)$ that is parallel to the given line has the following equation.

$y - (-1) = \frac{2}{3}(x - 2)$	Write in point-slope form.
$3(y + 1) = 2(x - 2)$	Multiply each side by 3.
$3y + 3 = 2x - 4$	Distributive Property
$y = \frac{2}{3}x - \frac{7}{3}$	Write in slope-intercept form.

b. Any line perpendicular to the given line must have a slope of $-\frac{3}{2}$ (because $-\frac{3}{2}$ is the negative reciprocal of $\frac{2}{3}$). So, the line through $(2, -1)$ that is perpendicular to the given line has the following equation.

$y - (-1) = -\frac{3}{2}(x - 2)$	Write in point-slope form.
$2(y + 1) = -3(x - 2)$	Multiply each side by 2.
$2y + 2 = -3x + 6$	Distributive Property
$y = -\frac{3}{2}x + 2$	Write in slope-intercept form.

✓CHECKPOINT   Now try Exercise 69.

Technology

On a graphing utility, lines will not appear to have the correct slope unless you use a viewing window that has a square setting. For instance, try graphing the lines in Example 4 using the standard setting $-10 \le x \le 10$ and $-10 \le y \le 10$. Then reset the viewing window with the square setting $-9 \le x \le 9$ and $-6 \le y \le 6$. On which setting do the lines $y = \frac{2}{3}x - \frac{5}{3}$ and $y = -\frac{3}{2}x + 2$ appear to be perpendicular?

Notice in Example 4 how the slope-intercept form is used to obtain information about the graph of a line, whereas the point-slope form is used to write the equation of a line.

Applications

In real-life problems, the slope of a line can be interpreted as either a *ratio* or a *rate*. If the *x*-axis and *y*-axis have the same unit of measure, then the slope has no units and is a **ratio.** If the *x*-axis and *y*-axis have different units of measure, then the slope is a **rate** or **rate of change.**

Example 5   Using Slope as a Ratio

The maximum recommended slope of a wheelchair ramp is $\frac{1}{12}$. A business is installing a wheelchair ramp that rises 22 inches over a horizontal length of 24 feet. Is the ramp steeper than recommended? (Source: *Americans with Disabilities Act Handbook*)

Solution

The horizontal length of the ramp is 24 feet or $12(24) = 288$ inches, as shown in Figure P.42. So, the slope of the ramp is

$$\text{Slope} = \frac{\text{vertical change}}{\text{horizontal change}} = \frac{22 \text{ in.}}{288 \text{ in.}} \approx 0.076.$$

Because $\frac{1}{12} \approx 0.083$, the slope of the ramp is not steeper than recommended.

22 in.

24 ft

FIGURE P.42

✓CHECKPOINT   Now try Exercise 97.

Example 6   Using Slope as a Rate of Change

A kitchen appliance manufacturing company determines that the total cost in dollars of producing *x* units of a blender is

$$C = 25x + 3500. \qquad \text{Cost equation}$$

Describe the practical significance of the *y*-intercept and slope of this line.

Solution

The *y*-intercept $(0, 3500)$ tells you that the cost of producing zero units is \$3500. This is the *fixed cost* of production—it includes costs that must be paid regardless of the number of units produced. The slope of $m = 25$ tells you that the cost of producing each unit is \$25, as shown in Figure P.43. Economists call the cost per unit the *marginal cost*. If the production increases by one unit, then the "margin," or extra amount of cost, is \$25. So, the cost increases at a rate of \$25 per unit.

✓CHECKPOINT   Now try Exercise 101.

Manufacturing

$C = 25x + 3500$

Marginal cost:
$m = \$25$

Fixed cost: \$3500

Cost (in dollars)

10,000
9,000
8,000
7,000
6,000
5,000
4,000
3,000
2,000
1,000

50   100   150

Number of units

FIGURE P.43   *Production cost*

Most business expenses can be deducted in the same year they occur. One exception is the cost of property that has a useful life of more than 1 year. Such costs must be *depreciated* (decreased in value) over the useful life of the property. If the *same amount* is depreciated each year, the procedure is called *linear* or *straight-line depreciation*. The *book value* is the difference between the original value and the total amount of depreciation accumulated to date.

Example 7   Straight-Line Depreciation

A college purchased exercise equipment worth $12,000 for the new campus fitness center. The equipment has a useful life of 8 years. The salvage value at the end of 8 years is $2000. Write a linear equation that describes the book value of the equipment each year.

Solution

Let V represent the value of the equipment at the end of year t. You can represent the initial value of the equipment by the data point $(0, 12{,}000)$ and the salvage value of the equipment by the data point $(8, 2000)$. The slope of the line is

$$m = \frac{2000 - 12{,}000}{8 - 0} = -\$1250$$

which represents the annual depreciation in *dollars per year*. Using the point-slope form, you can write the equation of the line as follows.

$$V - 12{,}000 = -1250(t - 0) \qquad \text{Write in point-slope form.}$$

$$V = -1250t + 12{,}000 \qquad \text{Write in slope-intercept form.}$$

The table shows the book value at the end of each year, and the graph of the equation is shown in Figure P.44.

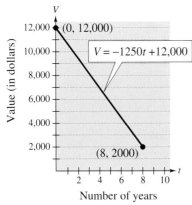

Useful Life of Equipment

$V = -1250t + 12{,}000$

FIGURE P.44   *Straight-line depreciation*

Year, t	Value, V
0	12,000
1	10,750
2	9,500
3	8,250
4	7,000
5	5,750
6	4,500
7	3,250
8	2,000

✔CHECKPOINT   Now try Exercise 107.

In many real-life applications, the two data points that determine the line are often given in a *disguised* form. Note how the data points are described in Example 7.

Example 8   Predicting Sales per Share

The sales per share for Starbucks Corporation were $6.97 in 2001 and $8.47 in 2002. Using only this information, write a linear equation that gives the sales per share in terms of the year. Then predict the sales per share for 2003.   (Source: Starbucks Corporation)

Solution

Let $t = 1$ represent 2001. Then the two given values are represented by the data points (1, 6.97) and (2, 8.47). The slope of the line through these points is

$$m = \frac{8.47 - 6.97}{2 - 1}$$

$$= 1.5.$$

Using the point-slope form, you can find the equation that relates the sales per share y and the year t to be

$$y - 6.97 = 1.5(t - 1) \qquad \text{Write in point-slope form.}$$

$$y = 1.5t + 5.47. \qquad \text{Write in slope-intercept form.}$$

According to this equation, the sales per share in 2003 was $y = 1.5(3) + 5.47 = \$9.97$, as shown in Figure P.45. (In this case, the prediction is quite good—the actual sales per share in 2003 was $10.35.)

✓**CHECKPOINT**   Now try Exercise 109.

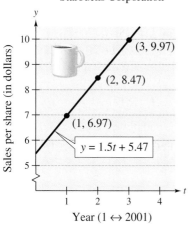

Starbucks Corporation

(3, 9.97)

(2, 8.47)

(1, 6.97)

$y = 1.5t + 5.47$

Sales per share (in dollars)

Year (1 ↔ 2001)

FIGURE P.45

The prediction method illustrated in Example 8 is called **linear extrapolation.** Note in Figure P.46 that an extrapolated point does not lie between the given points. When the estimated point lies between two given points, as shown in Figure P.47, the procedure is called **linear interpolation.**

Because the slope of a vertical line is not defined, its equation cannot be written in slope-intercept form. However, every line has an equation that can be written in the **general form**

$$Ax + By + C = 0 \qquad \text{General form}$$

where A and B are not both zero. For instance, the vertical line given by $x = a$ can be represented by the general form $x - a = 0$.

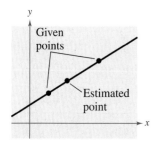

Given points

Estimated point

Linear extrapolation
FIGURE P.46

Given points

Estimated point

Linear interpolation
FIGURE P.47

Summary of Equations of Lines

1. General form: $Ax + By + C = 0$

2. Vertical line: $x = a$

3. Horizontal line: $y = b$

4. Slope-intercept form: $y = mx + b$

5. Point-slope form: $y - y_1 = m(x - x_1)$

6. Two-point form: $y - y_1 = \dfrac{y_2 - y_1}{x_2 - x_1}(x - x_1)$

P.4   Exercises

VOCABULARY CHECK:

In Exercises 1–6, fill in the blanks.

1. The simplest mathematical model for relating two variables is the _____ equation in two variables $y = mx + b$.

2. For a line, the ratio of the change in y to the change in x is called the _____ of the line.

3. Two lines are _____ if and only if their slopes are equal.

4. Two lines are _____ if and only if their slopes are negative reciprocals of each other.

5. When the x-axis and y-axis have different units of measure, the slope can be interpreted as a _____.

6. The prediction method _____ _____ is the method used to estimate a point on a line that does not lie between the given points.

7. Match each equation of a line with its form.

   (a) $Ax + By + C = 0$ (i) Vertical line

   (b) $x = a$ (ii) Slope-intercept form

   (c) $y = b$ (iii) General form

   (d) $y = mx + b$ (iv) Point-slope form

   (e) $y - y_1 = m(x - x_1)$ (v) Horizontal line

In Exercises 1 and 2, identify the line that has each slope.

1. (a) $m = \frac{2}{3}$

   (b) m is undefined.

   (c) $m = -2$

2. (a) $m = 0$

   (b) $m = -\frac{3}{4}$

   (c) $m = 1$

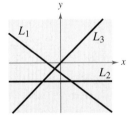

In Exercises 3 and 4, sketch the lines through the point with the indicated slopes on the same set of coordinate axes.

Point	Slopes
3. $(2, 3)$	(a) 0 (b) 1 (c) 2 (d) -3
4. $(-4, 1)$	(a) 3 (b) -3 (c) $\frac{1}{2}$ (d) Undefined

In Exercises 5–8, estimate the slope of the line.

5.

6.

7.

8.

In Exercises 9–20, find the slope and y-intercept (if possible) of the equation of the line. Sketch the line.

9. $y = 5x + 3$

10. $y = x - 10$

11. $y = -\frac{1}{2}x + 4$

12. $y = -\frac{3}{2}x + 6$

13. $5x - 2 = 0$

14. $3y + 5 = 0$

15. $7x + 6y = 30$

16. $2x + 3y = 9$

17. $y - 3 = 0$

18. $y + 4 = 0$

19. $x + 5 = 0$

20. $x - 2 = 0$

In Exercises 21–28, plot the points and find the slope of the line passing through the pair of points.

21. $(-3, -2), (1, 6)$

22. $(2, 4), (4, -4)$

23. $(-6, -1), (-6, 4)$

24. $(0, -10), (-4, 0)$

25. $\left(\frac{11}{2}, -\frac{4}{3}\right), \left(-\frac{3}{2}, -\frac{1}{3}\right)$

26. $\left(\frac{7}{8}, \frac{3}{4}\right), \left(\frac{5}{4}, -\frac{1}{4}\right)$

27. $(4.8, 3.1), (-5.2, 1.6)$

28. $(-1.75, -8.3), (2.25, -2.6)$

In Exercises 29–38, use the point on the line and the slope of the line to find three additional points through which the line passes. (There are many correct answers.)

Point	Slope
29. $(2, 1)$	$m = 0$
30. $(-4, 1)$	m is undefined.
31. $(5, -6)$	$m = 1$
32. $(10, -6)$	$m = -1$
33. $(-8, 1)$	m is undefined.
34. $(-3, -1)$	$m = 0$
35. $(-5, 4)$	$m = 2$
36. $(0, -9)$	$m = -2$
37. $(7, -2)$	$m = \frac{1}{2}$
38. $(-1, -6)$	$m = -\frac{1}{2}$

In Exercises 39–50, find the slope-intercept form of the equation of the line that passes through the given point and has the indicated slope. Sketch the line.

Point	Slope
39. $(0, -2)$	$m = 3$
40. $(0, 10)$	$m = -1$
41. $(-3, 6)$	$m = -2$
42. $(0, 0)$	$m = 4$
43. $(4, 0)$	$m = -\frac{1}{3}$
44. $(-2, -5)$	$m = \frac{3}{4}$
45. $(6, -1)$	m is undefined.
46. $(-10, 4)$	m is undefined.
47. $\left(4, \frac{5}{2}\right)$	$m = 0$
48. $\left(-\frac{1}{2}, \frac{3}{2}\right)$	$m = 0$
49. $(-5.1, 1.8)$	$m = 5$
50. $(2.3, -8.5)$	$m = -\frac{5}{2}$

In Exercises 51–64, find the slope-intercept form of the equation of the line passing through the points. Sketch the line.

51. $(5, -1), (-5, 5)$ **52.** $(4, 3), (-4, -4)$

53. $(-8, 1), (-8, 7)$ **54.** $(-1, 4), (6, 4)$

55. $\left(2, \frac{1}{2}\right), \left(\frac{1}{2}, \frac{5}{4}\right)$ **56.** $\left(1, 1\right), \left(6, -\frac{2}{3}\right)$

57. $\left(-\frac{1}{10}, -\frac{3}{5}\right), \left(\frac{9}{10}, -\frac{9}{5}\right)$ **58.** $\left(\frac{3}{4}, \frac{3}{2}\right), \left(-\frac{4}{3}, \frac{7}{4}\right)$

59. $(1, 0.6), (-2, -0.6)$

60. $(-8, 0.6), (2, -2.4)$

61. $(2, -1), \left(\frac{1}{3}, -1\right)$

62. $\left(\frac{1}{5}, -2\right), (-6, -2)$

63. $\left(\frac{7}{3}, -8\right), \left(\frac{7}{3}, 1\right)$

64. $(1.5, -2), (1.5, 0.2)$

In Exercises 65–68, determine whether the lines L_1 and L_2 passing through the pairs of points are parallel, perpendicular, or neither.

65. L_1: $(0, -1), (5, 9)$ **66.** L_1: $(-2, -1), (1, 5)$
 L_2: $(0, 3), (4, 1)$ L_2: $(1, 3), (5, -5)$

67. L_1: $(3, 6), (-6, 0)$ **68.** L_1: $(4, 8), (-4, 2)$
 L_2: $(0, -1), \left(5, \frac{7}{3}\right)$ L_2: $(3, -5), \left(-1, \frac{1}{3}\right)$

In Exercises 69–78, write the slope-intercept forms of the equations of the lines through the given point (a) parallel to the given line and (b) perpendicular to the given line.

Point	Line
69. $(2, 1)$	$4x - 2y = 3$
70. $(-3, 2)$	$x + y = 7$
71. $\left(-\frac{2}{3}, \frac{7}{8}\right)$	$3x + 4y = 7$
72. $\left(\frac{7}{8}, \frac{3}{4}\right)$	$5x + 3y = 0$
73. $(-1, 0)$	$y = -3$
74. $(4, -2)$	$y = 1$
75. $(2, 5)$	$x = 4$
76. $(-5, 1)$	$x = -2$
77. $(2.5, 6.8)$	$x - y = 4$
78. $(-3.9, -1.4)$	$6x + 2y = 9$

In Exercises 79–84, use the *intercept form* to find the equation of the line with the given intercepts. The intercept form of the equation of a line with intercepts $(a, 0)$ and $(0, b)$ is

$$\frac{x}{a} + \frac{y}{b} = 1, \ a \neq 0, \ b \neq 0.$$

79. x-intercept: $(2, 0)$ **80.** x-intercept: $(-3, 0)$
 y-intercept: $(0, 3)$ y-intercept: $(0, 4)$

81. x-intercept: $\left(-\frac{1}{6}, 0\right)$ **82.** x-intercept: $\left(\frac{2}{3}, 0\right)$
 y-intercept: $\left(0, -\frac{2}{3}\right)$ y-intercept: $(0, -2)$

83. Point on line: $(1, 2)$ **84.** Point on line: $(-3, 4)$
 x-intercept: $(c, 0)$ x-intercept: $(d, 0)$
 y-intercept: $(0, c)$, $c \neq 0$ y-intercept: $(0, d)$, $d \neq 0$

Graphical Interpretation In Exercises 85–88, identify any relationships that exist among the lines, and then use a graphing utility to graph the three equations in the same viewing window. Adjust the viewing window so that the slope appears visually correct—that is, so that parallel lines appear parallel and perpendicular lines appear to intersect at right angles.

85. (a) $y = 2x$ (b) $y = -2x$ (c) $y = \frac{1}{2}x$

86. (a) $y = \frac{2}{3}x$ (b) $y = -\frac{3}{2}x$ (c) $y = \frac{2}{3}x + 2$

87. (a) $y = -\frac{1}{2}x$ (b) $y = -\frac{1}{2}x + 3$ (c) $y = 2x - 4$

88. (a) $y = x - 8$ (b) $y = x + 1$ (c) $y = -x + 3$

In Exercises 89–92, find a relationship between x and y such that (x, y) is equidistant (the same distance) from the two points.

89. $(4, -1), (-2, 3)$

90. $(6, 5), (1, -8)$

91. $\left(3, \frac{5}{2}\right), (-7, 1)$

92. $\left(-\frac{1}{2}, -4\right), \left(\frac{7}{2}, \frac{5}{4}\right)$

93. *Sales*   The following are the slopes of lines representing annual sales y in terms of time x in years. Use the slopes to interpret any change in annual sales for a one-year increase in time.

(a) The line has a slope of $m = 135$.

(b) The line has a slope of $m = 0$.

(c) The line has a slope of $m = -40$.

94. *Revenue*   The following are the slopes of lines representing daily revenues y in terms of time x in days. Use the slopes to interpret any change in daily revenues for a one-day increase in time.

(a) The line has a slope of $m = 400$.

(b) The line has a slope of $m = 100$.

(c) The line has a slope of $m = 0$.

95. *Average Salary*   The graph shows the average salaries for senior high school principals from 1990 through 2002. (Source: Educational Research Service)

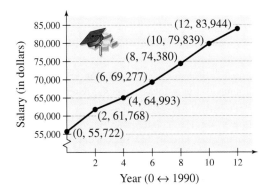

(a) Use the slopes to determine the time periods in which the average salary increased the greatest and the least.

(b) Find the slope of the line segment connecting the years 1990 and 2002.

(c) Interpret the meaning of the slope in part (b) in the context of the problem.

96. *Net Profit*   The graph shows the net profits (in millions) for Applebee's International, Inc. for the years 1994 through 2003.   (Source: Applebee's International, Inc.)

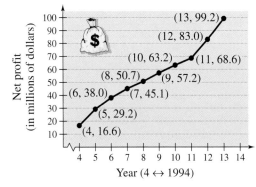

(a) Use the slopes to determine the years in which the net profit showed the greatest increase and the least increase.

(b) Find the slope of the line segment connecting the years 1994 and 2003.

(c) Interpret the meaning of the slope in part (b) in the context of the problem.

97. *Road Grade*   You are driving on a road that has a 6% uphill grade (see figure). This means that the slope of the road is $\frac{6}{100}$. Approximate the amount of vertical change in your position if you drive 200 feet.

98. *Road Grade*   From the top of a mountain road, a surveyor takes several horizontal measurements x and several vertical measurements y, as shown in the table (x and y are measured in feet).

x	300	600	900	1200	1500	1800	2100
y	-25	-50	-75	-100	-125	-150	-175

(a) Sketch a scatter plot of the data.

(b) Use a straightedge to sketch the line that you think best fits the data.

(c) Find an equation for the line you sketched in part (b).

(d) Interpret the meaning of the slope of the line in part (c) in the context of the problem.

(e) The surveyor needs to put up a road sign that indicates the steepness of the road. For instance, a surveyor would put up a sign that states "8% grade" on a road with a downhill grade that has a slope of $-\frac{8}{100}$. What should the sign state for the road in this problem?

Rate of Change **In Exercises 99 and 100, you are given the dollar value of a product in 2005 and the rate at which the value of the product is expected to change during the next 5 years. Use this information to write a linear equation that gives the dollar value V of the product in terms of the year t. (Let t = 5 represent 2005.)**

	2005 Value	*Rate*
99.	$2540	$125 decrease per year
100.	$156	$4.50 increase per year

Graphical Interpretation **In Exercises 101–104, match the description of the situation with its graph. Also determine the slope and y-intercept of each graph and interpret the slope and y-intercept in the context of the situation. [The graphs are labeled (a), (b), (c), and (d).]**

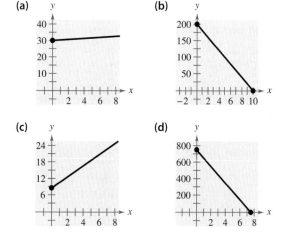

101. A person is paying $20 per week to a friend to repay a $200 loan.

102. An employee is paid $8.50 per hour plus $2 for each unit produced per hour.

103. A sales representative receives $30 per day for food plus $0.32 for each mile traveled.

104. A computer that was purchased for $750 depreciates $100 per year.

105. *Cash Flow per Share* The cash flow per share for the Timberland Co. was $0.18 in 1995 and $4.04 in 2003. Write a linear equation that gives the cash flow per share in terms of the year. Let t = 5 represent 1995. Then predict the cash flows for the years 2008 and 2010. (Source: The Timberland Co.)

106. *Number of Stores* In 1999 there were 4076 J.C. Penney stores and in 2003 there were 1078 stores. Write a linear equation that gives the number of stores in terms of the year. Let t = 9 represent 1999. Then predict the numbers of stores for the years 2008 and 2010. Are your answers reasonable? Explain. (Source: J.C. Penney Co.)

107. *Depreciation* A sub shop purchases a used pizza oven for $875. After 5 years, the oven will have to be replaced. Write a linear equation giving the value V of the equipment during the 5 years it will be in use.

108. *Depreciation* A school district purchases a high-volume printer, copier, and scanner for $25,000. After 10 years, the equipment will have to be replaced. Its value at that time is expected to be $2000. Write a linear equation giving the value V of the equipment during the 10 years it will be in use.

109. *College Enrollment* The Pennsylvania State University had enrollments of 40,571 students in 2000 and 41,289 students in 2004 at its main campus in University Park, Pennsylvania. (Source: Penn State Fact Book)

(a) Assuming the enrollment growth is linear, find a linear model that gives the enrollment in terms of the year t, where t = 0 corresponds to 2000.

(b) Use your model from part (a) to predict the enrollments in 2008 and 2010.

(c) What is the slope of your model? Explain its meaning in the context of the situation.

110. *College Enrollment* The University of Florida had enrollments of 36,531 students in 1990 and 48,673 students in 2003. (Source: University of Florida)

(a) What was the average annual change in enrollment from 1990 to 2003?

(b) Use the average annual change in enrollment to estimate the enrollments in 1994, 1998, and 2002.

(c) Write the equation of a line that represents the given data. What is its slope? Interpret the slope in the context of the problem.

(d) Using the results of parts (a)–(c), write a short paragraph discussing the concepts of *slope* and *average rate of change*.

111. *Sales* A discount outlet is offering a 15% discount on all items. Write a linear equation giving the sale price S for an item with a list price L.

112. *Hourly Wage* A microchip manufacturer pays its assembly line workers $11.50 per hour. In addition, workers receive a piecework rate of $0.75 per unit produced. Write a linear equation for the hourly wage W in terms of the number of units x produced per hour.

113. *Cost, Revenue, and Profit* A roofing contractor purchases a shingle delivery truck with a shingle elevator for $36,500. The vehicle requires an average expenditure of $5.25 per hour for fuel and maintenance, and the operator is paid $11.50 per hour.

(a) Write a linear equation giving the total cost C of operating this equipment for t hours. (Include the purchase cost of the equipment.)

(b) Assuming that customers are charged $27 per hour of machine use, write an equation for the revenue R derived from t hours of use.

(c) Use the formula for profit $(P = R - C)$ to write an equation for the profit derived from t hours of use.

(d) Use the result of part (c) to find the break-even point—that is, the number of hours this equipment must be used to yield a profit of 0 dollars.

114. *Rental Demand* A real estate office handles an apartment complex with 50 units. When the rent per unit is $580 per month, all 50 units are occupied. However, when the rent is $625 per month, the average number of occupied units drops to 47. Assume that the relationship between the monthly rent p and the demand x is linear.

(a) Write the equation of the line giving the demand x in terms of the rent p.

(b) Use this equation to predict the number of units occupied when the rent is $655.

(c) Predict the number of units occupied when the rent is $595.

115. *Geometry* The length and width of a rectangular garden are 15 meters and 10 meters, respectively. A walkway of width x surrounds the garden.

(a) Draw a diagram that gives a visual representation of the problem.

(b) Write the equation for the perimeter y of the walkway in terms of x.

 (c) Use a graphing utility to graph the equation for the perimeter.

 (d) Determine the slope of the graph in part (c). For each additional one-meter increase in the width of the walkway, determine the increase in its perimeter.

116. *Monthly Salary* A pharmaceutical salesperson receives a monthly salary of $2500 plus a commission of 7% of sales. Write a linear equation for the salesperson's monthly wage W in terms of monthly sales S.

117. *Business Costs* A sales representative of a company using a personal car receives $120 per day for lodging and meals plus $0.38 per mile driven. Write a linear equation giving the daily cost C to the company in terms of x, the number of miles driven.

118. *Sports* The median salaries (in thousands of dollars) for players on the Los Angeles Dodgers from 1996 to 2003 are shown in the scatter plot. Find the equation of the line that you think best fits these data. (Let y represent the median salary and let t represent the year, with $t = 6$ corresponding to 1996.) (Source: *USA TODAY*)

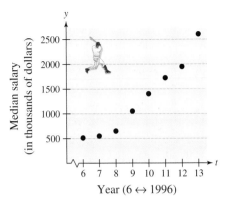

FIGURE FOR **118**

Model It

119. *Data Analysis: Cell Phone Suscribers* The numbers of cellular phone suscribers y (in millions) in the United States from 1990 through 2002, where x is the year, are shown as data points (x, y). (Source: Cellular Telecommunications & Internet Association)

(1990,	5.3)
(1991,	7.6)
(1992,	11.0)
(1993,	16.0)
(1994,	24.1)
(1995,	33.8)
(1996,	44.0)
(1997,	55.3)
(1998,	69.2)
(1999,	86.0)
(2000,	109.5)
(2001,	128.4)
(2002,	140.8)

(a) Sketch a scatter plot of the data. Let $x = 0$ correspond to 1990.

(b) Use a straightedge to sketch the line that you think best fits the data.

(c) Find the equation of the line from part (b). Explain the procedure you used.

(d) Write a short paragraph explaining the meanings of the slope and y-intercept of the line in terms of the data.

(e) Compare the values obtained using your model with the actual values.

(f) Use your model to estimate the number of cellular phone suscribers in 2008.

120. *Data Analysis: Average Scores* An instructor gives regular 20-point quizzes and 100-point exams in an algebra course. Average scores for six students, given as data points (x, y) where x is the average quiz score and y is the average test score, are $(18, 87)$, $(10, 55)$, $(19, 96)$, $(16, 79)$, $(13, 76)$, and $(15, 82)$. [*Note:* There are many correct answers for parts (b)–(d).]

(a) Sketch a scatter plot of the data.

(b) Use a straightedge to sketch the line that you think best fits the data.

(c) Find an equation for the line you sketched in part (b).

(d) Use the equation in part (c) to estimate the average test score for a person with an average quiz score of 17.

(e) The instructor adds 4 points to the average test score of each student in the class. Describe the changes in the positions of the plotted points and the change in the equation of the line.

Synthesis

True or False? **In Exercises 121 and 122, determine whether the statement is true or false. Justify your answer.**

121. A line with a slope of $-\frac{5}{7}$ is steeper than a line with a slope of $-\frac{6}{7}$.

122. The line through $(-8, 2)$ and $(-1, 4)$ and the line through $(0, -4)$ and $(-7, 7)$ are parallel.

123. Explain how you could show that the points $A\,(2, 3)$, $B\,(2, 9)$, and $C\,(4, 3)$ are the vertices of a right triangle.

124. Explain why the slope of a vertical line is said to be undefined.

125. With the information shown in the graphs, is it possible to determine the slope of each line? Is it possible that the lines could have the same slope? Explain.

(a)   (b)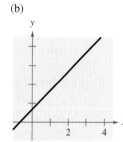

126. The slopes of two lines are -4 and $\frac{5}{2}$. Which is steeper? Explain.

127. The value V of a molding machine t years after it is purchased is

$$V = -4000t + 58{,}500, \quad 0 \le t \le 5.$$

Explain what the V-intercept and slope measure.

128. *Think About It* Is it possible for two lines with positive slopes to be perpendicular? Explain.

---

129. **Make a Decision** To work an extended application analyzing the numbers of bachelor's degrees earned by women in the United States from 1985 to 2002, visit this text's website at *college.hmco.com*. *(Data Source: U.S. National Center for Educational Statistics)*

P.5   Functions

What you should learn

- Determine whether relations between two variables are functions.
- Use function notation and evaluate functions.
- Find the domains of functions.
- Use functions to model and solve real-life problems.
- Evaluate difference quotients.

Why you should learn it

Functions can be used to model and solve real-life problems. For instance, in Exercise 100 on page 67, you will use a function to model the force of water against the face of a dam.

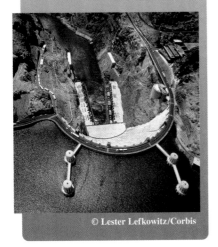

© Lester Lefkowitz/Corbis

Introduction to Functions

Many everyday phenomena involve two quantities that are related to each other by some rule of correspondence. The mathematical term for such a rule of correspondence is a **relation.** In mathematics, relations are often represented by mathematical equations and formulas. For instance, the simple interest I earned on $1000 for 1 year is related to the annual interest rate r by the formula $I = 1000r$.

The formula $I = 1000r$ represents a special kind of relation that matches each item from one set with *exactly one* item from a different set. Such a relation is called a **function.**

Definition of Function

A **function** f from a set A to a set B is a relation that assigns to each element x in the set A exactly one element y in the set B. The set A is the **domain** (or set of inputs) of the function f, and the set B contains the **range** (or set of outputs).

To help understand this definition, look at the function that relates the time of day to the temperature in Figure P.48.

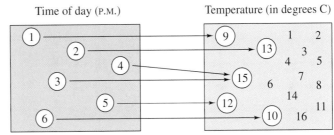

Set A is the domain.
Inputs: 1, 2, 3, 4, 5, 6

Set B contains the range.
Outputs: 9, 10, 12, 13, 15

FIGURE P.48

This function can be represented by the following ordered pairs, in which the first coordinate (x-value) is the input and the second coordinate (y-value) is the output.

$$\{(1, 9°), (2, 13°), (3, 15°), (4, 15°), (5, 12°), (6, 10°)\}$$

Characteristics of a Function from Set *A* to Set *B*

1. Each element in A must be matched with an element in B.

2. Some elements in B may not be matched with any element in A.

3. Two or more elements in A may be matched with the same element in B.

4. An element in A (the domain) cannot be matched with two different elements in B.

Functions are commonly represented in four ways.

Four Ways to Represent a Function

1. *Verbally* by a sentence that describes how the input variable is related to the output variable

2. *Numerically* by a table or a list of ordered pairs that matches input values with output values

3. *Graphically* by points on a graph in a coordinate plane in which the input values are represented by the horizontal axis and the output values are represented by the vertical axis

4. *Algebraically* by an equation in two variables

To determine whether or not a relation is a function, you must decide whether each input value is matched with exactly one output value. If any input value is matched with two or more output values, the relation is not a function.

Example 1 Testing for Functions

Determine whether the relation represents y as a function of x.

a. The input value x is the number of representatives from a state, and the output value y is the number of senators.

b.

Input, x	Output, y
2	11
2	10
3	8
4	5
5	1

c.

FIGURE P.49

Solution

a. This verbal description *does* describe y as a function of x. Regardless of the value of x, the value of y is always 2. Such functions are called *constant functions*.

b. This table *does not* describe y as a function of x. The input value 2 is matched with two different y-values.

c. The graph in Figure P.49 *does* describe y as a function of x. Each input value is matched with exactly one output value.

✓CHECKPOINT Now try Exercise 5.

Representing functions by sets of ordered pairs is common in *discrete mathematics*. In algebra, however, it is more common to represent functions by equations or formulas involving two variables. For instance, the equation

$$y = x^2 \qquad \text{\small y is a function of x.}$$

represents the variable y as a function of the variable x. In this equation, x is the **independent variable** and y is the **dependent variable.** The domain of the function is the set of all values taken on by the independent variable x, and the range of the function is the set of all values taken on by the dependent variable y.

Example 2   Testing for Functions Represented Algebraically

Which of the equations represent(s) y as a function of x?

a. $x^2 + y = 1$ **b.** $-x + y^2 = 1$

Solution

To determine whether y is a function of x, try to solve for y in terms of x.

a. Solving for y yields

$$x^2 + y = 1$$ Write original equation.

$$y = 1 - x^2.$$ Solve for y.

To each value of x there corresponds exactly one value of y. So, y is a function of x.

b. Solving for y yields

$$-x + y^2 = 1$$ Write original equation.

$$y^2 = 1 + x$$ Add x to each side.

$$y = \pm \sqrt{1 + x}.$$ Solve for y.

The $\pm$ indicates that to a given value of x there correspond two values of y. So, y is not a function of x.

✓CHECKPOINT   Now try Exercise 15.

Function Notation

When an equation is used to represent a function, it is convenient to name the function so that it can be referenced easily. For example, you know that the equation $y = 1 - x^2$ describes y as a function of x. Suppose you give this function the name "f." Then you can use the following **function notation.**

Input	Output	Equation
x	$f(x)$	$f(x) = 1 - x^2$

The symbol $f(x)$ is read as *the value of f at x* or simply *f of x.* The symbol $f(x)$ corresponds to the y-value for a given x. So, you can write $y = f(x)$. Keep in mind that f is the *name* of the function, whereas $f(x)$ is the *value* of the function at x. For instance, the function given by

$$f(x) = 3 - 2x$$

has *function values* denoted by $f(-1), f(0), f(2)$, and so on. To find these values, substitute the specified input values into the given equation.

For $x = -1$, $f(-1) = 3 - 2(-1) = 3 + 2 = 5.$

For $x = 0$, $f(0) = 3 - 2(0) = 3 - 0 = 3.$

For $x = 2$, $f(2) = 3 - 2(2) = 3 - 4 = -1.$

© Bettmann/Corbis

Historical Note
Leonhard Euler (1707–1783), a Swiss mathematician, is considered to have been the most prolific and productive mathematician in history. One of his greatest influences on mathematics was his use of symbols, or notation. The function notation $y = f(x)$ was introduced by Euler.

Understanding the concept of a function is essential. Be sure that students understand function notation. Frequently $f(x)$ is misinterpreted as "f times x" rather than "f of x."

Although f is often used as a convenient function name and x is often used as the independent variable, you can use other letters. For instance,

$$f(x) = x^2 - 4x + 7, \quad f(t) = t^2 - 4t + 7, \quad \text{and} \quad g(s) = s^2 - 4s + 7$$

all define the same function. In fact, the role of the independent variable is that of a "placeholder." Consequently, the function could be described by

$$f\left(\right) = \left(\right)^2 - 4\left(\right) + 7.$$

STUDY TIP

In Example 3, note that $g(x + 2)$ is not equal to $g(x) + g(2)$. In general, $g(u + v) \neq g(u) + g(v)$.

Students often have difficulty understanding how to evaluate piecewise-defined functions. You may want to use the following additional examples to demonstrate evaluation of piecewise-defined functions.

a. Evaluate at $x = 0, 1, 3$.

$$f(x) = \begin{cases} \dfrac{x}{2} + 1, & x \leq 1 \\ 3x + 2, & x > 1 \end{cases}$$

Solution

Because $x = 0$ is less than or equal to 1, use $f(x) = (x/2) + 1$ to obtain

$$f(0) = \frac{0}{2} + 1 = 1.$$

For $x = 1$, use $f(x) = (x/2) + 1$ to obtain

$$f(1) = \frac{1}{2} + 1 = 1\frac{1}{2}.$$

For $x = 3$, use $f(x) = 3x + 2$ to obtain

$$f(3) = 3(3) + 2 = 11.$$

b. Evaluate at $x = 0, 3, 5$.

$$f(x) = \begin{cases} x^2 + 3, & x < 2 \\ 7, & 2 \leq x \leq 4 \\ 2x - 1, & x > 4 \end{cases}$$

Solution

Because $x = 0$ is less than 2, use $f(x) = x^2 + 3$ to obtain

$$f(0) = 0^2 + 3 = 3.$$

For $x = 3$, use $f(x) = 7$ to obtain

$$f(3) = 7.$$

For $x = 5$, use $f(x) = 2x - 1$ to obtain

$$f(5) = 2(5) - 1 = 9.$$

Example 3 Evaluating a Function

Let $g(x) = -x^2 + 4x + 1$. Find each function value.

a. $g(2)$ **b.** $g(t)$ **c.** $g(x + 2)$

Solution

a. Replacing x with 2 in $g(x) = -x^2 + 4x + 1$ yields the following.

$$g(2) = -(2)^2 + 4(2) + 1 = -4 + 8 + 1 = 5$$

b. Replacing x with t yields the following.

$$g(t) = -(t)^2 + 4(t) + 1 = -t^2 + 4t + 1$$

c. Replacing x with $x + 2$ yields the following.

$$\begin{aligned} g(x + 2) &= -(x + 2)^2 + 4(x + 2) + 1 \\ &= -(x^2 + 4x + 4) + 4x + 8 + 1 \\ &= -x^2 - 4x - 4 + 4x + 8 + 1 \\ &= -x^2 + 5 \end{aligned}$$

✓**CHECKPOINT** Now try Exercise 29.

A function defined by two or more equations over a specified domain is called a **piecewise-defined function.**

Example 4 A Piecewise-Defined Function

Evaluate the function when $x = -1, 0$, and 1.

$$f(x) = \begin{cases} x^2 + 1, & x < 0 \\ x - 1, & x \geq 0 \end{cases}$$

Solution

Because $x = -1$ is less than 0, use $f(x) = x^2 + 1$ to obtain

$$f(-1) = (-1)^2 + 1 = 2.$$

For $x = 0$, use $f(x) = x - 1$ to obtain

$$f(0) = (0) - 1 = -1.$$

For $x = 1$, use $f(x) = x - 1$ to obtain

$$f(1) = (1) - 1 = 0.$$

✓**CHECKPOINT** Now try Exercise 35.

The Domain of a Function

The domain of a function can be described explicitly or it can be *implied* by the expression used to define the function. The **implied domain** is the set of all real numbers for which the expression is defined. For instance, the function given by

$$f(x) = \frac{1}{x^2 - 4}$$ Domain excludes x-values that result in division by zero.

has an implied domain that consists of all real x other than $x = \pm 2$. These two values are excluded from the domain because division by zero is undefined. Another common type of implied domain is that used to avoid even roots of negative numbers. For example, the function given by

$$f(x) = \sqrt{x}$$ Domain excludes x-values that result in even roots of negative numbers.

is defined only for $x \geq 0$. So, its implied domain is the interval $[0, \infty)$. In general, the domain of a function *excludes* values that would cause division by zero *or* that would result in the even root of a negative number.

Example 5   **Finding the Domain of a Function**

Find the domain of each function.

a. $f: \{(-3, 0), (-1, 4), (0, 2), (2, 2), (4, -1)\}$ **b.** $g(x) = \dfrac{1}{x + 5}$

c. Volume of a sphere: $V = \frac{4}{3}\pi r^3$ **d.** $h(x) = \sqrt{4 - x^2}$

Solution

a. The domain of f consists of all first coordinates in the set of ordered pairs.

Domain $= \{-3, -1, 0, 2, 4\}$

b. Excluding x-values that yield zero in the denominator, the domain of g is the set of all real numbers x except $x = -5$.

c. Because this function represents the volume of a sphere, the values of the radius r must be positive. So, the domain is the set of all real numbers r such that $r > 0$.

d. This function is defined only for x-values for which

$$4 - x^2 \geq 0.$$

By solving this inequality, you can conclude that $-2 \leq x \leq 2$. So, the domain is the interval $[-2, 2]$.

✓**CHECKPOINT**   Now try Exercise 59.

In Example 5(c), note that the domain of a function may be implied by the physical context. For instance, from the equation

$$V = \frac{4}{3}\pi r^3$$

you would have no reason to restrict r to positive values, but the physical context implies that a sphere cannot have a negative or zero radius.

$$\boxed{\frac{h}{r} = 4}$$

FIGURE P.50

Applications

Example 6 The Dimensions of a Container

You work in the marketing department of a soft-drink company and are experimenting with a new can for iced tea that is slightly narrower and taller than a standard can. For your experimental can, the ratio of the height to the radius is 4, as shown in Figure P.50.

a. Write the volume of the can as a function of the radius r.

b. Write the volume of the can as a function of the height h.

Solution

a. $V(r) = \pi r^2 h = \pi r^2 (4r) = 4\pi r^3$ Write V as a function of r.

b. $V(h) = \pi \left(\dfrac{h}{4}\right)^2 h = \dfrac{\pi h^3}{16}$ Write V as a function of h.

✓CHECKPOINT Now try Exercise 87.

Example 7 The Path of a Baseball

A baseball is hit at a point 3 feet above ground at a velocity of 100 feet per second and an angle of 45°. The path of the baseball is given by the function

$$f(x) = -0.0032x^2 + x + 3$$

where y and x are measured in feet, as shown in Figure P.51. Will the baseball clear a 10-foot fence located 300 feet from home plate?

FIGURE P.51

Solution

When $x = 300$, the height of the baseball is

$$f(300) = -0.0032(300)^2 + 300 + 3$$
$$= 15 \text{ feet.}$$

So, the baseball will clear the fence.

✓CHECKPOINT Now try Exercise 93.

In the equation in Example 7, the height of the baseball is a function of the distance from home plate.

Number of Alternative-Fueled
Vehicles in the U.S.

FIGURE P.52

Example 8   **Alternative-Fueled Vehicles**

The number V (in thousands) of alternative-fueled vehicles in the United States increased in a linear pattern from 1995 to 1999, as shown in Figure P.52. Then, in 2000, the number of vehicles took a jump and, until 2002, increased in a different linear pattern. These two patterns can be approximated by the function

$$V(t) = \begin{cases} 18.08t + 155.3 & 5 \leq t \leq 9 \\ 38.20t + 10.2, & 10 \leq t \leq 12 \end{cases}$$

where t represents the year, with $t = 5$ corresponding to 1995. Use this function to approximate the number of alternative-fueled vehicles for each year from 1995 to 2002. (Source: Science Applications International Corporation; Energy Information Administration)

Solution

From 1995 to 1999, use $V(t) = 18.08t + 155.3$.

$$\underbrace{245.7}_{1995} \quad \underbrace{263.8}_{1996} \quad \underbrace{281.9}_{1997} \quad \underbrace{299.9}_{1998} \quad \underbrace{318.0}_{1999}$$

From 2000 to 2002, use $V(t) = 38.20t + 10.2$.

$$\underbrace{392.2}_{2000} \quad \underbrace{430.4}_{2001} \quad \underbrace{468.6}_{2002}$$

✓CHECKPOINT   Now try Exercise 95.

Difference Quotients

One of the basic definitions in calculus employs the ratio

$$\frac{f(x + h) - f(x)}{h}, \quad h \neq 0.$$

This ratio is called a **difference quotient,** as illustrated in Example 9.

Example 9   **Evaluating a Difference Quotient**

For $f(x) = x^2 - 4x + 7$, find $\dfrac{f(x + h) - f(x)}{h}$.

Solution

$$\frac{f(x + h) - f(x)}{h} = \frac{[(x + h)^2 - 4(x + h) + 7] - (x^2 - 4x + 7)}{h}$$

$$= \frac{x^2 + 2xh + h^2 - 4x - 4h + 7 - x^2 + 4x - 7}{h}$$

$$= \frac{2xh + h^2 - 4h}{h} = \frac{h(2x + h - 4)}{h} = 2x + h - 4, \ h \neq 0$$

✓CHECKPOINT   Now try Exercise 79.

The symbol ∫ indicates an example or exercise that highlights algebraic techniques specifically used in calculus.

You may find it easier to calculate the difference quotient in Example 9 by first finding $f(x + h)$, and then substituting the resulting expression into the difference quotient, as follows.

$$f(x + h) = (x + h)^2 - 4(x + h) + 7 = x^2 + 2xh + h^2 - 4x - 4h + 7$$

$$\frac{f(x + h) - f(x)}{h} = \frac{(x^2 + 2xh + h^2 - 4x - 4h + 7) - (x^2 - 4x + 7)}{h}$$

$$= \frac{2xh + h^2 - 4h}{h} = \frac{h(2x + h - 4)}{h} = 2x + h - 4, \quad h \neq 0$$

Summary of Function Terminology

Function: A **function** is a relationship between two variables such that to each value of the independent variable there corresponds exactly one value of the dependent variable.

Function Notation: $y = f(x)$

   f is the *name* of the function.

   y is the **dependent variable.**

   x is the **independent variable.**

   $f(x)$ is the *value of the function at x.*

Domain: The **domain** of a function is the set of all values (inputs) of the independent variable for which the function is defined. If x is in the domain of f, f is said to be *defined* at x. If x is not in the domain of f, f is said to be *undefined* at x.

Range: The **range** of a function is the set of all values (outputs) assumed by the dependent variable (that is, the set of all function values).

Implied Domain: If f is defined by an algebraic expression and the domain is not specified, the **implied domain** consists of all real numbers for which the expression is defined.

*W*RITING ABOUT *M*ATHEMATICS

Everyday Functions   In groups of two or three, identify common real-life functions. Consider everyday activities, events, and expenses, such as long distance telephone calls and car insurance. Here are two examples.

a. The statement, "Your happiness is a function of the grade you receive in this course" *is not* a correct mathematical use of the word "function." The word "happiness" is ambiguous.

b. The statement, "Your federal income tax is a function of your adjusted gross income" *is* a correct mathematical use of the word "function." Once you have determined your adjusted gross income, your income tax can be determined.

Describe your functions in words. Avoid using ambiguous words. Can you find an example of a piecewise-defined function?

P.5 | Exercises

VOCABULARY CHECK: Fill in the blanks.

1. A relation that assigns to each element x from a set of inputs, or _____, exactly one element y in a set of outputs, or _____, is called a _____.

2. Functions are commonly represented in four different ways, _____, _____, _____, and _____.

3. For an equation that represents y as a function of x, the set of all values taken on by the _____ variable x is the domain, and the set of all values taken on by the _____ variable y is the range.

4. The function given by

$$f(x) = \begin{cases} 2x - 1, & x < 0 \\ x^2 + 4, & x \geq 0 \end{cases}$$

is an example of a _____ function.

5. If the domain of the function f is not given, then the set of values of the independent variable for which the expression is defined is called the _____ _____.

6. In calculus, one of the basic definitions is that of a _____ _____, given by $\dfrac{f(x + h) - f(x)}{h}$, $h \neq 0$.

In Exercises 1–4, is the relationship a function?

1. Domain   Range

-2 ⟶ 5
-1   6
0   7
1   8
2

2. Domain   Range

-2 ⟶ 3
-1   4
0   5
1
2

3. Domain   Range

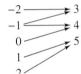

National League → Cubs, Pirates, Dodgers

American League → Orioles, Yankees, Twins

4. Domain (Year)   Range (Number of North Atlantic tropical storms and hurricanes)

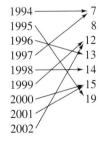

1994 ⟶ 7
1995   8
1996   12
1997   13
1998   14
1999   15
2000   19
2001
2002

In Exercises 5–8, does the table describe a function? Explain your reasoning.

5.

Input value	-2	-1	0	1	2
Output value	-8	-1	0	1	8

6.

Input value	0	1	2	1	0
Output value	-4	-2	0	2	4

7.

Input value	10	7	4	7	10
Output value	3	6	9	12	15

8.

Input value	0	3	9	12	15
Output value	3	3	3	3	3

In Exercises 9 and 10, which sets of ordered pairs represent functions from A to B? Explain.

9. $A = \{0, 1, 2, 3\}$ and $B = \{-2, -1, 0, 1, 2\}$

   (a) $\{(0, 1), (1, -2), (2, 0), (3, 2)\}$

   (b) $\{(0, -1), (2, 2), (1, -2), (3, 0), (1, 1)\}$

   (c) $\{(0, 0), (1, 0), (2, 0), (3, 0)\}$

   (d) $\{(0, 2), (3, 0), (1, 1)\}$

10. $A = \{a, b, c\}$ and $B = \{0, 1, 2, 3\}$

   (a) $\{(a, 1), (c, 2), (c, 3), (b, 3)\}$

   (b) $\{(a, 1), (b, 2), (c, 3)\}$

   (c) $\{(1, a), (0, a), (2, c), (3, b)\}$

   (d) $\{(c, 0), (b, 0), (a, 3)\}$

Circulation of Newspapers **In Exercises 11 and 12, use the graph, which shows the circulation (in millions) of daily newspapers in the United States.** (Source: Editor & Publisher Company)

11. Is the circulation of morning newspapers a function of the year? Is the circulation of evening newspapers a function of the year? Explain.

12. Let $f(x)$ represent the circulation of evening newspapers in year x. Find $f(1998)$.

In Exercises 13–24, determine whether the equation represents y as a function of x.

13. $x^2 + y^2 = 4$

14. $x = y^2$

15. $x^2 + y = 4$

16. $x + y^2 = 4$

17. $2x + 3y = 4$

18. $(x - 2)^2 + y^2 = 4$

19. $y^2 = x^2 - 1$

20. $y = \sqrt{x + 5}$

21. $y = |4 - x|$

22. $|y| = 4 - x$

23. $x = 14$

24. $y = -75$

In Exercises 25–38, evaluate the function at each specified value of the independent variable and simplify.

25. $f(x) = 2x - 3$

    (a) $f(1)$ (b) $f(-3)$ (c) $f(x - 1)$

26. $g(y) = 7 - 3y$

    (a) $g(0)$ (b) $g\left(\frac{7}{3}\right)$ (c) $g(s + 2)$

27. $V(r) = \frac{4}{3}\pi r^3$

    (a) $V(3)$ (b) $V\left(\frac{3}{2}\right)$ (c) $V(2r)$

28. $h(t) = t^2 - 2t$

    (a) $h(2)$ (b) $h(1.5)$ (c) $h(x + 2)$

29. $f(y) = 3 - \sqrt{y}$

    (a) $f(4)$ (b) $f(0.25)$ (c) $f(4x^2)$

30. $f(x) = \sqrt{x + 8} + 2$

    (a) $f(-8)$ (b) $f(1)$ (c) $f(x - 8)$

31. $q(x) = \dfrac{1}{x^2 - 9}$

    (a) $q(0)$ (b) $q(3)$ (c) $q(y + 3)$

32. $q(t) = \dfrac{2t^2 + 3}{t^2}$

    (a) $q(2)$ (b) $q(0)$ (c) $q(-x)$

33. $f(x) = \dfrac{|x|}{x}$

    (a) $f(2)$ (b) $f(-2)$ (c) $f(x - 1)$

34. $f(x) = |x| + 4$

    (a) $f(2)$ (b) $f(-2)$ (c) $f(x^2)$

35. $f(x) = \begin{cases} 2x + 1, & x < 0 \\ 2x + 2, & x \geq 0 \end{cases}$

    (a) $f(-1)$ (b) $f(0)$ (c) $f(2)$

36. $f(x) = \begin{cases} x^2 + 2, & x \leq 1 \\ 2x^2 + 2, & x > 1 \end{cases}$

    (a) $f(-2)$ (b) $f(1)$ (c) $f(2)$

37. $f(x) = \begin{cases} 3x - 1, & x < -1 \\ 4, & -1 \leq x \leq 1 \\ x^2, & x > 1 \end{cases}$

    (a) $f(-2)$ (b) $f\left(-\frac{1}{2}\right)$ (c) $f(3)$

38. $f(x) = \begin{cases} 4 - 5x, & x \leq -2 \\ 0, & -2 < x < 2 \\ x^2 + 1, & x > 2 \end{cases}$

    (a) $f(-3)$ (b) $f(4)$ (c) $f(-1)$

In Exercises 39–44, complete the table.

39. $f(x) = x^2 - 3$

x	-2	-1	0	1	2
$f(x)$					

40. $g(x) = \sqrt{x - 3}$

x	3	4	5	6	7
$g(x)$					

41. $h(t) = \frac{1}{2}|t + 3|$

t	-5	-4	-3	-2	-1
$h(t)$					

42. $f(s) = \dfrac{|s - 2|}{s - 2}$

s	0	1	$\frac{3}{2}$	$\frac{5}{2}$	4
$f(s)$					

43. $f(x) = \begin{cases} -\frac{1}{2}x + 4, & x \le 0 \\ (x - 2)^2, & x > 0 \end{cases}$

x	-2	-1	0	1	2
$f(x)$					

44. $f(x) = \begin{cases} 9 - x^2, & x < 3 \\ x - 3, & x \ge 3 \end{cases}$

x	1	2	3	4	5
$f(x)$					

In Exercises 45–52, find all real values of x such that $f(x) = 0$.

45. $f(x) = 15 - 3x$ **46.** $f(x) = 5x + 1$

47. $f(x) = \dfrac{3x - 4}{5}$ **48.** $f(x) = \dfrac{12 - x^2}{5}$

49. $f(x) = x^2 - 9$ **50.** $f(x) = x^2 - 8x + 15$

51. $f(x) = x^3 - x$ **52.** $f(x) = x^3 - x^2 - 4x + 4$

In Exercises 53–56, find the value(s) of x for which $f(x) = g(x)$.

53. $f(x) = x^2 + 2x + 1, \quad g(x) = 3x + 3$

54. $f(x) = x^4 - 2x^2, \quad g(x) = 2x^2$

55. $f(x) = \sqrt{3x} + 1, \quad g(x) = x + 1$

56. $f(x) = \sqrt{x} - 4, \quad g(x) = 2 - x$

In Exercises 57–70, find the domain of the function.

57. $f(x) = 5x^2 + 2x - 1$ **58.** $g(x) = 1 - 2x^2$

59. $h(t) = \dfrac{4}{t}$ **60.** $s(y) = \dfrac{3y}{y + 5}$

61. $g(y) = \sqrt{y - 10}$ **62.** $f(t) = \sqrt[3]{t + 4}$

63. $f(x) = \sqrt[4]{1 - x^2}$ **64.** $f(x) = \sqrt[4]{x^2 + 3x}$

65. $g(x) = \dfrac{1}{x} - \dfrac{3}{x + 2}$ **66.** $h(x) = \dfrac{10}{x^2 - 2x}$

67. $f(s) = \dfrac{\sqrt{s - 1}}{s - 4}$ **68.** $f(x) = \dfrac{\sqrt{x + 6}}{6 + x}$

69. $f(x) = \dfrac{x - 4}{\sqrt{x}}$ **70.** $f(x) = \dfrac{x - 5}{\sqrt{x^2 - 9}}$

In Exercises 71–74, assume that the domain of f is the set $A = \{-2, -1, 0, 1, 2\}$. Determine the set of ordered pairs that represents the function f.

71. $f(x) = x^2$ **72.** $f(x) = x^2 - 3$

73. $f(x) = |x| + 2$ **74.** $f(x) = |x + 1|$

Exploration   In Exercises 75–78, match the data with one of the following functions

$$f(x) = cx, \ g(x) = cx^2, \ h(x) = c\sqrt{|x|}, \ \text{and} \ r(x) = \frac{c}{x}$$

and determine the value of the constant c that will make the function fit the data in the table.

75.

x	-4	-1	0	1	4
y	-32	-2	0	-2	-32

76.

x	-4	-1	0	1	4
y	-1	$-\frac{1}{4}$	0	$\frac{1}{4}$	1

77.

x	-4	-1	0	1	4
y	-8	-32	Undef.	32	8

78.

x	-4	-1	0	1	4
y	6	3	0	3	6

In Exercises 79–86, find the difference quotient and simplify your answer.

79. $f(x) = x^2 - x + 1, \quad \dfrac{f(2 + h) - f(2)}{h}, h \ne 0$

80. $f(x) = 5x - x^2, \quad \dfrac{f(5 + h) - f(5)}{h}, h \ne 0$

81. $f(x) = x^3 + 3x, \quad \dfrac{f(x + h) - f(x)}{h}, h \ne 0$

82. $f(x) = 4x^2 - 2x, \quad \dfrac{f(x + h) - f(x)}{h}, h \ne 0$

83. $g(x) = \dfrac{1}{x^2}, \quad \dfrac{g(x) - g(3)}{x - 3}, x \ne 3$

84. $f(t) = \dfrac{1}{t - 2}, \quad \dfrac{f(t) - f(1)}{t - 1}, t \ne 1$

85. $f(x) = \sqrt{5x}, \quad \dfrac{f(x) - f(5)}{x - 5}, x \ne 5$

86. $f(x) = x^{2/3} + 1, \quad \dfrac{f(x) - f(8)}{x - 8}, x \ne 8$

87. *Geometry* Write the area A of a square as a function of its perimeter P.

88. *Geometry* Write the area A of a circle as a function of its circumference C.

The symbol ∯ indicates an example or exercise that highlights algebraic techniques specifically used in calculus.

89. *Maximum Volume* An open box of maximum volume is to be made from a square piece of material 24 centimeters on a side by cutting equal squares from the corners and turning up the sides (see figure).

(a) The table shows the volume V (in cubic centimeters) of the box for various heights x (in centimeters). Use the table to estimate the maximum volume.

Height, x	1	2	3	4	5	6
Volume, V	484	800	972	1024	980	864

(b) Plot the points (x, V) from the table in part (a). Does the relation defined by the ordered pairs represent V as a function of x?

(c) If V is a function of x, write the function and determine its domain.

90. *Maximum Profit* The cost per unit in the production of a portable CD player is $60. The manufacturer charges $90 per unit for orders of 100 or less. To encourage large orders, the manufacturer reduces the charge by $0.15 per CD player for each unit ordered in excess of 100 (for example, there would be a charge of $87 per CD player for an order size of 120).

(a) The table shows the profit P (in dollars) for various numbers of units ordered, x. Use the table to estimate the maximum profit.

Units, x	110	120	130	140
Profit, P	3135	3240	3315	3360

Units, x	150	160	170
Profit, P	3375	3360	3315

(b) Plot the points (x, P) from the table in part (a). Does the relation defined by the ordered pairs represent P as a function of x?

(c) If P is a function of x, write the function and determine its domain.

91. *Geometry* A right triangle is formed in the first quadrant by the x- and y-axes and a line through the point $(2, 1)$ (see figure). Write the area A of the triangle as a function of x, and determine the domain of the function.

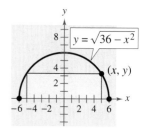

FIGURE FOR **91**   FIGURE FOR **92**

92. *Geometry* A rectangle is bounded by the x-axis and the semicircle $y = \sqrt{36 - x^2}$ (see figure). Write the area A of the rectangle as a function of x, and determine the domain of the function.

93. *Path of a Ball* The height y (in feet) of a baseball thrown by a child is

$$y = -\frac{1}{10}x^2 + 3x + 6$$

where x is the horizontal distance (in feet) from where the ball was thrown. Will the ball fly over the head of another child 30 feet away trying to catch the ball? (Assume that the child who is trying to catch the ball holds a baseball glove at a height of 5 feet.)

94. *Prescription Drugs* The amounts d (in billions of dollars) spent on prescription drugs in the United States from 1991 to 2002 (see figure) can be approximated by the model

$$d(t) = \begin{cases} 5.0t + 37, & 1 \le t \le 7 \\ 18.7t - 64, & 8 \le t \le 12 \end{cases}$$

where t represents the year, with $t = 1$ corresponding to 1991. Use this model to find the amount spent on prescription drugs in each year from 1991 to 2002. (Source: U.S. Centers for Medicare & Medicaid Services)

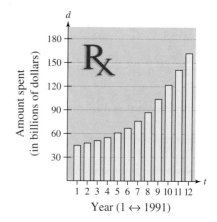

95. *Average Price* The average prices p (in thousands of dollars) of a new mobile home in the United States from 1990 to 2002 (see figure) can be approximated by the model

$$p(t) = \begin{cases} 0.182t^2 + 0.57t + 27.3, & 0 \leq t \leq 7 \\ 2.50t + 21.3, & 8 \leq t \leq 12 \end{cases}$$

where t represents the year, with $t = 0$ corresponding to 1990. Use this model to find the average price of a mobile home in each year from 1990 to 2002. (Source: U.S. Census Bureau)

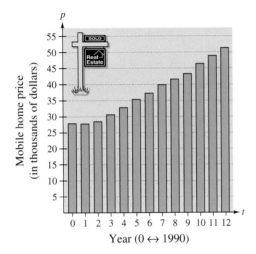

Year (0 ↔ 1990)

96. *Postal Regulations* A rectangular package to be sent by the U.S. Postal Service can have a maximum combined length and girth (perimeter of a cross section) of 108 inches (see figure).

(a) Write the volume V of the package as a function of x. What is the domain of the function?

 (b) Use a graphing utility to graph your function. Be sure to use an appropriate window setting.

(c) What dimensions will maximize the volume of the package? Explain your answer.

97. *Cost, Revenue, and Profit* A company produces a product for which the variable cost is $12.30 per unit and the fixed costs are $98,000. The product sells for $17.98. Let x be the number of units produced and sold.

(a) The total cost for a business is the sum of the variable cost and the fixed costs. Write the total cost C as a function of the number of units produced.

(b) Write the revenue R as a function of the number of units sold.

(c) Write the profit P as a function of the number of units sold. (*Note:* $P = R - C$)

98. *Average Cost* The inventor of a new game believes that the variable cost for producing the game is $0.95 per unit and the fixed costs are $6000. The inventor sells each game for $1.69. Let x be the number of games sold.

(a) The total cost for a business is the sum of the variable cost and the fixed costs. Write the total cost C as a function of the number of games sold.

(b) Write the average cost per unit $\overline{C} = C/x$ as a function of x.

99. *Transportation* For groups of 80 or more people, a charter bus company determines the rate per person according to the formula

Rate $= 8 - 0.05(n - 80)$, $n \geq 80$

where the rate is given in dollars and n is the number of people.

(a) Write the revenue R for the bus company as a function of n.

(b) Use the function in part (a) to complete the table. What can you conclude?

n	90	100	110	120	130	140	150
$R(n)$							

100. *Physics* The force F (in tons) of water against the face of a dam is estimated by the function $F(y) = 149.76\sqrt{10}\,y^{5/2}$, where y is the depth of the water (in feet).

(a) Complete the table. What can you conclude from the table?

y	5	10	20	30	40
$F(y)$					

(b) Use the table to approximate the depth at which the force against the dam is 1,000,000 tons.

(c) Find the depth at which the force against the dam is 1,000,000 tons algebraically.

101. *Height of a Balloon* A balloon carrying a transmitter ascends vertically from a point 3000 feet from the receiving station.

(a) Draw a diagram that gives a visual representation of the problem. Let h represent the height of the balloon and let d represent the distance between the balloon and the receiving station.

(b) Write the height of the balloon as a function of d. What is the domain of the function?

Model It

102. Wildlife The graph shows the numbers of threatened and endangered fish species in the world from 1996 through 2003. Let $f(t)$ represent the number of threatened and endangered fish species in the year t. (Source: U.S. Fish and Wildlife Service)

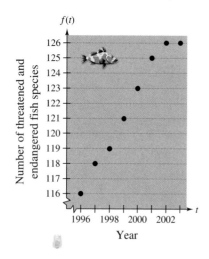

(a) Find $\dfrac{f(2003) - f(1996)}{2003 - 1996}$ and interpret the result in the context of the problem.

(b) Find a linear model for the data algebraically. Let N represent the number of threatened and endangered fish species and let $x = 6$ correspond to 1996.

(c) Use the model found in part (b) to complete the table.

x	6	7	8	9	10	11	12	13
N								

(d) Compare your results from part (c) with the actual data.

 (e) Use a graphing utility to find a linear model for the data. Let $x = 6$ correspond to 1996. How does the model you found in part (b) compare with the model given by the graphing utility?

Synthesis

True or False? **In Exercises 103 and 104, determine whether the statement is true or false. Justify your answer.**

103. The domain of the function given by $f(x) = x^4 - 1$ is $(-\infty, \infty)$, and the range of $f(x)$ is $(0, \infty)$.

104. The set of ordered pairs $\{(-8, -2), (-6, 0), (-4, 0), (-2, 2), (0, 4), (2, -2)\}$ represents a function.

105. Writing In your own words, explain the meanings of *domain* and *range*.

106. Think About It Consider $f(x) = \sqrt{x - 2}$ and $g(x) = \sqrt[3]{x - 2}$. Why are the domains of f and g different?

In Exercises 107 and 108, determine whether the statements use the word *function* in ways that are mathematically correct. Explain your reasoning.

107. (a) The sales tax on a purchased item is a function of the selling price.

(b) Your score on the next algebra exam is a function of the number of hours you study the night before the exam.

108. (a) The amount in your savings account is a function of your salary.

(b) The speed at which a free-falling baseball strikes the ground is a function of the height from which it was dropped.

P.6   **Analyzing Graphs of Functions**

What you should learn

- Use the Vertical Line Test for functions.
- Find the zeros of functions.
- Determine intervals on which functions are increasing or decreasing and determine relative maximum and relative minimum values of functions.
- Determine the average rate of change of a function.
- Identify even and odd functions.

Why you should learn it

Graphs of functions can help you visualize relationships between variables in real life. For instance, in Exercise 86 on page 79, you will use the graph of a function to represent visually the temperature for a city over a 24–hour period.

The Graph of a Function

In Section P.5, you studied functions from an algebraic point of view. In this section, you will study functions from a graphical perspective.

The **graph of a function** f is the collection of ordered pairs $(x, f(x))$ such that x is in the domain of f. As you study this section, remember that

$x =$ the directed distance from the y-axis

$y = f(x) =$ the directed distance from the x-axis

as shown in Figure P.53.

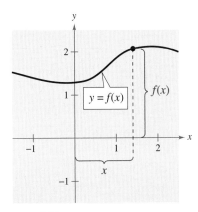

FIGURE P.53

Example 1   **Finding the Domain and Range of a Function**

Use the graph of the function f, shown in Figure P.54, to find (a) the domain of f, (b) the function values $f(-1)$ and $f(2)$, and (c) the range of f.

Solution

a. The closed dot at $(-1, 1)$ indicates that $x = -1$ is in the domain of f, whereas the open dot at $(5, 2)$ indicates that $x = 5$ is not in the domain. So, the domain of f is all x in the interval $[-1, 5)$.

b. Because $(-1, 1)$ is a point on the graph of f, it follows that $f(-1) = 1$. Similarly, because $(2, -3)$ is a point on the graph of f, it follows that $f(2) = -3$.

c. Because the graph does not extend below $f(2) = -3$ or above $f(0) = 3$, the range of f is the interval $[-3, 3]$.

✓*CHECKPOINT*   Now try Exercise 1.

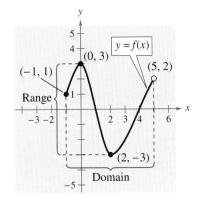

FIGURE P.54

The use of dots (open or closed) at the extreme left and right points of a graph indicates that the graph does not extend beyond these points. If no such dots are shown, assume that the graph extends beyond these points.

By the definition of a function, at most one y-value corresponds to a given x-value. This means that the graph of a function cannot have two or more different points with the same x-coordinate, and no two points on the graph of a function can be vertically above or below each other. It follows, then, that a vertical line can intersect the graph of a function at most once. This observation provides a convenient visual test called the **Vertical Line Test** for functions.

Vertical Line Test for Functions

A set of points in a coordinate plane is the graph of y as a function of x if and only if no *vertical* line intersects the graph at more than one point.

Example 2 Vertical Line Test for Functions

Use the Vertical Line Test to decide whether the graphs in Figure P.55 represent y as a function of x.

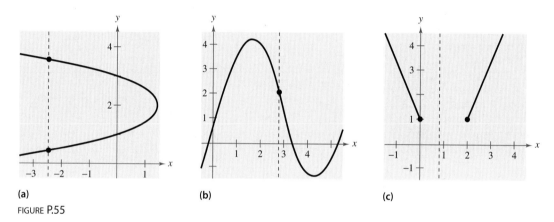

(a)   (b)   (c)

FIGURE P.55

Solution

a. This *is not* a graph of y as a function of x, because you can find a vertical line that intersects the graph twice. That is, for a particular input x, there is more than one output y.

b. This *is* a graph of y as a function of x, because every vertical line intersects the graph at most once. That is, for a particular input x, there is at most one output y.

c. This *is* a graph of y as a function of x. (Note that if a vertical line does not intersect the graph, it simply means that the function is undefined for that particular value of x.) That is, for a particular input x, there is at most one output y.

✓CHECKPOINT   Now try Exercise 9.

Zeros of a Function

If the graph of a function of x has an x-intercept at $(a, 0)$, then a is a **zero** of the function.

> ## Zeros of a Function
>
> The **zeros of a function** f of x are the x-values for which $f(x) = 0$.

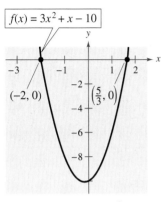

$f(x) = 3x^2 + x - 10$

$(-2, 0)$ $\left(\frac{5}{3}, 0\right)$

Zeros of f: $x = -2, x = \frac{5}{3}$

FIGURE P.56

Example 3   Finding the Zeros of a Function

Find the zeros of each function.

a. $f(x) = 3x^2 + x - 10$ **b.** $g(x) = \sqrt{10 - x^2}$ **c.** $h(t) = \dfrac{2t - 3}{t + 5}$

Solution

To find the zeros of a function, set the function equal to zero and solve for the independent variable.

a.
$$3x^2 + x - 10 = 0 \qquad \text{Set } f(x) \text{ equal to 0.}$$
$$(3x - 5)(x + 2) = 0 \qquad \text{Factor.}$$
$$3x - 5 = 0 \quad \Longrightarrow \quad x = \tfrac{5}{3} \qquad \text{Set 1st factor equal to 0.}$$
$$x + 2 = 0 \quad \Longrightarrow \quad x = -2 \qquad \text{Set 2nd factor equal to 0.}$$

The zeros of f are $x = \frac{5}{3}$ and $x = -2$. In Figure P.56, note that the graph of f has $\left(\frac{5}{3}, 0\right)$ and $(-2, 0)$ as its x-intercepts.

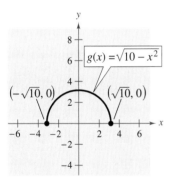

$g(x) = \sqrt{10 - x^2}$

$(-\sqrt{10}, 0)$ $(\sqrt{10}, 0)$

Zeros of g: $x = \pm\sqrt{10}$

FIGURE P.57

b.
$$\sqrt{10 - x^2} = 0 \qquad \text{Set } g(x) \text{ equal to 0.}$$
$$10 - x^2 = 0 \qquad \text{Square each side.}$$
$$10 = x^2 \qquad \text{Add } x^2 \text{ to each side.}$$
$$\pm\sqrt{10} = x \qquad \text{Extract square roots.}$$

The zeros of g are $x = -\sqrt{10}$ and $x = \sqrt{10}$. In Figure P.57, note that the graph of g has $\left(-\sqrt{10}, 0\right)$ and $\left(\sqrt{10}, 0\right)$ as its x-intercepts.

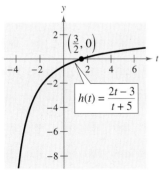

$\left(\frac{3}{2}, 0\right)$

$h(t) = \dfrac{2t - 3}{t + 5}$

Zero of h: $t = \frac{3}{2}$

FIGURE P.58

c.
$$\frac{2t - 3}{t + 5} = 0 \qquad \text{Set } h(t) \text{ equal to 0.}$$
$$2t - 3 = 0 \qquad \text{Multiply each side by } t + 5.$$
$$2t = 3 \qquad \text{Add 3 to each side.}$$
$$t = \frac{3}{2} \qquad \text{Divide each side by 2.}$$

The zero of h is $t = \frac{3}{2}$. In Figure P.58, note that the graph of h has $\left(\frac{3}{2}, 0\right)$ as its t-intercept.

✓**CHECKPOINT**   Now try Exercise 15.

Increasing and Decreasing Functions

The more you know about the graph of a function, the more you know about the function itself. Consider the graph shown in Figure P.59. As you move from *left to right*, this graph falls from $x = -2$ to $x = 0$, is constant from $x = 0$ to $x = 2$, and rises from $x = 2$ to $x = 4$.

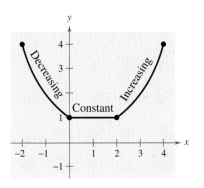

FIGURE P.59

> ### Increasing, Decreasing, and Constant Functions
>
> A function f is **increasing** on an interval if, for any x_1 and x_2 in the interval, $x_1 < x_2$ implies $f(x_1) < f(x_2)$.
>
> A function f is **decreasing** on an interval if, for any x_1 and x_2 in the interval, $x_1 < x_2$ implies $f(x_1) > f(x_2)$.
>
> A function f is **constant** on an interval if, for any x_1 and x_2 in the interval, $f(x_1) = f(x_2)$.

Example 4   Increasing and Decreasing Functions

Use the graphs in Figure P.60 to describe the increasing or decreasing behavior of each function.

Solution

a. This function is increasing over the entire real line.

b. This function is increasing on the interval $(-\infty, -1)$, decreasing on the interval $(-1, 1)$, and increasing on the interval $(1, \infty)$.

c. This function is increasing on the interval $(-\infty, 0)$, constant on the interval $(0, 2)$, and decreasing on the interval $(2, \infty)$.

(a)

(b)

(c)

FIGURE P.60

   Now try Exercise 33.

To help you decide whether a function is increasing, decreasing, or constant on an interval, you can evaluate the function for several values of *x*. However, calculus is needed to determine, for certain, all intervals on which a function is increasing, decreasing, or constant.

The points at which a function changes its increasing, decreasing, or constant behavior are helpful in determining the **relative minimum** or **relative maximum** values of the function.

Definitions of Relative Minimum and Relative Maximum

A function value $f(a)$ is called a **relative minimum** of f if there exists an interval (x_1, x_2) that contains a such that

$$x_1 < x < x_2 \quad \text{implies} \quad f(a) \leq f(x).$$

A function value $f(a)$ is called a **relative maximum** of f if there exists an interval (x_1, x_2) that contains a such that

$$x_1 < x < x_2 \quad \text{implies} \quad f(a) \geq f(x).$$

STUDY TIP

A relative minimum or relative maximum is also referred to as a *local* minimum or *local* maximum.

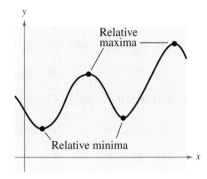

FIGURE P.61

Figure P.61 shows several different examples of relative minima and relative maxima. By writing a second-degree equation in standard form, $y = a(k - h)^2 + k$, you can find the *exact point* (h, k) at which it has a relative minimum or relative maximum. For the time being, however, you can use a graphing utility to find reasonable approximations of these points.

Example 5   Approximating a Relative Minimum

Use a graphing utility to approximate the relative minimum of the function given by $f(x) = 3x^2 - 4x - 2$.

Solution

The graph of f is shown in Figure P.62. By using the *zoom* and *trace* features or the *minimum* feature of a graphing utility, you can estimate that the function has a relative minimum at the point

$$(0.67, -3.33). \qquad \text{Relative minimum}$$

By writing this second-degree equation in standard form, $f(x) = 3\left(x - \frac{2}{3}\right)^2 - \frac{10}{3}$, you can determine that the exact point at which the relative minimum occurs is $\left(\frac{2}{3}, -\frac{10}{3}\right)$.

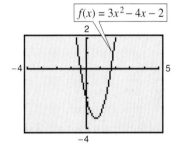

$f(x) = 3x^2 - 4x - 2$

FIGURE P.62

✓CHECKPOINT   Now try Exercise 49.

You can also use the *table* feature of a graphing utility to approximate numerically the relative minimum of the function in Example 5. Using a table that begins at 0.6 and increments the value of x by 0.01, you can approximate that the minimum of $f(x) = 3x^2 - 4x - 2$ occurs at the point $(0.67, -3.33)$.

Technology

If you use a graphing utility to estimate the x- and y-values of a relative minimum or relative maximum, the *zoom* feature will often produce graphs that are nearly flat. To overcome this problem, you can manually change the vertical setting of the viewing window. The graph will stretch vertically if the values of Ymin and Ymax are closer together.

FIGURE P.63

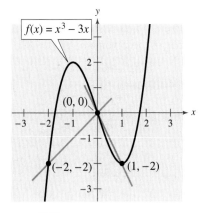

FIGURE P.64

Average Rate of Change

In Section P.4, you learned that the slope of a line can be interpreted as a *rate of change*. For a nonlinear graph whose slope changes at each point, the **average rate of change** between any two points $(x_1, f(x_1))$ and $(x_2, f(x_2))$ is the slope of the line through the two points (see Figure P.63). The line through the two points is called the **secant line,** and the slope of this line is denoted as m_{sec}.

$$\text{Average rate of change of } f \text{ from } x_1 \text{ to } x_2 = \frac{f(x_2) - f(x_1)}{x_2 - x_1}$$

$$= \frac{\text{change in } y}{\text{change in } x}$$

$$= m_{sec}$$

Example 6 Average Rate of Change of a Function

Find the average rates of change of $f(x) = x^3 - 3x$ (a) from $x_1 = -2$ to $x_2 = 0$ and (b) from $x_1 = 0$ to $x_2 = 1$ (see Figure P.64).

Solution

a. The average rate of change of f from $x_1 = -2$ to $x_2 = 0$ is

$$\frac{f(x_2) - f(x_1)}{x_2 - x_1} = \frac{f(0) - f(-2)}{0 - (-2)} = \frac{0 - (-2)}{2} = 1.$$ Secant line has positive slope.

b. The average rate of change of f from $x_1 = 0$ to $x_2 = 1$ is

$$\frac{f(x_2) - f(x_1)}{x_2 - x_1} = \frac{f(1) - f(0)}{1 - 0} = \frac{-2 - 0}{1} = -2.$$ Secant line has negative slope.

✓CHECKPOINT   Now try Exercise 63.

Example 7 Finding Average Speed

The distance s (in feet) a moving car is from a stoplight is given by the function $s(t) = 20t^{3/2}$, where t is the time (in seconds). Find the average speed of the car (a) from $t_1 = 0$ to $t_2 = 4$ seconds and (b) from $t_1 = 4$ to $t_2 = 9$ seconds.

Solution

a. The average speed of the car from $t_1 = 0$ to $t_2 = 4$ seconds is

$$\frac{s(t_2) - s(t_1)}{t_2 - t_1} = \frac{s(4) - s(0)}{4 - (0)} = \frac{160 - 0}{4} = 40 \text{ feet per second.}$$

b. The average speed of the car from $t_1 = 4$ to $t_2 = 9$ seconds is

$$\frac{s(t_2) - s(t_1)}{t_2 - t_1} = \frac{s(9) - s(4)}{9 - 4} = \frac{540 - 160}{5} = 76 \text{ feet per second.}$$

✓CHECKPOINT   Now try Exercise 89.

Even and Odd Functions

In Section P.3, you studied different types of symmetry of a graph. In the terminology of functions, a function is said to be **even** if its graph is symmetric with respect to the y-axis and to be **odd** if its graph is symmetric with respect to the origin. The symmetry tests in Section P.3 yield the following tests for even and odd functions.

Tests for Even and Odd Functions

A function $y = f(x)$ is **even** if, for each x in the domain of f,

$$f(-x) = f(x).$$

A function $y = f(x)$ is **odd** if, for each x in the domain of f,

$$f(-x) = -f(x).$$

Exploration

Graph each of the functions with a graphing utility. Determine whether the function is *even*, *odd*, or *neither*.

$$f(x) = x^2 - x^4$$

$$g(x) = 2x^3 + 1$$

$$h(x) = x^5 - 2x^3 + x$$

$$j(x) = 2 - x^6 - x^8$$

$$k(x) = x^5 - 2x^4 + x - 2$$

$$p(x) = x^9 + 3x^5 - x^3 + x$$

What do you notice about the equations of functions that are odd? What do you notice about the equations of functions that are even? Can you describe a way to identify a function as odd or even by inspecting the equation? Can you describe a way to identify a function as neither odd nor even by inspecting the equation?

Example 8   Even and Odd Functions

a. The function $g(x) = x^3 - x$ is odd because $g(-x) = -g(x)$, as follows.

$$g(-x) = (-x)^3 - (-x) \qquad \text{Substitute } -x \text{ for } x.$$

$$= -x^3 + x \qquad \text{Simplify.}$$

$$= -(x^3 - x) \qquad \text{Distributive Property}$$

$$= -g(x) \qquad \text{Test for odd function}$$

b. The function $h(x) = x^2 + 1$ is even because $h(-x) = h(x)$, as follows.

$$h(-x) = (-x)^2 + 1 \qquad \text{Substitute } -x \text{ for } x.$$

$$= x^2 + 1 \qquad \text{Simplify.}$$

$$= h(x) \qquad \text{Test for even function}$$

The graphs and symmetry of these two functions are shown in Figure P.65.

Additional Example

Is the function given by $f(x) = x^3 - 1$ even, odd, or neither?

Solution

Substituting $-x$ for x,

$f(-x) = (-x)^3 - 1 = -x^3 - 1.$

Because $f(x) = x^3 - 1$ and $f(-x) = -x^3 - 1$, $f(-x) \neq f(x)$ and $f(-x) \neq -f(x)$. The function is neither even nor odd.

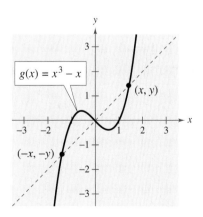

(a) **Symmetric to origin: Odd Function**

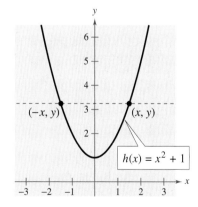

(b) **Symmetric to y-axis: Even Function**

FIGURE P.65

✓CHECKPOINT   Now try Exercise 71.

P.6 Exercises

VOCABULARY CHECK: Fill in the blanks.

1. The graph of a function f is the collection of _____ _____ or $(x, f(x))$ such that x is in the domain of f.

2. The _____ _____ _____ is used to determine whether the graph of an equation is a function of y in terms of x.

3. The _____ of a function f are the values of x for which $f(x) = 0$.

4. A function f is _____ on an interval if, for any x_1 and x_2 in the interval, $x_1 < x_2$ implies $f(x_1) > f(x_2)$.

5. A function value $f(a)$ is a relative _____ of f if there exists an interval (x_1, x_2) containing a such that $x_1 < x < x_2$ implies $f(a) \geq f(x)$.

6. The _____ _____ _____ _____ between any two points $(x_1, f(x_1))$ and $(x_2, f(x_2))$ is the slope of the line through the two points, and this line is called the _____ line.

7. A function f is _____ if for the each x in the domain of f, $f(-x) = -f(x)$.

8. A function f is _____ if its graph is symmetric with respect to the y-axis.

In Exercises 1–4, use the graph of the function to find the domain and range of f.

1.

2.

3.

4.

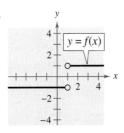

In Exercises 5–8, use the graph of the function to find the indicated function values.

5. (a) $f(-2)$ (b) $f(-1)$ **6.** (a) $f(-1)$ (b) $f(2)$
  (c) $f\left(\frac{1}{2}\right)$ (d) $f(1)$ (c) $f(0)$ (d) $f(1)$

7. (a) $f(-2)$ (b) $f(1)$ **8.** (a) $f(2)$ (b) $f(1)$
  (c) $f(0)$ (d) $f(2)$ (c) $f(3)$ (d) $f(-1)$

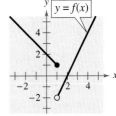

In Exercises 9–14, use the Vertical Line Test to determine whether y is a function of x. To print an enlarged copy of the graph, go to the website _www.mathgraphs.com_.

9. $y = \frac{1}{2}x^2$ **10.** $y = \frac{1}{4}x^3$

11. $x - y^2 = 1$ **12.** $x^2 + y^2 = 25$

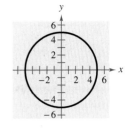

13. $x^2 = 2xy - 1$

14. $x = |y + 2|$

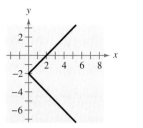

In Exercises 15–24, find the zeros of the function algebraically.

15. $f(x) = 2x^2 - 7x - 30$

16. $f(x) = 3x^2 + 22x - 16$

17. $f(x) = \dfrac{x}{9x^2 - 4}$

18. $f(x) = \dfrac{x^2 - 9x + 14}{4x}$

19. $f(x) = \frac{1}{2}x^3 - x$

20. $f(x) = x^3 - 4x^2 - 9x + 36$

21. $f(x) = 4x^3 - 24x^2 - x + 6$

22. $f(x) = 9x^4 - 25x^2$

23. $f(x) = \sqrt{2x - 1}$

24. $f(x) = \sqrt{3x + 2}$

 In Exercises 25–30, (a) use a graphing utility to graph the function and find the zeros of the function and (b) verify your results from part (a) algebraically.

25. $f(x) = 3 + \dfrac{5}{x}$

26. $f(x) = x(x - 7)$

27. $f(x) = \sqrt{2x + 11}$

28. $f(x) = \sqrt{3x - 14} - 8$

29. $f(x) = \dfrac{3x - 1}{x - 6}$

30. $f(x) = \dfrac{2x^2 - 9}{3 - x}$

In Exercises 31–38, determine the intervals over which the function is increasing, decreasing, or constant.

31. $f(x) = \frac{3}{2}x$

32. $f(x) = x^2 - 4x$

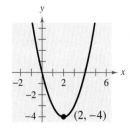

33. $f(x) = x^3 - 3x^2 + 2$

34. $f(x) = \sqrt{x^2 - 1}$

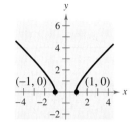

35. $f(x) = \begin{cases} x + 3, & x \le 0 \\ 3, & 0 < x \le 2 \\ 2x + 1, & x > 2 \end{cases}$

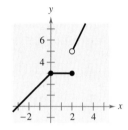

36. $f(x) = \begin{cases} 2x + 1, & x \le -1 \\ x^2 - 2, & x > -1 \end{cases}$

37. $f(x) = |x + 1| + |x - 1|$

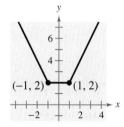

38. $f(x) = \dfrac{x^2 + x + 1}{x + 1}$

In Exercises 39–48, (a) use a graphing utility to graph the function and visually determine the intervals over which the function is increasing, decreasing, or constant, and (b) make a table of values to verify whether the function is increasing, decreasing, or constant over the intervals you identified in part (a).

39. $f(x) = 3$

40. $g(x) = x$

41. $g(s) = \dfrac{s^2}{4}$

42. $h(x) = x^2 - 4$

43. $f(t) = -t^4$

44. $f(x) = 3x^4 - 6x^2$

45. $f(x) = \sqrt{1-x}$

46. $f(x) = x\sqrt{x+3}$

47. $f(x) = x^{3/2}$

48. $f(x) = x^{2/3}$

 In Exercises 49–54, use a graphing utility to graph the function and approximate (to two decimal places) any relative minimum or relative maximum values.

49. $f(x) = (x-4)(x+2)$

50. $f(x) = 3x^2 - 2x - 5$

51. $f(x) = -x^2 + 3x - 2$

52. $f(x) = -2x^2 + 9x$

53. $f(x) = x(x-2)(x+3)$

54. $f(x) = x^3 - 3x^2 - x + 1$

In Exercises 55–62, graph the function and determine the interval(s) for which $f(x) \geq 0$.

55. $f(x) = 4 - x$

56. $f(x) = 4x + 2$

57. $f(x) = x^2 + x$

58. $f(x) = x^2 - 4x$

59. $f(x) = \sqrt{x-1}$

60. $f(x) = \sqrt{x+2}$

61. $f(x) = -(1 + |x|)$

62. $f(x) = \frac{1}{2}(2 + |x|)$

In Exercises 63–70, find the average rate of change of the function from x_1 to x_2.

Function	x-Values
63. $f(x) = -2x + 15$	$x_1 = 0, x_2 = 3$
64. $f(x) = 3x + 8$	$x_1 = 0, x_2 = 3$
65. $f(x) = x^2 + 12x - 4$	$x_1 = 1, x_2 = 5$
66. $f(x) = x^2 - 2x + 8$	$x_1 = 1, x_2 = 5$
67. $f(x) = x^3 - 3x^2 - x$	$x_1 = 1, x_2 = 3$
68. $f(x) = -x^3 + 6x^2 + x$	$x_1 = 1, x_2 = 6$
69. $f(x) = -\sqrt{x-2} + 5$	$x_1 = 3, x_2 = 11$
70. $f(x) = -\sqrt{x+1} + 3$	$x_1 = 3, x_2 = 8$

In Exercises 71–76, determine whether the function is even, odd, or neither. Then describe the symmetry.

71. $f(x) = x^6 - 2x^2 + 3$

72. $h(x) = x^3 - 5$

73. $g(x) = x^3 - 5x$

74. $f(x) = x\sqrt{1-x^2}$

75. $f(t) = t^2 + 2t - 3$

76. $g(s) = 4s^{2/3}$

In Exercises 77–80, write the height h of the rectangle as a function of x.

77.

78.

79.

80.

In Exercises 81–84, write the length L of the rectangle as a function of y.

81.

82.

83.

84.

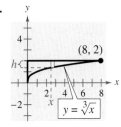 **85. *Electronics*** The number of lumens (time rate of flow of light) L from a fluorescent lamp can be approximated by the model

$$L = -0.294x^2 + 97.744x - 664.875, \quad 20 \leq x \leq 90$$

where x is the wattage of the lamp.

(a) Use a graphing utility to graph the function.

(b) Use the graph from part (a) to estimate the wattage necessary to obtain 2000 lumens.

Model It

86. Data Analysis: Temperature 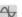 The table shows the temperature y (in degrees Fahrenheit) of a certain city over a 24-hour period. Let x represent the time of day, where $x = 0$ corresponds to 6 A.M.

Time, x	Temperature, y
0	34
2	50
4	60
6	64
8	63
10	59
12	53
14	46
16	40
18	36
20	34
22	37
24	45

A model that represents these data is given by

$$y = 0.026x^3 - 1.03x^2 + 10.2x + 34, \quad 0 \le x \le 24.$$

(a) Use a graphing utility to create a scatter plot of the data. Then graph the model in the same viewing window.

(b) How well does the model fit the data?

(c) Use the graph to approximate the times when the temperature was increasing and decreasing.

(d) Use the graph to approximate the maximum and minimum temperatures during this 24-hour period.

(e) Could this model be used to predict the temperature for the city during the next 24-hour period? Why or why not?

87. Coordinate Axis Scale Each function models the specified data for the years 1995 through 2005, with $t = 5$ corresponding to 1995. Estimate a reasonable scale for the vertical axis (e.g., hundreds, thousands, millions, etc.) of the graph and justify your answer. (There are many correct answers.)

(a) $f(t)$ represents the average salary of college professors.

(b) $f(t)$ represents the U.S. population.

(c) $f(t)$ represents the percent of the civilian work force that is unemployed.

88. Geometry Corners of equal size are cut from a square with sides of length 8 meters (see figure).

(a) Write the area A of the resulting figure as a function of x. Determine the domain of the function.

 (b) Use a graphing utility to graph the area function over its domain. Use the graph to find the range of the function.

(c) Identify the figure that would result if x were chosen to be the maximum value in the domain of the function. What would be the length of each side of the figure?

89. Digital Music Sales The estimated revenues r (in billions of dollars) from sales of digital music from 2002 to 2007 can be approximated by the model

$$r = 15.639t^3 - 104.75t^2 + 303.5t - 301, \quad 2 \le t \le 7$$

where t represents the year, with $t = 2$ corresponding to 2002. (Source: Fortune)

 (a) Use a graphing utility to graph the model.

(b) Find the average rate of change of the model from 2002 to 2007. Interpret your answer in the context of the problem.

90. Foreign College Students The numbers of foreign students F (in thousands) enrolled in colleges in the United States from 1992 to 2002 can be approximated by the model.

$$F = 0.004t^4 + 0.46t^2 + 431.6, \quad 2 \le t \le 12$$

where t represents the year, with $t = 2$ corresponding to 1992. (Source: Institute of International Education)

 (a) Use a graphing utility to graph the model.

 (b) Find the average rate of change of the model from 1992 to 2002. Interpret your answer in the context of the problem.

(c) Find the five-year time periods when the rate of change was the greatest and the least.

Physics   In Exercises 91–96, (a) use the position equation $s = -16t^2 + v_0 t + s_0$ to write a function that represents the situation, (b) use a graphing utility to graph the function, (c) find the average rate of change of the function from t_1 to t_2, (d) interpret your answer to part (c) in the context of the problem, (e) find the equation of the secant line through t_1 and t_2, and (f) graph the secant line in the same viewing window as your position function.

91. An object is thrown upward from a height of 6 feet at a velocity of 64 feet per second.

    $t_1 = 0, t_2 = 3$

92. An object is thrown upward from a height of 6.5 feet at a velocity of 72 feet per second.

    $t_1 = 0, t_2 = 4$

93. An object is thrown upward from ground level at a velocity of 120 feet per second.

    $t_1 = 3, t_2 = 5$

94. An object is thrown upward from ground level at a velocity of 96 feet per second.

    $t_1 = 2, t_2 = 5$

95. An object is dropped from a height of 120 feet.

    $t_1 = 0, t_2 = 2$

96. An object is dropped from a height of 80 feet.

    $t_1 = 1, t_2 = 2$

Synthesis

True or False?   In Exercises 97 and 98, determine whether the statement is true or false. Justify your answer.

97. A function with a square root cannot have a domain that is the set of real numbers.

98. It is possible for an odd function to have the interval $[0, \infty)$ as its domain.

99. If f is an even function, determine whether g is even, odd, or neither. Explain.

    (a) $g(x) = -f(x)$

    (b) $g(x) = f(-x)$

    (c) $g(x) = f(x) - 2$

    (d) $g(x) = f(x - 2)$

100. *Think About It*   Does the graph in Exercise 11 represent x as a function of y? Explain.

Think About It   In Exercises 101–104, find the coordinates of a second point on the graph of a function f if the given point is on the graph and the function is (a) even and (b) odd.

101. $\left(-\frac{3}{2}, 4\right)$

102. $\left(-\frac{5}{3}, -7\right)$

103. $(4, 9)$

104. $(5, -1)$

 105. *Writing*   Use a graphing utility to graph each function. Write a paragraph describing any similarities and differences you observe among the graphs.

(a) $y = x$

(b) $y = x^2$

(c) $y = x^3$

(d) $y = x^4$

(e) $y = x^5$

(f) $y = x^6$

 106. *Conjecture*   Use the results of Exercise 105 to make a conjecture about the graphs of $y = x^7$ and $y = x^8$. Use a graphing utility to graph the functions and compare the results with your conjecture.

P.7   A Library of Parent Functions

What you should learn

- Identify and graph linear and squaring functions.
- Identify and graph cubic, square root, and reciprocal functions.
- Identify and graph step and other piecewise-defined functions.
- Recognize graphs of parent functions.

Why you should learn it

Step functions can be used to model real-life situations. For instance, in Exercise 63 on page 87, you will use a step function to model the cost of sending an overnight package from Los Angeles to Miami.

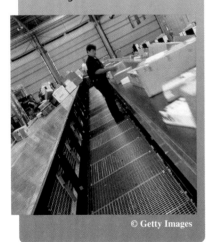

© Getty Images

Linear and Squaring Functions

One of the goals of this text is to enable you to recognize the basic shapes of the graphs of different types of functions. For instance, you know that the graph of the **linear function** $f(x) = ax + b$ is a line with slope $m = a$ and y-intercept at $(0, b)$. The graph of the linear function has the following characteristics.

- The domain of the function is the set of all real numbers.
- The range of the function is the set of all real numbers.
- The graph has an x-intercept of $(-b/m, 0)$ and a y-intercept of $(0, b)$.
- The graph is increasing if $m > 0$, decreasing if $m < 0$, and constant if $m = 0$.

Example 1   Writing a Linear Function

Write the linear function f for which $f(1) = 3$ and $f(4) = 0$.

Solution

To find the equation of the line that passes through $(x_1, y_1) = (1, 3)$ and $(x_2, y_2) = (4, 0)$, first find the slope of the line.

$$m = \frac{y_2 - y_1}{x_2 - x_1} = \frac{0 - 3}{4 - 1} = \frac{-3}{3} = -1$$

Next, use the point-slope form of the equation of a line.

$$y - y_1 = m(x - x_1) \qquad \text{Point-slope form}$$
$$y - 3 = -1(x - 1) \qquad \text{Substitute for } x_1, y_1, \text{ and } m.$$
$$y = -x + 4 \qquad \text{Simplify.}$$
$$f(x) = -x + 4 \qquad \text{Function notation}$$

The graph of this function is shown in Figure P.66.

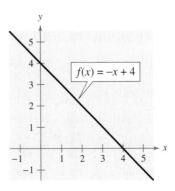

$f(x) = -x + 4$

FIGURE P.66

✓CHECKPOINT   Now try Exercise 1.

There are two special types of linear functions, the **constant function** and the **identity function.** A constant function has the form

$$f(x) = c$$

and has the domain of all real numbers with a range consisting of a single real number c. The graph of a constant function is a horizontal line, as shown in Figure P.67. The identity function has the form

$$f(x) = x.$$

Its domain and range are the set of all real numbers. The identity function has a slope of $m = 1$ and a y-intercept $(0, 0)$. The graph of the identity function is a line for which each x-coordinate equals the corresponding y-coordinate. The graph is always increasing, as shown in Figure P.68.

FIGURE P.67

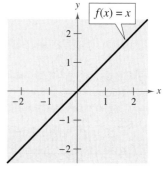

FIGURE P.68

The graph of the **squaring function**

$$f(x) = x^2$$

is a U-shaped curve with the following characteristics.

- The domain of the function is the set of all real numbers.
- The range of the function is the set of all nonnegative real numbers.
- The function is even.
- The graph has an intercept at $(0, 0)$.
- The graph is decreasing on the interval $(-\infty, 0)$ and increasing on the interval $(0, \infty)$.
- The graph is symmetric with respect to the y-axis.
- The graph has a relative minimum at $(0, 0)$.

The graph of the squaring function is shown in Figure P.69.

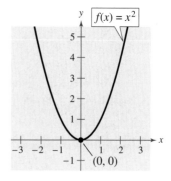

FIGURE P.69

Cubic, Square Root, and Reciprocal Functions

The basic characteristics of the graphs of the **cubic, square root,** and **reciprocal functions** are summarized below.

1. The graph of the cubic function $f(x) = x^3$ has the following characteristics.
   - The domain of the function is the set of all real numbers.
   - The range of the function is the set of all real numbers.
   - The function is odd.
   - The graph has an intercept at $(0, 0)$.
   - The graph is increasing on the interval $(-\infty, \infty)$.
   - The graph is symmetric with respect to the origin.

   The graph of the cubic function is shown in Figure P.70.

2. The graph of the *square root* function $f(x) = \sqrt{x}$ has the following characteristics.
   - The domain of the function is the set of all nonnegative real numbers.
   - The range of the function is the set of all nonnegative real numbers.
   - The graph has an intercept at $(0, 0)$.
   - The graph is increasing on the interval $(0, \infty)$.

   The graph of the square root function is shown in Figure P.71.

3. The graph of the reciprocal function $f(x) = \dfrac{1}{x}$ has the following characteristics.
   - The domain of the function is $(-\infty, 0) \cup (0, \infty)$.
   - The range of the function is $(-\infty, 0) \cup (0, \infty)$.
   - The function is odd.
   - The graph does not have any intercepts.
   - The graph is decreasing on the intervals $(-\infty, 0)$ and $(0, \infty)$.
   - The graph is symmetric with respect to the origin.

   The graph of the reciprocal function is shown in Figure P.72.

Cubic function
FIGURE P.70

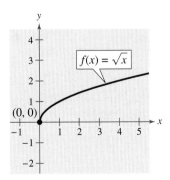

Square root function
FIGURE P.71

Reciprocal function
FIGURE P.72

FIGURE P.73

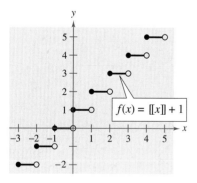

FIGURE P.74

Step and Piecewise-Defined Functions

Functions whose graphs resemble sets of stairsteps are known as **step functions.** The most famous of the step functions is the **greatest integer function,** which is denoted by $[\![x]\!]$ and defined as

$$f(x) = [\![x]\!] = \textit{the greatest integer less than or equal to } x.$$

Some values of the greatest integer function are as follows.

$$[\![-1]\!] = (\text{greatest integer} \leq -1) = -1$$

$$\left[\!\!\left[-\tfrac{1}{2}\right]\!\!\right] = \left(\text{greatest integer} \leq -\tfrac{1}{2}\right) = -1$$

$$\left[\!\!\left[\tfrac{1}{10}\right]\!\!\right] = \left(\text{greatest integer} \leq \tfrac{1}{10}\right) = 0$$

$$[\![1.5]\!] = (\text{greatest integer} \leq 1.5) = 1$$

The graph of the greatest integer function

$$f(x) = [\![x]\!]$$

has the following characteristics, as shown in Figure P.73.

- The domain of the function is the set of all real numbers.
- The range of the function is the set of all integers.
- The graph has a y-intercept at $(0, 0)$ and x-intercepts in the interval $[0, 1)$.
- The graph is constant between each pair of consecutive integers.
- The graph jumps vertically one unit at each integer value.

Example 2 **Evaluating a Step Function**

Evaluate the function when $x = -1, 2,$ and $\tfrac{3}{2}$.

$$f(x) = [\![x]\!] + 1$$

Solution

For $x = -1$, the greatest integer ≤ -1 is -1, so

$$f(-1) = [\![-1]\!] + 1 = -1 + 1 = 0.$$

For $x = 2$, the greatest integer ≤ 2 is 2, so

$$f(2) = [\![2]\!] + 1 = 2 + 1 = 3.$$

For $x = \tfrac{3}{2}$, the greatest integer $\leq \tfrac{3}{2}$ is 1, so

$$f\!\left(\tfrac{3}{2}\right) = \left[\!\!\left[\tfrac{3}{2}\right]\!\!\right] + 1 = 1 + 1 = 2.$$

You can verify your answers by examining the graph of $f(x) = [\![x]\!] + 1$ shown in Figure P.74.

✓CHECKPOINT   Now try Exercise 29.

Recall from Section P.5 that a piecewise-defined function is defined by two or more equations over a specified domain. To graph a piecewise-defined function, graph each equation separately over the specified domain, as shown in Example 3.

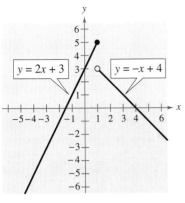

FIGURE P.75

Example 3   **Graphing a Piecewise-Defined Function**

Sketch the graph of

$$f(x) = \begin{cases} 2x + 3, & x \le 1 \\ -x + 4, & x > 1 \end{cases}.$$

Solution

This piecewise-defined function is composed of two linear functions. At $x = 1$ and to the left of $x = 1$ the graph is the line $y = 2x + 3$, and to the right of $x = 1$ the graph is the line $y = -x + 4$, as shown in Figure P.75. Notice that the point $(1, 5)$ is a solid dot and the point $(1, 3)$ is an open dot. This is because $f(1) = 2(1) + 3 = 5$.

✓CHECKPOINT   Now try Exercise 43.

Parent Functions

The eight graphs shown in Figure P.76 represent the most commonly used functions in algebra. Familiarity with the basic characteristics of these simple graphs will help you analyze the shapes of more complicated graphs—in particular, graphs obtained from these graphs by the rigid and nonrigid transformations studied in the next section.

(a) **Constant Function**

(b) **Identity Function**

(c) **Absolute Value Function**

(d) **Square Root Function**

(e) **Quadratic Function**

(f) **Cubic Function**

(g) **Reciprocal Function**

(h) **Greatest Integer Function**

FIGURE P.76

P.7 | Exercises

VOCABULARY CHECK: Match each function with its name.

1. $f(x) = [\![x]\!]$

2. $f(x) = x$

3. $f(x) = \dfrac{1}{x}$

4. $f(x) = x^2$

5. $f(x) = \sqrt{x}$

6. $f(x) = c$

7. $f(x) = |x|$

8. $f(x) = x^3$

9. $f(x) = ax + b$

(a) squaring function

(b) square root function

(c) cubic function

(d) linear function

(e) constant function

(f) absolute value function

(e) greatest integer function

(h) reciprocal function

(i) identity function

In Exercises 1–8, (a) write the linear function f such that it has the indicated function values and (b) sketch the graph of the function.

1. $f(1) = 4,\ f(0) = 6$

2. $f(-3) = -8,\ f(1) = 2$

3. $f(5) = -4,\ f(-2) = 17$

4. $f(3) = 9,\ f(-1) = -11$

5. $f(-5) = -1,\ f(5) = -1$

6. $f(-10) = 12,\ f(16) = -1$

7. $f\left(\tfrac{1}{2}\right) = -6,\ f(4) = -3$

8. $f\left(\tfrac{2}{3}\right) = -\tfrac{15}{2},\ f(-4) = -11$

 In Exercises 9–28, use a graphing utility to graph the function. Be sure to choose an appropriate viewing window.

9. $f(x) = -x - \tfrac{3}{4}$

10. $f(x) = 3x - \tfrac{5}{2}$

11. $f(x) = -\tfrac{1}{6}x - \tfrac{5}{2}$

12. $f(x) = \tfrac{5}{6} - \tfrac{2}{3}x$

13. $f(x) = x^2 - 2x$

14. $f(x) = -x^2 + 8x$

15. $h(x) = -x^2 + 4x + 12$

16. $g(x) = x^2 - 6x - 16$

17. $f(x) = x^3 - 1$

18. $f(x) = 8 - x^3$

19. $f(x) = (x - 1)^3 + 2$

20. $g(x) = 2(x + 3)^3 + 1$

21. $f(x) = 4\sqrt{x}$

22. $f(x) = 4 - 2\sqrt{x}$

23. $g(x) = 2 - \sqrt{x + 4}$

24. $h(x) = \sqrt{x + 2} + 3$

25. $f(x) = -\dfrac{1}{x}$

26. $f(x) = 4 + \dfrac{1}{x}$

27. $h(x) = \dfrac{1}{x + 2}$

28. $k(x) = \dfrac{1}{x - 3}$

In Exercises 29–36, evaluate the function for the indicated values.

29. $f(x) = [\![x]\!]$

(a) $f(2.1)$ (b) $f(2.9)$ (c) $f(-3.1)$ (d) $f\left(\tfrac{7}{2}\right)$

30. $g(x) = 2[\![x]\!]$

(a) $g(-3)$ (b) $g(0.25)$ (c) $g(9.5)$ (d) $g\left(\tfrac{11}{3}\right)$

31. $h(x) = [\![x + 3]\!]$

(a) $h(-2)$ (b) $h\left(\tfrac{1}{2}\right)$ (c) $h(4.2)$ (d) $h(-21.6)$

32. $f(x) = 4[\![x]\!] + 7$

(a) $f(0)$ (b) $f(-1.5)$ (c) $f(6)$ (d) $f\left(\tfrac{5}{3}\right)$

33. $h(x) = [\![3x - 1]\!]$

(a) $h(2.5)$ (b) $h(-3.2)$ (c) $h\left(\tfrac{7}{3}\right)$ (d) $h\left(-\tfrac{21}{3}\right)$

34. $k(x) = \left[\!\left[\tfrac{1}{2}x + 6\right]\!\right]$

(a) $k(5)$ (b) $k(-6.1)$ (c) $k(0.1)$ (d) $k(15)$

35. $g(x) = 3[\![x - 2]\!] + 5$

(a) $g(-2.7)$ (b) $g(-1)$ (c) $g(0.8)$ (d) $g(14.5)$

36. $g(x) = -7[\![x + 4]\!] + 6$

(a) $g\left(\tfrac{1}{8}\right)$ (b) $g(9)$ (c) $g(-4)$ (d) $g\left(\tfrac{3}{2}\right)$

In Exercises 37–42, sketch the graph of the function.

37. $g(x) = -[\![x]\!]$

38. $g(x) = 4[\![x]\!]$

39. $g(x) = [\![x]\!] - 2$

40. $g(x) = [\![x]\!] - 1$

41. $g(x) = [\![x + 1]\!]$

42. $g(x) = [\![x - 3]\!]$

In Exercises 43–50, graph the function.

43. $f(x) = \begin{cases} 2x + 3, & x < 0 \\ 3 - x, & x \ge 0 \end{cases}$

44. $g(x) = \begin{cases} x + 6, & x \le -4 \\ \tfrac{1}{2}x - 4, & x > -4 \end{cases}$

45. $f(x) = \begin{cases} \sqrt{4 + x}, & x < 0 \\ \sqrt{4 - x}, & x \ge 0 \end{cases}$

46. $f(x) = \begin{cases} 1 - (x - 1)^2, & x \le 2 \\ \sqrt{x - 2}, & x > 2 \end{cases}$

47. $f(x) = \begin{cases} x^2 + 5, & x \le 1 \\ -x^2 + 4x + 3, & x > 1 \end{cases}$

48. $h(x) = \begin{cases} 3 - x^2, & x < 0 \\ x^2 + 2, & x \geq 0 \end{cases}$

49. $h(x) = \begin{cases} 4 - x^2, & x < -2 \\ 3 + x, & -2 \leq x < 0 \\ x^2 + 1, & x \geq 0 \end{cases}$

50. $k(x) = \begin{cases} 2x + 1, & x \leq -1 \\ 2x^2 - 1, & -1 < x \leq 1 \\ 1 - x^2, & x > 1 \end{cases}$

In Exercises 51 and 52, (a) use a graphing utility to graph the function, (b) state the domain and range of the function, and (c) describe the pattern of the graph.

51. $s(x) = 2\left(\frac{1}{4}x - \left[\!\left[\frac{1}{4}x\right]\!\right]\right)$ **52.** $g(x) = 2\left(\frac{1}{4}x - \left[\!\left[\frac{1}{4}x\right]\!\right]\right)^2$

In Exercises 53–60, (a) identify the parent function and the transformed parent function shown in the graph, (b) write an equation for the function shown in the graph, and (c) use a graphing utility to verify your answers in parts (a) and (b).

53.

54.

55.

56.

57.

58.

59.

60.
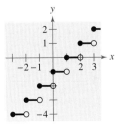

61. *Communications*   The cost of a telephone call between Denver and Boise is \$0.60 for the first minute and \$0.42 for each additional minute or portion of a minute. A model for the total cost C (in dollars) of the phone call is $C = 0.60 - 0.42[\![1 - t]\!]$, $t > 0$ where t is the length of the phone call in minutes.

  (a) Sketch the graph of the model.

  (b) Determine the cost of a call lasting 12 minutes and 30 seconds.

62. *Communications*   The cost of using a telephone calling card is \$1.05 for the first minute and \$0.38 for each additional minute or portion of a minute.

  (a) A customer needs a model for the cost C of using a calling card for a call lasting t minutes. Which of the following is the appropriate model? Explain.

$$C_1(t) = 1.05 + 0.38[\![t - 1]\!]$$
$$C_2(t) = 1.05 - 0.38[\![-(t - 1)]\!]$$

  (b) Graph the appropriate model. Determine the cost of a call lasting 18 minutes and 45 seconds.

63. *Delivery Charges*   The cost of sending an overnight package from Los Angeles to Miami is \$10.75 for a package weighing up to but not including 1 pound and \$3.95 for each additional pound or portion of a pound. A model for the total cost C (in dollars) of sending the package is $C = 10.75 + 3.95[\![x]\!]$, $x > 0$ where x is the weight in pounds.

  (a) Sketch a graph of the model.

  (b) Determine the cost of sending a package that weighs 10.33 pounds.

64. *Delivery Charges*   The cost of sending an overnight package from New York to Atlanta is \$9.80 for a package weighing up to but not including 1 pound and \$2.50 for each additional pound or portion of a pound.

  (a) Use the greatest integer function to create a model for the cost C of overnight delivery of a package weighing x pounds, $x > 0$.

  (b) Sketch the graph of the function.

65. *Wages*   A mechanic is paid \$12.00 per hour for regular time and time-and-a-half for overtime. The weekly wage function is given by

$$W(h) = \begin{cases} 12h, & 0 < h \leq 40 \\ 18(h - 40) + 480, & h > 40 \end{cases}$$

where h is the number of hours worked in a week.

  (a) Evaluate $W(30)$, $W(40)$, $W(45)$, and $W(50)$.

  (b) The company increased the regular work week to 45 hours. What is the new weekly wage function?

66. *Snowstorm* During a nine-hour snowstorm, it snows at a rate of 1 inch per hour for the first 2 hours, at a rate of 2 inches per hour for the next 6 hours, and at a rate of 0.5 inch per hour for the final hour. Write and graph a piecewise-defined function that gives the depth of the snow during the snowstorm. How many inches of snow accumulated from the storm?

Model It

67. *Revenue* The table shows the monthly revenue y (in thousands of dollars) of a landscaping business for each month of the year 2005, with $x = 1$ representing January.

Month, x	Revenue, y
1	5.2
2	5.6
3	6.6
4	8.3
5	11.5
6	15.8
7	12.8
8	10.1
9	8.6
10	6.9
11	4.5
12	2.7

A mathematical model that represents these data is

$$f(x) = \begin{cases} -1.97x + 26.3 \\ 0.505x^2 - 1.47x + 6.3 \end{cases}$$

(a) What is the domain of each part of the piecewise-defined function? How can you tell? Explain your reasoning.

(b) Sketch a graph of the model.

(c) Find $f(5)$ and $f(11)$, and interpret your results in the context of the problem.

(d) How do the values obtained from the model in part (b) compare with the actual data values?

68. *Fluid Flow* The intake pipe of a 100-gallon tank has a flow rate of 10 gallons per minute, and two drainpipes have flow rates of 5 gallons per minute each. The figure shows the volume V of fluid in the tank as a function of time t. Determine the combination of the input pipe and drain pipes in which the fluid is flowing in specific subintervals of the 1 hour of time shown on the graph. (There are many correct answers.)

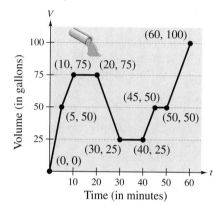

Synthesis

True or False? In Exercises 69 and 70, determine whether the statement is true or false. Justify your answer.

69. A piecewise-defined function will always have at least one x-intercept or at least one y-intercept.

70. $f(x) = \begin{cases} 2, & 1 \le x < 2 \\ 4, & 2 \le x < 3 \\ 6, & 3 \le x < 4 \end{cases}$

can be rewritten as $f(x) = 2[\![x]\!]$, $1 \le x < 4$.

Exploration In Exercises 71 and 72, write equations for the piecewise-defined function shown in the graph.

71.

72.

P.8 Transformations of Functions

What you should learn

- Use vertical and horizontal shifts to sketch graphs of functions.
- Use reflections to sketch graphs of functions.
- Use nonrigid transformations to sketch graphs of functions.

Why you should learn it

Knowing the graphs of common functions and knowing how to shift, reflect, and stretch graphs of functions can help you sketch a wide variety of simple functions by hand. This skill is useful in sketching graphs of functions that model real-life data, such as in Exercise 68 on page 98, where you are asked to sketch the graph of a function that models the amounts of mortgage debt outstanding from 1990 through 2002.

© Ken Fisher/Getty Images

Shifting Graphs

Many functions have graphs that are simple transformations of the parent graphs summarized in Section P.7. For example, you can obtain the graph of

$$h(x) = x^2 + 2$$

by shifting the graph of $f(x) = x^2$ *upward* two units, as shown in Figure P.77. In function notation, h and f are related as follows.

$$h(x) = x^2 + 2 = f(x) + 2 \qquad \text{Upward shift of two units}$$

Similarly, you can obtain the graph of

$$g(x) = (x - 2)^2$$

by shifting the graph of $f(x) = x^2$ to the *right* two units, as shown in Figure P.78. In this case, the functions g and f have the following relationship.

$$g(x) = (x - 2)^2 = f(x - 2) \qquad \text{Right shift of two units}$$

FIGURE P.77

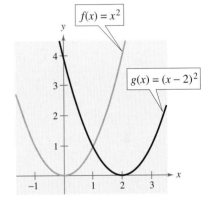

FIGURE P.78

The following list summarizes this discussion about horizontal and vertical shifts.

Vertical and Horizontal Shifts

Let c be a positive real number. **Vertical and horizontal shifts** in the graph of $y = f(x)$ are represented as follows.

1. Vertical shift c units *upward:* $h(x) = f(x) + c$

2. Vertical shift c units *downward:* $h(x) = f(x) - c$

3. Horizontal shift c units to the *right:* $h(x) = f(x - c)$

4. Horizontal shift c units to the *left:* $h(x) = f(x + c)$

You might also wish to illustrate simple transformations of functions numerically using tables to emphasize what happens to individual ordered pairs. For instance, if you have $f(x) = x^2, h(x) = x^2 + 2 = f(x) + 2,$ and $g(x) = (x - 2)^2 = f(x - 2),$ you can illustrate these transformations with the following tables.

x	$f(x)$	$h(x) = f(x) + 2$
-2	4	$4 + 2 = 6$
-1	1	$1 + 2 = 3$
0	0	$0 + 2 = 2$
1	1	$1 + 2 = 3$
2	4	$4 + 2 = 6$

x	$x - 2$	$g(x) = f(x - 2)$
0	$0 - 2 = -2$	4
1	$1 - 2 = -1$	1
2	$2 - 2 = 0$	0
3	$3 - 2 = 1$	1
4	$4 - 2 = 2$	4

Some graphs can be obtained from combinations of vertical and horizontal shifts, as demonstrated in Example 1(b). Vertical and horizontal shifts generate a *family of functions*, each with the same shape but at different locations in the plane.

Example 1   Shifts in the Graphs of a Function

Use the graph of $f(x) = x^3$ to sketch the graph of each function.

a. $g(x) = x^3 - 1$

b. $h(x) = (x + 2)^3 + 1$

Solution

a. Relative to the graph of $f(x) = x^3,$ the graph of $g(x) = x^3 - 1$ is a downward shift of one unit, as shown in Figure P.79.

b. Relative to the graph of $f(x) = x^3,$ the graph of $h(x) = (x + 2)^3 + 1$ involves a left shift of two units and an upward shift of one unit, as shown in Figure P.80.

FIGURE P.79

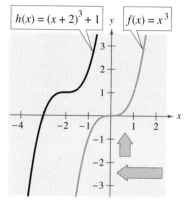

FIGURE P.80

✓CHECKPOINT   Now try Exercise 1.

In Figure P.80, notice that the same result is obtained if the vertical shift precedes the horizontal shift *or* if the horizontal shift precedes the vertical shift.

Exploration

Graphing utilities are ideal tools for exploring translations of functions. Graph $f, g,$ and h in same viewing window. Before looking at the graphs, try to predict how the graphs of g and h relate to the graph of f.

a. $f(x) = x^2,$ $g(x) = (x - 4)^2,$ $h(x) = (x - 4)^2 + 3$

b. $f(x) = x^2,$ $g(x) = (x + 1)^2,$ $h(x) = (x + 1)^2 - 2$

c. $f(x) = x^2,$ $g(x) = (x + 4)^2,$ $h(x) = (x + 4)^2 + 2$

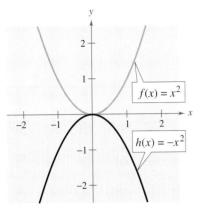

FIGURE P.81

Reflecting Graphs

The second common type of transformation is a **reflection.** For instance, if you consider the x-axis to be a mirror, the graph of

$$h(x) = -x^2$$

is the mirror image (or reflection) of the graph of

$$f(x) = x^2,$$

as shown in Figure P.81.

> **Reflections in the Coordinate Axes**
>
> **Reflections** in the coordinate axes of the graph of $y = f(x)$ are represented as follows.
>
> **1.** Reflection in the x-axis: $h(x) = -f(x)$
>
> **2.** Reflection in the y-axis: $h(x) = f(-x)$

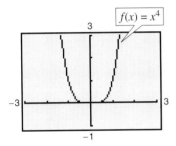

FIGURE P.82

Example 2   **Finding Equations from Graphs**

The graph of the function given by

$$f(x) = x^4$$

is shown in Figure P.82. Each of the graphs in Figure P.83 is a transformation of the graph of f. Find an equation for each of these functions.

(a)

(b)

FIGURE P.83

Solution

a. The graph of g is a reflection in the x-axis *followed by* an upward shift of two units of the graph of $f(x) = x^4$. So, the equation for g is

$$g(x) = -x^4 + 2.$$

b. The graph of h is a horizontal shift of three units to the right *followed by* a reflection in the x-axis of the graph of $f(x) = x^4$. So, the equation for h is

$$h(x) = -(x - 3)^4.$$

✓CHECKPOINT   Now try Exercise 9.

Exploration

Reverse the order of transformations in Example 2(a). Do you obtain the same graph? Do the same for Example 2(b). Do you obtain the same graph? Explain.

Example 3 **Reflections and Shifts**

Compare the graph of each function with the graph of $f(x) = \sqrt{x}$.

a. $g(x) = -\sqrt{x}$ **b.** $h(x) = \sqrt{-x}$ **c.** $k(x) = -\sqrt{x+2}$

Algebraic Solution

a. The graph of g is a reflection of the graph of f in the x-axis because

$$g(x) = -\sqrt{x}$$

$$= -f(x).$$

b. The graph of h is a reflection of the graph of f in the y-axis because

$$h(x) = \sqrt{-x}$$

$$= f(-x).$$

c. The graph of k is a left shift of two units followed by a reflection in the x-axis because

$$k(x) = -\sqrt{x+2}$$

$$= -f(x+2).$$

Graphical Solution

a. Graph f and g on the same set of coordinate axes. From the graph in Figure P.84, you can see that the graph of g is a reflection of the graph of f in the x-axis.

b. Graph f and h on the same set of coordinate axes. From the graph in Figure P.85, you can see that the graph of h is a reflection of the graph of f in the y-axis.

c. Graph f and k on the same set of coordinate axes. From the graph in Figure P.86, you can see that the graph of k is a left shift of two units of the graph of f, followed by a reflection in the x-axis.

FIGURE P.84   FIGURE P.85

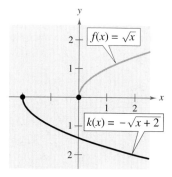

FIGURE P.86

✓CHECKPOINT   Now try Exercise 19.

Activities

1. How are the graphs of $f(x)$ and $g(x) = -f(x)$ related?

   Answer: They are reflections of each other in the x-axis.

2. Compare the graph of $f(x) = |x|$ with the graph of $g(x) = |x-9|$.

   Answer: $g(x)$ is $f(x)$ shifted to the right nine units.

When sketching the graphs of functions involving square roots, remember that the domain must be restricted to exclude negative numbers inside the radical. For instance, here are the domains of the functions in Example 3.

Domain of $g(x) = -\sqrt{x}$: $x \geq 0$

Domain of $h(x) = \sqrt{-x}$: $x \leq 0$

Domain of $k(x) = -\sqrt{x+2}$: $x \geq -2$

FIGURE P.87

FIGURE P.88

FIGURE P.89

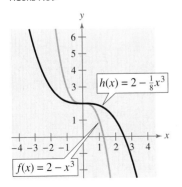

FIGURE P.90

Nonrigid Transformations

Horizontal shifts, vertical shifts, and reflections are **rigid transformations** because the basic shape of the graph is unchanged. These transformations change only the *position* of the graph in the coordinate plane. **Nonrigid transformations** are those that cause a *distortion*—a change in the shape of the original graph. For instance, a nonrigid transformation of the graph of $y = f(x)$ is represented by $g(x) = cf(x)$, where the transformation is a **vertical stretch** if $c > 1$ and a **vertical shrink** if $0 < c < 1$. Another nonrigid transformation of the graph of $y = f(x)$ is represented by $h(x) = f(cx)$, where the transformation is a **horizontal shrink** if $c > 1$ and a **horizontal stretch** if $0 < c < 1$.

Example 4   Nonrigid Transformations

Compare the graph of each function with the graph of $f(x) = |x|$.

a. $h(x) = 3|x|$ **b.** $g(x) = \frac{1}{3}|x|$

Solution

a. Relative to the graph of $f(x) = |x|$, the graph of

$$h(x) = 3|x| = 3f(x)$$

is a vertical stretch (each y-value is multiplied by 3) of the graph of f. (See Figure P.87.)

b. Similarly, the graph of

$$g(x) = \tfrac{1}{3}|x| = \tfrac{1}{3}f(x)$$

is a vertical shrink $\left(\text{each } y\text{-value is multiplied by } \tfrac{1}{3}\right)$ of the graph of f. (See Figure P.88.)

✓**CHECKPOINT**   Now try Exercise 23.

Example 5   Nonrigid Transformations

Compare the graph of each function with the graph of $f(x) = 2 - x^3$.

a. $g(x) = f(2x)$ **b.** $h(x) = f\!\left(\tfrac{1}{2}x\right)$

Solution

a. Relative to the graph of $f(x) = 2 - x^3$, the graph of

$$g(x) = f(2x) = 2 - (2x)^3 = 2 - 8x^3$$

is a horizontal shrink $(c > 1)$ of the graph of f. (See Figure P.89.)

b. Similarly, the graph of

$$h(x) = f\!\left(\tfrac{1}{2}x\right) = 2 - \left(\tfrac{1}{2}x\right)^3 = 2 - \tfrac{1}{8}x^3$$

is a horizontal stretch $(0 < c < 1)$ of the graph of f. (See Figure P.90.)

✓**CHECKPOINT**   Now try Exercise 27.

P.8 | Exercises

VOCABULARY CHECK:

In Exercises 1–5, fill in the blanks.

1. Horizontal shifts, vertical shifts, and reflections are called _____ transformations.

2. A reflection in the x-axis of $y = f(x)$ is represented by $h(x) =$ _____, while a reflection in the y-axis of $y = f(x)$ is represented by $h(x) =$ _____.

3. Transformations that cause a distortion in the shape of the graph of $y = f(x)$ are called _____ transformations.

4. A nonrigid transformation of $y = f(x)$ represented by $h(x) = f(cx)$ is a _____ _____ if $c > 1$ and a _____ _____ if $0 < c < 1$.

5. A nonrigid transformation of $y = f(x)$ represented by $g(x) = cf(x)$ is a _____ _____ if $c > 1$ and a _____ _____ if $0 < c < 1$.

6. Match the rigid transformation of $y = f(x)$ with the correct representation of the graph of h, where $c > 0$.

   (a) $h(x) = f(x) + c$ (i) A horizontal shift of f, c units to the right

   (b) $h(x) = f(x) - c$ (ii) A vertical shift of f, c units downward

   (c) $h(x) = f(x + c)$ (iii) A horizontal shift of f, c units to the left

   (d) $h(x) = f(x - c)$ (iv) A vertical shift of f, c units upward

---

1. For each function, sketch (on the same set of coordinate axes) a graph of each function for $c = -1, 1,$ and 3.

   (a) $f(x) = |x| + c$

   (b) $f(x) = |x - c|$

   (c) $f(x) = |x + 4| + c$

2. For each function, sketch (on the same set of coordinate axes) a graph of each function for $c = -3, -1, 1,$ and 3.

   (a) $f(x) = \sqrt{x} + c$

   (b) $f(x) = \sqrt{x - c}$

   (c) $f(x) = \sqrt{x - 3} + c$

3. For each function, sketch (on the same set of coordinate axes) a graph of each function for $c = -2, 0,$ and 2.

   (a) $f(x) = [\![x]\!] + c$

   (b) $f(x) = [\![x + c]\!]$

   (c) $f(x) = [\![x - 1]\!] + c$

4. For each function, sketch (on the same set of coordinate axes) a graph of each function for $c = -3, -1, 1,$ and 3.

   (a) $f(x) = \begin{cases} x^2 + c, & x < 0 \\ -x^2 + c, & x \geq 0 \end{cases}$

   (b) $f(x) = \begin{cases} (x + c)^2, & x < 0 \\ -(x + c)^2, & x \geq 0 \end{cases}$

In Exercises 5–8, use the graph of f to sketch each graph. To print an enlarged copy of the graph go to the website www.mathgraphs.com.

5. (a) $y = f(x) + 2$
   (b) $y = f(x - 2)$
   (c) $y = 2f(x)$
   (d) $y = -f(x)$
   (e) $y = f(x + 3)$
   (f) $y = f(-x)$
   (g) $y = f(\frac{1}{2}x)$

6. (a) $y = f(-x)$
   (b) $y = f(x) + 4$
   (c) $y = 2f(x)$
   (d) $y = -f(x - 4)$
   (e) $y = f(x) - 3$
   (f) $y = -f(x) - 1$
   (g) $y = f(2x)$

FIGURE FOR 5

FIGURE FOR 6

7. (a) $y = f(x) - 1$
   (b) $y = f(x - 1)$
   (c) $y = f(-x)$
   (d) $y = f(x + 1)$
   (e) $y = -f(x - 2)$
   (f) $y = \frac{1}{2}f(x)$
   (g) $y = f(2x)$

8. (a) $y = f(x - 5)$
   (b) $y = -f(x) + 3$
   (c) $y = \frac{1}{3}f(x)$
   (d) $y = -f(x + 1)$
   (e) $y = f(-x)$
   (f) $y = f(x) - 10$
   (g) $y = f(\frac{1}{3}x)$

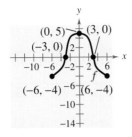

FIGURE FOR 7   FIGURE FOR 8

9. Use the graph of $f(x) = x^2$ to write an equation for each function whose graph is shown.

(a)

(b)

(c)

(d)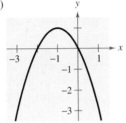

10. Use the graph of $f(x) = x^3$ to write an equation for each function whose graph is shown.

(a)

(b)

(c)

(d)

11. Use the graph of $f(x) = |x|$ to write an equation for each function whose graph is shown.

(a)

(b)

(c)

(d)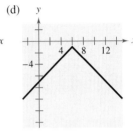

12. Use the graph of $f(x) = \sqrt{x}$ to write an equation for each function whose graph is shown.

(a)

(b)

(c)

(d)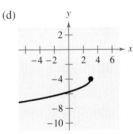

In Exercises 13–18, identify the parent function and the transformation shown in the graph. Write an equation for the function shown in the graph.

13.

14.

15.

16.

17.

18.

In Exercises 19–42, g is related to one of the parent functions described in this chapter. (a) Identify the parent function f. (b) Describe the sequence of tranformations from f to g. (c) Sketch the graph of g. (d) Use function notation to write g in terms of f.

19. $g(x) = 12 - x^2$

20. $g(x) = (x - 8)^2$

21. $g(x) = x^3 + 7$

22. $g(x) = -x^3 - 1$

23. $g(x) = \frac{2}{3}x^2 + 4$

24. $g(x) = 2(x - 7)^2$

25. $g(x) = 2 - (x + 5)^2$

26. $g(x) = -(x + 10)^2 + 5$

27. $g(x) = \sqrt{3x}$

28. $g(x) = \sqrt{\frac{1}{4}x}$

29. $g(x) = (x - 1)^3 + 2$

30. $g(x) = (x + 3)^3 - 10$

31. $g(x) = -|x| - 2$

32. $g(x) = 6 - |x + 5|$

33. $g(x) = -|x + 4| + 8$

34. $g(x) = |-x + 3| + 9$

35. $g(x) = 3 - [\![x]\!]$

36. $g(x) = 2[\![x + 5]\!]$

37. $g(x) = \sqrt{x - 9}$

38. $g(x) = \sqrt{x + 4} + 8$

39. $g(x) = \sqrt{7 - x} - 2$

40. $g(x) = -\sqrt{x + 1} - 6$

41. $g(x) = \sqrt{\frac{1}{2}x} - 4$

42. $g(x) = \sqrt{3x} + 1$

In Exercises 43–50, write an equation for the function that is described by the given characteristics.

43. The shape of $f(x) = x^2$, but moved two units to the right and eight units downward

44. The shape of $f(x) = x^2$, but moved three units to the left, seven units upward, and reflected in the x-axis

45. The shape of $f(x) = x^3$, but moved 13 units to the right

46. The shape of $f(x) = x^3$, but moved six units to the left, six units downward, and reflected in the y-axis

47. The shape of $f(x) = |x|$, but moved 10 units upward and reflected in the x-axis

48. The shape of $f(x) = |x|$, but moved one unit to the left and seven units downward

49. The shape of $f(x) = \sqrt{x}$, but moved six units to the left and reflected in both the x-axis and the y-axis

50. The shape of $f(x) = \sqrt{x}$, but moved nine units downward and reflected in both the x-axis and the y-axis

51. Use the graph of $f(x) = x^2$ to write an equation for each function whose graph is shown.

(a)

(b)

52. Use the graph of $f(x) = x^3$ to write an equation for each function whose graph is shown.

(a)

(b)

53. Use the graph of $f(x) = |x|$ to write an equation for each function whose graph is shown.

(a)

(b)

54. Use the graph of $f(x) = \sqrt{x}$ to write an equation for each function whose graph is shown.

(a)

(b)

In Exercises 55–60, identify the parent function and the transformation shown in the graph. Write an equation for the function shown in the graph. Then use a graphing utility to verify your answer.

55.

56.

57.

58.

59.

60.

 Graphical Analysis   In Exercises 61–64, use the viewing window shown to write a possible equation for the transformation of the parent function.

61.

62.

63.

64.

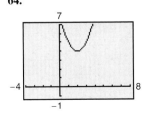

Graphical Reasoning   In Exercises 65 and 66, use the graph of *f* to sketch the graph of *g*. To print an enlarged copy of the graph, go to the website *www.mathgraphs.com*.

65.

(a) $g(x) = f(x) + 2$ (b) $g(x) = f(x) - 1$

(c) $g(x) = f(-x)$ (d) $g(x) = -2f(x)$

(e) $g(x) = f(4x)$ (f) $g(x) = f\left(\tfrac{1}{2}x\right)$

66.

(a) $g(x) = f(x) - 5$ (b) $g(x) = f(x) + \tfrac{1}{2}$

(c) $g(x) = f(-x)$ (d) $g(x) = -4f(x)$

(e) $g(x) = f(2x) + 1$ (f) $g(x) = f\left(\tfrac{1}{4}x\right) - 2$

Model It

67. *Fuel Use*   The amounts of fuel *F* (in billions of gallons) used by trucks from 1980 through 2002 can be approximated by the function

$$F = f(t) = 20.6 + 0.035t^2, \quad 0 \le t \le 22$$

where *t* represents the year, with $t = 0$ corresponding to 1980.   (Source: U.S. Federal Highway Administration)

(a) Describe the transformation of the parent function $f(x) = x^2$. Then sketch the graph over the specified domain.

(b) Find the average rate of change of the function from 1980 to 2002. Interpret your answer in the context of the problem.

(c) Rewrite the function so that $t = 0$ represents 1990. Explain how you got your answer.

(d) Use the model from part (c) to predict the amount of fuel used by trucks in 2010. Does your answer seem reasonable? Explain.

68. *Finance* The amounts M (in trillions of dollars) of mortgage debt outstanding in the United States from 1990 through 2002 can be approximated by the function

$$M = f(t) = 0.0054(t + 20.396)^2, \quad 0 \le t \le 12$$

where t represents the year, with $t = 0$ corresponding to 1990. (Source: Board of Governors of the Federal Reserve System)

(a) Describe the transformation of the parent function $f(x) = x^2$. Then sketch the graph over the specified domain.

(b) Rewrite the function so that $t = 0$ represents 2000. Explain how you got your answer.

Synthesis

True or False? In Exercises 69 and 70, determine whether the statement is true or false. Justify your answer.

69. The graphs of

$$f(x) = |x| + 6 \quad \text{and} \quad f(x) = |-x| + 6$$

are identical.

70. If the graph of the parent function $f(x) = x^2$ is moved six units to the right, three units upward, and reflected in the x-axis, then the point $(-2, 19)$ will lie on the graph of the transformation.

71. *Describing Profits* Management originally predicted that the profits from the sales of a new product would be approximated by the graph of the function f shown. The actual profits are shown by the function g along with a verbal description. Use the concepts of transformations of graphs to write g in terms of f.

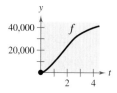

(a) The profits were only three-fourths as large as expected.

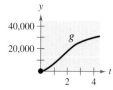

(b) The profits were consistently $10,000 greater than predicted.

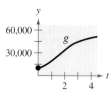

(c) There was a two-year delay in the introduction of the product. After sales began, profits grew as expected.

72. Explain why the graph of $y = -f(x)$ is a reflection of the graph of $y = f(x)$ about the x-axis.

73. The graph of $y = f(x)$ passes through the points $(0, 1)$, $(1, 2)$, and $(2, 3)$. Find the corresponding points on the graph of $y = f(x + 2) - 1$.

74. *Think About It* You can use either of two methods to graph a function: plotting points or translating a parent function as shown in this section. Which method of graphing do you prefer to use for each function? Explain.

(a) $f(x) = 3x^2 - 4x + 1$

(b) $f(x) = 2(x - 1)^2 - 6$

P.9   Combinations of Functions: Composite Functions

What you should learn

- Add, subtract, multiply, and divide functions.
- Find the composition of one function with another function.
- Use combinations and compositions of functions to model and solve real-life problems.

Why you should learn it

Compositions of functions can be used to model and solve real-life problems. For instance, in Exercise 68 on page 107, compositions of functions are used to determine the price of a new hybrid car.

© Jim West/The Image Works

Arithmetic Combinations of Functions

Just as two real numbers can be combined by the operations of addition, subtraction, multiplication, and division to form other real numbers, two *functions* can be combined to create new functions. For example, the functions given by $f(x) = 2x - 3$ and $g(x) = x^2 - 1$ can be combined to form the sum, difference, product, and quotient of f and g.

$$f(x) + g(x) = (2x - 3) + (x^2 - 1)$$
$$= x^2 + 2x - 4 \qquad \text{Sum}$$

$$f(x) - g(x) = (2x - 3) - (x^2 - 1)$$
$$= -x^2 + 2x - 2 \qquad \text{Difference}$$

$$f(x)g(x) = (2x - 3)(x^2 - 1)$$
$$= 2x^3 - 3x^2 - 2x + 3 \qquad \text{Product}$$

$$\frac{f(x)}{g(x)} = \frac{2x - 3}{x^2 - 1}, \quad x \neq \pm 1 \qquad \text{Quotient}$$

The domain of an **arithmetic combination** of functions f and g consists of all real numbers that are common to the domains of f and g. In the case of the quotient $f(x)/g(x)$, there is the further restriction that $g(x) \neq 0$.

Sum, Difference, Product, and Quotient of Functions

Let f and g be two functions with overlapping domains. Then, for all x common to both domains, the *sum*, *difference*, *product*, and *quotient* of f and g are defined as follows.

1. *Sum:* $(f + g)(x) = f(x) + g(x)$

2. *Difference:* $(f - g)(x) = f(x) - g(x)$

3. *Product:* $(fg)(x) = f(x) \cdot g(x)$

4. *Quotient:* $\left(\dfrac{f}{g}\right)(x) = \dfrac{f(x)}{g(x)}, \qquad g(x) \neq 0$

Example 1   Finding the Sum of Two Functions

Given $f(x) = 2x + 1$ and $g(x) = x^2 + 2x - 1$, find $(f + g)(x)$.

Solution

$$(f + g)(x) = f(x) + g(x) = (2x + 1) + (x^2 + 2x - 1) = x^2 + 4x$$

✓CHECKPOINT   Now try Exercise 5(a).

Example 2	**Finding the Difference of Two Functions**

Given $f(x) = 2x + 1$ and $g(x) = x^2 + 2x - 1$, find $(f - g)(x)$. Then evaluate the difference when $x = 2$.

Solution

The difference of f and g is

$$(f - g)(x) = f(x) - g(x)$$

$$= (2x + 1) - (x^2 + 2x - 1)$$

$$= -x^2 + 2.$$

When $x = 2$, the value of this difference is

$$(f - g)(2) = -(2)^2 + 2$$

$$= -2.$$

✓*CHECKPOINT*   Now try Exercise 5(b).

In Examples 1 and 2, both f and g have domains that consist of all real numbers. So, the domains of $(f + g)$ and $(f - g)$ are also the set of all real numbers. Remember that any restrictions on the domains of f and g must be considered when forming the sum, difference, product, or quotient of f and g.

Example 3	**Finding the Domains of Quotients of Functions**

Find $\left(\dfrac{f}{g}\right)(x)$ and $\left(\dfrac{g}{f}\right)(x)$ for the functions given by

$$f(x) = \sqrt{x} \quad \text{and} \quad g(x) = \sqrt{4 - x^2}.$$

Then find the domains of f/g and g/f.

Solution

The quotient of f and g is

$$\left(\frac{f}{g}\right)(x) = \frac{f(x)}{g(x)} = \frac{\sqrt{x}}{\sqrt{4 - x^2}}$$

and the quotient of g and f is

$$\left(\frac{g}{f}\right)(x) = \frac{g(x)}{f(x)} = \frac{\sqrt{4 - x^2}}{\sqrt{x}}.$$

The domain of f is $[0, \infty)$ and the domain of g is $[-2, 2]$. The intersection of these domains is $[0, 2]$. So, the domains of $\left(\dfrac{f}{g}\right)$ and $\left(\dfrac{g}{f}\right)$ are as follows.

$$\text{Domain of } \left(\frac{f}{g}\right): [0, 2) \qquad \text{Domain of } \left(\frac{g}{f}\right): (0, 2]$$

Note that the domain of (f/g) includes $x = 0$, but not $x = 2$, because $x = 2$ yields a zero in the denominator, whereas the domain of (g/f) includes $x = 2$, but not $x = 0$, because $x = 0$ yields a zero in the denominator.

✓*CHECKPOINT*   Now try Exercise 5(d).

Composition of Functions

Another way of combining two functions is to form the **composition** of one with the other. For instance, if $f(x) = x^2$ and $g(x) = x + 1,$ the composition of f with g is

$$f(g(x)) = f(x + 1)$$
$$= (x + 1)^2.$$

This composition is denoted as $(f \circ g)$ and reads as "f composed with g."

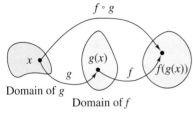

Domain of g
Domain of f

FIGURE P.91

> ### Definition of Composition of Two Functions
>
> The **composition** of the function f with the function g is
>
> $$(f \circ g)(x) = f(g(x)).$$
>
> The domain of $(f \circ g)$ is the set of all x in the domain of g such that $g(x)$ is in the domain of f. (See Figure P.91.)

Example 4   **Composition of Functions**

Given $f(x) = x + 2$ and $g(x) = 4 - x^2$, find the following.

a. $(f \circ g)(x)$ **b.** $(g \circ f)(x)$ **c.** $(g \circ f)(-2)$

Solution

a. The composition of f with g is as follows.

$$(f \circ g)(x) = f(g(x))$$ Definition of $f \circ g$

$$= f(4 - x^2)$$ Definition of $g(x)$

$$= (4 - x^2) + 2$$ Definition of $f(x)$

$$= -x^2 + 6$$ Simplify.

b. The composition of g with f is as follows.

$$(g \circ f)(x) = g(f(x))$$ Definition of $g \circ f$

$$= g(x + 2)$$ Definition of $f(x)$

$$= 4 - (x + 2)^2$$ Definition of $g(x)$

$$= 4 - (x^2 + 4x + 4)$$ Expand.

$$= -x^2 - 4x$$ Simplify.

Note that, in this case, $(f \circ g)(x) \neq (g \circ f)(x)$.

c. Using the result of part (b), you can write the following.

$$(g \circ f)(-2) = -(-2)^2 - 4(-2)$$ Substitute.

$$= -4 + 8$$ Simplify.

$$= 4$$ Simplify.

✓**CHECKPOINT**   Now try Exercise 31.

STUDY TIP

The following tables of values help illustrate the composition $(f \circ g)(x)$ given in Example 4.

x	0	1	2	3
$g(x)$	4	3	0	-5

$g(x)$	4	3	0	-5
$f(g(x))$	6	5	2	-3

x	0	1	2	3
$f(g(x))$	6	5	2	-3

Note that the first two tables can be combined (or "composed") to produce the values given in the third table.

Technology

You can use a graphing utility to determine the domain of a composition of functions. For the composition in Example 5, enter the function composition as

$$y = \left(\sqrt{9 - x^2}\right)^2 - 9.$$

You should obtain the graph shown below. Use the *trace* feature to determine that the x-coordinates of points on the graph extend from -3 to 3. So, the domain of $(f \circ g)(x)$ is $-3 \leq x \leq 3$.

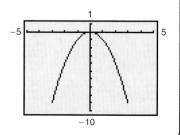

Activities

1. Given $f(x) = 3x^2 + 2$ and $g(x) = 2x$, find $f \circ g$.

    Answer: $(f \circ g)(x) = 12x^2 + 2$

2. Given the functions

    $f(x) = \dfrac{1}{x - 2}$ and $g(x) = \sqrt{x}$,

    find the composition of f with g. Then find the domain of the composition.

    Answer: $(f \circ g)(x) = \dfrac{1}{\sqrt{x} - 2}.$ The

    domain of $(f \circ g)$ is the set of all non-negative real numbers except $x = 4$.

3. Find two functions f and g such that $(f \circ g)(x) = h(x)$. (There are many correct answers.)

    a. $h(x) = \dfrac{1}{\sqrt{3x + 1}}$

        Answer: $f(x) = \dfrac{1}{\sqrt{x}}$ and

        $g(x) = 3x + 1$

    b. $h(x) = (2x + 3)^4$

        Answer: $f(x) = x^4$ and

        $g(x) = 2x + 3$

Example 5   Finding the Domain of a Composite Function

Given $f(x) = x^2 - 9$ and $g(x) = \sqrt{9 - x^2}$, find the composition $(f \circ g)(x)$. Then find the domain of $(f \circ g)$.

Solution

$$
\begin{aligned}
(f \circ g)(x) &= f(g(x)) \\
&= f\left(\sqrt{9 - x^2}\right) \\
&= \left(\sqrt{9 - x^2}\right)^2 - 9 \\
&= 9 - x^2 - 9 \\
&= -x^2
\end{aligned}
$$

From this, it might appear that the domain of the composition is the set of all real numbers. This, however is not true. Because the domain of f is the set of all real numbers and the domain of g is $-3 \leq x \leq 3$, the domain of $(f \circ g)$ is $-3 \leq x \leq 3$.

✓ **CHECKPOINT**   Now try Exercise 35.

In Examples 4 and 5, you formed the composition of two given functions. In calculus, it is also important to be able to identify two functions that make up a given composite function. For instance, the function h given by

$$h(x) = (3x - 5)^3$$

is the composition of f with g, where $f(x) = x^3$ and $g(x) = 3x - 5$. That is,

$$h(x) = (3x - 5)^3 = [g(x)]^3 = f(g(x)).$$

Basically, to "decompose" a composite function, look for an "inner" function and an "outer" function. In the function h above, $g(x) = 3x - 5$ is the inner function and $f(x) = x^3$ is the outer function.

Example 6   Decomposing a Composite Function

Write the function given by $h(x) = \dfrac{1}{(x - 2)^2}$ as a composition of two functions.

Solution

One way to write h as a composition of two functions is to take the inner function to be $g(x) = x - 2$ and the outer function to be

$$f(x) = \frac{1}{x^2} = x^{-2}.$$

Then you can write

$$h(x) = \frac{1}{(x - 2)^2} = (x - 2)^{-2} = f(x - 2) = f(g(x)).$$

✓ **CHECKPOINT**   Now try Exercise 47.

Application

Example 7 **Bacteria Count**

The number N of bacteria in a refrigerated food is given by

$$N(T) = 20T^2 - 80T + 500, \qquad 2 \le T \le 14$$

where T is the temperature of the food in degrees Celsius. When the food is removed from refrigeration, the temperature of the food is given by

$$T(t) = 4t + 2, \qquad 0 \le t \le 3$$

where t is the time in hours. (a) Find the composition $N(T(t))$ and interpret its meaning in context. (b) Find the time when the bacterial count reaches 2000.

Solution

a. $N(T(t)) = 20(4t + 2)^2 - 80(4t + 2) + 500$

$\qquad = 20(16t^2 + 16t + 4) - 320t - 160 + 500$

$\qquad = 320t^2 + 320t + 80 - 320t - 160 + 500$

$\qquad = 320t^2 + 420$

The composite function $N(T(t))$ represents the number of bacteria in the food as a function of the amount of time the food has been out of refrigeration.

b. The bacterial count will reach 2000 when $320t^2 + 420 = 2000$. Solve this equation to find that the count will reach 2000 when $t \approx 2.2$ hours. When you solve this equation, note that the negative value is rejected because it is not in the domain of the composite function.

✓*CHECKPOINT*   Now try Exercise 65.

Writing About Mathematics

To expand on this activity, you might consider asking your students to use the tables they created in parts (a) and (b), along with a table of values for x and $f(x)$, to demonstrate and explain how the tables can be manipulated to yield tables of values for $h(x)$ and $g(x)$.

𝒲RITING ABOUT 𝓜ATHEMATICS

Analyzing Arithmetic Combinations of Functions

a. Use the graphs of f and $(f + g)$ in Figure P.92 to make a table showing the values of $g(x)$ when $x = 1, 2, 3, 4, 5,$ and 6. Explain your reasoning.

b. Use the graphs of f and $(f - h)$ in Figure P.92 to make a table showing the values of $h(x)$ when $x = 1, 2, 3, 4, 5,$ and 6. Explain your reasoning.

FIGURE P.92

P.9 Exercises

VOCABULARY CHECK: Fill in the blanks.

1. Two functions f and g can be combined by the arithmetic operations of _____, _____, _____, and _____ to create new functions.

2. The _____ of the function f with g is $(f \circ g)(x) = f(g(x))$.

3. The domain of $(f \circ g)$ is all x in the domain of g such that _____ is in the domain of f.

4. To decompose a composite function, look for an _____ function and an _____ function.

In Exercises 1–4, use the graphs of f and g to graph $h(x) = (f + g)(x)$. To print an enlarged copy of the graph, go to the website www.mathgraphs.com.

1.

2.

3.

4.

In Exercises 5–12, find (a) $(f + g)(x)$, (b) $(f - g)(x)$, (c) $(fg)(x)$, and (d) $(f/g)(x)$. What is the domain of f/g?

5. $f(x) = x + 2$, $g(x) = x - 2$

6. $f(x) = 2x - 5$, $g(x) = 2 - x$

7. $f(x) = x^2$, $g(x) = 4x - 5$

8. $f(x) = 2x - 5$, $g(x) = 4$

9. $f(x) = x^2 + 6$, $g(x) = \sqrt{1 - x}$

10. $f(x) = \sqrt{x^2 - 4}$, $g(x) = \dfrac{x^2}{x^2 + 1}$

11. $f(x) = \dfrac{1}{x}$, $g(x) = \dfrac{1}{x^2}$

12. $f(x) = \dfrac{x}{x + 1}$, $g(x) = x^3$

In Exercises 13–24, evaluate the indicated function for $f(x) = x^2 + 1$ and $g(x) = x - 4$.

13. $(f + g)(2)$ **14.** $(f - g)(-1)$

15. $(f - g)(0)$ **16.** $(f + g)(1)$

17. $(f - g)(3t)$ **18.** $(f + g)(t - 2)$

19. $(fg)(6)$ **20.** $(fg)(-6)$

21. $\left(\dfrac{f}{g}\right)(5)$ **22.** $\left(\dfrac{f}{g}\right)(0)$

23. $\left(\dfrac{f}{g}\right)(-1) - g(3)$ **24.** $(fg)(5) + f(4)$

In Exercises 25–28, graph the functions f, g, and $f + g$ on the same set of coordinate axes.

25. $f(x) = \frac{1}{2}x$, $g(x) = x - 1$

26. $f(x) = \frac{1}{3}x$, $g(x) = -x + 4$

27. $f(x) = x^2$, $g(x) = -2x$

28. $f(x) = 4 - x^2$, $g(x) = x$

Graphical Reasoning **In Exercises 29 and 30, use a graphing utility to graph f, g, and $f + g$ in the same viewing window. Which function contributes most to the magnitude of the sum when $0 \le x \le 2$? Which function contributes most to the magnitude of the sum when $x > 6$?**

29. $f(x) = 3x$, $g(x) = -\dfrac{x^3}{10}$

30. $f(x) = \dfrac{x}{2}$, $g(x) = \sqrt{x}$

In Exercises 31–34, find (a) $f \circ g$, (b) $g \circ f$, and (c) $f \circ f$.

31. $f(x) = x^2$, $g(x) = x - 1$

32. $f(x) = 3x + 5$, $g(x) = 5 - x$

33. $f(x) = \sqrt[3]{x - 1}$, $g(x) = x^3 + 1$

34. $f(x) = x^3$, $g(x) = \dfrac{1}{x}$

In Exercises 35–42, find (a) $f \circ g$ and (b) $g \circ f$. Find the domain of each function and each composite function.

35. $f(x) = \sqrt{x + 4}$, $g(x) = x^2$

36. $f(x) = \sqrt[3]{x - 5}$, $g(x) = x^3 + 1$

37. $f(x) = x^2 + 1$, $g(x) = \sqrt{x}$

38. $f(x) = x^{2/3}$, $g(x) = x^6$

39. $f(x) = |x|$, $g(x) = x + 6$

40. $f(x) = |x - 4|$, $g(x) = 3 - x$

41. $f(x) = \dfrac{1}{x}$, $g(x) = x + 3$

42. $f(x) = \dfrac{3}{x^2 - 1}$, $g(x) = x + 1$

In Exercises 43–46, use the graphs of f and g to evaluate the functions.

43. (a) $(f + g)(3)$ (b) $(f/g)(2)$

44. (a) $(f - g)(1)$ (b) $(fg)(4)$

45. (a) $(f \circ g)(2)$ (b) $(g \circ f)(2)$

46. (a) $(f \circ g)(1)$ (b) $(g \circ f)(3)$

In Exercises 47–54, find two functions f and g such that $(f \circ g)(x) = h(x)$. (There are many correct answers.)

47. $h(x) = (2x + 1)^2$ 48. $h(x) = (1 - x)^3$

49. $h(x) = \sqrt[3]{x^2 - 4}$ 50. $h(x) = \sqrt{9 - x}$

51. $h(x) = \dfrac{1}{x + 2}$ 52. $h(x) = \dfrac{4}{(5x + 2)^2}$

53. $h(x) = \dfrac{-x^2 + 3}{4 - x^2}$ 54. $h(x) = \dfrac{27x^3 + 6x}{10 - 27x^3}$

55. **Stopping Distance** The research and development department of an automobile manufacturer has determined that when a driver is required to stop quickly to avoid an accident, the distance (in feet) the car travels during the driver's reaction time is given by $R(x) = \frac{3}{4}x$, where x is the speed of the car in miles per hour. The distance (in feet) traveled while the driver is braking is given by $B(x) = \frac{1}{15}x^2$. Find the function that represents the total stopping distance T. Graph the functions R, B, and T on the same set of coordinate axes for $0 \le x \le 60$.

56. **Sales**   From 2000 to 2005, the sales R_1 (in thousands of dollars) for one of two restaurants owned by the same parent company can be modeled by

$$R_1 = 480 - 8t - 0.8t^2, \qquad t = 0, 1, 2, 3, 4, 5$$

where $t = 0$ represents 2000. During the same six-year period, the sales R_2 (in thousands of dollars) for the second restaurant can be modeled by

$$R_2 = 254 + 0.78t, \qquad t = 0, 1, 2, 3, 4, 5.$$

(a) Write a function R_3 that represents the total sales of the two restaurants owned by the same parent company.

 (b) Use a graphing utility to graph R_1, R_2, and R_3 in the same viewing window.

57. **Vital Statistics**   Let $b(t)$ be the number of births in the United States in year t, and let $d(t)$ represent the number of deaths in the United States in year t, where $t = 0$ corresponds to 2000.

(a) If $p(t)$ is the population of the United States in year t, find the function $c(t)$ that represents the percent change in the population of the United States.

(b) Interpret the value of $c(5)$.

58. **Pets**   Let $d(t)$ be the number of dogs in the United States in year t, and let $c(t)$ be the number of cats in the United States in year t, where $t = 0$ corresponds to 2000.

(a) Find the function $p(t)$ that represents the total number of dogs and cats in the United States.

(b) Interpret the value of $p(5)$.

(c) Let $n(t)$ represent the population of the United States in year t, where $t = 0$ corresponds to 2000. Find and interpret

$$h(t) = \frac{p(t)}{n(t)}.$$

59. **Military Personnel**   The total numbers of Army personnel (in thousands) A and Navy personnel (in thousands) N from 1990 to 2002 can be approximated by the models

$$A(t) = 3.36t^2 - 59.8t + 735$$

and

$$N(t) = 1.95t^2 - 42.2t + 603$$

where t represents the year, with $t = 0$ corresponding to 1990.   (Source: Department of Defense)

(a) Find and interpret $(A + N)(t)$. Evaluate this function for $t = 4$, 8, and 12.

(b) Find and interpret $(A - N)(t)$. Evaluate this function for $t = 4$, 8, and 12.

60. Sales The sales of exercise equipment E (in millions of dollars) in the United States from 1997 to 2003 can be approximated by the function

$$E(t) = 25.95t^2 - 231.2t + 3356$$

and the U.S. population P (in millions) from 1997 to 2003 can be approximated by the function

$$P(t) = 3.02t + 252.0$$

where t represents the year, with $t = 7$ corresponding to 1997. (Source: National Sporting Goods Association, U.S. Census Bureau)

(a) Find and interpret $h(t) = \dfrac{E(t)}{P(t)}$.

(b) Evaluate the function in part (a) for $t = 7$, 10, and 12.

Model It

61. Health Care Costs The table shows the total amounts (in billions of dollars) spent on health services and supplies in the United States (including Puerto Rico) for the years 1995 through 2001. The variables y_1, y_2, and y_3 represent out-of-pocket payments, insurance premiums, and other types of payments, respectively. (Source: Centers for Medicare and Medicaid Services)

Year	y_1	y_2	y_3
1995	146.2	329.1	44.8
1996	152.0	344.1	48.1
1997	162.2	359.9	52.1
1998	175.2	382.0	55.6
1999	184.4	412.1	57.8
2000	194.7	449.0	57.4
2001	205.5	496.1	57.8

(a) Use the *regression* feature of a graphing utility to find a linear model for y_1 and quadratic models for y_2 and y_3. Let $t = 5$ represent 1995.

(b) Find $y_1 + y_2 + y_3$. What does this sum represent?

(c) Use a graphing utility to graph y_1, y_2, y_3, and $y_1 + y_2 + y_3$ in the same viewing window.

(d) Use the model from part (b) to estimate the total amounts spent on health services and supplies in the years 2008 and 2010.

---

62. Graphical Reasoning An electronically controlled thermostat in a home is programmed to lower the temperature automatically during the night. The temperature in the house T (in degrees Fahrenheit) is given in terms of t, the time in hours on a 24-hour clock (see figure).

(a) Explain why T is a function of t.

(b) Approximate $T(4)$ and $T(15)$.

(c) The thermostat is reprogrammed to produce a temperature H for which $H(t) = T(t - 1)$. How does this change the temperature?

(d) The thermostat is reprogrammed to produce a temperature H for which $H(t) = T(t) - 1$. How does this change the temperature?

(e) Write a piecewise-defined function that represents the graph.

63. Geometry A square concrete foundation is prepared as a base for a cylindrical tank (see figure).

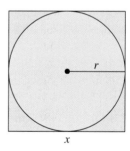

(a) Write the radius r of the tank as a function of the length x of the sides of the square.

(b) Write the area A of the circular base of the tank as a function of the radius r.

(c) Find and interpret $(A \circ r)(x)$.

64. *Physics*   A pebble is dropped into a calm pond, causing ripples in the form of concentric circles (see figure). The radius r (in feet) of the outer ripple is $r(t) = 0.6t$, where t is the time in seconds after the pebble strikes the water. The area A of the circle is given by the function $A(r) = \pi r^2$. Find and interpret $(A \circ r)(t)$.

65. *Bacteria Count*   The number N of bacteria in a refriger-ated food is given by

$$N(T) = 10T^2 - 20T + 600, \quad 1 \le T \le 20$$

where T is the temperature of the food in degrees Celsius. When the food is removed from refrigeration, the tempera-ture of the food is given by

$$T(t) = 3t + 2, \quad 0 \le t \le 6$$

where t is the time in hours.

(a) Find the composition $N(T(t))$ and interpret its meaning in context.

(b) Find the time when the bacterial count reaches 1500.

66. *Cost*   The weekly cost C of producing x units in a manu-facturing process is given by

$$C(x) = 60x + 750.$$

The number of units x produced in t hours is given by

$$x(t) = 50t.$$

(a) Find and interpret $(C \circ x)(t)$.

(b) Find the time that must elapse in order for the cost to increase to $15,000.

67. *Salary*   You are a sales representative for a clothing manufacturer. You are paid an annual salary, plus a bonus of 3% of your sales over $500,000. Consider the two functions given by

$$f(x) = x - 500{,}000 \quad \text{and} \quad g(x) = 0.03x.$$

If x is greater than $500,000, which of the following repre-sents your bonus? Explain your reasoning.

(a) $f(g(x))$ (b) $g(f(x))$

68. *Consumer Awareness*   The suggested retail price of a new hybrid car is p dollars. The dealership advertises a factory rebate of $2000 and a 10% discount.

(a) Write a function R in terms of p giving the cost of the hybrid car after receiving the rebate from the factory.

(b) Write a function S in terms of p giving the cost of the hybrid car after receiving the dealership discount.

(c) Form the composite functions $(R \circ S)(p)$ and $(S \circ R)(p)$ and interpret each.

(d) Find $(R \circ S)(20{,}500)$ and $(S \circ R)(20{,}500)$. Which yields the lower cost for the hybrid car? Explain.

Synthesis

True or False?   **In Exercises 69 and 70, determine whether the statement is true or false. Justify your answer.**

69. If $f(x) = x + 1$ and $g(x) = 6x$, then

$$(f \circ g)(x) = (g \circ f)(x).$$

70. If you are given two functions $f(x)$ and $g(x)$, you can calculate $(f \circ g)(x)$ if and only if the range of g is a subset of the domain of f.

71. *Proof*   Prove that the product of two odd functions is an even function, and that the product of two even functions is an even function.

72. *Conjecture*   Use examples to hypothesize whether the product of an odd function and an even function is even or odd. Then prove your hypothesis.

P.10 Inverse Functions

© Tim Boyle/Getty Images

Inverse Functions

Recall from Section P.5, that a function can be represented by a set of ordered pairs. For instance, the function $f(x) = x + 4$ from the set $A = \{1, 2, 3, 4\}$ to the set $B = \{5, 6, 7, 8\}$ can be written as follows.

$$f(x) = x + 4: \{(1, 5), (2, 6), (3, 7), (4, 8)\}$$

In this case, by interchanging the first and second coordinates of each of these ordered pairs, you can form the **inverse function** of f, which is denoted by f^{-1}. It is a function from the set B to the set A, and can be written as follows.

$$f^{-1}(x) = x - 4: \{(5, 1), (6, 2), (7, 3), (8, 4)\}$$

Note that the domain of f is equal to the range of f^{-1}, and vice versa, as shown in Figure P.93. Also note that the functions f and f^{-1} have the effect of "undoing" each other. In other words, when you form the composition of f with f^{-1} or the composition of f^{-1} with f, you obtain the identity function.

$$f(f^{-1}(x)) = f(x - 4) = (x - 4) + 4 = x$$
$$f^{-1}(f(x)) = f^{-1}(x + 4) = (x + 4) - 4 = x$$

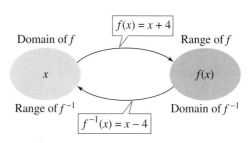

Domain of f $f(x) = x + 4$ Range of f

x $f(x)$

Range of f^{-1} $f^{-1}(x) = x - 4$ Domain of f^{-1}

FIGURE P.93

Example 1   Finding Inverse Functions Informally

Find the inverse function of $f(x) = 4x$. Then verify that both $f(f^{-1}(x))$ and $f^{-1}(f(x))$ are equal to the identity function.

Solution

The function f *multiplies* each input by 4. To "undo" this function, you need to *divide* each input by 4. So, the inverse function of $f(x) = 4x$ is

$$f^{-1}(x) = \frac{x}{4}.$$

You can verify that both $f(f^{-1}(x)) = x$ and $f^{-1}(f(x)) = x$ as follows.

$$f(f^{-1}(x)) = f\left(\frac{x}{4}\right) = 4\left(\frac{x}{4}\right) = x \qquad f^{-1}(f(x)) = f^{-1}(4x) = \frac{4x}{4} = x$$

✓CHECKPOINT   Now try Exercise 1.

Consider the functions given by

$$f(x) = x + 2$$

and

$$f^{-1}(x) = x - 2.$$

Evaluate $f(f^{-1}(x))$ and $f^{-1}(f(x))$ for the indicated values of x. What can you conclude about the functions?

x	-10	0	7	45
$f(f^{-1}(x))$				
$f^{-1}(f(x))$				

Definition of Inverse Function

Let f and g be two functions such that

$$f(g(x)) = x \qquad \text{for every } x \text{ in the domain of } g$$

and

$$g(f(x)) = x \qquad \text{for every } x \text{ in the domain of } f.$$

Under these conditions, the function g is the **inverse function** of the function f. The function g is denoted by f^{-1} (read "f-inverse"). So,

$$f(f^{-1}(x)) = x \qquad \text{and} \qquad f^{-1}(f(x)) = x.$$

The domain of f must be equal to the range of f^{-1}, and the range of f must be equal to the domain of f^{-1}.

Don't be confused by the use of -1 to denote the inverse function f^{-1}. In this text, whenever f^{-1} is written, it *always* refers to the inverse function of the function f and *not* to the reciprocal of $f(x)$.

If the function g is the inverse function of the function f, it must also be true that the function f is the inverse function of the function g. For this reason, you can say that the functions f and g are *inverse functions of each other*.

Example 2   Verifying Inverse Functions

Which of the functions is the inverse function of $f(x) = \dfrac{5}{x - 2}$?

$$g(x) = \frac{x - 2}{5} \qquad h(x) = \frac{5}{x} + 2$$

Additional Examples

Determine whether the two functions are inverse functions of each other.

a. $f(x) = 3x - 2, g(x) = \dfrac{1}{3}x + 2$

b. $f(x) = \dfrac{1}{4}x - 3, g(x) = 12 + 4x$

c. $f(x) = 2x + 4, g(x) = \dfrac{1}{2}x - 2$

Solution

The functions in parts (b) and (c) are inverse functions of each other because $f(g(x)) = x$ and $g(f(x)) = x$. The functions in part (a) are not inverse functions of each other.

Solution

By forming the composition of f with g, you have

$$f(g(x)) = f\left(\frac{x - 2}{5}\right)$$

$$= \frac{5}{\left(\dfrac{x - 2}{5}\right) - 2} \qquad \text{Substitute } \frac{x - 2}{5} \text{ for } x.$$

$$= \frac{25}{x - 12} \neq x.$$

Because this composition is not equal to the identity function x, it follows that g *is not* the inverse function of f. By forming the composition of f with h, you have

$$f(h(x)) = f\left(\frac{5}{x} + 2\right) = \frac{5}{\left(\dfrac{5}{x} + 2\right) - 2} = \frac{5}{\left(\dfrac{5}{x}\right)} = x.$$

So, it appears that h *is* the inverse function of f. You can confirm this by showing that the composition of h with f is also equal to the identity function.

✓**CHECKPOINT**   Now try Exercise 5.

FIGURE P.94

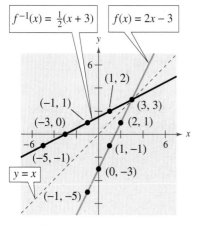

FIGURE P.95

The Graph of an Inverse Function

The graphs of a function f and its inverse function f^{-1} are related to each other in the following way. If the point (a, b) lies on the graph of f, then the point (b, a) must lie on the graph of f^{-1}, and vice versa. This means that the graph of f^{-1} is a *reflection* of the graph of f in the line $y = x$, as shown in Figure P.94.

Example 3   Finding Inverse Functions Graphically

Sketch the graphs of the inverse functions $f(x) = 2x - 3$ and $f^{-1}(x) = \frac{1}{2}(x + 3)$ on the same rectangular coordinate system and show that the graphs are reflections of each other in the line $y = x$.

Solution

The graphs of f and f^{-1} are shown in Figure P.95. It appears that the graphs are reflections of each other in the line $y = x$. You can further verify this reflective property by testing a few points on each graph. Note in the following list that if the point (a, b) is on the graph of f, the point (b, a) is on the graph of f^{-1}.

Graph of $f(x) = 2x - 3$	*Graph of $f^{-1}(x) = \frac{1}{2}(x + 3)$*
$(-1, -5)$	$(-5, -1)$
$(0, -3)$	$(-3, 0)$
$(1, -1)$	$(-1, 1)$
$(2, 1)$	$(1, 2)$
$(3, 3)$	$(3, 3)$

✓**CHECKPOINT**   Now try Exercise 15.

Example 4   Finding Inverse Functions Graphically

Sketch the graphs of the inverse functions $f(x) = x^2 \ (x \geq 0)$ and $f^{-1}(x) = \sqrt{x}$ on the same rectangular coordinate system and show that the graphs are reflections of each other in the line $y = x$.

Solution

The graphs of f and f^{-1} are shown in Figure P.96. It appears that the graphs are reflections of each other in the line $y = x$. You can further verify this reflective property by testing a few points on each graph. Note in the following list that if the point (a, b) is on the graph of f, the point (b, a) is on the graph of f^{-1}.

Graph of $f(x) = x^2, \quad x \geq 0$	*Graph of $f^{-1}(x) = \sqrt{x}$*
$(0, 0)$	$(0, 0)$
$(1, 1)$	$(1, 1)$
$(2, 4)$	$(4, 2)$
$(3, 9)$	$(9, 3)$

Try showing that $f(f^{-1}(x)) = x$ and $f^{-1}(f(x)) = x$.

✓**CHECKPOINT**   Now try Exercise 17.

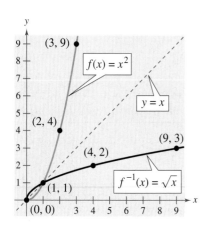

FIGURE P.96

One-to-One Functions

The reflective property of the graphs of inverse functions gives you a nice *geometric* test for determining whether a function has an inverse function. This test is called the **Horizontal Line Test** for inverse functions.

> ### Horizontal Line Test for Inverse Functions
>
> A function f has an inverse function if and only if no *horizontal* line intersects the graph of f at more than one point.

If no horizontal line intersects the graph of f at more than one point, then no y-value is matched with more than one x-value. This is the essential characteristic of what are called **one-to-one functions.**

> ### One-to-One Functions
>
> A function f is **one-to-one** if each value of the dependent variable corresponds to exactly one value of the independent variable. A function f has an inverse function if and only if f is one-to-one.

Consider the function given by $f(x) = x^2$. The table on the left is a table of values for $f(x) = x^2$. The table of values on the right is made up by interchanging the columns of the first table. The table on the right does not represent a function because the input $x = 4$ is matched with two different outputs: $y = -2$ and $y = 2$. So, $f(x) = x^2$ is not one-to-one and does not have an inverse function.

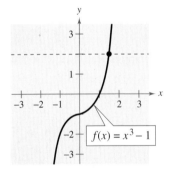

FIGURE P.97

x	$f(x) = x^2$
-2	4
-1	1
0	0
1	1
2	4
3	9

x	y
4	-2
1	-1
0	0
1	1
4	2
9	3

FIGURE P.98

Example 5   Applying the Horizontal Line Test

a. The graph of the function given by $f(x) = x^3 - 1$ is shown in Figure P.97. Because no horizontal line intersects the graph of f at more than one point, you can conclude that f *is* a one-to-one function and *does* have an inverse function.

b. The graph of the function given by $f(x) = x^2 - 1$ is shown in Figure P.98. Because it is possible to find a horizontal line that intersects the graph of f at more than one point, you can conclude that f *is not* a one-to-one function and *does not* have an inverse function.

✓CHECKPOINT   Now try Exercise 29.

Finding Inverse Functions Algebraically

For simple functions (such as the one in Example 1), you can find inverse functions by inspection. For more complicated functions, however, it is best to use the following guidelines. The key step in these guidelines is Step 3—interchanging the roles of x and y. This step corresponds to the fact that inverse functions have ordered pairs with the coordinates reversed.

Finding an Inverse Function

1. Use the Horizontal Line Test to decide whether f has an inverse function.

2. In the equation for $f(x)$, replace $f(x)$ by y.

3. Interchange the roles of x and y, and solve for y.

4. Replace y by $f^{-1}(x)$ in the new equation.

5. Verify that f and f^{-1} are inverse functions of each other by showing that the domain of f is equal to the range of f^{-1}, the range of f is equal to the domain of f^{-1}, and $f(f^{-1}(x)) = x$ and $f^{-1}(f(x)) = x$.

Example 6   **Finding an Inverse Function Algebraically**

Find the inverse function of

$$f(x) = \frac{5 - 3x}{2}.$$

Solution

The graph of f is a line, as shown in Figure P.99. This graph passes the Horizontal Line Test. So, you know that f is one-to-one and has an inverse function.

$$f(x) = \frac{5 - 3x}{2} \qquad \text{Write original function.}$$

$$y = \frac{5 - 3x}{2} \qquad \text{Replace } f(x) \text{ by } y.$$

$$x = \frac{5 - 3y}{2} \qquad \text{Interchange } x \text{ and } y.$$

$$2x = 5 - 3y \qquad \text{Multiply each side by 2.}$$

$$3y = 5 - 2x \qquad \text{Isolate the } y\text{-term.}$$

$$y = \frac{5 - 2x}{3} \qquad \text{Solve for } y.$$

$$f^{-1}(x) = \frac{5 - 2x}{3} \qquad \text{Replace } y \text{ by } f^{-1}(x).$$

Note that both f and f^{-1} have domains and ranges that consist of the entire set of real numbers. Check that $f(f^{-1}(x)) = x$ and $f^{-1}(f(x)) = x$.

✓**CHECKPOINT**   Now try Exercise 55.

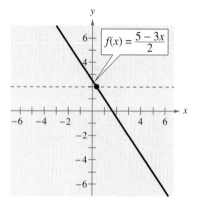

FIGURE P.99

Exploration

Restrict the domain of $f(x) = x^2 + 1$ to $x \geq 0$. Use a graphing utility to graph the function. Does the restricted function have an inverse function? Explain.

Example 7   **Finding an Inverse Function**

Find the inverse function of

$$f(x) = \sqrt[3]{x + 1}.$$

Solution

The graph of f is a curve, as shown in Figure P.100. Because this graph passes the Horizontal Line Test, you know that f is one-to-one and has an inverse function.

$f(x) = \sqrt[3]{x + 1}$	Write original function.
$y = \sqrt[3]{x + 1}$	Replace $f(x)$ by y.
$x = \sqrt[3]{y + 1}$	Interchange x and y.
$x^3 = y + 1$	Cube each side.
$x^3 - 1 = y$	Solve for y.
$x^3 - 1 = f^{-1}(x)$	Replace y by $f^{-1}(x)$.

Both f and f^{-1} have domains and ranges that consist of the entire set of real numbers. You can verify this result numerically as shown in the tables below.

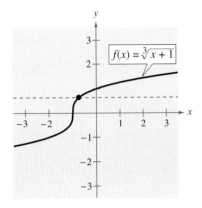

FIGURE P.100

x	$f(x)$
-28	-3
-9	-2
-2	-1
-1	0
0	1
7	2
26	3

x	$f^{-1}(x)$
-3	-28
-2	-9
-1	-2
0	-1
1	0
2	7
3	26

✓CHECKPOINT   Now try Exercise 61.

Activities

1. Given $f(x) = 5x - 7$, find $f^{-1}(x)$.

   Answer: $f^{-1}(x) = \dfrac{x + 7}{5}$

2. Show that f and g are inverse functions of each other by showing that $f(g(x)) = x$ and $g(f(x)) = x$.

   $f(x) = 3x^3 + 1$

   $g(x) = \sqrt[3]{\dfrac{x - 1}{3}}$

3. Describe the graphs of functions that have inverse functions and show how the graphs of a function and its inverse function are related.

𝒲RITING ABOUT 𝓜ATHEMATICS

The Existence of an Inverse Function Write a short paragraph describing why the following functions do or do not have inverse functions.

a. Let x represent the retail price of an item (in dollars), and let $f(x)$ represent the sales tax on the item. Assume that the sales tax is 6% of the retail price *and* that the sales tax is rounded to the nearest cent. Does this function have an inverse function? (*Hint:* Can you undo this function?

For instance, if you know that the sales tax is $0.12, can you determine exactly what the retail price is?)

b. Let x represent the temperature in degrees Celsius, and let $f(x)$ represent the temperature in degrees Fahrenheit. Does this function have an inverse function? (*Hint:* The formula for converting from degrees Celsius to degrees Fahrenheit is $F = \frac{9}{5}C + 32$.)

P.10 | Exercises

VOCABULARY CHECK: Fill in the blanks.

1. If the composite functions $f(g(x)) = x$ and $g(f(x)) = x$ then the function g is the _____ function of f.
2. The domain of f is the _____ of f^{-1}, and the _____ of f^{-1} is the range of f.
3. The graphs of f and f^{-1} are reflections of each other in the line _____.
4. A function f is _____ if each value of the dependent variable corresponds to exactly one value of the independent variable.
5. A graphical test for the existence of an inverse function of f is called the _____ Line Test.

In Exercises 1–8, find the inverse function of f informally. Verify that $f(f^{-1}(x)) = x$ and $f^{-1}(f(x)) = x$.

1. $f(x) = 6x$
2. $f(x) = \frac{1}{3}x$
3. $f(x) = x + 9$
4. $f(x) = x - 4$
5. $f(x) = 3x + 1$
6. $f(x) = \dfrac{x - 1}{5}$
7. $f(x) = \sqrt[3]{x}$
8. $f(x) = x^5$

In Exercises 9–12, match the graph of the function with the graph of its inverse function. [The graphs of the inverse functions are labeled (a), (b), (c), and (d).]

(a)

(b)

(c)

(d)

9.

10.

11.

12.

In Exercises 13–24, show that f and g are inverse functions (a) algebraically and (b) graphically.

13. $f(x) = 2x$, $g(x) = \dfrac{x}{2}$
14. $f(x) = x - 5$, $g(x) = x + 5$
15. $f(x) = 7x + 1$, $g(x) = \dfrac{x - 1}{7}$
16. $f(x) = 3 - 4x$, $g(x) = \dfrac{3 - x}{4}$
17. $f(x) = \dfrac{x^3}{8}$, $g(x) = \sqrt[3]{8x}$
18. $f(x) = \dfrac{1}{x}$, $g(x) = \dfrac{1}{x}$
19. $f(x) = \sqrt{x - 4}$, $g(x) = x^2 + 4$, $x \geq 0$
20. $f(x) = 1 - x^3$, $g(x) = \sqrt[3]{1 - x}$
21. $f(x) = 9 - x^2$, $x \geq 0$, $g(x) = \sqrt{9 - x}$, $x \leq 9$
22. $f(x) = \dfrac{1}{1 + x}$, $x \geq 0$, $g(x) = \dfrac{1 - x}{x}$, $0 < x \leq 1$
23. $f(x) = \dfrac{x - 1}{x + 5}$, $g(x) = -\dfrac{5x + 1}{x - 1}$
24. $f(x) = \dfrac{x + 3}{x - 2}$, $g(x) = \dfrac{2x + 3}{x - 1}$

In Exercises 25 and 26, does the function have an inverse function?

25.

x	-1	0	1	2	3	4
$f(x)$	-2	1	2	1	-2	-6

26.

x	-3	-2	-1	0	2	3
$f(x)$	10	6	4	1	-3	-10

In Exercises 27 and 28, use the table of values for $y = f(x)$ to complete a table for $y = f^{-1}(x)$.

27.

x	-2	-1	0	1	2	3
$f(x)$	-2	0	2	4	6	8

28.

x	-3	-2	-1	0	1	2
$f(x)$	-10	-7	-4	-1	2	5

In Exercises 29–32, does the function have an inverse function?

29.

30.

31.

32.
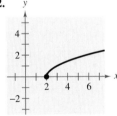

In Exercises 33–38, use a graphing utility to graph the function, and use the Horizontal Line Test to determine whether the function is one-to-one and so has an inverse function.

33. $g(x) = \dfrac{4 - x}{6}$

34. $f(x) = 10$

35. $h(x) = |x + 4| - |x - 4|$

36. $g(x) = (x + 5)^3$

37. $f(x) = -2x\sqrt{16 - x^2}$

38. $f(x) = \frac{1}{8}(x + 2)^2 - 1$

In Exercises 39–54, (a) find the inverse function of f, (b) graph both f and f^{-1} on the same set of coordinate axes, (c) describe the relationship between the graphs of f and f^{-1}, and (d) state the domain and range of f and f^{-1}.

39. $f(x) = 2x - 3$

40. $f(x) = 3x + 1$

41. $f(x) = x^5 - 2$

42. $f(x) = x^3 + 1$

43. $f(x) = \sqrt{x}$

44. $f(x) = x^2, \quad x \geq 0$

45. $f(x) = \sqrt{4 - x^2}, \quad 0 \leq x \leq 2$

46. $f(x) = x^2 - 2, \quad x \leq 0$

47. $f(x) = \dfrac{4}{x}$

48. $f(x) = -\dfrac{2}{x}$

49. $f(x) = \dfrac{x + 1}{x - 2}$

50. $f(x) = \dfrac{x - 3}{x + 2}$

51. $f(x) = \sqrt[3]{x - 1}$

52. $f(x) = x^{3/5}$

53. $f(x) = \dfrac{6x + 4}{4x + 5}$

54. $f(x) = \dfrac{8x - 4}{2x + 6}$

In Exercises 55–68, determine whether the function has an inverse function. If it does, find the inverse function.

55. $f(x) = x^4$

56. $f(x) = \dfrac{1}{x^2}$

57. $g(x) = \dfrac{x}{8}$

58. $f(x) = 3x + 5$

59. $p(x) = -4$

60. $f(x) = \dfrac{3x + 4}{5}$

61. $f(x) = (x + 3)^2, \quad x \geq -3$

62. $q(x) = (x - 5)^2$

63. $f(x) = \begin{cases} x + 3, & x < 0 \\ 6 - x, & x \geq 0 \end{cases}$

64. $f(x) = \begin{cases} -x, & x \leq 0 \\ x^2 - 3x, & x > 0 \end{cases}$

65. $h(x) = -\dfrac{4}{x^2}$

66. $f(x) = |x - 2|, \quad x \leq 2$

67. $f(x) = \sqrt{2x + 3}$

68. $f(x) = \sqrt{x - 2}$

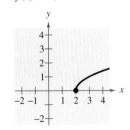

In **Exercises 69–74,** use the functions given by $f(x) = \frac{1}{8}x - 3$ and $g(x) = x^3$ to find the indicated value or function.

69. $(f^{-1} \circ g^{-1})(1)$

70. $(g^{-1} \circ f^{-1})(-3)$

71. $(f^{-1} \circ f^{-1})(6)$

72. $(g^{-1} \circ g^{-1})(-4)$

73. $(f \circ g)^{-1}$

74. $g^{-1} \circ f^{-1}$

In **Exercises 75–78,** use the functions given by $f(x) = x + 4$ and $g(x) = 2x - 5$ to find the specified function.

75. $g^{-1} \circ f^{-1}$

76. $f^{-1} \circ g^{-1}$

77. $(f \circ g)^{-1}$

78. $(g \circ f)^{-1}$

Model It

79. *U.S. Households* The numbers of households f (in thousands) in the United States from 1995 to 2003 are shown in the table. The time (in years) is given by t, with $t = 5$ corresponding to 1995. (Source: U.S. Census Bureau)

Year, t	Households, $f(t)$
5	98,990
6	99,627
7	101,018
8	102,528
9	103,874
10	104,705
11	108,209
12	109,297
13	111,278

(a) Find $f^{-1}(108,209)$.

(b) What does f^{-1} mean in the context of the problem?

 (c) Use the *regression* feature of a graphing utility to find a linear model for the data, $y = mx + b$. (Round m and b to two decimal places.)

 (d) Algebraically find the inverse function of the linear model in part (c).

 (e) Use the inverse function of the linear model you found in part (d) to approximate $f^{-1}(117,022)$.

(f) Use the inverse function of the linear model you found in part (d) to approximate $f^{-1}(108,209)$. How does this value compare with the original data shown in the table?

80. *Digital Camera Sales* The factory sales f (in millions of dollars) of digital cameras in the United States from 1998 through 2003 are shown in the table. The time (in years) is given by t, with $t = 8$ corresponding to 1998. (Source: Consumer Electronincs Association)

Year, t	Sales, $f(t)$
8	519
9	1209
10	1825
11	1972
12	2794
13	3421

(a) Does f^{-1} exist?

(b) If f^{-1} exists, what does it represent in the context of the problem?

(c) If f^{-1} exists, find $f^{-1}(1825)$.

(d) If the table was extended to 2004 and if the factory sales of digital cameras for that year was \$2794 million, would f^{-1} exist? Explain.

81. *Miles Traveled* The total numbers f (in billions) of miles traveled by motor vehicles in the United States from 1995 through 2002 are shown in the table. The time (in years) is given by t, with $t = 5$ corresponding to 1995. (Source: U.S. Federal Highway Administration)

Year, t	Miles traveled, $f(t)$
5	2423
6	2486
7	2562
8	2632
9	2691
10	2747
11	2797
12	2856

(a) Does f^{-1} exist?

(b) If f^{-1} exists, what does it mean in the context of the problem?

(c) If f^{-1} exists, find $f^{-1}(2632)$.

(d) If the table was extended to 2003 and if the total number of miles traveled by motor vehicles for that year was 2747 billion, would f^{-1} exist? Explain.

82. *Hourly Wage*   Your wage is $8.00 per hour plus $0.75 for each unit produced per hour. So, your hourly wage y in terms of the number of units produced is

$y = 8 + 0.75x.$

(a) Find the inverse function.

(b) What does each variable represent in the inverse function?

(c) Determine the number of units produced when your hourly wage is $22.25.

83. *Diesel Mechanics*   The function given by

$y = 0.03x^2 + 245.50, \qquad 0 < x < 100$

approximates the exhaust temperature y in degrees Fahrenheit, where x is the percent load for a diesel engine.

(a) Find the inverse function. What does each variable represent in the inverse function?

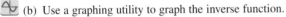 (b) Use a graphing utility to graph the inverse function.

(c) The exhaust temperature of the engine must not exceed 500 degrees Fahrenheit. What is the percent load interval?

84. *Cost*   You need a total of 50 pounds of two types of ground beef costing $1.25 and $1.60 per pound, respectively. A model for the total cost y of the two types of beef is

$y = 1.25x + 1.60(50 - x)$

where x is the number of pounds of the less expensive ground beef.

(a) Find the inverse function of the cost function. What does each variable represent in the inverse function?

(b) Use the context of the problem to determine the domain of the inverse function.

(c) Determine the number of pounds of the less expensive ground beef purchased when the total cost is $73.

Synthesis

True or False?   **In Exercises 85 and 86, determine whether the statement is true or false. Justify your answer.**

85. If f is an even function, f^{-1} exists.

86. If the inverse function of f exists and the graph of f has a y-intercept, the y-intercept of f is an x-intercept of f^{-1}.

87. *Proof*   Prove that if f and g are one-to-one functions, then $(f \circ g)^{-1}(x) = (g^{-1} \circ f^{-1})(x).$

88. *Proof*   Prove that if f is a one-to-one odd function, then f^{-1} is an odd function.

In Exercises 89–92, use the graph of the function f to create a table of values for the given points. Then create a second table that can be used to find f^{-1}, and sketch the graph of f^{-1} if possible.

89.

90.

91.

92.

93. *Think About It*   The function given by

$f(x) = k(2 - x - x^3)$

has an inverse function, and $f^{-1}(3) = -2$. Find k.

94. *Think About It*   The function given by

$f(x) = k(x^3 + 3x - 4)$

has an inverse function, and $f^{-1}(-5) = 2$. Find k.

P   Chapter Summary

What did you learn?

Section P.1	Review Exercises
☐ Represent and classify real numbers *(p. 2)*.	1, 2
☐ Order real numbers and use inequalities *(p. 3)*.	3–6
☐ Find the absolute values of real numbers and find the distance between two real numbers *(p. 5)*.	7–10
☐ Evaluate algebraic expressions *(p. 6)*.	11, 12
☐ Use the basic rules and properties of algebra *(p. 8)*.	13–22

Section P.2	
☐ Identify different types of equations *(p. 12)*.	23, 24
☐ Solve linear equations in one variable and equations that lead to linear equations *(p. 12)*.	25–32
☐ Solve quadratic equations by factoring, extracting square roots, completing the square, and using the Quadratic Formula *(p. 15)*.	33–42
☐ Solve polynomial equations of degree three or greater *(p. 19)*.	43–46
☐ Solve equations involving radicals *(p. 20)*.	47–52
☐ Solve equations with absolute values *(p. 21)*.	53–56

Section P.3	
☐ Plot points in the Cartesian plane *(p. 26)*.	57–60
☐ Use the Distance Formula to find the distance between two points *(p. 28)*.	61–64
☐ Use the Midpoint Formula to find the midpoint of a line segment *(p. 29)*.	61–64
☐ Use a coordinate plane to model and solve real-life problems *(p. 30)*.	65, 66
☐ Sketch graphs of equations *(p. 31)*.	67–70
☐ Find *x*- and *y*-intercepts of graphs of equations *(p. 32)*.	71, 72
☐ Use symmetry to sketch graphs of equations *(p. 33)*.	73–80
☐ Find equations of and sketch graphs of circles *(p. 35)*.	81–86

Section P.4	
☐ Use slope to graph linear equations in two variables *(p. 40)*.	87–94
☐ Find slopes of lines *(p. 42)*.	95–98
☐ Write linear equations in two variables *(p. 44)*.	99–106
☐ Use slope to identify parallel and perpendicular lines *(p. 45)*.	107, 108
☐ Use slope and linear equations in two variables to model and solve real-life problems *(p. 46)*.	109, 110

P   Review Exercises

P.1 In Exercises 1 and 2, determine which numbers in the set are (a) natural numbers, (b) whole numbers, (c) integers, (d) rational numbers, and (e) irrational numbers.

1. $\left\{11, -14, -\frac{8}{9}, \frac{5}{2}, \sqrt{6}, 0.4\right\}$
2. $\left\{\sqrt{15}, -22, -\frac{10}{3}, 0, 5.2, \frac{3}{7}\right\}$

In Exercises 3 and 4, use a calculator to find the decimal form of each rational number. If it is a nonterminating decimal, write the repeating pattern. Then plot the numbers on the real number line and place the appropriate inequality sign (< or >) between them.

3. (a) $\frac{5}{6}$ (b) $\frac{7}{8}$
4. (a) $\frac{9}{25}$ (b) $\frac{5}{7}$

In Exercises 5 and 6, give a verbal description of the subset of real numbers represented by the inequality, and sketch the subset on the real number line.

5. $x \le 7$
6. $x > 1$

In Exercises 7 and 8, find the distance between a and b.

7. $a = -92, b = 63$
8. $a = -112, b = -6$

In Exercises 9 and 10, use absolute value notation to describe the situation.

9. The distance between x and 7 is at least 4.
10. The distance between x and 25 is no more than 10.

In Exercises 11 and 12, evaluate the expression for each value of x.

Expression	*Values*	
11. $12x - 7$	(a) $x = 0$	(b) $x = -1$
12. $x^2 - 6x + 5$	(a) $x = -2$	(b) $x = 2$

In Exercises 13–16, identify the rule of algebra illustrated by the statement.

13. $2x + (3x - 10) = (2x + 3x) - 10$
14. $4(t + 2) = 4 \cdot t + 4 \cdot 2$
15. $0 + (a - 5) = a - 5$
16. $\frac{2}{y + 4} \cdot \frac{y + 4}{2} = 1, \quad y \ne -4$

In Exercises 17–22, perform the operation without using a calculator.

17. $|-3| + 4(-2) - 6$
18. $\frac{|-10|}{-10}$
19. $\frac{5}{18} \div \frac{10}{3}$
20. $(16 - 8) \div 4$
21. $6[4 - 2(6 + 8)]$
22. $-4[16 - 3(7 - 10)]$

P.2 In Exercises 23 and 24, determine whether the equation is an identity or a conditional equation.

23. $6 - (x - 2)^2 = 2 + 4x - x^2$
24. $3(x - 2) + 2x = 2(x + 3)$

In Exercises 25–32, solve the equation (if possible) and check your solution.

25. $3x - 2(x + 5) = 10$
26. $4x + 2(7 - x) = 5$
27. $4(x + 3) - 3 = 2(4 - 3x) - 4$
28. $\frac{1}{2}(x - 3) - 2(x + 1) = 5$
29. $\frac{x}{5} - 3 = \frac{x}{3} + 1$
30. $\frac{4x - 3}{6} + \frac{x}{4} = x - 2$
31. $\frac{18}{x} = \frac{10}{x - 4}$
32. $\frac{5}{x - 2} = \frac{13}{2x - 3}$

In Exercises 33–42, use any method to solve the quadratic equation.

33. $15 + x - 2x^2 = 0$
34. $2x^2 - x - 28 = 0$
35. $6 = 3x^2$
36. $16x^2 = 25$
37. $(x + 4)^2 = 18$
38. $(x - 8)^2 = 15$
39. $x^2 - 12x + 30 = 0$
40. $x^2 + 6x - 3 = 0$
41. $-2x^2 - 5x + 27 = 0$
42. $-20 - 3x + 3x^2 = 0$

In Exercises 43–56, find all solutions of the equation. Check your solutions in the original equation.

43. $5x^4 - 12x^3 = 0$
44. $4x^3 - 6x^2 = 0$
45. $x^4 - 5x^2 + 6 = 0$
46. $9x^4 + 27x^3 - 4x^2 - 12x = 0$
47. $\sqrt{x + 4} = 3$
48. $\sqrt{x - 2} - 8 = 0$
49. $\sqrt{2x + 3} + \sqrt{x - 2} = 2$
50. $5\sqrt{x} - \sqrt{x - 1} = 6$
51. $(x - 1)^{2/3} - 25 = 0$
52. $(x + 2)^{3/4} = 27$
53. $|x - 5| = 10$
54. $|2x + 3| = 7$
55. $|x^2 - 3| = 2x$
56. $|x^2 - 6| = x$

P.3 In Exercises 57 and 58, plot the points in the Cartesian plane.

57. $(2, 2), (0, -4), (-3, 6), (-1, -7)$
58. $(5, 0), (8, 1), (4, -2), (-3, -3)$

In Exercises 59 and 60, determine the quadrant(s) in which (x, y) is located so that the condition(s) is (are) satisfied.

59. $x > 0$ and $y = -2$ **60.** $y > 0$

In Exercises 61–64, (a) plot the points, (b) find the distance between the points, and (c) find the midpoint of the line segment joining the points.

61. $(-3, 8), (1, 5)$ **62.** $(-2, 6), (4, -3)$
63. $(5.6, 0), (0, 8.2)$ **64.** $(0, -1.2), (-3.6, 0)$

65. *Sales* The Cheesecake Factory had annual sales of $539.1 million in 2001 and $773.8 million in 2003. Use the Midpoint Formula to estimate the sales in 2002. (Source: The Cheesecake Factory, Inc.)

66. *Meteorology* The apparent temperature is a measure of relative discomfort to a person from heat and high humidity. The table shows the actual temperatures x (in degrees Fahrenheit) versus the apparent temperatures y (in degrees Fahrenheit) for a relative humidity of 75%.

x	70	75	80	85	90	95	100
y	70	77	85	95	109	130	150

(a) Sketch a scatter plot of the data shown in the table.

(b) Find the change in the apparent temperature when the actual temperature changes from 70°F to 100°F.

In Exercises 67–70, complete a table of values. Use the solution points to sketch the graph of the equation.

67. $y = 3x - 5$ **68.** $y = -\frac{1}{2}x + 2$
69. $y = x^2 - 3x$ **70.** $y = 2x^2 - x - 9$

In Exercises 71 and 72, find the x- and y-intercepts of the graph of the equation.

71. $y = (x - 3)^2 - 4$ **72.** $y = |x + 1| - 3$

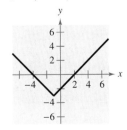

In Exercises 73–80, use the algebraic tests to check for symmetry with respect to both axes and the origin. Then sketch the graph of the equation.

73. $y = -4x + 1$ **74.** $y = 5x - 6$
75. $y = 5 - x^2$ **76.** $y = x^2 - 10$
77. $y = x^3 + 3$ **78.** $y = -6 - x^3$
79. $y = \sqrt{x + 5}$ **80.** $y = |x| + 9$

In Exercises 81–84, find the center and radius of the circle and sketch its graph.

81. $x^2 + y^2 = 9$ **82.** $x^2 + y^2 = 4$
83. $\left(x - \frac{1}{2}\right)^2 + (y + 1)^2 = 36$
84. $(x + 4)^2 + \left(y - \frac{3}{2}\right)^2 = 100$

85. Find the standard form of the equation of the circle for which the endpoints of a diameter are $(0, 0)$ and $(4, -6)$.

86. Find the standard form of the equation of the circle for which the endpoints of a diameter are $(-2, -3)$ and $(4, -10)$.

P.4 In Exercises 87–94, find the slope and y-intercept (if possible) of the equation of the line. Sketch the line.

87. $y = -2x - 7$ **88.** $y = 4x - 3$
89. $y = 6$ **90.** $x = -3$
91. $y = 3x + 13$ **92.** $y = -10x + 9$
93. $y = -\frac{5}{2}x - 1$ **94.** $y = \frac{5}{6}x + 5$

In Exercises 95–98, plot the points and find the slope of the line passing through the pair of points.

95. $(3, -4), (-7, 1)$ **96.** $(-1, 8), (6, 5)$
97. $(-4.5, 6), (2.1, 3)$ **98.** $(-3, 2), (8, 2)$

In Exercises 99–102, find the slope-intercept form of the equation of the line that passes through the given point and has the indicated slope. Sketch the line.

	Point	*Slope*
99.	$(0, -5)$	$m = \frac{3}{2}$
100.	$(-2, 6)$	$m = 0$
101.	$(10, -3)$	$m = -\frac{1}{2}$
102.	$(-8, 5)$	m is undefined.

In Exercises 103–106, find the slope-intercept form of the equation of the line passing through the points.

103. $(0, 0), (0, 10)$ **104.** $(2, 5), (-2, -1)$
105. $(-1, 4), (2, 0)$ **106.** $(11, -2), (6, -1)$

In Exercises 107 and 108, write the slope-intercept forms of the equations of the lines through the given point (a) parallel to the given line and (b) perpendicular to the given line.

Point	*Line*
107. $(3, -2)$	$5x - 4y = 8$
108. $(-8, 3)$	$2x + 3y = 5$

109. *Sales* During the second and third quarters of the year, a salvage yard had sales of $160,000 and $185,000, respectively. The growth of sales follows a linear pattern. Estimate sales during the fourth quarter.

110. *Inflation* The dollar value of a product in 2005 is $85, and the product is expected to increase in value at a rate of $3.75 per year.

(a) Write a linear equation that gives the dollar value V of the product in terms of the year t. (Let $t = 5$ represent 2005.)

 (b) Use a graphing utility to graph the equation found in part (a).

(c) Move the cursor along the graph of the sales model to estimate the dollar value of the product in 2010.

P.5 In Exercises 111 and 112, determine which of the sets of ordered pairs represents a function from A to B. Explain your reasoning.

111. $A = \{10, 20, 30, 40\}$ and $B = \{0, 2, 4, 6\}$

(a) $\{(20, 4), (40, 0), (20, 6), (30, 2)\}$

(b) $\{(10, 4), (20, 4), (30, 4), (40, 4)\}$

(c) $\{(40, 0), (30, 2), (20, 4), (10, 6)\}$

(d) $\{(20, 2), (10, 0), (40, 4)\}$

112. $A = \{u, v, w\}$ and $B = \{-2, -1, 0, 1, 2\}$

(a) $\{(v, -1), (u, 2), (w, 0), (u, -2)\}$

(b) $\{(u, -2), (v, 2), (w, 1)\}$

(c) $\{(u, 2), (v, 2), (w, 1), (w, 1)\}$

(d) $\{(w, -2), (v, 0), (w, 2)\}$

In Exercises 113–116, determine whether the equation represents y as a function of x.

113. $16x - y^4 = 0$ **114.** $2x - y - 3 = 0$
115. $y = \sqrt{1 - x}$ **116.** $|y| = x + 2$

In Exercises 117 and 118, evaluate the function at each specified value of the independent variable and simplify.

117. $h(x) = \begin{cases} 2x + 1, & x \le -1 \\ x^2 + 2, & x > -1 \end{cases}$

(a) $h(-2)$ (b) $h(-1)$ (c) $h(0)$ (d) $h(2)$

118. $f(x) = \dfrac{4}{x^2 + 1}$

(a) $f(1)$ (b) $f(-5)$ (c) $f(-t)$ (d) $f(0)$

In Exercises 119–124, find the domain of the function. Verify your result with a graph.

119. $f(x) = \sqrt{25 - x^2}$ **120.** $f(x) = 3x + 4$

121. $g(s) = \dfrac{5}{3s - 9}$ **122.** $f(x) = \sqrt{x^2 + 8x}$

123. $h(x) = \dfrac{x}{x^2 - x - 6}$ **124.** $h(t) = |t + 1|$

125. *Physics* The velocity of a ball projected upward from ground level is given by $v(t) = -32t + 48$, where t is the time in seconds and v is the velocity in feet per second.

(a) Find the velocity when $t = 1$.

(b) Find the time when the ball reaches its maximum height. [Hint: Find the time when $v(t) = 0$.]

(c) Find the velocity when $t = 2$.

126. *Total Cost* A hand tool manufacturer produces a product for which the variable cost is $5.35 per unit and the fixed costs are $16,000. The company sells the product for $8.20 and can sell all that it produces.

(a) Find the total cost as a function of x, the number of units produced.

(b) Find the profit as a function of x.

In Exercises 127 and 128, find the difference quotient and simplify your answer.

127. $f(x) = 2x^2 + 3x - 1$, $\dfrac{f(x + h) - f(x)}{h}$, $h \ne 0$

128. $f(x) = x^3 - 5x^2 + x$, $\dfrac{f(x + h) - f(x)}{h}$, $h \ne 0$

P.6 In Exercises 129–132, use the Vertical Line Test to determine whether y is a function of x. To print an enlarged copy of the graph, go to the website *www.mathgraphs.com*.

129. $y = (x - 3)^2$ **130.** $y = -\frac{3}{5}x^3 - 2x + 1$

131. $x - 4 = y^2$

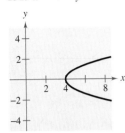

132. $x = -|4 - y|$

149. $f(x) = 2x\sqrt{x^2 + 3}$

150. $f(x) = \sqrt[5]{6x^2}$

P.7 In Exercises 151–154, write the linear function f such that it has the indicated function values. Then sketch the graph of the function.

151. $f(2) = -6,\ f(-1) = 3$

152. $f(0) = -5,\ f(4) = -8$

153. $f\left(-\frac{4}{5}\right) = 2,\ f\left(\frac{11}{5}\right) = 7$

154. $f(3.3) = 5.6,\ f(-4.7) = -1.4$

In Exercises 133–136, find the zeros of the function algebraically.

133. $f(x) = 3x^2 - 16x + 21$ **134.** $f(x) = 5x^2 + 4x - 1$

135. $f(x) = \dfrac{8x + 3}{11 - x}$

136. $f(x) = x^3 - x^2 - 25x + 25$

In Exercises 155–164, graph the function.

155. $f(x) = 3 - x^2$ **156.** $h(x) = x^3 - 2$

157. $f(x) = -\sqrt{x}$ **158.** $f(x) = \sqrt{x + 1}$

159. $g(x) = \dfrac{3}{x}$ **160.** $g(x) = \dfrac{1}{x + 5}$

161. $f(x) = [\![x]\!] - 2$ **162.** $g(x) = [\![x + 4]\!]$

163. $f(x) = \begin{cases} 5x - 3, & x \ge -1 \\ -4x + 5, & x < -1 \end{cases}$

164. $f(x) = \begin{cases} x^2 - 2, & x < -2 \\ 5, & -2 \le x \le 0 \\ 8x - 5, & x > 0 \end{cases}$

In Exercises 137 and 138, determine the intervals over which the function is increasing, decreasing, or constant.

137. $f(x) = |x| + |x + 1|$ **138.** $f(x) = (x^2 - 4)^2$

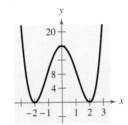

In Exercises 165 and 166, the figure shows the graph of a transformed parent function. Identify the parent function.

165.

166.

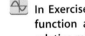 In Exercises 139–142, use a graphing utility to graph the function and approximate (to two decimal places) any relative minimum or relative maximum values.

139. $f(x) = -x^2 + 2x + 1$ **140.** $f(x) = x^4 - 4x^2 - 2$

141. $f(x) = x^3 - 6x^4$

142. $f(x) = x^3 - 4x^2 + x - 1$

P.8 In Exercises 167–180, h is related to one of the parent functions described in this chapter. (a) Identify the parent function f. (b) Describe the sequence of transformations from f to h. (c) Sketch the graph of h. (d) Use function notation to write h in terms of f.

167. $h(x) = x^2 - 9$ **168.** $h(x) = (x - 2)^3 + 2$

In Exercises 143–146, find the average rate of change of the function from x_1 to x_2.

Function	x-Values
143. $f(x) = -x^2 + 8x - 4$	$x_1 = 0, x_2 = 4$
144. $f(x) = x^3 + 12x - 2$	$x_1 = 0, x_2 = 4$
145. $f(x) = 2 - \sqrt{x + 1}$	$x_1 = 3, x_2 = 7$
146. $f(x) = 1 - \sqrt{x + 3}$	$x_1 = 1, x_2 = 6$

169. $h(x) = \sqrt{x - 7}$ **170.** $h(x) = |x + 3| - 5$

171. $h(x) = -(x + 3)^2 + 1$ **172.** $h(x) = -(x - 5)^3 - 5$

173. $h(x) = -[\![x]\!] + 6$ **174.** $h(x) = -\sqrt{x + 1} + 9$

175. $h(x) = -|-x + 4| + 6$ **176.** $h(x) = -(x + 1)^2 - 3$

177. $h(x) = 5[\![x - 9]\!]$ **178.** $h(x) = -\frac{1}{3}x^3$

179. $h(x) = -2\sqrt{x - 4}$ **180.** $h(x) = \frac{1}{2}|x| - 1$

In Exercises 147–150, determine whether the function is even, odd, or neither.

147. $f(x) = x^5 + 4x - 7$ **148.** $f(x) = x^4 - 20x^2$

P.9 In Exercises 181 and 182, find (a) $(f + g)(x)$, (b) $(f - g)(x)$, (c) $(fg)(x)$, and (d) $(f/g)(x)$. What is the domain of f/g?

181. $f(x) = x^2 + 3$, $g(x) = 2x - 1$
182. $f(x) = x^2 - 4$, $g(x) = \sqrt{3 - x}$

In Exercises 183 and 184, find (a) $f \circ g$ and (b) $g \circ f$. Find the domain of each function and each composite function.

183. $f(x) = \frac{1}{3}x - 3$, $g(x) = 3x + 1$
184. $f(x) = x^3 - 4$, $g(x) = \sqrt[3]{x + 7}$

In Exercises 185 and 186, find two functions f and g such that $(f \circ g)(x) = h(x)$. (There are many correct answers.)

185. $h(x) = (6x - 5)^3$
186. $h(x) = \sqrt[3]{x + 2}$

187. *Electronics Sales* The factory sales (in millions of dollars) for VCRs $v(t)$ and DVD players $d(t)$ from 1997 to 2003 can be approximated by the functions

$$v(t) = -31.86t^2 + 233.6t + 2594$$

and

$$d(t) = -4.18t^2 + 571.0t - 3706$$

where t represents the year, with $t = 7$ corresponding to 1997. (Source: Consumer Electronics Association)

(a) Find and interpret $(v + d)(t)$.

 (b) Use a graphing utility to graph $v(t)$, $d(t)$, and the function from part (a) in the same viewing window.

 (c) Find $(v + d)(10)$. Use the graph in part (b) to verify your result.

188. *Bacteria Count* The number N of bacteria in a refrigerated food is given by

$$N(T) = 25T^2 - 50T + 300, \quad 2 \le T \le 20$$

where T is the temperature of the food in degrees Celsius. When the food is removed from refrigeration, the temperature of the food is given by

$$T(t) = 2t + 1, \quad 0 \le t \le 9$$

where t is the time in hours (a) Find the composition $N(T(t))$ and interpret its meaning in context and (b) find the time when the bacterial count reaches 750.

P.10 In Exercises 189 and 190, find the inverse function of f informally. Verify that $f(f^{-1}(x)) = x$ and $f^{-1}(f(x)) = x$.

189. $f(x) = x - 7$
190. $f(x) = x + 5$

In Exercises 191 and 192, determine whether the function has an inverse function.

191.   **192.**

 In Exercises 193–196, use a graphing utility to graph the function, and use the Horizontal Line Test to determine whether the function is one-to-one and so has an inverse function.

193. $f(x) = 4 - \frac{1}{3}x$ **194.** $f(x) = (x - 1)^2$

195. $h(t) = \dfrac{2}{t - 3}$ **196.** $g(x) = \sqrt{x + 6}$

In Exercises 197–200, (a) find the inverse function of f, (b) graph both f and f^{-1} on the same set of coordinate axes, (c) describe the relationship between the graphs of f and f^{-1}, and (d) state the domain and range of f and f^{-1}.

197. $f(x) = \frac{1}{2}x - 3$ **198.** $f(x) = 5x - 7$
199. $f(x) = \sqrt{x + 1}$ **200.** $f(x) = x^3 + 2$

In Exercises 201 and 202, restrict the domain of the function f to an interval over which the function is increasing and determine f^{-1} over that interval.

201. $f(x) = 2(x - 4)^2$
202. $f(x) = |x - 2|$

Synthesis

True or False? In Exercises 203 and 204, determine whether the statement is true or false. Justify your answer.

203. Relative to the graph of $f(x) = \sqrt{x}$, the function given by $h(x) = -\sqrt{x + 9} - 13$ is shifted 9 units to the left and 13 units downward, then reflected in the x-axis.

204. If f and g are two inverse functions, then the domain of g is equal to the range of f.

205. *Writing* Explain why it is essential to check your solutions to radical, absolute value, and rational equations.

206. *Writing* Explain how to tell whether a relation between two variables is a function.

207. *Writing* Explain the difference between the Vertical Line Test and the Horizontal Line Test.

Chapter Test

Take this test as you would take a test in class. When you are finished, check your work against the answers given in the back of the book.

1. Place $<$ or $>$ between the real numbers $-\frac{10}{3}$ and $-|-4|$.

2. Find the distance between the real numbers -5.4 and $3\frac{3}{4}$.

3. Identify the rule of algebra illustrated by $(5 - x) + 0 = 5 - x$.

In Exercises 4–9, solve the equation (if possible).

4. $\frac{2}{3}(x - 1) + \frac{1}{4}x = 10$

5. $(x - 3)(x + 2) = 14$

6. $\dfrac{x - 2}{x + 2} + \dfrac{4}{x + 2} + 4 = 0$

7. $x^4 + x^2 - 6 = 0$

8. $2\sqrt{x} - \sqrt{2x + 1} = 1$

9. $|3x - 1| = 7$

10. Plot the points $(-2, 5)$ and $(6, 0)$. Find the coordinates of the midpoint of the line segment joining the points and the distance between the points.

In Exercises 11–13, check for symmetry with respect to both axes and the origin. Then sketch the graph of the equation. Identify any x- and y-intercepts.

11. $y = 4 - \frac{3}{4}x$

12. $y = 4 - (x - 2)^2$

13. $y = x - x^3$

14. Find the center and radius of the circle given by $(x - 3)^2 + y^2 = 9$. Then sketch its graph.

15. Find an equation of the line that passes through the point $(3, 8)$ and is (a) parallel to and (b) perpendicular to the line $-4x + 7y = -5$.

16. Evaluate the functions given by $f(x) = |x + 2| - 15$ at each specified value of the independent variable and simplify.

   (a) $f(-8)$ (b) $f(14)$ (c) $f(x - 6)$

In Exercises 17–19, (a) use a graphing utility to graph the function, (b) determine the domain of the function, (c) approximate the intervals over which the function is increasing, decreasing, or constant, and (d) determine whether the function is even, odd, or neither.

17. $f(x) = 2x^6 + 5x^4 - x^2$

18. $f(x) = 4x\sqrt{3 - x}$

19. $f(x) = |x + 5|$

In Exercises 20–22, (a) identify the parent function in the transformation, (b) describe the sequence of transformations from f to h, and (c) sketch the graph of h.

20. $h(x) = -\llbracket x \rrbracket$

21. $h(x) = -\sqrt{x + 5} + 8$

22. $h(x) = \frac{1}{4}|x + 1| - 3$

In Exercises 23 and 24, find (a) $(f + g)(x)$, (b) $(f - g)(x)$, (c) $(fg)(x)$, (d) $(f/g)(x)$, (e) $(f \circ g)(x)$, and (f) $(g \circ f)(x)$.

23. $f(x) = 3x^2 - 7, \quad g(x) = -x^2 - 4x + 5$

24. $f(x) = \dfrac{1}{x}, \quad g(x) = 2\sqrt{x}$

In Exercises 25–27, determine whether the function has an inverse function, and if so, find the inverse function.

25. $f(x) = x^3 + 8$

26. $f(x) = |x^2 - 3| + 6$

27. $f(x) = \dfrac{3x\sqrt{x}}{8}$

Proofs in Mathematics

What does the word *proof* mean to you? In mathematics, the word *proof* is used to mean simply a valid argument. When you are proving a statement or theorem, you must use facts, definitions, and accepted properties in a logical order. You can also use previously proved theorems in your proof. For instance, the Distance Formula is used in the proof of the Midpoint Formula below. There are several different proof methods, which you will see in later chapters.

> ## The Midpoint Formula   *(p. 29)*
> The midpoint of the line segment joining the points (x_1, y_1) and (x_2, y_2) is given by the Midpoint Formula
>
> $$\text{Midpoint} = \left(\frac{x_1 + x_2}{2}, \frac{y_1 + y_2}{2} \right).$$

The Cartesian Plane

The Cartesian plane was named after the French mathematician René Descartes (1596–1650). While Descartes was lying in bed, he noticed a fly buzzing around on the square ceiling tiles. He discovered that the position of the fly could be described by which ceiling tile the fly landed on. This led to the development of the Cartesian plane. Descartes felt that a coordinate plane could be used to facilitate description of the positions of objects.

Proof

Using the figure, you must show that $d_1 = d_2$ and $d_1 + d_2 = d_3$.

By the Distance Formula, you obtain

$$d_1 = \sqrt{\left(\frac{x_1 + x_2}{2} - x_1 \right)^2 + \left(\frac{y_1 + y_2}{2} - y_1 \right)^2}$$

$$= \frac{1}{2} \sqrt{(x_2 - x_1)^2 + (y_2 - y_1)^2}$$

$$d_2 = \sqrt{\left(x_2 - \frac{x_1 + x_2}{2} \right)^2 + \left(y_2 - \frac{y_1 + y_2}{2} \right)^2}$$

$$= \frac{1}{2} \sqrt{(x_2 - x_1)^2 + (y_2 - y_1)^2}$$

$$d_3 = \sqrt{(x_2 - x_1)^2 + (y_2 - y_1)^2}$$

So, it follows that $d_1 = d_2$ and $d_1 + d_2 = d_3$.

P.S. Problem Solving

This collection of thought-provoking and challenging exercises further explores and expands upon concepts learned in this chapter.

1. As a salesperson, you receive a monthly salary of $2000, plus a commission of 7% of sales. You are offered a new job at $2300 per month, plus a commission of 5% of sales.

   (a) Write a linear equation for your current monthly wage W_1 in terms of your monthly sales S.

   (b) Write a linear equation for the monthly wage W_2 of your new job offer in terms of the monthly sales S.

   (c) Use a graphing utility to graph both equations in the same viewing window. Find the point of intersection. What does it signify?

   (d) You think you can sell $20,000 per month. Should you change jobs? Explain.

2. For the numbers 2 through 9 on a telephone keypad (see figure), create two relations: one mapping numbers onto letters, and the other mapping letters onto numbers. Are both relations functions? Explain.

3. What can be said about the sum and difference of each of the following?

   (a) Two even functions   (b) Two odd functions

   (c) An odd function and an even function

4. The two functions given by

   $$f(x) = x \quad \text{and} \quad g(x) = -x$$

   are their own inverse functions. Graph each function and explain why this is true. Graph other linear functions that are their own inverse functions. Find a general formula for a family of linear functions that are their own inverse functions.

5. Prove that a function of the following form is even.

   $$y = a_{2n}x^{2n} + a_{2n-2}x^{2n-2} + \cdots + a_2x^2 + a_0$$

6. A miniature golf professional is trying to make a hole-in-one on the miniature golf green shown. A coordinate plane is placed over the golf green. The golf ball is at the point $(2.5, 2)$ and the hole is at the point $(9.5, 2)$. The professional wants to bank the ball off the side wall of the green at the point (x, y). Find the coordinates of the point (x, y). Then write an equation for the path of the ball.

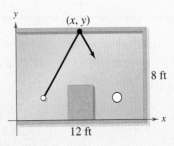

FIGURE FOR 6

7. At 2:00 P.M. on April 11, 1912, the *Titanic* left Cobh, Ireland, on her voyage to New York City. At 11:40 P.M. on April 14, the *Titanic* struck an iceberg and sank, having covered only about 2100 miles of the approximately 3400-mile trip.

   (a) What was the total length of the *Titantic's* voyage in hours?

   (b) What was the *Titanic's* average speed in miles per hour?

   (c) Write a function relating the *Titanic's* distance from New York City and the number of hours traveled. Find the domain and range of the function.

   (d) Graph the function from part (c).

8. Consider the function given by $f(x) = -x^2 + 4x - 3$. Find the average rate of change of the function from x_1 to x_2.

   (a) $x_1 = 1, x_2 = 2$ (b) $x_1 = 1, x_2 = 1.5$

   (c) $x_1 = 1, x_2 = 1.25$

   (d) $x_1 = 1, x_2 = 1.125$

   (e) $x_1 = 1, x_2 = 1.0625$

   (f) Does the average rate of change seem to be approaching one value? If so, what value?

   (g) Find the equation of the secant line through the points $(x_1, f(x_1))$ and $(x_2, f(x_2))$ for parts (a)–(e).

   (h) Find the equation of the line though the point $(1, f(1))$ using your answer from part (f) as the slope of the line.

9. Consider the functions given by $f(x) = 4x$ and $g(x) = x + 6$.

   (a) Find $(f \circ g)(x)$.

   (b) Find $(f \circ g)^{-1}(x)$.

   (c) Find $f^{-1}(x)$ and $g^{-1}(x)$.

   (d) Find $(g^{-1} \circ f^{-1})(x)$ and compare the result with that of part (b).

   (e) Repeat parts (a) through (d) for $f(x) = x^3 + 1$ and $g(x) = 2x$.

   (f) Write two one-to-one functions f and g, and repeat parts (a) through (d) for these functions.

   (g) Make a conjecture about $(f \circ g)^{-1}(x)$ and $(g^{-1} \circ f^{-1})(x)$.

127

10. You are in a boat 2 miles from the nearest point on the coast. You are to travel to a point Q, 3 miles down the coast and 1 mile inland (see figure). You can row at 2 miles per hour and walk at 4 miles per hour.

Not drawn to scale.

(a) Write the total time T of the trip as a function of x.

(b) Determine the domain of the function.

(c) Use a graphing utility to graph the function. Be sure to choose an appropriate viewing window.

(d) Use the *zoom* and *trace* features to find the value of x that minimizes T.

(e) Write a brief paragraph interpreting these values.

11. The **Heaviside function** $H(x)$ is widely used in engineering applications. (See figure.) To print an enlarged copy of the graph, go to the website *www.mathgraphs.com*.

$$H(x) = \begin{cases} 1, & x \geq 0 \\ 0, & x < 0 \end{cases}$$

Sketch the graph of each function by hand.

(a) $H(x) - 2$ (b) $H(x - 2)$ (c) $-H(x)$

(d) $H(-x)$ (e) $\frac{1}{2}H(x)$ (f) $-H(x - 2) + 2$

12. Let $f(x) = \dfrac{1}{1 - x}$.

(a) What are the domain and range of f?

(b) Find $f(f(x))$. What is the domain of this function?

(c) Find $f(f(f(x)))$. Is the graph a line? Why or why not?

13. Show that the Associative Property holds for compositions of functions—that is,

$$(f \circ (g \circ h))(x) = ((f \circ g) \circ h)(x).$$

14. Consider the graph of the function f shown in the figure. Use this graph to sketch the graph of each function. To print an enlarged copy of the graph, go to the website *www.mathgraphs.com*.

(a) $f(x + 1)$ (b) $f(x) + 1$ (c) $2f(x)$ (d) $f(-x)$

(e) $-f(x)$ (f) $|f(x)|$ (g) $f(|x|)$

15. Use the graphs of f and f^{-1} to complete each table of function values.

(a)

x	-4	-2	0	4
$(f(f^{-1}(x))$				

(b)

x	-3	-2	0	1
$(f + f^{-1})(x)$				

(c)

x	-3	-2	0	1
$(f \cdot f^{-1})(x)$				

(d)

x	-4	-3	0	4
$\|f^{-1}(x)\|$				

Trigonometry

1

Airport runways are named on the basis of the angles they form with due north, measured in a clockwise direction. These angles are called bearings and can be determined using trigonometry.

Rajs/Photonica/Getty Images

SELECTED APPLICATIONS

Trigonometric functions have many real-life applications. The applications listed below represent a small sample of the applications in this chapter.

What you should learn

- Describe angles.
- Use radian measure.
- Use degree measure.
- Use angles to model and solve real-life problems.

Why you should learn it

You can use angles to model and solve real-life problems. For instance, in Exercise 108 on page 141, you are asked to use angles to find the speed of a bicycle.

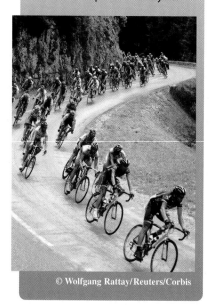

© Wolfgang Rattay/Reuters/Corbis

Angles

As derived from the Greek language, the word **trigonometry** means "measurement of triangles." Initially, trigonometry dealt with relationships among the sides and angles of triangles and was used in the development of astronomy, navigation, and surveying. With the development of calculus and the physical sciences in the 17th century, a different perspective arose—one that viewed the classic trigonometric relationships as *functions* with the set of real numbers as their domains. Consequently, the applications of trigonometry expanded to include a vast number of physical phenomena involving rotations and vibrations. These phenomena include sound waves, light rays, planetary orbits, vibrating strings, pendulums, and orbits of atomic particles.

The approach in this text incorporates *both* perspectives, starting with angles and their measure.

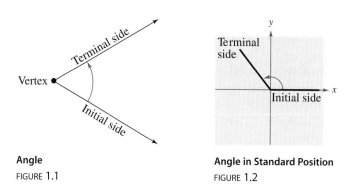

Angle

FIGURE 1.1

Angle in Standard Position

FIGURE 1.2

An **angle** is determined by rotating a ray (half-line) about its endpoint. The starting position of the ray is the **initial side** of the angle, and the position after rotation is the **terminal side,** as shown in Figure 1.1. The endpoint of the ray is the **vertex** of the angle. This perception of an angle fits a coordinate system in which the origin is the vertex and the initial side coincides with the positive *x*-axis. Such an angle is in **standard position,** as shown in Figure 1.2. **Positive angles** are generated by counterclockwise rotation, and **negative angles** by clockwise rotation, as shown in Figure 1.3. Angles are labeled with Greek letters α (alpha), β (beta), and θ (theta), as well as uppercase letters A, B, and C. In Figure 1.4, note that angles α and β have the same initial and terminal sides. Such angles are **coterminal.**

The *HM mathSpace*® CD-ROM and *Eduspace*® for this text contain additional resources related to the concepts discussed in this chapter.

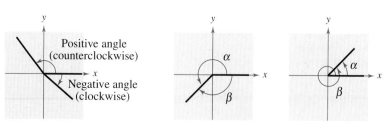

FIGURE 1.3

FIGURE 1.4   *Coterminal Angles*

Radian Measure

The **measure of an angle** is determined by the amount of rotation from the initial side to the terminal side. One way to measure angles is in *radians*. This type of measure is especially useful in calculus. To define a radian, you can use a **central angle** of a circle, one whose vertex is the center of the circle, as shown in Figure 1.5

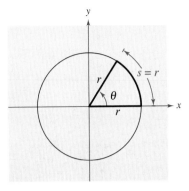

Arc length = radius when θ = 1 radian

FIGURE 1.5

> ### Definition of Radian
>
> One **radian** is the measure of a central angle θ that intercepts an arc s equal in length to the radius r of the circle. See Figure 1.5. Algebraically, this means that
>
> $$\theta = \frac{s}{r}$$
>
> where θ is measured in radians.

Because the circumference of a circle is $2\pi r$ units, it follows that a central angle of one full revolution (counterclockwise) corresponds to an arc length of

$$s = 2\pi r.$$

Moreover, because $2\pi \approx 6.28$, there are just over six radius lengths in a full circle, as shown in Figure 1.6. Because the units of measure for s and r are the same, the ratio s/r has no units—it is simply a real number.

Because the radian measure of an angle of one full revolution is 2π, you can obtain the following.

$$\frac{1}{2} \text{ revolution} = \frac{2\pi}{2} = \pi \text{ radians}$$

$$\frac{1}{4} \text{ revolution} = \frac{2\pi}{4} = \frac{\pi}{2} \text{ radians}$$

$$\frac{1}{6} \text{ revolution} = \frac{2\pi}{6} = \frac{\pi}{3} \text{ radians}$$

These and other common angles are shown in Figure 1.7.

FIGURE 1.6

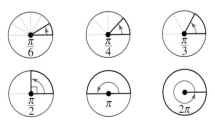

FIGURE 1.7

Recall that the four quadrants in a coordinate system are numbered I, II, III, and IV. Figure 1.8 on page 132 shows which angles between 0 and 2π lie in each of the four quadrants. Note that angles between 0 and $\pi/2$ are **acute** angles and angles between $\pi/2$ and π are **obtuse** angles.

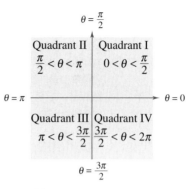

FIGURE 1.8

Two angles are coterminal if they have the same initial and terminal sides. For instance, the angles 0 and 2π are coterminal, as are the angles $\pi/6$ and $13\pi/6$. You can find an angle that is coterminal to a given angle θ by adding or subtracting 2π (one revolution), as demonstrated in Example 1. A given angle θ has infinitely many coterminal angles. For instance, $\theta = \pi/6$ is coterminal with

$$\frac{\pi}{6} + 2n\pi$$

where n is an integer.

Remind your students to work in radians.

Example 1 Sketching and Finding Coterminal Angles

a. For the positive angle $13\pi/6$, subtract 2π to obtain a coterminal angle

$$\frac{13\pi}{6} - 2\pi = \frac{\pi}{6}. \qquad \text{See Figure 1.9.}$$

b. For the positive angle $3\pi/4$, subtract 2π to obtain a coterminal angle

$$\frac{3\pi}{4} - 2\pi = -\frac{5\pi}{4}. \qquad \text{See Figure 1.10.}$$

c. For the negative angle $-2\pi/3$, add 2π to obtain a coterminal angle

$$-\frac{2\pi}{3} + 2\pi = \frac{4\pi}{3}. \qquad \text{See Figure 1.11.}$$

FIGURE 1.9

FIGURE 1.10

FIGURE 1.11

✓CHECKPOINT Now try Exercise 17.

You might point out that complementary and supplementary angles do not necessarily share a common side. For example, the acute angles of a right triangle are complementary because the sum of their measures is $\pi/2$.

Two positive angles α and β are **complementary** (complements of each other) if their sum is $\pi/2$. Two positive angles are **supplementary** (supplements of each other) if their sum is π. See Figure 1.12.

Complementary Angles

Supplementary Angles

FIGURE **1.12**

Example 2   Complementary and Supplementary Angles

If possible, find the complement and the supplement of (a) $2\pi/5$ and (b) $4\pi/5$.

Solution

a. The complement of $2\pi/5$ is

$$\frac{\pi}{2} - \frac{2\pi}{5} = \frac{5\pi}{10} - \frac{4\pi}{10} = \frac{\pi}{10}.$$

The supplement of $2\pi/5$ is

$$\pi - \frac{2\pi}{5} = \frac{5\pi}{5} - \frac{2\pi}{5} = \frac{3\pi}{5}.$$

b. Because $4\pi/5$ is greater than $\pi/2$, it has no complement. (Remember that complements are *positive* angles.) The supplement is

$$\pi - \frac{4\pi}{5} = \frac{5\pi}{5} - \frac{4\pi}{5} = \frac{\pi}{5}.$$

✓*CHECKPOINT*   Now try Exercise 21.

Degree Measure

A second way to measure angles is in terms of **degrees,** denoted by the symbol $°$. A measure of one degree ($1°$) is equivalent to a rotation of $\frac{1}{360}$ of a complete revolution about the vertex. To measure angles, it is convenient to mark degrees on the circumference of a circle, as shown in Figure 1.13. So, a full revolution (counterclockwise) corresponds to $360°$, a half revolution to $180°$, a quarter revolution to $90°$, and so on.

Because 2π radians corresponds to one complete revolution, degrees and radians are related by the equations

$$360° = 2\pi \text{ rad} \qquad \text{and} \qquad 180° = \pi \text{ rad}.$$

From the latter equation, you obtain

$$1° = \frac{\pi}{180} \text{ rad} \qquad \text{and} \qquad 1 \text{ rad} = \left(\frac{180°}{\pi}\right)$$

which lead to the conversion rules at the top of the next page.

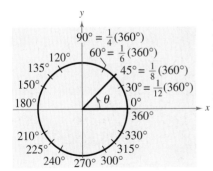
FIGURE **1.13**

Converting from degrees to radians and vice versa should help your students become familiar with radian measure.

Conversions Between Degrees and Radians

1. To convert degrees to radians, multiply degrees by $\dfrac{\pi \text{ rad}}{180°}$.

2. To convert radians to degrees, multiply radians by $\dfrac{180°}{\pi \text{ rad}}$.

To apply these two conversion rules, use the basic relationship $\pi \text{ rad} = 180°$. (See Figure 1.14.)

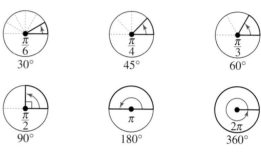

FIGURE 1.14

When no units of angle measure are specified, *radian measure is implied*. For instance, if you write $\theta = 2$, you imply that $\theta = 2$ radians.

Example 3 Converting from Degrees to Radians

a. $135° = (135 \text{ deg})\left(\dfrac{\pi \text{ rad}}{180 \text{ deg}}\right) = \dfrac{3\pi}{4}$ radians Multiply by $\pi/180$.

b. $540° = (540 \text{ deg})\left(\dfrac{\pi \text{ rad}}{180 \text{ deg}}\right) = 3\pi$ radians Multiply by $\pi/180$.

c. $-270° = (-270 \text{ deg})\left(\dfrac{\pi \text{ rad}}{180 \text{ deg}}\right) = -\dfrac{3\pi}{2}$ radians Multiply by $\pi/180$.

✓**CHECKPOINT** Now try Exercise 47.

Example 4 Converting from Radians to Degrees

a. $-\dfrac{\pi}{2} \text{ rad} = \left(-\dfrac{\pi}{2} \text{ rad}\right)\left(\dfrac{180 \text{ deg}}{\pi \text{ rad}}\right) = -90°$ Multiply by $180/\pi$.

b. $\dfrac{9\pi}{2} \text{ rad} = \left(\dfrac{9\pi}{2} \text{ rad}\right)\left(\dfrac{180 \text{ deg}}{\pi \text{ rad}}\right) = 810°$ Multiply by $180/\pi$.

c. $2 \text{ rad} = (2 \text{ rad})\left(\dfrac{180 \text{ deg}}{\pi \text{ rad}}\right) = \dfrac{360°}{\pi} \approx 114.59°$ Multiply by $180/\pi$.

✓**CHECKPOINT** Now try Exercise 51.

If you have a calculator with a "radian-to-degree" conversion key, try using it to verify the result shown in part (c) of Example 4.

Technology

With calculators it is convenient to use *decimal* degrees to denote fractional parts of degrees. Historically, however, fractional parts of degrees were expressed in *minutes* and *seconds*, using the prime (′) and double prime (″) notations, respectively. That is,

$1' = $ one minute $= \dfrac{1}{60}(1°)$

$1'' = $ one second $= \dfrac{1}{3600}(1°)$

Consequently, an angle of 64 degrees, 32 minutes, and 47 seconds is represented by $\theta = 64° \, 32' \, 47''$. Many calculators have special keys for converting an angle in degrees, minutes, and seconds (D° M′ S″) to decimal degree form, and vice versa.

Applications

The *radian measure* formula, $\theta = s/r$, can be used to measure arc length along a circle.

Arc Length

For a circle of radius r, a central angle θ intercepts an arc of length s given by

$$s = r\theta \qquad \text{Length of circular arc}$$

where θ is measured in radians. Note that if $r = 1$, then $s = \theta$, and the radian measure of θ equals the arc length.

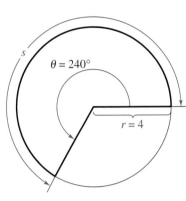

FIGURE 1.15

$\theta = 240°$

$r = 4$

Example 5   Finding Arc Length

A circle has a radius of 4 inches. Find the length of the arc intercepted by a central angle of $240°$, as shown in Figure 1.15.

Solution

To use the formula $s = r\theta$, first convert $240°$ to radian measure.

$$240° = (240 \text{ deg})\left(\frac{\pi \text{ rad}}{180 \text{ deg}}\right) = \frac{4\pi}{3} \text{ radians}$$

Then, using a radius of $r = 4$ inches, you can find the arc length to be

$$s = r\theta = 4\left(\frac{4\pi}{3}\right) = \frac{16\pi}{3} \approx 16.76 \text{ inches}.$$

Note that the units for $r\theta$ are determined by the units for r because θ is given in radian measure, which has no units.

✓CHECKPOINT   Now try Exercise 87.

The formula for the length of a circular arc can be used to analyze the motion of a particle moving at a *constant speed* along a circular path.

Because radian measure is so often used, you may want to encourage your students to be as familiar with the radian measure of angles as they are with degree measure. Measuring arc length along a circle is one of many applications in which radian measure is used.

Linear and Angular Speeds

Consider a particle moving at a constant speed along a circular arc of radius r. If s is the length of the arc traveled in time t, then the **linear speed** v of the particle is

$$\text{Linear speed } v = \frac{\text{arc length}}{\text{time}} = \frac{s}{t}.$$

Moreover, if θ is the angle (in radian measure) corresponding to the arc length s, then the **angular speed** ω (the lowercase Greek letter omega) of the particle is

$$\text{Angular speed } \omega = \frac{\text{central angle}}{\text{time}} = \frac{\theta}{t}.$$

STUDY TIP

Linear speed measures how fast the particle moves, and angular speed measures how fast the angle changes. By dividing the formula for arc length by t, you can establish a relationship between linear speed v and angular speed ω, as shown.

$$s = r\theta$$

$$\frac{s}{t} = \frac{r\theta}{t}$$

$$v = r\omega$$

FIGURE **1.16**

FIGURE **1.17**

Example 6 **Finding Linear Speed**

The second hand of a clock is 10.2 centimeters long, as shown in Figure 1.16. Find the linear speed of the tip of this second hand as it passes around the clock face.

Solution

In one revolution, the arc length traveled is

$$s = 2\pi r$$
$$= 2\pi(10.2) \qquad \text{Substitute for } r.$$
$$= 20.4\pi \text{ centimeters.}$$

The time required for the second hand to travel this distance is

$$t = 1 \text{ minute} = 60 \text{ seconds.}$$

So, the linear speed of the tip of the second hand is

$$\text{Linear speed} = \frac{s}{t}$$
$$= \frac{20.4\pi \text{ centimeters}}{60 \text{ seconds}}$$
$$\approx 1.068 \text{ centimeters per second.}$$

✔**CHECKPOINT** Now try Exercise 103.

Example 7 **Finding Angular and Linear Speeds**

A Ferris wheel with a 50-foot radius (see Figure 1.17) makes 1.5 revolutions per minute.

a. Find the angular speed of the Ferris wheel in radians per minute.

b. Find the linear speed of the Ferris wheel.

Solution

a. Because each revolution generates 2π radians, it follows that the wheel turns $(1.5)(2\pi) = 3\pi$ radians per minute. In other words, the angular speed is

$$\text{Angular speed} = \frac{\theta}{t}$$
$$= \frac{3\pi \text{ radians}}{1 \text{ minute}} = 3\pi \text{ radians per minute.}$$

b. The linear speed is

$$\text{Linear speed} = \frac{s}{t}$$
$$= \frac{r\theta}{t}$$
$$= \frac{50(3\pi) \text{ feet}}{1 \text{ minute}} \approx 471.2 \text{ feet per minute.}$$

✔**CHECKPOINT** Now try Exercise 105.

A **sector** of a circle is the region bounded by two radii of the circle and their intercepted arc (see Figure 1.18).

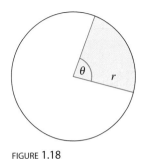

FIGURE 1.18

> ## Area of a Sector of a Circle
>
> For a circle of radius r, the area A of a sector of the circle with central angle θ is given by
>
> $$A = \frac{1}{2}r^2\theta$$
>
> where θ is measured in radians.

Example 8   Area of a Sector of a Circle

A sprinkler on a golf course fairway is set to spray water over a distance of 70 feet and rotates through an angle of 120° (see Figure 1.19). Find the area of the fairway watered by the sprinkler.

Solution

First convert 120° to radian measure as follows.

$$\theta = 120°$$

$$= (120 \text{ deg})\left(\frac{\pi \text{ rad}}{180 \text{ deg}}\right) \qquad \text{Multiply by } \pi/180.$$

$$= \frac{2\pi}{3} \text{ radians}$$

120°   70 ft

FIGURE 1.19

Then, using $\theta = 2\pi/3$ and $r = 70$, the area is

$$A = \frac{1}{2}r^2\theta \qquad \text{Formula for the area of a sector of a circle}$$

$$= \frac{1}{2}(70)^2\left(\frac{2\pi}{3}\right) \qquad \text{Substitute for } r \text{ and } \theta.$$

$$= \frac{4900\pi}{3} \qquad \text{Simplify.}$$

$$\approx 5131 \text{ square feet.} \qquad \text{Simplify.}$$

✓CHECKPOINT   Now try Exercise 107.

1.1 Exercises

The *HM mathSpace®* CD-ROM and *Eduspace®* for this text contain step-by-step solutions to all odd-numbered exercises. They also provide Tutorial Exercises for additional help.

VOCABULARY CHECK: Fill in the blanks.

1. _____ means "measurement of triangles."

2. An _____ is determined by rotating a ray about its endpoint.

3. Two angles that have the same initial and terminal sides are _____.

4. One _____ is the measure of a central angle that intercepts an arc equal to the radius of the circle.

5. Angles that measure between 0 and $\pi/2$ are _____ angles, and angles that measure between $\pi/2$ and π are _____ angles.

6. Two positive angles that have a sum of $\pi/2$ are _____ angles, whereas two positive angles that have a sum of π are _____ angles.

7. The angle measure that is equivalent to $\frac{1}{360}$ of a complete revolution about an angle's vertex is one _____.

8. The _____ speed of a particle is the ratio of the arc length traveled to the time traveled.

9. The _____ speed of a particle is the ratio of the change in the central angle to time.

10. The area of a sector of a circle with radius r and central angle θ, where θ is measured in radians, is given by the formula _____.

PREREQUISITE SKILLS REVIEW: Practice and review algebra skills needed for this section at **www.Eduspace.com**.

In Exercises 1–6, estimate the angle to the nearest one-half radian.

1.

2.

3.

4.

5.

6.

In Exercises 7–12, determine the quadrant in which each angle lies. (The angle measure is given in radians.)

7. (a) $\dfrac{\pi}{5}$ (b) $\dfrac{7\pi}{5}$

8. (a) $\dfrac{11\pi}{8}$ (b) $\dfrac{9\pi}{8}$

9. (a) $-\dfrac{\pi}{12}$ (b) -2

10. (a) -1 (b) $-\dfrac{11\pi}{9}$

11. (a) 3.5 (b) 2.25

12. (a) 6.02 (b) -4.25

In Exercises 13–16, sketch each angle in standard position.

13. (a) $\dfrac{5\pi}{4}$ (b) $-\dfrac{2\pi}{3}$ 14. (a) $-\dfrac{7\pi}{4}$ (b) $\dfrac{5\pi}{2}$

15. (a) $\dfrac{11\pi}{6}$ (b) -3 16. (a) 4 (b) 7π

In Exercises 17–20, determine two coterminal angles (one positive and one negative) for each angle. Give your answers in radians.

17. (a)   (b)

18. (a)   (b)

19. (a) $\theta = \dfrac{2\pi}{3}$ (b) $\theta = \dfrac{\pi}{12}$

20. (a) $\theta = -\dfrac{9\pi}{4}$ (b) $\theta = -\dfrac{2\pi}{15}$

In Exercises 21–24, find (if possible) the complement and supplement of each angle.

21. (a) $\dfrac{\pi}{3}$ (b) $\dfrac{3\pi}{4}$ **22.** (a) $\dfrac{\pi}{12}$ (b) $\dfrac{11\pi}{12}$

23. (a) 1   (b) 2   **24.** (a) 3   (b) 1.5

In Exercises 25–30, estimate the number of degrees in the angle.

25.

26.

27.

28.

29.

30.

In Exercises 31–34, determine the quadrant in which each angle lies.

31. (a) $130°$ (b) $285°$

32. (a) $8.3°$ (b) $257° \, 30'$

33. (a) $-132° \, 50'$ (b) $-336°$

34. (a) $-260°$ (b) $-3.4°$

In Exercises 35–38, sketch each angle in standard position.

35. (a) $30°$ (b) $150°$ **36.** (a) $-270°$ (b) $-120°$

37. (a) $405°$ (b) $480°$ **38.** (a) $-750°$ (b) $-600°$

In Exercises 39–42, determine two coterminal angles (one positive and one negative) for each angle. Give your answers in degrees.

39. (a)   (b) $\theta = -36°$

     $\theta = 45°$

40. (a)   (b) $\theta = -420°$

     $\theta = 120°$

41. (a) $\theta = 240°$ (b) $\theta = -180°$

42. (a) $\theta = -420°$ (b) $\theta = 230°$

In Exercises 43–46, find (if possible) the complement and supplement of each angle.

43. (a) $18°$ (b) $115°$ **44.** (a) $3°$ (b) $64°$

45. (a) $79°$ (b) $150°$ **46.** (a) $130°$ (b) $170°$

In Exercises 47–50, rewrite each angle in radian measure as a multiple of π. (Do not use a calculator.)

47. (a) $30°$ (b) $150°$ **48.** (a) $315°$ (b) $120°$

49. (a) $-20°$ (b) $-240°$ **50.** (a) $-270°$ (b) $144°$

In Exercises 51–54, rewrite each angle in degree measure. (Do not use a calculator.)

51. (a) $\dfrac{3\pi}{2}$ (b) $\dfrac{7\pi}{6}$ **52.** (a) $-\dfrac{7\pi}{12}$ (b) $\dfrac{\pi}{9}$

53. (a) $\dfrac{7\pi}{3}$ (b) $-\dfrac{11\pi}{30}$ **54.** (a) $\dfrac{11\pi}{6}$ (b) $\dfrac{34\pi}{15}$

In Exercises 55–62, convert the angle measure from degrees to radians. Round to three decimal places.

55. $115°$ **56.** $87.4°$

57. $-216.35°$ **58.** $-48.27°$

59. $532°$ **60.** $345°$

61. $-0.83°$ **62.** $0.54°$

In Exercises 63–70, convert the angle measure from radians to degrees. Round to three decimal places.

63. $\dfrac{\pi}{7}$ **64.** $\dfrac{5\pi}{11}$

65. $\dfrac{15\pi}{8}$ **66.** $\dfrac{13\pi}{2}$

67. -4.2π **68.** 4.8π

69. -2 **70.** -0.57

In Exercises 71–74, convert each angle measure to decimal degree form.

71. (a) $54° \, 45'$ (b) $-128° \, 30'$

72. (a) $245° \, 10'$ (b) $2° \, 12'$

73. (a) $85° \, 18' \, 30''$ (b) $330° \, 25''$

74. (a) $-135° \, 36''$ (b) $-408° \, 16' 20''$

In Exercises 75–78, convert each angle measure to D° M′ S″ form.

75. (a) $240.6°$ (b) $-145.8°$

76. (a) $-345.12°$ (b) $0.45°$

77. (a) $2.5°$ (b) $-3.58°$

78. (a) $-0.355°$ (b) $0.7865°$

In Exercises 79–82, find the angle in radians.

79.

80.

81.

82.

In Exercises 83–86, find the radian measure of the central angle of a circle of radius r that intercepts an arc of length s.

	Radius r	Arc Length s
83.	27 inches	6 inches
84.	14 feet	8 feet
85.	14.5 centimeters	25 centimeters
86.	80 kilometers	160 kilometers

In Exercises 87–90, find the length of the arc on a circle of radius r intercepted by a central angle θ.

	Radius r	Central Angle θ
87.	15 inches	180°
88.	9 feet	60°
89.	3 meters	1 radian
90.	20 centimeters	$\pi/4$ radian

In Exercises 91–94, find the area of the sector of the circle with radius r and central angle θ.

	Radius r	Central Angle θ
91.	4 inches	$\pi/3$
92.	12 millimeters	$\pi/4$
93.	2.5 feet	225°
94.	1.4 miles	330°

Distance Between Cities **In Exercises 95 and 96, find the distance between the cities. Assume that Earth is a sphere of radius 4000 miles and that the cities are on the same longitude (one city is due north of the other).**

	City	Latitude
95.	Dallas, Texas	32° 47′ 39″ N
	Omaha, Nebraska	41° 15′ 50″ N
96.	San Francisco, California	37° 47′ 36″ N
	Seattle, Washington	47° 37′ 18″ N

97. *Difference in Latitudes* Assuming that Earth is a sphere of radius 6378 kilometers, what is the difference in the latitudes of Syracuse, New York and Annapolis, Maryland, where Syracuse is 450 kilometers due north of Annapolis?

98. *Difference in Latitudes* Assuming that Earth is a sphere of radius 6378 kilometers, what is the difference in the latitudes of Lynchburg, Virginia and Myrtle Beach, South Carolina, where Lynchburg is 400 kilometers due north of Myrtle Beach?

99. *Instrumentation* The pointer on a voltmeter is 6 centimeters in length (see figure). Find the angle through which the pointer rotates when it moves 2.5 centimeters on the scale.

FIGURE FOR 99   FIGURE FOR 100

100. *Electric Hoist* An electric hoist is being used to lift a beam (see figure). The diameter of the drum on the hoist is 10 inches, and the beam must be raised 2 feet. Find the number of degrees through which the drum must rotate.

101. *Angular Speed* A car is moving at a rate of 65 miles per hour, and the diameter of its wheels is 2.5 feet.

(a) Find the number of revolutions per minute the wheels are rotating.

(b) Find the angular speed of the wheels in radians per minute.

102. *Angular Speed* A two-inch-diameter pulley on an electric motor that runs at 1700 revolutions per minute is connected by a belt to a four-inch-diameter pulley on a saw arbor.

(a) Find the angular speed (in radians per minute) of each pulley.

(b) Find the revolutions per minute of the saw.

103. *Linear and Angular Speeds* A $7\frac{1}{4}$-inch circular power saw rotates at 5200 revolutions per minute.

(a) Find the angular speed of the saw blade in radians per minute.

(b) Find the linear speed (in feet per minute) of one of the 24 cutting teeth as they contact the wood being cut.

104. *Linear and Angular Speeds* A carousel with a 50-foot diameter makes 4 revolutions per minute.

(a) Find the angular speed of the carousel in radians per minute.

(b) Find the linear speed of the platform rim of the carousel.

105. *Linear and Angular Speeds* The diameter of a DVD is approximately 12 centimeters. The drive motor of the DVD player is controlled to rotate precisely between 200 and 500 revolutions per minute, depending on what track is being read.

(a) Find an interval for the angular speed of a DVD as it rotates.

(b) Find an interval for the linear speed of a point on the outermost track as the DVD rotates.

106. *Area* A car's rear windshield wiper rotates 125°. The total length of the wiper mechanism is 25 inches and wipes the windshield over a distance of 14 inches. Find the area covered by the wiper.

107. *Area* A sprinkler system on a farm is set to spray water over a distance of 35 meters and to rotate through an angle of 140°. Draw a diagram that shows the region that can be irrigated with the sprinkler. Find the area of the region.

Model It

108. *Speed of a Bicycle* The radii of the pedal sprocket, the wheel sprocket, and the wheel of the bicycle in the figure are 4 inches, 2 inches, and 14 inches, respectively. A cyclist is pedaling at a rate of 1 revolution per second.

14 in.

4 in.

2 in.

(a) Find the speed of the bicycle in feet per second and miles per hour.

(b) Use your result from part (a) to write a function for the distance d (in miles) a cyclist travels in terms of the number n of revolutions of the pedal sprocket.

(c) Write a function for the distance d (in miles) a cyclist travels in terms of the time t (in seconds). Compare this function with part (b).

(d) Classify the types of functions you found in parts (b) and (c). Explain your reasoning.

Synthesis

True or False? In Exercises 109–111, determine whether the statement is true or false. Justify your answer.

109. A measurement of 4 radians corresponds to two complete revolutions from the initial side to the terminal side of an angle.

110. The difference between the measures of two coterminal angles is always a multiple of 360° if expressed in degrees and is always a multiple of 2π radians if expressed in radians.

111. An angle that measures $-1260°$ lies in Quadrant III.

112. *Writing* In your own words, explain the meanings of (a) an angle in standard position, (b) a negative angle, (c) coterminal angles, and (d) an obtuse angle.

113. *Think About It* A fan motor turns at a given angular speed. How does the speed of the tips of the blades change if a fan of greater diameter is installed on the motor? Explain.

114. *Think About It* Is a degree or a radian the larger unit of measure? Explain.

115. *Writing* If the radius of a circle is increasing and the magnitude of a central angle is held constant, how is the length of the intercepted arc changing? Explain your reasoning.

116. *Proof* Prove that the area of a circular sector of radius r with central angle θ is $A = \frac{1}{2}\theta r^2$, where θ is measured in radians.

Skills Review

In Exercises 117–124, simplify the radical expression.

117. $\dfrac{4}{4\sqrt{2}}$ **118.** $\dfrac{2}{\sqrt{3}}$

119. $\dfrac{2\sqrt{3}}{\sqrt{6}}$ **120.** $\dfrac{5\sqrt{5}}{2\sqrt{10}}$

121. $\sqrt{2^2 + 6^2}$

122. $\sqrt{18^2 - 12^2}$

123. $\sqrt{18^2 - 6^2}$

124. $\sqrt{17^2 - 9^2}$

In Exercises 125–128, sketch the graphs of $y = x^5$ and the specified transformation.

125. $f(x) = (x - 2)^5$

126. $f(x) = x^5 - 4$

127. $f(x) = 2 - x^5$

128. $f(x) = -(x + 3)^5$

1.2 Trigonometric Functions: The Unit Circle

What you should learn

- Identify a unit circle and describe its relationship to real numbers.
- Evaluate trigonometric functions using the unit circle.
- Use the domain and period to evaluate sine and cosine functions.
- Use a calculator to evaluate trigonometric functions.

Why you should learn it

Trigonometric functions are used to model the movement of an oscillating weight. For instance, in Exercise 57 on page 148, the displacement from equilibrium of an oscillating weight suspended by a spring is modeled as a function of time.

The Unit Circle

The two historical perspectives of trigonometry incorporate different methods for introducing the trigonometric functions. Our first introduction to these functions is based on the unit circle.

Consider the **unit circle** given by

$$x^2 + y^2 = 1 \qquad \text{Unit circle}$$

as shown in Figure 1.20.

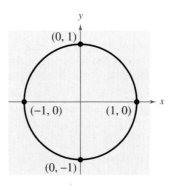

FIGURE 1.20

Imagine that the real number line is wrapped around this circle, with positive numbers corresponding to a counterclockwise wrapping and negative numbers corresponding to a clockwise wrapping, as shown in Figure 1.21.

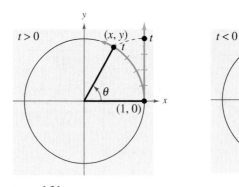

FIGURE 1.21

Try demonstrating the wrapping function by using a spool and thread. Demonstrate the real number t as the length of the thread. Then wrap the thread around the spool to show the correspondence between t and the point (x, y) on the circle. For example, if $t = \pi/2$, the point on the unit circle to which it corresponds is $(0, 1)$.

As the real number line is wrapped around the unit circle, each real number t corresponds to a point (x, y) on the circle. For example, the real number 0 corresponds to the point $(1, 0)$. Moreover, because the unit circle has a circumference of 2π, the real number 2π also corresponds to the point $(1, 0)$.

In general, each real number t also corresponds to a central angle θ (in standard position) whose radian measure is t. With this interpretation of t, the arc length formula $s = r\theta$ (with $r = 1$) indicates that the real number t is the length of the arc intercepted by the angle θ, given in radians.

The Trigonometric Functions

From the preceding discussion, it follows that the coordinates x and y are two functions of the real variable t. You can use these coordinates to define the six trigonometric functions of t.

sine   cosecant   cosine   secant   tangent   cotangent

These six functions are normally abbreviated sin, csc, cos, sec, tan, and cot, respectively.

> ### Definitions of Trigonometric Functions
>
> Let t be a real number and let (x, y) be the point on the unit circle corresponding to t.
>
> $$\sin t = y \qquad\qquad \cos t = x \qquad\qquad \tan t = \frac{y}{x}, \quad x \neq 0$$
>
> $$\csc t = \frac{1}{y}, \quad y \neq 0 \qquad \sec t = \frac{1}{x}, \quad x \neq 0 \qquad \cot t = \frac{x}{y}, \quad y \neq 0$$

In the definitions of the trigonometric functions, note that the tangent and secant are not defined when $x = 0$. For instance, because $t = \pi/2$ corresponds to $(x, y) = (0, 1)$, it follows that $\tan(\pi/2)$ and $\sec(\pi/2)$ are *undefined*. Similarly, the cotangent and cosecant are not defined when $y = 0$. For instance, because $t = 0$ corresponds to $(x, y) = (1, 0)$, cot 0 and csc 0 are *undefined*.

In Figure 1.22, the unit circle has been divided into eight equal arcs, corresponding to t-values of

$$0, \frac{\pi}{4}, \frac{\pi}{2}, \frac{3\pi}{4}, \pi, \frac{5\pi}{4}, \frac{3\pi}{2}, \frac{7\pi}{4}, \text{ and } 2\pi.$$

Similarly, in Figure 1.23, the unit circle has been divided into 12 equal arcs, corresponding to t-values of

$$0, \frac{\pi}{6}, \frac{\pi}{3}, \frac{\pi}{2}, \frac{2\pi}{3}, \frac{5\pi}{6}, \pi, \frac{7\pi}{6}, \frac{4\pi}{3}, \frac{3\pi}{2}, \frac{5\pi}{3}, \frac{11\pi}{6}, \text{ and } 2\pi.$$

To verify the points on the unit circle in Figure 1.22, note that $\left(\frac{\sqrt{2}}{2}, \frac{\sqrt{2}}{2}\right)$ also lies on the line $y = x$. So, substituting x for y in the equation of the unit circle produces the following.

$$x^2 + x^2 = 1 \implies 2x^2 = 1 \implies x^2 = \frac{1}{2} \implies x = \pm\frac{\sqrt{2}}{2}$$

Because the point is in the first quadrant, $x = \dfrac{\sqrt{2}}{2}$ and because $y = x$, you also have $y = \dfrac{\sqrt{2}}{2}$. You can use similar reasoning to verify the rest of the points in Figure 1.22 and the points in Figure 1.23.

Using the (x, y) coordinates in Figures 1.22 and 1.23, you can easily evaluate the trigonometric functions for common t-values. This procedure is demonstrated in Examples 1 and 2. You should study and learn these exact function values for common t-values because they will help you in later sections to perform calculations quickly and easily.

FIGURE **1.22**

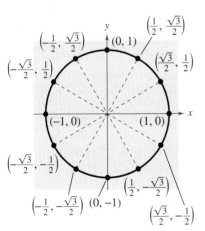

FIGURE **1.23**

Example 1   **Evaluating Trigonometric Functions**

Evaluate the six trigonometric functions at each real number.

a. $t = \dfrac{\pi}{6}$ **b.** $t = \dfrac{5\pi}{4}$ **c.** $t = 0$ **d.** $t = \pi$

Solution

For each t-value, begin by finding the corresponding point (x, y) on the unit circle. Then use the definitions of trigonometric functions listed on page 143.

a. $t = \dfrac{\pi}{6}$ corresponds to the point $(x, y) = \left(\dfrac{\sqrt{3}}{2}, \dfrac{1}{2}\right)$.

$$\sin\frac{\pi}{6} = y = \frac{1}{2} \qquad\qquad \csc\frac{\pi}{6} = \frac{1}{y} = \frac{1}{1/2} = 2$$

$$\cos\frac{\pi}{6} = x = \frac{\sqrt{3}}{2} \qquad\qquad \sec\frac{\pi}{6} = \frac{1}{x} = \frac{2}{\sqrt{3}} = \frac{2\sqrt{3}}{3}$$

$$\tan\frac{\pi}{6} = \frac{y}{x} = \frac{1/2}{\sqrt{3}/2} = \frac{1}{\sqrt{3}} = \frac{\sqrt{3}}{3} \qquad \cot\frac{\pi}{6} = \frac{x}{y} = \frac{\sqrt{3}/2}{1/2} = \sqrt{3}$$

b. $t = \dfrac{5\pi}{4}$ corresponds to the point $(x, y) = \left(-\dfrac{\sqrt{2}}{2}, -\dfrac{\sqrt{2}}{2}\right)$.

$$\sin\frac{5\pi}{4} = y = -\frac{\sqrt{2}}{2} \qquad\qquad \csc\frac{5\pi}{4} = \frac{1}{y} = -\frac{2}{\sqrt{2}} = -\sqrt{2}$$

$$\cos\frac{5\pi}{4} = x = -\frac{\sqrt{2}}{2} \qquad\qquad \sec\frac{5\pi}{4} = \frac{1}{x} = -\frac{2}{\sqrt{2}} = -\sqrt{2}$$

$$\tan\frac{5\pi}{4} = \frac{y}{x} = \frac{-\sqrt{2}/2}{-\sqrt{2}/2} = 1 \qquad \cot\frac{5\pi}{4} = \frac{x}{y} = \frac{-\sqrt{2}/2}{-\sqrt{2}/2} = 1$$

c. $t = 0$ corresponds to the point $(x, y) = (1, 0)$.

$$\sin 0 = y = 0 \qquad\qquad \csc 0 = \frac{1}{y} \text{ is undefined.}$$

$$\cos 0 = x = 1 \qquad\qquad \sec 0 = \frac{1}{x} = \frac{1}{1} = 1$$

$$\tan 0 = \frac{y}{x} = \frac{0}{1} = 0 \qquad\qquad \cot 0 = \frac{x}{y} \text{ is undefined.}$$

d. $t = \pi$ corresponds to the point $(x, y) = (-1, 0)$.

$$\sin \pi = y = 0 \qquad\qquad \csc \pi = \frac{1}{y} \text{ is undefined.}$$

$$\cos \pi = x = -1 \qquad\qquad \sec \pi = \frac{1}{x} = \frac{1}{-1} = -1$$

$$\tan \pi = \frac{y}{x} = \frac{0}{-1} = 0 \qquad\qquad \cot \pi = \frac{x}{y} \text{ is undefined.}$$

✓CHECKPOINT   Now try Exercise 23.

Additional Example

Evaluate the six trigonometric functions at $t = \dfrac{5\pi}{2}$.

Solution

Moving counterclockwise around the unit circle one and a quarter revolutions, you find that $t = 5\pi/2$ corresponds to the point $(x, y) = (0, 1)$.

$$\sin\frac{5\pi}{2} = y = 1 \qquad \csc\frac{5\pi}{2} = \frac{1}{y} = 1$$

$$\cos\frac{5\pi}{2} = x = 0$$

$$\sec\frac{5\pi}{2} = \frac{1}{x} \text{ is undefined.}$$

$$\tan\frac{5\pi}{2} = \frac{y}{x} \text{ is undefined.}$$

$$\cot\frac{5\pi}{2} = \frac{x}{y} = 0$$

With your graphing utility in *radian* and *parametric* modes, enter the equations

X1T = cos T   and   Y1T = sin T

and use the following settings.

Tmin = 0, Tmax = 6.3,
Tstep = 0.1
Xmin = -1.5, Xmax = 1.5,
Xscl = 1
Ymin = -1, Ymax = 1,
Yscl = 1

1. Graph the entered equations and describe the graph.

2. Use the *trace* feature to move the cursor around the graph. What do the *t*-values represent? What do the *x*- and *y*-values represent?

3. What are the least and greatest values of *x* and *y*?

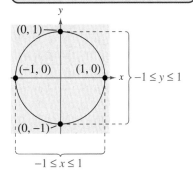

FIGURE 1.24

$$t = \frac{\pi}{2}, \frac{\pi}{2} + 2\pi, \frac{\pi}{2} + 4\pi, \dots$$

$$t = \frac{3\pi}{4}, \frac{3\pi}{4} + 2\pi, \dots \qquad t = \frac{\pi}{4}, \frac{\pi}{4} + 2\pi, \dots$$

$$t = \pi, 3\pi, \dots$$

$$t = 0, 2\pi, \dots$$

$$t = \frac{5\pi}{4}, \frac{5\pi}{4} + 2\pi, \dots$$

$$t = \frac{7\pi}{4}, \frac{7\pi}{4} + 2\pi, \frac{7\pi}{4} + 4\pi, \dots$$

$$t = \frac{3\pi}{2}, \frac{3\pi}{2} + 2\pi, \frac{3\pi}{2} + 4\pi, \dots$$

FIGURE 1.25

Example 2   **Evaluating Trigonometric Functions**

Evaluate the six trigonometric functions at $t = -\dfrac{\pi}{3}$.

Solution

Moving *clockwise* around the unit circle, it follows that $t = -\pi/3$ corresponds to the point $(x, y) = \left(1/2, -\sqrt{3}/2\right)$.

$$\sin\left(-\frac{\pi}{3}\right) = -\frac{\sqrt{3}}{2} \qquad\qquad \csc\left(-\frac{\pi}{3}\right) = -\frac{2}{\sqrt{3}} = -\frac{2\sqrt{3}}{3}$$

$$\cos\left(-\frac{\pi}{3}\right) = \frac{1}{2} \qquad\qquad \sec\left(-\frac{\pi}{3}\right) = 2$$

$$\tan\left(-\frac{\pi}{3}\right) = \frac{-\sqrt{3}/2}{1/2} = -\sqrt{3} \qquad \cot\left(-\frac{\pi}{3}\right) = \frac{1/2}{-\sqrt{3}/2} = -\frac{1}{\sqrt{3}} = -\frac{\sqrt{3}}{3}$$

✓CHECKPOINT   Now try Exercise 25.

Domain and Period of Sine and Cosine

The *domain* of the sine and cosine functions is the set of all real numbers. To determine the *range* of these two functions, consider the unit circle shown in Figure 1.24. Because $r = 1$, it follows that $\sin t = y$ and $\cos t = x$. Moreover, because (x, y) is on the unit circle, you know that $-1 \le y \le 1$ and $-1 \le x \le 1$. So, the values of sine and cosine also range between -1 and 1.

$$\begin{array}{ccc} -1 \le \ y \ \le 1 & & -1 \le \ x \ \le 1 \\ & \text{and} & \\ -1 \le \sin t \le 1 & & -1 \le \cos t \le 1 \end{array}$$

Adding 2π to each value of t in the interval $[0, 2\pi]$ completes a second revolution around the unit circle, as shown in Figure 1.25. The values of $\sin(t + 2\pi)$ and $\cos(t + 2\pi)$ correspond to those of $\sin t$ and $\cos t$. Similar results can be obtained for repeated revolutions (positive or negative) on the unit circle. This leads to the general result

$$\sin(t + 2\pi n) = \sin t$$

and

$$\cos(t + 2\pi n) = \cos t$$

for any integer n and real number t. Functions that behave in such a repetitive (or cyclic) manner are called **periodic.**

Definition of Periodic Function

A function f is **periodic** if there exists a positive real number c such that

$$f(t + c) = f(t)$$

for all t in the domain of f. The smallest number c for which f is periodic is called the **period** of f.

Recall from Section P.6 that a function f is *even* if $f(-t) = f(t)$, and is *odd* if $f(-t) = -f(t)$.

> ### Even and Odd Trigonometric Functions
>
> The cosine and secant functions are *even*.
>
> $$\cos(-t) = \cos t \qquad \sec(-t) = \sec t$$
>
> The sine, cosecant, tangent, and cotangent functions are *odd*.
>
> $$\sin(-t) = -\sin t \qquad \csc(-t) = -\csc t$$
>
> $$\tan(-t) = -\tan t \qquad \cot(-t) = -\cot t$$

STUDY TIP

From the definition of periodic function, it follows that the sine and cosine functions are periodic and have a period of 2π. The other four trigonometric functions are also periodic, and will be discussed further in Section 1.6.

Example 3 **Using the Period to Evaluate the Sine and Cosine**

a. Because $\dfrac{13\pi}{6} = 2\pi + \dfrac{\pi}{6}$, you have $\sin \dfrac{13\pi}{6} = \sin\left(2\pi + \dfrac{\pi}{6}\right) = \sin \dfrac{\pi}{6} = \dfrac{1}{2}$.

b. Because $-\dfrac{7\pi}{2} = -4\pi + \dfrac{\pi}{2}$, you have

$$\cos\left(-\dfrac{7\pi}{2}\right) = \cos\left(-4\pi + \dfrac{\pi}{2}\right) = \cos \dfrac{\pi}{2} = 0.$$

c. For $\sin t = \dfrac{4}{5}$, $\sin(-t) = -\dfrac{4}{5}$ because the sine function is odd.

✓**CHECKPOINT** Now try Exercise 31.

Evaluating Trigonometric Functions with a Calculator

When evaluating a trigonometric function with a calculator, you need to set the calculator to the desired *mode* of measurement (*degree* or *radian*).

Most calculators do not have keys for the cosecant, secant, and cotangent functions. To evaluate these functions, you can use the $\boxed{x^{-1}}$ key with their respective reciprocal functions sine, cosine, and tangent. For example, to evaluate $\csc(\pi/8)$, use the fact that

$$\csc \dfrac{\pi}{8} = \dfrac{1}{\sin(\pi/8)}$$

and enter the following keystroke sequence in *radian* mode.

$\boxed{(}\;\boxed{\text{SIN}}\;\boxed{(}\;\boxed{\pi}\;\boxed{\div}\;\boxed{8}\;\boxed{)}\;\boxed{)}\;\boxed{x^{-1}}\;\boxed{\text{ENTER}}$ Display 2.6131259

Technology

When evaluating trigonometric functions with a calculator, remember to enclose all fractional angle measures in parentheses. For instance, if you want to evaluate $\sin \theta$ for $\theta = \pi/6$, you should enter

$\boxed{\text{SIN}}\;\boxed{(}\;\boxed{\pi}\;\boxed{\div}\;\boxed{6}\;\boxed{)}\;\boxed{\text{ENTER}}$.

These keystrokes yield the correct value of 0.5. Note that some calculators automatically place a left parenthesis after trigonometric functions. Check the user's guide for your calculator for specific keystrokes on how to evaluate trigonometric functions.

Example 4 **Using a Calculator**

	Function	*Mode*	*Calculator Keystrokes*	*Display*
a.	$\sin \dfrac{2\pi}{3}$	Radian	$\boxed{\text{SIN}}\;\boxed{(}\;\boxed{2}\;\boxed{\pi}\;\boxed{\div}\;\boxed{3}\;\boxed{)}\;\boxed{\text{ENTER}}$	0.8660254
b.	$\cot 1.5$	Radian	$\boxed{(}\;\boxed{\text{TAN}}\;\boxed{(}\;\boxed{1.5}\;\boxed{)}\;\boxed{)}\;\boxed{x^{-1}}\;\boxed{\text{ENTER}}$	0.0709148

✓**CHECKPOINT** Now try Exercise 45.

1.2   Exercises

VOCABULARY CHECK: Fill in the blanks.

1. Each real number t corresponds to a point (x, y) on the _____ _____.
2. A function f is _____ if there exists a positive real number c such that $f(t + c) = f(t)$ for all t in the domain of f.
3. The smallest number c for which a function f is periodic is called the _____ of f.
4. A function f is _____ if $f(-t) = -f(t)$ and _____ if $f(-t) = f(t)$.

PREREQUISITE SKILLS REVIEW: Practice and review algebra skills needed for this section at **www.Eduspace.com.**

In Exercises 1–4, determine the exact values of the six trigonometric functions of the angle θ.

1.

2.

3.

4.
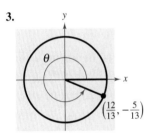

In Exercises 5–12, find the point (x, y) on the unit circle that corresponds to the real number t.

5. $t = \dfrac{\pi}{4}$

6. $t = \dfrac{\pi}{3}$

7. $t = \dfrac{7\pi}{6}$

8. $t = \dfrac{5\pi}{4}$

9. $t = \dfrac{4\pi}{3}$

10. $t = \dfrac{5\pi}{3}$

11. $t = \dfrac{3\pi}{2}$

12. $t = \pi$

In Exercises 13–22, evaluate (if possible) the sine, cosine, and tangent of the real number.

13. $t = \dfrac{\pi}{4}$

14. $t = \dfrac{\pi}{3}$

15. $t = -\dfrac{\pi}{6}$

16. $t = -\dfrac{\pi}{4}$

17. $t = -\dfrac{7\pi}{4}$

18. $t = -\dfrac{4\pi}{3}$

19. $t = \dfrac{11\pi}{6}$

20. $t = \dfrac{5\pi}{3}$

21. $t = -\dfrac{3\pi}{2}$

22. $t = -2\pi$

In Exercises 23–28, evaluate (if possible) the six trigonometric functions of the real number.

23. $t = \dfrac{3\pi}{4}$

24. $t = \dfrac{5\pi}{6}$

25. $t = -\dfrac{\pi}{2}$

26. $t = \dfrac{3\pi}{2}$

27. $t = \dfrac{4\pi}{3}$

28. $t = \dfrac{7\pi}{4}$

In Exercises 29–36, evaluate the trigonometric function using its period as an aid.

29. $\sin 5\pi$

30. $\cos 5\pi$

31. $\cos \dfrac{8\pi}{3}$

32. $\sin \dfrac{9\pi}{4}$

33. $\cos\left(-\dfrac{15\pi}{2}\right)$

34. $\sin \dfrac{19\pi}{6}$

35. $\sin\left(-\dfrac{9\pi}{4}\right)$

36. $\cos\left(-\dfrac{8\pi}{3}\right)$

In Exercises 37–42, use the value of the trigonometric function to evaluate the indicated functions.

37. $\sin t = \frac{1}{3}$
 (a) $\sin(-t)$
 (b) $\csc(-t)$

38. $\sin(-t) = \frac{3}{8}$
 (a) $\sin t$
 (b) $\csc t$

39. $\cos(-t) = -\frac{1}{5}$
 (a) $\cos t$
 (b) $\sec(-t)$

40. $\cos t = -\frac{3}{4}$
 (a) $\cos(-t)$
 (b) $\sec(-t)$

41. $\sin t = \frac{4}{5}$
 (a) $\sin(\pi - t)$
 (b) $\sin(t + \pi)$

42. $\cos t = \frac{4}{5}$
 (a) $\cos(\pi - t)$
 (b) $\cos(t + \pi)$

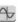 In Exercises 43–52, use a calculator to evaluate the trigonometric function. Round your answer to four decimal places. (Be sure the calculator is set in the correct angle mode.)

43. $\sin \dfrac{\pi}{4}$

44. $\tan \dfrac{\pi}{3}$

45. csc 1.3

46. cot 1

47. $\cos(-1.7)$

48. $\cos(-2.5)$

49. csc 0.8

50. sec 1.8

51. sec 22.8

52. $\sin(-0.9)$

Estimation In Exercises 53 and 54, use the figure and a straightedge to approximate the value of each trigonometric function. To print an enlarged copy of the graph, go to the website *www.mathgraphs.com.*

53. (a) sin 5   (b) cos 2

54. (a) sin 0.75   (b) cos 2.5

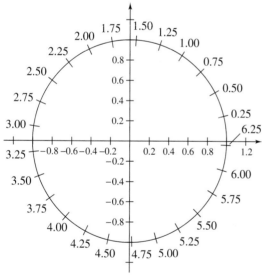

FIGURE FOR 53–56

Estimation In Exercises 55 and 56, use the figure and a straightedge to approximate the solution of each equation, where $0 \le t < 2\pi$. To print an enlarged copy of the graph, go to the website *www.mathgraphs.com.*

55. (a) sin $t = 0.25$ (b) cos $t = -0.25$

56. (a) sin $t = -0.75$ (b) cos $t = 0.75$

Model It

57. *Harmonic Motion* The displacement from equilibrium of an oscillating weight suspended by a spring is given by $y(t) = 3 \sin(\pi t/4)$, where y is the displacement (in feet) and t is the time (in seconds).

(a) Complete the table.

t	0	$\frac{1}{2}$	1	$\frac{3}{2}$	2
y					

 (b) Use the *table* feature of a graphing utility to determine when the displacement is maximum.

 (c) Use the *table* feature of a graphing utility to approximate the time t $(0 < t < 8)$ when the weight reaches equilibrium.

58. *Harmonic Motion* The displacement from equilibrium of an oscillating weight suspended by a spring is given by $y(t) = \frac{1}{4} \cos 6t$, where y is the displacement (in feet) and t is the time (in seconds). Find the displacement when (a) $t = 0$, (b) $t = \frac{1}{4}$, and (c) $t = \frac{1}{2}$.

Synthesis

True or False? In Exercises 59 and 60, determine whether the statement is true or false. Justify your answer.

59. Because $\sin(-t) = -\sin t$, it can be said that the sine of a negative angle is a negative number.

60. $\tan a = \tan(a - 6\pi)$

61. *Exploration* Let (x_1, y_1) and (x_2, y_2) be points on the unit circle corresponding to $t = t_1$ and $t = \pi - t_1$, respectively.

(a) Identify the symmetry of the points (x_1, y_1) and (x_2, y_2).

(b) Make a conjecture about any relationship between $\sin t_1$ and $\sin(\pi - t_1)$.

(c) Make a conjecture about any relationship between $\cos t_1$ and $\cos(\pi - t_1)$.

62. Use the unit circle to verify that the cosine and secant functions are even and that the sine, cosecant, tangent, and cotangent functions are odd.

Skills Review

In Exercises 63–66, find the inverse function f^{-1} of the one-to-one function f.

63. $f(x) = \frac{1}{2}(3x - 2)$ **64.** $f(x) = \frac{1}{4}x^3 + 1$

65. $f(x) = \sqrt{x^2 - 4}, \quad x \ge 2$ **66.** $f(x) = \dfrac{x + 2}{x - 4}$

1.3   Right Triangle Trigonometry

What you should learn

- Evaluate trigonometric functions of acute angles.
- Use the fundamental trigonometric identities.
- Use a calculator to evaluate trigonometric functions.
- Use trigonometric functions to model and solve real-life problems.

Why you should learn it

Trigonometric functions are often used to analyze real-life situations. For instance, in Exercise 71 on page 159, you can use trigonometric functions to find the height of a helium-filled balloon.

Joseph Sohm; Chromosohm

The Six Trigonometric Functions

Our second look at the trigonometric functions is from a *right triangle* perspective. Consider a right triangle, with one acute angle labeled θ, as shown in Figure 1.26. Relative to the angle θ, the three sides of the triangle are the **hypotenuse,** the **opposite side** (the side opposite the angle θ), and the **adjacent side** (the side adjacent to the angle θ).

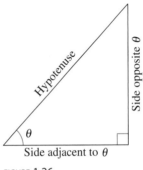

FIGURE 1.26

Using the lengths of these three sides, you can form six ratios that define the six trigonometric functions of the acute angle θ.

sine   cosecant   cosine   secant   tangent   cotangent

In the following definitions, it is important to see that $0° < \theta < 90°$ (θ lies in the first quadrant) and that for such angles the value of each trigonometric function is *positive*.

Right Triangle Definitions of Trigonometric Functions

Let θ be an *acute* angle of a right triangle. The six trigonometric functions of the angle θ are defined as follows. (Note that the functions in the second row are the *reciprocals* of the corresponding functions in the first row.)

$$\sin \theta = \frac{\text{opp}}{\text{hyp}} \qquad \cos \theta = \frac{\text{adj}}{\text{hyp}} \qquad \tan \theta = \frac{\text{opp}}{\text{adj}}$$

$$\csc \theta = \frac{\text{hyp}}{\text{opp}} \qquad \sec \theta = \frac{\text{hyp}}{\text{adj}} \qquad \cot \theta = \frac{\text{adj}}{\text{opp}}$$

The abbreviations opp, adj, and hyp represent the lengths of the three sides of a right triangle.

opp = the length of the side *opposite* θ

adj = the length of the side *adjacent to* θ

hyp = the length of the *hypotenuse*

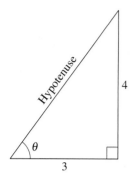

FIGURE 1.27

You may wish to review the Pythagorean Theorem before presenting the examples in this section.

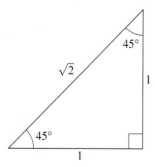

FIGURE 1.28

Example 1   **Evaluating Trigonometric Functions**

Use the triangle in Figure 1.27 to find the values of the six trigonometric functions of θ.

Solution

By the Pythagorean Theorem, $(\text{hyp})^2 = (\text{opp})^2 + (\text{adj})^2$, it follows that

$$\text{hyp} = \sqrt{4^2 + 3^2}$$
$$= \sqrt{25}$$
$$= 5.$$

So, the six trigonometric functions of θ are

$$\sin \theta = \frac{\text{opp}}{\text{hyp}} = \frac{4}{5} \qquad \csc \theta = \frac{\text{hyp}}{\text{opp}} = \frac{5}{4}$$

$$\cos \theta = \frac{\text{adj}}{\text{hyp}} = \frac{3}{5} \qquad \sec \theta = \frac{\text{hyp}}{\text{adj}} = \frac{5}{3}$$

$$\tan \theta = \frac{\text{opp}}{\text{adj}} = \frac{4}{3} \qquad \cot \theta = \frac{\text{adj}}{\text{opp}} = \frac{3}{4}.$$

✓**CHECKPOINT**   Now try Exercise 3.

In Example 1, you were given the lengths of two sides of the right triangle, but not the angle θ. Often, you will be asked to find the trigonometric functions of a *given* acute angle θ. To do this, construct a right triangle having θ as one of its angles.

Example 2   **Evaluating Trigonometric Functions of 45°**

Find the values of $\sin 45°$, $\cos 45°$, and $\tan 45°$.

Solution

Construct a right triangle having 45° as one of its acute angles, as shown in Figure 1.28. Choose the length of the adjacent side to be 1. From geometry, you know that the other acute angle is also 45°. So, the triangle is isosceles and the length of the opposite side is also 1. Using the Pythagorean Theorem, you find the length of the hypotenuse to be $\sqrt{2}$.

$$\sin 45° = \frac{\text{opp}}{\text{hyp}} = \frac{1}{\sqrt{2}} = \frac{\sqrt{2}}{2}$$

$$\cos 45° = \frac{\text{adj}}{\text{hyp}} = \frac{1}{\sqrt{2}} = \frac{\sqrt{2}}{2}$$

$$\tan 45° = \frac{\text{opp}}{\text{adj}} = \frac{1}{1} = 1$$

✓**CHECKPOINT**   Now try Exercise 17.

Consider having your students construct the triangle in Figure 1.29 with angles in the corresponding radian measures, then find the six trigonometric functions for each of the acute angles.

The triangles in Figures 1.27, 1.28, and 1.29 are useful problem-solving aids. Encourage your students to draw diagrams when they solve problems similar to those in Examples 1, 2, and 3.

Example 3   **Evaluating Trigonometric Functions of 30° and 60°**

Use the equilateral triangle shown in Figure 1.29 to find the values of sin 60°, cos 60°, sin 30°, and cos 30°.

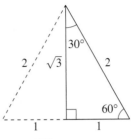

FIGURE **1.29**

Solution

Use the Pythagorean Theorem and the equilateral triangle in Figure 1.29 to verify the lengths of the sides shown in the figure. For $\theta = 60°$, you have adj $= 1$, opp $= \sqrt{3}$, and hyp $= 2$. So,

$$\sin 60° = \frac{\text{opp}}{\text{hyp}} = \frac{\sqrt{3}}{2} \quad \text{and} \quad \cos 60° = \frac{\text{adj}}{\text{hyp}} = \frac{1}{2}.$$

For $\theta = 30°$, adj $= \sqrt{3}$, opp $= 1$, and hyp $= 2$. So,

$$\sin 30° = \frac{\text{opp}}{\text{hyp}} = \frac{1}{2} \quad \text{and} \quad \cos 30° = \frac{\text{adj}}{\text{hyp}} = \frac{\sqrt{3}}{2}.$$

✓**CHECKPOINT**   Now try Exercise 19.

Sines, Cosines, and Tangents of Special Angles

$$\sin 30° = \sin \frac{\pi}{6} = \frac{1}{2} \qquad \cos 30° = \cos \frac{\pi}{6} = \frac{\sqrt{3}}{2} \qquad \tan 30° = \tan \frac{\pi}{6} = \frac{\sqrt{3}}{3}$$

$$\sin 45° = \sin \frac{\pi}{4} = \frac{\sqrt{2}}{2} \qquad \cos 45° = \cos \frac{\pi}{4} = \frac{\sqrt{2}}{2} \qquad \tan 45° = \tan \frac{\pi}{4} = 1$$

$$\sin 60° = \sin \frac{\pi}{3} = \frac{\sqrt{3}}{2} \qquad \cos 60° = \cos \frac{\pi}{3} = \frac{1}{2} \qquad \tan 60° = \tan \frac{\pi}{3} = \sqrt{3}$$

In the box, note that $\sin 30° = \frac{1}{2} = \cos 60°$. This occurs because 30° and 60° are complementary angles. In general, it can be shown from the right triangle definitions that *cofunctions of complementary angles are equal*. That is, if θ is an acute angle, the following relationships are true.

$$\sin(90° - \theta) = \cos \theta \qquad \cos(90° - \theta) = \sin \theta$$

$$\tan(90° - \theta) = \cot \theta \qquad \cot(90° - \theta) = \tan \theta$$

$$\sec(90° - \theta) = \csc \theta \qquad \csc(90° - \theta) = \sec \theta$$

Trigonometric Identities

In trigonometry, a great deal of time is spent studying relationships between trigonometric functions (identities).

These identities will be used many times in trigonometry and later in calculus. Encourage your students to learn them well.

Fundamental Trigonometric Identities

Reciprocal Identities

$$\sin \theta = \frac{1}{\csc \theta} \qquad \cos \theta = \frac{1}{\sec \theta} \qquad \tan \theta = \frac{1}{\cot \theta}$$

$$\csc \theta = \frac{1}{\sin \theta} \qquad \sec \theta = \frac{1}{\cos \theta} \qquad \cot \theta = \frac{1}{\tan \theta}$$

Quotient Identities

$$\tan \theta = \frac{\sin \theta}{\cos \theta} \qquad \cot \theta = \frac{\cos \theta}{\sin \theta}$$

Pythagorean Identities

$$\sin^2 \theta + \cos^2 \theta = 1 \qquad 1 + \tan^2 \theta = \sec^2 \theta$$

$$1 + \cot^2 \theta = \csc^2 \theta$$

Note that $\sin^2 \theta$ represents $(\sin \theta)^2$, $\cos^2 \theta$ represents $(\cos \theta)^2$, and so on.

Example 4   Applying Trigonometric Identities

Let θ be an acute angle such that $\sin \theta = 0.6$. Find the values of (a) $\cos \theta$ and (b) $\tan \theta$ using trigonometric identities.

Solution

a. To find the value of $\cos \theta$, use the Pythagorean identity

$$\sin^2 \theta + \cos^2 \theta = 1.$$

So, you have

$$(0.6)^2 + \cos^2 \theta = 1 \qquad \text{Substitute 0.6 for } \sin \theta.$$

$$\cos^2 \theta = 1 - (0.6)^2 = 0.64 \qquad \text{Subtract } (0.6)^2 \text{ from each side.}$$

$$\cos \theta = \sqrt{0.64} = 0.8. \qquad \text{Extract the positive square root.}$$

b. Now, knowing the sine and cosine of θ, you can find the tangent of θ to be

$$\tan \theta = \frac{\sin \theta}{\cos \theta}$$

$$= \frac{0.6}{0.8}$$

$$= 0.75.$$

Use the definitions of $\cos \theta$ and $\tan \theta$, and the triangle shown in Figure 1.30, to check these results.

FIGURE 1.30

✓CHECKPOINT   Now try Exercise 29.

| **Example 5** | **Applying Trigonometric Identities** |

Let θ be an acute angle such that $\tan \theta = 3$. Find the values of (a) $\cot \theta$ and (b) $\sec \theta$ using trigonometric identities.

Solution

a. $\cot \theta = \dfrac{1}{\tan \theta}$ Reciprocal identity

   $\cot \theta = \dfrac{1}{3}$

b. $\sec^2 \theta = 1 + \tan^2 \theta$ Pythagorean identity

   $\sec^2 \theta = 1 + 3^2$

   $\sec^2 \theta = 10$

   $\sec \theta = \sqrt{10}$

Use the definitions of $\cot \theta$ and $\sec \theta$, and the triangle shown in Figure 1.31, to check these results.

✓CHECKPOINT   Now try Exercise 31.

FIGURE **1.31**

Evaluating Trigonometric Functions with a Calculator

To use a calculator to evaluate trigonometric functions of angles measured in degrees, first set the calculator to *degree* mode and then proceed as demonstrated in Section 1.2. For instance, you can find values of $\cos 28°$ and $\sec 28°$ as follows.

	Function	*Mode*	*Calculator Keystrokes*	*Display*
a.	$\cos 28°$	Degree	$\boxed{\text{COS}}$ 28 $\boxed{\text{ENTER}}$	0.8829476
b.	$\sec 28°$	Degree	$\boxed{\text{(}}$ $\boxed{\text{COS}}$ $\boxed{\text{(}}$ 28 $\boxed{\text{)}}$ $\boxed{\text{)}}$ $\boxed{x^{-1}}$ $\boxed{\text{ENTER}}$	1.1325701

   Throughout this text, angles are assumed to be measured in radians unless noted otherwise. For example, $\sin 1$ means the sine of 1 radian and $\sin 1°$ means the sine of 1 degree.

| **Example 6** | **Using a Calculator** |

Use a calculator to evaluate $\sec(5°\,40'\,12'')$.

Solution

Begin by converting to decimal degree form. [Recall that $1' = \frac{1}{60}(1°)$ and $1'' = \frac{1}{3600}(1°)$].

$$5°\,40'\,12'' = 5° + \left(\frac{40}{60}\right)^° + \left(\frac{12}{3600}\right)^° = 5.67°$$

Then, use a calculator to evaluate $\sec 5.67°$.

Function	*Calculator Keystrokes*	*Display*
$\sec(5°\,40'\,12'') = \sec 5.67°$	$\boxed{\text{(}}$ $\boxed{\text{COS}}$ $\boxed{\text{(}}$ 5.67 $\boxed{\text{)}}$ $\boxed{\text{)}}$ $\boxed{x^{-1}}$ $\boxed{\text{ENTER}}$	1.0049166

✓CHECKPOINT   Now try Exercise 47.

One of the most common errors students make when they evaluate trigonometric functions with a calculator is not having their calculators set to the correct mode (*radian* vs. *degree*).

FIGURE 1.32

FIGURE 1.33

Applications Involving Right Triangles

Many applications of trigonometry involve a process called **solving right triangles.** In this type of application, you are usually given one side of a right triangle and one of the acute angles and are asked to find one of the other sides, *or* you are given two sides and are asked to find one of the acute angles.

In Example 7, the angle you are given is the **angle of elevation,** which represents the angle from the horizontal upward to an object. For objects that lie below the horizontal, it is common to use the term **angle of depression,** as shown in Figure 1.32.

Example 7 Using Trigonometry to Solve a Right Triangle

A surveyor is standing 115 feet from the base of the Washington Monument, as shown in Figure 1.33. The surveyor measures the angle of elevation to the top of the monument as 78.3°. How tall is the Washington Monument?

Solution

From Figure 1.33, you can see that

$$\tan 78.3° = \frac{\text{opp}}{\text{adj}} = \frac{y}{x}$$

where $x = 115$ and y is the height of the monument. So, the height of the Washington Monument is

$$y = x \tan 78.3° \approx 115(4.82882) \approx 555 \text{ feet.}$$

✔**CHECKPOINT** Now try Exercise 63.

Example 8 Using Trigonometry to Solve a Right Triangle

An historic lighthouse is 200 yards from a bike path along the edge of a lake. A walkway to the lighthouse is 400 yards long. Find the acute angle θ between the bike path and the walkway, as illustrated in Figure 1.34.

FIGURE 1.34

Solution

From Figure 1.34, you can see that the sine of the angle θ is

$$\sin \theta = \frac{\text{opp}}{\text{hyp}} = \frac{200}{400} = \frac{1}{2}.$$

Now you should recognize that $\theta = 30°$.

✔**CHECKPOINT** Now try Exercise 65.

By now you are able to recognize that $\theta = 30°$ is the acute angle that satisfies the equation $\sin \theta = \frac{1}{2}$. Suppose, however, that you were given the equation $\sin \theta = 0.6$ and were asked to find the acute angle θ. Because

$$\sin 30° = \frac{1}{2}$$

$$= 0.5000$$

and

$$\sin 45° = \frac{1}{\sqrt{2}}$$

$$\approx 0.7071$$

you might guess that θ lies somewhere between $30°$ and $45°$. In a later section, you will study a method by which a more precise value of θ can be determined.

Example 9   Solving a Right Triangle

Find the length c of the skateboard ramp shown in Figure 1.35.

FIGURE 1.35

Solution

From Figure 1.35, you can see that

$$\sin 18.4° = \frac{\text{opp}}{\text{hyp}}$$

$$= \frac{4}{c}.$$

So, the length of the skateboard ramp is

$$c = \frac{4}{\sin 18.4°}$$

$$\approx \frac{4}{0.3156}$$

$$\approx 12.7 \text{ feet}.$$

✓CHECKPOINT   Now try Exercise 67.

Activities

1. Use the right triangle shown to find each of the six trigonometric functions of the angle θ.

Answer:

$\sin \theta = \dfrac{2\sqrt{29}}{29}$, $\csc \theta = \dfrac{\sqrt{29}}{2}$,

$\cos \theta = \dfrac{5\sqrt{29}}{29}$, $\sec \theta = \dfrac{\sqrt{29}}{5}$,

$\tan \theta = \dfrac{2}{5}$, $\cot \theta = \dfrac{5}{2}$

2. A 10-foot ladder leans against the side of a house. The ladder makes an angle of $60°$ with the ground. How far up the side of the house does the ladder reach?

Answer: $5\sqrt{3} \approx 8.66$ feet

1.3 | Exercises

VOCABULARY CHECK:

1. Match the trigonometric function with its right triangle definition.

(a) Sine   (b) Cosine   (c) Tangent   (d) Cosecant   (e) Secant   (f) Cotangent

(i) $\dfrac{\text{hypotenuse}}{\text{adjacent}}$ (ii) $\dfrac{\text{adjacent}}{\text{opposite}}$ (iii) $\dfrac{\text{hypotenuse}}{\text{opposite}}$ (iv) $\dfrac{\text{adjacent}}{\text{hypotenuse}}$ (v) $\dfrac{\text{opposite}}{\text{hypotenuse}}$ (vi) $\dfrac{\text{opposite}}{\text{adjacent}}$

In Exercises 2 and 3, fill in the blanks.

2. Relative to the angle θ, the three sides of a right triangle are the _____ side, the _____ side, and the _____.

3. An angle that measures from the horizontal upward to an object is called the angle of _____, whereas an angle that measures from the horizontal downward to an object is called the angle of _____.

PREREQUISITE SKILLS REVIEW: Practice and review algebra skills needed for this section at **www.Eduspace.com.**

In Exercises 1–4, find the exact values of the six trigonometric functions of the angle θ shown in the figure. (Use the Pythagorean Theorem to find the third side of the triangle.)

1.

2.

3.

4.
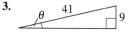

In Exercises 5–8, find the exact values of the six trigonometric functions of the angle θ for each of the two triangles. Explain why the function values are the same.

5.

6.

7.

8.

In Exercises 9–16, sketch a right triangle corresponding to the trigonometric function of the acute angle θ. Use the Pythagorean Theorem to determine the third side and then find the other five trigonometric functions of θ.

9. $\sin \theta = \dfrac{3}{4}$ **10.** $\cos \theta = \dfrac{5}{7}$

11. $\sec \theta = 2$ **12.** $\cot \theta = 5$

13. $\tan \theta = 3$ **14.** $\sec \theta = 6$

15. $\cot \theta = \dfrac{3}{2}$ **16.** $\csc \theta = \dfrac{17}{4}$

In Exercises 17–26, construct an appropriate triangle to complete the table. ($0 \le \theta \le 90°, 0 \le \theta \le \pi/2$)

	Function	θ (deg)	θ (rad)	Function Value
17.	sin	30°		
18.	cos	45°		
19.	tan		$\dfrac{\pi}{3}$	
20.	sec		$\dfrac{\pi}{4}$	
21.	cot			$\dfrac{\sqrt{3}}{3}$
22.	csc			$\sqrt{2}$
23.	cos		$\dfrac{\pi}{6}$	
24.	sin		$\dfrac{\pi}{4}$	
25.	cot			1
26.	tan			$\dfrac{\sqrt{3}}{3}$

In Exercises 27–32, use the given function value(s), and trigonometric identities (including the cofunction identities), to find the indicated trigonometric functions.

27. $\sin 60° = \dfrac{\sqrt{3}}{2}, \quad \cos 60° = \dfrac{1}{2}$

   (a) $\tan 60°$ (b) $\sin 30°$

   (c) $\cos 30°$ (d) $\cot 60°$

28. $\sin 30° = \dfrac{1}{2}, \quad \tan 30° = \dfrac{\sqrt{3}}{3}$

   (a) $\csc 30°$ (b) $\cot 60°$

   (c) $\cos 30°$ (d) $\cot 30°$

29. $\csc \theta = \dfrac{\sqrt{13}}{2}, \quad \sec \theta = \dfrac{\sqrt{13}}{3}$

   (a) $\sin \theta$ (b) $\cos \theta$

   (c) $\tan \theta$ (d) $\sec(90° - \theta)$

30. $\sec \theta = 5, \quad \tan \theta = 2\sqrt{6}$

   (a) $\cos \theta$ (b) $\cot \theta$

   (c) $\cot(90° - \theta)$ (d) $\sin \theta$

31. $\cos \alpha = \frac{1}{3}$

   (a) $\sec \alpha$ (b) $\sin \alpha$

   (c) $\cot \alpha$ (d) $\sin(90° - \alpha)$

32. $\tan \beta = 5$

   (a) $\cot \beta$ (b) $\cos \beta$

   (c) $\tan(90° - \beta)$ (d) $\csc \beta$

In Exercises 33–42, use trigonometric identities to transform the left side of the equation into the right side $(0 < \theta < \pi/2)$.

33. $\tan \theta \cot \theta = 1$

34. $\cos \theta \sec \theta = 1$

35. $\tan \alpha \cos \alpha = \sin \alpha$

36. $\cot \alpha \sin \alpha = \cos \alpha$

37. $(1 + \cos \theta)(1 - \cos \theta) = \sin^2 \theta$

38. $(1 + \sin \theta)(1 - \sin \theta) = \cos^2 \theta$

39. $(\sec \theta + \tan \theta)(\sec \theta - \tan \theta) = 1$

40. $\sin^2 \theta - \cos^2 \theta = 2 \sin^2 \theta - 1$

41. $\dfrac{\sin \theta}{\cos \theta} + \dfrac{\cos \theta}{\sin \theta} = \csc \theta \sec \theta$

42. $\dfrac{\tan \beta + \cot \beta}{\tan \beta} = \csc^2 \beta$

 In Exercises 43–52, use a calculator to evaluate each function. Round your answers to four decimal places. (Be sure the calculator is in the correct angle mode.)

43. (a) $\sin 10°$ (b) $\cos 80°$

44. (a) $\tan 23.5°$ (b) $\cot 66.5°$

45. (a) $\sin 16.35°$ (b) $\csc 16.35°$

46. (a) $\cos 16° \, 18'$ (b) $\sin 73° \, 56'$

47. (a) $\sec 42° \, 12'$ (b) $\csc 48° \, 7'$

48. (a) $\cos 4° \, 50' \, 15''$ (b) $\sec 4° \, 50' \, 15''$

49. (a) $\cot 11° \, 15'$ (b) $\tan 11° \, 15'$

50. (a) $\sec 56° \, 8 \, 10''$ (b) $\cos 56° \, 8 \, 10''$

51. (a) $\csc 32° \, 40' \, 3''$ (b) $\tan 44° \, 28 \, 16''$

52. (a) $\sec\left(\frac{9}{5} \cdot 20 + 32\right)°$ (b) $\cot\left(\frac{9}{5} \cdot 30 + 32\right)°$

In Exercises 53–58, find the values of θ in degrees $(0° < \theta < 90°)$ and radians $(0 < \theta < \pi/2)$ without the aid of a calculator.

53. (a) $\sin \theta = \dfrac{1}{2}$ (b) $\csc \theta = 2$

54. (a) $\cos \theta = \dfrac{\sqrt{2}}{2}$ (b) $\tan \theta = 1$

55. (a) $\sec \theta = 2$ (b) $\cot \theta = 1$

56. (a) $\tan \theta = \sqrt{3}$ (b) $\cos \theta = \dfrac{1}{2}$

57. (a) $\csc \theta = \dfrac{2\sqrt{3}}{3}$ (b) $\sin \theta = \dfrac{\sqrt{2}}{2}$

58. (a) $\cot \theta = \dfrac{\sqrt{3}}{3}$ (b) $\sec \theta = \sqrt{2}$

In Exercises 59–62, solve for x, y, or r as indicated.

59. Solve for x.   **60.** Solve for y.

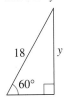

61. Solve for x.   **62.** Solve for r.

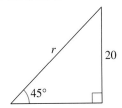

63. *Empire State Building*   You are standing 45 meters from the base of the Empire State Building. You estimate that the angle of elevation to the top of the 86th floor (the observatory) is 82°. If the total height of the building is another 123 meters above the 86th floor, what is the approximate height of the building? One of your friends is on the 86th floor. What is the distance between you and your friend?

64. Height   A six-foot person walks from the base of a broadcasting tower directly toward the tip of the shadow cast by the tower. When the person is 132 feet from the tower and 3 feet from the tip of the shadow, the person's shadow starts to appear beyond the tower's shadow.

(a) Draw a right triangle that gives a visual representation of the problem. Show the known quantities of the triangle and use a variable to indicate the height of the tower.

(b) Use a trigonometric function to write an equation involving the unknown quantity.

(c) What is the height of the tower?

65. Angle of Elevation   You are skiing down a mountain with a vertical height of 1500 feet. The distance from the top of the mountain to the base is 3000 feet. What is the angle of elevation from the base to the top of the mountain?

66. Width of a River   A biologist wants to know the width w of a river so in order to properly set instruments for studying the pollutants in the water. From point A, the biologist walks downstream 100 feet and sights to point C (see figure). From this sighting, it is determined that $\theta = 54°$. How wide is the river?

67. Length   A steel cable zip-line is being constructed for a competition on a reality television show. One end of the zip-line is attached to a platform on top of a 150-foot pole. The other end of the zip-line is attached to the top of a 5-foot stake. The angle of elevation to the platform is 23° (see figure).

(a) How long is the zip-line?

(b) How far is the stake from the pole?

(c) Contestants take an average of 6 seconds to reach the ground from the top of the zip-line. At what rate are contestants moving down the line? At what rate are they dropping vertically?

68. Height of a Mountain   In traveling across flat land, you notice a mountain directly in front of you. Its angle of elevation (to the peak) is 3.5°. After you drive 13 miles closer to the mountain, the angle of elevation is 9°. Approximate the height of the mountain.

69. Machine Shop Calculations   A steel plate has the form of one-fourth of a circle with a radius of 60 centimeters. Two two-centimeter holes are to be drilled in the plate positioned as shown in the figure. Find the coordinates of the center of each hole.

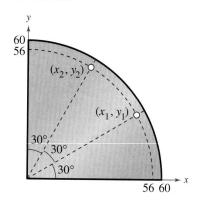

70. Machine Shop Calculations   A tapered shaft has a diameter of 5 centimeters at the small end and is 15 centimeters long (see figure). The taper is 3°. Find the diameter d of the large end of the shaft.

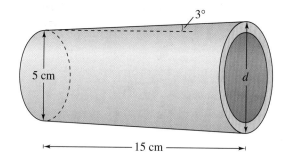

Model It

71. *Height*   A 20-meter line is used to tether a helium-filled balloon. Because of a breeze, the line makes an angle of approximately 85° with the ground.

(a) Draw a right triangle that gives a visual representation of the problem. Show the known quantities of the triangle and use a variable to indicate the height of the balloon.

(b) Use a trigonometric function to write an equation involving the unknown quantity.

(c) What is the height of the balloon?

(d) The breeze becomes stronger and the angle the balloon makes with the ground decreases. How does this affect the triangle you drew in part (a)?

(e) Complete the table, which shows the heights (in meters) of the balloon for decreasing angle measures θ.

Angle, θ	80°	70°	60°	50°
Height				

Angle, θ	40°	30°	20°	10°
Height				

(f) As the angle the balloon makes with the ground approaches 0°, how does this affect the height of the balloon? Draw a right triangle to explain your reasoning.

72. *Geometry*   Use a compass to sketch a quarter of a circle of radius 10 centimeters. Using a protractor, construct an angle of 20° in standard position (see figure). Drop a perpendicular line from the point of intersection of the terminal side of the angle and the arc of the circle. By actual measurement, calculate the coordinates (x, y) of the point of intersection and use these measurements to approximate the six trigonometric functions of a 20° angle.

Synthesis

True or False?   In Exercises 73–78, determine whether the statement is true or false. Justify your answer.

73. $\sin 60° \csc 60° = 1$

74. $\sec 30° = \csc 60°$

75. $\sin 45° + \cos 45° = 1$

76. $\cot^2 10° - \csc^2 10° = -1$

77. $\dfrac{\sin 60°}{\sin 30°} = \sin 2°$

78. $\tan[(5°)^2] = \tan^2(5°)$

79. *Writing*   In right triangle trigonometry, explain why $\sin 30° = \frac{1}{2}$ regardless of the size of the triangle.

80. *Think About It*   You are given only the value $\tan \theta$. Is it possible to find the value of $\sec \theta$ without finding the measure of θ? Explain.

81. *Exploration*

(a) Complete the table.

θ	0.1	0.2	0.3	0.4	0.5
$\sin \theta$					

(b) Is θ or $\sin \theta$ greater for θ in the interval $(0, 0.5]$?

(c) As θ approaches 0, how do θ and $\sin \theta$ compare? Explain.

82. *Exploration*

(a) Complete the table.

θ	0°	18°	36°	54°	72°	90°
$\sin \theta$						
$\cos \theta$						

(b) Discuss the behavior of the sine function for θ in the range from 0° to 90°.

(c) Discuss the behavior of the cosine function for θ in the range from 0° to 90°.

(d) Use the definitions of the sine and cosine functions to explain the results of parts (b) and (c).

Skills Review

In Exercises 83–86, perform the operations and simplify.

83. $\dfrac{x^2 - 6x}{x^2 + 4x - 12} \cdot \dfrac{x^2 + 12x + 36}{x^2 - 36}$

84. $\dfrac{2t^2 + 5t - 12}{9 - 4t^2} \div \dfrac{t^2 - 16}{4t^2 + 12t + 9}$

85. $\dfrac{3}{x + 2} - \dfrac{2}{x - 2} + \dfrac{x}{x^2 + 4x + 4}$

86. $\dfrac{\left(\dfrac{3}{x} - \dfrac{1}{4}\right)}{\left(\dfrac{12}{x} - 1\right)}$

1.4 Trigonometric Functions of Any Angle

What you should learn

- Evaluate trigonometric functions of any angle.
- Use reference angles to evaluate trigonometric functions.
- Evaluate trigonometric functions of real numbers.

Why you should learn it

You can use trigonometric functions to model and solve real-life problems. For instance, in Exercise 87 on page 167, you can use trigonometric functions to model the monthly normal temperatures in New York City and Fairbanks, Alaska.

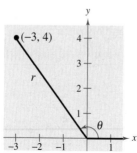

Introduction

In Section 1.3, the definitions of trigonometric functions were restricted to acute angles. In this section, the definitions are extended to cover *any* angle. If θ is an *acute* angle, these definitions coincide with those given in the preceding section.

Definitions of Trigonometric Functions of Any Angle

Let θ be an angle in standard position with (x, y) a point on the terminal side of θ and $r = \sqrt{x^2 + y^2} \neq 0$.

$$\sin \theta = \frac{y}{r} \qquad \cos \theta = \frac{x}{r}$$

$$\tan \theta = \frac{y}{x}, \quad x \neq 0 \qquad \cot \theta = \frac{x}{y}, \quad y \neq 0$$

$$\sec \theta = \frac{r}{x}, \quad x \neq 0 \qquad \csc \theta = \frac{r}{y}, \quad y \neq 0$$

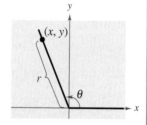

Because $r = \sqrt{x^2 + y^2}$ *cannot* be zero, it follows that the sine and cosine functions are defined for any real value of θ. However, if $x = 0$, the tangent and secant of θ are undefined. For example, the tangent of $90°$ is undefined. Similarly, if $y = 0$, the cotangent and cosecant of θ are undefined.

Example 1   Evaluating Trigonometric Functions

Let $(-3, 4)$ be a point on the terminal side of θ. Find the sine, cosine, and tangent of θ.

Solution

Referring to Figure 1.36, you can see that $x = -3$, $y = 4$, and

$$r = \sqrt{x^2 + y^2} = \sqrt{(-3)^2 + 4^2} = \sqrt{25} = 5.$$

So, you have the following.

$$\sin \theta = \frac{y}{r} = \frac{4}{5}$$

$$\cos \theta = \frac{x}{r} = -\frac{3}{5}$$

$$\tan \theta = \frac{y}{x} = -\frac{4}{3}$$

✓**CHECKPOINT**   Now try Exercise 1.

FIGURE **1.36**

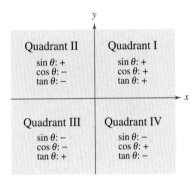

FIGURE 1.37

The *signs* of the trigonometric functions in the four quadrants can be determined easily from the definitions of the functions. For instance, because $\cos \theta = x/r$, it follows that $\cos \theta$ is positive wherever $x > 0$, which is in Quadrants I and IV. (Remember, r is always positive.) In a similar manner, you can verify the results shown in Figure 1.37.

Example 2　Evaluating Trigonometric Functions

Given $\tan \theta = -\frac{5}{4}$ and $\cos \theta > 0$, find $\sin \theta$ and $\sec \theta$.

Solution

Note that θ lies in Quadrant IV because that is the only quadrant in which the tangent is negative and the cosine is positive. Moreover, using

$$\tan \theta = \frac{y}{x} = -\frac{5}{4}$$

and the fact that y is negative in Quadrant IV, you can let $y = -5$ and $x = 4$. So, $r = \sqrt{16 + 25} = \sqrt{41}$ and you have

$$\sin \theta = \frac{y}{r} = \frac{-5}{\sqrt{41}}$$

$$\approx -0.7809$$

$$\sec \theta = \frac{r}{x} = \frac{\sqrt{41}}{4}$$

$$\approx 1.6008.$$

✓CHECKPOINT　Now try Exercise 17.

Example 3　Trigonometric Functions of Quadrant Angles

Evaluate the cosine and tangent functions at the four quadrant angles 0, $\dfrac{\pi}{2}$, π, and $\dfrac{3\pi}{2}$.

Solution

To begin, choose a point on the terminal side of each angle, as shown in Figure 1.38. For each of the four points, $r = 1$, and you have the following.

$$\cos 0 = \frac{x}{r} = \frac{1}{1} = 1 \qquad \tan 0 = \frac{y}{x} = \frac{0}{1} = 0 \qquad (x, y) = (1, 0)$$

$$\cos \frac{\pi}{2} = \frac{x}{r} = \frac{0}{1} = 0 \qquad \tan \frac{\pi}{2} = \frac{y}{x} = \frac{1}{0} \Rightarrow \text{undefined} \qquad (x, y) = (0, 1)$$

$$\cos \pi = \frac{x}{r} = \frac{-1}{1} = -1 \qquad \tan \pi = \frac{y}{x} = \frac{0}{-1} = 0 \qquad (x, y) = (-1, 0)$$

$$\cos \frac{3\pi}{2} = \frac{x}{r} = \frac{0}{1} = 0 \qquad \tan \frac{3\pi}{2} = \frac{y}{x} = \frac{-1}{0} \Rightarrow \text{undefined} \qquad (x, y) = (0, -1)$$

✓CHECKPOINT　Now try Exercise 29.

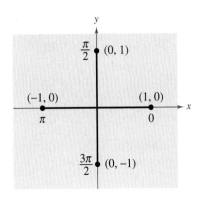

FIGURE 1.38

Reference Angles

Sketching several angles with their reference angles may help reinforce the fact that the reference angle is the acute angle formed with the horizontal.

The values of the trigonometric functions of angles greater than 90° (or less than 0°) can be determined from their values at corresponding acute angles called **reference angles.**

> ## Definition of Reference Angle
>
> Let θ be an angle in standard position. Its **reference angle** is the acute angle θ' formed by the terminal side of θ and the horizontal axis.

Figure 1.39 shows the reference angles for θ in Quadrants II, III, and IV.

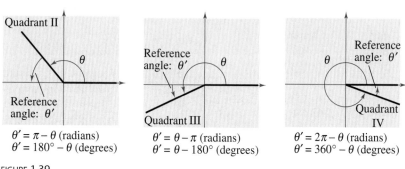

$\theta' = \pi - \theta$ (radians)
$\theta' = 180° - \theta$ (degrees)

$\theta' = \theta - \pi$ (radians)
$\theta' = \theta - 180°$ (degrees)

$\theta' = 2\pi - \theta$ (radians)
$\theta' = 360° - \theta$ (degrees)

FIGURE 1.39

Example 4   **Finding Reference Angles**

Find the reference angle θ'.

a. $\theta = 300°$ **b.** $\theta = 2.3$ **c.** $\theta = -135°$

Solution

a. Because 300° lies in Quadrant IV, the angle it makes with the x-axis is

$$\theta' = 360° - 300°$$
$$= 60°. \qquad \text{Degrees}$$

Figure 1.40 shows the angle $\theta = 300°$ and its reference angle $\theta' = 60°$.

b. Because 2.3 lies between $\pi/2 \approx 1.5708$ and $\pi \approx 3.1416$, it follows that it is in Quadrant II and its reference angle is

$$\theta' = \pi - 2.3$$
$$\approx 0.8416. \qquad \text{Radians}$$

Figure 1.41 shows the angle $\theta = 2.3$ and its reference angle $\theta' = \pi - 2.3$.

c. First, determine that $-135°$ is coterminal with 225°, which lies in Quadrant III. So, the reference angle is

$$\theta' = 225° - 180°$$
$$= 45°. \qquad \text{Degrees}$$

Figure 1.42 shows the angle $\theta = -135°$ and its reference angle $\theta' = 45°$.

✓CHECKPOINT   Now try Exercise 37.

FIGURE 1.40

FIGURE 1.41

FIGURE 1.42

Trigonometric Functions of Real Numbers

To see how a reference angle is used to evaluate a trigonometric function, consider the point (x, y) on the terminal side of θ, as shown in Figure 1.43. By definition, you know that

$$\sin \theta = \frac{y}{r} \quad \text{and} \quad \tan \theta = \frac{y}{x}.$$

For the right triangle with acute angle θ' and sides of lengths $|x|$ and $|y|$, you have

$$\sin \theta' = \frac{\text{opp}}{\text{hyp}} = \frac{|y|}{r}$$

and

$$\tan \theta' = \frac{\text{opp}}{\text{adj}} = \frac{|y|}{|x|}.$$

So, it follows that $\sin \theta$ and $\sin \theta'$ are equal, *except possibly in sign*. The same is true for $\tan \theta$ and $\tan \theta'$ *and* for the other four trigonometric functions. In all cases, the sign of the function value can be determined by the quadrant in which θ lies.

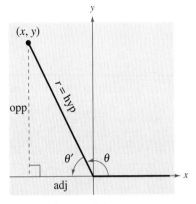

opp $= |y|$, adj $= |x|$

FIGURE 1.43

Evaluating Trigonometric Functions of Any Angle

To find the value of a trigonometric function of any angle θ:

1. Determine the function value for the associated reference angle θ'.

2. Depending on the quadrant in which θ lies, affix the appropriate sign to the function value.

By using reference angles and the special angles discussed in the preceding section, you can greatly extend the scope of *exact* trigonometric values. For instance, knowing the function values of 30° means that you know the function values of all angles for which 30° is a reference angle. For convenience, the table below shows the exact values of the trigonometric functions of special angles and quadrant angles.

Trigonometric Values of Common Angles

θ (degrees)	0°	30°	45°	60°	90°	180°	270°
θ (radians)	0	$\dfrac{\pi}{6}$	$\dfrac{\pi}{4}$	$\dfrac{\pi}{3}$	$\dfrac{\pi}{2}$	π	$\dfrac{3\pi}{2}$
$\sin \theta$	0	$\dfrac{1}{2}$	$\dfrac{\sqrt{2}}{2}$	$\dfrac{\sqrt{3}}{2}$	1	0	-1
$\cos \theta$	1	$\dfrac{\sqrt{3}}{2}$	$\dfrac{\sqrt{2}}{2}$	$\dfrac{1}{2}$	0	-1	0
$\tan \theta$	0	$\dfrac{\sqrt{3}}{3}$	1	$\sqrt{3}$	Undef.	0	Undef.

STUDY TIP

Learning the table of values at the right is worth the effort because doing so will increase both your efficiency and your confidence. Here is a pattern for the sine function that may help you remember the values.

θ	0°	30°	45°	60°	90°
$\sin \theta$	$\dfrac{\sqrt{0}}{2}$	$\dfrac{\sqrt{1}}{2}$	$\dfrac{\sqrt{2}}{2}$	$\dfrac{\sqrt{3}}{2}$	$\dfrac{\sqrt{4}}{2}$

Reverse the order to get cosine values of the same angles.

Example 5   Using Reference Angles

Evaluate each trigonometric function.

a. $\cos \dfrac{4\pi}{3}$ **b.** $\tan(-210°)$ **c.** $\csc \dfrac{11\pi}{4}$

Solution

a. Because $\theta = 4\pi/3$ lies in Quadrant III, the reference angle is $\theta' = (4\pi/3) - \pi = \pi/3$, as shown in Figure 1.44. Moreover, the cosine is negative in Quadrant III, so

$$\cos \frac{4\pi}{3} = (-) \cos \frac{\pi}{3}$$

$$= -\frac{1}{2}.$$

b. Because $-210° + 360° = 150°$, it follows that $-210°$ is coterminal with the second-quadrant angle $150°$. So, the reference angle is $\theta' = 180° - 150° = 30°$, as shown in Figure 1.45. Finally, because the tangent is negative in Quadrant II, you have

$$\tan(-210°) = (-) \tan 30°$$

$$= -\frac{\sqrt{3}}{3}.$$

c. Because $(11\pi/4) - 2\pi = 3\pi/4$, it follows that $11\pi/4$ is coterminal with the second-quadrant angle $3\pi/4$. So, the reference angle is $\theta' = \pi - (3\pi/4) = \pi/4$, as shown in Figure 1.46. Because the cosecant is positive in Quadrant II, you have

$$\csc \frac{11\pi}{4} = (+) \csc \frac{\pi}{4}$$

$$= \frac{1}{\sin(\pi/4)}$$

$$= \sqrt{2}.$$

FIGURE 1.44

FIGURE 1.45

FIGURE 1.46

✓ CHECKPOINT   Now try Exercise 51.

Example 6   Using Trigonometric Identities

Let θ be an angle in Quadrant II such that $\sin \theta = \frac{1}{3}$. Find (a) $\cos \theta$ and (b) $\tan \theta$ by using trigonometric identities.

Solution

a. Using the Pythagorean identity $\sin^2 \theta + \cos^2 \theta = 1$, you obtain

$$\left(\frac{1}{3}\right)^2 + \cos^2 \theta = 1 \qquad \text{Substitute } \tfrac{1}{3} \text{ for } \sin \theta.$$

$$\cos^2 \theta = 1 - \frac{1}{9} = \frac{8}{9}.$$

Because $\cos \theta < 0$ in Quadrant II, you can use the negative root to obtain

$$\cos \theta = -\frac{\sqrt{8}}{\sqrt{9}}$$

$$= -\frac{2\sqrt{2}}{3}.$$

b. Using the trigonometric identity $\tan \theta = \dfrac{\sin \theta}{\cos \theta}$, you obtain

$$\tan \theta = \frac{1/3}{-2\sqrt{2}/3} \qquad \text{Substitute for } \sin \theta \text{ and } \cos \theta.$$

$$= -\frac{1}{2\sqrt{2}}$$

$$= -\frac{\sqrt{2}}{4}.$$

✔**CHECKPOINT**   Now try Exercise 59.

You can use a calculator to evaluate trigonometric functions, as shown in the next example.

Example 7   Using a Calculator

Use a calculator to evaluate each trigonometric function.

a. $\cot 410°$ **b.** $\sin(-7)$ **c.** $\sec \dfrac{\pi}{9}$

Solution

Function	Mode	Calculator Keystrokes	Display
a. $\cot 410°$	Degree	(TAN (410)) x^{-1} ENTER	0.8390996
b. $\sin(-7)$	Radian	SIN ((−) 7) ENTER	−0.6569866
c. $\sec \dfrac{\pi}{9}$	Radian	(COS (π ÷ 9)) x^{-1} ENTER	1.0641778

✔**CHECKPOINT**   Now try Exercise 69.

Students often have difficulty determining angles, especially when the functions given are csc, sec, and/or cot. Have your students rewrite the expression in terms of sin, cos, or tan, whichever is applicable, before evaluating.

Activities

1. Determine the exact values of the six trigonometric functions of the angle in standard position whose terminal side contains the point $(-3, -7)$.

   Answer:

   $\sin \theta = -\dfrac{7}{\sqrt{58}} \qquad \csc \theta = -\dfrac{\sqrt{58}}{7}$

   $\cos \theta = -\dfrac{3}{\sqrt{58}} \qquad \sec \theta = -\dfrac{\sqrt{58}}{3}$

   $\tan \theta = \dfrac{7}{3} \qquad \cot \theta = \dfrac{3}{7}$

2. For the angle $\theta = -135°$, find the reference angle θ', and sketch θ and θ' in standard position.

   Answer: $\theta' = 45°$

3. Find two values of $\theta, 0 \leq \theta < 2\pi$, that satisfy the equation $\tan \theta = -1$. Do not use your calculator.

   Answer: $\theta = \dfrac{3\pi}{4}, \dfrac{7\pi}{4}$

1.4 | Exercises

VOCABULARY CHECK:

In Exercises 1–6, let θ be an angle in standard position, with (x, y) a point on the terminal side of θ and $r\sqrt{x^2 + y^2} \neq 0$.

1. $\sin \theta =$ _____

2. $\dfrac{r}{y} =$ _____

3. $\tan \theta =$ _____

4. $\sec \theta =$ _____

5. $\dfrac{x}{r} =$ _____

6. $\dfrac{x}{y} =$ _____

7. The acute positive angle that is formed by the terminal side of the angle θ and the horizontal axis is called the _____ angle of θ and is denoted by θ'.

PREREQUISITE SKILLS REVIEW: Practice and review algebra skills needed for this section at **www.Eduspace.com.**

In Exercises 1–4, determine the exact values of the six trigonometric functions of the angle θ.

1. (a)

(b)

2. (a)

(b)

3. (a)

(b)

4. (a)

(b)
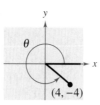

In Exercises 5–10, the point is on the terminal side of an angle in standard position. Determine the exact values of the six trigonometric functions of the angle.

5. $(7, 24)$

6. $(8, 15)$

7. $(-4, 10)$

8. $(-5, -2)$

9. $(-3.5, 6.8)$

10. $\left(3\frac{1}{2}, -7\frac{3}{4}\right)$

In Exercises 11–14, state the quadrant in which θ lies.

11. $\sin \theta < 0$ and $\cos \theta < 0$

12. $\sin \theta > 0$ and $\cos \theta > 0$

13. $\sin \theta > 0$ and $\tan \theta < 0$

14. $\sec \theta > 0$ and $\cot \theta < 0$

In Exercises 15–24, find the values of the six trigonometric functions of θ with the given constraint.

Function Value	*Constraint*
15. $\sin \theta = \frac{3}{5}$	θ lies in Quadrant II.
16. $\cos \theta = -\frac{4}{5}$	θ lies in Quadrant III.
17. $\tan \theta = -\frac{15}{8}$	$\sin \theta < 0$
18. $\cos \theta = \frac{8}{17}$	$\tan \theta < 0$
19. $\cot \theta = -3$	$\cos \theta > 0$
20. $\csc \theta = 4$	$\cot \theta < 0$
21. $\sec \theta = -2$	$\sin \theta > 0$
22. $\sin \theta = 0$	$\sec \theta = -1$
23. $\cot \theta$ is undefined.	$\pi/2 \leq \theta \leq 3\pi/2$
24. $\tan \theta$ is undefined.	$\pi \leq \theta \leq 2\pi$

In Exercises 25–28, the terminal side of θ lies on the given line in the specified quadrant. Find the values of the six trigonometric functions of θ by finding a point on the line.

Line	*Quadrant*
25. $y = -x$	II
26. $y = \frac{1}{3}x$	III
27. $2x - y = 0$	III
28. $4x + 3y = 0$	IV

In Exercises 29–36, evaluate the trigonometric function of the quadrant angle.

29. $\sin \pi$

30. $\csc \dfrac{3\pi}{2}$

31. $\sec \dfrac{3\pi}{2}$

32. $\sec \pi$

33. $\sin \dfrac{\pi}{2}$

34. $\cot \pi$

35. $\csc \pi$

36. $\cot \dfrac{\pi}{2}$

In Exercises 37–44, find the reference angle θ', and sketch θ and θ' in standard position.

37. $\theta = 203°$

38. $\theta = 309°$

39. $\theta = -245°$

40. $\theta = -145°$

41. $\theta = \dfrac{2\pi}{3}$

42. $\theta = \dfrac{7\pi}{4}$

43. $\theta = 3.5$

44. $\theta = \dfrac{11\pi}{3}$

In Exercises 45–58, evaluate the sine, cosine, and tangent of the angle without using a calculator.

45. $225°$

46. $300°$

47. $750°$

48. $-405°$

49. $-150°$

50. $-840°$

51. $\dfrac{4\pi}{3}$

52. $\dfrac{\pi}{4}$

53. $-\dfrac{\pi}{6}$

54. $-\dfrac{\pi}{2}$

55. $\dfrac{11\pi}{4}$

56. $\dfrac{10\pi}{3}$

57. $-\dfrac{3\pi}{2}$

58. $-\dfrac{25\pi}{4}$

In Exercises 59–64, find the indicated trigonometric value in the specified quadrant.

	Function	Quadrant	Trigonometric Value
59.	$\sin \theta = -\frac{3}{5}$	IV	$\cos \theta$
60.	$\cot \theta = -3$	II	$\sin \theta$
61.	$\tan \theta = \frac{3}{2}$	III	$\sec \theta$
62.	$\csc \theta = -2$	IV	$\cot \theta$
63.	$\cos \theta = \frac{5}{8}$	I	$\sec \theta$
64.	$\sec \theta = -\frac{9}{4}$	III	$\tan \theta$

In Exercises 65–80, use a calculator to evaluate the trigonometric function. Round your answer to four decimal places. (Be sure the calculator is set in the correct angle mode.)

65. $\sin 10°$

66. $\sec 225°$

67. $\cos(-110°)$

68. $\csc(-330°)$

69. $\tan 304°$

70. $\cot 178°$

71. $\sec 72°$

72. $\tan(-188°)$

73. $\tan 4.5$

74. $\cot 1.35$

75. $\tan \dfrac{\pi}{9}$

76. $\tan\left(-\dfrac{\pi}{9}\right)$

77. $\sin(-0.65)$

78. $\sec 0.29$

79. $\cot\left(-\dfrac{11\pi}{8}\right)$

80. $\csc\left(-\dfrac{15\pi}{14}\right)$

In Exercises 81–86, find two solutions of the equation. Give your answers in degrees ($0° \le \theta < 360°$) and in radians ($0 \le \theta < 2\pi$). Do not use a calculator.

81. (a) $\sin \theta = \frac{1}{2}$ (b) $\sin \theta = -\frac{1}{2}$

82. (a) $\cos \theta = \dfrac{\sqrt{2}}{2}$ (b) $\cos \theta = -\dfrac{\sqrt{2}}{2}$

83. (a) $\csc \theta = \dfrac{2\sqrt{3}}{2}$ (b) $\cot \theta = -1$

84. (a) $\sec \theta = 2$ (b) $\sec \theta = -2$

85. (a) $\tan \theta = 1$ (b) $\cot \theta = -\sqrt{3}$

86. (a) $\sin \theta = \dfrac{\sqrt{3}}{2}$ (b) $\sin \theta = -\dfrac{\sqrt{3}}{2}$

Model It

87. *Data Analysis: Meteorology*   The table shows the monthly normal temperatures (in degrees Fahrenheit) for selected months for New York City (N) and Fairbanks, Alaska (F). (Source: National Climatic Data Center)

Month	New York City, N	Fairbanks, F
January	33	−10
April	52	32
July	77	62
October	58	24
December	38	−6

(a) Use the *regression* feature of a graphing utility to find a model of the form

$$y = a \sin(bt + c) + d$$

for each city. Let t represent the month, with $t = 1$ corresponding to January.

Model It (continued)

(b) Use the models from part (a) to find the monthly normal temperatures for the two cities in February, March, May, June, August, September, and November.

(c) Compare the models for the two cities.

88. Sales A company that produces snowboards, which are seasonal products, forecasts monthly sales over the next 2 years to be

$$S = 23.1 + 0.442t + 4.3 \cos \frac{\pi t}{6}$$

where S is measured in thousands of units and t is the time in months, with $t = 1$ representing January 2006. Predict sales for each of the following months.

(a) February 2006   (b) February 2007

(c) June 2006   (d) June 2007

Path of a Projectile In Exercises 89 and 90, use the following information. The horizontal distance d (in feet) traveled by a projectile with an initial speed of v feet per second is modeled by

$$d = \frac{v^2}{32} \sin 2\theta.$$

where θ is the angle at which the projectile is launched.

89. Find the horizontal distance traveled by a golf ball that is hit with an initial speed of 100 feet per second when the ball is hit at an angle of (a) $\theta = 30°$, (b) $\theta = 50°$, and (c) $\theta = 60°$.

90. Find the horizontal distance traveled by a model rocket that is launched with an initial speed of 120 feet per second when the model rocket is launched at an angle of (a) $\theta = 60°$, (b) $\theta = 70°$, and (c) $\theta = 80°$.

91. Harmonic Motion The displacement from equilibrium of an oscillating weight suspended by a spring and subject to the damping effect of friction is given by

$$y(t) = 2 \cos 6t$$

where y is the displacement (in centimeters) and t is the time (in seconds). Find the displacement when (a) $t = 0$, (b) $t = \frac{1}{4}$, and (c) $t = \frac{1}{2}$.

92. Distance An airplane, flying at an altitude of 6 miles, is on a flight path that passes directly over an observer (see figure). If θ is the angle of elevation from the observer to the plane, find the distance d from the observer to the plane when (a) $\theta = 30°$, (b) $\theta = 90°$, and (c) $\theta = 120°$.

FIGURE FOR 92

Synthesis

True or False? In Exercises 93 and 94, determine whether the statement is true or false. Justify your answer.

93. In each of the four quadrants, the signs of the secant function and sine function will be the same.

94. To find the reference angle for an angle θ (given in degrees), find the integer n such that $0 \le 360°n - \theta \le 360°$. The difference $360°n - \theta$ is the reference angle.

95. Writing Consider an angle in standard position with $r = 12$ centimeters, as shown in the figure. Write a short paragraph describing the changes in the values of x, y, $\sin \theta$, $\cos \theta$, and $\tan \theta$ as θ increases continuously from $0°$ to $90°$.

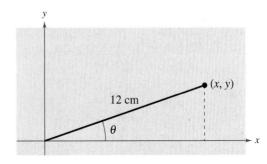

96. Writing Explain how reference angles are used to find the trigonometric functions of obtuse angles.

Skills Review

In Exercises 97–104, graph the function. Identify the domain and any intercepts of the function.

97. $y = x - 8$ **98.** $y = 6 - 7x$

99. $y = x^2 + 3x - 4$ **100.** $y = 2x^2 - 5x$

101. $f(x) = x^3 + 8$ **102.** $g(x) = x^4 + 2x^2 - 3$

103. $g(x) = \sqrt{x + 5}$ **104.** $f(x) = \sqrt{4x - 1}$

1.5   Graphs of Sine and Cosine Functions

What you should learn

- Sketch the graphs of basic sine and cosine functions.
- Use amplitude and period to help sketch the graphs of sine and cosine functions.
- Sketch translations of the graphs of sine and cosine functions.
- Use sine and cosine functions to model real-life data.

Why you should learn it

Sine and cosine functions are often used in scientific calculations. For instance, in Exercise 73 on page 178, you can use a trigonometric function to model the airflow of your respiratory cycle.

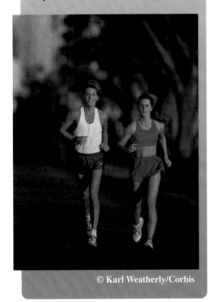

© Karl Weatherly/Corbis

Basic Sine and Cosine Curves

In this section, you will study techniques for sketching the graphs of the sine and cosine functions. The graph of the sine function is a **sine curve.** In Figure 1.47, the black portion of the graph represents one period of the function and is called **one cycle** of the sine curve. The gray portion of the graph indicates that the basic sine curve repeats indefinitely in the positive and negative directions. The graph of the cosine function is shown in Figure 1.48.

Recall from Section 1.2 that the domain of the sine and cosine functions is the set of all real numbers. Moreover, the range of each function is the interval $[-1, 1]$, and each function has a period of 2π. Do you see how this information is consistent with the basic graphs shown in Figures 1.47 and 1.48?

FIGURE 1.47

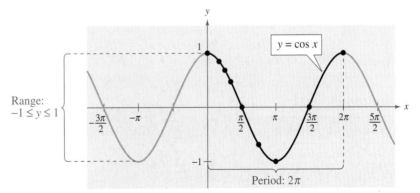

FIGURE 1.48

Note in Figures 1.47 and 1.48 that the sine curve is symmetric with respect to the *origin*, whereas the cosine curve is symmetric with respect to the *y-axis*. These properties of symmetry follow from the fact that the sine function is odd and the cosine function is even.

To sketch the graphs of the basic sine and cosine functions by hand, it helps to note five **key points** in one period of each graph: the *intercepts*, *maximum points*, and *minimum points* (see Figure 1.49).

 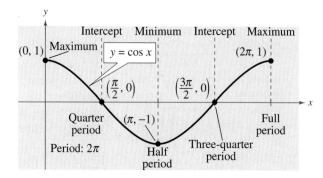

FIGURE 1.49

Example 1   Using Key Points to Sketch a Sine Curve

Sketch the graph of $y = 2 \sin x$ on the interval $[-\pi, 4\pi]$.

Solution

Note that

$$y = 2 \sin x = 2(\sin x)$$

indicates that the y-values for the key points will have twice the magnitude of those on the graph of $y = \sin x$. Divide the period 2π into four equal parts to get the key points for $y = 2 \sin x$.

Intercept	*Maximum*	*Intercept*	*Minimum*		*Intercept*
$(0, 0),$	$\left(\dfrac{\pi}{2}, 2\right),$	$(\pi, 0),$	$\left(\dfrac{3\pi}{2}, -2\right),$	and	$(2\pi, 0)$

By connecting these key points with a smooth curve and extending the curve in both directions over the interval $[-\pi, 4\pi]$, you obtain the graph shown in Figure 1.50.

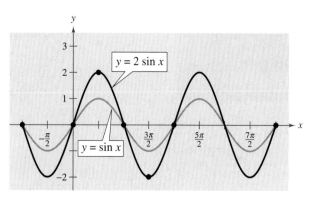

FIGURE 1.50

Now try Exercise 35.

Technology

When using a graphing utility to graph trigonometric functions, pay special attention to the viewing window you use. For instance, try graphing $y = [\sin(10x)]/10$ in the standard viewing window in *radian* mode. What do you observe? Use the *zoom* feature to find a viewing window that displays a good view of the graph.

Amplitude and Period

In the remainder of this section you will study the graphic effect of each of the constants a, b, c, and d in equations of the forms

$$y = d + a \sin(bx - c)$$

and

$$y = d + a \cos(bx - c).$$

A quick review of the transformations you studied in Section P.8 should help in this investigation.

The constant factor a in $y = a \sin x$ acts as a *scaling factor*—a *vertical stretch* or *vertical shrink* of the basic sine curve. If $|a| > 1$, the basic sine curve is stretched, and if $|a| < 1$, the basic sine curve is shrunk. The result is that the graph of $y = a \sin x$ ranges between $-a$ and a instead of between -1 and 1. The absolute value of a is the **amplitude** of the function $y = a \sin x$. The range of the function $y = a \sin x$ for $a > 0$ is $-a \leq y \leq a$.

Definition of Amplitude of Sine and Cosine Curves

The **amplitude** of $y = a \sin x$ and $y = a \cos x$ represents half the distance between the maximum and minimum values of the function and is given by

$$\text{Amplitude} = |a|.$$

FIGURE **1.51**

Example 2   Scaling: Vertical Shrinking and Stretching

On the same coordinate axes, sketch the graph of each function.

a. $y = \dfrac{1}{2} \cos x$ **b.** $y = 3 \cos x$

Solution

a. Because the amplitude of $y = \frac{1}{2} \cos x$ is $\frac{1}{2}$, the maximum value is $\frac{1}{2}$ and the minimum value is $-\frac{1}{2}$. Divide one cycle, $0 \leq x \leq 2\pi$, into four equal parts to get the key points

Maximum	Intercept	Minimum	Intercept		Maximum
$\left(0, \dfrac{1}{2}\right),$	$\left(\dfrac{\pi}{2}, 0\right),$	$\left(\pi, -\dfrac{1}{2}\right),$	$\left(\dfrac{3\pi}{2}, 0\right),$	and	$\left(2\pi, \dfrac{1}{2}\right).$

b. A similar analysis shows that the amplitude of $y = 3 \cos x$ is 3, and the key points are

Maximum	Intercept	Minimum	Intercept		Maximum
$(0, 3),$	$\left(\dfrac{\pi}{2}, 0\right),$	$(\pi, -3),$	$\left(\dfrac{3\pi}{2}, 0\right),$	and	$(2\pi, 3).$

The graphs of these two functions are shown in Figure 1.51. Notice that the graph of $y = \frac{1}{2} \cos x$ is a vertical *shrink* of the graph of $y = \cos x$ and the graph of $y = 3 \cos x$ is a vertical *stretch* of the graph of $y = \cos x$.

✓CHECKPOINT   Now try Exercise 37.

Exploration

Sketch the graph of $y = \cos bx$ for $b = \frac{1}{2}$, 2, and 3. How does the value of b affect the graph? How many complete cycles occur between 0 and 2π for each value of b?

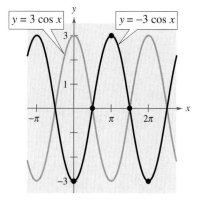

You know from Section P.8 that the graph of $y = -f(x)$ is a **reflection** in the x-axis of the graph of $y = f(x)$. For instance, the graph of $y = -3 \cos x$ is a reflection of the graph of $y = 3 \cos x$, as shown in Figure 1.52.

Because $y = a \sin x$ completes one cycle from $x = 0$ to $x = 2\pi$, it follows that $y = a \sin bx$ completes one cycle from $x = 0$ to $x = 2\pi/b$.

Period of Sine and Cosine Functions

Let b be a positive real number. The **period** of $y = a \sin bx$ and $y = a \cos bx$ is given by

$$\text{Period} = \frac{2\pi}{b}.$$

Note that if $0 < b < 1$, the period of $y = a \sin bx$ is greater than 2π and represents a *horizontal stretching* of the graph of $y = a \sin x$. Similarly, if $b > 1$, the period of $y = a \sin bx$ is less than 2π and represents a *horizontal shrinking* of the graph of $y = a \sin x$. If b is negative, the identities $\sin(-x) = -\sin x$ and $\cos(-x) = \cos x$ are used to rewrite the function.

Example 3   Scaling: Horizontal Stretching

Sketch the graph of $y = \sin \dfrac{x}{2}$.

Solution

The amplitude is 1. Moreover, because $b = \frac{1}{2}$, the period is

$$\frac{2\pi}{b} = \frac{2\pi}{\frac{1}{2}} = 4\pi. \qquad \text{Substitute for } b.$$

Now, divide the period-interval $[0, 4\pi]$ into four equal parts with the values π, 2π, and 3π to obtain the key points on the graph.

Intercept	Maximum	Intercept	Minimum		Intercept
$(0, 0)$,	$(\pi, 1)$,	$(2\pi, 0)$,	$(3\pi, -1)$,	and	$(4\pi, 0)$

The graph is shown in Figure 1.53.

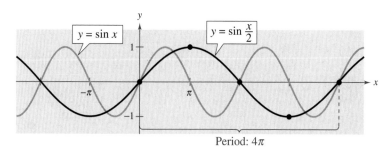

Period: 4π

✓CHECKPOINT   Now try Exercise 39.

Sketch the graph of

$$y = \sin(x - c)$$

where $c = -\pi/4, 0,$ and $\pi/4$. How does the value of c affect the graph?

STUDY TIP

In general, to divide a period-interval into four equal parts, successively add "period/4," starting with the left endpoint of the interval. For instance, for the period-interval $[-\pi/6, \pi/2]$ of length $2\pi/3$, you would successively add

$$\frac{2\pi/3}{4} = \frac{\pi}{6}$$

to get $-\pi/6, 0, \pi/6, \pi/3,$ and $\pi/2$ as the x-values for the key points on the graph.

Translations of Sine and Cosine Curves

The constant c in the general equations

$$y = a \sin(bx - c) \qquad \text{and} \qquad y = a \cos(bx - c)$$

creates a *horizontal translation* (shift) of the basic sine and cosine curves. Comparing $y = a \sin bx$ with $y = a \sin(bx - c)$, you find that the graph of $y = a \sin(bx - c)$ completes one cycle from $bx - c = 0$ to $bx - c = 2\pi$. By solving for x, you can find the interval for one cycle to be

Left endpoint   Right endpoint

$$\frac{c}{b} \le x \le \frac{c}{b} + \frac{2\pi}{b}.$$

Period

This implies that the period of $y = a \sin(bx - c)$ is $2\pi/b$, and the graph of $y = a \sin bx$ is shifted by an amount c/b. The number c/b is the **phase shift.**

Graphs of Sine and Cosine Functions

The graphs of $y = a \sin(bx - c)$ and $y = a \cos(bx - c)$ have the following characteristics. (Assume $b > 0$.)

$$\text{Amplitude} = |a| \qquad \text{Period} = \frac{2\pi}{b}$$

The left and right endpoints of a one-cycle interval can be determined by solving the equations $bx - c = 0$ and $bx - c = 2\pi$.

Horizontal Translation

Example 4

Sketch the graph of $y = \dfrac{1}{2} \sin\left(x - \dfrac{\pi}{3}\right)$.

Solution

The amplitude is $\frac{1}{2}$ and the period is 2π. By solving the equations

$$x - \frac{\pi}{3} = 0 \quad \Longrightarrow \quad x = \frac{\pi}{3}$$

and

$$x - \frac{\pi}{3} = 2\pi \quad \Longrightarrow \quad x = \frac{7\pi}{3}$$

you see that the interval $[\pi/3, 7\pi/3]$ corresponds to one cycle of the graph. Dividing this interval into four equal parts produces the key points

Intercept	Maximum	Intercept	Minimum		Intercept
$\left(\dfrac{\pi}{3}, 0\right),$	$\left(\dfrac{5\pi}{6}, \dfrac{1}{2}\right),$	$\left(\dfrac{4\pi}{3}, 0\right),$	$\left(\dfrac{11\pi}{6}, -\dfrac{1}{2}\right),$	and	$\left(\dfrac{7\pi}{3}, 0\right).$

The graph is shown in Figure 1.54.

✔CHECKPOINT   Now try Exercise 45.

Period: 2π

FIGURE **1.54**

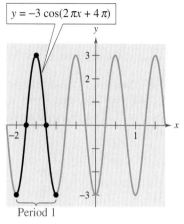

$y = -3\cos(2\pi x + 4\pi)$

Period 1

FIGURE **1.55**

Activities

1. Describe the relationship between the graphs of $f(x) = \sin x$ and $g(x) = 3\sin(2x + 1)$.

   Answer: The amplitude of the basic sine curve is 1, whereas the amplitude of g is 3. The period of the basic sine curve is 2π, whereas the period of g is π. Lastly, the graph of g has a phase shift $\frac{1}{2}$ unit to the left of the graph of $f(x) = \sin x$.

2. Determine the amplitude, period, and phase shift of

   $y = \frac{1}{2}\cos(\pi x - 1)$.

   Answer: Amplitude: $\frac{1}{2}$;

   period: 2;

   phase shift $\frac{1}{\pi}$

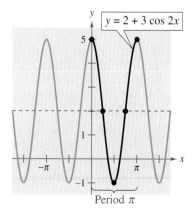

$y = 2 + 3\cos 2x$

Period π

FIGURE **1.56**

Example 5   Horizontal Translation

Sketch the graph of

$$y = -3\cos(2\pi x + 4\pi).$$

Solution

The amplitude is 3 and the period is $2\pi/2\pi = 1$. By solving the equations

$$2\pi x + 4\pi = 0$$
$$2\pi x = -4\pi$$
$$x = -2$$

and

$$2\pi x + 4\pi = 2\pi$$
$$2\pi x = -2\pi$$
$$x = -1$$

you see that the interval $[-2, -1]$ corresponds to one cycle of the graph. Dividing this interval into four equal parts produces the key points

Minimum	*Intercept*	*Maximum*	*Intercept*		*Minimum*
$(-2, -3),$	$\left(-\frac{7}{4}, 0\right),$	$\left(-\frac{3}{2}, 3\right),$	$\left(-\frac{5}{4}, 0\right),$	and	$(-1, -3).$

The graph is shown in Figure 1.55.

✓CHECKPOINT   Now try Exercise 47.

The final type of transformation is the *vertical translation* caused by the constant d in the equations

$$y = d + a\sin(bx - c)$$

and

$$y = d + a\cos(bx - c).$$

The shift is d units upward for $d > 0$ and d units downward for $d < 0$. In other words, the graph oscillates about the horizontal line $y = d$ instead of about the x-axis.

Example 6   Vertical Translation

Sketch the graph of

$$y = 2 + 3\cos 2x.$$

Solution

The amplitude is 3 and the period is π. The key points over the interval $[0, \pi]$ are

$$(0, 5), \quad \left(\frac{\pi}{4}, 2\right), \quad \left(\frac{\pi}{2}, -1\right), \quad \left(\frac{3\pi}{4}, 2\right), \quad \text{and} \quad (\pi, 5).$$

The graph is shown in Figure 1.56. Compared with the graph of $f(x) = 3\cos 2x$, the graph of $y - 2 + 3\cos 2x$ is shifted upward two units.

✓CHECKPOINT   Now try Exercise 53.

Mathematical Modeling

Sine and cosine functions can be used to model many real-life situations, including electric currents, musical tones, radio waves, tides, and weather patterns.

Time, t	Depth, y
Midnight	3.4
2 A.M.	8.7
4 A.M.	11.3
6 A.M.	9.1
8 A.M.	3.8
10 A.M.	0.1
Noon	1.2

Example 7   **Finding a Trigonometric Model**

Throughout the day, the depth of water at the end of a dock in Bar Harbor, Maine varies with the tides. The table shows the depths (in feet) at various times during the morning.   (Source: Nautical Software, Inc.)

a. Use a trigonometric function to model the data.

b. Find the depths at 9 A.M. and 3 P.M.

c. A boat needs at least 10 feet of water to moor at the dock. During what times in the afternoon can it safely dock?

Solution

a. Begin by graphing the data, as shown in Figure 1.57. You can use either a sine or cosine model. Suppose you use a cosine model of the form

$$y = a \cos(bt - c) + d.$$

The difference between the maximum height and the minimum height of the graph is twice the amplitude of the function. So, the amplitude is

$$a = \frac{1}{2}[(\text{maximum depth}) - (\text{minimum depth})] = \frac{1}{2}(11.3 - 0.1) = 5.6.$$

The cosine function completes one half of a cycle between the times at which the maximum and minimum depths occur. So, the period is

$$p = 2[(\text{time of min. depth}) - (\text{time of max. depth})] = 2(10 - 4) = 12$$

which implies that $b = 2\pi/p \approx 0.524$. Because high tide occurs 4 hours after midnight, consider the left endpoint to be $c/b = 4$, so $c \approx 2.094$. Moreover, because the average depth is $\frac{1}{2}(11.3 + 0.1) = 5.7$, it follows that $d = 5.7$. So, you can model the depth with the function given by

$$y = 5.6 \cos(0.524t - 2.094) + 5.7.$$

b. The depths at 9 A.M. and 3 P.M. are as follows.

$$y = 5.6 \cos(0.524 \cdot 9 - 2.094) + 5.7$$

$$\approx 0.84 \text{ foot} \qquad\qquad 9 \text{ A.M.}$$

$$y = 5.6 \cos(0.524 \cdot 15 - 2.094) + 5.7$$

$$\approx 10.57 \text{ feet} \qquad\qquad 3 \text{ P.M.}$$

c. To find out when the depth y is at least 10 feet, you can graph the model with the line $y = 10$ using a graphing utility, as shown in Figure 1.58. Using the *intersect* feature, you can determine that the depth is at least 10 feet between 2:42 P.M. ($t \approx 14.7$) and 5:18 P.M. ($t \approx 17.3$).

✓*CHECKPOINT*   Now try Exercise 77.

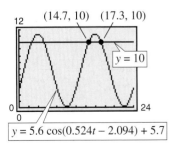

Changing Tides

FIGURE 1.57

(14.7, 10)   (17.3, 10)
$y = 10$
$y = 5.6 \cos(0.524t - 2.094) + 5.7$

FIGURE 1.58

1.5 | Exercises

VOCABULARY CHECK: Fill in the blanks.

1. One period of a sine or cosine function function is called one _____ of the sine curve or cosine curve.

2. The _____ of a sine or cosine curve represents half the distance between the maximum and minimum values of the function.

3. The period of a sine or cosine function is given by _____.

4. For the function given by $y = a \sin(bx - c)$, $\frac{c}{b}$ represents the _____ _____ of the graph of the function.

5. For the function given by $y = d + a \cos(bx - c)$, d represents a _____ _____ of the graph of the function.

PREREQUISITE SKILLS REVIEW: Practice and review algebra skills needed for this section at **www.Eduspace.com.**

In Exercises 1–14, find the period and amplitude.

1. $y = 3 \sin 2x$

2. $y = 2 \cos 3x$

3. $y = \frac{5}{2} \cos \frac{x}{2}$

4. $y = -3 \sin \frac{x}{3}$

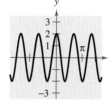

5. $y = \frac{1}{2} \sin \frac{\pi x}{3}$

6. $y = \frac{3}{2} \cos \frac{\pi x}{2}$

7. $y = -2 \sin x$

8. $y = -\cos \frac{2x}{3}$

9. $y = 3 \sin 10x$

10. $y = \frac{1}{3} \sin 8x$

11. $y = \frac{1}{2} \cos \frac{2x}{3}$

12. $y = \frac{5}{2} \cos \frac{x}{4}$

13. $y = \frac{1}{4} \sin 2\pi x$

14. $y = \frac{2}{3} \cos \frac{\pi x}{10}$

In Exercises 15–22, describe the relationship between the graphs of f and g. Consider amplitude, period, and shifts.

15. $f(x) = \sin x$
$g(x) = \sin(x - \pi)$

16. $f(x) = \cos x$
$g(x) = \cos(x + \pi)$

17. $f(x) = \cos 2x$
$g(x) = -\cos 2x$

18. $f(x) = \sin 3x$
$g(x) = \sin(-3x)$

19. $f(x) = \cos x$
$g(x) = \cos 2x$

20. $f(x) = \sin x$
$g(x) = \sin 3x$

21. $f(x) = \sin 2x$
$g(x) = 3 + \sin 2x$

22. $f(x) = \cos 4x$
$g(x) = -2 + \cos 4x$

In Exercises 23–26, describe the relationship between the graphs of f and g. Consider amplitude, period, and shifts.

23.

24.

25.

26.

In Exercises 27–34, graph f and g on the same set of coordinate axes. (Include two full periods.)

27. $f(x) = -2 \sin x$
$g(x) = 4 \sin x$

28. $f(x) = \sin x$
$g(x) = \sin \dfrac{x}{3}$

29. $f(x) = \cos x$
$g(x) = 1 + \cos x$

30. $f(x) = 2 \cos 2x$
$g(x) = -\cos 4x$

31. $f(x) = -\dfrac{1}{2} \sin \dfrac{x}{2}$
$g(x) = 3 - \dfrac{1}{2} \sin \dfrac{x}{2}$

32. $f(x) = 4 \sin \pi x$
$g(x) = 4 \sin \pi x - 3$

33. $f(x) = 2 \cos x$
$g(x) = 2 \cos(x + \pi)$

34. $f(x) = -\cos x$
$g(x) = -\cos(x - \pi)$

In Exercises 35–56, sketch the graph of the function. (Include two full periods.)

35. $y = 3 \sin x$

36. $y = \tfrac{1}{4} \sin x$

37. $y = \tfrac{1}{3} \cos x$

38. $y = 4 \cos x$

39. $y = \cos \dfrac{x}{2}$

40. $y = \sin 4x$

41. $y = \cos 2\pi x$

42. $y = \sin \dfrac{\pi x}{4}$

43. $y = -\sin \dfrac{2\pi x}{3}$

44. $y = -10 \cos \dfrac{\pi x}{6}$

45. $y = \sin\left(x - \dfrac{\pi}{4}\right)$

46. $y = \sin(x - \pi)$

47. $y = 3 \cos(x + \pi)$

48. $y = 4 \cos\left(x + \dfrac{\pi}{4}\right)$

49. $y = 2 - \sin \dfrac{2\pi x}{3}$

50. $y = -3 + 5 \cos \dfrac{\pi t}{12}$

51. $y = 2 + \tfrac{1}{10} \cos 60\pi x$

52. $y = 2 \cos x - 3$

53. $y = 3 \cos(x + \pi) - 3$

54. $y = 4 \cos\left(x + \dfrac{\pi}{4}\right) + 4$

55. $y = \dfrac{2}{3} \cos\left(\dfrac{x}{2} - \dfrac{\pi}{4}\right)$

56. $y = -3 \cos(6x + \pi)$

 In Exercises 57–62, use a graphing utility to graph the function. Include two full periods. Be sure to choose an appropriate viewing window.

57. $y = -2 \sin(4x + \pi)$

58. $y = -4 \sin\left(\dfrac{2}{3}x - \dfrac{\pi}{3}\right)$

59. $y = \cos\left(2\pi x - \dfrac{\pi}{2}\right) + 1$

60. $y = 3 \cos\left(\dfrac{\pi x}{2} + \dfrac{\pi}{2}\right) - 2$

61. $y = -0.1 \sin\left(\dfrac{\pi x}{10} + \pi\right)$

62. $y = \tfrac{1}{100} \sin 120\pi t$

Graphical Reasoning   **In Exercises 63–66, find a and d for the function $f(x) = a \cos x + d$ such that the graph of f matches the figure.**

63.

64.

65.

66.

Graphical Reasoning   **In Exercises 67–70, find a, b, and c for the function $f(x) = a \sin(bx - c)$ such that the graph of f matches the figure.**

67.

68.

69.

70.

 In Exercises 71 and 72, use a graphing utility to graph y_1 and y_2 in the interval $[-2\pi, 2\pi]$. Use the graphs to find real numbers x such that $y_1 = y_2$.

71. $y_1 = \sin x$
$y_2 = -\tfrac{1}{2}$

72. $y_1 = \cos x$
$y_2 = -1$

73. *Respiratory Cycle* For a person at rest, the velocity v (in liters per second) of air flow during a respiratory cycle (the time from the beginning of one breath to the beginning of the next) is given by $v = 0.85 \sin \frac{\pi t}{3}$, where t is the time (in seconds). (Inhalation occurs when $v > 0$, and exhalation occurs when $v < 0$.)

(a) Find the time for one full respiratory cycle.

(b) Find the number of cycles per minute.

(c) Sketch the graph of the velocity function.

74. *Respiratory Cycle* After exercising for a few minutes, a person has a respiratory cycle for which the velocity of air flow is approximated by $v = 1.75 \sin \frac{\pi t}{2}$, where t is the time (in seconds). (Inhalation occurs when $v > 0$, and exhalation occurs when $v < 0$.)

(a) Find the time for one full respiratory cycle.

(b) Find the number of cycles per minute.

(c) Sketch the graph of the velocity function.

75. *Data Analysis: Meteorology* The table shows the maximum daily high temperatures for Tallahassee T and Chicago C (in degrees Fahrenheit) for month t, with $t = 1$ corresponding to January. (Source: National Climatic Data Center)

Month, t	Tallahassee, T	Chicago, C
1	63.8	29.6
2	67.4	34.7
3	74.0	46.1
4	80.0	58.0
5	86.5	69.9
6	90.9	79.2
7	92.0	83.5
8	91.5	81.2
9	88.5	73.9
10	81.2	62.1
11	72.9	47.1
12	65.8	34.4

(a) A model for the temperature in Tallahassee is given by

$$T(t) = 77.90 + 14.10 \cos\left(\frac{\pi t}{6} - 3.67\right).$$

Find a trigonometric model for Chicago.

(b) Use a graphing utility to graph the data points and the model for the temperatures in Tallahassee. How well does the model fit the data?

(c) Use a graphing utility to graph the data points and the model for the temperatures in Chicago. How well does the model fit the data?

(d) Use the models to estimate the average maximum temperature in each city. Which term of the models did you use? Explain.

(e) What is the period of each model? Are the periods what you expected? Explain.

(f) Which city has the greater variability in temperature throughout the year? Which factor of the models determines this variability? Explain.

76. *Health* The function given by $P = 100 - 20 \cos \frac{5\pi t}{3}$ approximates the blood pressure P (in millimeters) of mercury at time t (in seconds) for a person at rest.

(a) Find the period of the function.

(b) Find the number of heartbeats per minute.

77. *Piano Tuning* When tuning a piano, a technician strikes a tuning fork for the A above middle C and sets up a wave motion that can be approximated by $y = 0.001 \sin 880\pi t$, where t is the time (in seconds).

(a) What is the period of the function?

(b) The frequency f is given by $f = 1/p$. What is the frequency of the note?

Model It

78. *Data Analysis: Astronomy* The percent y of the moon's face that is illuminated on day x of the year 2007, where $x = 1$ represents January 1, is shown in the table. (Source: U.S. Naval Observatory)

x	y
3	1.0
11	0.5
19	0.0
26	0.5
32	1.0
40	0.5

(a) Create a scatter plot of the data.

(b) Find a trigonometric model that fits the data.

(c) Add the graph of your model in part (b) to the scatter plot. How well does the model fit the data?

(d) What is the period of the model?

(e) Estimate the moon's percent illumination on March 12, 2007.

79. Fuel Consumption The daily consumption C (in gallons) of diesel fuel on a farm is modeled by

$$C = 30.3 + 21.6 \sin\left(\frac{2\pi t}{365} + 10.9\right)$$

where t is the time (in days), with $t = 1$ corresponding to January 1.

(a) What is the period of the model? Is it what you expected? Explain.

(b) What is the average daily fuel consumption? Which term of the model did you use? Explain.

 (c) Use a graphing utility to graph the model. Use the graph to approximate the time of the year when consumption exceeds 40 gallons per day.

80. Ferris Wheel A Ferris wheel is built such that the height h (in feet) above ground of a seat on the wheel at time t (in seconds) can be modeled by

$$h(t) = 53 + 50 \sin\left(\frac{\pi}{10}t - \frac{\pi}{2}\right).$$

(a) Find the period of the model. What does the period tell you about the ride?

(b) Find the amplitude of the model. What does the amplitude tell you about the ride?

 (c) Use a graphing utility to graph one cycle of the model.

Synthesis

True or False? **In Exercises 81–83, determine whether the statement is true or false. Justify your answer.**

81. The graph of the function given by $f(x) = \sin(x + 2\pi)$ translates the graph of $f(x) = \sin x$ exactly one period to the right so that the two graphs look identical.

82. The function given by $y = \frac{1}{2}\cos 2x$ has an amplitude that is twice that of the function given by $y = \cos x$.

83. The graph of $y = -\cos x$ is a reflection of the graph of $y = \sin(x + \pi/2)$ in the x-axis.

 84. Writing Use a graphing utility to graph the function given by $y = d + a\sin(bx - c)$, for several different values of a, b, c, and d. Write a paragraph describing the changes in the graph corresponding to changes in each constant.

Conjecture **In Exercises 85 and 86, graph f and g on the same set of coordinate axes. Include two full periods. Make a conjecture about the functions.**

85. $f(x) = \sin x$, $g(x) = \cos\left(x - \frac{\pi}{2}\right)$

86. $f(x) = \sin x$, $g(x) = -\cos\left(x + \frac{\pi}{2}\right)$

 87. Exploration Using calculus, it can be shown that the sine and cosine functions can be approximated by the polynomials

$$\sin x \approx x - \frac{x^3}{3!} + \frac{x^5}{5!} \quad \text{and} \quad \cos x \approx 1 - \frac{x^2}{2!} + \frac{x^4}{4!}$$

where x is in radians.

(a) Use a graphing utility to graph the sine function and its polynomial approximation in the same viewing window. How do the graphs compare?

(b) Use a graphing utility to graph the cosine function and its polynomial approximation in the same viewing window. How do the graphs compare?

(c) Study the patterns in the polynomial approximations of the sine and cosine functions and predict the next term in each. Then repeat parts (a) and (b). How did the accuracy of the approximations change when an additional term was added?

 88. Exploration Use the polynomial approximations for the sine and cosine functions in Exercise 87 to approximate the following function values. Compare the results with those given by a calculator. Is the error in the approximation the same in each case? Explain.

(a) $\sin\dfrac{1}{2}$ (b) $\sin 1$ (c) $\sin\dfrac{\pi}{6}$

(d) $\cos(-0.5)$ (e) $\cos 1$ (f) $\cos\dfrac{\pi}{4}$

Skills Review

In Exercises 89–92, identify the rule of algebra illustrated by the statement.

89. $(7 - x)14 = 7 \cdot 14 - x \cdot 14$

90. $3x + 2y = 2y + 3x$

91. $0 + \dfrac{1}{x^2} = \dfrac{1}{x^2}$

92. $(2x^2 + x) + 8 = 2x^2 + (x + 8)$

In Exercises 93–96, find the slope-intercept form of the equation of the line passing through the points. Then sketch the line.

93. $(-3, 5), (2, -1)$ **94.** $(1, 6), (2, 1)$

95. $(-6, -1), (4, 5)$ **96.** $(0, -3), (8, 0)$

97. Make a Decision To work an extended application analyzing the normal daily maximum temperature and normal precipitation in Honolulu, Hawaii, visit this text's website at *college.hmco.com.* *(Data Source: NOAA)*

1.6 Graphs of Other Trigonometric Functions

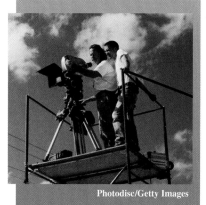

Photodisc/Getty Images

Graph of the Tangent Function

Recall that the tangent function is odd. That is, $\tan(-x) = -\tan x$. Consequently, the graph of $y = \tan x$ is symmetric with respect to the origin. You also know from the identity $\tan x = \sin x / \cos x$ that the tangent is undefined for values at which $\cos x = 0$. Two such values are $x = \pm \pi/2 \approx \pm 1.5708$.

x	$-\dfrac{\pi}{2}$	-1.57	-1.5	$-\dfrac{\pi}{4}$	0	$\dfrac{\pi}{4}$	1.5	1.57	$\dfrac{\pi}{2}$
$\tan x$	Undef.	-1255.8	-14.1	-1	0	1	14.1	1255.8	Undef.

As indicated in the table, $\tan x$ increases without bound as x approaches $\pi/2$ from the left, and decreases without bound as x approaches $-\pi/2$ from the right. So, the graph of $y = \tan x$ has *vertical asymptotes* at $x = \pi/2$ and $x = -\pi/2$, as shown in Figure 1.59. Moreover, because the period of the tangent function is π, vertical asymptotes also occur when $x = \pi/2 + n\pi$, where n is an integer. The domain of the tangent function is the set of all real numbers other than $x = \pi/2 + n\pi$, and the range is the set of all real numbers.

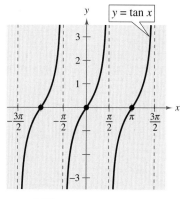

$y = \tan x$

PERIOD: π
DOMAIN: ALL $x \neq \dfrac{\pi}{2} + n\pi$
RANGE: $(-\infty, \infty)$
VERTICAL ASYMPTOTES: $x = \dfrac{\pi}{2} + n\pi$

FIGURE 1.59

Sketching the graph of $y = a \tan(bx - c)$ is similar to sketching the graph of $y = a \sin(bx - c)$ in that you locate key points that identify the intercepts and asymptotes. Two consecutive vertical asymptotes can be found by solving the equations

$$bx - c = -\frac{\pi}{2} \quad \text{and} \quad bx - c = \frac{\pi}{2}.$$

The midpoint between two consecutive vertical asymptotes is an x-intercept of the graph. The period of the function $y = a \tan(bx - c)$ is the distance between two consecutive vertical asymptotes. The amplitude of a tangent function is not defined. After plotting the asymptotes and the x-intercept, plot a few additional points between the two asymptotes and sketch one cycle. Finally, sketch one or two additional cycles to the left and right.

Consider reviewing period, range, and domain for all six trigonometric functions, especially emphasizing the difference between the periods of the tangent and cotangent functions.

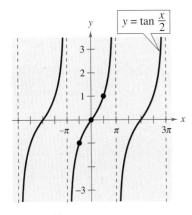

FIGURE 1.60

Example 1   Sketching the Graph of a Tangent Function

Sketch the graph of $y = \tan \dfrac{x}{2}$.

Solution

By solving the equations

$$\frac{x}{2} = -\frac{\pi}{2} \quad \text{and} \quad \frac{x}{2} = \frac{\pi}{2}$$

$$x = -\pi \qquad\qquad x = \pi$$

you can see that two consecutive vertical asymptotes occur at $x = -\pi$ and $x = \pi$. Between these two asymptotes, plot a few points, including the x-intercept, as shown in the table. Three cycles of the graph are shown in Figure 1.60.

x	$-\pi$	$-\dfrac{\pi}{2}$	0	$\dfrac{\pi}{2}$	π
$\tan \dfrac{x}{2}$	Undef.	-1	0	1	Undef.

✓CHECKPOINT   Now try Exercise 7.

Example 2   Sketching the Graph of a Tangent Function

Sketch the graph of $y = -3 \tan 2x$.

Solution

By solving the equations

$$2x = -\frac{\pi}{2} \quad \text{and} \quad 2x = \frac{\pi}{2}$$

$$x = -\frac{\pi}{4} \qquad\qquad x = \frac{\pi}{4}$$

you can see that two consecutive vertical asymptotes occur at $x = -\pi/4$ and $x = \pi/4$. Between these two asymptotes, plot a few points, including the x-intercept, as shown in the table. Three cycles of the graph are shown in Figure 1.61.

x	$-\dfrac{\pi}{4}$	$-\dfrac{\pi}{8}$	0	$\dfrac{\pi}{8}$	$\dfrac{\pi}{4}$
$-3 \tan 2x$	Undef.	3	0	-3	Undef.

FIGURE 1.61

✓CHECKPOINT   Now try Exercise 9.

By comparing the graphs in Examples 1 and 2, you can see that the graph of $y = a \tan(bx - c)$ increases between consecutive vertical asymptotes when $a > 0$, and decreases between consecutive vertical asymptotes when $a < 0$. In other words, the graph for $a < 0$ is a reflection in the x-axis of the graph for $a > 0$.

Graph of the Cotangent Function

The graph of the cotangent function is similar to the graph of the tangent function. It also has a period of π. However, from the identity

$$y = \cot x = \frac{\cos x}{\sin x}$$

you can see that the cotangent function has vertical asymptotes when $\sin x$ is zero, which occurs at $x = n\pi$, where n is an integer. The graph of the cotangent function is shown in Figure 1.62. Note that two consecutive vertical asymptotes of the graph of $y = a \cot(bx - c)$ can be found by solving the equations $bx - c = 0$ and $bx - c = \pi$.

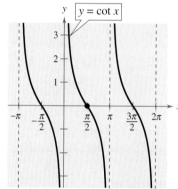

PERIOD: π
DOMAIN: ALL $x \neq n\pi$
RANGE: $(-\infty, \infty)$
VERTICAL ASYMPTOTES: $x = n\pi$

FIGURE 1.62

> **Technology**
>
> Some graphing utilities have difficulty graphing trigonometric functions that have vertical asymptotes. Your graphing utility may connect parts of the graphs of tangent, cotangent, secant, and cosecant functions that are not supposed to be connected. To eliminate this problem, change the mode of the graphing utility to *dot* mode.

Example 3   **Sketching the Graph of a Cotangent Function**

Sketch the graph of $y = 2 \cot \dfrac{x}{3}$.

Solution

By solving the equations

$$\frac{x}{3} = 0 \qquad \text{and} \qquad \frac{x}{3} = \pi$$
$$x = 0 \qquad\qquad\qquad x = 3\pi$$

you can see that two consecutive vertical asymptotes occur at $x = 0$ and $x = 3\pi$. Between these two asymptotes, plot a few points, including the x-intercept, as shown in the table. Three cycles of the graph are shown in Figure 1.63. Note that the period is 3π, the distance between consecutive asymptotes.

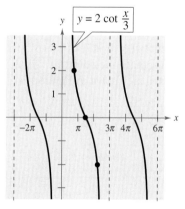

FIGURE 1.63

x	0	$\dfrac{3\pi}{4}$	$\dfrac{3\pi}{2}$	$\dfrac{9\pi}{4}$	3π
$2 \cot \dfrac{x}{3}$	Undef.	2	0	-2	Undef.

✔CHECKPOINT   Now try Exercise 19.

Graphs of the Reciprocal Functions

Using a graphing utility to graph cosecant and secant functions can help reinforce the need to use the reciprocal functions.

The graphs of the two remaining trigonometric functions can be obtained from the graphs of the sine and cosine functions using the reciprocal identities

$$\csc x = \frac{1}{\sin x} \quad \text{and} \quad \sec x = \frac{1}{\cos x}.$$

For instance, at a given value of x, the y-coordinate of sec x is the reciprocal of the y-coordinate of cos x. Of course, when cos $x = 0$, the reciprocal does not exist. Near such values of x, the behavior of the secant function is similar to that of the tangent function. In other words, the graphs of

$$\tan x = \frac{\sin x}{\cos x} \quad \text{and} \quad \sec x = \frac{1}{\cos x}$$

have vertical asymptotes at $x = \pi/2 + n\pi$, where n is an integer, and the cosine is zero at these x-values. Similarly,

$$\cot x = \frac{\cos x}{\sin x} \quad \text{and} \quad \csc x = \frac{1}{\sin x}$$

have vertical asymptotes where sin $x = 0$—that is, at $x = n\pi$.

To sketch the graph of a secant or cosecant function, you should first make a sketch of its reciprocal function. For instance, to sketch the graph of $y = \csc x$, first sketch the graph of $y = \sin x$. Then take reciprocals of the y-coordinates to obtain points on the graph of $y = \csc x$. This procedure is used to obtain the graphs shown in Figure 1.64.

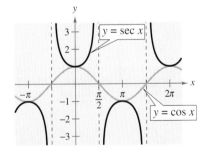

PERIOD: 2π
DOMAIN: ALL $x \neq n\pi$
RANGE: $(-\infty, -1] \cup [1, \infty)$
VERTICAL ASYMPTOTES: $x = n\pi$
SYMMETRY: ORIGIN
FIGURE 1.64

PERIOD: 2π
DOMAIN: ALL $x \neq \frac{\pi}{2} + n\pi$
RANGE: $(-\infty, -1] \cup [1, \infty)$
VERTICAL ASYMPTOTES: $x = \frac{\pi}{2} + n\pi$
SYMMETRY: y-AXIS

In comparing the graphs of the cosecant and secant functions with those of the sine and cosine functions, note that the "hills" and "valleys" are interchanged. For example, a hill (or maximum point) on the sine curve corresponds to a valley (a relative minimum) on the cosecant curve, and a valley (or minimum point) on the sine curve corresponds to a hill (a relative maximum) on the cosecant curve, as shown in Figure 1.65. Additionally, x-intercepts of the sine and cosine functions become vertical asymptotes of the cosecant and secant functions, respectively (see Figure 1.65).

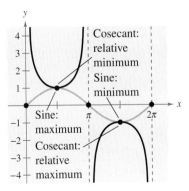

FIGURE 1.65

Example 4 Sketching the Graph of a Cosecant Function

Sketch the graph of $y = 2 \csc\left(x + \dfrac{\pi}{4}\right)$.

Solution

Begin by sketching the graph of

$$y = 2 \sin\left(x + \frac{\pi}{4}\right).$$

For this function, the amplitude is 2 and the period is 2π. By solving the equations

$$x + \frac{\pi}{4} = 0 \qquad \text{and} \qquad x + \frac{\pi}{4} = 2\pi$$

$$x = -\frac{\pi}{4} \qquad\qquad\qquad x = \frac{7\pi}{4}$$

you can see that one cycle of the sine function corresponds to the interval from $x = -\pi/4$ to $x = 7\pi/4$. The graph of this sine function is represented by the gray curve in Figure 1.66. Because the sine function is zero at the midpoint and endpoints of this interval, the corresponding cosecant function

$$y = 2 \csc\left(x + \frac{\pi}{4}\right)$$

$$= 2\left(\frac{1}{\sin[x + (\pi/4)]}\right)$$

has vertical asymptotes at $x = -\pi/4$, $x = 3\pi/4$, $x = 7\pi/4$, etc. The graph of the cosecant function is represented by the black curve in Figure 1.66.

✓**CHECKPOINT** Now try Exercise 25.

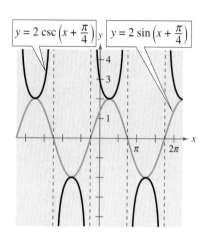

$y = 2 \csc\left(x + \dfrac{\pi}{4}\right)$ $y = 2 \sin\left(x + \dfrac{\pi}{4}\right)$

FIGURE **1.66**

Example 5 Sketching the Graph of a Secant Function

Sketch the graph of $y = \sec 2x$.

Solution

Begin by sketching the graph of $y = \cos 2x$, as indicated by the gray curve in Figure 1.67. Then, form the graph of $y = \sec 2x$ as the black curve in the figure. Note that the x-intercepts of $y = \cos 2x$

$$\left(-\frac{\pi}{4}, 0\right), \qquad \left(\frac{\pi}{4}, 0\right), \qquad \left(\frac{3\pi}{4}, 0\right), \ldots$$

correspond to the vertical asymptotes

$$x = -\frac{\pi}{4}, \qquad x = \frac{\pi}{4}, \qquad x = \frac{3\pi}{4}, \ldots$$

of the graph of $y = \sec 2x$. Moreover, notice that the period of $y = \cos 2x$ and $y = \sec 2x$ is π.

✓**CHECKPOINT** Now try Exercise 27.

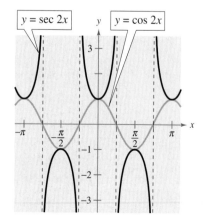

$y = \sec 2x$ $y = \cos 2x$

FIGURE **1.67**

Damped Trigonometric Graphs

A *product* of two functions can be graphed using properties of the individual functions. For instance, consider the function

$$f(x) = x \sin x$$

as the product of the functions $y = x$ and $y = \sin x$. Using properties of absolute value and the fact that $|\sin x| \le 1$, you have $0 \le |x||\sin x| \le |x|$. Consequently,

$$-|x| \le x \sin x \le |x|$$

which means that the graph of $f(x) = x \sin x$ lies between the lines $y = -x$ and $y = x$. Furthermore, because

$$f(x) = x \sin x = \pm x \qquad \text{at} \qquad x = \frac{\pi}{2} + n\pi$$

and

$$f(x) = x \sin x = 0 \qquad \text{at} \qquad x = n\pi$$

the graph of f touches the line $y = -x$ or the line $y = x$ at $x = \pi/2 + n\pi$ and has x-intercepts at $x = n\pi$. A sketch of f is shown in Figure 1.68. In the function $f(x) = x \sin x$, the factor x is called the **damping factor.**

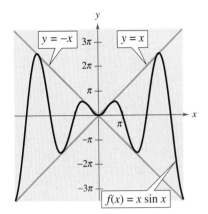

FIGURE 1.68

Example 6   Damped Sine Wave

Sketch the graph of

$$f(x) = x^2 \sin 3x.$$

Solution

Consider $f(x)$ as the product of the two functions

$$y = x^2 \qquad \text{and} \qquad y = \sin 3x$$

each of which has the set of real numbers as its domain. For any real number x, you know that $x^2 \ge 0$ and $|\sin 3x| \le 1$. So, $x^2 |\sin 3x| \le x^2$, which means that

$$-x^2 \le x^2 \sin 3x \le x^2.$$

Furthermore, because

$$f(x) = x^2 \sin 3x = \pm x^2 \qquad \text{at} \qquad x = \frac{\pi}{6} + \frac{n\pi}{3}$$

and

$$f(x) = x^2 \sin 3x = 0 \qquad \text{at} \qquad x = \frac{n\pi}{3}$$

the graph of f touches the curves $y = -x^2$ and $y = x^2$ at $x = \pi/6 + n\pi/3$ and has intercepts at $x = n\pi/3$. A sketch is shown in Figure 1.69.

FIGURE 1.69

 CHECKPOINT   Now try Exercise 29.

Figure 1.70 summarizes the characteristics of the six basic trigonometric functions.

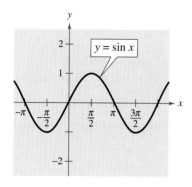

DOMAIN: ALL REALS
RANGE: $[-1, 1]$
PERIOD: 2π

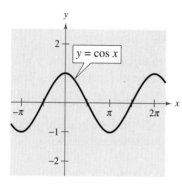

DOMAIN: ALL REALS
RANGE: $[-1, 1]$
PERIOD: 2π

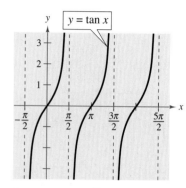

DOMAIN: ALL $x \neq \frac{\pi}{2} + n\pi$
RANGE: $(-\infty, \infty)$
PERIOD: π

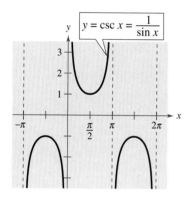

DOMAIN: ALL $x \neq n\pi$
RANGE: $(-\infty, -1] \cup [1, \infty)$
PERIOD: 2π

FIGURE 1.70

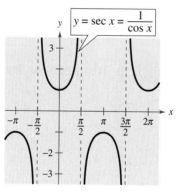

DOMAIN: ALL $x \neq \frac{\pi}{2} + n\pi$
RANGE: $(-\infty, -1] \cup [1, \infty)$
PERIOD: 2π

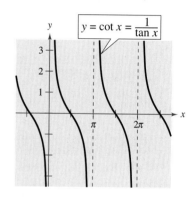

DOMAIN: ALL $x \neq n\pi$
RANGE: $(-\infty, \infty)$
PERIOD: π

𝒲RITING ABOUT 𝓜ATHEMATICS

Combining Trigonometric Functions Recall from Section P.9 that functions can be combined arithmetically. This also applies to trigonometric functions. For each of the functions

$$h(x) = x + \sin x \quad \text{and} \quad h(x) = \cos x - \sin 3x$$

(a) identify two simpler functions f and g that comprise the combination, (b) use a table to show how to obtain the numerical values of $h(x)$ from the numerical values of $f(x)$ and $g(x)$, and (c) use graphs of f and g to show how h may be formed.

Can you find functions

$$f(x) = d + a\sin(bx + c) \quad \text{and} \quad g(x) = d + a\cos(bx + c)$$

such that $f(x) + g(x) = 0$ for all x?

1.6 | Exercises

VOCABULARY CHECK: Fill in the blanks.

1. The graphs of the tangent, cotangent, secant, and cosecant functions all have _____ asymptotes.

2. To sketch the graph of a secant or cosecant function, first make a sketch of its corresponding _____ function.

3. For the functions given by $f(x) = g(x) \cdot \sin x$, $g(x)$ is called the _____ factor of the function $f(x)$.

4. The period of $y = \tan x$ is _____.

5. The domain of $y = \cot x$ is all real numbers such that _____.

6. The range of $y = \sec x$ is _____.

7. The period of $y = \csc x$ is _____.

PREREQUISITE SKILLS REVIEW: Practice and review algebra skills needed for this section at **www.Eduspace.com.**

In Exercises 1–6, match the function with its graph. State the period of the function. [The graphs are labeled (a), (b), (c), (d), (e), and (f).]

(a)

(b)

(c)

(d)

(e)

(f)

1. $y = \sec 2x$

2. $y = \tan \dfrac{x}{2}$

3. $y = \dfrac{1}{2} \cot \pi x$

4. $y = -\csc x$

5. $y = \dfrac{1}{2} \sec \dfrac{\pi x}{2}$

6. $y = -2 \sec \dfrac{\pi x}{2}$

In Exercises 7–30, sketch the graph of the function. Include two full periods.

7. $y = \dfrac{1}{3} \tan x$

8. $y = \dfrac{1}{4} \tan x$

9. $y = \tan 3x$

10. $y = -3 \tan \pi x$

11. $y = -\dfrac{1}{2} \sec x$

12. $y = \dfrac{1}{4} \sec x$

13. $y = \csc \pi x$

14. $y = 3 \csc 4x$

15. $y = \sec \pi x - 1$

16. $y = -2 \sec 4x + 2$

17. $y = \csc \dfrac{x}{2}$

18. $y = \csc \dfrac{x}{3}$

19. $y = \cot \dfrac{x}{2}$

20. $y = 3 \cot \dfrac{\pi x}{2}$

21. $y = \dfrac{1}{2} \sec 2x$

22. $y = -\dfrac{1}{2} \tan x$

23. $y = \tan \dfrac{\pi x}{4}$

24. $y = \tan(x + \pi)$

25. $y = \csc(\pi - x)$

26. $y = \csc(2x - \pi)$

27. $y = 2 \sec(x + \pi)$

28. $y = -\sec \pi x + 1$

29. $y = \dfrac{1}{4} \csc\left(x + \dfrac{\pi}{4}\right)$

30. $y = 2 \cot\left(x + \dfrac{\pi}{2}\right)$

In Exercises 31–40, use a graphing utility to graph the function. Include two full periods.

31. $y = \tan \dfrac{x}{3}$

32. $y = -\tan 2x$

33. $y = -2 \sec 4x$

34. $y = \sec \pi x$

35. $y = \tan\left(x - \dfrac{\pi}{4}\right)$

36. $y = \dfrac{1}{4} \cot\left(x - \dfrac{\pi}{2}\right)$

37. $y = -\csc(4x - \pi)$

38. $y = 2 \sec(2x - \pi)$

39. $y = 0.1 \tan\left(\dfrac{\pi x}{4} + \dfrac{\pi}{4}\right)$

40. $y = \dfrac{1}{3} \sec\left(\dfrac{\pi x}{2} + \dfrac{\pi}{2}\right)$

In Exercises 41–48, use a graph to solve the equation on the interval $[-2\pi, 2\pi]$.

41. $\tan x = 1$

42. $\tan x = \sqrt{3}$

43. $\cot x = -\dfrac{\sqrt{3}}{3}$

44. $\cot x = 1$

45. $\sec x = -2$

46. $\sec x = 2$

47. $\csc x = \sqrt{2}$

48. $\csc x = -\dfrac{2\sqrt{3}}{3}$

In Exercises 49 and 50, use the graph of the function to determine whether the function is even, odd, or neither.

49. $f(x) = \sec x$

50. $f(x) = \tan x$

51. *Graphical Reasoning* Consider the functions given by

$$f(x) = 2 \sin x \quad \text{and} \quad g(x) = \frac{1}{2} \csc x$$

on the interval $(0, \pi)$.

(a) Graph f and g in the same coordinate plane.

(b) Approximate the interval in which $f > g$.

(c) Describe the behavior of each of the functions as x approaches π. How is the behavior of g related to the behavior of f as x approaches π?

 52. *Graphical Reasoning* Consider the functions given by

$$f(x) = \tan \frac{\pi x}{2} \quad \text{and} \quad g(x) = \frac{1}{2} \sec \frac{\pi x}{2}$$

on the interval $(-1, 1)$.

(a) Use a graphing utility to graph f and g in the same viewing window.

(b) Approximate the interval in which $f < g$.

(c) Approximate the interval in which $2f < 2g$. How does the result compare with that of part (b)? Explain.

 In Exercises 53–56, use a graphing utility to graph the two equations in the same viewing window. Use the graphs to determine whether the expressions are equivalent. Verify the results algebraically.

53. $y_1 = \sin x \csc x, \quad y_2 = 1$

54. $y_1 = \sin x \sec x, \quad y_2 = \tan x$

55. $y_1 = \dfrac{\cos x}{\sin x}, \quad y_2 = \cot x$

56. $y_1 = \sec^2 x - 1, \quad y_2 = \tan^2 x$

In Exercises 57–60, match the function with its graph. Describe the behavior of the function as x approaches zero. [The graphs are labeled (a), (b), (c), and (d).]

(a)

(b)

(c)

(d)

57. $f(x) = |x \cos x|$

58. $f(x) = x \sin x$

59. $g(x) = |x| \sin x$

60. $g(x) = |x| \cos x$

Conjecture In Exercises 61–64, graph the functions f and g. Use the graphs to make a conjecture about the relationship between the functions.

61. $f(x) = \sin x + \cos\left(x + \dfrac{\pi}{2}\right), \quad g(x) = 0$

62. $f(x) = \sin x - \cos\left(x + \dfrac{\pi}{2}\right), \quad g(x) = 2 \sin x$

63. $f(x) = \sin^2 x, \quad g(x) = \frac{1}{2}(1 - \cos 2x)$

64. $f(x) = \cos^2 \dfrac{\pi x}{2}, \quad g(x) = \dfrac{1}{2}(1 + \cos \pi x)$

 In Exercises 65–68, use a graphing utility to graph the function and the damping factor of the function in the same viewing window. Describe the behavior of the function as x increases without bound.

65. $g(x) = x \cos \pi x$

66. $f(x) = x^2 \cos x$

67. $f(x) = x^3 \sin x$

68. $h(x) = x^3 \cos x$

Exploration In Exercises 69–74, use a graphing utility to graph the function. Describe the behavior of the function as x approaches zero.

69. $y = \dfrac{6}{x} + \cos x, \quad x > 0$

70. $y = \dfrac{4}{x} + \sin 2x, \quad x > 0$

71. $g(x) = \dfrac{\sin x}{x}$ **72.** $f(x) = \dfrac{1 - \cos x}{x}$

73. $f(x) = \sin \dfrac{1}{x}$ **74.** $h(x) = x \sin \dfrac{1}{x}$

75. Distance   A plane flying at an altitude of 7 miles above a radar antenna will pass directly over the radar antenna (see figure). Let d be the ground distance from the antenna to the point directly under the plane and let x be the angle of elevation to the plane from the antenna. (d is positive as the plane approaches the antenna.) Write d as a function of x and graph the function over the interval $0 < x < \pi$.

7 mi

Not drawn to scale

76. Television Coverage   A television camera is on a reviewing platform 27 meters from the street on which a parade will be passing from left to right (see figure). Write the distance d from the camera to a particular unit in the parade as a function of the angle x, and graph the function over the interval $-\pi/2 < x < \pi/2$. (Consider x as negative when a unit in the parade approaches from the left.)

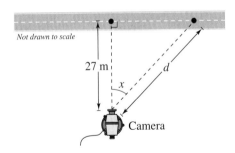

Not drawn to scale

27 m

d

x

Camera

Model It

77. Predator-Prey Model   The population C of coyotes (a predator) at time t (in months) in a region is estimated to be

$$C = 5000 + 2000 \sin \frac{\pi t}{12}$$

and the population R of rabbits (its prey) is estimated to be

---

Model It (continued)

$$R = 25{,}000 + 15{,}000 \cos \frac{\pi t}{12}.$$

(a) Use a graphing utility to graph both models in the same viewing window. Use the window setting $0 \le t \le 100$.

(b) Use the graphs of the models in part (a) to explain the oscillations in the size of each population.

(c) The cycles of each population follow a periodic pattern. Find the period of each model and describe several factors that could be contributing to the cyclical patterns.

78. Sales   The projected monthly sales S (in thousands of units) of lawn mowers (a seasonal product) are modeled by $S = 74 + 3t - 40 \cos(\pi t/6)$, where t is the time (in months), with $t = 1$ corresponding to January. Graph the sales function over 1 year.

79. Meterology   The normal monthly high temperatures H (in degrees Fahrenheit) for Erie, Pennsylvania are approximated by

$$H(t) = 54.33 - 20.38 \cos \frac{\pi t}{6} - 15.69 \sin \frac{\pi t}{6}$$

and the normal monthly low temperatures L are approximated by

$$L(t) = 39.36 - 15.70 \cos \frac{\pi t}{6} - 14.16 \sin \frac{\pi t}{6}$$

where t is the time (in months), with $t = 1$ corresponding to January (see figure). (Source: National Oceanic and Atmospheric Administration)

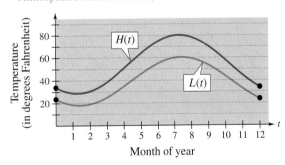

(a) What is the period of each function?

(b) During what part of the year is the difference between the normal high and normal low temperatures greatest? When is it smallest?

(c) The sun is northernmost in the sky around June 21, but the graph shows the warmest temperatures at a later date. Approximate the lag time of the temperatures relative to the position of the sun.

80. *Harmonic Motion* An object weighing W pounds is suspended from the ceiling by a steel spring (see figure). The weight is pulled downward (positive direction) from its equilibrium position and released. The resulting motion of the weight is described by the function

$$y = \frac{4}{t} \cos 4t, \quad t > 0$$

where y is the distance (in feet) and t is the time (in seconds).

Equilibrium

 (a) Use a graphing utility to graph the function.

(b) Describe the behavior of the displacement function for increasing values of time t.

Synthesis

True or False? In Exercises 81 and 82, determine whether the statement is true or false. Justify your answer.

81. The graph of $y = \csc x$ can be obtained on a calculator by graphing the reciprocal of $y = \sin x$.

82. The graph of $y = \sec x$ can be obtained on a calculator by graphing a translation of the reciprocal of $y = \sin x$.

83. *Writing* Describe the behavior of $f(x) = \tan x$ as x approaches $\pi/2$ from the left and from the right.

84. *Writing* Describe the behavior of $f(x) = \csc x$ as x approaches π from the left and from the right.

85. *Exploration* Consider the function given by

$$f(x) = x - \cos x.$$

 (a) Use a graphing utility to graph the function and verify that there exists a zero between 0 and 1. Use the graph to approximate the zero.

(b) Starting with $x_0 = 1$, generate a sequence $x_1, x_2, x_3, \ldots$, where $x_n = \cos(x_{n-1})$. For example,

$$x_0 = 1$$
$$x_1 = \cos(x_0)$$
$$x_2 = \cos(x_1)$$
$$x_3 = \cos(x_2)$$
$$\vdots$$

What value does the sequence approach?

 86. *Approximation* Using calculus, it can be shown that the tangent function can be approximated by the polynomial

$$\tan x \approx x + \frac{2x^3}{3!} + \frac{16x^5}{5!}$$

where x is in radians. Use a graphing utility to graph the tangent function and its polynomial approximation in the same viewing window. How do the graphs compare?

 87. *Approximation* Using calculus, it can be shown that the secant function can be approximated by the polynomial

$$\sec x \approx 1 + \frac{x^2}{2!} + \frac{5x^4}{4!}$$

where x is in radians. Use a graphing utility to graph the secant function and its polynomial approximation in the same viewing window. How do the graphs compare?

88. *Pattern Recognition*

 (a) Use a graphing utility to graph each function.

$$y_1 = \frac{4}{\pi}\left(\sin \pi x + \frac{1}{3}\sin 3\pi x\right)$$
$$y_2 = \frac{4}{\pi}\left(\sin \pi x + \frac{1}{3}\sin 3\pi x + \frac{1}{5}\sin 5\pi x\right)$$

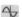 (b) Identify the pattern started in part (a) and find a function y_3 that continues the pattern one more term. Use a graphing utility to graph y_3.

(c) The graphs in parts (a) and (b) approximate the periodic function in the figure. Find a function y_4 that is a better approximation.

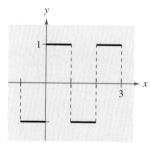

Skills Review

In Exercises 89–96, solve the equation by any convenient method.

89. $x^2 = 64$

90. $(x - 5)^2 = 8$

91. $4x^2 - 12x + 9 = 0$

92. $9x^2 + 12x + 3 = 0$

93. $x^2 - 6x + 4 = 0$

94. $2x^2 - 4x - 6 = 0$

95. $50 + 5x = 3x^2$

96. $2x^2 + 4x - 9 = 2(x - 1)^2$

1.7   Inverse Trigonometric Functions

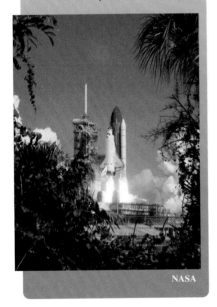

NASA

Inverse Sine Function

Recall from Section P.10 that, for a function to have an inverse function, it must be one-to-one—that is, it must pass the Horizontal Line Test. From Figure 1.71, you can see that $y = \sin x$ does not pass the test because different values of x yield the same y-value.

sin x has an inverse function
on this interval.

FIGURE 1.71

However, if you restrict the domain to the interval $-\pi/2 \le x \le \pi/2$ (corresponding to the black portion of the graph in Figure 1.71), the following properties hold.

1. On the interval $[-\pi/2, \pi/2]$, the function $y = \sin x$ is increasing.
2. On the interval $[-\pi/2, \pi/2]$, $y = \sin x$ takes on its full range of values, $-1 \le \sin x \le 1$.
3. On the interval $[-\pi/2, \pi/2]$, $y = \sin x$ is one-to-one.

So, on the restricted domain $-\pi/2 \le x \le \pi/2$, $y = \sin x$ has a unique inverse function called the **inverse sine function.** It is denoted by

$$y = \arcsin x \qquad \text{or} \qquad y = \sin^{-1} x.$$

The notation $\sin^{-1} x$ is consistent with the inverse function notation $f^{-1}(x)$. The arcsin x notation (read as "the arcsine of x") comes from the association of a central angle with its intercepted *arc length* on a unit circle. So, arcsin x means the angle (or arc) whose sine is x. Both notations, arcsin x and $\sin^{-1} x$, are commonly used in mathematics, so remember that $\sin^{-1} x$ denotes the *inverse* sine function rather than $1/\sin x$. The values of arcsin x lie in the interval $-\pi/2 \le \arcsin x \le \pi/2$. The graph of $y = \arcsin x$ is shown in Example 2.

> ### Definition of Inverse Sine Function
>
> The **inverse sine function** is defined by
>
> $$y = \arcsin x \qquad \text{if and only if} \qquad \sin y = x$$
>
> where $-1 \le x \le 1$ and $-\pi/2 \le y \le \pi/2$. The domain of $y = \arcsin x$ is $[-1, 1]$, and the range is $[-\pi/2, \pi/2]$.

STUDY TIP

As with the trigonometric functions, much of the work with the inverse trigonometric functions can be done by *exact* calculations rather than by calculator approximations. Exact calculations help to increase your understanding of the inverse functions by relating them to the right triangle definitions of the trigonometric functions.

Example 1   Evaluating the Inverse Sine Function

If possible, find the exact value.

a. $\arcsin\left(-\dfrac{1}{2}\right)$ **b.** $\sin^{-1}\dfrac{\sqrt{3}}{2}$ **c.** $\sin^{-1}2$

Solution

a. Because $\sin\left(-\dfrac{\pi}{6}\right) = -\dfrac{1}{2}$ for $-\dfrac{\pi}{2} \le y \le \dfrac{\pi}{2}$, it follows that

$$\arcsin\left(-\dfrac{1}{2}\right) = -\dfrac{\pi}{6}.$$ Angle whose sine is $-\tfrac{1}{2}$

b. Because $\sin\dfrac{\pi}{3} = \dfrac{\sqrt{3}}{2}$ for $-\dfrac{\pi}{2} \le y \le \dfrac{\pi}{2}$, it follows that

$$\sin^{-1}\dfrac{\sqrt{3}}{2} = \dfrac{\pi}{3}.$$ Angle whose sine is $\sqrt{3}/2$

c. It is not possible to evaluate $y = \sin^{-1}x$ when $x = 2$ because there is no angle whose sine is 2. Remember that the domain of the inverse sine function is $[-1, 1]$.

✓**CHECKPOINT**   Now try Exercise 1.

You may wish to illustrate the reflections of $y = \sin x$ and $y = \arcsin x$ about the line $y = x$. Consider using a graphing utility to do this.

Example 2   Graphing the Arcsine Function

Sketch a graph of

$$y = \arcsin x.$$

Solution

By definition, the equations $y = \arcsin x$ and $\sin y = x$ are equivalent for $-\pi/2 \le y \le \pi/2$. So, their graphs are the same. From the interval $[-\pi/2, \pi/2]$, you can assign values to y in the second equation to make a table of values. Then plot the points and draw a smooth curve through the points.

y	$-\dfrac{\pi}{2}$	$-\dfrac{\pi}{4}$	$-\dfrac{\pi}{6}$	0	$\dfrac{\pi}{6}$	$\dfrac{\pi}{4}$	$\dfrac{\pi}{2}$
$x = \sin y$	-1	$-\dfrac{\sqrt{2}}{2}$	$-\dfrac{1}{2}$	0	$\dfrac{1}{2}$	$\dfrac{\sqrt{2}}{2}$	1

The resulting graph for $y = \arcsin x$ is shown in Figure 1.72. Note that it is the reflection (in the line $y = x$) of the black portion of the graph in Figure 1.71. Be sure you see that Figure 1.72 shows the *entire* graph of the inverse sine function. Remember that the domain of $y = \arcsin x$ is the closed interval $[-1, 1]$ and the range is the closed interval $[-\pi/2, \pi/2]$.

✓**CHECKPOINT**   Now try Exercise 17.

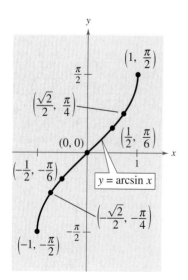

FIGURE 1.72

Other Inverse Trigonometric Functions

The cosine function is decreasing and one-to-one on the interval $0 \le x \le \pi$, as shown in Figure 1.73.

cos x has an inverse function
on this interval.

FIGURE **1.73**

Consequently, on this interval the cosine function has an inverse function—the **inverse cosine function**—denoted by

$$y = \arccos x \quad \text{or} \quad y = \cos^{-1} x.$$

Similarly, you can define an **inverse tangent function** by restricting the domain of $y = \tan x$ to the interval $(-\pi/2, \pi/2)$. The following list summarizes the definitions of the three most common inverse trigonometric functions. The remaining three are defined in Exercises 101–103.

> You may need to point out to your students that the range for each of these functions is different. Students should know these ranges well to ensure that their answers are within the correct range. Referencing the graphs of the inverse trigonometric functions may also be helpful.

Definitions of the Inverse Trigonometric Functions

Function	Domain	Range
$y = \arcsin x$ if and only if $\sin y = x$	$-1 \le x \le 1$	$-\dfrac{\pi}{2} \le y \le \dfrac{\pi}{2}$
$y = \arccos x$ if and only if $\cos y = x$	$-1 \le x \le 1$	$0 \le y \le \pi$
$y = \arctan x$ if and only if $\tan y = x$	$-\infty < x < \infty$	$-\dfrac{\pi}{2} < y < \dfrac{\pi}{2}$

The graphs of these three inverse trigonometric functions are shown in Figure 1.74.

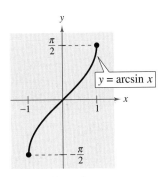

DOMAIN: $[-1, 1]$
RANGE: $\left[-\dfrac{\pi}{2}, \dfrac{\pi}{2}\right]$
FIGURE **1.74**

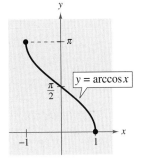

DOMAIN: $[-1, 1]$
RANGE: $[0, \pi]$

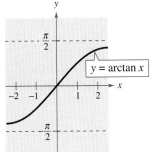

DOMAIN: $(-\infty, \infty)$
RANGE: $\left(-\dfrac{\pi}{2}, \dfrac{\pi}{2}\right)$

Example 3 Evaluating Inverse Trigonometric Functions

Find the exact value.

a. $\arccos \dfrac{\sqrt{2}}{2}$ **b.** $\cos^{-1}(-1)$

c. $\arctan 0$ **d.** $\tan^{-1}(-1)$

Solution

a. Because $\cos(\pi/4) = \sqrt{2}/2$, and $\pi/4$ lies in $[0, \pi]$, it follows that

$$\arccos \frac{\sqrt{2}}{2} = \frac{\pi}{4}.$$ Angle whose cosine is $\sqrt{2}/2$

b. Because $\cos \pi = -1$, and π lies in $[0, \pi]$, it follows that

$$\cos^{-1}(-1) = \pi.$$ Angle whose cosine is -1

c. Because $\tan 0 = 0$, and 0 lies in $(-\pi/2, \pi/2)$, it follows that

$$\arctan 0 = 0.$$ Angle whose tangent is 0

d. Because $\tan(-\pi/4) = -1$, and $-\pi/4$ lies in $(-\pi/2, \pi/2)$, it follows that

$$\tan^{-1}(-1) = -\frac{\pi}{4}.$$ Angle whose tangent is -1

✓CHECKPOINT   Now try Exercise 11.

Example 4 Calculators and Inverse Trigonometric Functions

Use a calculator to approximate the value (if possible).

a. $\arctan(-8.45)$

b. $\sin^{-1} 0.2447$

c. $\arccos 2$

Solution

Function	Mode	Calculator Keystrokes
a. $\arctan(-8.45)$	Radian	[TAN⁻¹] [(] [(-)] 8.45 [)] [ENTER]

From the display, it follows that $\arctan(-8.45) \approx -1.453001$.

b. $\sin^{-1} 0.2447$	Radian	[SIN⁻¹] [(] 0.2447 [)] [ENTER]

From the display, it follows that $\sin^{-1} 0.2447 \approx 0.2472103$.

c. $\arccos 2$	Radian	[COS⁻¹] [(] 2 [)] [ENTER]

In *real number* mode, the calculator should display an *error message* because the domain of the inverse cosine function is $[-1, 1]$.

✓CHECKPOINT   Now try Exercise 25.

In Example 4, if you had set the calculator to *degree* mode, the displays would have been in degrees rather than radians. This convention is peculiar to calculators. By definition, the values of inverse trigonometric functions are *always in radians*.

STUDY TIP

It is important to remember that the domain of the inverse sine function and the inverse cosine function is $[-1, 1]$, as indicated in Example 4(c).

Compositions of Functions

Recall from Section P.10 that for all x in the domains of f and f^{-1}, inverse functions have the properties

$$f(f^{-1}(x)) = x \qquad \text{and} \qquad f^{-1}(f(x)) = x.$$

Inverse Properties of Trigonometric Functions

If $-1 \leq x \leq 1$ and $-\pi/2 \leq y \leq \pi/2$, then

$$\sin(\arcsin x) = x \qquad \text{and} \qquad \arcsin(\sin y) = y.$$

If $-1 \leq x \leq 1$ and $0 \leq y \leq \pi$, then

$$\cos(\arccos x) = x \qquad \text{and} \qquad \arccos(\cos y) = y.$$

If x is a real number and $-\pi/2 < y < \pi/2$, then

$$\tan(\arctan x) = x \qquad \text{and} \qquad \arctan(\tan y) = y.$$

Activities

1. Evaluate

$$\arccos\left(-\frac{\sqrt{3}}{2}\right).$$

Answer: $\frac{5\pi}{6}$

2. Use a calculator to evaluate arctan 3.2.

Answer: 1.268

3. Write an algebraic expression that is equivalent to $\sin(\arctan 3x)$.

Answer: $\dfrac{3x}{\sqrt{1 + 9x^2}}$

Keep in mind that these inverse properties do not apply for arbitrary values of x and y. For instance,

$$\arcsin\left(\sin\frac{3\pi}{2}\right) = \arcsin(-1) = -\frac{\pi}{2} \neq \frac{3\pi}{2}.$$

In other words, the property

$$\arcsin(\sin y) = y$$

is not valid for values of y outside the interval $\left[-\pi/2, \pi/2\right]$.

Example 5   Using Inverse Properties

If possible, find the exact value.

a. $\tan[\arctan(-5)]$ **b.** $\arcsin\left(\sin\dfrac{5\pi}{3}\right)$ **c.** $\cos(\cos^{-1}\pi)$

Solution

a. Because -5 lies in the domain of the arctan function, the inverse property applies, and you have

$$\tan[\arctan(-5)] = -5.$$

b. In this case, $5\pi/3$ does not lie within the range of the arcsine function, $-\pi/2 \leq y \leq \pi/2$. However, $5\pi/3$ is coterminal with

$$\frac{5\pi}{3} - 2\pi = -\frac{\pi}{3}$$

which does lie in the range of the arcsine function, and you have

$$\arcsin\left(\sin\frac{5\pi}{3}\right) = \arcsin\left[\sin\left(-\frac{\pi}{3}\right)\right] = -\frac{\pi}{3}.$$

c. The expression $\cos(\cos^{-1}\pi)$ is not defined because $\cos^{-1}\pi$ is not defined. Remember that the domain of the inverse cosine function is $[-1, 1]$.

✓**CHECKPOINT**   Now try Exercise 43.

Example 6 shows how to use right triangles to find exact values of compositions of inverse functions. Then, Example 7 shows how to use right triangles to convert a trigonometric expression into an algebraic expression. This conversion technique is used frequently in calculus.

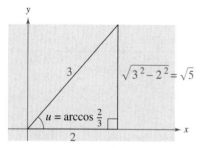

Angle whose cosine is $\frac{2}{3}$
FIGURE **1.75**

Example 6   Evaluating Compositions of Functions

Find the exact value.

a. $\tan\left(\arccos\dfrac{2}{3}\right)$ **b.** $\cos\left[\arcsin\left(-\dfrac{3}{5}\right)\right]$

Solution

a. If you let $u = \arccos\frac{2}{3}$, then $\cos u = \frac{2}{3}$. Because $\cos u$ is positive, u is a *first*-quadrant angle. You can sketch and label angle u as shown in Figure 1.75. Consequently,

$$\tan\left(\arccos\frac{2}{3}\right) = \tan u = \frac{\text{opp}}{\text{adj}} = \frac{\sqrt{5}}{2}.$$

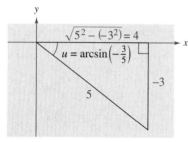

Angle whose sine is $-\frac{3}{5}$
FIGURE **1.76**

b. If you let $u = \arcsin\left(-\frac{3}{5}\right)$, then $\sin u = -\frac{3}{5}$. Because $\sin u$ is negative, u is a *fourth*-quadrant angle. You can sketch and label angle u as shown in Figure 1.76. Consequently,

$$\cos\left[\arcsin\left(-\frac{3}{5}\right)\right] = \cos u = \frac{\text{adj}}{\text{hyp}} = \frac{4}{5}.$$

✓CHECKPOINT   Now try Exercise 51.

Example 7   Some Problems from Calculus

Write each of the following as an algebraic expression in x.

a. $\sin(\arccos 3x), \quad 0 \le x \le \dfrac{1}{3}$ **b.** $\cot(\arccos 3x), \quad 0 \le x < \dfrac{1}{3}$

Solution

If you let $u = \arccos 3x$, then $\cos u = 3x$, where $-1 \le 3x \le 1$. Because

$$\cos u = \frac{\text{adj}}{\text{hyp}} = \frac{3x}{1}$$

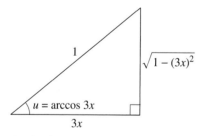

Angle whose cosine is $3x$
FIGURE **1.77**

you can sketch a right triangle with acute angle u, as shown in Figure 1.77. From this triangle, you can easily convert each expression to algebraic form.

a. $\sin(\arccos 3x) = \sin u = \dfrac{\text{opp}}{\text{hyp}} = \sqrt{1 - 9x^2}, \quad 0 \le x \le \dfrac{1}{3}$

b. $\cot(\arccos 3x) = \cot u = \dfrac{\text{adj}}{\text{opp}} = \dfrac{3x}{\sqrt{1 - 9x^2}}, \quad 0 \le x < \dfrac{1}{3}$

✓CHECKPOINT   Now try Exercise 59.

You may want to review with students how to rationalize the denominator of a fractional expression.

In Example 7, similar arguments can be made for x-values lying in the interval $\left[-\frac{1}{3}, 0\right]$.

1.7 │ Exercises

VOCABULARY CHECK: Fill in the blanks.

Function	Alternative Notation	Domain	Range
1. $y = \arcsin x$	_____	_____	$-\dfrac{\pi}{2} \le y \le \dfrac{\pi}{2}$
2. _____	$y = \cos^{-1} x$	$-1 \le x \le 1$	_____
3. $y = \arctan x$	_____	_____	_____

PREREQUISITE SKILLS REVIEW: Practice and review algebra skills needed for this section at **www.Eduspace.com**.

In Exercises 1–16, evaluate the expression without using a calculator.

1. $\arcsin \frac{1}{2}$ **2.** $\arcsin 0$

3. $\arccos \frac{1}{2}$ **4.** $\arccos 0$

5. $\arctan \dfrac{\sqrt{3}}{3}$ **6.** $\arctan(-1)$

7. $\cos^{-1}\left(-\dfrac{\sqrt{3}}{2}\right)$ **8.** $\sin^{-1}\left(-\dfrac{\sqrt{2}}{2}\right)$

9. $\arctan\left(-\sqrt{3}\right)$ **10.** $\arctan \sqrt{3}$

11. $\arccos\left(-\dfrac{1}{2}\right)$ **12.** $\arcsin \dfrac{\sqrt{2}}{2}$

13. $\sin^{-1} \dfrac{\sqrt{3}}{2}$ **14.** $\tan^{-1}\left(-\dfrac{\sqrt{3}}{3}\right)$

15. $\tan^{-1} 0$ **16.** $\cos^{-1} 1$

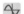 In Exercises 17 and 18, use a graphing utility to graph f, g, and $y = x$ in the same viewing window to verify geometrically that g is the inverse function of f. (Be sure to restrict the domain of f properly.)

17. $f(x) = \sin x$, $g(x) = \arcsin x$
18. $f(x) = \tan x$, $g(x) = \arctan x$

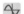 In Exercises 19–34, use a calculator to evaluate the expression. Round your result to two decimal places.

19. $\arccos 0.28$ **20.** $\arcsin 0.45$

21. $\arcsin(-0.75)$ **22.** $\arccos(-0.7)$

23. $\arctan(-3)$ **24.** $\arctan 15$

25. $\sin^{-1} 0.31$ **26.** $\cos^{-1} 0.26$

27. $\arccos(-0.41)$ **28.** $\arcsin(-0.125)$

29. $\arctan 0.92$ **30.** $\arctan 2.8$

31. $\arcsin \frac{3}{4}$ **32.** $\arccos\left(-\frac{1}{3}\right)$

33. $\tan^{-1} \frac{7}{2}$ **34.** $\tan^{-1}\left(-\frac{95}{7}\right)$

In Exercises 35 and 36, determine the missing coordinates of the points on the graph of the function.

35.

36.

In Exercises 37–42, use an inverse trigonometric function to write θ as a function of x.

37.

38.

39.

40.

41.

42.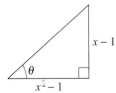

In Exercises 43–48, use the properties of inverse trigonometric functions to evaluate the expression.

43. $\sin(\arcsin 0.3)$ **44.** $\tan(\arctan 25)$

45. $\cos[\arccos(-0.1)]$ **46.** $\sin[\arcsin(-0.2)]$

47. $\arcsin(\sin 3\pi)$ **48.** $\arccos\left(\cos \dfrac{7\pi}{2}\right)$

In Exercises 49–58, find the exact value of the expression. (*Hint:* Sketch a right triangle.)

49. $\sin\left(\arctan \frac{3}{4}\right)$

50. $\sec\left(\arcsin \frac{4}{5}\right)$

51. $\cos(\tan^{-1} 2)$

52. $\sin\left(\cos^{-1} \frac{\sqrt{5}}{5}\right)$

53. $\cos\left(\arcsin \frac{5}{13}\right)$

54. $\csc\left[\arctan\left(-\frac{5}{12}\right)\right]$

55. $\sec\left[\arctan\left(-\frac{3}{5}\right)\right]$

56. $\tan\left[\arcsin\left(-\frac{3}{4}\right)\right]$

57. $\sin\left[\arccos\left(-\frac{2}{3}\right)\right]$

58. $\cot\left(\arctan \frac{5}{8}\right)$

In Exercises 59–68, write an algebraic expression that is equivalent to the expression. (*Hint:* Sketch a right triangle, as demonstrated in Example 7.)

59. $\cot(\arctan x)$

60. $\sin(\arctan x)$

61. $\cos(\arcsin 2x)$

62. $\sec(\arctan 3x)$

63. $\sin(\arccos x)$

64. $\sec[\arcsin(x - 1)]$

65. $\tan\left(\arccos \frac{x}{3}\right)$

66. $\cot\left(\arctan \frac{1}{x}\right)$

67. $\csc\left(\arctan \frac{x}{\sqrt{2}}\right)$

68. $\cos\left(\arcsin \frac{x - h}{r}\right)$

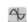 In Exercises 69 and 70, use a graphing utility to graph f and g in the same viewing window to verify that the two functions are equal. Explain why they are equal. Identify any asymptotes of the graphs.

69. $f(x) = \sin(\arctan 2x), \quad g(x) = \frac{2x}{\sqrt{1 + 4x^2}}$

70. $f(x) = \tan\left(\arccos \frac{x}{2}\right), \quad g(x) = \frac{\sqrt{4 - x^2}}{x}$

In Exercises 71–74, fill in the blank.

71. $\arctan \frac{9}{x} = \arcsin(\quad), \quad x \neq 0$

72. $\arcsin \frac{\sqrt{36 - x^2}}{6} = \arccos(\quad), \quad 0 \leq x \leq 6$

73. $\arccos \frac{3}{\sqrt{x^2 - 2x + 10}} = \arcsin(\quad)$

74. $\arccos \frac{x - 2}{2} = \arctan(\quad), \quad |x - 2| \leq 2$

In Exercises 75 and 76, sketch a graph of the function and compare the graph of g with the graph of $f(x) = \arcsin x$.

75. $g(x) = \arcsin(x - 1)$

76. $g(x) = \arcsin \frac{x}{2}$

In Exercises 77–82, sketch a graph of the function.

77. $y = 2 \arccos x$

78. $g(t) = \arccos(t + 2)$

79. $f(x) = \arctan 2x$

80. $f(x) = \frac{\pi}{2} + \arctan x$

81. $h(v) = \tan(\arccos v)$

82. $f(x) = \arccos \frac{x}{4}$

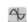 In Exercises 83–88, use a graphing utility to graph the function.

83. $f(x) = 2 \arccos(2x)$

84. $f(x) = \pi \arcsin(4x)$

85. $f(x) = \arctan(2x - 3)$

86. $f(x) = -3 + \arctan(\pi x)$

87. $f(x) = \pi - \sin^{-1}\left(\frac{2}{3}\right)$

88. $f(x) = \frac{\pi}{2} + \cos^{-1}\left(\frac{1}{\pi}\right)$

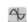 In Exercises 89 and 90, write the function in terms of the sine function by using the identity

$$A \cos \omega t + B \sin \omega t = \sqrt{A^2 + B^2} \sin\left(\omega t + \arctan \frac{A}{B}\right).$$

Use a graphing utility to graph both forms of the function. What does the graph imply?

89. $f(t) = 3 \cos 2t + 3 \sin 2t$

90. $f(t) = 4 \cos \pi t + 3 \sin \pi t$

91. *Docking a Boat*   A boat is pulled in by means of a winch located on a dock 5 feet above the deck of the boat (see figure). Let θ be the angle of elevation from the boat to the winch and let s be the length of the rope from the winch to the boat.

(a) Write θ as a function of s.

(b) Find θ when $s = 40$ feet and $s = 20$ feet.

92. *Photography*   A television camera at ground level is filming the lift-off of a space shuttle at a point 750 meters from the launch pad (see figure). Let θ be the angle of elevation to the shuttle and let s be the height of the shuttle.

(a) Write θ as a function of s.

(b) Find θ when $s = 300$ meters and $s = 1200$ meters.

Model It

93. *Photography*   A photographer is taking a picture of a three-foot-tall painting hung in an art gallery. The camera lens is 1 foot below the lower edge of the painting (see figure). The angle β subtended by the camera lens x feet from the painting is

$$\beta = \arctan \frac{3x}{x^2 + 4}, \quad x > 0.$$

(a) Use a graphing utility to graph β as a function of x.

(b) Move the cursor along the graph to approximate the distance from the picture when β is maximum.

(c) Identify the asymptote of the graph and discuss its meaning in the context of the problem.

94. *Granular Angle of Repose*   Different types of granular substances naturally settle at different angles when stored in cone-shaped piles. This angle θ is called the *angle of repose* (see figure). When rock salt is stored in a cone-shaped pile 11 feet high, the diameter of the pile's base is about 34 feet. (Source: Bulk-Store Structures, Inc.)

(a) Find the angle of repose for rock salt.

(b) How tall is a pile of rock salt that has a base diameter of 40 feet?

95. *Granular Angle of Repose*   When whole corn is stored in a cone-shaped pile 20 feet high, the diameter of the pile's base is about 82 feet.

(a) Find the angle of repose for whole corn.

(b) How tall is a pile of corn that has a base diameter of 100 feet?

96. *Angle of Elevation*   An airplane flies at an altitude of 6 miles toward a point directly over an observer. Consider θ and x as shown in the figure.

(a) Write θ as a function of x.

(b) Find θ when $x = 7$ miles and $x = 1$ mile.

97. *Security Patrol*   A security car with its spotlight on is parked 20 meters from a warehouse. Consider θ and x as shown in the figure.

(a) Write θ as a function of x.

(b) Find θ when $x = 5$ meters and $x = 12$ meters.

Synthesis

True or False?   **In Exercises 98–100, determine whether the statement is true or false. Justify your answer.**

98. $\sin \dfrac{5\pi}{6} = \dfrac{1}{2}$ ⟹ $\arcsin \dfrac{1}{2} = \dfrac{5\pi}{6}$

99. $\tan \dfrac{5\pi}{4} = 1$ ⟹ $\arctan 1 = \dfrac{5\pi}{4}$

100. $\arctan x = \dfrac{\arcsin x}{\arccos x}$

101. Define the inverse cotangent function by restricting the domain of the cotangent function to the interval $(0, \pi)$, and sketch its graph.

102. Define the inverse secant function by restricting the domain of the secant function to the intervals $[0, \pi/2)$ and $(\pi/2, \pi]$, and sketch its graph.

103. Define the inverse cosecant function by restricting the domain of the cosecant function to the intervals $[-\pi/2, 0)$ and $(0, \pi/2]$, and sketch its graph.

104. Use the results of Exercises 101–103 to evaluate each expression without using a calculator.

(a) $\operatorname{arcsec} \sqrt{2}$ (b) $\operatorname{arcsec} 1$

(c) $\operatorname{arccot}\left(-\sqrt{3}\right)$ (d) $\operatorname{arccsc} 2$

105. *Area*   In calculus, it is shown that the area of the region bounded by the graphs of $y = 0$, $y = 1/(x^2 + 1)$, $x = a$, and $x = b$ is given by

$$\text{Area} = \arctan b - \arctan a$$

(see figure). Find the area for the following values of a and b.

(a) $a = 0, b = 1$ (b) $a = -1, b = 1$

(c) $a = 0, b = 3$ (d) $a = -1, b = 3$

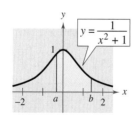

106. *Think About It*   Use a graphing utility to graph the functions

$$f(x) = \sqrt{x} \text{ and } g(x) = 6 \arctan x.$$

For $x > 0$, it appears that $g > f$. Explain why you know that there exists a positive real number a such that $g < f$ for $x > a$. Approximate the number a.

107. *Think About It*   Consider the functions given by

$$f(x) = \sin x \quad \text{and} \quad f^{-1}(x) = \arcsin x.$$

(a) Use a graphing utility to graph the composite functions $f \circ f^{-1}$ and $f^{-1} \circ f$.

(b) Explain why the graphs in part (a) are not the graph of the line $y = x$. Why do the graphs of $f \circ f^{-1}$ and $f^{-1} \circ f$ differ?

108. *Proof*   Prove each identity.

(a) $\arcsin(-x) = -\arcsin x$

(b) $\arctan(-x) = -\arctan x$

(c) $\arctan x + \arctan \dfrac{1}{x} = \dfrac{\pi}{2}, \quad x > 0$

(d) $\arcsin x + \arccos x = \dfrac{\pi}{2}$

(e) $\arcsin x = \arctan \dfrac{x}{\sqrt{1 - x^2}}$

Skills Review

In Exercises 109–112, sketch a right triangle corresponding to the trigonometric function of the acute angle θ. Use the Pythagorean Theorem to determine the third side. Then find the other five trigonometric functions of θ.

109. $\sin \theta = \frac{3}{4}$

110. $\tan \theta = 2$

111. $\cos \theta = \frac{5}{6}$

112. $\sec \theta = 3$

113. *Partnership Costs*   A group of people agree to share equally in the cost of a $250,000 endowment to a college. If they could find two more people to join the group, each person's share of the cost would decrease by $6250. How many people are presently in the group?

114. *Speed*   A boat travels at a speed of 18 miles per hour in still water. It travels 35 miles upstream and then returns to the starting point in a total of 4 hours. Find the speed of the current.

1.8   Applications and Models

What you should learn

- Solve real-life problems involving right triangles.
- Solve real-life problems involving directional bearings.
- Solve real-life problems involving harmonic motion.

Why you should learn it

Right triangles often occur in real-life situations. For instance, in Exercise 62 on page 210, right triangles are used to determine the shortest grain elevator for a grain storage bin on a farm.

Applications Involving Right Triangles

In this section, the three angles of a right triangle are denoted by the letters A, B, and C (where C is the right angle), and the lengths of the sides opposite these angles by the letters a, b, and c (where c is the hypotenuse).

Example 1   **Solving a Right Triangle**

Solve the right triangle shown in Figure 1.78 for all unknown sides and angles.

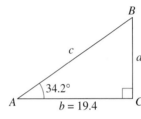

FIGURE 1.78

Solution

Because $C = 90°$, it follows that $A + B = 90°$ and $B = 90° - 34.2° = 55.8°$. To solve for a, use the fact that

$$\tan A = \frac{\text{opp}}{\text{adj}} = \frac{a}{b} \implies a = b \tan A.$$

So, $a = 19.4 \tan 34.2° \approx 13.18$. Similarly, to solve for c, use the fact that

$$\cos A = \frac{\text{adj}}{\text{hyp}} = \frac{b}{c} \implies c = \frac{b}{\cos A}.$$

So, $c = \dfrac{19.4}{\cos 34.2°} \approx 23.46$.

✓CHECKPOINT   Now try Exercise 1.

Example 2   **Finding a Side of a Right Triangle**

A safety regulation states that the maximum angle of elevation for a rescue ladder is 72°. A fire department's longest ladder is 110 feet. What is the maximum safe rescue height?

Solution

A sketch is shown in Figure 1.79. From the equation $\sin A = a/c$, it follows that

$$a = c \sin A = 110 \sin 72° \approx 104.6.$$

So, the maximum safe rescue height is about 104.6 feet above the height of the fire truck.

✓CHECKPOINT   Now try Exercise 15.

FIGURE 1.79

FIGURE **1.80**

| Example 3 | **Finding a Side of a Right Triangle**

At a point 200 feet from the base of a building, the angle of elevation to the *bottom* of a smokestack is 35°, whereas the angle of elevation to the *top* is 53°, as shown in Figure 1.80. Find the height s of the smokestack alone.

Solution

Note from Figure 1.80 that this problem involves two right triangles. For the smaller right triangle, use the fact that

$$\tan 35° = \frac{a}{200}$$

to conclude that the height of the building is

$$a = 200 \tan 35°.$$

For the larger right triangle, use the equation

$$\tan 53° = \frac{a + s}{200}$$

to conclude that $a + s = 200 \tan 53°$. So, the height of the smokestack is

$$s = 200 \tan 53° - a$$

$$= 200 \tan 53° - 200 \tan 35°$$

$$\approx 125.4 \text{ feet.}$$

✓CHECKPOINT   Now try Exercise 19.

FIGURE **1.81**

| Example 4 | **Finding an Acute Angle of a Right Triangle**

A swimming pool is 20 meters long and 12 meters wide. The bottom of the pool is slanted so that the water depth is 1.3 meters at the shallow end and 4 meters at the deep end, as shown in Figure 1.81. Find the angle of depression of the bottom of the pool.

Solution

Using the tangent function, you can see that

$$\tan A = \frac{\text{opp}}{\text{adj}}$$

$$= \frac{2.7}{20}$$

$$= 0.135.$$

So, the angle of depression is

$$A = \arctan 0.135$$

$$\approx 0.13419 \text{ radian}$$

$$\approx 7.69°.$$

✓CHECKPOINT   Now try Exercise 25.

Trigonometry and Bearings

In surveying and navigation, directions are generally given in terms of **bearings.** A bearing measures the acute angle that a path or line of sight makes with a fixed north-south line, as shown in Figure 1.82. For instance, the bearing S 35° E in Figure 1.82 means 35 degrees east of south.

FIGURE 1.82

Example 5   Finding Directions in Terms of Bearings

A ship leaves port at noon and heads due west at 20 knots, or 20 nautical miles (nm) per hour. At 2 P.M. the ship changes course to N 54° W, as shown in Figure 1.83. Find the ship's bearing and distance from the port of departure at 3 P.M.

FIGURE 1.83

Solution

For triangle BCD, you have $B = 90° - 54° = 36°$. The two sides of this triangle can be determined to be

$$b = 20 \sin 36° \qquad \text{and} \qquad d = 20 \cos 36°.$$

For triangle ACD, you can find angle A as follows.

$$\tan A = \frac{b}{d + 40} = \frac{20 \sin 36°}{20 \cos 36° + 40} \approx 0.2092494$$

$$A \approx \arctan 0.2092494 \approx 0.2062732 \text{ radian} \approx 11.82°$$

The angle with the north-south line is $90° - 11.82° = 78.18°$. So, the bearing of the ship is N 78.18° W. Finally, from triangle ACD, you have $\sin A = b/c$, which yields

$$c = \frac{b}{\sin A} = \frac{20 \sin 36°}{\sin 11.82°}$$

$$\approx 57.4 \text{ nautical miles.} \qquad \text{Distance from port}$$

✓CHECKPOINT   Now try Exercise 31.

Harmonic Motion

The periodic nature of the trigonometric functions is useful for describing the motion of a point on an object that vibrates, oscillates, rotates, or is moved by wave motion.

For example, consider a ball that is bobbing up and down on the end of a spring, as shown in Figure 1.84. Suppose that 10 centimeters is the maximum distance the ball moves vertically upward or downward from its equilibrium (at rest) position. Suppose further that the time it takes for the ball to move from its maximum displacement above zero to its maximum displacement below zero and back again is $t = 4$ seconds. Assuming the ideal conditions of perfect elasticity and no friction or air resistance, the ball would continue to move up and down in a uniform and regular manner.

Equilibrium   Maximum negative   Maximum positive
displacement   displacement

FIGURE 1.84

From this spring you can conclude that the period (time for one complete cycle) of the motion is

Period = 4 seconds

its amplitude (maximum displacement from equilibrium) is

Amplitude = 10 centimeters

and its **frequency** (number of cycles per second) is

Frequency = $\dfrac{1}{4}$ cycle per second.

Motion of this nature can be described by a sine or cosine function, and is called **simple harmonic motion.**

Definition of Simple Harmonic Motion

A point that moves on a coordinate line is said to be in **simple harmonic motion** if its distance d from the origin at time t is given by either

$$d = a \sin \omega t \quad \text{or} \quad d = a \cos \omega t$$

where a and ω are real numbers such that $\omega > 0$. The motion has amplitude $|a|$, period $2\pi/\omega$, and frequency $\omega/(2\pi)$.

Example 6   Simple Harmonic Motion

Write the equation for the simple harmonic motion of the ball described in Figure 1.84, where the period is 4 seconds. What is the frequency of this harmonic motion?

Solution

Because the spring is at equilibrium ($d = 0$) when $t = 0$, you use the equation

$$d = a \sin \omega t.$$

Moreover, because the maximum displacement from zero is 10 and the period is 4, you have

$$\text{Amplitude} = |a| = 10$$

$$\text{Period} = \frac{2\pi}{\omega} = 4 \quad \implies \quad \omega = \frac{\pi}{2}.$$

Consequently, the equation of motion is

$$d = 10 \sin \frac{\pi}{2} t.$$

Note that the choice of $a = 10$ or $a = -10$ depends on whether the ball initially moves up or down. The frequency is

$$\text{Frequency} = \frac{\omega}{2\pi}$$

$$= \frac{\pi/2}{2\pi}$$

$$= \frac{1}{4} \text{ cycle per second.}$$

✓**CHECKPOINT**   Now try Exercise 51.

One illustration of the relationship between sine waves and harmonic motion can be seen in the wave motion resulting when a stone is dropped into a calm pool of water. The waves move outward in roughly the shape of sine (or cosine) waves, as shown in Figure 1.85. As an example, suppose you are fishing and your fishing bob is attached so that it does not move horizontally. As the waves move outward from the dropped stone, your fishing bob will move up and down in simple harmonic motion, as shown in Figure 1.86.

FIGURE **1.85**

FIGURE **1.86**

Example 7   **Simple Harmonic Motion**

Given the equation for simple harmonic motion

$$d = 6 \cos \frac{3\pi}{4} t$$

find (a) the maximum displacement, (b) the frequency, (c) the value of d when $t = 4$, and (d) the least positive value of t for which $d = 0$.

Algebraic Solution

The given equation has the form $d = a \cos \omega t$, with $a = 6$ and $\omega = 3\pi/4$.

a. The maximum displacement (from the point of equilibrium) is given by the amplitude. So, the maximum displacement is 6.

b. Frequency $= \dfrac{\omega}{2\pi}$

$$= \frac{3\pi/4}{2\pi} = \frac{3}{8} \text{ cycle per unit of time}$$

c. $d = 6 \cos\left[\dfrac{3\pi}{4}(4) \right]$

$$= 6 \cos 3\pi$$

$$= 6(-1)$$

$$= -6$$

d. To find the least positive value of t for which $d = 0$, solve the equation

$$d = 6 \cos \frac{3\pi}{4} t = 0.$$

First divide each side by 6 to obtain

$$\cos \frac{3\pi}{4} t = 0.$$

This equation is satisfied when

$$\frac{3\pi}{4} t = \frac{\pi}{2}, \frac{3\pi}{2}, \frac{5\pi}{2}, \ldots.$$

Multiply these values by $4/(3\pi)$ to obtain

$$t = \frac{2}{3}, 2, \frac{10}{3}, \ldots.$$

So, the least positive value of t is $t = \frac{2}{3}$.

✓CHECKPOINT   Now try Exercise 55.

Graphical Solution

Use a graphing utility set in *radian* mode to graph

$$y = 6 \cos \frac{3\pi}{4} x.$$

a. Use the *maximum* feature of the graphing utility to estimate that the maximum displacement from the point of equilibrium $y = 0$ is 6, as shown in Figure 1.87.

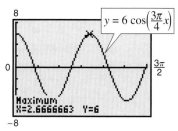

FIGURE 1.87

b. The period is the time for the graph to complete one cycle, which is $x \approx 2.667$. You can estimate the frequency as follows.

$$\text{Frequency} \approx \frac{1}{2.667} \approx 0.375 \text{ cycle per unit of time}$$

c. Use the *trace* feature to estimate that the value of y when $x = 4$ is $y = -6$, as shown in Figure 1.88.

d. Use the *zero* or *root* feature to estimate that the least positive value of x for which $y = 0$ is $x \approx 0.6667$, as shown in Figure 1.89.

FIGURE 1.88

FIGURE 1.89

1.8 | Exercises

VOCABULARY CHECK: Fill in the blanks.

1. An angle that measures from the horizontal upward to an object is called the angle of _____, whereas an angle that measures from the horizontal downward to an object is called the angle of _____.

2. A _____ measures the acute angle a path or line of sight makes with a fixed north-south line.

3. A point that moves on a coordinate line is said to be in simple _____ _____ if its distance d from the origin at time t is given by either $d = a \sin \omega t$ or $d = a \cos \omega t$.

PREREQUISITE SKILLS REVIEW: Practice and review algebra skills needed for this section at **www.Eduspace.com.**

In Exercises 1–10, solve the right triangle shown in the figure. Round your answers to two decimal places.

1. $A = 20°$, $b = 10$	**2.** $B = 54°$, $c = 15$
3. $B = 71°$, $b = 24$	**4.** $A = 8.4°$, $a = 40.5$
5. $a = 6$, $b = 10$	**6.** $a = 25$, $c = 35$
7. $b = 16$, $c = 52$	**8.** $b = 1.32$, $c = 9.45$
9. $A = 12°15'$, $c = 430.5$	
10. $B = 65°12'$, $a = 14.2$	

FIGURE FOR 1–10   FIGURE FOR 11–14

In Exercises 11–14, find the altitude of the isosceles triangle shown in the figure. Round your answers to two decimal places.

11. $\theta = 52°$, $b = 4$ inches
12. $\theta = 18°$, $b = 10$ meters
13. $\theta = 41°$, $b = 46$ inches
14. $\theta = 27°$, $b = 11$ feet

15. **Length**   The sun is 25° above the horizon. Find the length of a shadow cast by a silo that is 50 feet tall (see figure).

16. **Length**   The sun is 20° above the horizon. Find the length of a shadow cast by a building that is 600 feet tall.

17. **Height**   A ladder 20 feet long leans against the side of a house. Find the height from the top of the ladder to the ground if the angle of elevation of the ladder is 80°.

18. **Height**   The length of a shadow of a tree is 125 feet when the angle of elevation of the sun is 33°. Approximate the height of the tree.

19. **Height**   From a point 50 feet in front of a church, the angles of elevation to the base of the steeple and the top of the steeple are 35° and 47° 40′, respectively.

    (a) Draw right triangles that give a visual representation of the problem. Label the known and unknown quantities.

    (b) Use a trigonometric function to write an equation involving the unknown quantity.

    (c) Find the height of the steeple.

20. **Height**   You are standing 100 feet from the base of a platform from which people are bungee jumping. The angle of elevation from your position to the top of the platform from which they jump is 51°. From what height are the people jumping?

21. **Depth**   The sonar of a navy cruiser detects a submarine that is 4000 feet from the cruiser. The angle between the water line and the submarine is 34° (see figure). How deep is the submarine?

22. **Angle of Elevation**   An engineer erects a 75-foot cellular telephone tower. Find the angle of elevation to the top of the tower at a point on level ground 50 feet from its base.

23. *Angle of Elevation*   The height of an outdoor basketball backboard is $12\frac{1}{2}$ feet, and the backboard casts a shadow $17\frac{1}{3}$ feet long.

(a) Draw a right triangle that gives a visual representation of the problem. Label the known and unknown quantities.

(b) Use a trigonometric function to write an equation involving the unknown quantity.

(c) Find the angle of elevation of the sun.

24. *Angle of Depression*   A Global Positioning System satellite orbits 12,500 miles above Earth's surface (see figure). Find the angle of depression from the satellite to the horizon. Assume the radius of Earth is 4000 miles.

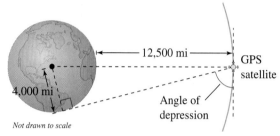

Not drawn to scale

25. *Angle of Depression*   A cellular telephone tower that is 150 feet tall is placed on top of a mountain that is 1200 feet above sea level. What is the angle of depression from the top of the tower to a cell phone user who is 5 horizontal miles away and 400 feet above sea level?

26. *Airplane Ascent*   During takeoff, an airplane's angle of ascent is 18° and its speed is 275 feet per second.

(a) Find the plane's altitude after 1 minute.

(b) How long will it take the plane to climb to an altitude of 10,000 feet?

27. *Mountain Descent*   A sign on a roadway at the top of a mountain indicates that for the next 4 miles the grade is 10.5° (see figure). Find the change in elevation over that distance for a car descending the mountain.

Not drawn to scale

28. *Mountain Descent*   A roadway sign at the top of a mountain indicates that for the next 4 miles the grade is 12%. Find the angle of the grade and the change in elevation over the 4 miles for a car descending the mountain.

29. *Navigation*   An airplane flying at 600 miles per hour has a bearing of 52°. After flying for 1.5 hours, how far north and how far east will the plane have traveled from its point of departure?

30. *Navigation*   A jet leaves Reno, Nevada and is headed toward Miami, Florida at a bearing of 100°. The distance between the two cities is approximately 2472 miles.

(a) How far north and how far west is Reno relative to Miami?

(b) If the jet is to return directly to Reno from Miami, at what bearing should it travel?

31. *Navigation*   A ship leaves port at noon and has a bearing of S 29° W. The ship sails at 20 knots.

(a) How many nautical miles south and how many nautical miles west will the ship have traveled by 6:00 P.M.?

(b) At 6:00 P.M., the ship changes course to due west. Find the ship's bearing and distance from the port of departure at 7:00 P.M.

32. *Navigation*   A privately owned yacht leaves a dock in Myrtle Beach, South Carolina and heads toward Freeport in the Bahamas at a bearing of S 1.4° E. The yacht averages a speed of 20 knots over the 428 nautical-mile trip.

(a) How long will it take the yacht to make the trip?

(b) How far east and south is the yacht after 12 hours?

(c) If a plane leaves Myrtle Beach to fly to Freeport, what bearing should be taken?

33. *Surveying*   A surveyor wants to find the distance across a swamp (see figure). The bearing from A to B is N 32° W. The surveyor walks 50 meters from A, and at the point C the bearing to B is N 68° W. Find (a) the bearing from A to C and (b) the distance from A to B.

34. *Location of a Fire*   Two fire towers are 30 kilometers apart, where tower A is due west of tower B. A fire is spotted from the towers, and the bearings from A and B are E 14° N and W 34° N, respectively (see figure). Find the distance d of the fire from the line segment AB.

Not drawn to scale

35. *Navigation*   A ship is 45 miles east and 30 miles south of port. The captain wants to sail directly to port. What bearing should be taken?

36. *Navigation*   An airplane is 160 miles north and 85 miles east of an airport. The pilot wants to fly directly to the airport. What bearing should be taken?

37. *Distance*   An observer in a lighthouse 350 feet above sea level observes two ships directly offshore. The angles of depression to the ships are 4° and 6.5° (see figure). How far apart are the ships?

38. *Distance*   A passenger in an airplane at an altitude of 10 kilometers sees two towns directly to the east of the plane. The angles of depression to the towns are 28° and 55° (see figure). How far apart are the towns?

39. *Altitude*   A plane is observed approaching your home and you assume that its speed is 550 miles per hour. The angle of elevation of the plane is 16° at one time and 57° one minute later. Approximate the altitude of the plane.

40. *Height*   While traveling across flat land, you notice a mountain directly in front of you. The angle of elevation to the peak is 2.5°. After you drive 17 miles closer to the mountain, the angle of elevation is 9°. Approximate the height of the mountain.

Geometry   **In Exercises 41 and 42, find the angle α between two nonvertical lines L_1 and L_2. The angle α satisfies the equation**

$$\tan \alpha = \left| \frac{m_2 - m_1}{1 + m_2 m_1} \right|$$

where m_1 and m_2 are the slopes of L_1 and L_2, respectively. (Assume that $m_1 m_2 \neq -1$.)

41. L_1: $3x - 2y = 5$
     L_2: $x + y = 1$

42. L_1: $2x - y = 8$
     L_2: $x - 5y = -4$

43. *Geometry*   Determine the angle between the diagonal of a cube and the diagonal of its base, as shown in the figure.

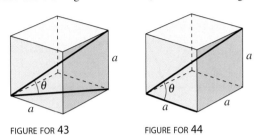

FIGURE FOR **43**   FIGURE FOR **44**

44. *Geometry*   Determine the angle between the diagonal of a cube and its edge, as shown in the figure.

45. *Geometry*   Find the length of the sides of a regular pentagon inscribed in a circle of radius 25 inches.

46. *Geometry*   Find the length of the sides of a regular hexagon inscribed in a circle of radius 25 inches.

47. *Hardware*   Write the distance y across the flat sides of a hexagonal nut as a function of r, as shown in the figure.

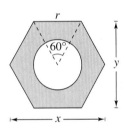

48. *Bolt Holes*   The figure shows a circular piece of sheet metal that has a diameter of 40 centimeters and contains 12 equally spaced bolt holes. Determine the straight-line distance between the centers of consecutive bolt holes.

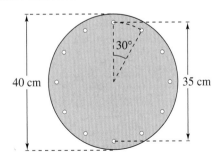

Trusses **In Exercises 49 and 50, find the lengths of all the unknown members of the truss.**

49.

50.

Harmonic Motion **In Exercises 51–54, find a model for simple harmonic motion satisfying the specified conditions.**

Displacement ($t = 0$)	Amplitude	Period
51. 0	4 centimeters	2 seconds
52. 0	3 meters	6 seconds
53. 3 inches	3 inches	1.5 seconds
54. 2 feet	2 feet	10 seconds

Harmonic Motion **In Exercises 55–58, for the simple harmonic motion described by the trigonometric function, find (a) the maximum displacement, (b) the frequency, (c) the value of d when $t = 5$, and (d) the least positive value of t for which $d = 0$. Use a graphing utility to verify your results.**

55. $d = 4 \cos 8\pi t$

56. $d = \frac{1}{2} \cos 20\pi t$

57. $d = \frac{1}{16} \sin 120\pi t$

58. $d = \frac{1}{64} \sin 792\pi t$

59. *Tuning Fork* A point on the end of a tuning fork moves in simple harmonic motion described by $d = a \sin \omega t$. Find ω given that the tuning fork for middle C has a frequency of 264 vibrations per second.

60. *Wave Motion* A buoy oscillates in simple harmonic motion as waves go past. It is noted that the buoy moves a total of 3.5 feet from its low point to its high point (see figure), and that it returns to its high point every 10 seconds. Write an equation that describes the motion of the buoy if its high point is at $t = 0$.

FIGURE FOR **60**

61. *Oscillation of a Spring* A ball that is bobbing up and down on the end of a spring has a maximum displacement of 3 inches. Its motion (in ideal conditions) is modeled by $y = \frac{1}{4} \cos 16t$ ($t > 0$), where y is measured in feet and t is the time in seconds.

(a) Graph the function.

(b) What is the period of the oscillations?

(c) Determine the first time the weight passes the point of equilibrium ($y = 0$).

Model It

62. *Numerical and Graphical Analysis* A two-meter-high fence is 3 meters from the side of a grain storage bin. A grain elevator must reach from ground level outside the fence to the storage bin (see figure). The objective is to determine the shortest elevator that meets the constraints.

(a) Complete four rows of the table.

θ	L_1	L_2	$L_1 + L_2$
0.1	$\dfrac{2}{\sin 0.1}$	$\dfrac{3}{\cos 0.1}$	23.0
0.2	$\dfrac{2}{\sin 0.2}$	$\dfrac{3}{\cos 0.2}$	13.1

Model It (continued)

 (b) Use a graphing utility to generate additional rows of the table. Use the table to estimate the minimum length of the elevator.

(c) Write the length $L_1 + L_2$ as a function of θ.

 (d) Use a graphing utility to graph the function. Use the graph to estimate the minimum length. How does your estimate compare with that of part (b)?

63. *Numerical and Graphical Analysis* The cross section of an irrigation canal is an isosceles trapezoid of which three of the sides are 8 feet long (see figure). The objective is to find the angle θ that maximizes the area of the cross section. [*Hint:* The area of a trapezoid is $(h/2)(b_1 + b_2)$.]

(a) Complete seven additional rows of the table.

Base 1	Base 2	Altitude	Area
8	$8 + 16 \cos 10°$	$8 \sin 10°$	22.1
8	$8 + 16 \cos 20°$	$8 \sin 20°$	42.5

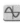 (b) Use a graphing utility to generate additional rows of the table. Use the table to estimate the maximum cross-sectional area.

(c) Write the area A as a function of θ.

 (d) Use a graphing utility to graph the function. Use the graph to estimate the maximum cross-sectional area. How does your estimate compare with that of part (b)?

64. *Data Analysis* The table shows the average sales S (in millions of dollars) of an outerwear manufacturer for each month t, where $t = 1$ represents January.

Time, t	1	2	3	4	5	6
Sales, s	13.46	11.15	8.00	4.85	2.54	1.70

Time, t	7	8	9	10	11	12
Sales, s	2.54	4.85	8.00	11.15	13.46	14.3

(a) Create a scatter plot of the data.

(b) Find a trigonometric model that fits the data. Graph the model with your scatter plot. How well does the model fit the data?

(c) What is the period of the model? Do you think it is reasonable given the context? Explain your reasoning.

(d) Interpret the meaning of the model's amplitude in the context of the problem.

Synthesis

True or False? In Exercises 65 and 66, determine whether the statement is true or false. Justify your answer.

65. The Leaning Tower of Pisa is not vertical, but if you know the exact angle of elevation θ to the 191-foot tower when you stand near it, then you can determine the exact distance to the tower d by using the formula

$$\tan \theta = \frac{191}{d}.$$

66. For the harmonic motion of a ball bobbing up and down on the end of a spring, one period can be described as the length of one coil of the spring.

67. *Writing* Is it true that N 24° E means 24 degrees north of east? Explain.

68. *Writing* Explain the difference between bearings used in nautical navigation and bearings used in air navigation.

Skills Review

In Exercises 69–72, write the slope-intercept form of the equation of the line with the specified characteristics. Then sketch the line.

69. $m = 4$, passes through $(-1, 2)$

70. $m = -\frac{1}{2}$, passes through $\left(\frac{1}{3}, 0\right)$

71. Passes through $(-2, 6)$ and $(3, 2)$

72. Passes through $\left(\frac{1}{4}, -\frac{2}{3}\right)$ and $\left(-\frac{1}{2}, \frac{1}{3}\right)$

1 Chapter Summary

What did you learn?

1   **Review Exercises**

1.1 In Exercises 1 and 2, estimate the angle to the nearest one-half radian.

1.

2.

In Exercises 3–10, (a) sketch the angle in standard position, (b) determine the quadrant in which the angle lies, and (c) determine one positive and one negative coterminal angle.

3. $\dfrac{11\pi}{4}$ **4.** $\dfrac{2\pi}{9}$

5. $-\dfrac{4\pi}{3}$ **6.** $-\dfrac{23\pi}{3}$

7. $70°$ **8.** $280°$

9. $-110°$ **10.** $-405°$

In Exercises 11–14, convert the angle measure from degrees to radians. Round your answer to three decimal places.

11. $480°$ **12.** $-127.5°$

13. $-33°\,45'$ **14.** $196°\,77'$

In Exercises 15–18, convert the angle measure from radians to degrees. Round your answer to three decimal places.

15. $\dfrac{5\pi}{7}$ **16.** $-\dfrac{11\pi}{6}$

17. -3.5 **18.** 5.7

19. *Arc Length*   Find the length of the arc on a circle with a radius of 20 inches intercepted by a central angle of $138°$.

20. *Arc Length*   Find the length of the arc on a circle with a radius of 11 meters intercepted by a central angle of $60°$.

21. *Phonograph*   Compact discs have all but replaced phonograph records. Phonograph records are vinyl discs that rotate on a turntable. A typical record album is 12 inches in diameter and plays at $33\frac{1}{3}$ revolutions per minute.

    (a) What is the angular speed of a record album?

    (b) What is the linear speed of the outer edge of a record album?

22. *Bicycle*   At what speed is a bicyclist traveling when his 27-inch-diameter tires are rotating at an angular speed of 5π radians per second?

23. *Circular Sector*   Find the area of the sector of a circle with a radius of 18 inches and central angle $\theta = 120°$.

24. *Circular Sector*   Find the area of the sector of a circle with a radius of 6.5 millimeters and central angle $\theta = 5\pi/6$.

1.2 In Exercises 25–28, find the point (x, y) on the unit circle that corresponds to the real number t.

25. $t = \dfrac{2\pi}{3}$ **26.** $t = \dfrac{3\pi}{4}$

27. $t = \dfrac{5\pi}{6}$ **28.** $t = -\dfrac{4\pi}{3}$

In Exercises 29–32, evaluate (if possible) the six trigonometric functions of the real number.

29. $t = \dfrac{7\pi}{6}$ **30.** $t = \dfrac{\pi}{4}$

31. $t = -\dfrac{2\pi}{3}$ **32.** $t = 2\pi$

In Exercises 33–36, evaluate the trigonometric function using its period as an aid.

33. $\sin\dfrac{11\pi}{4}$ **34.** $\cos 4\pi$

35. $\sin\left(-\dfrac{17\pi}{6}\right)$ **36.** $\cos\left(-\dfrac{13\pi}{3}\right)$

 In Exercises 37–40, use a calculator to evaluate the trigonometric function. Round your answer to four decimal places.

37. $\tan 33$ **38.** $\csc 10.5$

39. $\sec\dfrac{12\pi}{5}$ **40.** $\sin\left(-\dfrac{\pi}{9}\right)$

1.3 In Exercises 41–44, find the exact values of the six trigonometric functions of the angle θ shown in the figure.

41.

42.

43.

44.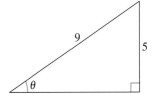

In Exercises 45–48, use the given function value and trigonometric identities (including the cofunction identities) to find the indicated trigonometric functions.

45. $\sin \theta = \frac{1}{3}$ (a) $\csc \theta$ (b) $\cos \theta$
                  (c) $\sec \theta$ (d) $\tan \theta$

46. $\tan \theta = 4$ (a) $\cot \theta$ (b) $\sec \theta$
                  (c) $\cos \theta$ (d) $\csc \theta$

47. $\csc \theta = 4$ (a) $\sin \theta$ (b) $\cos \theta$
                  (c) $\sec \theta$ (d) $\tan \theta$

48. $\csc \theta = 5$ (a) $\sin \theta$ (b) $\cot \theta$
                  (c) $\tan \theta$ (d) $\sec(90° - \theta)$

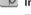 In Exercises 49–54, use a calculator to evaluate the trigonometric function. Round your answer to four decimal places.

49. $\tan 33°$

50. $\csc 11°$

51. $\sin 34.2°$

52. $\sec 79.3°$

53. $\cot 15° 14'$

54. $\cos 78° 11' 58''$

55. *Railroad Grade* A train travels 3.5 kilometers on a straight track with a grade of $1° 10'$ (see figure). What is the vertical rise of the train in that distance?

Not drawn to scale

56. *Guy Wire* A guy wire runs from the ground to the top of a 25-foot telephone pole. The angle formed between the wire and the ground is 52°. How far from the base of the pole is the wire attached to the ground?

1.4 In Exercises 57–64, the point is on the terminal side of an angle θ in standard position. Determine the exact values of the six trigonometric functions of the angle θ.

57. $(12, 16)$

58. $(3, -4)$

59. $\left(\frac{2}{3}, \frac{5}{2}\right)$

60. $\left(-\frac{10}{3}, -\frac{2}{3}\right)$

61. $(-0.5, 4.5)$

62. $(0.3, 0.4)$

63. $(x, 4x), \ x > 0$

64. $(-2x, -3x), \ x > 0$

In Exercises 65–70, find the values of the six trigonometric functions of θ.

Function Value	Constraint
65. $\sec \theta = \frac{6}{5}$	$\tan \theta < 0$
66. $\csc \theta = \frac{3}{2}$	$\cos \theta < 0$
67. $\sin \theta = \frac{3}{8}$	$\cos \theta < 0$
68. $\tan \theta = \frac{5}{4}$	$\cos \theta < 0$
69. $\cos \theta = -\frac{2}{5}$	$\sin \theta > 0$
70. $\sin \theta = -\frac{2}{4}$	$\cos \theta > 0$

In Exercises 71–74, find the reference angle θ', and sketch θ and θ' in standard position.

71. $\theta = 264°$ **72.** $\theta = 635°$

73. $\theta = -\dfrac{6\pi}{5}$ **74.** $\theta = \dfrac{17\pi}{3}$

In Exercises 75–82, evaluate the sine, cosine, and tangent of the angle without using a calculator.

75. $\dfrac{\pi}{3}$ **76.** $\dfrac{\pi}{4}$

77. $-\dfrac{7\pi}{3}$ **78.** $-\dfrac{5\pi}{4}$

79. $495°$ **80.** $-150°$

81. $-240°$ **82.** $315°$

 In Exercises 83–88, use a calculator to evaluate the trigonometric function. Round your answer to four decimal places.

83. $\sin 4$ **84.** $\tan 3$

85. $\sin(-3.2)$ **86.** $\cot(-4.8)$

85. $\sin \dfrac{12\pi}{5}$ **88.** $\tan\left(-\dfrac{25\pi}{7}\right)$

1.5 In Exercises 89–96, sketch the graph of the function. Include two full periods.

89. $y = \sin x$ **90.** $y = \cos x$

91. $f(x) = 5 \sin \dfrac{2x}{5}$ **92.** $f(x) = 8 \cos\left(-\dfrac{x}{4}\right)$

93. $y = 2 + \sin x$ **94.** $y = -4 - \cos \pi x$

95. $g(t) = \frac{5}{2} \sin(t - \pi)$ **96.** $g(t) = 3 \cos(t + \pi)$

97. *Sound Waves* Sound waves can be modeled by sine functions of the form $y = a \sin bx$, where x is measured in seconds.

(a) Write an equation of a sound wave whose amplitude is 2 and whose period is $\frac{1}{264}$ second.

(b) What is the frequency of the sound wave described in part (a)?

98. *Data Analysis: Meteorology* The times S of sunset (Greenwich Mean Time) at 40° north latitude on the 15th of each month are: 1(16:59), 2(17:35), 3(18:06), 4(18:38), 5(19:08), 6(19:30), 7(19:28), 8(18:57), 9(18:09), 10(17:21), 11(16:44), 12(16:36). The month is represented by t, with $t = 1$ corresponding to January. A model (in which minutes have been converted to the decimal parts of an hour) for the data is

$$S(t) = 18.09 + 1.41 \sin\left(\frac{\pi t}{6} + 4.60\right).$$

 (a) Use a graphing utility to graph the data points and the model in the same viewing window.

(b) What is the period of the model? Is it what you expected? Explain.

(c) What is the amplitude of the model? What does it represent in the model? Explain.

1.6 In Exercises 99–106, sketch a graph of the function. Include two full periods.

99. $f(x) = \tan x$

100. $f(t) = \tan\left(t - \frac{\pi}{4}\right)$

101. $f(x) = \cot x$

102. $g(t) = 2 \cot 2t$

103. $f(x) = \sec x$

104. $h(t) = \sec\left(t - \frac{\pi}{4}\right)$

105. $f(x) = \csc x$

106. $f(t) = 3 \csc\left(2t + \frac{\pi}{4}\right)$

 In Exercises 107 and 108, use a graphing utility to graph the function and the damping factor of the function in the same viewing window. Describe the behavior of the function as x increases without bound.

107. $f(x) = x \cos x$

108. $g(x) = x^4 \cos x$

1.7 In Exercises 109–114, evaluate the expression. If necessary, round your answer to two decimal places.

109. $\arcsin\left(-\frac{1}{2}\right)$

110. $\arcsin(-1)$

111. $\arcsin 0.4$

112. $\arcsin 0.213$

113. $\sin^{-1}(-0.44)$

114. $\sin^{-1} 0.89$

In Exercises 115–118, evaluate the expression without the aid of a calculator.

115. $\arccos \dfrac{\sqrt{3}}{2}$

116. $\arccos \dfrac{\sqrt{2}}{2}$

117. $\cos^{-1}(-1)$

118. $\cos^{-1} \dfrac{\sqrt{3}}{2}$

 In Exercises 119–122, use a calculator to evaluate the expression. Round your answer to two decimal places.

119. $\arccos 0.324$

120. $\arccos(-0.888)$

121. $\tan^{-1}(-1.5)$

122. $\tan^{-1} 8.2$

In Exercises 123–126, use a graphing utility to graph the function.

123. $f(x) = 2 \arcsin x$

124. $f(x) = 3 \arccos x$

125. $f(x) = \arctan \dfrac{x}{2}$

126. $f(x) = -\arcsin 2x$

In Exercises 127–130, find the exact value of the expression.

127. $\cos\left(\arctan \frac{3}{4}\right)$

128. $\tan\left(\arccos \frac{3}{5}\right)$

129. $\sec\left(\arctan \frac{12}{5}\right)$

130. $\cot\left[\arcsin\left(-\frac{12}{13}\right)\right]$

In Exercises 131 and 132, write an algebraic expression that is equivalent to the expression.

131. $\tan\left(\arccos \dfrac{x}{2}\right)$

132. $\sec[\arcsin(x - 1)]$

1.8 **133.** *Angle of Elevation* The height of a radio transmission tower is 70 meters, and it casts a shadow of length 30 meters (see figure). Find the angle of elevation of the sun.

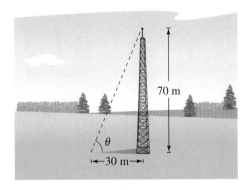

134. *Height* Your football has landed at the edge of the roof of your school building. When you are 25 feet from the base of the building, the angle of elevation to your football is 21°. How high off the ground is your football?

135. *Distance* From city A to city B, a plane flies 650 miles at a bearing of 48°. From city B to city C, the plane flies 810 miles at a bearing of 115°. Find the distance from city A to city C and the bearing from city A to city C.

136. *Wave Motion* Your fishing bobber oscillates in simple harmonic motion from the waves in the lake where you fish. Your bobber moves a total of 1.5 inches from its high point to its low point and returns to its high point every 3 seconds. Write an equation modeling the motion of your bobber if it is at its high point at time $t = 0$.

Synthesis

True or False? In Exercises 137–140, determine whether the statement is true or false. Justify your answer.

137. The tangent function is often useful for modeling simple harmonic motion.

138. The inverse sine function $y = \arcsin x$ cannot be defined as a function over any interval that is greater than the interval defined as $-\pi/2 \le y \le \pi/2$.

139. $y = \sin \theta$ is not a function because $\sin 30° = \sin 150°$.

140. Because $\tan 3\pi/4 = -1$, $\arctan(-1) = 3\pi/4$.

In Exercises 141–144, match the function $y = a \sin bx$ with its graph. Base your selection solely on your interpretation of the constants a and b. Explain your reasoning. [The graphs are labeled (a), (b), (c), and (d).]

(a)

(b)

(c)

(d)

141. $y = 3 \sin x$

142. $y = -3 \sin x$

143. $y = 2 \sin \pi x$

144. $y = 2 \sin \dfrac{x}{2}$

145. *Writing* Describe the behavior of $f(\theta) = \sec \theta$ at the zeros of $g(\theta) = \cos \theta$. Explain your reasoning.

146. *Conjecture*

(a) Use a graphing utility to complete the table.

θ	0.1	0.4	0.7	1.0	1.3
$\tan\left(\theta - \dfrac{\pi}{2}\right)$					
$-\cot \theta$					

(b) Make a conjecture about the relationship between $\tan\left(\theta - \dfrac{\pi}{2}\right)$ and $-\cot \theta$.

147. *Writing* When graphing the sine and cosine functions, determining the amplitude is part of the analysis. Explain why this is not true for the other four trigonometric functions.

148. *Graphical Reasoning* The formulas for the area of a circular sector and arc length are $A = \frac{1}{2}r^2\theta$ and $s = r\theta$, respectively. (r is the radius and θ is the angle measured in radians.)

(a) For $\theta = 0.8$, write the area and arc length as functions of r. What is the domain of each function? Use a graphing utility to graph the functions. Use the graphs to determine which function changes more rapidly as r increases. Explain.

(b) For $r = 10$ centimeters, write the area and arc length as functions of θ. What is the domain of each function? Use a graphing utility to graph and identify the functions.

149. *Writing* Describe a real-life application that can be represented by a simple harmonic motion model and is different from any that you've seen in this chapter. Explain which function you would use to model your application and why. Explain how you would determine the amplitude, period, and frequency of the model for your application.

1 | Chapter Test

Take this test as you would take a test in class. When you are finished, check your work against the answers given in the back of the book.

$(-2, 6)$

FIGURE FOR 4

1. Consider an angle that measures $\dfrac{5\pi}{4}$ radians.

   (a) Sketch the angle in standard position.

   (b) Determine two coterminal angles (one positive and one negative).

   (c) Convert the angle to degree measure.

2. A truck is moving at a rate of 90 kilometers per hour, and the diameter of its wheels is 1 meter. Find the angular speed of the wheels in radians per minute.

3. A water sprinkler sprays water on a lawn over a distance of 25 feet and rotates through an angle of 130°. Find the area of the lawn watered by the sprinkler.

4. Find the exact values of the six trigonometric functions of the angle θ shown in the figure.

5. Given that $\tan \theta = \frac{3}{2}$, find the other five trigonometric functions of θ.

6. Determine the reference angle θ' of the angle $\theta = 290°$ and sketch θ and θ' in standard position.

7. Determine the quadrant in which θ lies if $\sec \theta < 0$ and $\tan \theta > 0$.

8. Find two exact values of θ in degrees $(0 \le \theta < 360°)$ if $\cos \theta = -\sqrt{3}/2$. (Do not use a calculator.)

9. Use a calculator to approximate two values of θ in radians $(0 \le \theta < 2\pi)$ if $\csc \theta = 1.030$. Round the results to two decimal places.

In Exercises 10 and 11, find the remaining five trigonometric functions of θ satisfying the conditions.

10. $\cos \theta = \frac{3}{5}, \quad \tan \theta < 0$

11. $\sec \theta = -\frac{17}{8}, \quad \sin \theta > 0$

In Exercises 12 and 13, sketch the graph of the function. (Include two full periods.)

12. $g(x) = -2 \sin\left(x - \dfrac{\pi}{4}\right)$

13. $f(\alpha) = \dfrac{1}{2} \tan 2\alpha$

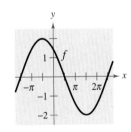

FIGURE FOR 16

In Exercises 14 and 15, use a graphing utility to graph the function. If the function is periodic, find its period.

14. $y = \sin 2\pi x + 2 \cos \pi x$

15. $y = 6t \cos(0.25t), \quad 0 \le t \le 32$

16. Find a, b, and c for the function $f(x) = a \sin(bx + c)$ such that the graph of f matches the figure.

17. Find the exact value of $\tan\left(\arccos \frac{2}{3}\right)$ without the aid of a calculator.

18. Graph the function $f(x) = 2 \arcsin\left(\frac{1}{2}x\right)$.

19. A plane is 80 miles south and 95 miles east of Cleveland Hopkins International Airport. What bearing should be taken to fly directly to the airport?

20. Write the equation for the simple harmonic motion of a ball on a spring that starts at its lowest point of 6 inches below equilibrium, bounces to its maximum height of 6 inches above equilibrium, and returns to its lowest point in a total of 2 seconds.

Proofs in Mathematics

The Pythagorean Theorem

The Pythagorean Theorem is one of the most famous theorems in mathematics. More than 100 different proofs now exist. James A. Garfield, the twentieth president of the United States, developed a proof of the Pythagorean Theorem in 1876. His proof, shown below, involved the fact that a trapezoid can be formed from two congruent right triangles and an isosceles right triangle.

The Pythagorean Theorem

In a right triangle, the sum of the squares of the lengths of the legs is equal to the square of the length of the hypotenuse, where a and b are the legs and c is the hypotenuse.

$$a^2 + b^2 = c^2$$

Proof

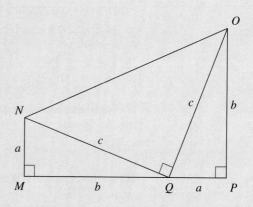

$$\text{Area of trapezoid } MNOP = \text{Area of } \triangle MNQ + \text{Area of } \triangle PQO + \text{Area of } \triangle NOQ$$

$$\frac{1}{2}(a + b)(a + b) = \frac{1}{2}ab + \frac{1}{2}ab + \frac{1}{2}c^2$$

$$\frac{1}{2}(a + b)(a + b) = ab + \frac{1}{2}c^2$$

$$(a + b)(a + b) = 2ab + c^2$$

$$a^2 + 2ab + b^2 = 2ab + c^2$$

$$a^2 + b^2 = c^2$$

This collection of thought-provoking and challenging exercises further explores and expands upon concepts learned in this chapter.

1. The restaurant at the top of the Space Needle in Seattle, Washington is circular and has a radius of 47.25 feet. The dining part of the restaurant revolves, making about one complete revolution every 48 minutes. A dinner party was seated at the edge of the revolving restaurant at 6:45 P.M. and was finished at 8:57 P.M.

(a) Find the angle through which the dinner party rotated.

(b) Find the distance the party traveled during dinner.

2. A bicycle's gear ratio is the number of times the freewheel turns for every one turn of the chainwheel (see figure). The table shows the numbers of teeth in the freewheel and chainwheel for the first five gears of an 18-speed touring bicycle. The chainwheel completes one rotation for each gear. Find the angle through which the freewheel turns for each gear. Give your answers in both degrees and radians.

Gear number	Number of teeth in freewheel	Number of teeth in chainwheel
1	32	24
2	26	24
3	22	24
4	32	40
5	19	24

Freewheel

Chainwheel

3. A surveyor in a helicopter is trying to determine the width of an island, as shown in the figure.

127°
39°
3000 ft
d
x
w
Not drawn to scale

(a) What is the shortest distance d the helicopter would have to travel to land on the island?

(b) What is the horizontal distance x that the helicopter would have to travel before it would be directly over the nearer end of the island?

(c) Find the width w of the island. Explain how you obtained your answer.

4. Use the figure below.

(a) Explain why $\triangle ABC$, $\triangle ADE$, and $\triangle AFG$ are similar triangles.

(b) What does similarity imply about the ratios

$$\frac{BC}{AB}, \frac{DE}{AD}, \text{ and } \frac{FG}{AF}?$$

(c) Does the value of sin A depend on which triangle from part (a) is used to calculate it? Would the value of sin A change if it were found using a different right triangle that was similar to the three given triangles?

(d) Do your conclusions from part (c) apply to the other five trigonometric functions? Explain.

5. Use a graphing utility to graph h, and use the graph to decide whether h is even, odd, or neither.

(a) $h(x) = \cos^2 x$

(b) $h(x) = \sin^2 x$

6. If f is an even function and g is an odd function, use the results of Exercise 5 to make a conjecture about h, where

(a) $h(x) = [f(x)]^2$

(b) $h(x) = [g(x)]^2$.

7. The model for the height h (in feet) of a Ferris wheel car is

$$h = 50 + 50 \sin 8\pi t$$

where t is the time (in minutes). (The Ferris wheel has a radius of 50 feet.) This model yields a height of 50 feet when $t = 0$. Alter the model so that the height of the car is 1 foot when $t = 0$.

8. The pressure P (in millimeters of mercury) against the walls of the blood vessels of a patient is modeled by

$$P = 100 - 20 \cos\left(\frac{8\pi}{3}t\right)$$

where t is time (in seconds).

 (a) Use a graphing utility to graph the model.

(b) What is the period of the model? What does the period tell you about this situation?

(c) What is the amplitude of the model? What does it tell you about this situation?

(d) If one cycle of this model is equivalent to one heartbeat, what is the pulse of this patient?

(e) If a physician wants this patient's pulse rate to be 64 beats per minute or less, what should the period be? What should the coefficient of t be?

9. A popular theory that attempts to explain the ups and downs of everyday life states that each of us has three cycles, called biorhythms, which begin at birth. These three cycles can be modeled by sine waves.

Physical (23 days): $P = \sin\dfrac{2\pi t}{23}, \quad t \geq 0$

Emotional (28 days): $E = \sin\dfrac{2\pi t}{28}, \quad t \geq 0$

Intellectual (33 days): $I = \sin\dfrac{2\pi t}{33}, \quad t \geq 0$

where t is the number of days since birth. Consider a person who was born on July 20, 1986.

 (a) Use a graphing utility to graph the three models in the same viewing window for $7300 \leq t \leq 7380$.

(b) Describe the person's biorhythms during the month of September 2006.

(c) Calculate the person's three energy levels on September 22, 2006.

 10. (a) Use a graphing utility to graph the functions given by

$$f(x) = 2 \cos 2x + 3 \sin 3x$$

and

$$g(x) = 2 \cos 2x + 3 \sin 4x.$$

(b) Use the graphs from part (a) to find the period of each function.

(c) If α and β are positive integers, is the function given by

$$h(x) = A \cos \alpha x + B \sin \beta x$$

periodic? Explain your reasoning.

11. Two trigonometric functions f and g have periods of 2, and their graphs intersect at $x = 5.35$.

(a) Give one smaller and one larger positive value of x at which the functions have the same value.

(b) Determine one negative value of x at which the graphs intersect.

(c) Is it true that $f(13.35) = g(-4.65)$? Explain your reasoning.

12. The function f is periodic, with period c. So, $f(t + c) = f(t)$. Are the following equal? Explain.

(a) $f(t - 2c) = f(t)$ (b) $f\left(t + \frac{1}{2}c\right) = f\left(\frac{1}{2}t\right)$

(c) $f\left(\frac{1}{2}(t + c)\right) = f\left(\frac{1}{2}t\right)$

13. If you stand in shallow water and look at an object below the surface of the water, the object will look farther away from you than it really is. This is because when light rays pass between air and water, the water refracts, or bends, the light rays. The index of refraction for water is 1.333. This is the ratio of the sine of θ_1 and the sine of θ_2 (see figure).

(a) You are standing in water that is 2 feet deep and are looking at a rock at angle $\theta_1 = 60°$ (measured from a line perpendicular to the surface of the water). Find θ_2.

(b) Find the distances x and y.

(c) Find the distance d between where the rock is and where it appears to be.

(d) What happens to d as you move closer to the rock? Explain your reasoning.

14. In calculus, it can be shown that the arctangent function can be approximated by the polynomial

$$\arctan x \approx x - \frac{x^3}{3} + \frac{x^5}{5} - \frac{x^7}{7}$$

where x is in radians.

(a) Use a graphing utility to graph the arctangent function and its polynomial approximation in the same viewing window. How do the graphs compare?

(b) Study the pattern in the polynomial approximation of the arctangent function and guess the next term. Then repeat part (a). How does the accuracy of the approximation change when additional terms are added?

Analytic Trigonometry

2

Concepts of trigonometry can be used to model the height above ground of a seat on a Ferris wheel.

© Patrick Ward/Corbis

SELECTED APPLICATIONS

Trigonometric equations and identities have many real-life applications. The applications listed below represent a small sample of the applications in this chapter.

- Friction,
  Exercise 95, page 229

- Shadow Length,
  Exercise 56, page 236

- Ferris Wheel,
  Exercise 75, page 246

- Data Analysis: Unemployment Rate,
  Exercise 76, page 246

- Harmonic Motion,
  Exercise 75, page 253

- Mach Number,
  Exercise 121, page 265

- Projectile Motion,
  Exercise 101, page 269

- Ocean Depth,
  Exercise 10, page 276

2.1 Using Fundamental Identities

What you should learn

- Recognize and write the fundamental trigonometric identities.
- Use the fundamental trigonometric identities to evaluate trigonometric functions, simplify trigonometric expressions, and rewrite trigonometric expressions.

Why you should learn it

Fundamental trigonometric identities can be used to simplify trigonometric expressions. For instance, in Exercise 95 on page 229, you can use trigonometric identities to simplify an expression for the coefficient of friction.

Introduction

In Chapter 1, you studied the basic definitions, properties, graphs, and applications of the individual trigonometric functions. In this chapter, you will learn how to use the fundamental identities to do the following.

1. Evaluate trigonometric functions.
2. Simplify trigonometric expressions.
3. Develop additional trigonometric identities.
4. Solve trigonometric equations.

Fundamental Trigonometric Identities

Reciprocal Identities

$$\sin u = \frac{1}{\csc u} \qquad \cos u = \frac{1}{\sec u} \qquad \tan u = \frac{1}{\cot u}$$

$$\csc u = \frac{1}{\sin u} \qquad \sec u = \frac{1}{\cos u} \qquad \cot u = \frac{1}{\tan u}$$

Quotient Identities

$$\tan u = \frac{\sin u}{\cos u} \qquad \cot u = \frac{\cos u}{\sin u}$$

Pythagorean Identities

$$\sin^2 u + \cos^2 u = 1 \qquad 1 + \tan^2 u = \sec^2 u \qquad 1 + \cot^2 u = \csc^2 u$$

Cofunction Identities

$$\sin\left(\frac{\pi}{2} - u\right) = \cos u \qquad \cos\left(\frac{\pi}{2} - u\right) = \sin u$$

$$\tan\left(\frac{\pi}{2} - u\right) = \cot u \qquad \cot\left(\frac{\pi}{2} - u\right) = \tan u$$

$$\sec\left(\frac{\pi}{2} - u\right) = \csc u \qquad \csc\left(\frac{\pi}{2} - u\right) = \sec u$$

Even/Odd Identities

$$\sin(-u) = -\sin u \qquad \cos(-u) = \cos u \qquad \tan(-u) = -\tan u$$

$$\csc(-u) = -\csc u \qquad \sec(-u) = \sec u \qquad \cot(-u) = -\cot u$$

Pythagorean identities are sometimes used in radical form such as

$$\sin u = \pm\sqrt{1 - \cos^2 u}$$

or

$$\tan u = \pm\sqrt{\sec^2 u - 1}$$

where the sign depends on the choice of u.

The *HM mathSpace®* CD-ROM and *Eduspace®* for this text contain additional resources related to the concepts discussed in this chapter.

Using the Fundamental Identities

One common use of trigonometric identities is to use given values of trigonometric functions to evaluate other trigonometric functions.

Example 1 **Using Identities to Evaluate a Function**

Use the values $\sec u = -\frac{3}{2}$ and $\tan u > 0$ to find the values of all six trigonometric functions.

Solution

Using a reciprocal identity, you have

$$\cos u = \frac{1}{\sec u} = \frac{1}{-3/2} = -\frac{2}{3}.$$

Using a Pythagorean identity, you have

$$\sin^2 u = 1 - \cos^2 u \qquad\qquad \text{Pythagorean identity}$$

$$= 1 - \left(-\frac{2}{3}\right)^2 \qquad\qquad \text{Substitute } -\tfrac{2}{3} \text{ for } \cos u.$$

$$= 1 - \frac{4}{9} = \frac{5}{9}. \qquad\qquad \text{Simplify.}$$

Because $\sec u < 0$ and $\tan u > 0$, it follows that u lies in Quadrant III. Moreover, because $\sin u$ is negative when u is in Quadrant III, you can choose the negative root and obtain $\sin u = -\sqrt{5}/3$. Now, knowing the values of the sine and cosine, you can find the values of all six trigonometric functions.

$$\sin u = -\frac{\sqrt{5}}{3} \qquad\qquad \csc u = \frac{1}{\sin u} = -\frac{3}{\sqrt{5}} = -\frac{3\sqrt{5}}{5}$$

$$\cos u = -\frac{2}{3} \qquad\qquad \sec u = \frac{1}{\cos u} = -\frac{3}{2}$$

$$\tan u = \frac{\sin u}{\cos u} = \frac{-\sqrt{5}/3}{-2/3} = \frac{\sqrt{5}}{2} \qquad \cot u = \frac{1}{\tan u} = \frac{2}{\sqrt{5}} = \frac{2\sqrt{5}}{5}$$

✓**CHECKPOINT**   Now try Exercise 11.

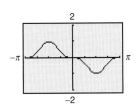

Example 2 **Simplifying a Trigonometric Expression**

Simplify $\sin x \cos^2 x - \sin x$.

Solution

First factor out a common monomial factor and then use a fundamental identity.

$$\sin x \cos^2 x - \sin x = \sin x(\cos^2 x - 1) \qquad \text{Factor out common monomial factor.}$$

$$= -\sin x(1 - \cos^2 x) \qquad \text{Factor out } -1.$$

$$= -\sin x(\sin^2 x) \qquad \text{Pythagorean identity}$$

$$= -\sin^3 x \qquad \text{Multiply.}$$

✓**CHECKPOINT**   Now try Exercise 45.

When factoring trigonometric expressions, it is helpful to find a special polynomial factoring form that fits the expression, as shown in Example 3.

Example 3   Factoring Trigonometric Expressions

Factor each expression.

a. $\sec^2 \theta - 1$ **b.** $4 \tan^2 \theta + \tan \theta - 3$

Solution

a. Here you have the difference of two squares, which factors as

$$\sec^2 \theta - 1 = (\sec \theta - 1)(\sec \theta + 1).$$

b. This expression has the polynomial form $ax^2 + bx + c$, and it factors as

$$4 \tan^2 \theta + \tan \theta - 3 = (4 \tan \theta - 3)(\tan \theta + 1).$$

✔CHECKPOINT   Now try Exercise 47.

On occasion, factoring or simplifying can best be done by first rewriting the expression in terms of just *one* trigonometric function or in terms of *sine and cosine only*. These strategies are illustrated in Examples 4 and 5, respectively.

Example 4   Factoring a Trigonometric Expression

Factor $\csc^2 x - \cot x - 3$.

Solution

Use the identity $\csc^2 x = 1 + \cot^2 x$ to rewrite the expression in terms of the cotangent.

$$\csc^2 x - \cot x - 3 = (1 + \cot^2 x) - \cot x - 3 \qquad \text{Pythagorean identity}$$

$$= \cot^2 x - \cot x - 2 \qquad \text{Combine like terms.}$$

$$= (\cot x - 2)(\cot x + 1) \qquad \text{Factor.}$$

✔CHECKPOINT   Now try Exercise 51.

Example 5   Simplifying a Trigonometric Expression

Simplify $\sin t + \cot t \cos t$.

Solution

Begin by rewriting $\cot t$ in terms of sine and cosine.

$$\sin t + \cot t \cos t = \sin t + \left(\frac{\cos t}{\sin t}\right) \cos t \qquad \text{Quotient identity}$$

$$= \frac{\sin^2 t + \cos^2 t}{\sin t} \qquad \text{Add fractions.}$$

$$= \frac{1}{\sin t} \qquad \text{Pythagorean identity}$$

$$= \csc t \qquad \text{Reciprocal identity}$$

✔CHECKPOINT   Now try Exercise 57.

STUDY TIP

Remember that when adding rational expressions, you must first find the least common denominator (LCD). In Example 5, the LCD is $\sin t$.

Example 6   **Adding Trigonometric Expressions**

Perform the addition and simplify.

$$\frac{\sin\theta}{1+\cos\theta}+\frac{\cos\theta}{\sin\theta}$$

Solution

$$\frac{\sin\theta}{1+\cos\theta}+\frac{\cos\theta}{\sin\theta}=\frac{(\sin\theta)(\sin\theta)+(\cos\theta)(1+\cos\theta)}{(1+\cos\theta)(\sin\theta)}$$

$$=\frac{\sin^2\theta+\cos^2\theta+\cos\theta}{(1+\cos\theta)(\sin\theta)}\qquad\text{Multiply.}$$

$$=\frac{1+\cos\theta}{(1+\cos\theta)(\sin\theta)}\qquad\begin{array}{l}\text{Pythagorean identity:}\\ \sin^2\theta+\cos^2\theta=1\end{array}$$

$$=\frac{1}{\sin\theta}\qquad\text{Divide out common factor.}$$

$$=\csc\theta\qquad\text{Reciprocal identity}$$

✔*CHECKPOINT*   Now try Exercise 61.

The last two examples in this section involve techniques for rewriting expressions in forms that are used in calculus.

Example 7   **Rewriting a Trigonometric Expression**

Rewrite $\dfrac{1}{1+\sin x}$ so that it is *not* in fractional form.

Solution

From the Pythagorean identity $\cos^2 x=1-\sin^2 x=(1-\sin x)(1+\sin x)$, you can see that multiplying both the numerator and the denominator by $(1-\sin x)$ will produce a monomial denominator.

$$\frac{1}{1+\sin x}=\frac{1}{1+\sin x}\cdot\frac{1-\sin x}{1-\sin x}\qquad\begin{array}{l}\text{Multiply numerator and}\\ \text{denominator by }(1-\sin x).\end{array}$$

$$=\frac{1-\sin x}{1-\sin^2 x}\qquad\text{Multiply.}$$

$$=\frac{1-\sin x}{\cos^2 x}\qquad\text{Pythagorean identity}$$

$$=\frac{1}{\cos^2 x}-\frac{\sin x}{\cos^2 x}\qquad\text{Write as separate fractions.}$$

$$=\frac{1}{\cos^2 x}-\frac{\sin x}{\cos x}\cdot\frac{1}{\cos x}\qquad\text{Product of fractions}$$

$$=\sec^2 x-\tan x\sec x\qquad\text{Reciprocal and quotient identities}$$

✔*CHECKPOINT*   Now try Exercise 65.

Example 8 Trigonometric Substitution

Use the substitution $x = 2 \tan \theta$, $0 < \theta < \pi/2$, to write

$$\sqrt{4 + x^2}$$

as a trigonometric function of θ.

Solution

Begin by letting $x = 2 \tan \theta$. Then, you can obtain

$$
\begin{aligned}
\sqrt{4 + x^2} &= \sqrt{4 + (2 \tan \theta)^2} && \text{Substitute } 2 \tan \theta \text{ for } x. \\
&= \sqrt{4 + 4 \tan^2 \theta} && \text{Rule of exponents} \\
&= \sqrt{4(1 + \tan^2 \theta)} && \text{Factor.} \\
&= \sqrt{4 \sec^2 \theta} && \text{Pythagorean identity} \\
&= 2 \sec \theta. && \sec \theta > 0 \text{ for } 0 < \theta < \pi/2
\end{aligned}
$$

✓**CHECKPOINT** Now try Exercise 77.

Figure 2.1 shows the right triangle illustration of the trigonometric substitution $x = 2 \tan \theta$ in Example 8. You can use this triangle to check the solution of Example 8. For $0 < \theta < \pi/2$, you have

$$\text{opp} = x, \quad \text{adj} = 2, \quad \text{and} \quad \text{hyp} = \sqrt{4 + x^2}.$$

With these expressions, you can write the following.

$$\sec \theta = \frac{\text{hyp}}{\text{adj}}$$

$$\sec \theta = \frac{\sqrt{4 + x^2}}{2}$$

$$2 \sec \theta = \sqrt{4 + x^2}$$

So, the solution checks.

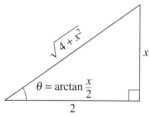

Angle whose tangent is $\dfrac{x}{2}$.

FIGURE **2.1**

| **2.1** | **Exercises** | The *HM mathSpace*® CD-ROM and *Eduspace*® for this text contain step-by-step solutions to all odd-numbered exercises. They also provide Tutorial Exercises for additional help. |

VOCABULARY CHECK: Fill in the blank to complete the trigonometric identity.

1. $\dfrac{\sin u}{\cos u} = $ _____

2. $\dfrac{1}{\sec u} = $ _____

3. $\dfrac{1}{\tan u} = $ _____

4. $\dfrac{1}{\sin u} = $ _____

5. $1 + $ _____ $= \csc^2 u$

6. $1 + \tan^2 u = $ _____

7. $\sin\left(\dfrac{\pi}{2} - u\right) = $ _____

8. $\sec\left(\dfrac{\pi}{2} - u\right) = $ _____

9. $\cos(-u) = $ _____

10. $\tan(-u) = $ _____

PREREQUISITE SKILLS REVIEW: Practice and review algebra skills needed for this section at **www.Eduspace.com.**

In Exercises 1–14, use the given values to evaluate (if possible) all six trigonometric functions.

1. $\sin x = \dfrac{\sqrt{3}}{2}, \quad \cos x = -\dfrac{1}{2}$

2. $\tan x = \dfrac{\sqrt{3}}{3}, \quad \cos x = -\dfrac{\sqrt{3}}{2}$

3. $\sec \theta = \sqrt{2}, \quad \sin \theta = -\dfrac{\sqrt{2}}{2}$

4. $\csc \theta = \dfrac{5}{3}, \quad \tan \theta = \dfrac{3}{4}$

5. $\tan x = \dfrac{5}{12}, \quad \sec x = -\dfrac{13}{12}$

6. $\cot \phi = -3, \quad \sin \phi = \dfrac{\sqrt{10}}{10}$

7. $\sec \phi = \dfrac{3}{2}, \quad \csc \phi = -\dfrac{3\sqrt{5}}{5}$

8. $\cos\left(\dfrac{\pi}{2} - x\right) = \dfrac{3}{5}, \quad \cos x = \dfrac{4}{5}$

9. $\sin(-x) = -\dfrac{1}{3}, \quad \tan x = -\dfrac{\sqrt{2}}{4}$

10. $\sec x = 4, \quad \sin x > 0$

11. $\tan \theta = 2, \quad \sin \theta < 0$

12. $\csc \theta = -5, \quad \cos \theta < 0$

13. $\sin \theta = -1, \quad \cot \theta = 0$

14. $\tan \theta$ is undefined, $\quad \sin \theta > 0$

In Exercises 15–20, match the trigonometric expression with one of the following.

(a) $\sec x$ (b) -1 (c) $\cot x$

(d) 1 (e) $-\tan x$ (f) $\sin x$

15. $\sec x \cos x$

16. $\tan x \csc x$

17. $\cot^2 x - \csc^2 x$

18. $(1 - \cos^2 x)(\csc x)$

19. $\dfrac{\sin(-x)}{\cos(-x)}$

20. $\dfrac{\sin[(\pi/2) - x]}{\cos[(\pi/2) - x]}$

In Exercises 21–26, match the trigonometric expression with one of the following.

(a) $\csc x$ (b) $\tan x$ (c) $\sin^2 x$

(d) $\sin x \tan x$ (e) $\sec^2 x$ (f) $\sec^2 x + \tan^2 x$

21. $\sin x \sec x$

22. $\cos^2 x(\sec^2 x - 1)$

23. $\sec^4 x - \tan^4 x$

24. $\cot x \sec x$

25. $\dfrac{\sec^2 x - 1}{\sin^2 x}$

26. $\dfrac{\cos^2[(\pi/2) - x]}{\cos x}$

In Exercises 27–44, use the fundamental identities to simplify the expression. There is more than one correct form of each answer.

27. $\cot \theta \sec \theta$

28. $\cos \beta \tan \beta$

29. $\sin \phi(\csc \phi - \sin \phi)$

30. $\sec^2 x(1 - \sin^2 x)$

31. $\dfrac{\cot x}{\csc x}$

32. $\dfrac{\csc \theta}{\sec \theta}$

33. $\dfrac{1 - \sin^2 x}{\csc^2 x - 1}$

34. $\dfrac{1}{\tan^2 x + 1}$

35. $\sec \alpha \cdot \dfrac{\sin \alpha}{\tan \alpha}$

36. $\dfrac{\tan^2 \theta}{\sec^2 \theta}$

37. $\cos\left(\dfrac{\pi}{2} - x\right)\sec x$

38. $\cot\left(\dfrac{\pi}{2} - x\right)\cos x$

39. $\dfrac{\cos^2 y}{1 - \sin y}$

40. $\cos t(1 + \tan^2 t)$

41. $\sin \beta \tan \beta + \cos \beta$

42. $\csc \phi \tan \phi + \sec \phi$

43. $\cot u \sin u + \tan u \cos u$

44. $\sin \theta \sec \theta + \cos \theta \csc \theta$

In Exercises 45–56, factor the expression and use the fundamental identities to simplify. There is more than one correct form of each answer.

45. $\tan^2 x - \tan^2 x \sin^2 x$

46. $\sin^2 x \csc^2 x - \sin^2 x$

47. $\sin^2 x \sec^2 x - \sin^2 x$

48. $\cos^2 x + \cos^2 x \tan^2 x$

49. $\dfrac{\sec^2 x - 1}{\sec x - 1}$

50. $\dfrac{\cos^2 x - 4}{\cos x - 2}$

51. $\tan^4 x + 2 \tan^2 x + 1$

52. $1 - 2 \cos^2 x + \cos^4 x$

53. $\sin^4 x - \cos^4 x$

54. $\sec^4 x - \tan^4 x$

55. $\csc^3 x - \csc^2 x - \csc x + 1$

56. $\sec^3 x - \sec^2 x - \sec x + 1$

In Exercises 57–60, perform the multiplication and use the fundamental identities to simplify. There is more than one correct form of each answer.

57. $(\sin x + \cos x)^2$

58. $(\cot x + \csc x)(\cot x - \csc x)$

59. $(2 \csc x + 2)(2 \csc x - 2)$

60. $(3 - 3 \sin x)(3 + 3 \sin x)$

In Exercises 61–64, perform the addition or subtraction and use the fundamental identities to simplify. There is more than one correct form of each answer.

61. $\dfrac{1}{1 + \cos x} + \dfrac{1}{1 - \cos x}$

62. $\dfrac{1}{\sec x + 1} - \dfrac{1}{\sec x - 1}$

63. $\dfrac{\cos x}{1 + \sin x} + \dfrac{1 + \sin x}{\cos x}$

64. $\tan x - \dfrac{\sec^2 x}{\tan x}$

 In Exercises 65–68, rewrite the expression so that it is not in fractional form. There is more than one correct form of each answer.

65. $\dfrac{\sin^2 y}{1 - \cos y}$

66. $\dfrac{5}{\tan x + \sec x}$

67. $\dfrac{3}{\sec x - \tan x}$

68. $\dfrac{\tan^2 x}{\csc x + 1}$

 Numerical and Graphical Analysis In Exercises 69–72, use a graphing utility to complete the table and graph the functions. Make a conjecture about y_1 and y_2.

x	0.2	0.4	0.6	0.8	1.0	1.2	1.4
y_1							
y_2							

69. $y_1 = \cos\left(\dfrac{\pi}{2} - x\right), \quad y_2 = \sin x$

70. $y_1 = \sec x - \cos x, \quad y_2 = \sin x \tan x$

71. $y_1 = \dfrac{\cos x}{1 - \sin x}, \quad y_2 = \dfrac{1 + \sin x}{\cos x}$

72. $y_1 = \sec^4 x - \sec^2 x, \quad y_2 = \tan^2 x + \tan^4 x$

 In Exercises 73–76, use a graphing utility to determine which of the six trigonometric functions is equal to the expression. Verify your answer algebraically.

73. $\cos x \cot x + \sin x$

74. $\sec x \csc x - \tan x$

75. $\dfrac{1}{\sin x}\left(\dfrac{1}{\cos x} - \cos x\right)$

76. $\dfrac{1}{2}\left(\dfrac{1 + \sin \theta}{\cos \theta} + \dfrac{\cos \theta}{1 + \sin \theta}\right)$

In Exercises 77–82, use the trigonometric substitution to write the algebraic expression as a trigonometric function of θ, where $0 < \theta < \pi/2$.

77. $\sqrt{9 - x^2}, \quad x = 3 \cos \theta$

78. $\sqrt{64 - 16x^2}, \quad x = 2 \cos \theta$

79. $\sqrt{x^2 - 9}, \quad x = 3 \sec \theta$

80. $\sqrt{x^2 - 4}, \quad x = 2 \sec \theta$

81. $\sqrt{x^2 + 25}, \quad x = 5 \tan \theta$

82. $\sqrt{x^2 + 100}, \quad x = 10 \tan \theta$

In Exercises 83–86, use the trigonometric substitution to write the algebraic equation as a trigonometric function of θ, where $-\pi/2 < \theta < \pi/2$. Then find $\sin \theta$ and $\cos \theta$.

83. $3 = \sqrt{9 - x^2}, \quad x = 3 \sin \theta$

84. $3 = \sqrt{36 - x^2}, \quad x = 6 \sin \theta$

85. $2\sqrt{2} = \sqrt{16 - 4x^2}, \quad x = 2 \cos \theta$

86. $-5\sqrt{3} = \sqrt{100 - x^2}, \quad x = 10 \cos \theta$

In Exercises 87–90, use a graphing utility to solve the equation for θ, where $0 \le \theta < 2\pi$.

87. $\sin \theta = \sqrt{1 - \cos^2 \theta}$

88. $\cos \theta = -\sqrt{1 - \sin^2 \theta}$

89. $\sec \theta = \sqrt{1 + \tan^2 \theta}$

90. $\csc \theta = \sqrt{1 + \cot^2 \theta}$

In Exercises 91–94, use a calculator to demonstrate the identity for each value of θ.

91. $\csc^2 \theta - \cot^2 \theta = 1$

    (a) $\theta = 132°$,   (b) $\theta = \dfrac{2\pi}{7}$

92. $\tan^2 \theta + 1 = \sec^2 \theta$

    (a) $\theta = 346°$,   (b) $\theta = 3.1$

93. $\cos\left(\dfrac{\pi}{2} - \theta\right) = \sin \theta$

    (a) $\theta = 80°$,   (b) $\theta = 0.8$

94. $\sin(-\theta) = -\sin \theta$

    (a) $\theta = 250°$,   (b) $\theta = \frac{1}{2}$

95. *Friction* The forces acting on an object weighing W units on an inclined plane positioned at an angle of θ with the horizontal (see figure) are modeled by

$$\mu W \cos \theta = W \sin \theta$$

where μ is the coefficient of friction. Solve the equation for μ and simplify the result.

96. *Rate of Change* The rate of change of the function

$$f(x) = -\csc x - \sin x$$

is given by the expression

$$\csc x \cot x - \cos x.$$

Show that this expression can also be written as $\cos x \cot^2 x.$

Synthesis

True or False? **In Exercises 97 and 98, determine whether the statement is true or false. Justify your answer.**

97. The even and odd trigonometric identities are helpful for determining whether the value of a trigonometric function is positive or negative.

98. A cofunction identity can be used to transform a tangent function so that it can be represented by a cosecant function.

In Exercises 99–102, fill in the blanks. (*Note:* The notation $x \to c^+$ indicates that x approaches c from the right and $x \to c$ indicates that x approaches c from the left.)

99. As $x \to \dfrac{\pi^-}{2}$, $\sin x \to$ _____ and $\csc x \to$ _____.

100. As $x \to 0^+$, $\cos x \to$ _____ and $\sec x \to$ _____.

101. As $x \to \dfrac{\pi^-}{2}$, $\tan x \to$ _____ and $\cot x \to$ _____.

102. As $x \to \pi^+$, $\sin x \to$ _____ and $\csc x \to$ _____.

In Exercises 103–108, determine whether or not the equation is an identity, and give a reason for your answer.

103. $\cos \theta = \sqrt{1 - \sin^2 \theta}$ **104.** $\cot \theta = \sqrt{\csc^2 \theta + 1}$

105. $\dfrac{(\sin k\theta)}{(\cos k\theta)} = \tan \theta$, k is a constant.

106. $\dfrac{1}{(5 \cos \theta)} = 5 \sec \theta$

107. $\sin \theta \csc \theta = 1$ **108.** $\csc^2 \theta = 1$

109. Use the definitions of sine and cosine to derive the Pythagorean identity $\sin^2 \theta + \cos^2 \theta = 1$.

110. *Writing* Use the Pythagorean identity

$$\sin^2 \theta + \cos^2 \theta = 1$$

to derive the other Pythagorean identities, $1 + \tan^2 \theta = \sec^2 \theta$ and $1 + \cot^2 \theta = \csc^2 \theta$. Discuss how to remember these identities and other fundamental identities.

Skills Review

In Exercises 111 and 112, perform the operation and simplify.

111. $\left(\sqrt{x} + 5\right)\left(\sqrt{x} - 5\right)$ **112.** $\left(2\sqrt{z} + 3\right)^2$

In Exercises 113–116, perform the addition or subtraction and simplify.

113. $\dfrac{1}{x + 5} + \dfrac{x}{x - 8}$ **114.** $\dfrac{6x}{x - 4} - \dfrac{3}{4 - x}$

115. $\dfrac{2x}{x^2 - 4} - \dfrac{7}{x + 4}$ **116.** $\dfrac{x}{x^2 - 25} + \dfrac{x^2}{x - 5}$

In Exercises 117–120, sketch the graph of the function. (Include two full periods.)

117. $f(x) = \dfrac{1}{2} \sin \pi x$ **118.** $f(x) = -2 \tan \dfrac{\pi x}{2}$

119. $f(x) = \dfrac{1}{2} \sec\left(x + \dfrac{\pi}{4}\right)$ **120.** $f(x) = \dfrac{3}{2} \cos(x - \pi) + 3$

2.2 Verifying Trigonometric Identities

What you should learn

• Verify trigonometric identities.

Why you should learn it

You can use trigonometric identities to rewrite trigonometric equations that model real-life situations. For instance, in Exercise 56 on page 236, you can use trigonometric identities to simplify the equation that models the length of a shadow cast by a gnomon (a device used to tell time).

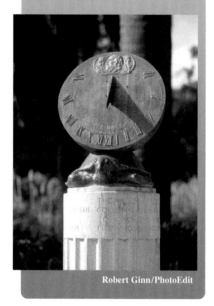

Robert Ginn/PhotoEdit

You may want to review the distinctions among expressions, equations, and identities. Have your students look at some algebraic identities and conditional equations before starting this section. It is important for them to understand what it means to verify an identity and not try to solve it as an equation.

Introduction

In this section, you will study techniques for verifying trigonometric identities. In the next section, you will study techniques for solving trigonometric equations. The key to verifying identities *and* solving equations is the ability to use the fundamental identities and the rules of algebra to rewrite trigonometric expressions.

Remember that a *conditional equation* is an equation that is true for only some of the values in its domain. For example, the conditional equation

$$\sin x = 0 \qquad \text{Conditional equation}$$

is true only for $x = n\pi$, where n is an integer. When you find these values, you are *solving* the equation.

On the other hand, an equation that is true for all real values in the domain of the variable is an *identity*. For example, the familiar equation

$$\sin^2 x = 1 - \cos^2 x \qquad \text{Identity}$$

is true for all real numbers x. So, it is an identity.

Verifying Trigonometric Identities

Although there are similarities, verifying that a trigonometric equation is an identity is quite different from solving an equation. There is no well-defined set of rules to follow in verifying trigonometric identities, and the process is best learned by practice.

Guidelines for Verifying Trigonometric Identities

1. Work with one side of the equation at a time. It is often better to work with the more complicated side first.

2. Look for opportunities to factor an expression, add fractions, square a binomial, or create a monomial denominator.

3. Look for opportunities to use the fundamental identities. Note which functions are in the final expression you want. Sines and cosines pair up well, as do secants and tangents, and cosecants and cotangents.

4. If the preceding guidelines do not help, try converting all terms to sines and cosines.

5. Always try *something*. Even paths that lead to dead ends provide insights.

Verifying trigonometric identities is a useful process if you need to convert a trigonometric expression into a form that is more useful algebraically. When you verify an identity, you cannot *assume* that the two sides of the equation are equal because you are trying to verify that they *are* equal. As a result, when verifying identities, you cannot use operations such as adding the same quantity to each side of the equation or cross multiplication.

Example 1   Verifying a Trigonometric Identity

Verify the identity $\dfrac{\sec^2 \theta - 1}{\sec^2 \theta} = \sin^2 \theta.$

Solution

Because the left side is more complicated, start with it.

$$\frac{\sec^2 \theta - 1}{\sec^2 \theta} = \frac{(\tan^2 \theta + 1) - 1}{\sec^2 \theta} \qquad \text{Pythagorean identity}$$

$$= \frac{\tan^2 \theta}{\sec^2 \theta} \qquad \text{Simplify.}$$

$$= \tan^2 \theta(\cos^2 \theta) \qquad \text{Reciprocal identity}$$

$$= \frac{\sin^2 \theta}{(\cos^2 \theta)} (\cos^2 \theta) \qquad \text{Quotient identity}$$

$$= \sin^2 \theta \qquad \text{Simplify.}$$

Notice how the identity is verified. You start with the left side of the equation (the more complicated side) and use the fundamental trigonometric identities to simplify it until you obtain the right side.

✓**CHECKPOINT**   Now try Exercise 5.

There is more than one way to verify an identity. Here is another way to verify the identity in Example 1.

$$\frac{\sec^2 \theta - 1}{\sec^2 \theta} = \frac{\sec^2 \theta}{\sec^2 \theta} - \frac{1}{\sec^2 \theta} \qquad \text{Rewrite as the difference of fractions.}$$

$$= 1 - \cos^2 \theta \qquad \text{Reciprocal identity}$$

$$= \sin^2 \theta \qquad \text{Pythagorean identity}$$

Example 2   Combining Fractions Before Using Identities

Verify the identity $\dfrac{1}{1 - \sin \alpha} + \dfrac{1}{1 + \sin \alpha} = 2 \sec^2 \alpha.$

Solution

$$\frac{1}{1 - \sin \alpha} + \frac{1}{1 + \sin \alpha} = \frac{1 + \sin \alpha + 1 - \sin \alpha}{(1 - \sin \alpha)(1 + \sin \alpha)} \qquad \text{Add fractions.}$$

$$= \frac{2}{1 - \sin^2 \alpha} \qquad \text{Simplify.}$$

$$= \frac{2}{\cos^2 \alpha} \qquad \text{Pythagorean identity}$$

$$= 2 \sec^2 \alpha \qquad \text{Reciprocal identity}$$

✓**CHECKPOINT**   Now try Exercise 19.

STUDY TIP

Remember that an identity is only true for all real values in the domain of the variable. For instance, in Example 1 the identity is not true when $\theta = \pi/2$ because $\sec^2 \theta$ is not defined when $\theta = \pi/2$.

Encourage your students to identify the reasoning behind each solution step in the examples of this section while covering the comment lines. This will help students to recognize and remember the fundamental trigonometric identities.

Example 3 Verifying Trigonometric Identity

Verify the identity $(\tan^2 x + 1)(\cos^2 x - 1) = -\tan^2 x$.

Algebraic Solution

By applying identities before multiplying, you obtain the following.

$(\tan^2 x + 1)(\cos^2 x - 1) = (\sec^2 x)(-\sin^2 x)$ Pythagorean identities

$$= -\frac{\sin^2 x}{\cos^2 x}$$ Reciprocal identity

$$= -\left(\frac{\sin x}{\cos x}\right)^2$$ Rule of exponents

$$= -\tan^2 x$$ Quotient identity

✓CHECKPOINT   Now try Exercise 39.

Numerical Solution

Use the *table* feature of a graphing utility set in *radian* mode to create a table that shows the values of $y_1 = (\tan^2 x + 1)(\cos^2 x - 1)$ and $y_2 = -\tan^2 x$ for different values of x, as shown in Figure 2.2. From the table you can see that the values of y_1 and y_2 appear to be identical, so $(\tan^2 x + 1)(\cos^2 x - 1) = -\tan^2 x$ appears to be an identity.

X	Y1	Y2
-.5	-.2984	-.2984
-.25	-.0652	-.0652
0	0	0
.25	-.0652	-.0652
.5	-.2984	-.2984
.75	-.8679	-.8679
1	-2.426	-2.426

X=-.5

FIGURE 2.2

Example 4 Converting to Sines and Cosines

Verify the identity $\tan x + \cot x = \sec x \csc x$.

Solution

Try converting the left side into sines and cosines.

$$\tan x + \cot x = \frac{\sin x}{\cos x} + \frac{\cos x}{\sin x}$$ Quotient identities

$$= \frac{\sin^2 x + \cos^2 x}{\cos x \sin x}$$ Add fractions.

$$= \frac{1}{\cos x \sin x}$$ Pythagorean identity

$$= \frac{1}{\cos x} \cdot \frac{1}{\sin x} = \sec x \csc x$$ Reciprocal identities

✓CHECKPOINT   Now try Exercise 29.

> **STUDY TIP**
>
> Although a graphing utility can be useful in helping to verify an identity, you must use algebraic techniques to produce a *valid* proof.

> **STUDY TIP**
>
> As shown at the right, $\csc^2 x(1 + \cos x)$ is considered a simplified form of $1/(1 - \cos x)$ because the expression does not contain any fractions.

Recall from algebra that *rationalizing the denominator* using conjugates is, on occasion, a powerful simplification technique. A related form of this technique, shown below, works for simplifying trigonometric expressions as well.

$$\frac{1}{1 - \cos x} = \frac{1}{1 - \cos x}\left(\frac{1 + \cos x}{1 + \cos x}\right) = \frac{1 + \cos x}{1 - \cos^2 x} = \frac{1 + \cos x}{\sin^2 x}$$

$$= \csc^2 x(1 + \cos x)$$

This technique is demonstrated in the next example.

Example 5   Verifying Trigonometric Identities

Verify the identity $\sec y + \tan y = \dfrac{\cos y}{1 - \sin y}$.

Solution

Begin with the *right* side, because you can create a monomial denominator by multiplying the numerator and denominator by $1 + \sin y$.

$$\frac{\cos y}{1 - \sin y} = \frac{\cos y}{1 - \sin y}\left(\frac{1 + \sin y}{1 + \sin y}\right) \qquad \text{Multiply numerator and denominator by } 1 + \sin y.$$

$$= \frac{\cos y + \cos y \sin y}{1 - \sin^2 y} \qquad \text{Multiply.}$$

$$= \frac{\cos y + \cos y \sin y}{\cos^2 y} \qquad \text{Pythagorean identity}$$

$$= \frac{\cos y}{\cos^2 y} + \frac{\cos y \sin y}{\cos^2 y} \qquad \text{Write as separate fractions.}$$

$$= \frac{1}{\cos y} + \frac{\sin y}{\cos y} \qquad \text{Simplify.}$$

$$= \sec y + \tan y \qquad \text{Identities}$$

✓CHECKPOINT   Now try Exercise 33.

In Examples 1 through 5, you have been verifying trigonometric identities by working with one side of the equation and converting to the form given on the other side. On occasion, it is practical to work with each side *separately*, to obtain one common form equivalent to both sides. This is illustrated in Example 6.

Example 6   Working with Each Side Separately

Verify the identity $\dfrac{\cot^2 \theta}{1 + \csc \theta} = \dfrac{1 - \sin \theta}{\sin \theta}$.

Solution

Working with the left side, you have

$$\frac{\cot^2 \theta}{1 + \csc \theta} = \frac{\csc^2 \theta - 1}{1 + \csc \theta} \qquad \text{Pythagorean identity}$$

$$= \frac{(\csc \theta - 1)(\csc \theta + 1)}{1 + \csc \theta} \qquad \text{Factor.}$$

$$= \csc \theta - 1. \qquad \text{Simplify.}$$

Now, simplifying the right side, you have

$$\frac{1 - \sin \theta}{\sin \theta} = \frac{1}{\sin \theta} - \frac{\sin \theta}{\sin \theta} \qquad \text{Write as separate fractions.}$$

$$= \csc \theta - 1. \qquad \text{Reciprocal identity}$$

The identity is verified because both sides are equal to $\csc \theta - 1$.

✓CHECKPOINT   Now try Exercise 47.

In Example 7, powers of trigonometric functions are rewritten as more complicated sums of products of trigonometric functions. This is a common procedure used in calculus.

Example 7   Three Examples from Calculus

Verify each identity.

a. $\tan^4 x = \tan^2 x \sec^2 x - \tan^2 x$

b. $\sin^3 x \cos^4 x = (\cos^4 x - \cos^6 x) \sin x$

c. $\csc^4 x \cot x = \csc^2 x(\cot x + \cot^3 x)$

Solution

a. $\tan^4 x = (\tan^2 x)(\tan^2 x)$ Write as separate factors.

$\qquad\quad = \tan^2 x(\sec^2 x - 1)$ Pythagorean identity

$\qquad\quad = \tan^2 x \sec^2 x - \tan^2 x$ Multiply.

b. $\sin^3 x \cos^4 x = \sin^2 x \cos^4 x \sin x$ Write as separate factors.

$\qquad\qquad\quad = (1 - \cos^2 x)\cos^4 x \sin x$ Pythagorean identity

$\qquad\qquad\quad = (\cos^4 x - \cos^6 x) \sin x$ Multiply.

c. $\csc^4 x \cot x = \csc^2 x \csc^2 x \cot x$ Write as separate factors.

$\qquad\qquad\quad = \csc^2 x(1 + \cot^2 x) \cot x$ Pythagorean identity

$\qquad\qquad\quad = \csc^2 x(\cot x + \cot^3 x)$ Multiply.

✓**CHECKPOINT**   Now try Exercise 49.

𝒲RITING ABOUT 𝓜ATHEMATICS

Error Analysis   You are tutoring a student in trigonometry. One of the homework problems your student encounters asks whether the following statement is an identity.

$$\tan^2 x \sin^2 x \overset{?}{=} \frac{5}{6} \tan^2 x$$

Your student does not attempt to verify the equivalence algebraically, but mistakenly uses only a graphical approach. Using range settings of

Xmin = -3π	Ymin = -20
Xmax = 3π	Ymax = 20
Xscl = $\pi/2$	Yscl = 1

your student graphs both sides of the expression on a graphing utility and concludes that the statement is an identity.

What is wrong with your student's reasoning? Explain. Discuss the limitations of verifying identities graphically.

2.2 | Exercises

VOCABULARY CHECK:

In Exercises 1 and 2, fill in the blanks.

1. An equation that is true for all real values in its domain is called an _____.

2. An equation that is true for only some values in its domain is called a _____ _____.

In Exercises 3–8, fill in the blank to complete the trigonometric identity.

3. $\dfrac{1}{\cot u} = $ _____

4. $\dfrac{\cos u}{\sin u} = $ _____

5. $\sin^2 u + $ _____ $= 1$

6. $\cos\left(\dfrac{\pi}{2} - u\right) = $ _____

7. $\csc(-u) = $ _____

8. $\sec(-u) = $ _____

PREREQUISITE SKILLS REVIEW: Practice and review algebra skills needed for this section at **www.Eduspace.com.**

In Exercises 1–38, verify the identity.

1. $\sin t \csc t = 1$

2. $\sec y \cos y = 1$

3. $(1 + \sin \alpha)(1 - \sin \alpha) = \cos^2 \alpha$

4. $\cot^2 y(\sec^2 y - 1) = 1$

5. $\cos^2 \beta - \sin^2 \beta = 1 - 2\sin^2 \beta$

6. $\cos^2 \beta - \sin^2 \beta = 2\cos^2 \beta - 1$

7. $\sin^2 \alpha - \sin^4 \alpha = \cos^2 \alpha - \cos^4 \alpha$

8. $\cos x + \sin x \tan x = \sec x$

9. $\dfrac{\csc^2 \theta}{\cot \theta} = \csc \theta \sec \theta$

10. $\dfrac{\cot^3 t}{\csc t} = \cos t(\csc^2 t - 1)$

11. $\dfrac{\cot^2 t}{\csc t} = \csc t - \sin t$

12. $\dfrac{1}{\tan \beta} + \tan \beta = \dfrac{\sec^2 \beta}{\tan \beta}$

13. $\sin^{1/2} x \cos x - \sin^{5/2} x \cos x = \cos^3 x \sqrt{\sin x}$

14. $\sec^6 x(\sec x \tan x) - \sec^4 x(\sec x \tan x) = \sec^5 x \tan^3 x$

15. $\dfrac{1}{\sec x \tan x} = \csc x - \sin x$

16. $\dfrac{\sec \theta - 1}{1 - \cos \theta} = \sec \theta$

17. $\csc x - \sin x = \cos x \cot x$

18. $\sec x - \cos x = \sin x \tan x$

19. $\dfrac{1}{\tan x} + \dfrac{1}{\cot x} = \tan x + \cot x$

20. $\dfrac{1}{\sin x} - \dfrac{1}{\csc x} = \csc x - \sin x$

21. $\dfrac{\cos \theta \cot \theta}{1 - \sin \theta} - 1 = \csc \theta$

22. $\dfrac{1 + \sin \theta}{\cos \theta} + \dfrac{\cos \theta}{1 + \sin \theta} = 2 \sec \theta$

23. $\dfrac{1}{\sin x + 1} + \dfrac{1}{\csc x + 1} = 1$

24. $\cos x - \dfrac{\cos x}{1 - \tan x} = \dfrac{\sin x \cos x}{\sin x - \cos x}$

25. $\tan\left(\dfrac{\pi}{2} - \theta\right)\tan \theta = 1$

26. $\dfrac{\cos[(\pi/2) - x]}{\sin[(\pi/2) - x]} = \tan x$

27. $\dfrac{\csc(-x)}{\sec(-x)} = -\cot x$

28. $(1 + \sin y)[1 + \sin(-y)] = \cos^2 y$

29. $\dfrac{\tan x \cot x}{\cos x} = \sec x$

30. $\dfrac{\tan x + \tan y}{1 - \tan x \tan y} = \dfrac{\cot x + \cot y}{\cot x \cot y - 1}$

31. $\dfrac{\tan x + \cot y}{\tan x \cot y} = \tan y + \cot x$

32. $\dfrac{\cos x - \cos y}{\sin x + \sin y} + \dfrac{\sin x - \sin y}{\cos x + \cos y} = 0$

33. $\sqrt{\dfrac{1 + \sin \theta}{1 - \sin \theta}} = \dfrac{1 + \sin \theta}{|\cos \theta|}$

34. $\sqrt{\dfrac{1 - \cos \theta}{1 + \cos \theta}} = \dfrac{1 - \cos \theta}{|\sin \theta|}$

35. $\cos^2 \beta + \cos^2\left(\dfrac{\pi}{2} - \beta\right) = 1$

36. $\sec^2 y - \cot^2\left(\dfrac{\pi}{2} - y\right) = 1$

37. $\sin t \csc\left(\dfrac{\pi}{2} - t\right) = \tan t$

38. $\sec^2\left(\dfrac{\pi}{2} - x\right) - 1 = \cot^2 x$

In Exercises 39–46, (a) use a graphing utility to graph each side of the equation to determine whether the equation is an identity, (b) use the *table* feature of a graphing utility to determine whether the equation is an identity, and (c) confirm the results of parts (a) and (b) algebraically.

39. $2 \sec^2 x - 2 \sec^2 x \sin^2 x - \sin^2 x - \cos^2 x = 1$

40. $\csc x(\csc x - \sin x) + \dfrac{\sin x - \cos x}{\sin x} + \cot x = \csc^2 x$

41. $2 + \cos^2 x - 3 \cos^4 x = \sin^2 x(3 + 2 \cos^2 x)$

42. $\tan^4 x + \tan^2 x - 3 = \sec^2 x(4 \tan^2 x - 3)$

43. $\csc^4 x - 2 \csc^2 x + 1 = \cot^4 x$

44. $(\sin^4 \beta - 2 \sin^2 \beta + 1) \cos \beta = \cos^5 \beta$

45. $\dfrac{\cos x}{1 - \sin x} = \dfrac{1 - \sin x}{\cos x}$ **46.** $\dfrac{\cot \alpha}{\csc \alpha + 1} = \dfrac{\csc \alpha + 1}{\cot \alpha}$

In Exercises 47–50, verify the identity.

47. $\tan^5 x = \tan^3 x \sec^2 x - \tan^3 x$

48. $\sec^4 x \tan^2 x = (\tan^2 x + \tan^4 x) \sec^2 x$

49. $\cos^3 x \sin^2 x = (\sin^2 x - \sin^4 x) \cos x$

50. $\sin^4 x + \cos^4 x = 1 - 2 \cos^2 x + 2 \cos^4 x$

In Exercises 51–54, use the cofunction identities to evaluate the expression without the aid of a calculator.

51. $\sin^2 25° + \sin^2 65°$ **52.** $\cos^2 55° + \cos^2 35°$

53. $\cos^2 20° + \cos^2 52° + \cos^2 38° + \cos^2 70°$

54. $\sin^2 12° + \sin^2 40° + \sin^2 50° + \sin^2 78°$

55. *Rate of Change*   The rate of change of the function $f(x) = \sin x + \csc x$ with respect to change in the variable x is given by the expression $\cos x - \csc x \cot x$. Show that the expression for the rate of change can also be $-\cos x \cot^2 x$.

Model It

56. *Shadow Length*   The length s of a shadow cast by a vertical gnomon (a device used to tell time) of height h when the angle of the sun above the horizon is θ (see figure) can be modeled by the equation

$$s = \frac{h \sin(90° - \theta)}{\sin \theta}.$$

Model It (continued)

(a) Verify that the equation for s is equal to $h \cot \theta$.

 (b) Use a graphing utility to complete the table. Let $h = 5$ feet.

θ	10°	20°	30°	40°	50°
s					

θ	60°	70°	80°	90°
s				

(c) Use your table from part (b) to determine the angles of the sun for which the length of the shadow is the greatest and the least.

(d) Based on your results from part (c), what time of day do you think it is when the angle of the sun above the horizon is 90°?

Synthesis

True or False?   In Exercises 57 and 58, determine whether the statement is true or false. Justify your answer.

57. The equation $\sin^2 \theta + \cos^2 \theta = 1 + \tan^2 \theta$ is an identity, because $\sin^2(0) + \cos^2(0) = 1$ and $1 + \tan^2(0) = 1$.

58. The equation $1 + \tan^2 \theta = 1 + \cot^2 \theta$ is *not* an identity, because it is true that $1 + \tan^2(\pi/6) = 1\frac{1}{3}$, and $1 + \cot^2(\pi/6) = 4$.

Think About It   In Exercises 59 and 60, explain why the equation is not an identity and find one value of the variable for which the equation is not true.

59. $\sin \theta = \sqrt{1 - \cos^2 \theta}$

60. $\tan \theta = \sqrt{\sec^2 \theta - 1}$

Skills Review

In Exercises 61–64, use the Quadratic Formula to solve the quadratic equation.

61. $x^2 + 6x - 12 = 0$

62. $x^2 + 5x - 7 = 0$

63. $3x^2 - 6x - 12 = 0$

64. $8x^2 - 4x - 3 = 0$

2.3　Solving Trigonometric Equations

Tom Stillo/Index Stock Imagery

Introduction

To solve a trigonometric equation, use standard algebraic techniques such as collecting like terms and factoring. Your preliminary goal in solving a trigonometric equation is to *isolate* the trigonometric function involved in the equation. For example, to solve the equation $2 \sin x = 1$, divide each side by 2 to obtain

$$\sin x = \frac{1}{2}.$$

To solve for x, note in Figure 2.3 that the equation $\sin x = \frac{1}{2}$ has solutions $x = \pi/6$ and $x = 5\pi/6$ in the interval $[0, 2\pi)$. Moreover, because $\sin x$ has a period of 2π, there are infinitely many other solutions, which can be written as

$$x = \frac{\pi}{6} + 2n\pi \qquad \text{and} \qquad x = \frac{5\pi}{6} + 2n\pi \qquad \text{General solution}$$

where n is an integer, as shown in Figure 2.3.

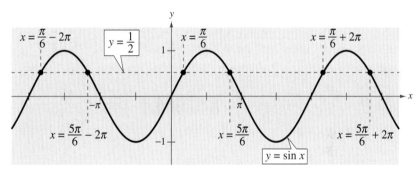

FIGURE 2.3

Another way to show that the equation $\sin x = \frac{1}{2}$ has infinitely many solutions is indicated in Figure 2.4. Any angles that are coterminal with $\pi/6$ or $5\pi/6$ will also be solutions of the equation.

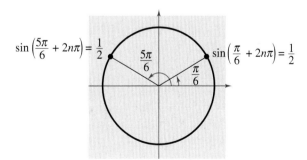

FIGURE 2.4

When solving trigonometric equations, you should write your answer(s) using exact values rather than decimal approximations.

Example 1 **Collecting Like Terms**

Solve $\sin x + \sqrt{2} = -\sin x$.

Solution

Begin by rewriting the equation so that $\sin x$ is isolated on one side of the equation.

$$\sin x + \sqrt{2} = -\sin x \qquad \text{Write original equation.}$$

$$\sin x + \sin x + \sqrt{2} = 0 \qquad \text{Add } \sin x \text{ to each side.}$$

$$\sin x + \sin x = -\sqrt{2} \qquad \text{Subtract } \sqrt{2} \text{ from each side.}$$

$$2 \sin x = -\sqrt{2} \qquad \text{Combine like terms.}$$

$$\sin x = -\frac{\sqrt{2}}{2} \qquad \text{Divide each side by 2.}$$

Because $\sin x$ has a period of 2π, first find all solutions in the interval $[0, 2\pi)$. These solutions are $x = 5\pi/4$ and $x = 7\pi/4$. Finally, add multiples of 2π to each of these solutions to get the general form

$$x = \frac{5\pi}{4} + 2n\pi \qquad \text{and} \qquad x = \frac{7\pi}{4} + 2n\pi \qquad \text{General solution}$$

where n is an integer.

✔CHECKPOINT Now try Exercise 7.

Example 2 **Extracting Square Roots**

Solve $3 \tan^2 x - 1 = 0$.

Solution

Begin by rewriting the equation so that $\tan x$ is isolated on one side of the equation.

$$3 \tan^2 x - 1 = 0 \qquad \text{Write original equation.}$$

$$3 \tan^2 x = 1 \qquad \text{Add 1 to each side.}$$

$$\tan^2 x = \frac{1}{3} \qquad \text{Divide each side by 3.}$$

$$\tan x = \pm\frac{1}{\sqrt{3}} = \pm\frac{\sqrt{3}}{3} \qquad \text{Extract square roots.}$$

Because $\tan x$ has a period of π, first find all solutions in the interval $[0, \pi)$. These solutions are $x = \pi/6$ and $x = 5\pi/6$. Finally, add multiples of π to each of these solutions to get the general form

$$x = \frac{\pi}{6} + n\pi \qquad \text{and} \qquad x = \frac{5\pi}{6} + n\pi \qquad \text{General solution}$$

where n is an integer.

✔CHECKPOINT Now try Exercise 11.

The equations in Examples 1 and 2 involved only one trigonometric function. When two or more functions occur in the same equation, collect all terms on one side and try to separate the functions by factoring or by using appropriate identities. This may produce factors that yield no solutions, as illustrated in Example 3.

Example 3   Factoring

Solve $\cot x \cos^2 x = 2 \cot x$.

Solution

Begin by rewriting the equation so that all terms are collected on one side of the equation.

$$\cot x \cos^2 x = 2 \cot x \qquad \text{Write original equation.}$$

$$\cot x \cos^2 x - 2 \cot x = 0 \qquad \text{Subtract 2 cot } x \text{ from each side.}$$

$$\cot x(\cos^2 x - 2) = 0 \qquad \text{Factor.}$$

By setting each of these factors equal to zero, you obtain

$$\cot x = 0 \qquad \text{and} \qquad \cos^2 x - 2 = 0$$

$$x = \frac{\pi}{2} \qquad\qquad \cos^2 x = 2$$

$$\cos x = \pm \sqrt{2}.$$

The equation $\cot x = 0$ has the solution $x = \pi/2$ [in the interval $(0, \pi)$]. No solution is obtained for $\cos x = \pm \sqrt{2}$ because $\pm \sqrt{2}$ are outside the range of the cosine function. Because $\cot x$ has a period of π, the general form of the solution is obtained by adding multiples of π to $x = \pi/2$, to get

$$x = \frac{\pi}{2} + n\pi \qquad \text{General solution}$$

where n is an integer. You can confirm this graphically by sketching the graph of $y = \cot x \cos^2 x - 2 \cot x$, as shown in Figure 2.5. From the graph you can see that the x-intercepts occur at $-3\pi/2$, $-\pi/2$, $\pi/2$, $3\pi/2$, and so on. These x-intercepts correspond to the solutions of $\cot x \cos^2 x - 2 \cot x = 0$.

$$y = \cot x \cos^2 x - 2 \cot x$$

FIGURE 2.5

✔CHECKPOINT   Now try Exercise 15.

Equations of Quadratic Type

Many trigonometric equations are of quadratic type $ax^2 + bx + c = 0$. Here are a couple of examples.

Quadratic in sin x	*Quadratic in sec x*
$2 \sin^2 x - \sin x - 1 = 0$	$\sec^2 x - 3 \sec x - 2 = 0$
$2(\sin x)^2 - \sin x - 1 = 0$	$(\sec x)^2 - 3(\sec x) - 2 = 0$

To solve equations of this type, factor the quadratic or, if this is not possible, use the Quadratic Formula.

Example 4   **Factoring an Equation of Quadratic Type**

Find all solutions of $2 \sin^2 x - \sin x - 1 = 0$ in the interval $[0, 2\pi)$.

Algebraic Solution

Begin by treating the equation as a quadratic in $\sin x$ and factoring.

$$2 \sin^2 x - \sin x - 1 = 0 \quad \text{Write original equation.}$$

$$(2 \sin x + 1)(\sin x - 1) = 0 \quad \text{Factor.}$$

Setting each factor equal to zero, you obtain the following solutions in the interval $[0, 2\pi)$.

$$2 \sin x + 1 = 0 \qquad \text{and} \qquad \sin x - 1 = 0$$

$$\sin x = -\frac{1}{2} \qquad\qquad\qquad \sin x = 1$$

$$x = \frac{7\pi}{6}, \frac{11\pi}{6} \qquad\qquad\qquad x = \frac{\pi}{2}$$

✔CHECKPOINT   Now try Exercise 29.

Graphical Solution

Use a graphing utility set in *radian* mode to graph $y = 2 \sin^2 x - \sin x - 1$ for $0 \le x < 2\pi$, as shown in Figure 2.6. Use the *zero* or *root* feature or the *zoom* and *trace* features to approximate the *x*-intercepts to be

$$x \approx 1.571 \approx \frac{\pi}{2}, x \approx 3.665 \approx \frac{7\pi}{6}, \text{ and } x \approx 5.760 \approx \frac{11\pi}{6}.$$

These values are the approximate solutions of $2 \sin^2 x - \sin x - 1 = 0$ in the interval $[0, 2\pi)$.

FIGURE 2.6

Example 5   **Rewriting with a Single Trigonometric Function**

Solve $2 \sin^2 x + 3 \cos x - 3 = 0$.

Solution

The algebraic model can be difficult for students to visualize. Encourage them to write or visualize the algebraic model, rewrite the corresponding trigonometric equation in terms of a single trigonometric function, and then solve for x.

This equation contains both sine and cosine functions. You can rewrite the equation so that it has only cosine functions by using the identity $\sin^2 x = 1 - \cos^2 x$.

$$2 \sin^2 x + 3 \cos x - 3 = 0 \qquad \text{Write original equation.}$$

$$2(1 - \cos^2 x) + 3 \cos x - 3 = 0 \qquad \text{Pythagorean identity}$$

$$2 \cos^2 x - 3 \cos x + 1 = 0 \qquad \text{Multiply each side by } -1.$$

$$(2 \cos x - 1)(\cos x - 1) = 0 \qquad \text{Factor.}$$

Set each factor equal to zero to find the solutions in the interval $[0, 2\pi)$.

$$2 \cos x - 1 = 0 \quad\Longrightarrow\quad \cos x = \frac{1}{2} \quad\Longrightarrow\quad x = \frac{\pi}{3}, \frac{5\pi}{3}$$

$$\cos x - 1 = 0 \quad\Longrightarrow\quad \cos x = 1 \quad\Longrightarrow\quad x = 0$$

Because $\cos x$ has a period of 2π, the general form of the solution is obtained by adding multiples of 2π to get

$$x = 2n\pi, \quad x = \frac{\pi}{3} + 2n\pi, \quad x = \frac{5\pi}{3} + 2n\pi \qquad \text{General solution}$$

where n is an integer.

✔CHECKPOINT   Now try Exercise 31.

Sometimes you must square each side of an equation to obtain a quadratic, as demonstrated in the next example. Because this procedure can introduce extraneous solutions, you should check any solutions in the original equation to see whether they are valid or extraneous.

Example 6   Squaring and Converting to Quadratic Type

Find all solutions of $\cos x + 1 = \sin x$ in the interval $[0, 2\pi)$.

Solution

It is not clear how to rewrite this equation in terms of a single trigonometric function. Notice what happens when you square each side of the equation.

$\cos x + 1 = \sin x$	Write original equation.
$\cos^2 x + 2 \cos x + 1 = \sin^2 x$	Square each side.
$\cos^2 x + 2 \cos x + 1 = 1 - \cos^2 x$	Pythagorean identity
$\cos^2 x + \cos^2 x + 2 \cos x + 1 - 1 = 0$	Rewrite equation.
$2 \cos^2 x + 2 \cos x = 0$	Combine like terms.
$2 \cos x(\cos x + 1) = 0$	Factor.

Setting each factor equal to zero produces

$$2 \cos x = 0 \qquad \text{and} \qquad \cos x + 1 = 0$$

$$\cos x = 0 \qquad\qquad\qquad \cos x = -1$$

$$x = \frac{\pi}{2}, \frac{3\pi}{2} \qquad\qquad\qquad x = \pi.$$

Because you squared the original equation, check for extraneous solutions.

Check $x = \pi/2$

$$\cos \frac{\pi}{2} + 1 \overset{?}{=} \sin \frac{\pi}{2} \qquad \text{Substitute } \pi/2 \text{ for } x.$$

$$0 + 1 = 1 \qquad \text{Solution checks. } \checkmark$$

Check $x = 3\pi/2$

$$\cos \frac{3\pi}{2} + 1 \overset{?}{=} \sin \frac{3\pi}{2} \qquad \text{Substitute } 3\pi/2 \text{ for } x.$$

$$0 + 1 \ne -1 \qquad \text{Solution does not check.}$$

Check $x = \pi$

$$\cos \pi + 1 \overset{?}{=} \sin \pi \qquad \text{Substitute } \pi \text{ for } x.$$

$$-1 + 1 = 0 \qquad \text{Solution checks. } \checkmark$$

Of the three possible solutions, $x = 3\pi/2$ is extraneous. So, in the interval $[0, 2\pi)$, the only two solutions are $x = \pi/2$ and $x = \pi$.

✓CHECKPOINT   Now try Exercise 33.

STUDY TIP

You square each side of the equation in Example 6 because the squares of the sine and cosine functions are related by a Pythagorean identity. The same is true for the squares of the secant and tangent functions and the cosecant and cotangent functions.

Exploration

Use a graphing utility to confirm the solutions found in Example 6 in two different ways. Do both methods produce the same x-values? Which method do you prefer? Why?

1. Graph both sides of the equation and find the x-coordinates of the points at which the graphs intersect.

    Left side: $y = \cos x + 1$

    Right side: $y = \sin x$

2. Graph the equation

    $$y = \cos x + 1 - \sin x$$

    and find the x-intercepts of the graph.

Functions Involving Multiple Angles

The next two examples involve trigonometric functions of multiple angles of the forms sin ku and cos ku. To solve equations of these forms, first solve the equation for ku, then divide your result by k.

Example 7 Functions of Multiple Angles

Solve $2 \cos 3t - 1 = 0$.

Solution

$$2 \cos 3t - 1 = 0 \qquad \text{Write original equation.}$$

$$2 \cos 3t = 1 \qquad \text{Add 1 to each side.}$$

$$\cos 3t = \frac{1}{2} \qquad \text{Divide each side by 2.}$$

In the interval $[0, 2\pi)$, you know that $3t = \pi/3$ and $3t = 5\pi/3$ are the only solutions, so, in general, you have

$$3t = \frac{\pi}{3} + 2n\pi \qquad \text{and} \qquad 3t = \frac{5\pi}{3} + 2n\pi.$$

Dividing these results by 3, you obtain the general solution

$$t = \frac{\pi}{9} + \frac{2n\pi}{3} \qquad \text{and} \qquad t = \frac{5\pi}{9} + \frac{2n\pi}{3} \qquad \text{General solution}$$

where n is an integer.

✓**CHECKPOINT** Now try Exercise 35.

Additional Examples
Find all solutions for $0 \leq x < 2\pi$.
a. $2 \sin^2 2x = 1$
   Solution:
   $x = \dfrac{\pi}{8}, \dfrac{3\pi}{8}, \dfrac{5\pi}{8}, \dfrac{7\pi}{8}, \dfrac{9\pi}{8}, \dfrac{11\pi}{8}, \dfrac{13\pi}{8}, \dfrac{15\pi}{8}$
b. $\tan 4x = 1$
   Solution:
   $x = \dfrac{\pi}{16}, \dfrac{5\pi}{16}, \dfrac{9\pi}{16}, \dfrac{13\pi}{16}, \dfrac{17\pi}{16}, \dfrac{21\pi}{16}, \dfrac{25\pi}{16}, \dfrac{29\pi}{16}$

Example 8 Functions of Multiple Angles

Solve $3 \tan \dfrac{x}{2} + 3 = 0$.

Solution

$$3 \tan \frac{x}{2} + 3 = 0 \qquad \text{Write original equation.}$$

$$3 \tan \frac{x}{2} = -3 \qquad \text{Subtract 3 from each side.}$$

$$\tan \frac{x}{2} = -1 \qquad \text{Divide each side by 3.}$$

In the interval $[0, \pi)$, you know that $x/2 = 3\pi/4$ is the only solution, so, in general, you have

$$\frac{x}{2} = \frac{3\pi}{4} + n\pi.$$

Multiplying this result by 2, you obtain the general solution

$$x = \frac{3\pi}{2} + 2n\pi \qquad \text{General solution}$$

where n is an integer.

Consider having your students check their solutions with a graphing utility. This will help to reinforce the fact that there are many solutions, and will demonstrate why.

✓**CHECKPOINT** Now try Exercise 39.

Using Inverse Functions

In the next example, you will see how inverse trigonometric functions can be used to solve an equation.

Example 9   Using Inverse Functions

Solve $\sec^2 x - 2 \tan x = 4$.

Solution

$$\sec^2 x - 2 \tan x = 4 \qquad \text{Write original equation.}$$

$$1 + \tan^2 x - 2 \tan x - 4 = 0 \qquad \text{Pythagorean identity}$$

$$\tan^2 x - 2 \tan x - 3 = 0 \qquad \text{Combine like terms.}$$

$$(\tan x - 3)(\tan x + 1) = 0 \qquad \text{Factor.}$$

Setting each factor equal to zero, you obtain two solutions in the interval $(-\pi/2, \pi/2)$. [Recall that the range of the inverse tangent function is $(-\pi/2, \pi/2)$.]

$\tan x - 3 = 0$	and	$\tan x + 1 = 0$
$\tan x = 3$		$\tan x = -1$
$x = \arctan 3$		$x = -\dfrac{\pi}{4}$

Finally, because $\tan x$ has a period of π, you obtain the general solution by adding multiples of π

$$x = \arctan 3 + n\pi \qquad \text{and} \qquad x = -\frac{\pi}{4} + n\pi \qquad \text{General solution}$$

where n is an integer. You can use a calculator to approximate the value of arctan 3.

✓CHECKPOINT   Now try Exercise 59.

Activities

1. Find all solutions of the equation without using a calculator.

    $\tan^2 x = 1$

    Answer: $x = \dfrac{\pi}{4} + n\pi, \dfrac{3\pi}{4} + n\pi$

2. Solve for x, $0 \le x < 2\pi$, without using a calculator.

    $\sin^2 3x - 2 \sin 3x + 1 = 0$

    Answer: $x = \dfrac{\pi}{6}, \dfrac{5\pi}{6}, \dfrac{3\pi}{2}$

3. Solve for x, $0 \le x < 2\pi$.

    $6 \cos^2 x - \cos x - 2 = 0$

    Answer: $x = 0.841, \dfrac{2\pi}{3}, \dfrac{4\pi}{3}, 5.442$

$\mathcal{W}$RITING ABOUT $\mathcal{M}$ATHEMATICS

Equations with No Solutions   One of the following equations has solutions and the other two do not. Which two equations do not have solutions?

a. $\sin^2 x - 5 \sin x + 6 = 0$

b. $\sin^2 x - 4 \sin x + 6 = 0$

c. $\sin^2 x - 5 \sin x - 6 = 0$

Find conditions involving the constants b and c that will guarantee that the equation

$$\sin^2 x + b \sin x + c = 0$$

has at least one solution on some interval of length 2π.

2.3 Exercises

VOCABULARY CHECK: Fill in the blanks.

1. The equation $2 \sin \theta + 1 = 0$ has the solutions $\theta = \dfrac{7\pi}{6} + 2n\pi$ and $\theta = \dfrac{11\pi}{6} + 2n\pi$, which are called _____ solutions.

2. The equation $2 \tan^2 x - 3 \tan x + 1 = 0$ is a trigonometric equation that is of _____ type.

3. A solution to an equation that does not satisfy the original equation is called an _____ solution.

PREREQUISITE SKILLS REVIEW: Practice and review algebra skills needed for this section at **www.Eduspace.com.**

In Exercises 1–6, verify that the x-values are solutions of the equation.

1. $2 \cos x - 1 = 0$

(a) $x = \dfrac{\pi}{3}$ (b) $x = \dfrac{5\pi}{3}$

2. $\sec x - 2 = 0$

(a) $x = \dfrac{\pi}{3}$ (b) $x = \dfrac{5\pi}{3}$

3. $3 \tan^2 2x - 1 = 0$

(a) $x = \dfrac{\pi}{12}$ (b) $x = \dfrac{5\pi}{12}$

4. $2 \cos^2 4x - 1 = 0$

(a) $x = \dfrac{\pi}{16}$ (b) $x = \dfrac{3\pi}{16}$

5. $2 \sin^2 x - \sin x - 1 = 0$

(a) $x = \dfrac{\pi}{2}$ (b) $x = \dfrac{7\pi}{6}$

6. $\csc^4 x - 4 \csc^2 x = 0$

(a) $x = \dfrac{\pi}{6}$ (b) $x = \dfrac{5\pi}{6}$

In Exercises 7–20, solve the equation.

7. $2 \cos x + 1 = 0$ **8.** $2 \sin x + 1 = 0$

9. $\sqrt{3} \csc x - 2 = 0$ **10.** $\tan x + \sqrt{3} = 0$

11. $3 \sec^2 x - 4 = 0$ **12.** $3 \cot^2 x - 1 = 0$

13. $\sin x(\sin x + 1) = 0$

14. $(3 \tan^2 x - 1)(\tan^2 x - 3) = 0$

15. $4 \cos^2 x - 1 = 0$ **16.** $\sin^2 x = 3 \cos^2 x$

17. $2 \sin^2 2x = 1$ **18.** $\tan^2 3x = 3$

19. $\tan 3x(\tan x - 1) = 0$ **20.** $\cos 2x(2 \cos x + 1) = 0$

In Exercises 21–34, find all solutions of the equation in the interval $[0, 2\pi)$.

21. $\cos^3 x = \cos x$ **22.** $\sec^2 x - 1 = 0$

23. $3 \tan^3 x = \tan x$ **24.** $2 \sin^2 x = 2 + \cos x$

25. $\sec^2 x - \sec x = 2$ **26.** $\sec x \csc x = 2 \csc x$

27. $2 \sin x + \csc x = 0$ **28.** $\sec x + \tan x = 1$

29. $2 \cos^2 x + \cos x - 1 = 0$

30. $2 \sin^2 x + 3 \sin x + 1 = 0$

31. $2 \sec^2 x + \tan^2 x - 3 = 0$

32. $\cos x + \sin x \tan x = 2$

33. $\csc x + \cot x = 1$

34. $\sin x - 2 = \cos x - 2$

In Exercises 35–40, solve the multiple-angle equation.

35. $\cos 2x = \dfrac{1}{2}$ **36.** $\sin 2x = -\dfrac{\sqrt{3}}{2}$

37. $\tan 3x = 1$ **38.** $\sec 4x = 2$

39. $\cos \dfrac{x}{2} = \dfrac{\sqrt{2}}{2}$ **40.** $\sin \dfrac{x}{2} = -\dfrac{\sqrt{3}}{2}$

In Exercises 41–44, find the x-intercepts of the graph.

41. $y = \sin \dfrac{\pi x}{2} + 1$ **42.** $y = \sin \pi x + \cos \pi x$

43. $y = \tan^2 \left(\dfrac{\pi x}{6} \right) - 3$ **44.** $y = \sec^4 \left(\dfrac{\pi x}{8} \right) - 4$

In Exercises 45–54, use a graphing utility to approximate the solutions (to three decimal places) of the equation in the interval $[0, 2\pi)$.

45. $2 \sin x + \cos x = 0$

46. $4 \sin^3 x + 2 \sin^2 x - 2 \sin x - 1 = 0$

47. $\dfrac{1 + \sin x}{\cos x} + \dfrac{\cos x}{1 + \sin x} = 4$

48. $\dfrac{\cos x \cot x}{1 - \sin x} = 3$

49. $x \tan x - 1 = 0$ **50.** $x \cos x - 1 = 0$

51. $\sec^2 x + 0.5 \tan x - 1 = 0$

52. $\csc^2 x + 0.5 \cot x - 5 = 0$

53. $2 \tan^2 x + 7 \tan x - 15 = 0$

54. $6 \sin^2 x - 7 \sin x + 2 = 0$

In Exercises 55–58, use the Quadratic Formula to solve the equation in the interval $[0, 2\pi)$. Then use a graphing utility to approximate the angle x.

55. $12 \sin^2 x - 13 \sin x + 3 = 0$

56. $3 \tan^2 x + 4 \tan x - 4 = 0$

57. $\tan^2 x + 3 \tan x + 1 = 0$

58. $4 \cos^2 x - 4 \cos x - 1 = 0$

In Exercises 59–62, use inverse functions where needed to find all solutions of the equation in the interval $[0, 2\pi)$.

59. $\tan^2 x - 6 \tan x + 5 = 0$

60. $\sec^2 x + \tan x - 3 = 0$

61. $2 \cos^2 x - 5 \cos x + 2 = 0$

62. $2 \sin^2 x - 7 \sin x + 3 = 0$

In Exercises 63 and 64, (a) use a graphing utility to graph the function and approximate the maximum and minimum points on the graph in the interval $[0, 2\pi)$, and (b) solve the trigonometric equation and demonstrate that its solutions are the x-coordinates of the maximum and minimum points of f. (Calculus is required to find the trigonometric equation.)

Function	Trigonometric Equation
63. $f(x) = \sin x + \cos x$	$\cos x - \sin x = 0$
64. $f(x) = 2 \sin x + \cos 2x$	$2 \cos x - 4 \sin x \cos x = 0$

Fixed Point **In Exercises 65 and 66, find the smallest positive fixed point of the function f. [A *fixed point* of a function f is a real number c such that $f(c) = c$.]**

65. $f(x) = \tan \dfrac{\pi x}{4}$ **66.** $f(x) = \cos x$

67. *Graphical Reasoning* Consider the function given by

$$f(x) = \cos \frac{1}{x}$$

and its graph shown in the figure.

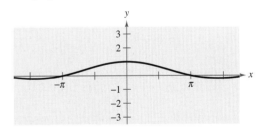

(a) What is the domain of the function?

(b) Identify any symmetry and any asymptotes of the graph.

(c) Describe the behavior of the function as $x \to 0$.

(d) How many solutions does the equation

$$\cos \frac{1}{x} = 0$$

have in the interval $[-1, 1]$? Find the solutions.

(e) Does the equation $\cos(1/x) = 0$ have a greatest solution? If so, approximate the solution. If not, explain why.

68. *Graphical Reasoning* Consider the function given by

$$f(x) = \frac{\sin x}{x}$$

and its graph shown in the figure.

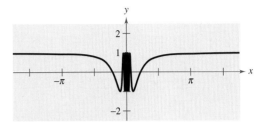

(a) What is the domain of the function?

(b) Identify any symmetry and any asymptotes of the graph.

(c) Describe the behavior of the function as $x \to 0$.

(d) How many solutions does the equation

$$\frac{\sin x}{x} = 0$$

have in the interval $[-8, 8]$? Find the solutions.

69. *Harmonic Motion* A weight is oscillating on the end of a spring (see figure). The position of the weight relative to the point of equilibrium is given by $y = \frac{1}{12}(\cos 8t - 3 \sin 8t)$, where y is the displacement (in meters) and t is the time (in seconds). Find the times when the weight is at the point of equilibrium $(y = 0)$ for $0 \le t \le 1$.

70. *Damped Harmonic Motion* The displacement from equilibrium of a weight oscillating on the end of a spring is given by $y = 1.56t^{-1/2}\cos 1.9t$, where y is the displacement (in feet) and t is the time (in seconds). Use a graphing utility to graph the displacement function for $0 \le t \le 10$. Find the time beyond which the displacement does not exceed 1 foot from equilibrium.

71. *Sales* The monthly sales S (in thousands of units) of a seasonal product are approximated by

$$S = 74.50 + 43.75 \sin \frac{\pi t}{6}$$

where t is the time (in months), with $t = 1$ corresponding to January. Determine the months when sales exceed 100,000 units.

72. *Sales* The monthly sales S (in hundreds of units) of skiing equipment at a sports store are approximated by

$$S = 58.3 + 32.5 \cos \frac{\pi t}{6}$$

where t is the time (in months), with $t = 1$ corresponding to January. Determine the months when sales exceed 7500 units.

73. *Projectile Motion* A batted baseball leaves the bat at an angle of θ with the horizontal and an initial velocity of $v_0 = 100$ feet per second. The ball is caught by an outfielder 300 feet from home plate (see figure). Find θ if the range r of a projectile is given by $r = \frac{1}{32}v_0^2 \sin 2\theta$.

Not drawn to scale

74. *Projectile Motion* A sharpshooter intends to hit a target at a distance of 1000 yards with a gun that has a muzzle velocity of 1200 feet per second (see figure). Neglecting air resistance, determine the gun's minimum angle of elevation θ if the range r is given by

$$r = \frac{1}{32}v_0^2 \sin 2\theta.$$

Not drawn to scale

75. *Ferris Wheel* A Ferris wheel is built such that the height h (in feet) above ground of a seat on the wheel at time t (in minutes) can be modeled by

$$h(t) = 53 + 50 \sin\left(\frac{\pi}{16}t - \frac{\pi}{2}\right).$$

The wheel makes one revolution every 32 seconds. The ride begins when $t = 0$.

(a) During the first 32 seconds of the ride, when will a person on the Ferris wheel be 53 feet above ground?

(b) When will a person be at the top of the Ferris wheel for the first time during the ride? If the ride lasts 160 seconds, how many times will a person be at the top of the ride, and at what times?

Model It

76. *Data Analysis: Unemployment Rate* The table shows the unemployment rates r in the United States for selected years from 1990 through 2004. The time t is measured in years, with $t = 0$ corresponding to 1990. (Source: U.S. Bureau of Labor Statistics)

Time, t	Rate, r	Time, t	Rate, r
0	5.6	8	4.5
2	7.5	10	4.0
4	6.1	12	5.8
6	5.4	14	5.5

(a) Create a scatter plot of the data.

Model It (continued)

(b) Which of the following models best represents the data? Explain your reasoning.

    (1) $r = 1.24 \sin(0.47t + 0.40) + 5.45$

    (2) $r = 1.24 \sin(0.47t - 0.01) + 5.45$

    (3) $r = \sin(0.10t + 5.61) + 4.80$

    (4) $r = 896 \sin(0.57t - 2.05) + 6.48$

(c) What term in the model gives the average unemployment rate? What is the rate?

(d) Economists study the lengths of business cycles such as cycles of unemployment rates. Based on this short span of time, use the model to find the length of this cycle.

(e) Use the model to estimate the next time the unemployment rate will be 5% or less.

---

77. *Geometry* The area of a rectangle (see figure) inscribed in one arc of the graph of $y = \cos x$ is given by

$$A = 2x \cos x, \quad 0 < x < \frac{\pi}{2}.$$

 (a) Use a graphing utility to graph the area function, and approximate the area of the largest inscribed rectangle.

(b) Determine the values of x for which $A \geq 1$.

78. *Quadratic Approximation* Consider the function given by $f(x) = 3 \sin(0.6x - 2)$.

(a) Approximate the zero of the function in the interval $[0, 6]$.

 (b) A quadratic approximation agreeing with f at $x = 5$ is $g(x) = -0.45x^2 + 5.52x - 13.70$. Use a graphing utility to graph f and g in the same viewing window. Describe the result.

(c) Use the Quadratic Formula to find the zeros of g. Compare the zero in the interval $[0, 6]$ with the result of part (a).

Synthesis

True or False?   In Exercises 79 and 80, determine whether the statement is true or false. Justify your answer.

79. The equation $2 \sin 4t - 1 = 0$ has four times the number of solutions in the interval $[0, 2\pi)$ as the equation $2 \sin t - 1 = 0$.

80. If you correctly solve a trigonometric equation to the statement $\sin x = 3.4$, then you can finish solving the equation by using an inverse function.

In Exercises 81 and 82, use the graph to approximate the number of points of intersection of the graphs of y_1 and y_2.

81. $y_1 = 2 \sin x$
$y_2 = 3x + 1$

82. $y_1 = 2 \sin x$
$y_2 = \frac{1}{2}x + 1$

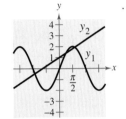

Skills Review

In Exercises 83 and 84, solve triangle ABC by finding all missing angle measures and side lengths.

83.

84.

In Exercises 85–88, use reference angles to find the exact values of the sine, cosine, and tangent of the angle with the given measure.

85. $390°$ **86.** $600°$

87. $-1845°$ **88.** $-1410°$

89. *Angle of Depression* Find the angle of depression from the top of a lighthouse 250 feet above water level to the water line of a ship 2 miles offshore.

90. *Height* From a point 100 feet in front of a public library, the angles of elevation to the base of the flagpole and the top of the pole are $28°$ and $39° \, 45'$, respectively. The flagpole is mounted on the front of the library's roof. Find the height of the flagpole.

---

91. Make a Decision To work an extended application analyzing the normal daily high temperatures in Phoenix and in Seattle, visit this text's website at *college.hmco.com*. *(Data Source: NOAA)*

2.4 Sum and Difference Formulas

What you should learn

- Use sum and difference formulas to evaluate trigonometric functions, verify identities, and solve trigonometric equations.

Why you should learn it

You can use identities to rewrite trigonometric expressions. For instance, in Exercise 75 on page 253, you can use an identity to rewrite a trigonometric expression in a form that helps you analyze a harmonic motion equation.

Richard Megna/Fundamental Photographs

Using Sum and Difference Formulas

In this and the following section, you will study the uses of several trigonometric identities and formulas.

Sum and Difference Formulas

$$\sin(u + v) = \sin u \cos v + \cos u \sin v$$

$$\sin(u - v) = \sin u \cos v - \cos u \sin v$$

$$\cos(u + v) = \cos u \cos v - \sin u \sin v$$

$$\cos(u - v) = \cos u \cos v + \sin u \sin v$$

$$\tan(u + v) = \frac{\tan u + \tan v}{1 - \tan u \tan v}$$

$$\tan(u - v) = \frac{\tan u - \tan v}{1 + \tan u \tan v}$$

For a proof of the sum and difference formulas, see Proofs in Mathematics on page 272.

Exploration

Use a graphing utility to graph $y_1 = \cos(x + 2)$ and $y_2 = \cos x + \cos 2$ in the same viewing window. What can you conclude about the graphs? Is it true that $\cos(x + 2) = \cos x + \cos 2$?

Use a graphing utility to graph $y_1 = \sin(x + 4)$ and $y_2 = \sin x + \sin 4$ in the same viewing window. What can you conclude about the graphs? Is it true that $\sin(x + 4) = \sin x + \sin 4$?

Examples 1 and 2 show how **sum and difference formulas** can be used to find exact values of trigonometric functions involving sums or differences of special angles.

Example 1 **Evaluating a Trigonometric Function**

Find the exact value of $\cos 75°$.

Solution

To find the *exact* value of $\cos 75°$, use the fact that $75° = 30° + 45°$. Consequently, the formula for $\cos(u + v)$ yields

$$\cos 75° = \cos(30° + 45°)$$

$$= \cos 30° \cos 45° - \sin 30° \sin 45°$$

$$= \frac{\sqrt{3}}{2}\left(\frac{\sqrt{2}}{2}\right) - \frac{1}{2}\left(\frac{\sqrt{2}}{2}\right) = \frac{\sqrt{6} - \sqrt{2}}{4}.$$

Try checking this result on your calculator. You will find that $\cos 75° \approx 0.259$.

✓CHECKPOINT Now try Exercise 1.

Historical Note
Hipparchus, considered the most eminent of Greek astronomers, was born about 160 B.C. in Nicaea. He was credited with the invention of trigonometry. He also derived the sum and difference formulas for $\sin(A \pm B)$ and $\cos(A \pm B)$.

Example 2   **Evaluating a Trigonometric Expression**

Find the exact value of $\sin \dfrac{\pi}{12}$.

Solution

Using the fact that

$$\frac{\pi}{12} = \frac{\pi}{3} - \frac{\pi}{4}$$

together with the formula for $\sin(u - v)$, you obtain

$$\sin \frac{\pi}{12} = \sin\left(\frac{\pi}{3} - \frac{\pi}{4}\right)$$

$$= \sin \frac{\pi}{3} \cos \frac{\pi}{4} - \cos \frac{\pi}{3} \sin \frac{\pi}{4}$$

$$= \frac{\sqrt{3}}{2}\left(\frac{\sqrt{2}}{2}\right) - \frac{1}{2}\left(\frac{\sqrt{2}}{2}\right)$$

$$= \frac{\sqrt{6} - \sqrt{2}}{4}.$$

✓*CHECKPOINT*   Now try Exercise 3.

Example 3   **Evaluating a Trigonometric Expression**

Find the exact value of $\sin 42° \cos 12° - \cos 42° \sin 12°$.

Solution

Recognizing that this expression fits the formula for $\sin(u - v)$, you can write

$$\sin 42° \cos 12° - \cos 42° \sin 12° = \sin(42° - 12°)$$

$$= \sin 30°$$

$$= \tfrac{1}{2}.$$

✓*CHECKPOINT*   Now try Exercise 31.

Example 4   **An Application of a Sum Formula**

Write $\cos(\arctan 1 + \arccos x)$ as an algebraic expression.

Solution

This expression fits the formula for $\cos(u + v)$. Angles $u = \arctan 1$ and $v = \arccos x$ are shown in Figure 2.7. So

$$\cos(u + v) = \cos(\arctan 1) \cos(\arccos x) - \sin(\arctan 1) \sin(\arccos x)$$

$$= \frac{1}{\sqrt{2}} \cdot x - \frac{1}{\sqrt{2}} \cdot \sqrt{1 - x^2}$$

$$= \frac{x - \sqrt{1 - x^2}}{\sqrt{2}}.$$

✓*CHECKPOINT*   Now try Exercise 51.

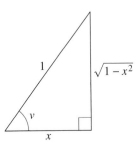

FIGURE 2.7

Example 5 shows how to use a difference formula to prove the cofunction identity

$$\cos\left(\frac{\pi}{2} - x\right) = \sin x.$$

Example 5 Proving a Cofunction Identity

Prove the cofunction identity $\cos\left(\frac{\pi}{2} - x\right) = \sin x.$

Solution

Using the formula for $\cos(u - v)$, you have

$$\cos\left(\frac{\pi}{2} - x\right) = \cos\frac{\pi}{2}\cos x + \sin\frac{\pi}{2}\sin x$$

$$= (0)(\cos x) + (1)(\sin x) = \sin x.$$

✓CHECKPOINT   Now try Exercise 55.

Sum and difference formulas can be used to rewrite expressions such as

$$\sin\left(\theta + \frac{n\pi}{2}\right) \quad \text{and} \quad \cos\left(\theta + \frac{n\pi}{2}\right), \quad \text{where } n \text{ is an integer}$$

as expressions involving only $\sin\theta$ or $\cos\theta$. The resulting formulas are called **reduction formulas.**

Example 6 Deriving Reduction Formulas

Simplify each expression.

a. $\cos\left(\theta - \frac{3\pi}{2}\right)$ **b.** $\tan(\theta + 3\pi)$

Solution

a. Using the formula for $\cos(u - v)$, you have

$$\cos\left(\theta - \frac{3\pi}{2}\right) = \cos\theta\cos\frac{3\pi}{2} + \sin\theta\sin\frac{3\pi}{2}$$

$$= (\cos\theta)(0) + (\sin\theta)(-1)$$

$$= -\sin\theta.$$

b. Using the formula for $\tan(u + v)$, you have

$$\tan(\theta + 3\pi) = \frac{\tan\theta + \tan 3\pi}{1 - \tan\theta\tan 3\pi}$$

$$= \frac{\tan\theta + 0}{1 - (\tan\theta)(0)}$$

$$= \tan\theta.$$

✓CHECKPOINT   Now try Exercise 65.

Activities

1. Use the sum and difference formulas to find the exact value of $\cos 15°$.

   Answer: $\dfrac{\sqrt{6} + \sqrt{2}}{4}$

2. Rewrite the expression using the sum and difference formulas.

   $\dfrac{\tan 40° + \tan 10°}{1 - \tan 40°\tan 10°}$

   Answer: $\tan(40° + 10°) = \tan 50°$

3. Verify the identity

   $\sin\left(\dfrac{\pi}{2} - \theta\right) = \cos\theta.$

   Answer:

   $\sin\left(\dfrac{\pi}{2} - \theta\right) = \sin\dfrac{\pi}{2}\cos\theta - \cos\dfrac{\pi}{2}\sin\theta$

   $= (1)\cos\theta - (0)\sin\theta$

   $= \cos\theta$

Example 7   **Solving a Trigonometric Equation**

Find all solutions of $\sin\left(x + \dfrac{\pi}{4}\right) + \sin\left(x - \dfrac{\pi}{4}\right) = -1$ in the interval $[0, 2\pi)$.

Solution

Using sum and difference formulas, rewrite the equation as

$$\sin x \cos \frac{\pi}{4} + \cos x \sin \frac{\pi}{4} + \sin x \cos \frac{\pi}{4} - \cos x \sin \frac{\pi}{4} = -1$$

$$2 \sin x \cos \frac{\pi}{4} = -1$$

$$2(\sin x)\left(\frac{\sqrt{2}}{2}\right) = -1$$

$$\sin x = -\frac{1}{\sqrt{2}}$$

$$\sin x = -\frac{\sqrt{2}}{2}.$$

So, the only solutions in the interval $[0, 2\pi)$ are

$$x = \frac{5\pi}{4} \qquad \text{and} \qquad x = \frac{7\pi}{4}.$$

You can confirm this graphically by sketching the graph of

$$y = \sin\left(x + \frac{\pi}{4}\right) + \sin\left(x - \frac{\pi}{4}\right) + 1 \text{ for } 0 \leq x < 2\pi,$$

as shown in Figure 2.8. From the graph you can see that the x-intercepts are $5\pi/4$ and $7\pi/4$.

✔CHECKPOINT   Now try Exercise 69.

$$y = \sin\left(x + \frac{\pi}{4}\right) + \sin\left(x - \frac{\pi}{4}\right) + 1$$

FIGURE 2.8

The next example was taken from calculus. It is used to derive the derivative of the sine function.

Example 8   **An Application from Calculus**

Verify that

$$\frac{\sin(x + h) - \sin x}{h} = (\cos x)\left(\frac{\sin h}{h}\right) - (\sin x)\left(\frac{1 - \cos h}{h}\right)$$

where $h \neq 0$.

Solution

Using the formula for $\sin(u + v)$, you have

$$\frac{\sin(x + h) - \sin x}{h} = \frac{\sin x \cos h + \cos x \sin h - \sin x}{h}$$

$$= \frac{\cos x \sin h - \sin x(1 - \cos h)}{h}$$

$$= (\cos x)\left(\frac{\sin h}{h}\right) - (\sin x)\left(\frac{1 - \cos h}{h}\right).$$

✔CHECKPOINT   Now try Exercise 91.

2.4 Exercises

VOCABULARY CHECK: Fill in the blank to complete the trigonometric identity.

1. $\sin(u - v) =$ _____

2. $\cos(u + v) =$ _____

3. $\tan(u + v) =$ _____

4. $\sin(u + v) =$ _____

5. $\cos(u - v) =$ _____

6. $\tan(u - v) =$ _____

PREREQUISITE SKILLS REVIEW: Practice and review algebra skills needed for this section at **www.Eduspace.com.**

In Exercises 1–6, find the exact value of each expression.

1. (a) $\cos(120° + 45°)$ (b) $\cos 120° + \cos 45°$

2. (a) $\sin(135° - 30°)$ (b) $\sin 135° - \cos 30°$

3. (a) $\cos\left(\dfrac{\pi}{4} + \dfrac{\pi}{3}\right)$ (b) $\cos\dfrac{\pi}{4} + \cos\dfrac{\pi}{3}$

4. (a) $\sin\left(\dfrac{3\pi}{4} + \dfrac{5\pi}{6}\right)$ (b) $\sin\dfrac{3\pi}{4} + \sin\dfrac{5\pi}{6}$

5. (a) $\sin\left(\dfrac{7\pi}{6} - \dfrac{\pi}{3}\right)$ (b) $\sin\dfrac{7\pi}{6} - \sin\dfrac{\pi}{3}$

6. (a) $\sin(315° - 60°)$ (b) $\sin 315° - \sin 60°$

In Exercises 7–22, find the exact values of the sine, cosine, and tangent of the angle by using a sum or difference formula.

7. $105° = 60° + 45°$

8. $165° = 135° + 30°$

9. $195° = 225° - 30°$

10. $255° = 300° - 45°$

11. $\dfrac{11\pi}{12} = \dfrac{3\pi}{4} + \dfrac{\pi}{6}$

12. $\dfrac{7\pi}{12} = \dfrac{\pi}{3} + \dfrac{\pi}{4}$

13. $\dfrac{17\pi}{12} = \dfrac{9\pi}{4} - \dfrac{5\pi}{6}$

14. $-\dfrac{\pi}{12} = \dfrac{\pi}{6} - \dfrac{\pi}{4}$

15. $285°$

16. $-105°$

17. $-165°$

18. $15°$

19. $\dfrac{13\pi}{12}$

20. $-\dfrac{7\pi}{12}$

21. $-\dfrac{13\pi}{12}$

22. $\dfrac{5\pi}{12}$

In Exercises 23–30, write the expression as the sine, cosine, or tangent of an angle.

23. $\cos 25° \cos 15° - \sin 25° \sin 15°$

24. $\sin 140° \cos 50° + \cos 140° \sin 50°$

25. $\dfrac{\tan 325° - \tan 86°}{1 + \tan 325° \tan 86°}$

26. $\dfrac{\tan 140° - \tan 60°}{1 + \tan 140° \tan 60°}$

27. $\sin 3 \cos 1.2 - \cos 3 \sin 1.2$

28. $\cos\dfrac{\pi}{7} \cos\dfrac{\pi}{5} - \sin\dfrac{\pi}{7} \sin\dfrac{\pi}{5}$

29. $\dfrac{\tan 2x + \tan x}{1 - \tan 2x \tan x}$

30. $\cos 3x \cos 2y + \sin 3x \sin 2y$

In Exercises 31–36, find the exact value of the expression.

31. $\sin 330° \cos 30° - \cos 330° \sin 30°$

32. $\cos 15° \cos 60° + \sin 15° \sin 60°$

33. $\sin\dfrac{\pi}{12} \cos\dfrac{\pi}{4} + \cos\dfrac{\pi}{12} \sin\dfrac{\pi}{4}$

34. $\cos\dfrac{\pi}{16} \cos\dfrac{3\pi}{16} - \sin\dfrac{\pi}{16} \sin\dfrac{3\pi}{16}$

35. $\dfrac{\tan 25° + \tan 110°}{1 - \tan 25° \tan 110°}$

36. $\dfrac{\tan(5\pi/4) - \tan(\pi/12)}{1 + \tan(5\pi/4) \tan(\pi/12)}$

In Exercises 37–44, find the exact value of the trigonometric function given that $\sin u = \dfrac{5}{13}$ and $\cos v = -\dfrac{3}{5}$. (Both u and v are in Quadrant II.)

37. $\sin(u + v)$

38. $\cos(u - v)$

39. $\cos(u + v)$

40. $\sin(v - u)$

41. $\tan(u + v)$

42. $\csc(u - v)$

43. $\sec(v - u)$

44. $\cot(u + v)$

In Exercises 45–50, find the exact value of the trigonometric function given that $\sin u = -\dfrac{7}{25}$ and $\cos v = -\dfrac{4}{5}$. (Both u and v are in Quadrant III.)

45. $\cos(u + v)$

46. $\sin(u + v)$

47. $\tan(u - v)$

48. $\cot(v - u)$

49. $\sec(u + v)$

50. $\cos(u - v)$

In Exercises 51–54, write the trigonometric expression as an algebraic expression.

51. $\sin(\arcsin x + \arccos x)$ **52.** $\sin(\arctan 2x - \arccos x)$

53. $\cos(\arccos x + \arcsin x)$

54. $\cos(\arccos x - \arctan x)$

In Exercises 55–64, verify the identity.

55. $\sin(3\pi - x) = \sin x$ **56.** $\sin\left(\dfrac{\pi}{2} + x\right) = \cos x$

57. $\sin\left(\dfrac{\pi}{6} + x\right) = \dfrac{1}{2}(\cos x + \sqrt{3}\sin x)$

58. $\cos\left(\dfrac{5\pi}{4} - x\right) = -\dfrac{\sqrt{2}}{2}(\cos x + \sin x)$

59. $\cos(\pi - \theta) + \sin\left(\dfrac{\pi}{2} + \theta\right) = 0$

60. $\tan\left(\dfrac{\pi}{4} - \theta\right) = \dfrac{1 - \tan\theta}{1 + \tan\theta}$

61. $\cos(x + y)\cos(x - y) = \cos^2 x - \sin^2 y$

62. $\sin(x + y)\sin(x - y) = \sin^2 x - \sin^2 y$

63. $\sin(x + y) + \sin(x - y) = 2\sin x \cos y$

64. $\cos(x + y) + \cos(x - y) = 2\cos x \cos y$

In Exercises 65–68, simplify the expression algebraically and use a graphing utility to confirm your answer graphically.

65. $\cos\left(\dfrac{3\pi}{2} - x\right)$ **66.** $\cos(\pi + x)$

67. $\sin\left(\dfrac{3\pi}{2} + \theta\right)$ **68.** $\tan(\pi + \theta)$

In Exercises 69–72, find all solutions of the equation in the interval $[0, 2\pi)$.

69. $\sin\left(x + \dfrac{\pi}{3}\right) + \sin\left(x - \dfrac{\pi}{3}\right) = 1$

70. $\sin\left(x + \dfrac{\pi}{6}\right) - \sin\left(x - \dfrac{\pi}{6}\right) = \dfrac{1}{2}$

71. $\cos\left(x + \dfrac{\pi}{4}\right) - \cos\left(x - \dfrac{\pi}{4}\right) = 1$

72. $\tan(x + \pi) + 2\sin(x + \pi) = 0$

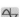 **In Exercises 73 and 74, use a graphing utility to approximate the solutions in the interval $[0, 2\pi)$.**

73. $\cos\left(x + \dfrac{\pi}{4}\right) + \cos\left(x - \dfrac{\pi}{4}\right) = 1$

74. $\tan(x + \pi) - \cos\left(x + \dfrac{\pi}{2}\right) = 0$

Model It

75. *Harmonic Motion*   A weight is attached to a spring suspended vertically from a ceiling. When a driving force is applied to the system, the weight moves vertically from its equilibrium position, and this motion is modeled by

$$y = \frac{1}{3}\sin 2t + \frac{1}{4}\cos 2t$$

where y is the distance from equilibrium (in feet) and t is the time (in seconds).

(a) Use the identity

$$a \sin B\theta + b \cos B\theta = \sqrt{a^2 + b^2}\,\sin(B\theta + C)$$

where $C = \arctan(b/a)$, $a > 0$, to write the model in the form

$$y = \sqrt{a^2 + b^2}\,\sin(Bt + C).$$

(b) Find the amplitude of the oscillations of the weight.

(c) Find the frequency of the oscillations of the weight.

76. *Standing Waves*   The equation of a standing wave is obtained by adding the displacements of two waves traveling in opposite directions (see figure). Assume that each of the waves has amplitude A, period T, and wavelength λ. If the models for these waves are

$$y_1 = A \cos 2\pi\left(\frac{t}{T} - \frac{x}{\lambda}\right) \quad \text{and} \quad y_2 = A \cos 2\pi\left(\frac{t}{T} + \frac{x}{\lambda}\right)$$

show that

$$y_1 + y_2 = 2A \cos\frac{2\pi t}{T}\cos\frac{2\pi x}{\lambda}.$$

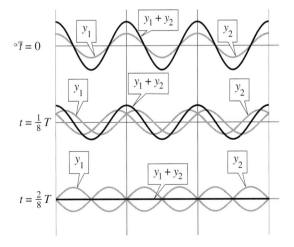

Synthesis

True or False?   **In Exercises 77–80, determine whether the statement is true or false. Justify your answer.**

77. $\sin(u \pm v) = \sin u \pm \sin v$

78. $\cos(u \pm v) = \cos u \pm \cos v$

79. $\cos\left(x - \dfrac{\pi}{2}\right) = -\sin x$ **80.** $\sin\left(x - \dfrac{\pi}{2}\right) = -\cos x$

In Exercises 81–84, verify the identity.

81. $\cos(n\pi + \theta) = (-1)^n \cos\theta,$ n is an integer

82. $\sin(n\pi + \theta) = (-1)^n \sin\theta,$ n is an integer

83. $a\sin B\theta + b\cos B\theta = \sqrt{a^2 + b^2}\sin(B\theta + C),$
where $C = \arctan(b/a)$ and $a > 0$

84. $a\sin B\theta + b\cos B\theta = \sqrt{a^2 + b^2}\cos(B\theta - C),$
where $C = \arctan(a/b)$ and $b > 0$

In Exercises 85–88, use the formulas given in Exercises 83 and 84 to write the trigonometric expression in the following forms.

(a) $\sqrt{a^2 + b^2}\sin(B\theta + C)$ **(b)** $\sqrt{a^2 + b^2}\cos(B\theta - C)$

85. $\sin\theta + \cos\theta$ **86.** $3\sin 2\theta + 4\cos 2\theta$

87. $12\sin 3\theta + 5\cos 3\theta$ **88.** $\sin 2\theta - \cos 2\theta$

In Exercises 89 and 90, use the formulas given in Exercises 83 and 84 to write the trigonometric expression in the form $a\sin B\theta + b\cos B\theta$.

89. $2\sin\left(\theta + \dfrac{\pi}{2}\right)$ **90.** $5\cos\left(\theta + \dfrac{3\pi}{4}\right)$

91. Verify the following identity used in calculus.

$$\dfrac{\cos(x + h) - \cos x}{h}$$
$$= \dfrac{\cos x(\cos h - 1)}{h} - \dfrac{\sin x \sin h}{h}$$

92. ***Exploration***   Let $x = \pi/6$ in the identity in Exercise 91 and define the functions f and g as follows.
$$f(h) = \dfrac{\cos(\pi/6 + h) - \cos(\pi/6)}{h}$$
$$g(h) = \cos\dfrac{\pi}{6}\left(\dfrac{\cos h - 1}{h}\right) - \sin\dfrac{\pi}{6}\left(\dfrac{\sin h}{h}\right)$$

(a) What are the domains of the functions f and g?

 (b) Use a graphing utility to complete the table.

h	0.01	0.02	0.05	0.1	0.2	0.5
$f(h)$						
$g(h)$						

 (c) Use a graphing utility to graph the functions f and g.

(d) Use the table and the graphs to make a conjecture about the values of the functions f and g as $h \to 0$.

In Exercises 93 and 94, use the figure, which shows two lines whose equations are

$$y_1 = m_1x + b_1 \quad \text{and} \quad y_2 = m_2x + b_2.$$

Assume that both lines have positive slopes. Derive a formula for the angle between the two lines. Then use your formula to find the angle between the given pair of lines.

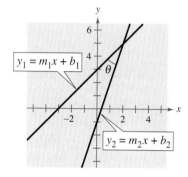

93. $y = x$ and $y = \sqrt{3}x$

94. $y = x$ and $y = \dfrac{1}{\sqrt{3}}x$

 95. ***Conjecture***   Consider the function given by
$$f(\theta) = \sin^2\left(\theta + \dfrac{\pi}{4}\right) + \sin^2\left(\theta - \dfrac{\pi}{4}\right).$$
Use a graphing utility to graph the function and use the graph to create an identity. Prove your conjecture.

96. ***Proof***

(a) Write a proof of the formula for $\sin(u + v)$.

(b) Write a proof of the formula for $\sin(u - v)$.

Skills Review

In Exercises 97–100, find the inverse function of f. Verify that $f(f^{-1}(x)) = x$ and $f^{-1}(f(x)) = x$.

97. $f(x) = 5(x - 3)$

98. $f(x) = \dfrac{7 - x}{8}$

99. $f(x) = x^2 - 8$

100. $f(x) = \sqrt{x - 16}$

What you should learn

- Use multiple-angle formulas to rewrite and evaluate trigonometric functions.
- Use power-reducing formulas to rewrite and evaluate trigonometric functions.
- Use half-angle formulas to rewrite and evaluate trigonometric functions.
- Use product-to-sum and sum-to-product formulas to rewrite and evaluate trigonometric functions.
- Use trigonometric formulas to rewrite real-life models.

Why you should learn it

You can use a variety of trigonometric formulas to rewrite trigonometric functions in more convenient forms. For instance, in Exercise 119 on page 265, you can use a double-angle formula to determine at what angle an athlete must throw a javelin.

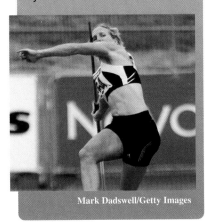

Mark Dadswell/Getty Images

Consider having your students prove the double-angle formulas. Double-angle formulas are often used in calculus.

Multiple-Angle Formulas

In this section, you will study four other categories of trigonometric identities.

1. The first category involves *functions of multiple angles* such as $\sin ku$ and $\cos ku$.

2. The second category involves *squares of trigonometric functions* such as $\sin^2 u$.

3. The third category involves *functions of half-angles* such as $\sin(u/2)$.

4. The fourth category involves *products of trigonometric functions* such as $\sin u \cos v$.

You should learn the **double-angle formulas** because they are used often in trigonometry and calculus. For proofs of the formulas, see Proofs in Mathematics on page 273.

Double-Angle Formulas

$$\sin 2u = 2 \sin u \cos u \qquad \cos 2u = \cos^2 u - \sin^2 u$$

$$\tan 2u = \frac{2 \tan u}{1 - \tan^2 u} \qquad\qquad = 2 \cos^2 u - 1$$

$$= 1 - 2 \sin^2 u$$

| **Example 1** | **Solving a Multiple-Angle Equation** |

Solve $2 \cos x + \sin 2x = 0$.

Solution

Begin by rewriting the equation so that it involves functions of x (rather than $2x$). Then factor and solve as usual.

$$2 \cos x + \sin 2x = 0 \qquad \text{Write original equation.}$$

$$2 \cos x + 2 \sin x \cos x = 0 \qquad \text{Double-angle formula}$$

$$2 \cos x(1 + \sin x) = 0 \qquad \text{Factor.}$$

$$2 \cos x = 0 \quad \text{and} \quad 1 + \sin x = 0 \qquad \text{Set factors equal to zero.}$$

$$x = \frac{\pi}{2}, \frac{3\pi}{2} \qquad\qquad x = \frac{3\pi}{2} \qquad \text{Solutions in } [0, 2\pi)$$

So, the general solution is

$$x = \frac{\pi}{2} + 2n\pi \qquad \text{and} \qquad x = \frac{3\pi}{2} + 2n\pi$$

where n is an integer. Try verifying these solutions graphically.

✓CHECKPOINT   Now try Exercise 9.

Example 2 Using Double-Angle Formulas to Analyze Graphs

Use a double-angle formula to rewrite the equation

$$y = 4\cos^2 x - 2.$$

Then sketch the graph of the equation over the interval $[0, 2\pi]$.

Solution

Using the double-angle formula for $\cos 2u$, you can rewrite the original equation as

$y = 4\cos^2 x - 2$	Write original equation.
$\quad = 2(2\cos^2 x - 1)$	Factor.
$\quad = 2\cos 2x.$	Use double-angle formula.

Using the techniques discussed in Section 1.5, you can recognize that the graph of this function has an amplitude of 2 and a period of π. The key points in the interval $[0, \pi]$ are as follows.

Maximum	Intercept	Minimum	Intercept	Maximum
$(0, 2)$	$\left(\dfrac{\pi}{4}, 0\right)$	$\left(\dfrac{\pi}{2}, -2\right)$	$\left(\dfrac{3\pi}{4}, 0\right)$	$(\pi, 2)$

Two cycles of the graph are shown in Figure 2.9.

✓CHECKPOINT Now try Exercise 21.

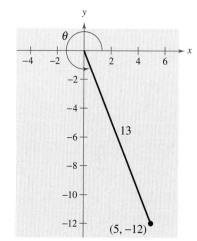

FIGURE 2.9

Example 3 Evaluating Functions Involving Double Angles

Use the following to find $\sin 2\theta$, $\cos 2\theta$, and $\tan 2\theta$.

$$\cos\theta = \frac{5}{13}, \qquad \frac{3\pi}{2} < \theta < 2\pi$$

Solution

From Figure 2.10, you can see that $\sin\theta = y/r = -12/13$. Consequently, using each of the double-angle formulas, you can write

$$\sin 2\theta = 2\sin\theta\cos\theta = 2\left(-\frac{12}{13}\right)\left(\frac{5}{13}\right) = -\frac{120}{169}$$

$$\cos 2\theta = 2\cos^2\theta - 1 = 2\left(\frac{25}{169}\right) - 1 = -\frac{119}{169}$$

$$\tan 2\theta = \frac{\sin 2\theta}{\cos 2\theta} = \frac{120}{119}.$$

✓CHECKPOINT Now try Exercise 23.

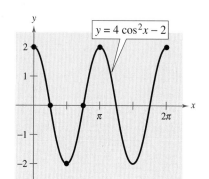

FIGURE 2.10

The double-angle formulas are not restricted to angles 2θ and θ. Other *double* combinations, such as 4θ and 2θ or 6θ and 3θ, are also valid. Here are two examples.

$$\sin 4\theta = 2\sin 2\theta\cos 2\theta \qquad \text{and} \qquad \cos 6\theta = \cos^2 3\theta - \sin^2 3\theta$$

By using double-angle formulas together with the sum formulas given in the preceding section, you can form other multiple-angle formulas.

Example 4 **Deriving a Triple-Angle Formula**

$$\sin 3x = \sin(2x + x)$$

$$= \sin 2x \cos x + \cos 2x \sin x$$

$$= 2 \sin x \cos x \cos x + (1 - 2 \sin^2 x)\sin x$$

$$= 2 \sin x \cos^2 x + \sin x - 2 \sin^3 x$$

$$= 2 \sin x(1 - \sin^2 x) + \sin x - 2 \sin^3 x$$

$$= 2 \sin x - 2 \sin^3 x + \sin x - 2 \sin^3 x$$

$$= 3 \sin x - 4 \sin^3 x$$

✓CHECKPOINT   Now try Exercise 97.

Power-Reducing Formulas

The double-angle formulas can be used to obtain the following **power-reducing formulas.** Example 5 shows a typical power reduction that is used in calculus.

Power-reducing formulas are often used in calculus.

> **Power-Reducing Formulas**
>
> $$\sin^2 u = \frac{1 - \cos 2u}{2} \qquad \cos^2 u = \frac{1 + \cos 2u}{2} \qquad \tan^2 u = \frac{1 - \cos 2u}{1 + \cos 2u}$$

For a proof of the power-reducing formulas, see Proofs in Mathematics on page 273.

Example 5 **Reducing a Power**

Rewrite $\sin^4 x$ as a sum of first powers of the cosines of multiple angles.

Solution

Note the repeated use of power-reducing formulas.

$$\sin^4 x = (\sin^2 x)^2 \qquad \text{Property of exponents}$$

$$= \left(\frac{1 - \cos 2x}{2}\right)^2 \qquad \text{Power-reducing formula}$$

$$= \frac{1}{4}(1 - 2 \cos 2x + \cos^2 2x) \qquad \text{Expand.}$$

$$= \frac{1}{4}\left(1 - 2 \cos 2x + \frac{1 + \cos 4x}{2}\right) \qquad \text{Power-reducing formula}$$

$$= \frac{1}{4} - \frac{1}{2} \cos 2x + \frac{1}{8} + \frac{1}{8} \cos 4x \qquad \text{Distributive Property}$$

$$= \frac{1}{8}(3 - 4 \cos 2x + \cos 4x) \qquad \text{Factor out common factor.}$$

✓CHECKPOINT   Now try Exercise 29.

Half-Angle Formulas

You can derive some useful alternative forms of the power-reducing formulas by replacing u with $u/2$. The results are called **half-angle formulas.**

Half-Angle Formulas

$$\sin \frac{u}{2} = \pm \sqrt{\frac{1 - \cos u}{2}}$$

$$\cos \frac{u}{2} = \pm \sqrt{\frac{1 + \cos u}{2}}$$

$$\tan \frac{u}{2} = \frac{1 - \cos u}{\sin u} = \frac{\sin u}{1 + \cos u}$$

The signs of $\sin \dfrac{u}{2}$ and $\cos \dfrac{u}{2}$ depend on the quadrant in which $\dfrac{u}{2}$ lies.

Example 6 Using a Half-Angle Formula

Find the exact value of $\sin 105°$.

Solution

Begin by noting that $105°$ is half of $210°$. Then, using the half-angle formula for $\sin(u/2)$ and the fact that $105°$ lies in Quadrant II, you have

$$\sin 105° = \sqrt{\frac{1 - \cos 210°}{2}}$$

$$= \sqrt{\frac{1 - (-\cos 30°)}{2}}$$

$$= \sqrt{\frac{1 + \left(\sqrt{3}/2\right)}{2}}$$

$$= \frac{\sqrt{2 + \sqrt{3}}}{2}.$$

The positive square root is chosen because $\sin \theta$ is positive in Quadrant II.

✓**CHECKPOINT** Now try Exercise 41.

Use your calculator to verify the result obtained in Example 6. That is, evaluate $\sin 105°$ and $\left(\sqrt{2 + \sqrt{3}}\right)/2$.

$$\sin 105° \approx 0.9659258$$

$$\frac{\sqrt{2 + \sqrt{3}}}{2} \approx 0.9659258$$

You can see that both values are approximately 0.9659258.

STUDY TIP

To find the exact value of a trigonometric function with an angle measure in D°M′S″ form using a half-angle formula, first convert the angle measure to decimal degree form. Then multiply the resulting angle measure by 2.

Example 7   **Solving a Trigonometric Equation**

Find all solutions of $2 - \sin^2 x = 2 \cos^2 \dfrac{x}{2}$ in the interval $[0, 2\pi)$.

Algebraic Solution

$$2 - \sin^2 x = 2 \cos^2 \frac{x}{2} \qquad \text{Write original equation.}$$

$$2 - \sin^2 x = 2 \left(\pm \sqrt{\frac{1 + \cos x}{2}} \right)^2 \qquad \text{Half-angle formula}$$

$$2 - \sin^2 x = 2 \left(\frac{1 + \cos x}{2} \right) \qquad \text{Simplify.}$$

$$2 - \sin^2 x = 1 + \cos x \qquad \text{Simplify.}$$

$$2 - (1 - \cos^2 x) = 1 + \cos x \qquad \text{Pythagorean identity}$$

$$\cos^2 x - \cos x = 0 \qquad \text{Simplify.}$$

$$\cos x (\cos x - 1) = 0 \qquad \text{Factor.}$$

By setting the factors $\cos x$ and $\cos x - 1$ equal to zero, you find that the solutions in the interval $[0, 2\pi)$ are

$$x = \frac{\pi}{2}, \qquad x = \frac{3\pi}{2}, \qquad \text{and} \qquad x = 0.$$

✓**CHECKPOINT**   Now try Exercise 59.

Graphical Solution

Use a graphing utility set in *radian* mode to graph $y = 2 - \sin^2 x - 2 \cos^2(x/2)$, as shown in Figure 2.11. Use the *zero* or *root* feature or the *zoom* and *trace* features to approximate the x-intercepts in the interval $[0, 2\pi)$ to be

$$x = 0, \, x \approx 1.571 \approx \frac{\pi}{2}, \text{ and } x \approx 4.712 \approx \frac{3\pi}{2}.$$

These values are the approximate solutions of $2 - \sin^2 x - 2 \cos^2(x/2) = 0$ in the interval $[0, 2\pi)$.

FIGURE **2.11**

Product-to-Sum Formulas

A common error is to write

$$2 \cos^2 \frac{x}{2}$$

as $\cos^2 x$, rather than to use the correct identity

$$2 \cos^2 \frac{x}{2} = 2 \left(\frac{1 + \cos x}{2} \right).$$

Each of the following **product-to-sum formulas** is easily verified using the sum and difference formulas discussed in the preceding section.

Product-to-Sum Formulas

$$\sin u \sin v = \frac{1}{2} [\cos(u - v) - \cos(u + v)]$$

$$\cos u \cos v = \frac{1}{2} [\cos(u - v) + \cos(u + v)]$$

$$\sin u \cos v = \frac{1}{2} [\sin(u + v) + \sin(u - v)]$$

$$\cos u \sin v = \frac{1}{2} [\sin(u + v) - \sin(u - v)]$$

Product-to-sum formulas are used in calculus to evaluate integrals involving the products of sines and cosines of two different angles.

Example 8 **Writing Products as Sums**

Rewrite the product $\cos 5x \sin 4x$ as a sum or difference.

Solution

Using the appropriate product-to-sum formula, you obtain

$$\cos 5x \sin 4x = \tfrac{1}{2}[\sin(5x + 4x) - \sin(5x - 4x)]$$

$$= \tfrac{1}{2} \sin 9x - \tfrac{1}{2} \sin x.$$

✓CHECKPOINT   Now try Exercise 67.

Occasionally, it is useful to reverse the procedure and write a sum of trigonometric functions as a product. This can be accomplished with the following **sum-to-product formulas.**

Activities

1. Find the exact values of $\sin 2u$, $\cos 2u$, and $\tan 2u$ given $\cot u = -5$, $\dfrac{3\pi}{2} < u < 2\pi$.

   Answer: $\sin 2u = -\dfrac{5}{13}$

   $\cos 2u = \dfrac{12}{13}$

   $\tan 2u = -\dfrac{5}{12}$

2. Find all solutions of the equation in the interval $[0, 2\pi)$.

   $4 \sin x \cos x = 1$

   Answer: $x = \dfrac{\pi}{12}, \dfrac{5\pi}{12}, \dfrac{13\pi}{12}, \dfrac{17\pi}{12}$

3. Rewrite as a sum: $2 \sin \theta \sin \phi$.

   Answer: $\cos(\theta - \phi) - \cos(\theta + \phi)$

Sum-to-Product Formulas

$$\sin u + \sin v = 2 \sin\left(\frac{u + v}{2}\right) \cos\left(\frac{u - v}{2}\right)$$

$$\sin u - \sin v = 2 \cos\left(\frac{u + v}{2}\right) \sin\left(\frac{u - v}{2}\right)$$

$$\cos u + \cos v = 2 \cos\left(\frac{u + v}{2}\right) \cos\left(\frac{u - v}{2}\right)$$

$$\cos u - \cos v = -2 \sin\left(\frac{u + v}{2}\right) \sin\left(\frac{u - v}{2}\right)$$

For a proof of the sum-to-product formulas, see Proofs in Mathematics on page 274.

Example 9 **Using a Sum-to-Product Formula**

Find the exact value of $\cos 195° + \cos 105°$.

Solution

Using the appropriate sum-to-product formula, you obtain

$$\cos 195° + \cos 105° = 2 \cos\left(\frac{195° + 105°}{2}\right) \cos\left(\frac{195° - 105°}{2}\right)$$

$$= 2 \cos 150° \cos 45°$$

$$= 2\left(-\frac{\sqrt{3}}{2}\right)\left(\frac{\sqrt{2}}{2}\right)$$

$$= -\frac{\sqrt{6}}{2}.$$

✓CHECKPOINT   Now try Exercise 83.

Example 10   **Solving a Trigonometric Equation**

Solve $\sin 5x + \sin 3x = 0$.

Solution

$$\sin 5x + \sin 3x = 0 \qquad \text{Write original equation.}$$

$$2 \sin\left(\frac{5x + 3x}{2}\right) \cos\left(\frac{5x - 3x}{2}\right) = 0 \qquad \text{Sum-to-product formula}$$

$$2 \sin 4x \cos x = 0 \qquad \text{Simplify.}$$

By setting the factor $2 \sin 4x$ equal to zero, you can find that the solutions in the interval $[0, 2\pi)$ are

$$x = 0, \frac{\pi}{4}, \frac{\pi}{2}, \frac{3\pi}{4}, \pi, \frac{5\pi}{4}, \frac{3\pi}{2}, \frac{7\pi}{4}.$$

The equation $\cos x = 0$ yields no additional solutions, and you can conclude that the solutions are of the form

$$x = \frac{n\pi}{4}$$

where n is an integer. You can confirm this graphically by sketching the graph of $y = \sin 5x + \sin 3x$, as shown in Figure 2.12. From the graph you can see that the x-intercepts occur at multiples of $\pi/4$.

✓CHECKPOINT   Now try Exercise 87.

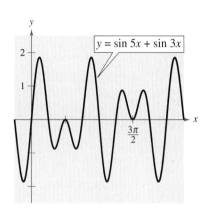

$y = \sin 5x + \sin 3x$

FIGURE **2.12**

Example 11   **Verifying a Trigonometric Identity**

Verify the identity

$$\frac{\sin t + \sin 3t}{\cos t + \cos 3t} = \tan 2t.$$

Solution

Using appropriate sum-to-product formulas, you have

$$\frac{\sin t + \sin 3t}{\cos t + \cos 3t} = \frac{2 \sin\left(\dfrac{t + 3t}{2}\right) \cos\left(\dfrac{t - 3t}{2}\right)}{2 \cos\left(\dfrac{t + 3t}{2}\right) \cos\left(\dfrac{t - 3t}{2}\right)}$$

$$= \frac{2 \sin(2t)\cos(-t)}{2 \cos(2t)\cos(-t)}$$

$$= \frac{\sin 2t}{\cos 2t}$$

$$= \tan 2t.$$

✓CHECKPOINT   Now try Exercise 105.

Application

 Example 12   Projectile Motion

Ignoring air resistance, the range of a projectile fired at an angle θ with the horizontal and with an initial velocity of v_0 feet per second is given by

$$r = \frac{1}{16}v_0^2 \sin\theta \cos\theta$$

where r is the horizontal distance (in feet) that the projectile will travel. A place kicker for a football team can kick a football from ground level with an initial velocity of 80 feet per second (see Figure 2.13).

FIGURE 2.13

a. Write the projectile motion model in a simpler form.

b. At what angle must the player kick the football so that the football travels 200 feet?

c. For what angle is the horizontal distance the football travels a maximum?

Solution

a. You can use a double-angle formula to rewrite the projectile motion model as

$$r = \frac{1}{32}v_0^2(2\sin\theta\cos\theta)$$ Rewrite original projectile motion model.

$$= \frac{1}{32}v_0^2 \sin 2\theta.$$ Rewrite model using a double-angle formula.

b. $r = \dfrac{1}{32}v_0^2 \sin 2\theta$ Write projectile motion model.

$200 = \dfrac{1}{32}(80)^2 \sin 2\theta$ Substitute 200 for r and 80 for v_0.

$200 = 200 \sin 2\theta$ Simplify.

$1 = \sin 2\theta$ Divide each side by 200.

You know that $2\theta = \pi/2$, so dividing this result by 2 produces $\theta = \pi/4$. Because $\pi/4 = 45°$, you can conclude that the player must kick the football at an angle of $45°$ so that the football will travel 200 feet.

c. From the model $r = 200 \sin 2\theta$ you can see that the amplitude is 200. So the maximum range is $r = 200$ feet. From part (b), you know that this corresponds to an angle of $45°$. Therefore, kicking the football at an angle of $45°$ will produce a maximum horizontal distance of 200 feet.

✓CHECKPOINT   Now try Exercise 119.

𝒲RITING ABOUT 𝓜ATHEMATICS

Deriving an Area Formula   Describe how you can use a double-angle formula or a half-angle formula to derive a formula for the area of an isosceles triangle. Use a labeled sketch to illustrate your derivation. Then write two examples that show how your formula can be used.

2.5	**Exercises**

VOCABULARY CHECK: Fill in the blank to complete the trigonometric formula.

1. $\sin 2u =$ _____

2. $\dfrac{1 + \cos 2u}{2} =$ _____

3. $\cos 2u =$ _____

4. $\dfrac{1 - \cos 2u}{1 + \cos 2u} =$ _____

5. $\sin \dfrac{u}{2} =$ _____

6. $\tan \dfrac{u}{2} =$ _____

7. $\cos u \cos v =$ _____

8. $\sin u \cos v =$ _____

9. $\sin u + \sin v =$ _____

10. $\cos u - \cos v =$ _____

PREREQUISITE SKILLS REVIEW: Practice and review algebra skills needed for this section at **www.Eduspace.com.**

In Exercises 1–8, use the figure to find the exact value of the trigonometric function.

1. $\sin \theta$

2. $\tan \theta$

3. $\cos 2\theta$

4. $\sin 2\theta$

5. $\tan 2\theta$

6. $\sec 2\theta$

7. $\csc 2\theta$

8. $\cot 2\theta$

In Exercises 9–18, find the exact solutions of the equation in the interval $[0, 2\pi)$.

9. $\sin 2x - \sin x = 0$

10. $\sin 2x + \cos x = 0$

11. $4 \sin x \cos x = 1$

12. $\sin 2x \sin x = \cos x$

13. $\cos 2x - \cos x = 0$

14. $\cos 2x + \sin x = 0$

15. $\tan 2x - \cot x = 0$

16. $\tan 2x - 2 \cos x = 0$

17. $\sin 4x = -2 \sin 2x$

18. $(\sin 2x + \cos 2x)^2 = 1$

In Exercises 19–22, use a double-angle formula to rewrite the expression.

19. $6 \sin x \cos x$

20. $6 \cos^2 x - 3$

21. $4 - 8 \sin^2 x$

22. $(\cos x + \sin x)(\cos x - \sin x)$

In Exercises 23–28, find the exact values of sin 2u, cos 2u, and tan 2u using the double-angle formulas.

23. $\sin u = -\dfrac{4}{5}, \quad \pi < u < \dfrac{3\pi}{2}$

24. $\cos u = -\dfrac{2}{3}, \quad \dfrac{\pi}{2} < u < \pi$

25. $\tan u = \dfrac{3}{4}, \quad 0 < u < \dfrac{\pi}{2}$

26. $\cot u = -4, \quad \dfrac{3\pi}{2} < u < 2\pi$

27. $\sec u = -\dfrac{5}{2}, \quad \dfrac{\pi}{2} < u < \pi$

28. $\csc u = 3, \quad \dfrac{\pi}{2} < u < \pi$

In Exercises 29–34, use the power-reducing formulas to rewrite the expression in terms of the first power of the cosine.

29. $\cos^4 x$

30. $\sin^8 x$

31. $\sin^2 x \cos^2 x$

32. $\sin^4 x \cos^4 x$

33. $\sin^2 x \cos^4 x$

34. $\sin^4 x \cos^2 x$

In Exercises 35–40, use the figure to find the exact value of the trigonometric function.

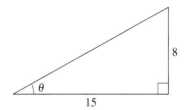

35. $\cos \dfrac{\theta}{2}$

36. $\sin \dfrac{\theta}{2}$

37. $\tan \dfrac{\theta}{2}$

38. $\sec \dfrac{\theta}{2}$

39. $\csc \dfrac{\theta}{2}$

40. $\cot \dfrac{\theta}{2}$

In Exercises 41–48, use the half-angle formulas to determine the exact values of the sine, cosine, and tangent of the angle.

41. $75°$

42. $165°$

43. $112° \, 30'$

44. $67° \, 30'$

45. $\dfrac{\pi}{8}$

46. $\dfrac{\pi}{12}$

47. $\dfrac{3\pi}{8}$

48. $\dfrac{7\pi}{12}$

In Exercises 49–54, find the exact values of $\sin(u/2)$, $\cos(u/2)$, and $\tan(u/2)$ using the half-angle formulas.

49. $\sin u = \dfrac{5}{13}, \quad \dfrac{\pi}{2} < u < \pi$

50. $\cos u = \dfrac{3}{5}, \quad 0 < u < \dfrac{\pi}{2}$

51. $\tan u = -\dfrac{5}{8}, \quad \dfrac{3\pi}{2} < u < 2\pi$

52. $\cot u = 3, \quad \pi < u < \dfrac{3\pi}{2}$

53. $\csc u = -\dfrac{5}{3}, \quad \pi < u < \dfrac{3\pi}{2}$

54. $\sec u = -\dfrac{7}{2}, \quad \dfrac{\pi}{2} < u < \pi$

In Exercises 55–58, use the half-angle formulas to simplify the expression.

55. $\sqrt{\dfrac{1 - \cos 6x}{2}}$

56. $\sqrt{\dfrac{1 + \cos 4x}{2}}$

57. $-\sqrt{\dfrac{1 - \cos 8x}{1 + \cos 8x}}$

58. $-\sqrt{\dfrac{1 - \cos(x - 1)}{2}}$

 In Exercises 59–62, find all solutions of the equation in the interval $[0, 2\pi)$. Use a graphing utility to graph the equation and verify the solutions.

59. $\sin\dfrac{x}{2} + \cos x = 0$

60. $\sin\dfrac{x}{2} + \cos x - 1 = 0$

61. $\cos\dfrac{x}{2} - \sin x = 0$

62. $\tan\dfrac{x}{2} - \sin x = 0$

In Exercises 63–74, use the product-to-sum formulas to write the product as a sum or difference.

63. $6 \sin \dfrac{\pi}{4} \cos \dfrac{\pi}{4}$

64. $4 \cos \dfrac{\pi}{3} \sin \dfrac{5\pi}{6}$

65. $10 \cos 75° \cos 15°$

66. $6 \sin 45° \cos 15°$

67. $\cos 4\theta \sin 6\theta$

68. $3 \sin 2\alpha \sin 3\alpha$

69. $5 \cos(-5\beta) \cos 3\beta$

70. $\cos 2\theta \cos 4\theta$

71. $\sin(x + y) \sin(x - y)$

72. $\sin(x + y) \cos(x - y)$

73. $\cos(\theta - \pi) \sin(\theta + \pi)$

74. $\sin(\theta + \pi) \sin(\theta - \pi)$

In Exercises 75–82, use the sum-to-product formulas to write the sum or difference as a product.

75. $\sin 5\theta - \sin 3\theta$

76. $\sin 3\theta + \sin \theta$

77. $\cos 6x + \cos 2x$

78. $\sin x + \sin 5x$

79. $\sin(\alpha + \beta) - \sin(\alpha - \beta)$

80. $\cos(\phi + 2\pi) + \cos \phi$

81. $\cos\left(\theta + \dfrac{\pi}{2}\right) - \cos\left(\theta - \dfrac{\pi}{2}\right)$

82. $\sin\left(x + \dfrac{\pi}{2}\right) + \sin\left(x - \dfrac{\pi}{2}\right)$

In Exercises 83–86, use the sum-to-product formulas to find the exact value of the expression.

83. $\sin 60° + \sin 30°$

84. $\cos 120° + \cos 30°$

85. $\cos \dfrac{3\pi}{4} - \cos \dfrac{\pi}{4}$

86. $\sin \dfrac{5\pi}{4} - \sin \dfrac{3\pi}{4}$

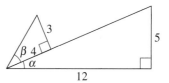 In Exercises 87–90, find all solutions of the equation in the interval $[0, 2\pi)$. Use a graphing utility to graph the equation and verify the solutions.

87. $\sin 6x + \sin 2x = 0$

88. $\cos 2x - \cos 6x = 0$

89. $\dfrac{\cos 2x}{\sin 3x - \sin x} - 1 = 0$

90. $\sin^2 3x - \sin^2 x = 0$

In Exercises 91–94, use the figure and trigonometric identities to find the exact value of the trigonometric function in two ways.

91. $\sin^2 \alpha$

92. $\cos^2 \alpha$

93. $\sin \alpha \cos \beta$

94. $\cos \alpha \sin \beta$

In Exercises 95–110, verify the identity.

95. $\csc 2\theta = \dfrac{\csc \theta}{2 \cos \theta}$

96. $\sec 2\theta = \dfrac{\sec^2 \theta}{2 - \sec^2 \theta}$

97. $\cos^2 2\alpha - \sin^2 2\alpha = \cos 4\alpha$

98. $\cos^4 x - \sin^4 x = \cos 2x$

99. $(\sin x + \cos x)^2 = 1 + \sin 2x$

100. $\sin \dfrac{\alpha}{3} \cos \dfrac{\alpha}{3} = \dfrac{1}{2} \sin \dfrac{2\alpha}{3}$

101. $1 + \cos 10y = 2 \cos^2 5y$

102. $\dfrac{\cos 3\beta}{\cos \beta} = 1 - 4 \sin^2 \beta$

103. $\sec \dfrac{u}{2} = \pm \sqrt{\dfrac{2 \tan u}{\tan u + \sin u}}$

104. $\tan \dfrac{u}{2} = \csc u - \cot u$

105. $\dfrac{\sin x \pm \sin y}{\cos x + \cos y} = \tan \dfrac{x \pm y}{2}$

106. $\dfrac{\sin x + \sin y}{\cos x - \cos y} = -\cot \dfrac{x - y}{2}$

107. $\dfrac{\cos 4x + \cos 2x}{\sin 4x + \sin 2x} = \cot 3x$

108. $\dfrac{\cos t + \cos 3t}{\sin 3t - \sin t} = \cot t$

109. $\sin\left(\dfrac{\pi}{6} + x\right) + \sin\left(\dfrac{\pi}{6} - x\right) = \cos x$

110. $\cos\left(\dfrac{\pi}{3} + x\right) + \cos\left(\dfrac{\pi}{3} - x\right) = \cos x$

In Exercises 111–114, use a graphing utility to verify the identity. Confirm that it is an identity algebraically.

111. $\cos 3\beta = \cos^3 \beta - 3 \sin^2 \beta \cos \beta$

112. $\sin 4\beta = 4 \sin \beta \cos \beta(1 - 2 \sin^2 \beta)$

113. $(\cos 4x - \cos 2x)/(2 \sin 3x) = -\sin x$

114. $(\cos 3x - \cos x)/(\sin 3x - \sin x) = -\tan 2x$

In Exercises 115 and 116, graph the function by hand in the interval $[0, 2\pi]$ by using the power-reducing formulas.

115. $f(x) = \sin^2 x$ **116.** $f(x) = \cos^2 x$

In Exercises 117 and 118, write the trigonometric expression as an algebraic expression.

117. $\sin(2 \arcsin x)$ **118.** $\cos(2 \arccos x)$

119. *Projectile Motion*   The range of a projectile fired at an angle θ with the horizontal and with an initial velocity of v_0 feet per second is

$$r = \dfrac{1}{32} v_0^2 \sin 2\theta$$

where r is measured in feet. An athlete throws a javelin at 75 feet per second. At what angle must the athlete throw the javelin so that the javelin travels 130 feet?

120. *Geometry*   The length of each of the two equal sides of an isosceles triangle is 10 meters (see figure). The angle between the two sides is θ.

(a) Write the area of the triangle as a function of $\theta/2$.

(b) Write the area of the triangle as a function of θ. Determine the value of θ such that the area is a maximum.

Model It

121. *Mach Number*   The mach number M of an airplane is the ratio of its speed to the speed of sound. When an airplane travels faster than the speed of sound, the sound waves form a cone behind the airplane (see figure). The mach number is related to the apex angle θ of the cone by

$$\sin \dfrac{\theta}{2} = \dfrac{1}{M}.$$

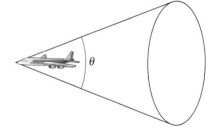

(a) Find the angle θ that corresponds to a mach number of 1.

(b) Find the angle θ that corresponds to a mach number of 4.5.

(c) The speed of sound is about 760 miles per hour. Determine the speed of an object with the mach numbers from parts (a) and (b).

(d) Rewrite the equation in terms of θ.

122. ***Railroad Track*** When two railroad tracks merge, the overlapping portions of the tracks are in the shapes of circular arcs (see figure). The radius of each arc r (in feet) and the angle θ are related by

$$\frac{x}{2} = 2r \sin^2 \frac{\theta}{2}.$$

Write a formula for x in terms of $\cos \theta$.

Synthesis

True or False? In Exercises 123 and 124, determine whether the statement is true or false. Justify your answer.

123. Because the sine function is an odd function, for a negative number u, $\sin 2u = -2 \sin u \cos u$.

124. $\sin \dfrac{u}{2} = -\sqrt{\dfrac{1 - \cos u}{2}}$ when u is in the second quadrant.

 In Exercises 125 and 126, (a) use a graphing utility to graph the function and approximate the maximum and minimum points on the graph in the interval $[0, 2\pi)$ and (b) solve the trigonometric equation and verify that its solutions are the x-coordinates of the maximum and minimum points of f. (Calculus is required to find the trigonometric equation.)

Function	*Trigonometric Equation*
125. $f(x) = 4 \sin \dfrac{x}{2} + \cos x$	$2 \cos \dfrac{x}{2} - \sin x = 0$
126. $f(x) = \cos 2x - 2 \sin x$	$-2 \cos x(2 \sin x + 1) = 0$

127. ***Exploration*** Consider the function given by

$$f(x) = \sin^4 x + \cos^4 x.$$

(a) Use the power-reducing formulas to write the function in terms of cosine to the first power.

(b) Determine another way of rewriting the function. Use a graphing utility to rule out incorrectly rewritten functions.

(c) Add a trigonometric term to the function so that it becomes a perfect square trinomial. Rewrite the function as a perfect square trinomial minus the term that you added. Use a graphing utility to rule out incorrectly rewritten functions.

(d) Rewrite the result of part (c) in terms of the sine of a double angle. Use a graphing utility to rule out incorrectly rewritten functions.

(e) When you rewrite a trigonometric expression, the result may not be the same as a friend's. Does this mean that one of you is wrong? Explain.

128. ***Conjecture*** Consider the function given by

$$f(x) = 2 \sin x\left(2 \cos^2 \frac{x}{2} - 1\right).$$

(a) Use a graphing utility to graph the function.

(b) Make a conjecture about the function that is an identity with f.

(c) Verify your conjecture analytically.

Skills Review

In Exercises 129–132, (a) plot the points, (b) find the distance between the points, and (c) find the midpoint of the line segment connecting the points.

129. $(5, 2), (-1, 4)$

130. $(-4, -3), (6, 10)$

131. $\left(0, \frac{1}{2}\right), \left(\frac{4}{3}, \frac{5}{2}\right)$

132. $\left(\frac{1}{3}, \frac{2}{3}\right), \left(-1, -\frac{3}{2}\right)$

In Exercises 133–136, find (if possible) the complement and supplement of each angle.

133. (a) $55°$ (b) $162°$

134. (a) $109°$ (b) $78°$

135. (a) $\dfrac{\pi}{18}$ (b) $\dfrac{9\pi}{20}$

136. (a) 0.95 (b) 2.76

137. ***Profit*** The total profit for a car manufacturer in October was 16% higher than it was in September. The total profit for the 2 months was $507,600. Find the profit for each month.

138. ***Mixture Problem*** A 55-gallon barrel contains a mixture with a concentration of 30%. How much of this mixture must be withdrawn and replaced by 100% concentrate to bring the mixture up to 50% concentration?

139. ***Distance*** A baseball diamond has the shape of a square in which the distance between each of the consecutive bases is 90 feet. Approximate the straight-line distance from home plate to second base.

2 Chapter Summary

What did you learn?

2   Review Exercises

In Exercises 1–6, name the trigonometric function that is equivalent to the expression.

1. $\dfrac{1}{\cos x}$

2. $\dfrac{1}{\sin x}$

3. $\dfrac{1}{\sec x}$

4. $\dfrac{1}{\tan x}$

5. $\dfrac{\cos x}{\sin x}$

6. $\sqrt{1 + \tan^2 x}$

In Exercises 7–10, use the given values and trigonometric identities to evaluate (if possible) all six trigonometric functions.

7. $\sin x = \frac{3}{5}, \quad \cos x = \frac{4}{5}$

8. $\tan \theta = \dfrac{2}{3}, \quad \sec \theta = \dfrac{\sqrt{13}}{3}$

9. $\sin\left(\dfrac{\pi}{2} - x\right) = \dfrac{\sqrt{2}}{2}, \quad \sin x = -\dfrac{\sqrt{2}}{2}$

10. $\csc\left(\dfrac{\pi}{2} - \theta\right) = 9, \quad \sin \theta = \dfrac{4\sqrt{5}}{9}$

In Exercises 11–22, use the fundamental trigonometric identities to simplify the expression.

11. $\dfrac{1}{\cot^2 x + 1}$

12. $\dfrac{\tan \theta}{1 - \cos^2 \theta}$

13. $\tan^2 x(\csc^2 x - 1)$

14. $\cot^2 x(\sin^2 x)$

15. $\dfrac{\sin\left(\dfrac{\pi}{2} - \theta\right)}{\sin \theta}$

16. $\dfrac{\cot\left(\dfrac{\pi}{2} - u\right)}{\cos u}$

17. $\cos^2 x + \cos^2 x \cot^2 x$

18. $\tan^2 \theta \csc^2 \theta - \tan^2 \theta$

19. $(\tan x + 1)^2 \cos x$

20. $(\sec x - \tan x)^2$

21. $\dfrac{1}{\csc \theta + 1} - \dfrac{1}{\csc \theta - 1}$

22. $\dfrac{\cos^2 x}{1 - \sin x}$

23. *Rate of Change* The rate of change of the function $f(x) = \csc x - \cot x$ is given by the expression $\csc^2 x - \csc x \cot x$. Show that this expression can also be written as

$$\dfrac{1 - \cos x}{\sin^2 x}.$$

24. *Rate of Change* The rate of change of the function $f(x) = 2\sqrt{\sin x}$ is given by the expression $\sin^{-1/2} x \cos x$. Show that this expression can also be written as $\cot x \sqrt{\sin x}$.

In Exercises 25–32, verify the identity.

25. $\cos x(\tan^2 x + 1) = \sec x$

26. $\sec^2 x \cot x - \cot x = \tan x$

27. $\cos\left(x + \dfrac{\pi}{2}\right) = -\sin x$

28. $\cot\left(\dfrac{\pi}{2} - x\right) = \tan x$

29. $\dfrac{1}{\tan \theta \csc \theta} = \cos \theta$

30. $\dfrac{1}{\tan x \csc x \sin x} = \cot x$

31. $\sin^5 x \cos^2 x = (\cos^2 x - 2\cos^4 x + \cos^6 x)\sin x$

32. $\cos^3 x \sin^2 x = (\sin^2 x - \sin^4 x)\cos x$

In Exercises 33–38, solve the equation.

33. $\sin x = \sqrt{3} - \sin x$

34. $4 \cos \theta = 1 + 2 \cos \theta$

35. $3\sqrt{3} \tan u = 3$

36. $\frac{1}{2} \sec x - 1 = 0$

37. $3 \csc^2 x = 4$

38. $4 \tan^2 u - 1 = \tan^2 u$

In Exercises 39–46, find all solutions of the equation in the interval $[0, 2\pi)$.

39. $2 \cos^2 x - \cos x = 1$

40. $2 \sin^2 x - 3 \sin x = -1$

41. $\cos^2 x + \sin x = 1$

42. $\sin^2 x + 2 \cos x = 2$

43. $2 \sin 2x - \sqrt{2} = 0$

44. $\sqrt{3} \tan 3x = 0$

45. $\cos 4x(\cos x - 1) = 0$

46. $3 \csc^2 5x = -4$

In Exercises 47–50, use inverse functions where needed to find all solutions of the equation in the interval $[0, 2\pi)$.

47. $\sin^2 x - 2 \sin x = 0$

48. $2 \cos^2 x + 3 \cos x = 0$

49. $\tan^2 \theta + \tan \theta - 12 = 0$

50. $\sec^2 x + 6 \tan x + 4 = 0$

In Exercises 51–54, find the exact values of the sine, cosine, and tangent of the angle by using a sum or difference formula.

51. $285° = 315° - 30°$

52. $345° = 300° + 45°$

53. $\dfrac{25\pi}{12} = \dfrac{11\pi}{6} + \dfrac{\pi}{4}$

54. $\dfrac{19\pi}{12} = \dfrac{11\pi}{6} - \dfrac{\pi}{4}$

In Exercises 55–58, write the expression as the sine, cosine, or tangent of an angle.

55. $\sin 60° \cos 45° - \cos 60° \sin 45°$

56. $\cos 45° \cos 120° - \sin 45° \sin 120°$

57. $\dfrac{\tan 25° + \tan 10°}{1 - \tan 25° \tan 10°}$

58. $\dfrac{\tan 68° - \tan 115°}{1 + \tan 68° \tan 115°}$

In Exercises 59–64, find the exact value of the trigonometric function given that $\sin u = \frac{3}{4}$ and $\cos v = -\frac{5}{13}$. (Both u and v are in Quadrant II.)

59. $\sin(u + v)$

60. $\tan(u + v)$

61. $\cos(u - v)$

62. $\sin(u - v)$

63. $\cos(u + v)$

64. $\tan(u - v)$

In Exercises 65–70, verify the identity.

65. $\cos\left(x + \dfrac{\pi}{2}\right) = -\sin x$

66. $\sin\left(x - \dfrac{3\pi}{2}\right) = \cos x$

67. $\cot\left(\dfrac{\pi}{2} - x\right) = \tan x$

68. $\sin(\pi - x) = \sin x$

69. $\cos 3x = 4\cos^3 x - 3\cos x$

70. $\dfrac{\sin(\alpha + \beta)}{\cos \alpha \cos \beta} = \tan \alpha + \tan\beta$

In Exercises 71–74, find all solutions of the equation in the interval $[0, 2\pi)$.

71. $\sin\left(x + \dfrac{\pi}{4}\right) - \sin\left(x - \dfrac{\pi}{4}\right) = 1$

72. $\cos\left(x + \dfrac{\pi}{6}\right) - \cos\left(x - \dfrac{\pi}{6}\right) = 1$

73. $\sin\left(x + \dfrac{\pi}{2}\right) - \sin\left(x - \dfrac{\pi}{2}\right) = \sqrt{3}$

74. $\cos\left(x + \dfrac{3\pi}{4}\right) - \cos\left(x - \dfrac{3\pi}{4}\right) = 0$

2.5 In Exercises 75 and 76, find the exact values of $\sin 2u$, $\cos 2u$, and $\tan 2u$ using the double-angle formulas.

75. $\sin u = -\dfrac{4}{5}, \quad \pi < u < \dfrac{3\pi}{2}$

76. $\cos u = -\dfrac{2}{\sqrt{5}}, \quad \dfrac{\pi}{2} < u < \pi$

In Exercises 77 and 78, use double-angle formulas to verify the identity algebraically and use a graphing utility to confirm your result graphically.

77. $\sin 4x = 8\cos^3 x \sin x - 4\cos x \sin x$

78. $\tan^2 x = \dfrac{1 - \cos 2x}{1 + \cos 2x}$

In Exercises 79–82, use the power-reducing formulas to rewrite the expression in terms of the first power of the cosine.

79. $\tan^2 2x$

80. $\cos^2 3x$

81. $\sin^2 x \tan^2 x$

82. $\cos^2 x \tan^2 x$

In Exercises 83–86, use the half-angle formulas to determine the exact values of the sine, cosine, and tangent of the angle.

83. $-75°$

84. $15°$

85. $\dfrac{19\pi}{12}$

86. $-\dfrac{17\pi}{12}$

In Exercises 87–90, find the exact values of $\sin(u/2)$, $\cos(u/2)$, and $\tan(u/2)$ using the half-angle formulas.

87. $\sin u = \frac{3}{5}, \ 0 < u < \pi/2$

88. $\tan u = \frac{5}{8}, \ \pi < u < 3\pi/2$

89. $\cos u = -\frac{2}{7}, \ \pi/2 < u < \pi$

90. $\sec u = -6, \ \pi/2 < u < \pi$

In Exercises 91 and 92, use the half-angle formulas to simplify the expression.

91. $-\sqrt{\dfrac{1 + \cos 10x}{2}}$

92. $\dfrac{\sin 6x}{1 + \cos 6x}$

In Exercises 93–96, use the product-to-sum formulas to write the product as a sum or difference.

93. $\cos \dfrac{\pi}{6} \sin \dfrac{\pi}{6}$

94. $6 \sin 15° \sin 45°$

95. $\cos 5\theta \cos 3\theta$

96. $4 \sin 3\alpha \cos 2\alpha$

In Exercises 97–100, use the sum-to-product formulas to write the sum or difference as a product.

97. $\sin 4\theta - \sin 2\theta$

98. $\cos 3\theta + \cos 2\theta$

99. $\cos\left(x + \dfrac{\pi}{6}\right) - \cos\left(x - \dfrac{\pi}{6}\right)$

100. $\sin\left(x + \dfrac{\pi}{4}\right) - \sin\left(x - \dfrac{\pi}{4}\right)$

101. *Projectile Motion* A baseball leaves the hand of the person at first base at an angle of θ with the horizontal and at an initial velocity of $v_0 = 80$ feet per second. The ball is caught by the person at second base 100 feet away. Find θ if the range r of a projectile is

$$r = \dfrac{1}{32} v_0^2 \sin 2\theta.$$

102. *Geometry* A trough for feeding cattle is 4 meters long and its cross sections are isosceles triangles with the two equal sides being $\frac{1}{2}$ meter (see figure). The angle between the two sides is θ.

4 m

$\frac{1}{2}$ m θ

$\frac{1}{2}$ m

(a) Write the trough's volume as a function of $\dfrac{\theta}{2}$.

(b) Write the volume of the trough as a function of θ and determine the value of θ such that the volume is maximum.

Harmonic Motion **In Exercises 103–106, use the following information. A weight is attached to a spring suspended vertically from a ceiling. When a driving force is applied to the system, the weight moves vertically from its equilibrium position, and this motion is described by the model**

$$y = 1.5 \sin 8t - 0.5 \cos 8t$$

where y is the distance from equilibrium (in feet) and t is the time (in seconds).

 103. Use a graphing utility to graph the model.

104. Write the model in the form

$$y = \sqrt{a^2 + b^2}\,\sin(Bt + C).$$

105. Find the amplitude of the oscillations of the weight.

106. Find the frequency of the oscillations of the weight.

Synthesis

True or False? **In Exercises 107–110, determine whether the statement is true or false. Justify your answer.**

107. If $\dfrac{\pi}{2} < \theta < \pi$, then $\cos\dfrac{\theta}{2} < 0$.

108. $\sin(x + y) = \sin x + \sin y$

109. $4 \sin(-x)\cos(-x) = -2 \sin 2x$

110. $4 \sin 45° \cos 15° = 1 + \sqrt{3}$

111. List the reciprocal identities, quotient identities, and Pythagorean identities from memory.

112. *Think About It* If a trigonometric equation has an infinite number of solutions, is it true that the equation is an identity? Explain.

113. *Think About It* Explain why you know from observation that the equation $a \sin x - b = 0$ has no solution if $|a| < |b|$.

114. *Surface Area* The surface area of a honeycomb is given by the equation

$$S = 6hs + \frac{3}{2}s^2\left(\frac{\sqrt{3} - \cos \theta}{\sin \theta}\right), \quad 0 < \theta \le 90°$$

where $h = 2.4$ inches, $s = 0.75$ inch, and θ is the angle shown in the figure.

θ

$h = 2.4$ in.

$s = 0.75$ in.

(a) For what value(s) of θ is the surface area 12 square inches?

(b) What value of θ gives the minimum surface area?

In Exercises 115 and 116, use the graphs of y_1 and y_2 to determine how to change one function to form the identity $y_1 = y_2$.

115. $y_1 = \sec^2\left(\dfrac{\pi}{2} - x\right)$

$y_2 = \cot^2 x$

116. $y_1 = \dfrac{\cos 3x}{\cos x}$

$y_2 = (2 \sin x)^2$

 In Exercises 117 and 118, use the *zero* or *root* feature of a graphing utility to approximate the solutions of the equation.

117. $y = \sqrt{x + 3} + 4 \cos x$

118. $y = 2 - \dfrac{1}{2}x^2 + 3 \sin \dfrac{\pi x}{2}$

2 Chapter Test

Take this test as you would take a test in class. When you are finished, check your work against the answers given in the back of the book.

1. If $\tan \theta = \frac{3}{2}$ and $\cos \theta < 0$, use the fundamental identities to evaluate the other five trigonometric functions of θ.

2. Use the fundamental identities to simplify $\csc^2 \beta(1 - \cos^2 \beta)$.

3. Factor and simplify $\dfrac{\sec^4 x - \tan^4 x}{\sec^2 x + \tan^2 x}$. 4. Add and simplify $\dfrac{\cos \theta}{\sin \theta} + \dfrac{\sin \theta}{\cos \theta}$.

5. Determine the values of θ, $0 \le \theta < 2\pi$, for which $\tan \theta = -\sqrt{\sec^2 \theta - 1}$ is true.

6. Use a graphing utility to graph the functions $y_1 = \cos x + \sin x \tan x$ and $y_2 = \sec x$. Make a conjecture about y_1 and y_2. Verify the result algebraically.

In Exercises 7–12, verify the identity.

7. $\sin \theta \sec \theta = \tan \theta$

8. $\sec^2 x \tan^2 x + \sec^2 x = \sec^4 x$

9. $\dfrac{\csc \alpha + \sec \alpha}{\sin \alpha + \cos \alpha} = \cot \alpha + \tan \alpha$

10. $\cos\left(x + \dfrac{\pi}{2}\right) = -\sin x$

11. $\sin(n\pi + \theta) = (-1)^n \sin \theta$, n is an integer.

12. $(\sin x + \cos x)^2 = 1 + \sin 2x$

13. Rewrite $\sin^4 x \tan^2 x$ in terms of the first power of the cosine.

14. Use a half-angle formula to simplify the expression $\dfrac{\sin 4\theta}{1 + \cos 4\theta}$.

15. Write $4 \cos 2\theta \sin 4\theta$ as a sum or difference.

16. Write $\sin 3\theta - \sin 4\theta$ as a product.

In Exercises 17–20, find all solutions of the equation in the interval $[0, 2\pi)$.

17. $\tan^2 x + \tan x = 0$

18. $\sin 2\alpha - \cos \alpha = 0$

19. $4 \cos^2 x - 3 = 0$

20. $\csc^2 x - \csc x - 2 = 0$

21. Use a graphing utility to approximate the solutions of the equation $3 \cos x - x = 0$ accurate to three decimal places.

22. Find the exact value of $\cos 105°$ using the fact that $105° = 135° - 30°$.

23. Use the figure to find the exact values of $\sin 2u$, $\cos 2u$, and $\tan 2u$.

24. Cheyenne, Wyoming has a latitude of 41°N. At this latitude, the position of the sun at sunrise can be modeled by

$$D = 31 \sin\left(\frac{2\pi}{365}t - 1.4\right)$$

where t is the time (in days) and $t = 1$ represents January 1. In this model, D represents the number of degrees north or south of due east that the sun rises. Use a graphing utility to determine the days on which the sun is more than 20° north of due east at sunrise.

25. The heights h (in feet) of two people in different seats on a Ferris wheel can be modeled by

$$h_1 = 28 \cos 10t + 38 \quad \text{and} \quad h_2 = 28 \cos\left[10\left(t - \frac{\pi}{6}\right)\right] + 38, \ 0 \le t \le 2$$

where t is the time (in minutes). When are the two people at the same height?

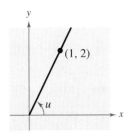

$(1, 2)$

u

FIGURE FOR **23**

Proofs in Mathematics

Sum and Difference Formulas *(p. 248)*

$$\sin(u + v) = \sin u \cos v + \cos u \sin v$$

$$\sin(u - v) = \sin u \cos v - \cos u \sin v$$

$$\cos(u + v) = \cos u \cos v - \sin u \sin v$$

$$\cos(u - v) = \cos u \cos v + \sin u \sin v$$

$$\tan(u + v) = \frac{\tan u + \tan v}{1 - \tan u \tan v}$$

$$\tan(u - v) = \frac{\tan u - \tan v}{1 + \tan u \tan v}$$

Proof

You can use the figures at the left for the proofs of the formulas for $\cos(u \pm v)$. In the top figure, let A be the point $(1, 0)$ and then use u and v to locate the points $B = (x_1, y_1)$, $C = (x_2, y_2)$, and $D = (x_3, y_3)$ on the unit circle. So, $x_i^2 + y_i^2 = 1$ for $i = 1, 2,$ and 3. For convenience, assume that $0 < v < u < 2\pi$. In the bottom figure, note that arcs AC and BD have the same length. So, line segments AC and BD are also equal in length, which implies that

$$\sqrt{(x_2 - 1)^2 + (y_2 - 0)^2} = \sqrt{(x_3 - x_1)^2 + (y_3 - y_1)^2}$$

$$x_2^2 - 2x_2 + 1 + y_2^2 = x_3^2 - 2x_1x_3 + x_1^2 + y_3^2 - 2y_1y_3 + y_1^2$$

$$(x_2^2 + y_2^2) + 1 - 2x_2 = (x_3^2 + y_3^2) + (x_1^2 + y_1^2) - 2x_1x_3 - 2y_1y_3$$

$$1 + 1 - 2x_2 = 1 + 1 - 2x_1x_3 - 2y_1y_3$$

$$x_2 = x_3x_1 + y_3y_1.$$

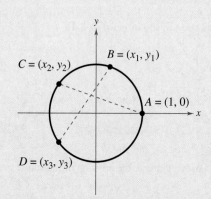

Finally, by substituting the values $x_2 = \cos(u - v)$, $x_3 = \cos u$, $x_1 = \cos v$, $y_3 = \sin u$, and $y_1 = \sin v$, you obtain $\cos(u - v) = \cos u \cos v + \sin u \sin v$. The formula for $\cos(u + v)$ can be established by considering $u + v = u - (-v)$ and using the formula just derived to obtain

$$\cos(u + v) = \cos[u - (-v)] = \cos u \cos(-v) + \sin u \sin(-v)$$

$$= \cos u \cos v - \sin u \sin v.$$

You can use the sum and difference formulas for sine and cosine to prove the formulas for $\tan(u \pm v)$.

$$\tan(u \pm v) = \frac{\sin(u \pm v)}{\cos(u \pm v)} \qquad \text{Quotient identity}$$

$$= \frac{\sin u \cos v \pm \cos u \sin v}{\cos u \cos v \mp \sin u \sin v} \qquad \text{Sum and difference formulas}$$

$$= \frac{\dfrac{\sin u \cos v \pm \cos u \sin v}{\cos u \cos v}}{\dfrac{\cos u \cos v \mp \sin u \sin v}{\cos u \cos v}} \qquad \text{Divide numerator and denominator by } \cos u \cos v.$$

$$= \cfrac{\dfrac{\sin u \cos v}{\cos u \cos v} \pm \dfrac{\cos u \sin v}{\cos u \cos v}}{\dfrac{\cos u \cos v}{\cos u \cos v} \mp \dfrac{\sin u \sin v}{\cos u \cos v}}$$

Write as separate fractions.

$$= \cfrac{\dfrac{\sin u}{\cos u} \pm \dfrac{\sin v}{\cos v}}{1 \mp \dfrac{\sin u}{\cos u} \cdot \dfrac{\sin v}{\cos v}}$$

Product of fractions

$$= \frac{\tan u \pm \tan v}{1 \mp \tan u \tan v}$$

Quotient identity

Double-Angle Formulas   (p. 255)

$$\sin 2u = 2 \sin u \cos u \qquad \cos 2u = \cos^2 u - \sin^2 u$$

$$\tan 2u = \frac{2 \tan u}{1 - \tan^2 u} \qquad\qquad = 2 \cos^2 u - 1 = 1 - 2 \sin^2 u$$

Proof

To prove all three formulas, let $v = u$ in the corresponding sum formulas.

$$\sin 2u = \sin(u + u) = \sin u \cos u + \cos u \sin u = 2 \sin u \cos u$$

$$\cos 2u = \cos(u + u) = \cos u \cos u - \sin u \sin u = \cos^2 u - \sin^2 u$$

$$\tan 2u = \tan(u + u) = \frac{\tan u + \tan u}{1 - \tan u \tan u} = \frac{2 \tan u}{1 - \tan^2 u}$$

Power-Reducing Formulas   (p. 257)

$$\sin^2 u = \frac{1 - \cos 2u}{2} \qquad \cos^2 u = \frac{1 + \cos 2u}{2} \qquad \tan^2 u = \frac{1 - \cos 2u}{1 + \cos 2u}$$

Proof

To prove the first formula, solve for $\sin^2 u$ in the double-angle formula $\cos 2u = 1 - 2 \sin^2 u$, as follows.

$$\cos 2u = 1 - 2 \sin^2 u \qquad\qquad \text{Write double-angle formula.}$$

$$2 \sin^2 u = 1 - \cos 2u \qquad\qquad \text{Subtract } \cos 2u \text{ from and add } 2 \sin^2 u \text{ to each side.}$$

$$\sin^2 u = \frac{1 - \cos 2u}{2} \qquad\qquad \text{Divide each side by 2.}$$

In a similar way you can prove the second formula, by solving for $\cos^2 u$ in the double-angle formula

$$\cos 2u = 2\cos^2 u - 1.$$

To prove the third formula, use a quotient identity, as follows.

$$\tan^2 u = \frac{\sin^2 u}{\cos^2 u}$$

$$= \frac{\dfrac{1 - \cos 2u}{2}}{\dfrac{1 + \cos 2u}{2}}$$

$$= \frac{1 - \cos 2u}{1 + \cos 2u}$$

Sum-to-Product Formulas *(p. 260)*

$$\sin u + \sin v = 2\sin\left(\frac{u + v}{2}\right)\cos\left(\frac{u - v}{2}\right)$$

$$\sin u - \sin v = 2\cos\left(\frac{u + v}{2}\right)\sin\left(\frac{u - v}{2}\right)$$

$$\cos u + \cos v = 2\cos\left(\frac{u + v}{2}\right)\cos\left(\frac{u - v}{2}\right)$$

$$\cos u - \cos v = -2\sin\left(\frac{u + v}{2}\right)\sin\left(\frac{u - v}{2}\right)$$

Proof

To prove the first formula, let $x = u + v$ and $y = u - v$. Then substitute $u = (x + y)/2$ and $v = (x - y)/2$ in the product-to-sum formula.

$$\sin u \cos v = \frac{1}{2}\left[\sin(u + v) + \sin(u - v)\right]$$

$$\sin\left(\frac{x + y}{2}\right)\cos\left(\frac{x - y}{2}\right) = \frac{1}{2}(\sin x + \sin y)$$

$$2\sin\left(\frac{x + y}{2}\right)\cos\left(\frac{x - y}{2}\right) = \sin x + \sin y$$

The other sum-to-product formulas can be proved in a similar manner.

This collection of thought-provoking and challenging exercises further explores and expands upon concepts learned in this chapter.

1. (a) Write each of the other trigonometric functions of θ in terms of $\sin \theta$.

   (b) Write each of the other trigonometric functions of θ in terms of $\cos \theta$.

2. Verify that for all integers n,

$$\cos\left[\frac{(2n+1)\pi}{2}\right] = 0.$$

3. Verify that for all integers n,

$$\sin\left[\frac{(12n+1)\pi}{6}\right] = \frac{1}{2}.$$

 4. A particular sound wave is modeled by

$$p(t) = \frac{1}{4\pi}\left(p_1(t) + 30p_2(t) + p_3(t) + p_5(t) + 30p_6(t)\right)$$

where $p_n(t) = \frac{1}{n}\sin(524n\pi t)$, and t is the time (in seconds).

   (a) Find the sine components $p_n(t)$ and use a graphing utility to graph each component. Then verify the graph of p that is shown.

   (b) Find the period of each sine component of p. Is p periodic? If so, what is its period?

   (c) Use the *zero* or *root* feature or the *zoom* and *trace* features of a graphing utility to find the t-intercepts of the graph of p over one cycle.

   (d) Use the *maximum* and *minimum* features of a graphing utility to approximate the absolute maximum and absolute minimum values of p over one cycle.

5. Three squares of side s are placed side by side (see figure). Make a conjecture about the relationship between the sum $u + v$ and w. Prove your conjecture by using the identity for the tangent of the sum of two angles.

FIGURE FOR 5

6. The path traveled by an object (neglecting air resistance) that is projected at an initial height of h_0 feet, an initial velocity of v_0 feet per second, and an initial angle θ is given by

$$y = -\frac{16}{v_0^2 \cos^2 \theta}x^2 + (\tan \theta)x + h_0$$

where x and y are measured in feet. Find a formula for the maximum height of an object projected from ground level at velocity v_0 and angle θ. To do this, find half of the horizontal distance

$$\frac{1}{32}v_0^2 \sin 2\theta$$

and then substitute it for x in the general model for the path of a projectile (where $h_0 = 0$).

7. Use the figure to derive the formulas for

$$\sin\frac{\theta}{2}, \cos\frac{\theta}{2}, \text{ and } \tan\frac{\theta}{2}$$

where θ is an acute angle.

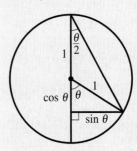

8. The force F (in pounds) on a person's back when he or she bends over at an angle θ is modeled by

$$F = \frac{0.6W \sin(\theta + 90°)}{\sin 12°}$$

where W is the person's weight (in pounds).

   (a) Simplify the model.

    (b) Use a graphing utility to graph the model, where $W = 185$ and $0° < \theta < 90°$.

   (c) At what angle is the force a maximum? At what angle is the force a minimum?

9. The number of hours of daylight that occur at any location on Earth depends on the time of year and the latitude of the location. The following equations model the numbers of hours of daylight in Seward, Alaska (60° latitude) and New Orleans, Louisiana (30° latitude).

$$D = 12.2 - 6.4 \cos\left[\frac{\pi(t + 0.2)}{182.6}\right] \qquad \text{Seward}$$

$$D = 12.2 - 1.9 \cos\left[\frac{\pi(t + 0.2)}{182.6}\right] \qquad \text{New Orleans}$$

In these models, D represents the number of hours of daylight and t represents the day, with $t = 0$ corresponding to January 1.

 (a) Use a graphing utility to graph both models in the same viewing window. Use a viewing window of $0 \le t \le 365$.

(b) Find the days of the year on which both cities receive the same amount of daylight. What are these days called?

(c) Which city has the greater variation in the number of daylight hours? Which constant in each model would you use to determine the difference between the greatest and least numbers of hours of daylight?

(d) Determine the period of each model.

10. The tide, or depth of the ocean near the shore, changes throughout the day. The water depth d (in feet) of a bay can be modeled by

$$d = 35 - 28 \cos \frac{\pi}{6.2} t$$

where t is the time in hours, with $t = 0$ corresponding to 12:00 A.M.

(a) Algebraically find the times at which the high and low tides occur.

(b) Algebraically find the time(s) at which the water depth is 3.5 feet.

 (c) Use a graphing utility to verify your results from parts (a) and (b).

11. Find the solution of each inequality in the interval $[0, 2\pi]$.

(a) $\sin x \ge 0.5$ (b) $\cos x \le -0.5$

(c) $\tan x < \sin x$ (d) $\cos x \ge \sin x$

12. The index of refraction n of a transparent material is the ratio of the speed of light in a vacuum to the speed of light in the material. Some common materials and their indices are air (1.00), water (1.33), and glass (1.50). Triangular prisms are often used to measure the index of refraction based on the formula

$$n = \frac{\sin\left(\dfrac{\theta}{2} + \dfrac{\alpha}{2}\right)}{\sin \dfrac{\theta}{2}}.$$

For the prism shown in the figure, $\alpha = 60°$.

(a) Write the index of refraction as a function of $\cot(\theta/2)$.

(b) Find θ for a prism made of glass.

13. (a) Write a sum formula for $\sin(u + v + w)$.

(b) Write a sum formula for $\tan(u + v + w)$.

14. (a) Derive a formula for $\cos 3\theta$.

(b) Derive a formula for $\cos 4\theta$.

15. The heights h (in inches) of pistons 1 and 2 in an automobile engine can be modeled by

$$h_1 = 3.75 \sin 733t + 7.5$$

and

$$h_2 = 3.75 \sin 733\left(t + \frac{4\pi}{3}\right) + 7.5$$

where t is measured in seconds.

 (a) Use a graphing utility to graph the heights of these two pistons in the same viewing window for $0 \le t \le 1$.

(b) How often are the pistons at the same height?

Additional Topics in Trigonometry

3

The work done by a force, such as pushing and pulling objects, can be calculated using vector operations.

© Amy Etra/PhotoEdit

SELECTED APPLICATIONS

Triangles and vectors have many real-life applications. The applications listed below represent a small sample of the applications in this chapter.

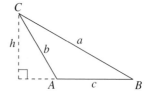

3.1 Law of Sines

What you should learn

- Use the Law of Sines to solve oblique triangles (AAS or ASA).
- Use the Law of Sines to solve oblique triangles (SSA).
- Find the areas of oblique triangles.
- Use the Law of Sines to model and solve real-life problems.

Why you should learn it

You can use the Law of Sines to solve real-life problems involving oblique triangles. For instance, in Exercise 44 on page 286, you can use the Law of Sines to determine the length of the shadow of the Leaning Tower of Pisa.

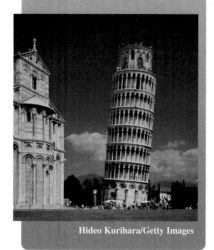

Hideo Kurihara/Getty Images

Introduction

In Chapter 1, you studied techniques for solving right triangles. In this section and the next, you will solve **oblique triangles**—triangles that have no right angles. As standard notation, the angles of a triangle are labeled A, B, and C, and their opposite sides are labeled a, b, and c, as shown in Figure 3.1.

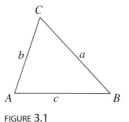

FIGURE 3.1

To solve an oblique triangle, you need to know the measure of at least one side and any two other parts of the triangle—either two sides, two angles, or one angle and one side. This breaks down into the following four cases.

1. Two angles and any side (AAS or ASA)
2. Two sides and an angle opposite one of them (SSA)
3. Three sides (SSS)
4. Two sides and their included angle (SAS)

The first two cases can be solved using the **Law of Sines,** whereas the last two cases require the Law of Cosines (see Section 3.2).

Law of Sines

If ABC is a triangle with sides a, b, and c, then

$$\frac{a}{\sin A} = \frac{b}{\sin B} = \frac{c}{\sin C}.$$

A is acute.   A is obtuse.

The Law of Sines can also be written in the reciprocal form

$$\frac{\sin A}{a} = \frac{\sin B}{b} = \frac{\sin C}{c}.$$

For a proof of the Law of Sines, see Proofs in Mathematics on page 325.

The *HM mathSpace*® CD-ROM and *Eduspace*® for this text contain additional resources related to the concepts discussed in this chapter.

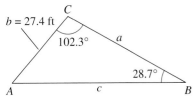

FIGURE **3.2**

STUDY TIP

When solving triangles, a careful sketch is useful as a quick test for the feasibility of an answer. Remember that the longest side lies opposite the largest angle, and the shortest side lies opposite the smallest angle.

| **Example 1** | **Given Two Angles and One Side—AAS** |

For the triangle in Figure 3.2, $C = 102.3°$, $B = 28.7°$, and $b = 27.4$ feet. Find the remaining angle and sides.

Solution

The third angle of the triangle is

$$A = 180° - B - C$$
$$= 180° - 28.7° - 102.3°$$
$$= 49.0°.$$

By the Law of Sines, you have

$$\frac{a}{\sin A} = \frac{b}{\sin B} = \frac{c}{\sin C}.$$

Using $b = 27.4$ produces

$$a = \frac{b}{\sin B}(\sin A) = \frac{27.4}{\sin 28.7°}(\sin 49.0°) \approx 43.06 \text{ feet}$$

and

$$c = \frac{b}{\sin B}(\sin C) = \frac{27.4}{\sin 28.7°}(\sin 102.3°) \approx 55.75 \text{ feet}.$$

✓CHECKPOINT　　Now try Exercise 1.

| **Example 2** | **Given Two Angles and One Side—ASA** | |

A pole tilts *toward* the sun at an 8° angle from the vertical, and it casts a 22-foot shadow. The angle of elevation from the tip of the shadow to the top of the pole is 43°. How tall is the pole?

Solution

From Figure 3.3, note that $A = 43°$ and $B = 90° + 8° = 98°$. So, the third angle is

$$C = 180° - A - B$$
$$= 180° - 43° - 98°$$
$$= 39°.$$

By the Law of Sines, you have

$$\frac{a}{\sin A} = \frac{c}{\sin C}.$$

Because $c = 22$ feet, the length of the pole is

$$a = \frac{c}{\sin C}(\sin A) = \frac{22}{\sin 39°}(\sin 43°) \approx 23.84 \text{ feet}.$$

✓CHECKPOINT　　Now try Exercise 35.

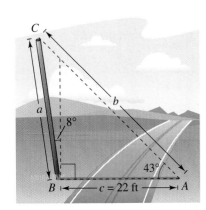

FIGURE **3.3**

For practice, try reworking Example 2 for a pole that tilts *away from* the sun under the same conditions.

The Ambiguous Case (SSA)

In Examples 1 and 2 you saw that two angles and one side determine a unique triangle. However, if two sides and one opposite angle are given, three possible situations can occur: (1) no such triangle exists, (2) one such triangle exists, or (3) two distinct triangles may satisfy the conditions.

The Ambiguous Case (SSA)

Consider a triangle in which you are given a, b, and A.   ($h = b \sin A$)

	A is acute.	A is acute.	A is acute.	A is acute.	A is obtuse.	A is obtuse.
Sketch						
Necessary condition	$a < h$	$a = h$	$a > b$	$h < a < b$	$a \leq b$	$a > b$
Triangles possible	None	One	One	Two	None	One

$b = 12$ in. $a = 22$ in.

C

$42°$

A c B

One solution: $a > b$

FIGURE **3.4**

Example 3   Single-Solution Case—SSA

For the triangle in Figure 3.4, $a = 22$ inches, $b = 12$ inches, and $A = 42°$. Find the remaining side and angles.

Solution

By the Law of Sines, you have

$$\frac{\sin B}{b} = \frac{\sin A}{a} \qquad \text{Reciprocal form}$$

$$\sin B = b\left(\frac{\sin A}{a}\right) \qquad \text{Multiply each side by } b.$$

$$\sin B = 12\left(\frac{\sin 42°}{22}\right) \qquad \text{Substitute for } A, a, \text{ and } b.$$

$$B \approx 21.41°. \qquad B \text{ is acute.}$$

Now, you can determine that

$$C \approx 180° - 42° - 21.41° = 116.59°.$$

Then, the remaining side is

$$\frac{c}{\sin C} = \frac{a}{\sin A}$$

$$c = \frac{a}{\sin A}(\sin C) = \frac{22}{\sin 42°}(\sin 116.59°) \approx 29.40 \text{ inches.}$$

Encourage your students to sketch the triangle, keeping in mind that the longest side lies opposite the largest angle of the triangle. For practice, suggest that students also find h.

✓**CHECKPOINT**   Now try Exercise 19.

No solution: a < h

FIGURE **3.5**

Example 4   No-Solution Case—SSA

Show that there is no triangle for which $a = 15$, $b = 25$, and $A = 85°$.

Solution

Begin by making the sketch shown in Figure 3.5. From this figure it appears that no triangle is formed. You can verify this using the Law of Sines.

$$\frac{\sin B}{b} = \frac{\sin A}{a} \qquad \text{Reciprocal form}$$

$$\sin B = b\left(\frac{\sin A}{a}\right) \qquad \text{Multiply each side by } b.$$

$$\sin B = 25\left(\frac{\sin 85°}{15}\right) \approx 1.660 > 1$$

This contradicts the fact that $|\sin B| \leq 1$. So, no triangle can be formed having sides $a = 15$ and $b = 25$ and an angle of $A = 85°$.

✓**CHECKPOINT**   Now try Exercise 21.

Example 5   Two-Solution Case—SSA

Find two triangles for which $a = 12$ meters, $b = 31$ meters, and $A = 20.5°$.

Solution

By the Law of Sines, you have

$$\frac{\sin B}{b} = \frac{\sin A}{a} \qquad \text{Reciprocal form}$$

$$\sin B = b\left(\frac{\sin A}{a}\right) = 31\left(\frac{\sin 20.5°}{12}\right) \approx 0.9047.$$

There are two angles $B_1 \approx 64.8°$ and $B_2 \approx 180° - 64.8° = 115.2°$ between $0°$ and $180°$ whose sine is 0.9047. For $B_1 \approx 64.8°$, you obtain

$$C \approx 180° - 20.5° - 64.8° = 94.7°$$

$$c = \frac{a}{\sin A}(\sin C) = \frac{12}{\sin 20.5°}(\sin 94.7°) \approx 34.15 \text{ meters.}$$

For $B_2 \approx 115.2°$, you obtain

$$C \approx 180° - 20.5° - 115.2° = 44.3°$$

$$c = \frac{a}{\sin A}(\sin C) = \frac{12}{\sin 20.5°}(\sin 44.3°) \approx 23.93 \text{ meters.}$$

The resulting triangles are shown in Figure 3.6.

FIGURE **3.6**

✓**CHECKPOINT**   Now try Exercise 23.

Activities

1. Use the given information to find (if possible) the remaining side and angles of the oblique triangle. If two solutions exist, find both.

   $A = 58°, \ a = 20, \ c = 10$

   Answer: $B = 97°, \ C = 25°,$
   $\qquad\qquad b = 23.4$

2. Use the given information to find (if possible) the remaining side and angles of the oblique triangle. If two solutions exist, find both.

   $B = 78°, \ b = 207, \ c = 210$

   Answer: Two solutions
   $A = 19.1°, \ a = 69.2, \ C = 82.9°$
   $A = 4.9°, \ a = 18.1, \ C = 97.1°$

3. Find the area of the triangle with $B = 120°, a = 32,$ and $c = 50$.

   Answer: Area = 692.8 square units

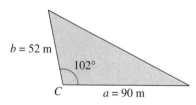

FIGURE 3.8

Area of an Oblique Triangle

The procedure used to prove the Law of Sines leads to a simple formula for the area of an oblique triangle. Referring to Figure 3.7, note that each triangle has a height of $h = b \sin A$. Consequently, the area of each triangle is

$$\text{Area} = \frac{1}{2}(\text{base})(\text{height}) = \frac{1}{2}(c)(b \sin A) = \frac{1}{2}bc \sin A.$$

By similar arguments, you can develop the formulas

$$\text{Area} = \frac{1}{2}ab \sin C = \frac{1}{2}ac \sin B.$$

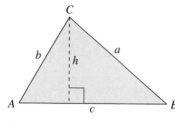

A is acute

FIGURE 3.7

A is obtuse

Area of an Oblique Triangle

The area of any triangle is one-half the product of the lengths of two sides times the sine of their included angle. That is,

$$\text{Area} = \frac{1}{2}bc \sin A = \frac{1}{2}ab \sin C = \frac{1}{2}ac \sin B.$$

Note that if angle A is 90°, the formula gives the area for a right triangle:

$$\text{Area} = \frac{1}{2}bc \,(\sin 90°) = \frac{1}{2}bc = \frac{1}{2}(\text{base})(\text{height}). \qquad \sin 90° = 1$$

Similar results are obtained for angles C and B equal to 90°.

Example 6 **Finding the Area of a Triangular Lot**

Find the area of a triangular lot having two sides of lengths 90 meters and 52 meters and an included angle of 102°.

Solution

Consider $a = 90$ meters, $b = 52$ meters, and angle $C = 102°$, as shown in Figure 3.8. Then, the area of the triangle is

$$\text{Area} = \frac{1}{2}ab \sin C = \frac{1}{2}(90)(52)(\sin 102°) \approx 2289 \text{ square meters.}$$

✓**CHECKPOINT** Now try Exercise 29.

FIGURE **3.9**

FIGURE **3.10**

Application

Example 7   An Application of the Law of Sines

The course for a boat race starts at point A in Figure 3.9 and proceeds in the direction S 52° W to point B, then in the direction S 40° E to point C, and finally back to A. Point C lies 8 kilometers directly south of point A. Approximate the total distance of the race course.

Solution

Because lines BD and AC are parallel, it follows that $\angle BCA \cong \angle DBC$. Consequently, triangle ABC has the measures shown in Figure 3.10. For angle B, you have $B = 180° - 52° - 40° = 88°$. Using the Law of Sines

$$\frac{a}{\sin 52°} = \frac{b}{\sin 88°} = \frac{c}{\sin 40°}$$

you can let $b = 8$ and obtain

$$a = \frac{8}{\sin 88°}(\sin 52°) \approx 6.308$$

and

$$c = \frac{8}{\sin 88°}(\sin 40°) \approx 5.145.$$

The total length of the course is approximately

$$\text{Length} \approx 8 + 6.308 + 5.145$$

$$= 19.453 \text{ kilometers.}$$

✓CHECKPOINT   Now try Exercise 39.

Alternative Writing About Mathematics: Error Analysis

You are a math instructor, and one of your students hands in the following solution. Discuss what is wrong with your student's solution. How could you help the student avoid making a similar mistake in the future? Use a diagram to illustrate your explanation.

Find side c in a triangle that has $a = 5.8$, $b = 7$, and $C = 82°$.

Student's Solution

Because this is an SSA situation, I can use the Law of Sines.

$$\frac{5.8}{7} = \frac{c}{\sin 82°}$$

$$(\sin 82°)\frac{5.8}{7} = 0.821$$

WRITING ABOUT **M**ATHEMATICS

Using the Law of Sines   In this section, you have been using the Law of Sines to solve *oblique* triangles. Can the Law of Sines also be used to solve a right triangle? If so, write a short paragraph explaining how to use the Law of Sines to solve each triangle. Is there an easier way to solve these triangles?

a. (AAS)

b. (ASA)

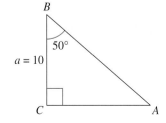

3.1 Exercises

The *HM mathSpace®* CD-ROM and *Eduspace®* for this text contain step-by-step solutions to all odd-numbered exercises. They also provide Tutorial Exercises for additional help.

Exercises with no solution: 20, 21, 24
Exercises with two solutions: 6, 23

VOCABULARY CHECK: Fill in the blanks.

1. An _____ triangle is a triangle that has no right angle.

2. For triangle *ABC*, the Law of Sines is given by $\dfrac{a}{\sin A} = $ _____ $= \dfrac{c}{\sin C}$.

3. The area of an oblique triangle is given by $\frac{1}{2}bc \sin A = \frac{1}{2}ab \sin C = $ _____ .

PREREQUISITE SKILLS REVIEW: Practice and review algebra skills needed for this section at **www.Eduspace.com.**

In Exercises 1–18, use the Law of Sines to solve the triangle. Round your answers to two decimal places.

1.

2.

3.

4.

5. $A = 36°, \quad a = 8, \quad b = 5$

6. $A = 60°, \quad a = 9, \quad c = 10$

7. $A = 102.4°, \quad C = 16.7°, \quad a = 21.6$

8. $A = 24.3°, \quad C = 54.6°, \quad c = 2.68$

9. $A = 83° \, 20', \quad C = 54.6°, \quad c = 18.1$

10. $A = 5° \, 40', \quad B = 8° \, 15', \quad b = 4.8$

11. $B = 15° \, 30', \quad a = 4.5, \quad b = 6.8$

12. $B = 2° \, 45', \quad b = 6.2, \quad c = 5.8$

13. $C = 145°, \quad b = 4, \quad c = 14$

14. $A = 100°, \quad a = 125, \quad c = 10$

15. $A = 110° \, 15', \quad a = 48, \quad b = 16$

16. $C = 85° \, 20', \quad a = 35, \quad c = 50$

17. $A = 55°, \quad B = 42°, \quad c = \frac{3}{4}$

18. $B = 28°, \quad C = 104°, \quad a = 3\frac{5}{8}$

In Exercises 19–24, use the Law of Sines to solve (if possible) the triangle. If two solutions exist, find both. Round your answers to two decimal places.

19. $A = 110°, \quad a = 125, \quad b = 100$

20. $A = 110°, \quad a = 125, \quad b = 200$

21. $A = 76°, \quad a = 18, \quad b = 20$

22. $A = 76°, \quad a = 34, \quad b = 21$

23. $A = 58°, \quad a = 11.4, \quad b = 12.8$

24. $A = 58°, \quad a = 4.5, \quad b = 12.8$

In Exercises 25–28, find values for *b* such that the triangle has (a) one solution, (b) two solutions, and (c) no solution.

25. $A = 36°, \quad a = 5$

26. $A = 60°, \quad a = 10$

27. $A = 10°, \quad a = 10.8$

28. $A = 88°, \quad a = 315.6$

In Exercises 29–34, find the area of the triangle having the indicated angle and sides.

29. $C = 120°, \quad a = 4, \quad b = 6$

30. $B = 130°, \quad a = 62, \quad c = 20$

31. $A = 43° \, 45', \quad b = 57, \quad c = 85$

32. $A = 5° \, 15', \quad b = 4.5, \quad c = 22$

33. $B = 72° \, 30', \quad a = 105, \quad c = 64$

34. $C = 84° \, 30', \quad a = 16, \quad b = 20$

35. *Height* Because of prevailing winds, a tree grew so that it was leaning 4° from the vertical. At a point 35 meters from the tree, the angle of elevation to the top of the tree is 23° (see figure). Find the height *h* of the tree.

36. *Height* A flagpole at a right angle to the horizontal is located on a slope that makes an angle of 12° with the horizontal. The flagpole's shadow is 16 meters long and points directly up the slope. The angle of elevation from the tip of the shadow to the sun is 20°.

(a) Draw a triangle that represents the problem. Show the known quantities on the triangle and use a variable to indicate the height of the flagpole.

(b) Write an equation involving the unknown quantity.

(c) Find the height of the flagpole.

37. *Angle of Elevation* A 10-meter telephone pole casts a 17-meter shadow directly down a slope when the angle of elevation of the sun is 42° (see figure). Find θ, the angle of elevation of the ground.

38. *Flight Path* A plane flies 500 kilometers with a bearing of 316° from Naples to Elgin (see figure). The plane then flies 720 kilometers from Elgin to Canton. Find the bearing of the flight from Elgin to Canton.

39. *Bridge Design* A bridge is to be built across a small lake from a gazebo to a dock (see figure). The bearing from the gazebo to the dock is S 41° W. From a tree 100 meters from the gazebo, the bearings to the gazebo and the dock are S 74° E and S 28° E, respectively. Find the distance from the gazebo to the dock.

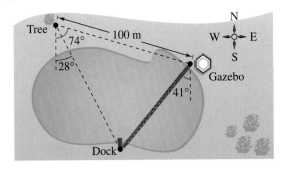

40. *Railroad Track Design* The circular arc of a railroad curve has a chord of length 3000 feet and a central angle of 40°.

(a) Draw a diagram that visually represents the problem. Show the known quantities on the diagram and use the variables *r* and *s* to represent the radius of the arc and the length of the arc, respectively.

(b) Find the radius *r* of the circular arc.

(c) Find the length *s* of the circular arc.

41. *Glide Path* A pilot has just started on the glide path for landing at an airport with a runway of length 9000 feet. The angles of depression from the plane to the ends of the runway are 17.5° and 18.8°.

(a) Draw a diagram that visually represents the problem.

(b) Find the air distance the plane must travel until touching down on the near end of the runway.

(c) Find the ground distance the plane must travel until touching down.

(d) Find the altitude of the plane when the pilot begins the descent.

42. *Locating a Fire* The bearing from the Pine Knob fire tower to the Colt Station fire tower is N 65° E, and the two towers are 30 kilometers apart. A fire spotted by rangers in each tower has a bearing of N 80° E from Pine Knob and S 70° E from Colt Station (see figure). Find the distance of the fire from each tower.

43. *Distance* A boat is sailing due east parallel to the shoreline at a speed of 10 miles per hour. At a given time, the bearing to the lighthouse is S 70° E, and 15 minutes later the bearing is S 63° E (see figure). The lighthouse is located at the shoreline. What is the distance from the boat to the shoreline?

Model It

44. *Shadow Length* The Leaning Tower of Pisa in Italy is characterized by its tilt. The tower leans because it was built on a layer of unstable soil—clay, sand, and water. The tower is approximately 58.36 meters tall from its foundation (see figure). The top of the tower leans about 5.45 meters off center.

Not drawn to scale

(a) Find the angle of lean α of the tower.

(b) Write β as a function of d and θ, where θ is the angle of elevation to the sun.

(c) Use the Law of Sines to write an equation for the length d of the shadow cast by the tower.

(d) Use a graphing utility to complete the table.

θ	10°	20°	30°	40°	50°	60°
d						

Synthesis

True or False? **In Exercises 45 and 46, determine whether the statement is true or false. Justify your answer.**

45. If a triangle contains an obtuse angle, then it must be oblique.

46. Two angles and one side of a triangle do not necessarily determine a unique triangle.

47. *Graphical and Numerical Analysis* In the figure, α and β are positive angles.

(a) Write α as a function of β.

(b) Use a graphing utility to graph the function. Determine its domain and range.

(c) Use the result of part (a) to write c as a function of β.

(d) Use a graphing utility to graph the function in part (c). Determine its domain and range.

(e) Complete the table. What can you infer?

β	0.4	0.8	1.2	1.6	2.0	2.4	2.8
α							
c							

FIGURE FOR 47   FIGURE FOR 48

48. *Graphical Analysis*

(a) Write the area A of the shaded region in the figure as a function of θ.

(b) Use a graphing utility to graph the area function.

(c) Determine the domain of the area function. Explain how the area of the region and the domain of the function would change if the eight-centimeter line segment were decreased in length.

Skills Review

In Exercises 49–52, use the fundamental trigonometric identities to simplify the expression.

49. $\sin x \cot x$

50. $\tan x \cos x \sec x$

51. $1 - \sin^2\left(\dfrac{\pi}{2} - x\right)$

52. $1 + \cot^2\left(\dfrac{\pi}{2} - x\right)$

3.2   Law of Cosines

What you should learn

- Use the Law of Cosines to solve oblique triangles (SSS or SAS).
- Use the Law of Cosines to model and solve real-life problems.
- Use Heron's Area Formula to find the area of a triangle.

Why you should learn it

You can use the Law of Cosines to solve real-life problems involving oblique triangles. For instance, in Exercise 31 on page 292, you can use the Law of Cosines to approximate the length of a marsh.

© Roger Ressmeyer/Corbis

Introduction

Two cases remain in the list of conditions needed to solve an oblique triangle—SSS and SAS. If you are given three sides (SSS), or two sides and their included angle (SAS), none of the ratios in the Law of Sines would be complete. In such cases, you can use the **Law of Cosines.**

Law of Cosines

Standard Form	*Alternative Form*
$a^2 = b^2 + c^2 - 2bc \cos A$	$\cos A = \dfrac{b^2 + c^2 - a^2}{2bc}$
$b^2 = a^2 + c^2 - 2ac \cos B$	$\cos B = \dfrac{a^2 + c^2 - b^2}{2ac}$
$c^2 = a^2 + b^2 - 2ab \cos C$	$\cos C = \dfrac{a^2 + b^2 - c^2}{2ab}$

For a proof of the Law of Cosines, see Proofs in Mathematics on page 326.

Example 1   **Three Sides of a Triangle—SSS**

Find the three angles of the triangle in Figure 3.11.

FIGURE **3.11**

Solution

It is a good idea first to find the angle opposite the longest side—side b in this case. Using the alternative form of the Law of Cosines, you find that

$$\cos B = \frac{a^2 + c^2 - b^2}{2ac} = \frac{8^2 + 14^2 - 19^2}{2(8)(14)} \approx -0.45089.$$

Because $\cos B$ is negative, you know that B is an *obtuse* angle given by $B \approx 116.80°$. At this point, it is simpler to use the Law of Sines to determine A.

$$\sin A = a\left(\frac{\sin B}{b}\right) \approx 8\left(\frac{\sin 116.80°}{19}\right) \approx 0.37583$$

Because B is obtuse, A must be acute, because a triangle can have, at most, one obtuse angle. So, $A \approx 22.08°$ and $C \approx 180° - 22.08° - 116.80° = 41.12°$.

✓CHECKPOINT   Now try Exercise 1.

In cases where the Law of Cosines must be used, encourage your students to solve for the largest angle first, then finish the problem using either the Law of Sines or the Law of Cosines.

Do you see why it was wise to find the largest angle *first* in Example 1? Knowing the cosine of an angle, you can determine whether the angle is acute or obtuse. That is,

$$\cos \theta > 0 \quad \text{for} \quad 0° < \theta < 90° \qquad \text{Acute}$$

$$\cos \theta < 0 \quad \text{for} \quad 90° < \theta < 180°. \qquad \text{Obtuse}$$

So, in Example 1, once you found that angle B was obtuse, you knew that angles A and C were both acute. If the largest angle is acute, the remaining two angles are acute also.

Example 2   Two Sides and the Included Angle—SAS

Find the remaining angles and side of the triangle in Figure 3.12.

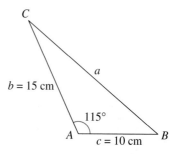

FIGURE **3.**12

Solution

Use the Law of Cosines to find the unknown side a in the figure.

$$a^2 = b^2 + c^2 - 2bc \cos A$$

$$a^2 = 15^2 + 10^2 - 2(15)(10) \cos 115°$$

$$a^2 \approx 451.79$$

$$a \approx 21.26$$

Because $a \approx 21.26$ centimeters, you now know the ratio $\sin A / a$ and you can use the reciprocal form of the Law of Sines to solve for B.

$$\frac{\sin B}{b} = \frac{\sin A}{a}$$

$$\sin B = b\left(\frac{\sin A}{a}\right)$$

$$= 15\left(\frac{\sin 115°}{21.26}\right)$$

$$\approx 0.63945$$

So, $B = \arcsin 0.63945 \approx 39.75°$ and $C \approx 180° - 115° - 39.75° = 25.25°$.

✓CHECKPOINT   Now try Exercise 3.

Applications

Example 3   An Application of the Law of Cosines

The pitcher's mound on a women's softball field is 43 feet from home plate and the distance between the bases is 60 feet, as shown in Figure 3.13. (The pitcher's mound is not halfway between home plate and second base.) How far is the pitcher's mound from first base?

Solution

In triangle HPF, $H = 45°$ (line HP bisects the right angle at H), $f = 43$, and $p = 60$. Using the Law of Cosines for this SAS case, you have

$$h^2 = f^2 + p^2 - 2fp \cos H$$

$$= 43^2 + 60^2 - 2(43)(60) \cos 45° \approx 1800.3$$

So, the approximate distance from the pitcher's mound to first base is

$$h \approx \sqrt{1800.3} \approx 42.43 \text{ feet.}$$

✓CHECKPOINT   Now try Exercise 31.

FIGURE 3.13

Example 4   An Application of the Law of Cosines

A ship travels 60 miles due east, then adjusts its course northward, as shown in Figure 3.14. After traveling 80 miles in that direction, the ship is 139 miles from its point of departure. Describe the bearing from point B to point C.

FIGURE 3.14

Solution

You have $a = 80$, $b = 139$, and $c = 60$; so, using the alternative form of the Law of Cosines, you have

$$\cos B = \frac{a^2 + c^2 - b^2}{2ac}$$

$$= \frac{80^2 + 60^2 - 139^2}{2(80)(60)}$$

$$\approx -0.97094.$$

So, $B \approx \arccos(-0.97094) \approx 166.15°$, and thus the bearing measured from due north from point B to point C is $166.15° - 90° = 76.15°$, or N 76.15° E.

✓CHECKPOINT   Now try Exercise 37.

Historical Note

Heron of Alexandria (c. 100 B.C.) was a Greek geometer and inventor. His works describe how to find the areas of triangles, quadrilaterals, regular polygons having 3 to 12 sides, and circles as well as the surface areas and volumes of three-dimensional objects.

Heron's Area Formula

The Law of Cosines can be used to establish the following formula for the area of a triangle. This formula is called **Heron's Area Formula** after the Greek mathematician Heron (c. 100 B.C.).

> ## Heron's Area Formula
>
> Given any triangle with sides of lengths a, b, and c, the area of the triangle is
>
> $$\text{Area} = \sqrt{s(s-a)(s-b)(s-c)}$$
>
> where $s = (a+b+c)/2$.

For a proof of Heron's Area Formula, see Proofs in Mathematics on page 327.

Activities

1. Determine whether the Law of Sines or the Law of Cosines is needed to solve each of the triangles.
    a. $A = 15°, B = 58°, c = 94$
    b. $a = 96, b = 43, A = 105°$
    c. $a = 24, b = 16, c = 29$
    d. $a = 15, c = 42, B = 49°$
    Answer: a. Law of Sines,
    b. Law of Sines, c. Law of Cosines,
    d. Law of Cosines
2. Solve the triangle: $a = 31, b = 52$, $c = 28$.
    Answer: $A = 29.8°, B = 123.5°$,
    $C = 26.7°$
3. Use Heron's Area Formula to find the area of a triangle with sides of lengths $a = 31, b = 52$, and $c = 28$.
    Answer: Area ≈ 361.8 square units

Example 5   Using Heron's Area Formula

Find the area of a triangle having sides of lengths $a = 43$ meters, $b = 53$ meters, and $c = 72$ meters.

Solution

Because $s = (a+b+c)/2 = 168/2 = 84$, Heron's Area Formula yields

$$\text{Area} = \sqrt{s(s-a)(s-b)(s-c)}$$

$$= \sqrt{84(41)(31)(12)} \approx 1131.89 \text{ square meters.}$$

✓CHECKPOINT   Now try Exercise 47.

You have now studied three different formulas for the area of a triangle.

Standard Formula	Area $= \frac{1}{2}bh$
Oblique Triangle	Area $= \frac{1}{2}bc\sin A = \frac{1}{2}ab\sin C = \frac{1}{2}ac\sin B$
Heron's Area Formula	Area $= \sqrt{s(s-a)(s-b)(s-c)}$

Writing About Mathematics
Suggestion:

You may want to ask students to draw the diagram of a triangle (labeling whatever sides and/or angles are necessary) that is well-suited to each area formula. Then ask students to exchange triangles with a partner to find the area of each.

𝒲RITING ABOUT 𝓜ATHEMATICS

The Area of a Triangle   Use the most appropriate formula to find the area of each triangle below. Show your work and give your reasons for choosing each formula.

a.

b.

c.

d.
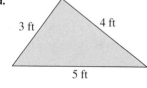

3.2 **Exercises**

VOCABULARY CHECK: Fill in the blanks.

1. If you are given three sides of a triangle, you would use the Law of _____ to find the three angles of the triangle.

2. The standard form of the Law of Cosines for $\cos B = \dfrac{a^2 + c^2 - b^2}{2ac}$ is _____ .

3. The Law of Cosines can be used to establish a formula for finding the area of a triangle called _____ _____ Formula.

PREREQUISITE SKILLS REVIEW: Practice and review algebra skills needed for this section at **www.Eduspace.com**.

In Exercises 1–16, use the Law of Cosines to solve the triangle. Round your answers to two decimal places.

1.

2.

3.

4.

5. $a = 11$, $b = 14$, $c = 20$

6. $a = 55$, $b = 25$, $c = 72$

7. $a = 75.4$, $b = 52$, $c = 52$

8. $a = 1.42$, $b = 0.75$, $c = 1.25$

9. $A = 135°$, $b = 4$, $c = 9$

10. $A = 55°$, $b = 3$, $c = 10$

11. $B = 10°\ 35'$, $a = 40$, $c = 30$

12. $B = 75°\ 20'$, $a = 6.2$, $c = 9.5$

13. $B = 125°\ 40'$, $a = 32$, $c = 32$

14. $C = 15°\ 15'$, $a = 6.25$, $b = 2.15$

15. $C = 43°$, $a = \frac{4}{9}$, $b = \frac{7}{9}$

16. $C = 103°$, $a = \frac{3}{8}$, $b = \frac{3}{4}$

In Exercises 17–22, complete the table by solving the parallelogram shown in the figure. (The lengths of the diagonals are given by c and d.)

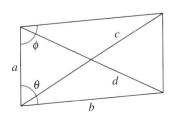

	a	b	c	d	θ	ϕ
17.	5	8			45°	
18.	25	35				120°
19.	10	14	20			
20.	40	60		80		
21.	15		25	20		
22.		25	50	35		

In Exercises 23–28, use Heron's Area Formula to find the area of the triangle.

23. $a = 5$, $b = 7$, $c = 10$

24. $a = 12$, $b = 15$, $c = 9$

25. $a = 2.5$, $b = 10.2$, $c = 9$

26. $a = 75.4$, $b = 52$, $c = 52$

27. $a = 12.32$, $b = 8.46$, $c = 15.05$

28. $a = 3.05$, $b = 0.75$, $c = 2.45$

29. *Navigation* A boat race runs along a triangular course marked by buoys A, B, and C. The race starts with the boats headed west for 3700 meters. The other two sides of the course lie to the north of the first side, and their lengths are 1700 meters and 3000 meters. Draw a figure that gives a visual representation of the problem, and find the bearings for the last two legs of the race.

30. *Navigation* A plane flies 810 miles from Franklin to Centerville with a bearing of 75°. Then it flies 648 miles from Centerville to Rosemount with a bearing of 32°. Draw a figure that visually represents the problem, and find the straight-line distance and bearing from Franklin to Rosemount.

31. *Surveying* To approximate the length of a marsh, a surveyor walks 250 meters from point *A* to point *B*, then turns 75° and walks 220 meters to point *C* (see figure). Approximate the length *AC* of the marsh.

32. *Surveying* A triangular parcel of land has 115 meters of frontage, and the other boundaries have lengths of 76 meters and 92 meters. What angles does the frontage make with the two other boundaries?

33. *Surveying* A triangular parcel of ground has sides of lengths 725 feet, 650 feet, and 575 feet. Find the measure of the largest angle.

34. *Streetlight Design* Determine the angle θ in the design of the streetlight shown in the figure.

35. *Distance* Two ships leave a port at 9 A.M. One travels at a bearing of N 53° W at 12 miles per hour, and the other travels at a bearing of S 67° W at 16 miles per hour. Approximate how far apart they are at noon that day.

36. *Length* A 100-foot vertical tower is to be erected on the side of a hill that makes a 6° angle with the horizontal (see figure). Find the length of each of the two guy wires that will be anchored 75 feet uphill and downhill from the base of the tower.

37. *Navigation* On a map, Orlando is 178 millimeters due south of Niagara Falls, Denver is 273 millimeters from Orlando, and Denver is 235 millimeters from Niagara Falls (see figure).

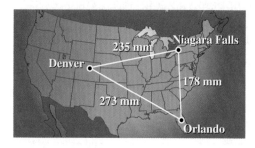

(a) Find the bearing of Denver from Orlando.

(b) Find the bearing of Denver from Niagara Falls.

38. *Navigation* On a map, Minneapolis is 165 millimeters due west of Albany, Phoenix is 216 millimeters from Minneapolis, and Phoenix is 368 millimeters from Albany (see figure).

(a) Find the bearing of Minneapolis from Phoenix.

(b) Find the bearing of Albany from Phoenix.

39. *Baseball* On a baseball diamond with 90-foot sides, the pitcher's mound is 60.5 feet from home plate. How far is it from the pitcher's mound to third base?

40. *Baseball* The baseball player in center field is playing approximately 330 feet from the television camera that is behind home plate. A batter hits a fly ball that goes to the wall 420 feet from the camera (see figure). The camera turns 8° to follow the play. Approximately how far does the center fielder have to run to make the catch?

41. *Aircraft Tracking*   To determine the distance between two aircraft, a tracking station continuously determines the distance to each aircraft and the angle A between them (see figure). Determine the distance a between the planes when $A = 42°$, $b = 35$ miles, and $c = 20$ miles.

42. *Aircraft Tracking*   Use the figure for Exercise 41 to determine the distance a between the planes when $A = 11°$, $b = 20$ miles, and $c = 20$ miles.

43. *Trusses* Q is the midpoint of the line segment $\overline{PR}$ in the truss rafter shown in the figure. What are the lengths of the line segments $\overline{PQ}$, $\overline{QS}$, and $\overline{RS}$?

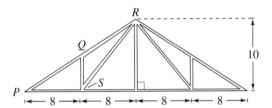

Model It

44. *Engine Design*   An engine has a seven-inch connecting rod fastened to a crank (see figure).

(a) Use the Law of Cosines to write an equation giving the relationship between x and θ.

(b) Write x as a function of θ. (Select the sign that yields positive values of x.)

 (c) Use a graphing utility to graph the function in part (b).

 (d) Use the graph in part (c) to determine the maximum distance the piston moves in one cycle.

---

45. *Paper Manufacturing*   In a process with continuous paper, the paper passes across three rollers of radii 3 inches, 4 inches, and 6 inches (see figure). The centers of the three-inch and six-inch rollers are d inches apart, and the length of the arc in contact with the paper on the four-inch roller is s inches. Complete the table.

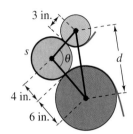

d (inches)	9	10	12	13	14	15	16
θ (degrees)							
s (inches)							

46. *Awning Design*   A retractable awning above a patio door lowers at an angle of 50° from the exterior wall at a height of 10 feet above the ground (see figure). No direct sunlight is to enter the door when the angle of elevation of the sun is greater than 70°. What is the length x of the awning?

47. *Geometry*   The lengths of the sides of a triangular parcel of land are approximately 200 feet, 500 feet, and 600 feet. Approximate the area of the parcel.

48. *Geometry*   A parking lot has the shape of a parallelogram (see figure). The lengths of two adjacent sides are 70 meters and 100 meters. The angle between the two sides is 70°. What is the area of the parking lot?

49. **Geometry** You want to buy a triangular lot measuring 510 yards by 840 yards by 1120 yards. The price of the land is $2000 per acre. How much does the land cost? (*Hint:* 1 acre = 4840 square yards)

50. **Geometry** You want to buy a triangular lot measuring 1350 feet by 1860 feet by 2490 feet. The price of the land is $2200 per acre. How much does the land cost? (*Hint:* 1 acre = 43,560 square feet)

Synthesis

True or False? In Exercises 51–53, determine whether the statement is true or false. Justify your answer.

51. In Heron's Area Formula, s is the average of the lengths of the three sides of the triangle.

52. In addition to SSS and SAS, the Law of Cosines can be used to solve triangles with SSA conditions.

53. A triangle with side lengths of 10 centimeters, 16 centimeters, and 5 centimeters can be solved using the Law of Cosines.

54. **Circumscribed and Inscribed Circles** Let R and r be the radii of the circumscribed and inscribed circles of a triangle ABC, respectively (see figure), and let

$$s = \frac{a + b + c}{2}.$$

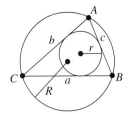

(a) Prove that $2R = \dfrac{a}{\sin A} = \dfrac{b}{\sin B} = \dfrac{c}{\sin C}$.

(b) Prove that $r = \sqrt{\dfrac{(s - a)(s - b)(s - c)}{s}}$.

Circumscribed and Inscribed Circles In Exercises 55 and 56, use the results of Exercise 54.

55. Given a triangle with

$a = 25$, $b = 55$, and $c = 72$

find the areas of (a) the triangle, (b) the circumscribed circle, and (c) the inscribed circle.

56. Find the length of the largest circular running track that can be built on a triangular piece of property with sides of lengths 200 feet, 250 feet, and 325 feet.

57. **Proof** Use the Law of Cosines to prove that

$$\frac{1}{2} bc(1 + \cos A) = \frac{a + b + c}{2} \cdot \frac{-a + b + c}{2}.$$

58. **Proof** Use the Law of Cosines to prove that

$$\frac{1}{2} bc(1 - \cos A) = \frac{a - b + c}{2} \cdot \frac{a + b - c}{2}.$$

Skills Review

In Exercises 59–64, evaluate the expression without using a calculator.

59. $\arcsin(-1)$

60. $\arccos 0$

61. $\arctan \sqrt{3}$

62. $\arctan\left(-\sqrt{3}\right)$

63. $\arcsin\left(-\dfrac{\sqrt{3}}{2}\right)$

64. $\arccos\left(-\dfrac{\sqrt{3}}{2}\right)$

In Exercises 65–68, write an algebraic expression that is equivalent to the expression.

65. $\sec(\arcsin 2x)$

66. $\tan(\arccos 3x)$

67. $\cot[\arctan(x - 2)]$

68. $\cos\left(\arcsin \dfrac{x - 1}{2}\right)$

In Exercises 69–72, use trigonometric substitution to write the algebraic equation as a trigonometric function of θ, where $-\pi/2 < \theta < \pi/2$. Then find $\sec \theta$ and $\csc \theta$.

69. $5 = \sqrt{25 - x^2}$, $x = 5 \sin \theta$

70. $-\sqrt{2} = \sqrt{4 - x^2}$, $x = 2 \cos \theta$

71. $-\sqrt{3} = \sqrt{x^2 - 9}$, $x = 3 \sec \theta$

72. $12 = \sqrt{36 + x^2}$, $x = 6 \tan \theta$

In Exercises 73 and 74, write the sum or difference as a product.

73. $\cos \dfrac{5\pi}{6} - \cos \dfrac{\pi}{3}$

74. $\sin\left(x - \dfrac{\pi}{2}\right) - \sin\left(x + \dfrac{\pi}{2}\right)$

3.3   Vectors in the Plane

What you should learn

- Represent vectors as directed line segments.
- Write the component forms of vectors.
- Perform basic vector operations and represent them graphically.
- Write vectors as linear combinations of unit vectors.
- Find the direction angles of vectors.
- Use vectors to model and solve real-life problems.

Why you should learn it

You can use vectors to model and solve real-life problems involving magnitude and direction. For instance, in Exercise 84 on page 307, you can use vectors to determine the true direction of a commercial jet.

Introduction

Quantities such as force and velocity involve both *magnitude* and *direction* and cannot be completely characterized by a single real number. To represent such a quantity, you can use a **directed line segment,** as shown in Figure 3.15. The directed line segment $\overrightarrow{PQ}$ has **initial point** P and **terminal point** Q. Its **magnitude** (or length) is denoted by $\|\overrightarrow{PQ}\|$ and can be found using the Distance Formula.

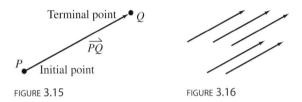

FIGURE 3.15   FIGURE 3.16

Two directed line segments that have the same magnitude and direction are equivalent. For example, the directed line segments in Figure 3.16 are all equivalent. The set of all directed line segments that are equivalent to the directed line segment $\overrightarrow{PQ}$ is a **vector v in the plane,** written $\mathbf{v} = \overrightarrow{PQ}$. Vectors are denoted by lowercase, boldface letters such as $\mathbf{u}$, $\mathbf{v}$, and $\mathbf{w}$.

Example 1   Vector Representation by Directed Line Segments

Let $\mathbf{u}$ be represented by the directed line segment from $P = (0, 0)$ to $Q = (3, 2)$, and let $\mathbf{v}$ be represented by the directed line segment from $R = (1, 2)$ to $S = (4, 4)$, as shown in Figure 3.17. Show that $\mathbf{u} = \mathbf{v}$.

FIGURE 3.17

Solution

From the Distance Formula, it follows that $\overrightarrow{PQ}$ and $\overrightarrow{RS}$ have the *same magnitude*.

$$\|\overrightarrow{PQ}\| = \sqrt{(3 - 0)^2 + (2 - 0)^2} = \sqrt{13}$$

$$\|\overrightarrow{RS}\| = \sqrt{(4 - 1)^2 + (4 - 2)^2} = \sqrt{13}$$

Moreover, both line segments have the *same direction* because they are both directed toward the upper right on lines having a slope of $\frac{2}{3}$. So, $\overrightarrow{PQ}$ and $\overrightarrow{RS}$ have the same magnitude and direction, and it follows that $\mathbf{u} = \mathbf{v}$.

✔CHECKPOINT   Now try Exercise 1.

Component Form of a Vector

The directed line segment whose initial point is the origin is often the most convenient representative of a set of equivalent directed line segments. This representative of the vector **v** is in **standard position.**

A vector whose initial point is the origin $(0, 0)$ can be uniquely represented by the coordinates of its terminal point (v_1, v_2). This is the **component form of a vector v,** written as

$$\mathbf{v} = \langle v_1, v_2 \rangle.$$

Help students see that $\mathbf{v} = \langle 1, 3 \rangle$ can be thought of as a vector with initial point $(0, 0)$ and terminal point $(1, 3)$, as well as a vector with initial point $(0, -1)$ and terminal point $(1, 2)$, and so on.

The coordinates v_1 and v_2 are the *components* of **v**. If both the initial point and the terminal point lie at the origin, **v** is the **zero vector** and is denoted by $\mathbf{0} = \langle 0, 0 \rangle$.

Component Form of a Vector

The component form of the vector with initial point $P = (p_1, p_2)$ and terminal point $Q = (q_1, q_2)$ is given by

$$\overrightarrow{PQ} = \langle q_1 - p_1, q_2 - p_2 \rangle = \langle v_1, v_2 \rangle = \mathbf{v}.$$

The **magnitude** (or length) of **v** is given by

$$\|\mathbf{v}\| = \sqrt{(q_1 - p_1)^2 + (q_2 - p_2)^2} = \sqrt{v_1^2 + v_2^2}.$$

If $\|\mathbf{v}\| = 1$, **v** is a **unit vector.** Moreover, $\|\mathbf{v}\| = 0$ if and only if **v** is the zero vector **0**.

Two vectors $\mathbf{u} = \langle u_1, u_2 \rangle$ and $\mathbf{v} = \langle v_1, v_2 \rangle$ are *equal* if and only if $u_1 = v_1$ and $u_2 = v_2$. For instance, in Example 1, the vector **u** from $P = (0, 0)$ to $Q = (3, 2)$ is

$$\mathbf{u} = \overrightarrow{PQ} = \langle 3 - 0, 2 - 0 \rangle = \langle 3, 2 \rangle$$

and the vector **v** from $R = (1, 2)$ to $S = (4, 4)$ is

$$\mathbf{v} = \overrightarrow{RS} = \langle 4 - 1, 4 - 2 \rangle = \langle 3, 2 \rangle.$$

Example 2　Finding the Component Form of a Vector

Find the component form and magnitude of the vector **v** that has initial point $(4, -7)$ and terminal point $(-1, 5)$.

Solution

Let $P = (4, -7) = (p_1, p_2)$ and let $Q = (-1, 5) = (q_1, q_2)$, as shown in Figure 3.18. Then, the components of $\mathbf{v} = \langle v_1, v_2 \rangle$ are

$$v_1 = q_1 - p_1 = -1 - 4 = -5$$
$$v_2 = q_2 - p_2 = 5 - (-7) = 12.$$

So, $\mathbf{v} = \langle -5, 12 \rangle$ and the magnitude of **v** is

$$\|\mathbf{v}\| = \sqrt{(-5)^2 + 12^2}$$
$$= \sqrt{169} = 13.$$

✓*CHECKPOINT*　Now try Exercise 9.

FIGURE 3.18

FIGURE 3.19

Vector Operations

The two basic vector operations are **scalar multiplication** and **vector addition.** In operations with vectors, numbers are usually referred to as **scalars.** In this text, scalars will always be real numbers. Geometrically, the product of a vector **v** and a scalar k is the vector that is $|k|$ times as long as **v**. If k is positive, $k\mathbf{v}$ has the same direction as **v**, and if k is negative, $k\mathbf{v}$ has the direction opposite that of **v**, as shown in Figure 3.19.

To add two vectors geometrically, position them (without changing their lengths or directions) so that the initial point of one coincides with the terminal point of the other. The sum $\mathbf{u} + \mathbf{v}$ is formed by joining the initial point of the second vector **v** with the terminal point of the first vector **u**, as shown in Figure 3.20. This technique is called the **parallelogram law** for vector addition because the vector $\mathbf{u} + \mathbf{v}$, often called the **resultant** of vector addition, is the diagonal of a parallelogram having **u** and **v** as its adjacent sides.

 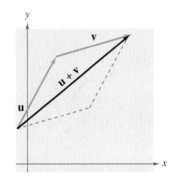

FIGURE 3.20

Definitions of Vector Addition and Scalar Multiplication

Let $\mathbf{u} = \langle u_1, u_2 \rangle$ and $\mathbf{v} = \langle v_1, v_2 \rangle$ be vectors and let k be a scalar (a real number). Then the *sum* of **u** and **v** is the vector

$$\mathbf{u} + \mathbf{v} = \langle u_1 + v_1, u_2 + v_2 \rangle \qquad \text{Sum}$$

and the *scalar multiple* of k times **u** is the vector

$$k\mathbf{u} = k\langle u_1, u_2 \rangle = \langle ku_1, ku_2 \rangle. \qquad \text{Scalar multiple}$$

The graphical representation of the difference of two vectors may not be obvious to your students. You may want to go over this carefully.

$\mathbf{u} - \mathbf{v} = \mathbf{u} + (-\mathbf{v})$

FIGURE 3.21

The **negative** of $\mathbf{v} = \langle v_1, v_2 \rangle$ is

$$-\mathbf{v} = (-1)\mathbf{v}$$
$$= \langle -v_1, -v_2 \rangle \qquad \text{Negative}$$

and the **difference** of **u** and **v** is

$$\mathbf{u} - \mathbf{v} = \mathbf{u} + (-\mathbf{v}) \qquad \text{Add } (-\mathbf{v}). \text{ See Figure 3.21.}$$
$$= \langle u_1 - v_1, u_2 - v_2 \rangle. \qquad \text{Difference}$$

To represent $\mathbf{u} - \mathbf{v}$ geometrically, you can use directed line segments with the *same* initial point. The difference $\mathbf{u} - \mathbf{v}$ is the vector from the terminal point of **v** to the terminal point of **u**, which is equal to $\mathbf{u} + (-\mathbf{v})$, as shown in Figure 3.21.

The component definitions of vector addition and scalar multiplication are illustrated in Example 3. In this example, notice that each of the vector operations can be interpreted geometrically.

Example 3   Vector Operations

Let $\mathbf{v} = \langle -2, 5 \rangle$ and $\mathbf{w} = \langle 3, 4 \rangle$, and find each of the following vectors.

a. $2\mathbf{v}$ **b.** $\mathbf{w} - \mathbf{v}$ **c.** $\mathbf{v} + 2\mathbf{w}$

Solution

a. Because $\mathbf{v} = \langle -2, 5 \rangle$, you have

$$2\mathbf{v} = 2\langle -2, 5 \rangle$$
$$= \langle 2(-2), 2(5) \rangle$$
$$= \langle -4, 10 \rangle.$$

A sketch of $2\mathbf{v}$ is shown in Figure 3.22.

b. The difference of $\mathbf{w}$ and $\mathbf{v}$ is

$$\mathbf{w} - \mathbf{v} = \langle 3 - (-2), 4 - 5 \rangle$$
$$= \langle 5, -1 \rangle.$$

A sketch of $\mathbf{w} - \mathbf{v}$ is shown in Figure 3.23. Note that the figure shows the vector difference $\mathbf{w} - \mathbf{v}$ as the sum $\mathbf{w} + (-\mathbf{v})$.

c. The sum of $\mathbf{v}$ and $2\mathbf{w}$ is

$$\mathbf{v} + 2\mathbf{w} = \langle -2, 5 \rangle + 2\langle 3, 4 \rangle$$
$$= \langle -2, 5 \rangle + \langle 2(3), 2(4) \rangle$$
$$= \langle -2, 5 \rangle + \langle 6, 8 \rangle$$
$$= \langle -2 + 6, 5 + 8 \rangle$$
$$= \langle 4, 13 \rangle.$$

A sketch of $\mathbf{v} + 2\mathbf{w}$ is shown in Figure 3.24.

FIGURE 3.22

FIGURE 3.23

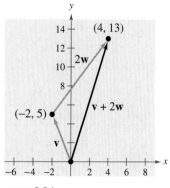

FIGURE 3.24

✓CHECKPOINT   Now try Exercise 21.

Vector addition and scalar multiplication share many of the properties of ordinary arithmetic.

Properties of Vector Addition and Scalar Multiplication

Let $\mathbf{u}$, $\mathbf{v}$, and $\mathbf{w}$ be vectors and let c and d be scalars. Then the following properties are true.

1. $\mathbf{u} + \mathbf{v} = \mathbf{v} + \mathbf{u}$
2. $(\mathbf{u} + \mathbf{v}) + \mathbf{w} = \mathbf{u} + (\mathbf{v} + \mathbf{w})$
3. $\mathbf{u} + \mathbf{0} = \mathbf{u}$
4. $\mathbf{u} + (-\mathbf{u}) = \mathbf{0}$
5. $c(d\mathbf{u}) = (cd)\mathbf{u}$
6. $(c + d)\mathbf{u} = c\mathbf{u} + d\mathbf{u}$
7. $c(\mathbf{u} + \mathbf{v}) = c\mathbf{u} + c\mathbf{v}$
8. $1(\mathbf{u}) = \mathbf{u}, 0(\mathbf{u}) = \mathbf{0}$
9. $\|c\mathbf{v}\| = |c|\,\|\mathbf{v}\|$

Property 9 can be stated as follows: the magnitude of the vector $c\mathbf{v}$ is the absolute value of c times the magnitude of $\mathbf{v}$.

Unit Vectors

In many applications of vectors, it is useful to find a unit vector that has the same direction as a given nonzero vector $\mathbf{v}$. To do this, you can divide $\mathbf{v}$ by its magnitude to obtain

$$\mathbf{u} = \text{unit vector} = \frac{\mathbf{v}}{\|\mathbf{v}\|} = \left(\frac{1}{\|\mathbf{v}\|}\right)\mathbf{v}. \qquad \text{Unit vector in direction of } \mathbf{v}$$

Note that $\mathbf{u}$ is a scalar multiple of $\mathbf{v}$. The vector $\mathbf{u}$ has a magnitude of 1 and the same direction as $\mathbf{v}$. The vector $\mathbf{u}$ is called a **unit vector in the direction of v**.

Example 4   Finding a Unit Vector

Find a unit vector in the direction of $\mathbf{v} = \langle -2, 5 \rangle$ and verify that the result has a magnitude of 1.

Solution

The unit vector in the direction of $\mathbf{v}$ is

$$\frac{\mathbf{v}}{\|\mathbf{v}\|} = \frac{\langle -2, 5 \rangle}{\sqrt{(-2)^2 + (5)^2}}$$

$$= \frac{1}{\sqrt{29}}\langle -2, 5 \rangle$$

$$= \left\langle \frac{-2}{\sqrt{29}}, \frac{5}{\sqrt{29}} \right\rangle.$$

This vector has a magnitude of 1 because

$$\sqrt{\left(\frac{-2}{\sqrt{29}}\right)^2 + \left(\frac{5}{\sqrt{29}}\right)^2} = \sqrt{\frac{4}{29} + \frac{25}{29}} = \sqrt{\frac{29}{29}} = 1.$$

✓CHECKPOINT   Now try Exercise 31.

The Granger Collection

Historical Note

William Rowan Hamilton (1805–1865), an Irish mathematician, did some of the earliest work with vectors. Hamilton spent many years developing a system of vector-like quantities called quaternions. Although Hamilton was convinced of the benefits of quaternions, the operations he defined did not produce good models for physical phenomena. It wasn't until the latter half of the nineteenth century that the Scottish physicist James Maxwell (1831–1879) restructured Hamilton's quaternions in a form useful for representing physical quantities such as force, velocity, and acceleration.

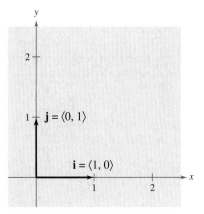

FIGURE **3.25**

The unit vectors $\langle 1, 0 \rangle$ and $\langle 0, 1 \rangle$ are called the **standard unit vectors** and are denoted by

$$\mathbf{i} = \langle 1, 0 \rangle \qquad \text{and} \qquad \mathbf{j} = \langle 0, 1 \rangle$$

as shown in Figure 3.25. (Note that the lowercase letter $\mathbf{i}$ is written in boldface to distinguish it from the imaginary number $i = \sqrt{-1}$.) These vectors can be used to represent any vector $\mathbf{v} = \langle v_1, v_2 \rangle$, as follows.

$$\mathbf{v} = \langle v_1, v_2 \rangle$$
$$= v_1 \langle 1, 0 \rangle + v_2 \langle 0, 1 \rangle$$
$$= v_1 \mathbf{i} + v_2 \mathbf{j}$$

The scalars v_1 and v_2 are called the **horizontal** and **vertical components of v,** respectively. The vector sum

$$v_1 \mathbf{i} + v_2 \mathbf{j}$$

is called a **linear combination** of the vectors $\mathbf{i}$ and $\mathbf{j}$. Any vector in the plane can be written as a linear combination of the standard unit vectors $\mathbf{i}$ and $\mathbf{j}$.

Example 5   Writing a Linear Combination of Unit Vectors

Let $\mathbf{u}$ be the vector with initial point $(2, -5)$ and terminal point $(-1, 3)$. Write $\mathbf{u}$ as a linear combination of the standard unit vectors $\mathbf{i}$ and $\mathbf{j}$.

Solution

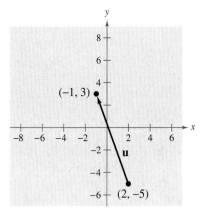

FIGURE **3.26**

Begin by writing the component form of the vector $\mathbf{u}$.

$$\mathbf{u} = \langle -1 - 2, 3 - (-5) \rangle$$
$$= \langle -3, 8 \rangle$$
$$= -3\mathbf{i} + 8\mathbf{j}$$

This result is shown graphically in Figure 3.26.

✓*CHECKPOINT*   Now try Exercise 43.

Example 6   Vector Operations

Let $\mathbf{u} = -3\mathbf{i} + 8\mathbf{j}$ and let $\mathbf{v} = 2\mathbf{i} - \mathbf{j}$. Find $2\mathbf{u} - 3\mathbf{v}$.

Solution

You could solve this problem by converting $\mathbf{u}$ and $\mathbf{v}$ to component form. This, however, is not necessary. It is just as easy to perform the operations in unit vector form.

$$2\mathbf{u} - 3\mathbf{v} = 2(-3\mathbf{i} + 8\mathbf{j}) - 3(2\mathbf{i} - \mathbf{j})$$
$$= -6\mathbf{i} + 16\mathbf{j} - 6\mathbf{i} + 3\mathbf{j}$$
$$= -12\mathbf{i} + 19\mathbf{j}$$

✓*CHECKPOINT*   Now try Exercise 49.

Direction Angles

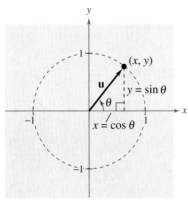

FIGURE 3.27 $\|\mathbf{u}\| = 1$

If **u** is a *unit vector* such that θ is the angle (measured counterclockwise) from the positive x-axis to **u**, the terminal point of **u** lies on the unit circle and you have

$$\mathbf{u} = \langle x, y \rangle = \langle \cos \theta, \sin \theta \rangle = (\cos \theta)\mathbf{i} + (\sin \theta)\mathbf{j}$$

as shown in Figure 3.27. The angle θ is the **direction angle** of the vector **u**.

Suppose that **u** is a unit vector with direction angle θ. If $\mathbf{v} = a\mathbf{i} + b\mathbf{j}$ is any vector that makes an angle θ with the positive x-axis, it has the same direction as **u** and you can write

$$\mathbf{v} = \| \mathbf{v} \| \langle \cos \theta, \sin \theta \rangle$$
$$= \| \mathbf{v} \| (\cos \theta)\mathbf{i} + \| \mathbf{v} \| (\sin \theta)\mathbf{j}.$$

Because $\mathbf{v} = a\mathbf{i} + b\mathbf{j} = \| \mathbf{v} \| (\cos \theta) \mathbf{i} + \| \mathbf{v} \| (\sin \theta) \mathbf{j}$, it follows that the direction angle θ for **v** is determined from

$$\tan \theta = \frac{\sin \theta}{\cos \theta} \qquad \text{Quotient identity}$$

$$= \frac{\| \mathbf{v} \| \sin \theta}{\| \mathbf{v} \| \cos \theta} \qquad \text{Multiply numerator and denominator by } \| \mathbf{v} \|.$$

$$= \frac{b}{a}. \qquad \text{Simplify.}$$

Example 7   Finding Direction Angles of Vectors

Find the direction angle of each vector.

a. $\mathbf{u} = 3\mathbf{i} + 3\mathbf{j}$

b. $\mathbf{v} = 3\mathbf{i} - 4\mathbf{j}$

Solution

a. The direction angle is

$$\tan \theta = \frac{b}{a} = \frac{3}{3} = 1.$$

So, $\theta = 45°$, as shown in Figure 3.28.

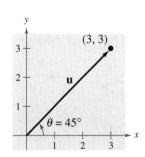

FIGURE 3.28

b. The direction angle is

$$\tan \theta = \frac{b}{a} = \frac{-4}{3}.$$

Moreover, because $\mathbf{v} = 3\mathbf{i} - 4\mathbf{j}$ lies in Quadrant IV, θ lies in Quadrant IV and its reference angle is

$$\theta = \left| \arctan\left(-\frac{4}{3} \right) \right| \approx |-53.13°| = 53.13°.$$

So, it follows that $\theta \approx 360° - 53.13° = 306.87°$, as shown in Figure 3.29.

FIGURE 3.29

✔CHECKPOINT   Now try Exercise 55.

Applications of Vectors

FIGURE **3.30**

Example 8 **Finding the Component Form of a Vector**

Find the component form of the vector that represents the velocity of an airplane descending at a speed of 100 miles per hour at an angle 30° below the horizontal, as shown in Figure 3.30.

Solution

The velocity vector **v** has a magnitude of 100 and a direction angle of $\theta = 210°$.

$$\mathbf{v} = \|\mathbf{v}\| (\cos \theta)\mathbf{i} + \|\mathbf{v}\| (\sin \theta)\mathbf{j}$$

$$= 100(\cos 210°)\mathbf{i} + 100(\sin 210°)\mathbf{j}$$

$$= 100\left(-\frac{\sqrt{3}}{2}\right)\mathbf{i} + 100\left(-\frac{1}{2}\right)\mathbf{j}$$

$$= -50\sqrt{3}\,\mathbf{i} - 50\mathbf{j}$$

$$= \langle -50\sqrt{3}, -50 \rangle$$

You can check that **v** has a magnitude of 100, as follows.

$$\|\mathbf{v}\| = \sqrt{\left(-50\sqrt{3}\right)^2 + (-50)^2}$$

$$= \sqrt{7500 + 2500}$$

$$= \sqrt{10,000} = 100$$

✓*CHECKPOINT* Now try Exercise 77.

Example 9 **Using Vectors to Determine Weight**

A force of 600 pounds is required to pull a boat and trailer up a ramp inclined at 15° from the horizontal. Find the combined weight of the boat and trailer.

Solution

Based on Figure 3.31, you can make the following observations.

$$\|\overrightarrow{BA}\| = \text{force of gravity} = \text{combined weight of boat and trailer}$$

$$\|\overrightarrow{BC}\| = \text{force against ramp}$$

$$\|\overrightarrow{AC}\| = \text{force required to move boat up ramp} = 600 \text{ pounds}$$

By construction, triangles *BWD* and *ABC* are similar. So, angle *ABC* is 15°, and so in triangle *ABC* you have

$$\sin 15° = \frac{\|\overrightarrow{AC}\|}{\|\overrightarrow{BA}\|} = \frac{600}{\|\overrightarrow{BA}\|}$$

$$\|\overrightarrow{BA}\| = \frac{600}{\sin 15°} \approx 2318.$$

Consequently, the combined weight is approximately 2318 pounds. (In Figure 3.31, note that $\overrightarrow{AC}$ is parallel to the ramp.)

✓*CHECKPOINT* Now try Exercise 81.

FIGURE **3.31**

Example 10   **Using Vectors to Find Speed and Direction**

An airplane is traveling at a speed of 500 miles per hour with a bearing of 330° at a fixed altitude with a negligible wind velocity as shown in Figure 3.32(a). When the airplane reaches a certain point, it encounters a wind with a velocity of 70 miles per hour in the direction N 45° E, as shown in Figure 3.32(b). What are the resultant speed and direction of the airplane?

(a)

(b)

FIGURE 3.32

Solution

Using Figure 3.32, the velocity of the airplane (alone) is

$$\mathbf{v}_1 = 500\langle\cos 120°, \sin 120°\rangle$$
$$= \langle-250, 250\sqrt{3}\rangle$$

and the velocity of the wind is

$$\mathbf{v}_2 = 70\langle\cos 45°, \sin 45°\rangle$$
$$= \langle35\sqrt{2}, 35\sqrt{2}\rangle.$$

So, the velocity of the airplane (in the wind) is

$$\mathbf{v} = \mathbf{v}_1 + \mathbf{v}_2$$
$$= \langle-250 + 35\sqrt{2}, 250\sqrt{3} + 35\sqrt{2}\rangle$$
$$\approx \langle-200.5, 482.5\rangle$$

and the resultant speed of the airplane is

$$\|\mathbf{v}\| = \sqrt{(-200.5)^2 + (482.5)^2}$$
$$\approx 522.5 \text{ miles per hour.}$$

Finally, if θ is the direction angle of the flight path, you have

$$\tan \theta = \frac{482.5}{-200.5}$$
$$\approx -2.4065$$

which implies that

$$\theta \approx 180° + \arctan(-2.4065) \approx 180° - 67.4° = 112.6°.$$

So, the true direction of the airplane is 337.4°.

✓CHECKPOINT   Now try Exercise 83.

3.3 | Exercises

VOCABULARY CHECK: Fill in the blanks.

1. A _____ _____ _____ can be used to represent a quantity that involves both magnitude and direction.

2. The directed line segment $\overrightarrow{PQ}$ has _____ point P and _____ point Q.

3. The _____ of the directed line segment $\overrightarrow{PQ}$ is denoted by $\|PQ\|$.

4. The set of all directed line segments that are equivalent to a given directed line segment $\overrightarrow{PQ}$ is a _____ **v** in the plane.

5. The directed line segment whose initial point is the origin is said to be in _____ _____ .

6. A vector that has a magnitude of 1 is called a _____ _____ .

7. The two basic vector operations are scalar _____ and vector _____ .

8. The vector **u** + **v** is called the _____ of vector addition.

9. The vector sum $v_1\mathbf{i} + v_2\mathbf{j}$ is called a _____ _____ of the vectors **i** and **j**, and the scalars v_1 and v_2 are called the _____ and _____ components of **v**, respectively.

PREREQUISITE SKILLS REVIEW: Practice and review algebra skills needed for this section at **www.Eduspace.com.**

In Exercises 1 and 2, show that **u** = **v**.

1.

2.
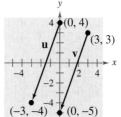

In Exercises 3–14, find the component form and the magnitude of the vector **v**.

3.

4.

5.

6.

7.

8.

Initial Point	Terminal Point
9. $(-1, 5)$	$(15, 12)$
10. $(1, 11)$	$(9, 3)$
11. $(-3, -5)$	$(5, 1)$
12. $(-3, 11)$	$(9, 40)$
13. $(1, 3)$	$(-8, -9)$
14. $(-2, 7)$	$(5, -17)$

In Exercises 15–20, use the figure to sketch a graph of the specified vector. To print an enlarged copy of the graph, go to the website, *www.mathgraphs.com.*

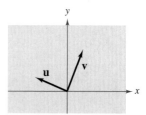

15. $-\mathbf{v}$ **16.** $5\mathbf{v}$

17. $\mathbf{u} + \mathbf{v}$ **18.** $\mathbf{u} - \mathbf{v}$

19. $\mathbf{u} + 2\mathbf{v}$ **20.** $\mathbf{v} - \frac{1}{2}\mathbf{u}$

In Exercises 21–28, find (a) $\mathbf{u} + \mathbf{v}$, (b) $\mathbf{u} - \mathbf{v}$, and (c) $2\mathbf{u} - 3\mathbf{v}$. Then sketch the resultant vector.

21. $\mathbf{u} = \langle 2, 1 \rangle$, $\mathbf{v} = \langle 1, 3 \rangle$
22. $\mathbf{u} = \langle 2, 3 \rangle$, $\mathbf{v} = \langle 4, 0 \rangle$
23. $\mathbf{u} = \langle -5, 3 \rangle$, $\mathbf{v} = \langle 0, 0 \rangle$
24. $\mathbf{u} = \langle 0, 0 \rangle$, $\mathbf{v} = \langle 2, 1 \rangle$
25. $\mathbf{u} = \mathbf{i} + \mathbf{j}$, $\mathbf{v} = 2\mathbf{i} - 3\mathbf{j}$
26. $\mathbf{u} = -2\mathbf{i} + \mathbf{j}$, $\mathbf{v} = -\mathbf{i} + 2\mathbf{j}$
27. $\mathbf{u} = 2\mathbf{i}$, $\mathbf{v} = \mathbf{j}$
28. $\mathbf{u} = 3\mathbf{j}$, $\mathbf{v} = 2\mathbf{i}$

In Exercises 29–38, find a unit vector in the direction of the given vector.

29. $\mathbf{u} = \langle 3, 0 \rangle$
30. $\mathbf{u} = \langle 0, -2 \rangle$
31. $\mathbf{v} = \langle -2, 2 \rangle$
32. $\mathbf{v} = \langle 5, -12 \rangle$
33. $\mathbf{v} = 6\mathbf{i} - 2\mathbf{j}$
34. $\mathbf{v} = \mathbf{i} + \mathbf{j}$
35. $\mathbf{w} = 4\mathbf{j}$
36. $\mathbf{w} = -6\mathbf{i}$
37. $\mathbf{w} = \mathbf{i} - 2\mathbf{j}$
38. $\mathbf{w} = 7\mathbf{j} - 3\mathbf{i}$

In Exercises 39–42, find the vector $\mathbf{v}$ with the given magnitude and the same direction as $\mathbf{u}$.

	Magnitude	Direction
39.	$\|\mathbf{v}\| = 5$	$\mathbf{u} = \langle 3, 3 \rangle$
40.	$\|\mathbf{v}\| = 6$	$\mathbf{u} = \langle -3, 3 \rangle$
41.	$\|\mathbf{v}\| = 9$	$\mathbf{u} = \langle 2, 5 \rangle$
42.	$\|\mathbf{v}\| = 10$	$\mathbf{u} = \langle -10, 0 \rangle$

In Exercises 43–46, the initial and terminal points of a vector are given. Write a linear combination of the standard unit vectors $\mathbf{i}$ and $\mathbf{j}$.

	Initial Point	Terminal Point
43.	$(-3, 1)$	$(4, 5)$
44.	$(0, -2)$	$(3, 6)$
45.	$(-1, -5)$	$(2, 3)$
46.	$(-6, 4)$	$(0, 1)$

In Exercises 47–52, find the component form of $\mathbf{v}$ and sketch the specified vector operations geometrically, where $\mathbf{u} = 2\mathbf{i} - \mathbf{j}$ and $\mathbf{w} = \mathbf{i} + 2\mathbf{j}$.

47. $\mathbf{v} = \frac{3}{2}\mathbf{u}$
48. $\mathbf{v} = \frac{3}{4}\mathbf{w}$
49. $\mathbf{v} = \mathbf{u} + 2\mathbf{w}$
50. $\mathbf{v} = -\mathbf{u} + \mathbf{w}$
51. $\mathbf{v} = \frac{1}{2}(3\mathbf{u} + \mathbf{w})$
52. $\mathbf{v} = \mathbf{u} - 2\mathbf{w}$

In Exercises 53–56, find the magnitude and direction angle of the vector $\mathbf{v}$.

53. $\mathbf{v} = 3(\cos 60°\mathbf{i} + \sin 60°\mathbf{j})$
54. $\mathbf{v} = 8(\cos 135°\mathbf{i} + \sin 135°\mathbf{j})$
55. $\mathbf{v} = 6\mathbf{i} - 6\mathbf{j}$
56. $\mathbf{v} = -5\mathbf{i} + 4\mathbf{j}$

In Exercises 57–64, find the component form of $\mathbf{v}$ given its magnitude and the angle it makes with the positive x-axis. Sketch $\mathbf{v}$.

	Magnitude	Angle
57.	$\|\mathbf{v}\| = 3$	$\theta = 0°$
58.	$\|\mathbf{v}\| = 1$	$\theta = 45°$
59.	$\|\mathbf{v}\| = \frac{7}{2}$	$\theta = 150°$
60.	$\|\mathbf{v}\| = \frac{5}{2}$	$\theta = 45°$
61.	$\|\mathbf{v}\| = 3\sqrt{2}$	$\theta = 150°$
62.	$\|\mathbf{v}\| = 4\sqrt{3}$	$\theta = 90°$
63.	$\|\mathbf{v}\| = 2$	$\mathbf{v}$ in the direction $\mathbf{i} + 3\mathbf{j}$
64.	$\|\mathbf{v}\| = 3$	$\mathbf{v}$ in the direction $3\mathbf{i} + 4\mathbf{j}$

In Exercises 65–68, find the component form of the sum of $\mathbf{u}$ and $\mathbf{v}$ with direction angles θ_u and θ_v.

	Magnitude	Angle
65.	$\|\mathbf{u}\| = 5$	$\theta_{\mathbf{u}} = 0°$
	$\|\mathbf{v}\| = 5$	$\theta_{\mathbf{v}} = 90°$
66.	$\|\mathbf{u}\| = 4$	$\theta_{\mathbf{u}} = 60°$
	$\|\mathbf{v}\| = 4$	$\theta_{\mathbf{v}} = 90°$
67.	$\|\mathbf{u}\| = 20$	$\theta_{\mathbf{u}} = 45°$
	$\|\mathbf{v}\| = 50$	$\theta_{\mathbf{v}} = 180°$
68.	$\|\mathbf{u}\| = 50$	$\theta_{\mathbf{u}} = 30°$
	$\|\mathbf{v}\| = 30$	$\theta_{\mathbf{v}} = 110°$

In Exercises 69 and 70, use the Law of Cosines to find the angle α between the vectors. (Assume $0° \le \alpha \le 180°$.)

69. $\mathbf{v} = \mathbf{i} + \mathbf{j}$, $\mathbf{w} = 2\mathbf{i} - 2\mathbf{j}$
70. $\mathbf{v} = \mathbf{i} + 2\mathbf{j}$, $\mathbf{w} = 2\mathbf{i} - \mathbf{j}$

Resultant Force   In Exercises 71 and 72, find the angle between the forces given the magnitude of their resultant. (*Hint:* Write force 1 as a vector in the direction of the positive x-axis and force 2 as a vector at an angle θ with the positive x-axis.)

	Force 1	Force 2	Resultant Force
71.	45 pounds	60 pounds	90 pounds
72.	3000 pounds	1000 pounds	3750 pounds

73. *Resultant Force* Forces with magnitudes of 125 newtons and 300 newtons act on a hook (see figure). The angle between the two forces is 45°. Find the direction and magnitude of the resultant of these forces.

74. *Resultant Force* Forces with magnitudes of 2000 newtons and 900 newtons act on a machine part at angles of 30° and −45°, respectively, with the *x*-axis (see figure). Find the direction and magnitude of the resultant of these forces.

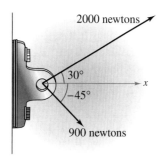

75. *Resultant Force* Three forces with magnitudes of 75 pounds, 100 pounds, and 125 pounds act on an object at angles of 30°, 45°, and 120°, respectively, with the positive *x*-axis. Find the direction and magnitude of the resultant of these forces.

76. *Resultant Force* Three forces with magnitudes of 70 pounds, 40 pounds, and 60 pounds act on an object at angles of −30°, 445°, and 135°, respectively, with the positive *x*-axis. Find the direction and magnitude of the resultant of these forces.

77. *Velocity* A ball is thrown with an initial velocity of 70 feet per second, at an angle of 35° with the horizontal (see figure). Find the vertical and horizontal components of the velocity.

78. *Velocity* A gun with a muzzle velocity of 1200 feet per second is fired at an angle of 6° with the horizontal. Find the vertical and horizontal components of the velocity.

Cable Tension In Exercises 79 and 80, use the figure to determine the tension in each cable supporting the load.

81. *Tow Line Tension* A loaded barge is being towed by two tugboats, and the magnitude of the resultant is 6000 pounds directed along the axis of the barge (see figure). Find the tension in the tow lines if they each make an 18° angle with the axis of the barge.

82. *Rope Tension* To carry a 100-pound cylindrical weight, two people lift on the ends of short ropes that are tied to an eyelet on the top center of the cylinder. Each rope makes a 20° angle with the vertical. Draw a figure that gives a visual representation of the problem, and find the tension in the ropes.

83. *Navigation* An airplane is flying in the direction of 148°, with an airspeed of 875 kilometers per hour. Because of the wind, its groundspeed and direction are 800 kilometers per hour and 140°, respectively (see figure). Find the direction and speed of the wind.

Model It

84. *Navigation*   A commercial jet is flying from Miami to Seattle. The jet's velocity with respect to the air is 580 miles per hour, and its bearing is 332°. The wind, at the altitude of the plane, is blowing from the southwest with a velocity of 60 miles per hour.

(a) Draw a figure that gives a visual representation of the problem.

(b) Write the velocity of the wind as a vector in component form.

(c) Write the velocity of the jet relative to the air in component form.

(d) What is the speed of the jet with respect to the ground?

(e) What is the true direction of the jet?

85. *Work*   A heavy implement is pulled 30 feet across a floor, using a force of 100 pounds. The force is exerted at an angle of 50° above the horizontal (see figure). Find the work done. (Use the formula for work, $W = FD$, where F is the component of the force in the direction of motion and D is the distance.)

FIGURE FOR 85   FIGURE FOR 86

86. *Rope Tension*   A tetherball weighing 1 pound is pulled outward from the pole by a horizontal force **u** until the rope makes a 45° angle with the pole (see figure). Determine the resulting tension in the rope and the magnitude of **u**.

Synthesis

True or False?   In Exercises 87 and 88, decide whether the statement is true or false. Justify your answer.

87. If **u** and **v** have the same magnitude and direction, then **u** = **v**.

88. If **u** = $a\mathbf{i} + b\mathbf{j}$ is a unit vector, then $a^2 + b^2 = 1$.

89. *Think About It*   Consider two forces of equal magnitude acting on a point.

(a) If the magnitude of the resultant is the sum of the magnitudes of the two forces, make a conjecture about the angle between the forces.

(b) If the resultant of the forces is **0**, make a conjecture about the angle between the forces.

(c) Can the magnitude of the resultant be greater than the sum of the magnitudes of the two forces? Explain.

90. *Graphical Reasoning*   Consider two forces

$$\mathbf{F}_1 = \langle 10, 0 \rangle \text{ and } \mathbf{F}_2 = 5\langle \cos \theta, \sin \theta \rangle.$$

(a) Find $\|\mathbf{F}_1 + \mathbf{F}_2\|$ as a function of θ.

(b) Use a graphing utility to graph the function in part (a) for $0 \le \theta < 2\pi$.

(c) Use the graph in part (b) to determine the range of the function. What is its maximum, and for what value of θ does it occur? What is its minimum, and for what value of θ does it occur?

(d) Explain why the magnitude of the resultant is never 0.

91. *Proof*   Prove that $(\cos \theta)\mathbf{i} + (\sin \theta)\mathbf{j}$ is a unit vector for any value of θ.

92. *Technology*   Write a program for your graphing utility that graphs two vectors and their difference given the vectors in component form.

In Exercises 93 and 94, use the program in Exercise 92 to find the difference of the vectors shown in the figure.

93.

94.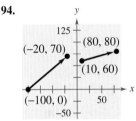

Skills Review

In Exercises 95–98, use the trigonometric substitution to write the algebraic expression as a trigonometric function of θ, where $0 < \theta < \pi/2$.

95. $\sqrt{x^2 - 64}$, $x = 8 \sec \theta$

96. $\sqrt{64 - x^2}$, $x = 8 \sin \theta$

97. $\sqrt{x^2 + 36}$, $x = 6 \tan \theta$

98. $\sqrt{(x^2 - 25)^3}$, $x = 5 \sec \theta$

In Exercises 99–102, solve the equation.

99. $\cos x(\cos x + 1) = 0$

100. $\sin x(2 \sin x + \sqrt{2}) = 0$

101. $3 \sec x \sin x - 2\sqrt{3} \sin x = 0$

102. $\cos x \csc x + \cos x\sqrt{2} = 0$

3.4 Vectors and Dot Products

Why you should learn it

You can use the dot product of two vectors to solve real-life problems involving two vector quantities. For instance, in Exercise 68 on page 316, you can use the dot product to find the force necessary to keep a sport utility vehicle from rolling down a hill.

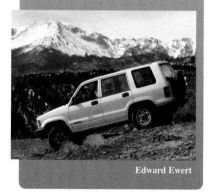

Edward Ewert

The Dot Product of Two Vectors

So far you have studied two vector operations—vector addition and multiplication by a scalar—each of which yields another vector. In this section, you will study a third vector operation, the **dot product.** This product yields a scalar, rather than a vector.

Definition of the Dot Product

The **dot product** of $\mathbf{u} = \langle u_1, u_2 \rangle$ and $\mathbf{v} = \langle v_1, v_2 \rangle$ is

$$\mathbf{u} \cdot \mathbf{v} = u_1 v_1 + u_2 v_2.$$

Properties of the Dot Product

Let $\mathbf{u}$, $\mathbf{v}$, and $\mathbf{w}$ be vectors in the plane or in space and let c be a scalar.

1. $\mathbf{u} \cdot \mathbf{v} = \mathbf{v} \cdot \mathbf{u}$

2. $\mathbf{0} \cdot \mathbf{v} = 0$

3. $\mathbf{u} \cdot (\mathbf{v} + \mathbf{w}) = \mathbf{u} \cdot \mathbf{v} + \mathbf{u} \cdot \mathbf{w}$

4. $\mathbf{v} \cdot \mathbf{v} = \|\mathbf{v}\|^2$

5. $c(\mathbf{u} \cdot \mathbf{v}) = c\mathbf{u} \cdot \mathbf{v} = \mathbf{u} \cdot c\mathbf{v}$

For proofs of the properties of the dot product, see Proofs in Mathematics on page 328.

Example 1 Finding Dot Products

Find each dot product.

a. $\langle 4, 5 \rangle \cdot \langle 2, 3 \rangle$ **b.** $\langle 2, -1 \rangle \cdot \langle 1, 2 \rangle$ **c.** $\langle 0, 3 \rangle \cdot \langle 4, -2 \rangle$

Solution

a. $\langle 4, 5 \rangle \cdot \langle 2, 3 \rangle = 4(2) + 5(3)$
$= 8 + 15$
$= 23$

b. $\langle 2, -1 \rangle \cdot \langle 1, 2 \rangle = 2(1) + (-1)(2) = 2 - 2 = 0$

c. $\langle 0, 3 \rangle \cdot \langle 4, -2 \rangle = 0(4) + 3(-2) = 0 - 6 = -6$

✓CHECKPOINT   Now try Exercise 1.

In Example 1, be sure you see that the dot product of two vectors is a scalar (a real number), not a vector. Moreover, notice that the dot product can be positive, zero, or negative.

Example 2   Using Properties of Dot Products

Let $\mathbf{u} = \langle -1, 3 \rangle$, $\mathbf{v} = \langle 2, -4 \rangle$, and $\mathbf{w} = \langle 1, -2 \rangle$. Find each dot product.

a. $(\mathbf{u} \cdot \mathbf{v})\mathbf{w}$ **b.** $\mathbf{u} \cdot 2\mathbf{v}$

Solution

Begin by finding the dot product of $\mathbf{u}$ and $\mathbf{v}$.

$$\mathbf{u} \cdot \mathbf{v} = \langle -1, 3 \rangle \cdot \langle 2, -4 \rangle$$
$$= (-1)(2) + 3(-4)$$
$$= -14$$

a. $(\mathbf{u} \cdot \mathbf{v})\mathbf{w} = -14\langle 1, -2 \rangle$
$$= \langle -14, 28 \rangle$$

b. $\mathbf{u} \cdot 2\mathbf{v} = 2(\mathbf{u} \cdot \mathbf{v})$
$$= 2(-14)$$
$$= -28$$

Notice that the product in part (a) is a vector, whereas the product in part (b) is a scalar. Can you see why?

✓*CHECKPOINT*   Now try Exercise 11.

Example 3   Dot Product and Magnitude

The dot product of $\mathbf{u}$ with itself is 5. What is the magnitude of $\mathbf{u}$?

Solution

Because $\|\mathbf{u}\|^2 = \mathbf{u} \cdot \mathbf{u}$ and $\mathbf{u} \cdot \mathbf{u} = 5$, it follows that

$$\|\mathbf{u}\| = \sqrt{\mathbf{u} \cdot \mathbf{u}}$$
$$= \sqrt{5}.$$

✓*CHECKPOINT*   Now try Exercise 19.

The Angle Between Two Vectors

The **angle between two nonzero vectors** is the angle θ, $0 \le \theta \le \pi$, between their respective standard position vectors, as shown in Figure 3.33. This angle can be found using the dot product. (Note that the angle between the zero vector and another vector is not defined.)

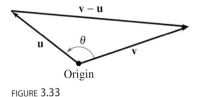

FIGURE **3.33**

> ### Angle Between Two Vectors
>
> If θ is the angle between two nonzero vectors $\mathbf{u}$ and $\mathbf{v}$, then
>
> $$\cos \theta = \frac{\mathbf{u} \cdot \mathbf{v}}{\|\mathbf{u}\| \, \|\mathbf{v}\|}.$$

For a proof of the angle between two vectors, see Proofs in Mathematics on page 328.

| Example 4 | Finding the Angle Between Two Vectors |

Find the angle between $\mathbf{u} = \langle 4, 3 \rangle$ and $\mathbf{v} = \langle 3, 5 \rangle$.

Solution

$$\cos \theta = \frac{\mathbf{u} \cdot \mathbf{v}}{\|\mathbf{u}\| \, \|\mathbf{v}\|}$$

$$= \frac{\langle 4, 3 \rangle \cdot \langle 3, 5 \rangle}{\|\langle 4, 3 \rangle\| \, \|\langle 3, 5 \rangle\|}$$

$$= \frac{27}{5\sqrt{34}}$$

This implies that the angle between the two vectors is

$$\theta = \arccos \frac{27}{5\sqrt{34}} \approx 22.2°$$

as shown in Figure 3.34.

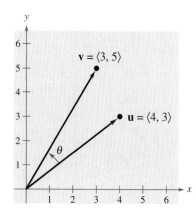

FIGURE 3.34

✓**CHECKPOINT** Now try Exercise 29.

Rewriting the expression for the angle between two vectors in the form

$$\mathbf{u} \cdot \mathbf{v} = \|\mathbf{u}\| \, \|\mathbf{v}\| \cos \theta \qquad \text{Alternative form of dot product}$$

produces an alternative way to calculate the dot product. From this form, you can see that because $\|\mathbf{u}\|$ and $\|\mathbf{v}\|$ are always positive, $\mathbf{u} \cdot \mathbf{v}$ and $\cos \theta$ will always have the same sign. Figure 3.35 shows the five possible orientations of two vectors.

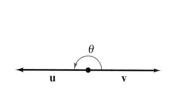

$\theta = \pi$
$\cos \theta = -1$
Opposite Direction
FIGURE 3.35

$\dfrac{\pi}{2} < \theta < \pi$
$-1 < \cos \theta < 0$
Obtuse Angle

$\theta = \dfrac{\pi}{2}$
$\cos \theta = 0$
90° Angle

$0 < \theta < \dfrac{\pi}{2}$
$0 < \cos \theta < 1$
Acute Angle

$\theta = 0$
$\cos \theta = 1$
Same Direction

Definition of Orthogonal Vectors

The vectors $\mathbf{u}$ and $\mathbf{v}$ are **orthogonal** if $\mathbf{u} \cdot \mathbf{v} = 0$.

The terms *orthogonal* and *perpendicular* mean essentially the same thing—meeting at right angles. Even though the angle between the zero vector and another vector is not defined, it is convenient to extend the definition of orthogonality to include the zero vector. In other words, the zero vector is orthogonal to every vector $\mathbf{u}$, because $\mathbf{0} \cdot \mathbf{u} = 0$.

Example 5   **Determining Orthogonal Vectors**

Are the vectors $\mathbf{u} = \langle 2, -3 \rangle$ and $\mathbf{v} = \langle 6, 4 \rangle$ orthogonal?

Solution

Begin by finding the dot product of the two vectors.

$$\mathbf{u} \cdot \mathbf{v} = \langle 2, -3 \rangle \cdot \langle 6, 4 \rangle = 2(6) + (-3)(4) = 0$$

Because the dot product is 0, the two vectors are orthogonal (see Figure 3.36).

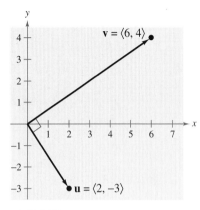

FIGURE 3.36

✓CHECKPOINT   Now try Exercise 47.

Finding Vector Components

You have already seen applications in which two vectors are added to produce a resultant vector. Many applications in physics and engineering pose the reverse problem—decomposing a given vector into the sum of two **vector components.**

Consider a boat on an inclined ramp, as shown in Figure 3.37. The force **F** due to gravity pulls the boat *down* the ramp and *against* the ramp. These two orthogonal forces, $\mathbf{w}_1$ and $\mathbf{w}_2$, are vector components of **F**. That is,

$$\mathbf{F} = \mathbf{w}_1 + \mathbf{w}_2. \qquad \text{Vector components of } \mathbf{F}$$

The negative of component $\mathbf{w}_1$ represents the force needed to keep the boat from rolling down the ramp, whereas $\mathbf{w}_2$ represents the force that the tires must withstand against the ramp. A procedure for finding $\mathbf{w}_1$ and $\mathbf{w}_2$ is shown on the following page.

FIGURE 3.37

Definition of Vector Components

Let $\mathbf{u}$ and $\mathbf{v}$ be nonzero vectors such that

$$\mathbf{u} = \mathbf{w}_1 + \mathbf{w}_2$$

where $\mathbf{w}_1$ and $\mathbf{w}_2$ are orthogonal and $\mathbf{w}_1$ is parallel to (or a scalar multiple of) $\mathbf{v}$, as shown in Figure 3.38. The vectors $\mathbf{w}_1$ and $\mathbf{w}_2$ are called **vector components** of $\mathbf{u}$. The vector $\mathbf{w}_1$ is the **projection** of $\mathbf{u}$ onto $\mathbf{v}$ and is denoted by

$$\mathbf{w}_1 = \text{proj}_{\mathbf{v}}\mathbf{u}.$$

The vector $\mathbf{w}_2$ is given by $\mathbf{w}_2 = \mathbf{u} - \mathbf{w}_1$.

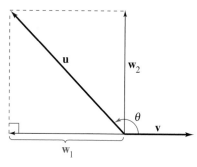

θ is acute. θ is obtuse.

FIGURE 3.38

From the definition of vector components, you can see that it is easy to find the component $\mathbf{w}_2$ once you have found the projection of $\mathbf{u}$ onto $\mathbf{v}$. To find the projection, you can use the dot product, as follows.

$$\mathbf{u} = \mathbf{w}_1 + \mathbf{w}_2 = c\mathbf{v} + \mathbf{w}_2 \qquad \text{$\mathbf{w}_1$ is a scalar multiple of $\mathbf{v}$.}$$

$$\mathbf{u} \cdot \mathbf{v} = (c\mathbf{v} + \mathbf{w}_2) \cdot \mathbf{v} \qquad \text{Take dot product of each side with $\mathbf{v}$.}$$

$$= c\mathbf{v} \cdot \mathbf{v} + \mathbf{w}_2 \cdot \mathbf{v}$$

$$= c\|\mathbf{v}\|^2 + 0 \qquad \text{$\mathbf{w}_2$ and $\mathbf{v}$ are orthogonal.}$$

So,

$$c = \frac{\mathbf{u} \cdot \mathbf{v}}{\|\mathbf{v}\|^2}$$

and

$$\mathbf{w}_1 = \text{proj}_{\mathbf{v}} \mathbf{u} = c\mathbf{v} = \frac{\mathbf{u} \cdot \mathbf{v}}{\|\mathbf{v}\|^2} \mathbf{v}.$$

Projection of u onto v

Let $\mathbf{u}$ and $\mathbf{v}$ be nonzero vectors. The projection of $\mathbf{u}$ onto $\mathbf{v}$ is

$$\text{proj}_{\mathbf{v}}\mathbf{u} = \left(\frac{\mathbf{u} \cdot \mathbf{v}}{\|\mathbf{v}\|^2}\right)\mathbf{v}.$$

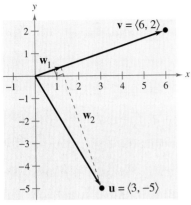

FIGURE **3.39**

Example 6   **Decomposing a Vector into Components**

Find the projection of $\mathbf{u} = \langle 3, -5 \rangle$ onto $\mathbf{v} = \langle 6, 2 \rangle$. Then write $\mathbf{u}$ as the sum of two orthogonal vectors, one of which is $\text{proj}_{\mathbf{v}}\mathbf{u}$.

Solution

The projection of $\mathbf{u}$ onto $\mathbf{v}$ is

$$\mathbf{w}_1 = \text{proj}_{\mathbf{v}}\mathbf{u} = \left(\frac{\mathbf{u} \cdot \mathbf{v}}{\|\mathbf{v}\|^2} \right)\mathbf{v} = \left(\frac{8}{40} \right)\langle 6, 2 \rangle = \left\langle \frac{6}{5}, \frac{2}{5} \right\rangle$$

as shown in Figure 3.39. The other component, $\mathbf{w}_2$, is

$$\mathbf{w}_2 = \mathbf{u} - \mathbf{w}_1 = \langle 3, -5 \rangle - \left\langle \frac{6}{5}, \frac{2}{5} \right\rangle = \left\langle \frac{9}{5}, -\frac{27}{5} \right\rangle.$$

So,

$$\mathbf{u} = \mathbf{w}_1 + \mathbf{w}_2 = \left\langle \frac{6}{5}, \frac{2}{5} \right\rangle + \left\langle \frac{9}{5}, -\frac{27}{5} \right\rangle = \langle 3, -5 \rangle.$$

✔**CHECKPOINT**   Now try Exercise 53.

Example 7   **Finding a Force**

A 200-pound cart sits on a ramp inclined at $30°$, as shown in Figure 3.40. What force is required to keep the cart from rolling down the ramp?

Solution

Because the force due to gravity is vertical and downward, you can represent the gravitational force by the vector

$$\mathbf{F} = -200\mathbf{j}. \qquad \text{Force due to gravity}$$

To find the force required to keep the cart from rolling down the ramp, project $\mathbf{F}$ onto a unit vector $\mathbf{v}$ in the direction of the ramp, as follows.

$$\mathbf{v} = (\cos 30°)\mathbf{i} + (\sin 30°)\mathbf{j} = \frac{\sqrt{3}}{2}\mathbf{i} + \frac{1}{2}\mathbf{j} \qquad \text{Unit vector along ramp}$$

Therefore, the projection of $\mathbf{F}$ onto $\mathbf{v}$ is

$$\begin{aligned}
\mathbf{w}_1 &= \text{proj}_{\mathbf{v}}\mathbf{F} \\
&= \left(\frac{\mathbf{F} \cdot \mathbf{v}}{\|\mathbf{v}\|^2} \right)\mathbf{v} \\
&= (\mathbf{F} \cdot \mathbf{v})\mathbf{v} \\
&= (-200)\left(\frac{1}{2} \right)\mathbf{v} \\
&= -100\left(\frac{\sqrt{3}}{2}\mathbf{i} + \frac{1}{2}\mathbf{j} \right).
\end{aligned}$$

The magnitude of this force is 100, and so a force of 100 pounds is required to keep the cart from rolling down the ramp.

✔**CHECKPOINT**   Now try Exercise 67.

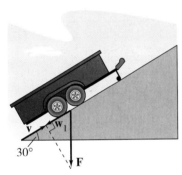

FIGURE **3.40**

Work

The work W done by a *constant* force $\mathbf{F}$ acting along the line of motion of an object is given by

$$W = (\text{magnitude of force})(\text{distance}) = \|\mathbf{F}\|\,\|\overrightarrow{PQ}\|$$

as shown in Figure 3.41. If the constant force $\mathbf{F}$ is not directed along the line of motion, as shown in Figure 3.42, the work W done by the force is given by

$$W = \|\text{proj}_{\overrightarrow{PQ}}\mathbf{F}\|\,\|\overrightarrow{PQ}\| \qquad \text{Projection form for work}$$

$$= (\cos\theta)\|\mathbf{F}\|\,\|\overrightarrow{PQ}\| \qquad \|\text{proj}_{\overrightarrow{PQ}}\mathbf{F}\| = (\cos\theta)\|\mathbf{F}\|$$

$$= \mathbf{F}\cdot\overrightarrow{PQ}. \qquad \text{Alternative form of dot product}$$

Force acts along the line of motion.
FIGURE 3.41

Force acts at angle θ with the line of motion.
FIGURE 3.42

This notion of work is summarized in the following definition.

Definition of Work

The **work** W done by a constant force $\mathbf{F}$ as its point of application moves along the vector $\overrightarrow{PQ}$ is given by either of the following.

1. $W = \|\text{proj}_{\overrightarrow{PQ}}\mathbf{F}\|\,\|\overrightarrow{PQ}\|$ Projection form

2. $W = \mathbf{F}\cdot\overrightarrow{PQ}$ Dot product form

Example 8   **Finding Work**

To close a sliding door, a person pulls on a rope with a constant force of 50 pounds at a constant angle of 60°, as shown in Figure 3.43. Find the work done in moving the door 12 feet to its closed position.

Solution

Using a projection, you can calculate the work as follows.

$$W = \|\text{proj}_{\overrightarrow{PQ}}\mathbf{F}\|\,\|\overrightarrow{PQ}\| \qquad \text{Projection form for work}$$

$$= (\cos 60°)\|\mathbf{F}\|\,\|\overrightarrow{PQ}\|$$

$$= \frac{1}{2}(50)(12) = 300 \text{ foot-pounds}$$

So, the work done is 300 foot-pounds. You can verify this result by finding the vectors $\mathbf{F}$ and $\overrightarrow{PQ}$ and calculating their dot product.

FIGURE 3.43

✓CHECKPOINT   Now try Exercise 69.

3.4   Exercises

VOCABULARY CHECK: Fill in the blanks.

1. The _____ _____ of two vectors yields a scalar, rather than a vector.
2. If θ is the angle between two nonzero vectors **u** and **v**, then $\cos \theta =$ _____ .
3. The vectors **u** and **v** are _____ if $\mathbf{u} \cdot \mathbf{v} = 0$.
4. The projection of **u** onto **v** is given by $\text{proj}_{\mathbf{v}}\mathbf{u} =$ _____ .
5. The work W done by a constant force **F** as its point of application moves along the vector $\overrightarrow{PQ}$
   is given by $W =$ _____ or $W =$ _____ .

PREREQUISITE SKILLS REVIEW: Practice and review algebra skills needed for this section at **www.Eduspace.com.**

In Exercises 1–8, find the dot product of **u** and **v**.

1. $\mathbf{u} = \langle 6, 1 \rangle$
$\mathbf{v} = \langle -2, 3 \rangle$

2. $\mathbf{u} = \langle 5, 12 \rangle$
$\mathbf{v} = \langle -3, 2 \rangle$

3. $\mathbf{u} = \langle -4, 1 \rangle$
$\mathbf{v} = \langle 2, -3 \rangle$

4. $\mathbf{u} = \langle -2, 5 \rangle$
$\mathbf{v} = \langle -1, -2 \rangle$

5. $\mathbf{u} = 4\mathbf{i} - 2\mathbf{j}$
$\mathbf{v} = \mathbf{i} - \mathbf{j}$

6. $\mathbf{u} = 3\mathbf{i} + 4\mathbf{j}$
$\mathbf{v} = 7\mathbf{i} - 2\mathbf{j}$

7. $\mathbf{u} = 3\mathbf{i} + 2\mathbf{j}$
$\mathbf{v} = -2\mathbf{i} - 3\mathbf{j}$

8. $\mathbf{u} = \mathbf{i} - 2\mathbf{j}$
$\mathbf{v} = -2\mathbf{i} + \mathbf{j}$

In Exercises 9–18, use the vectors $\mathbf{u} = \langle 2, 2 \rangle$, $\mathbf{v} = \langle -3, 4 \rangle$, and $\mathbf{w} = \langle 1, -2 \rangle$ to find the indicated quantity. State whether the result is a vector or a scalar.

9. $\mathbf{u} \cdot \mathbf{u}$

10. $3\mathbf{u} \cdot \mathbf{v}$

11. $(\mathbf{u} \cdot \mathbf{v})\mathbf{v}$

12. $(\mathbf{v} \cdot \mathbf{u})\mathbf{w}$

13. $(3\mathbf{w} \cdot \mathbf{v})\mathbf{u}$

14. $(\mathbf{u} \cdot 2\mathbf{v})\mathbf{w}$

15. $\|\mathbf{w}\| - 1$

16. $2 - \|\mathbf{u}\|$

17. $(\mathbf{u} \cdot \mathbf{v}) - (\mathbf{u} \cdot \mathbf{w})$

18. $(\mathbf{v} \cdot \mathbf{u}) - (\mathbf{w} \cdot \mathbf{v})$

In Exercises 19–24, use the dot product to find the magnitude of **u**.

19. $\mathbf{u} = \langle -5, 12 \rangle$

20. $\mathbf{u} = \langle 2, -4 \rangle$

21. $\mathbf{u} = 20\mathbf{i} + 25\mathbf{j}$

22. $\mathbf{u} = 12\mathbf{i} - 16\mathbf{j}$

23. $\mathbf{u} = 6\mathbf{j}$

24. $\mathbf{u} = -21\mathbf{i}$

In Exercises 25–34, find the angle θ between the vectors.

25. $\mathbf{u} = \langle 1, 0 \rangle$
$\mathbf{v} = \langle 0, -2 \rangle$

26. $\mathbf{u} = \langle 3, 2 \rangle$
$\mathbf{v} = \langle 4, 0 \rangle$

27. $\mathbf{u} = 3\mathbf{i} + 4\mathbf{j}$
$\mathbf{v} = -2\mathbf{j}$

28. $\mathbf{u} = 2\mathbf{i} - 3\mathbf{j}$
$\mathbf{v} = \mathbf{i} - 2\mathbf{j}$

29. $\mathbf{u} = 2\mathbf{i} - \mathbf{j}$
$\mathbf{v} = 6\mathbf{i} + 4\mathbf{j}$

30. $\mathbf{u} = -6\mathbf{i} - 3\mathbf{j}$
$\mathbf{v} = -8\mathbf{i} + 4\mathbf{j}$

31. $\mathbf{u} = 5\mathbf{i} + 5\mathbf{j}$
$\mathbf{v} = -6\mathbf{i} + 6\mathbf{j}$

32. $\mathbf{u} = 2\mathbf{i} - 3\mathbf{j}$
$\mathbf{v} = 4\mathbf{i} + 3\mathbf{j}$

33. $\mathbf{u} = \cos\left(\dfrac{\pi}{3}\right)\mathbf{i} + \sin\left(\dfrac{\pi}{3}\right)\mathbf{j}$
$\mathbf{v} = \cos\left(\dfrac{3\pi}{4}\right)\mathbf{i} + \sin\left(\dfrac{3\pi}{4}\right)\mathbf{j}$

34. $\mathbf{u} = \cos\left(\dfrac{\pi}{4}\right)\mathbf{i} + \sin\left(\dfrac{\pi}{4}\right)\mathbf{j}$
$\mathbf{v} = \cos\left(\dfrac{\pi}{2}\right)\mathbf{i} + \sin\left(\dfrac{\pi}{2}\right)\mathbf{j}$

In Exercises 35–38, graph the vectors and find the degree measure of the angle θ between the vectors.

35. $\mathbf{u} = 3\mathbf{i} + 4\mathbf{j}$
$\mathbf{v} = -7\mathbf{i} + 5\mathbf{j}$

36. $\mathbf{u} = 6\mathbf{i} + 3\mathbf{j}$
$\mathbf{v} = -4\mathbf{i} + 4\mathbf{j}$

37. $\mathbf{u} = 5\mathbf{i} + 5\mathbf{j}$
$\mathbf{v} = -8\mathbf{i} + 8\mathbf{j}$

38. $\mathbf{u} = 2\mathbf{i} - 3\mathbf{j}$
$\mathbf{v} = 8\mathbf{i} + 3\mathbf{j}$

In Exercises 39–42, use vectors to find the interior angles of the triangle with the given vertices.

39. $(1, 2), (3, 4), (2, 5)$

40. $(-3, -4), (1, 7), (8, 2)$

41. $(-3, 0), (2, 2), (0, 6)$

42. $(-3, 5), (-1, 9), (7, 9)$

In Exercises 43–46, find $\mathbf{u} \cdot \mathbf{v}$, where θ is the angle between **u** and **v**.

43. $\|\mathbf{u}\| = 4$, $\|\mathbf{v}\| = 10$, $\theta = \dfrac{2\pi}{3}$

44. $\|\mathbf{u}\| = 100$, $\|\mathbf{v}\| = 250$, $\theta = \dfrac{\pi}{6}$

45. $\|\mathbf{u}\| = 9$, $\|\mathbf{v}\| = 36$, $\theta = \dfrac{3\pi}{4}$

46. $\|\mathbf{u}\| = 4$, $\|\mathbf{v}\| = 12$, $\theta = \dfrac{\pi}{3}$

In Exercises 47–52, determine whether **u** and **v** are orthogonal, parallel, or neither.

47. $\mathbf{u} = \langle -12, 30 \rangle$
$\mathbf{v} = \left\langle \frac{1}{2}, -\frac{5}{4} \right\rangle$

48. $\mathbf{u} = \langle 3, 15 \rangle$
$\mathbf{v} = \langle -1, 5 \rangle$

49. $\mathbf{u} = \frac{1}{4}(3\mathbf{i} - \mathbf{j})$
$\mathbf{v} = 5\mathbf{i} + 6\mathbf{j}$

50. $\mathbf{u} = \mathbf{i}$
$\mathbf{v} = -2\mathbf{i} + 2\mathbf{j}$

51. $\mathbf{u} = 2\mathbf{i} - 2\mathbf{j}$
$\mathbf{v} = -\mathbf{i} - \mathbf{j}$

52. $\mathbf{u} = \langle \cos\theta, \sin\theta \rangle$
$\mathbf{v} = \langle \sin\theta, -\cos\theta \rangle$

In Exercises 53–56, find the projection of **u** onto **v**. Then write **u** as the sum of two orthogonal vectors, one of which is proj$_\mathbf{v}$**u**.

53. $\mathbf{u} = \langle 2, 2 \rangle$
$\mathbf{v} = \langle 6, 1 \rangle$

54. $\mathbf{u} = \langle 4, 2 \rangle$
$\mathbf{v} = \langle 1, -2 \rangle$

55. $\mathbf{u} = \langle 0, 3 \rangle$
$\mathbf{v} = \langle 2, 15 \rangle$

56. $\mathbf{u} = \langle -3, -2 \rangle$
$\mathbf{v} = \langle -4, -1 \rangle$

In Exercises 57 and 58, use the graph to determine mentally the projection of **u** onto **v**. (The coordinates of the terminal points of the vectors in standard position are given.) Use the formula for the projection of **u** onto **v** to verify your result.

57.

58.

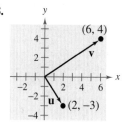

In Exercises 59–62, find two vectors in opposite directions that are orthogonal to the vector **u**. (There are many correct answers.)

59. $\mathbf{u} = \langle 3, 5 \rangle$

60. $\mathbf{u} = \langle -8, 3 \rangle$

61. $\mathbf{u} = \frac{1}{2}\mathbf{i} - \frac{2}{3}\mathbf{j}$

62. $\mathbf{u} = -\frac{5}{2}\mathbf{i} - 3\mathbf{j}$

Work In Exercises 63 and 64, find the work done in moving a particle from P to Q if the magnitude and direction of the force are given by **v**.

63. $P = (0, 0)$, $Q = (4, 7)$, $\mathbf{v} = \langle 1, 4 \rangle$

64. $P = (1, 3)$, $Q = (-3, 5)$, $\mathbf{v} = -2\mathbf{i} + 3\mathbf{j}$

65. ***Revenue*** The vector $\mathbf{u} = \langle 1650, 3200 \rangle$ gives the numbers of units of two types of baking pans produced by a company. The vector $\mathbf{v} = \langle 15.25, 10.50 \rangle$ gives the prices (in dollars) of the two types of pans, respectively.

(a) Find the dot product $\mathbf{u} \cdot \mathbf{v}$ and interpret the result in the context of the problem.

(b) Identify the vector operation used to increase the prices by 5%.

66. ***Revenue*** The vector $\mathbf{u} = \langle 3240, 2450 \rangle$ gives the numbers of hamburgers and hot dogs, respectively, sold at a fast-food stand in one month. The vector $\mathbf{v} = \langle 1.75, 1.25 \rangle$ gives the prices (in dollars) of the food items.

(a) Find the dot product $\mathbf{u} \cdot \mathbf{v}$ and interpret the result in the context of the problem.

(b) Identify the vector operation used to increase the prices by 2.5%.

Model It

67. ***Braking Load*** A truck with a gross weight of 30,000 pounds is parked on a slope of $d°$ (see figure). Assume that the only force to overcome is the force of gravity.

Weight = 30,000 lb

(a) Find the force required to keep the truck from rolling down the hill in terms of the slope d.

(b) Use a graphing utility to complete the table.

d	0°	1°	2°	3°	4°	5°
Force						

d	6°	7°	8°	9°	10°
Force					

(c) Find the force perpendicular to the hill when $d = 5°$.

68. ***Braking Load*** A sport utility vehicle with a gross weight of 5400 pounds is parked on a slope of 10°. Assume that the only force to overcome is the force of gravity. Find the force required to keep the vehicle from rolling down the hill. Find the force perpendicular to the hill.

69. **Work**   Determine the work done by a person lifting a 25-kilogram (245-newton) bag of sugar.

70. **Work**   Determine the work done by a crane lifting a 2400-pound car 5 feet.

71. **Work**   A force of 45 pounds exerted at an angle of 30° above the horizontal is required to slide a table across a floor (see figure). The table is dragged 20 feet. Determine the work done in sliding the table.

45 lb

30°

|← 20 ft →|

72. **Work**   A tractor pulls a log 800 meters, and the tension in the cable connecting the tractor and log is approximately 1600 kilograms (15,691 newtons). The direction of the force is 35° above the horizontal. Approximate the work done in pulling the log.

73. **Work**   One of the events in a local strongman contest is to pull a cement block 100 feet. One competitor pulls the block by exerting a force of 250 pounds on a rope attached to the block at an angle of 30° with the horizontal (see figure). Find the work done in pulling the block.

30°

|←100 ft→|

Not drawn to scale

74. **Work**   A toy wagon is pulled by exerting a force of 25 pounds on a handle that makes a 20° angle with the horizontal (see figure). Find the work done in pulling the wagon 50 feet.

20°

Synthesis

True or False?   In Exercises 75 and 76, determine whether the statement is true or false. Justify your answer.

75. The work W done by a constant force $\mathbf{F}$ acting along the line of motion of an object is represented by a vector.

76. A sliding door moves along the line of vector $\overrightarrow{PQ}$. If a force is applied to the door along a vector that is orthogonal to $\overrightarrow{PQ}$, then no work is done.

77. **Think About It**   What is known about θ, the angle between two nonzero vectors $\mathbf{u}$ and $\mathbf{v}$, under each condition?

   (a) $\mathbf{u} \cdot \mathbf{v} = 0$ (b) $\mathbf{u} \cdot \mathbf{v} > 0$ (c) $\mathbf{u} \cdot \mathbf{v} < 0$

78. **Think About It**   What can be said about the vectors $\mathbf{u}$ and $\mathbf{v}$ under each condition?

   (a) The projection of $\mathbf{u}$ onto $\mathbf{v}$ equals $\mathbf{u}$.

   (b) The projection of $\mathbf{u}$ onto $\mathbf{v}$ equals $\mathbf{0}$.

79. **Proof**   Use vectors to prove that the diagonals of a rhombus are perpendicular.

80. **Proof**   Prove the following.

   $$\|\mathbf{u} - \mathbf{v}\|^2 = \|\mathbf{u}\|^2 + \|\mathbf{v}\|^2 - 2\mathbf{u} \cdot \mathbf{v}$$

Skills Review

In Exercises 81–84, find all solutions of the equation in the interval $[0, 2\pi)$.

81. $\sin 2x - \sqrt{3} \sin x = 0$

82. $\sin 2x + \sqrt{2} \cos x = 0$

83. $2 \tan x = \tan 2x$

84. $\cos 2x - 3 \sin x = 2$

In Exercises 85–88, find the exact value of the trigonometric function given that $\sin u = -\frac{12}{13}$ and $\cos v = \frac{24}{25}$. (Both u and v are in Quadrant IV.)

85. $\sin(u - v)$

86. $\sin(u + v)$

87. $\cos(v - u)$

88. $\tan(u - v)$

3 | Chapter Summary

What did you learn?

Section 3.1

Review Exercises

☐ Use the Law of Sines to solve oblique triangles (AAS, ASA, or SSA) *(p. 278, 280)*.   1–12

☐ Find areas of oblique triangles *(p. 282)*.   13–16

☐ Use the Law of Sines to model and solve real-life problems *(p. 283)*.   17–20

Section 3.2

☐ Use the Law of Cosines to solve oblique triangles (SSS or SAS) *(p. 287)*.   21–28

☐ Use the Law of Cosines to model and solve real-life problems *(p. 289)*.   29–32

☐ Use Heron's Area Formula to find areas of triangles *(p. 290)*.   33–36

Section 3.3

☐ Represent vectors as directed line segments *(p. 295)*.   37, 38

☐ Write the component forms of vectors *(p. 296)*.   39–44

☐ Perform basic vector operations and represent vectors graphically *(p. 297)*.   45–56

☐ Write vectors as linear combinations of unit vectors *(p. 299)*.   57–62

☐ Find the direction angles of vectors *(p. 301)*.   63–68

☐ Use vectors to model and solve real-life problems *(p. 302)*.   69–72

Section 3.4

☐ Find the dot product of two vectors and use the properties of the dot product *(p. 308)*.   73–80

☐ Find the angle between two vectors and determine whether two vectors are orthogonal *(p. 309)*.   81–88

☐ Write vectors as sums of two vector components *(p. 311)*.   89–92

☐ Use vectors to find the work done by a force *(p. 314)*.   93–96

3 **Review Exercises**

3.1 In Exercises 1–12, use the Law of Sines to solve (if possible) the triangle. If two solutions exist, find both. Round your answers to two decimal places.

1.

2.

3. $B = 72°$, $C = 82°$, $b = 54$

4. $B = 10°$, $C = 20°$, $c = 33$

5. $A = 16°$, $B = 98°$, $c = 8.4$

6. $A = 95°$, $B = 45°$, $c = 104.8$

7. $A = 24°$, $C = 48°$, $b = 27.5$

8. $B = 64°$, $C = 36°$, $a = 367$

9. $B = 150°$, $b = 30$, $c = 10$

10. $B = 150°$, $a = 10$, $b = 3$

11. $A = 75°$, $a = 51.2$, $b = 33.7$

12. $B = 25°$, $a = 6.2$, $b = 4$

In Exercises 13–16, find the area of the triangle having the indicated angle and sides.

13. $A = 27°$, $b = 5$, $c = 7$

14. $B = 80°$, $a = 4$, $c = 8$

15. $C = 123°$, $a = 16$, $b = 5$

16. $A = 11°$, $b = 22$, $c = 21$

17. *Height* From a certain distance, the angle of elevation to the top of a building is 17°. At a point 50 meters closer to the building, the angle of elevation is 31°. Approximate the height of the building.

18. *Geometry* Find the length of the side w of the parallelogram.

19. *Height* A tree stands on a hillside of slope 28° from the horizontal. From a point 75 feet down the hill, the angle of elevation to the top of the tree is 45° (see figure). Find the height of the tree.

FIGURE FOR 19

20. *River Width* A surveyor finds that a tree on the opposite bank of a river, flowing due east, has a bearing of N 22° 30′ E from a certain point and a bearing of N 15° W from a point 400 feet downstream. Find the width of the river.

3.2 In Exercises 21–28, use the Law of Cosines to solve the triangle. Round your answers to two decimal places.

21. $a = 5$, $b = 8$, $c = 10$

22. $a = 80$, $b = 60$, $c = 100$

23. $a = 2.5$, $b = 5.0$, $c = 4.5$

24. $a = 16.4$, $b = 8.8$, $c = 12.2$

25. $B = 110°$, $a = 4$, $c = 4$

26. $B = 150°$, $a = 10$, $c = 20$

27. $C = 43°$, $a = 22.5$, $b = 31.4$

28. $A = 62°$, $b = 11.34$, $c = 19.52$

29. *Geometry* The lengths of the diagonals of a parallelogram are 10 feet and 16 feet. Find the lengths of the sides of the parallelogram if the diagonals intersect at an angle of 28°.

30. *Geometry* The lengths of the diagonals of a parallelogram are 30 meters and 40 meters. Find the lengths of the sides of the parallelogram if the diagonals intersect at an angle of 34°.

31. *Surveying* To approximate the length of a marsh, a surveyor walks 425 meters from point A to point B. Then the surveyor turns 65° and walks 300 meters to point C (see figure). Approximate the length AC of the marsh.

32. *Navigation* Two planes leave Raleigh-Durham Airport at approximately the same time. One is flying 425 miles per hour at a bearing of 355°, and the other is flying 530 miles per hour at a bearing of 67°. Draw a figure that gives a visual representation of the problem and determine the distance between the planes after they have flown for 2 hours.

In Exercises 33–36, use Heron's Area Formula to find the area of the triangle.

33. $a = 4$, $b = 5$, $c = 7$
34. $a = 15$, $b = 8$, $c = 10$
35. $a = 12.3$, $b = 15.8$, $c = 3.7$
36. $a = 38.1$, $b = 26.7$, $c = 19.4$

3.3 **In Exercises 37 and 38, show that u = v.**

37.
38.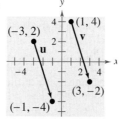

In Exercises 39–44, find the component form of the vector v satisfying the conditions.

39.
40.

41. Initial point: $(0, 10)$; terminal point: $(7, 3)$
42. Initial point: $(1, 5)$; terminal point: $(15, 9)$
43. $\|v\| = 8$, $\theta = 120°$
44. $\|v\| = \frac{1}{2}$, $\theta = 225°$

In Exercises 45–52, find (a) u + v, (b) u − v, (c) 3u, and (d) 2v + 5u.

45. $u = \langle -1, -3 \rangle$, $v = \langle -3, 6 \rangle$
46. $u = \langle 4, 5 \rangle$, $v = \langle 0, -1 \rangle$
47. $u = \langle -5, 2 \rangle$, $v = \langle 4, 4 \rangle$
48. $u = \langle 1, -8 \rangle$, $v = \langle 3, -2 \rangle$
49. $u = 2i - j$, $v = 5i + 3j$
50. $u = -7i - 3j$, $v = 4i - j$
51. $u = 4i$, $v = -i + 6j$
52. $u = -6j$, $v = i + j$

In Exercises 53–56, find the component form of w and sketch the specified vector operations geometrically, where u = 6i − 5j and v = 1 − i + 3j.

53. $w = 2u + v$
54. $w = 4u - 5v$
55. $w = 3v$
56. $w = \frac{1}{2}v$

In Exercises 57–60, write vector u as a linear combination of the standard unit vectors i and j.

57. $u = \langle -3, 4 \rangle$
58. $u = \langle -6, -8 \rangle$
59. u has initial point $(3, 4)$ and terminal point $(9, 8)$.
60. u has initial point $(-2, 7)$ and terminal point $(5, -9)$.

In Exercises 61 and 62, write the vector v in the form $\|v\|(\cos \theta i + \sin \theta j)$.

61. $v = -10i + 10j$
62. $v = 4i - j$

In Exercises 63–68, find the magnitude and the direction angle of the vector v.

63. $v = 7(\cos 60°i + \sin 60°j)$
64. $v = 3(\cos 150°i + \sin 150°j)$
65. $v = 5i + 4j$
66. $v = -4i + 7j$
67. $v = -3i - 3j$
68. $v = 8i - j$

69. *Resultant Force* Forces with magnitudes of 85 pounds and 50 pounds act on a single point. The angle between the forces is 15°. Describe the resultant force.

70. *Rope Tension* A 180-pound weight is supported by two ropes, as shown in the figure. Find the tension in each rope.

71. *Navigation* An airplane has an airspeed of 430 miles per hour at a bearing of 135°. The wind velocity is 35 miles per hour in the direction of N 30° E. Find the resultant speed and direction of the airplane.

72. *Navigation* An airplane has an airspeed of 724 kilometers per hour at a bearing of 30°. The wind velocity is 32 kilometers per hour from the west. Find the resultant speed and direction of the airplane.

3.4 In Exercises 73–76, find the dot product of **u** and **v**.

73. $\mathbf{u} = \langle 6, 7 \rangle$
$\quad \mathbf{v} = \langle -3, 9 \rangle$

74. $\mathbf{u} = \langle -7, 12 \rangle$
$\quad \mathbf{v} = \langle -4, -14 \rangle$

75. $\mathbf{u} = 3\mathbf{i} + 7\mathbf{j}$
$\quad \mathbf{v} = 11\mathbf{i} - 5\mathbf{j}$

76. $\mathbf{u} = -7\mathbf{i} + 2\mathbf{j}$
$\quad \mathbf{v} = 16\mathbf{i} - 12\mathbf{j}$

In Exercises 77–80, use the vectors $\mathbf{u} = \langle -3, 4 \rangle$ and $\mathbf{v} = \langle 2, 1 \rangle$ to find the indicated quantity. State whether the result is a vector or a scalar.

77. $2\mathbf{u} \cdot \mathbf{u}$

78. $\|\mathbf{v}\|^2$

79. $\mathbf{u}(\mathbf{u} \cdot \mathbf{v})$

80. $3\mathbf{u} \cdot \mathbf{v}$

In Exercises 81–84, find the angle θ between the vectors.

81. $\mathbf{u} = \cos\dfrac{7\pi}{4}\mathbf{i} + \sin\dfrac{7\pi}{4}\mathbf{j}$
$\quad \mathbf{v} = \cos\dfrac{5\pi}{6}\mathbf{i} + \sin\dfrac{5\pi}{6}\mathbf{j}$

82. $\mathbf{u} = \cos 45°\mathbf{i} + \sin 45°\mathbf{j}$
$\quad \mathbf{v} = \cos 300°\mathbf{i} + \sin 300°\mathbf{j}$

83. $\mathbf{u} = \langle 2\sqrt{2}, -4 \rangle, \quad \mathbf{v} = \langle -\sqrt{2}, 1 \rangle$

84. $\mathbf{u} = \langle 3, \sqrt{3} \rangle, \quad \mathbf{v} = \langle 4, 3\sqrt{3} \rangle$

In Exercises 85–88, determine whether **u** and **v** are orthogonal, parallel, or neither.

85. $\mathbf{u} = \langle -3, 8 \rangle$
$\quad \mathbf{v} = \langle 8, 3 \rangle$

86. $\mathbf{u} = \langle \frac{1}{4}, -\frac{1}{2} \rangle$
$\quad \mathbf{v} = \langle -2, 4 \rangle$

87. $\mathbf{u} = -\mathbf{i}$
$\quad \mathbf{v} = \mathbf{i} + 2\mathbf{j}$

88. $\mathbf{u} = -2\mathbf{i} + \mathbf{j}$
$\quad \mathbf{v} = 3\mathbf{i} + 6\mathbf{j}$

In Exercises 89–92, find the projection of **u** onto **v**. Then write **u** as the sum of two orthogonal vectors, one of which is $\text{proj}_\mathbf{v}\mathbf{u}$.

89. $\mathbf{u} = \langle -4, 3 \rangle, \ \mathbf{v} = \langle -8, -2 \rangle$

90. $\mathbf{u} = \langle 5, 6 \rangle, \ \mathbf{v} = \langle 10, 0 \rangle$

91. $\mathbf{u} = \langle 2, 7 \rangle, \ \mathbf{v} = \langle 1, -1 \rangle$

92. $\mathbf{u} = \langle -3, 5 \rangle, \ \mathbf{v} = \langle -5, 2 \rangle$

Work In Exercises 93 and 94, find the work done in moving a particle from P to Q if the magnitude and direction of the force are given by **v**.

93. $P = (5, 3), Q = (8, 9), \mathbf{v} = \langle 2, 7 \rangle$

94. $P = (-2, -9), Q = (-12, 8), \mathbf{v} = 3\mathbf{i} - 6\mathbf{j}$

95. **Work** Determine the work done by a crane lifting an 18,000-pound truck 48 inches.

96. **Work** A mover exerts a horizontal force of 25 pounds on a crate as it is pushed up a ramp that is 12 feet long and inclined at an angle of 20° above the horizontal. Find the work done in pushing the crate.

Synthesis

True or False? In Exercises 97–100, determine whether the statement is true or false. Justify your answer.

97. The Law of Sines is true if one of the angles in the triangle is a right angle.

98. When the Law of Sines is used, the solution is always unique.

99. If **u** is a unit vector in the direction of **v**, then $\mathbf{v} = \|\mathbf{v}\| \mathbf{u}$.

100. If $\mathbf{v} = a\mathbf{i} + b\mathbf{j} = \mathbf{0}$, then $a = -b$.

101. State the Law of Sines from memory.

102. State the Law of Cosines from memory.

103. What characterizes a vector in the plane?

104. Which vectors in the figure appear to be equivalent?

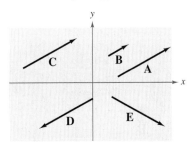

105. The vectors **u** and **v** have the same magnitudes in the two figures. In which figure will the magnitude of the sum be greater? Give a reason for your answer.

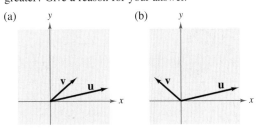

(a)   (b)

106. Give a geometric description of the scalar multiple $k\mathbf{u}$ of the vector **u**, for $k > 0$ and for $k < 0$.

107. Give a geometric description of the sum of the vectors **u** and **v**.

3 Chapter Test

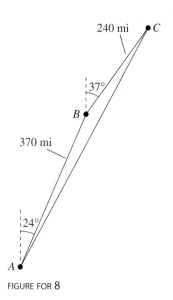

240 mi

C

37°

B

370 mi

24°

A

FIGURE FOR **8**

Take this test as you would take a test in class. When you are finished, check your work against the answers given in the back of the book.

In Exercises 1–6, use the information to solve the triangle. If two solutions exist, find both solutions. Round your answers to two decimal places.

1. $A = 24°$, $B = 68°$, $a = 12.2$
2. $B = 104°$, $C = 33°$, $a = 18.1$
3. $A = 24°$, $a = 11.2$, $b = 13.4$
4. $a = 4.0$, $b = 7.3$, $c = 12.4$
5. $B = 100°$, $a = 15$, $b = 23$
6. $C = 123°$, $a = 41$, $b = 57$

7. A triangular parcel of land has borders of lengths 60 meters, 70 meters, and 82 meters. Find the area of the parcel of land.

8. An airplane flies 370 miles from point A to point B with a bearing of 24°. It then flies 240 miles from point B to point C with a bearing of 37° (see figure). Find the distance and bearing from point A to point C.

In Exercises 9 and 10, find the component form of the vector **v** satisfying the given conditions.

9. Initial point of **v**: $(-3, 7)$; terminal point of **v**: $(11, -16)$
10. Magnitude of **v**: $\|\mathbf{v}\| = 12$; direction of **v**: $\mathbf{u} = \langle 3, -5 \rangle$

In Exercises 11–13, $\mathbf{u} = \langle 3, 5 \rangle$ and $\mathbf{v} = \langle -7, 1 \rangle$. Find the resultant vector and sketch its graph.

11. $\mathbf{u} + \mathbf{v}$ 12. $\mathbf{u} - \mathbf{v}$ 13. $5\mathbf{u} - 3\mathbf{v}$

14. Find a unit vector in the direction of $\mathbf{u} = \langle 4, -3 \rangle$.

15. Forces with magnitudes of 250 pounds and 130 pounds act on an object at angles of 45° and −60°, respectively, with the *x*-axis. Find the direction and magnitude of the resultant of these forces.

16. Find the angle between the vectors $\mathbf{u} = \langle -1, 5 \rangle$ and $\mathbf{v} = \langle 3, -2 \rangle$.

17. Are the vectors $\mathbf{u} = \langle 6, 10 \rangle$ and $\mathbf{v} = \langle 2, 3 \rangle$ orthogonal?

18. Find the projection of $\mathbf{u} = \langle 6, 7 \rangle$ onto $\mathbf{v} = \langle -5, -1 \rangle$. Then write **u** as the sum of two orthogonal vectors.

19. A 500-pound motorcycle is headed up a hill inclined at 12°. What force is required to keep the motorcycle from rolling down the hill when stopped at a red light?

3   Cumulative Test for Chapters 1–3

Take this test to review the material from earlier chapters. When you are finished, check your work against the answers given in the back of the book.

1. Consider the angle $\theta = -120°$.

   (a) Sketch the angle in standard position.

   (b) Determine a coterminal angle in the interval $[0°, 360°)$.

   (c) Convert the angle to radian measure.

   (d) Find the reference angle θ'.

   (e) Find the exact values of the six trigonometric functions of θ.

2. Convert the angle $\theta = 2.35$ radians to degrees. Round the answer to one decimal place.

3. Find $\cos \theta$ if $\tan \theta = -\frac{4}{3}$ and $\sin \theta < 0$.

In Exercises 4–6, sketch the graph of the function. (Include two full periods.)

4. $f(x) = 3 - 2 \sin \pi x$ 5. $g(x) = \frac{1}{2} \tan\left(x - \frac{\pi}{2}\right)$ 6. $h(x) = -\sec(x + \pi)$

FIGURE FOR **7**

7. Find a, b, and c such that the graph of the function $h(x) = a \cos(bx + c)$ matches the graph in the figure.

8. Sketch the graph of the function $f(x) = \frac{1}{2}x \sin x$ over the interval $-3\pi \le x \le 3\pi$.

In Exercises 9 and 10, find the exact value of the expression without using a calculator.

9. $\tan(\arctan 6.7)$ 10. $\tan\left(\arcsin \frac{3}{5}\right)$

11. Write an algebraic expression equivalent to $\sin(\arccos 2x)$.

12. Use the fundamental identities to simplify: $\cos\left(\frac{\pi}{2} - x\right) \csc x$.

13. Subtract and simplify: $\dfrac{\sin \theta - 1}{\cos \theta} - \dfrac{\cos \theta}{\sin \theta - 1}$.

In Exercises 14–16, verify the identity.

14. $\cot^2 \alpha(\sec^2 \alpha - 1) = 1$

15. $\sin(x + y) \sin(x - y) = \sin^2 x - \sin^2 y$

16. $\sin^2 x \cos^2 x = \frac{1}{8}(1 - \cos 4x)$

In Exercises 17 and 18, find all solutions of the equation in the interval $[0, 2\pi)$.

17. $2 \cos^2 \beta - \cos \beta = 0$

18. $3 \tan \theta - \cot \theta = 0$

19. Use the Quadratic Formula to solve the equation in the interval $[0, 2\pi)$: $\sin^2 x + 2 \sin x + 1 = 0$.

20. Given that $\sin u = \frac{12}{13}$, $\cos v = \frac{3}{5}$, and angles u and v are both in Quadrant I, find $\tan(u - v)$.

21. If $\tan \theta = \frac{1}{2}$, find the exact value of $\tan(2\theta)$.

22. If $\tan \theta = \dfrac{4}{3}$, find the exact value of $\sin \dfrac{\theta}{2}$.

23. Write the product $5 \sin \dfrac{3\pi}{4} \cdot \cos \dfrac{7\pi}{4}$ as a sum or difference.

24. Write $\cos 8x + \cos 4x$ as a product.

In Exercises 25–28, use the information to solve the triangle shown in the figure. Round your answers to two decimal places.

25. $A = 30°$, $a = 9$, $b = 8$

26. $A = 30°$, $b = 8$, $c = 10$

27. $A = 30°$, $C = 90°$, $b = 10$

28. $a = 4$, $b = 8$, $c = 9$

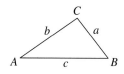

FIGURE FOR 25–28

29. Two sides of a triangle have lengths 7 inches and 12 inches. Their included angle measures 60°. Find the area of the triangle.

30. Find the area of a triangle with sides of lengths 11 inches, 16 inches, and 17 inches.

31. Write the vector $\mathbf{u} = \langle 3, 5 \rangle$ as a linear combination of the standard unit vectors $\mathbf{i}$ and $\mathbf{j}$.

32. Find a unit vector in the direction of $\mathbf{v} = \mathbf{i} + \mathbf{j}$.

33. Find $\mathbf{u} \cdot \mathbf{v}$ for $\mathbf{u} = 3\mathbf{i} + 4\mathbf{j}$ and $\mathbf{v} = \mathbf{i} - 2\mathbf{j}$.

34. Find the projection of $\mathbf{u} = \langle 8, -2 \rangle$ onto $\mathbf{v} = \langle 1, 5 \rangle$. Then write $\mathbf{u}$ as the sum of two orthogonal vectors.

35. A ceiling fan with 21-inch blades makes 63 revolutions per minute. Find the angular speed of the fan in radians per minute. Find the linear speed of the tips of the blades in inches per minute.

36. Find the area of the sector of a circle with a radius of 8 yards and a central angle of 114°.

37. From a point 200 feet from a flagpole, the angles of elevation to the bottom and top of the flag are 16° 45′ and 18°, respectively. Approximate the height of the flag to the nearest foot.

38. To determine the angle of elevation of a star in the sky, you get the star in your line of vision with the backboard of a basketball hoop that is 5 feet higher than your eyes (see figure). Your horizontal distance from the backboard is 12 feet. What is the angle of elevation of the star?

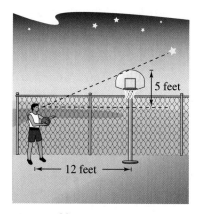

FIGURE FOR 38

39. Write a model for a particle in simple harmonic motion with a displacement of 4 inches and a period of 8 seconds.

40. An airplane's velocity with respect to the air is 500 kilometers per hour, with a bearing of 30°. The wind at the altitude of the plane has a velocity of 50 kilometers per hour with a bearing of N 60° E. What is the true direction of the plane, and what is its speed relative to the ground?

41. A force of 85 pounds exerted at an angle of 60° above the horizontal is required to slide an object across a floor. The object is dragged 10 feet. Determine the work done in sliding the object.

Proofs in Mathematics

Law of Tangents

Besides the Law of Sines and the Law of Cosines, there is also a Law of Tangents, which was developed by Francois Viète (1540–1603). The Law of Tangents follows from the Law of Sines and the sum-to-product formulas for sine and is defined as follows.

$$\frac{a + b}{a - b} = \frac{\tan[(A + B)/2]}{\tan[(A - B)/2]}$$

The Law of Tangents can be used to solve a triangle when two sides and the included angle are given (SAS). Before calculators were invented, the Law of Tangents was used to solve the SAS case instead of the Law of Cosines, because computation with a table of tangent values was easier.

A is acute.

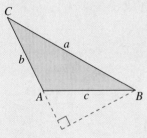

A is obtuse.

Law of Sines *(p. 278)*

If ABC is a triangle with sides a, b, and c, then

$$\frac{a}{\sin A} = \frac{b}{\sin B} = \frac{c}{\sin C}.$$

A is acute.

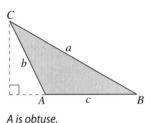

A is obtuse.

Proof

Let h be the altitude of either triangle found in the figure above. Then you have

$$\sin A = \frac{h}{b} \qquad \text{or} \qquad h = b \sin A$$

$$\sin B = \frac{h}{a} \qquad \text{or} \qquad h = a \sin B.$$

Equating these two values of h, you have

$$a \sin B = b \sin A \qquad \text{or} \qquad \frac{a}{\sin A} = \frac{b}{\sin B}.$$

Note that $\sin A \neq 0$ and $\sin B \neq 0$ because no angle of a triangle can have a measure of $0°$ or $180°$. In a similar manner, construct an altitude from vertex B to side AC (extended in the obtuse triangle), as shown at the left. Then you have

$$\sin A = \frac{h}{c} \qquad \text{or} \qquad h = c \sin A$$

$$\sin C = \frac{h}{a} \qquad \text{or} \qquad h = a \sin C.$$

Equating these two values of h, you have

$$a \sin C = c \sin A \qquad \text{or} \qquad \frac{a}{\sin A} = \frac{c}{\sin C}.$$

By the Transitive Property of Equality you know that

$$\frac{a}{\sin A} = \frac{b}{\sin B} = \frac{c}{\sin C}.$$

So, the Law of Sines is established.

Standard Form	*Alternative Form*
$a^2 = b^2 + c^2 - 2bc \cos A$	$\cos A = \dfrac{b^2 + c^2 - a^2}{2bc}$
$b^2 = a^2 + c^2 - 2ac \cos B$	$\cos B = \dfrac{a^2 + c^2 - b^2}{2ac}$
$c^2 = a^2 + b^2 - 2ab \cos C$	$\cos C = \dfrac{a^2 + b^2 - c^2}{2ab}$

Proof

To prove the first formula, consider the top triangle at the left, which has three acute angles. Note that vertex B has coordinates $(c, 0)$. Furthermore, C has coordinates (x, y), where $x = b \cos A$ and $y = b \sin A$. Because a is the distance from vertex C to vertex B, it follows that

$a = \sqrt{(x - c)^2 + (y - 0)^2}$	Distance Formula
$a^2 = (x - c)^2 + (y - 0)^2$	Square each side.
$a^2 = (b \cos A - c)^2 + (b \sin A)^2$	Substitute for x and y.
$a^2 = b^2 \cos^2 A - 2bc \cos A + c^2 + b^2 \sin^2 A$	Expand.
$a^2 = b^2(\sin^2 A + \cos^2 A) + c^2 - 2bc \cos A$	Factor out b^2.
$a^2 = b^2 + c^2 - 2bc \cos A.$	$\sin^2 A + \cos^2 A = 1$

To prove the second formula, consider the bottom triangle at the left, which also has three acute angles. Note that vertex A has coordinates $(c, 0)$. Furthermore, C has coordinates (x, y), where $x = a \cos B$ and $y = a \sin B$. Because b is the distance from vertex C to vertex A, it follows that

$b = \sqrt{(x - c)^2 + (y - 0)^2}$	Distance Formula
$b^2 = (x - c)^2 + (y - 0)^2$	Square each side.
$b^2 = (a \cos B - c)^2 + (a \sin B)^2$	Substitute for x and y.
$b^2 = a^2 \cos^2 B - 2ac \cos B + c^2 + a^2 \sin^2 B$	Expand.
$b^2 = a^2(\sin^2 B + \cos^2 B) + c^2 - 2ac \cos B$	Factor out a^2.
$b^2 = a^2 + c^2 - 2ac \cos B.$	$\sin^2 B + \cos^2 B = 1$

A similar argument is used to establish the third formula.

Heron's Area Formula   *(p. 290)*

Given any triangle with sides of lengths a, b, and c, the area of the triangle is

$$\text{Area} = \sqrt{s(s - a)(s - b)(s - c)}$$

where $s = \dfrac{(a + b + c)}{2}$.

Proof

From Section 3.1, you know that

$$\text{Area} = \frac{1}{2} bc \sin A \qquad \text{Formula for the area of an oblique triangle}$$

$$(\text{Area})^2 = \frac{1}{4} b^2 c^2 \sin^2 A \qquad \text{Square each side.}$$

$$\text{Area} = \sqrt{\frac{1}{4} b^2 c^2 \sin^2 A} \qquad \text{Take the square root of each side.}$$

$$= \sqrt{\frac{1}{4} b^2 c^2 (1 - \cos^2 A)} \qquad \text{Pythagorean Identity}$$

$$= \sqrt{\left[\frac{1}{2} bc(1 + \cos A)\right]\left[\frac{1}{2} bc(1 - \cos A)\right]}. \qquad \text{Factor.}$$

Using the Law of Cosines, you can show that

$$\frac{1}{2} bc(1 + \cos A) = \frac{a + b + c}{2} \cdot \frac{-a + b + c}{2}$$

and

$$\frac{1}{2} bc(1 - \cos A) = \frac{a - b + c}{2} \cdot \frac{a + b - c}{2}.$$

Letting $s = (a + b + c)/2$, these two equations can be rewritten as

$$\frac{1}{2} bc(1 + \cos A) = s(s - a)$$

and

$$\frac{1}{2} bc(1 - \cos A) = (s - b)(s - c).$$

By substituting into the last formula for area, you can conclude that

$$\text{Area} = \sqrt{s(s - a)(s - b)(s - c)}.$$

Properties of the Dot Product *(p. 308)*

Let **u**, **v**, and **w** be vectors in the plane or in space and let c be a scalar.

1. $\mathbf{u} \cdot \mathbf{v} = \mathbf{v} \cdot \mathbf{u}$　　　　　　　**2.** $\mathbf{0} \cdot \mathbf{v} = 0$

3. $\mathbf{u} \cdot (\mathbf{v} + \mathbf{w}) = \mathbf{u} \cdot \mathbf{v} + \mathbf{u} \cdot \mathbf{w}$　　**4.** $\mathbf{v} \cdot \mathbf{v} = \|\mathbf{v}\|^2$

5. $c(\mathbf{u} \cdot \mathbf{v}) = c\mathbf{u} \cdot \mathbf{v} = \mathbf{u} \cdot c\mathbf{v}$

Proof

Let $\mathbf{u} = \langle u_1, u_2 \rangle$, $\mathbf{v} = \langle v_1, v_2 \rangle$, $\mathbf{w} = \langle w_1, w_2 \rangle$, $\mathbf{0} = \langle 0, 0 \rangle$, and let c be a scalar.

1. $\mathbf{u} \cdot \mathbf{v} = u_1 v_1 + u_2 v_2 = v_1 u_1 + v_2 u_2 = \mathbf{v} \cdot \mathbf{u}$

2. $\mathbf{0} \cdot \mathbf{v} = 0 \cdot v_1 + 0 \cdot v_2 = 0$

3. $\mathbf{u} \cdot (\mathbf{v} + \mathbf{w}) = \mathbf{u} \cdot \langle v_1 + w_1, v_2 + w_2 \rangle$
$$= u_1(v_1 + w_1) + u_2(v_2 + w_2)$$
$$= u_1 v_1 + u_1 w_1 + u_2 v_2 + u_2 w_2$$
$$= (u_1 v_1 + u_2 v_2) + (u_1 w_1 + u_2 w_2) = \mathbf{u} \cdot \mathbf{v} + \mathbf{u} \cdot \mathbf{w}$$

4. $\mathbf{v} \cdot \mathbf{v} = v_1^2 + v_2^2 = \left(\sqrt{v_1^2 + v_2^2} \right)^2 = \|\mathbf{v}\|^2$

5. $c(\mathbf{u} \cdot \mathbf{v}) = c(\langle u_1, u_2 \rangle \cdot \langle v_1, v_2 \rangle)$
$$= c(u_1 v_1 + u_2 v_2)$$
$$= (cu_1)v_1 + (cu_2)v_2$$
$$= \langle cu_1, cu_2 \rangle \cdot \langle v_1, v_2 \rangle$$
$$= c\mathbf{u} \cdot \mathbf{v}$$

Angle Between Two Vectors *(p. 309)*

If θ is the angle between two nonzero vectors **u** and **v**, then $\cos \theta = \dfrac{\mathbf{u} \cdot \mathbf{v}}{\|\mathbf{u}\| \, \|\mathbf{v}\|}$.

Proof

Consider the triangle determined by vectors **u**, **v**, and $\mathbf{v} - \mathbf{u}$, as shown in the figure. By the Law of Cosines, you can write

$$\|\mathbf{v} - \mathbf{u}\|^2 = \|\mathbf{u}\|^2 + \|\mathbf{v}\|^2 - 2\|\mathbf{u}\| \, \|\mathbf{v}\| \cos \theta$$

$$(\mathbf{v} - \mathbf{u}) \cdot (\mathbf{v} - \mathbf{u}) = \|\mathbf{u}\|^2 + \|\mathbf{v}\|^2 - 2\|\mathbf{u}\| \, \|\mathbf{v}\| \cos \theta$$

$$(\mathbf{v} - \mathbf{u}) \cdot \mathbf{v} - (\mathbf{v} - \mathbf{u}) \cdot \mathbf{u} = \|\mathbf{u}\|^2 + \|\mathbf{v}\|^2 - 2\|\mathbf{u}\| \, \|\mathbf{v}\| \cos \theta$$

$$\mathbf{v} \cdot \mathbf{v} - \mathbf{u} \cdot \mathbf{v} - \mathbf{v} \cdot \mathbf{u} + \mathbf{u} \cdot \mathbf{u} = \|\mathbf{u}\|^2 + \|\mathbf{v}\|^2 - 2\|\mathbf{u}\| \, \|\mathbf{v}\| \cos \theta$$

$$\|\mathbf{v}\|^2 - 2\mathbf{u} \cdot \mathbf{v} + \|\mathbf{u}\|^2 = \|\mathbf{u}\|^2 + \|\mathbf{v}\|^2 - 2\|\mathbf{u}\| \, \|\mathbf{v}\| \cos \theta$$

$$\cos \theta = \frac{\mathbf{u} \cdot \mathbf{v}}{\|\mathbf{u}\| \, \|\mathbf{v}\|}.$$

Problem Solving

This collection of thought-provoking and challenging exercises further explores and expands upon concepts learned in this chapter.

1. In the figure, a beam of light is directed at the blue mirror, reflected to the red mirror, and then reflected back to the blue mirror. Find the distance PT that the light travels from the red mirror back to the blue mirror.

2. A triathlete sets a course to swim S 25° E from a point on shore to a buoy $\frac{3}{4}$ mile away. After swimming 300 yards through a strong current, the triathlete is off course at a bearing of S 35° E. Find the bearing and distance the triathlete needs to swim to correct her course.

3. A hiking party is lost in a national park. Two ranger stations have received an emergency SOS signal from the party. Station B is 75 miles due east of station A. The bearing from station A to the signal is S 60° E and the bearing from station B to the signal is S 75° W.

   (a) Draw a diagram that gives a visual representation of the problem.

   (b) Find the distance from each station to the SOS signal.

   (c) A rescue party is in the park 20 miles from station A at a bearing of S 80° E. Find the distance and the bearing the rescue party must travel to reach the lost hiking party.

4. You are seeding a triangular courtyard. One side of the courtyard is 52 feet long and another side is 46 feet long. The angle opposite the 52-foot side is 65°.

   (a) Draw a diagram that gives a visual representation of the problem.

   (b) How long is the third side of the courtyard?

   (c) One bag of grass covers an area of 50 square feet. How many bags of grass will you need to cover the courtyard?

5. For each pair of vectors, find the following.

   (i) $\|\mathbf{u}\|$ (ii) $\|\mathbf{v}\|$ (iii) $\|\mathbf{u} + \mathbf{v}\|$

   (iv) $\left\|\dfrac{\mathbf{u}}{\|\mathbf{u}\|}\right\|$ (v) $\left\|\dfrac{\mathbf{v}}{\|\mathbf{v}\|}\right\|$ (vi) $\left\|\dfrac{\mathbf{u} + \mathbf{v}}{\|\mathbf{u} + \mathbf{v}\|}\right\|$

   (a) $\mathbf{u} = \langle 1, -1 \rangle$ (b) $\mathbf{u} = \langle 0, 1 \rangle$
        $\mathbf{v} = \langle -1, 2 \rangle$ $\mathbf{v} = \langle 3, -3 \rangle$

   (c) $\mathbf{u} = \langle 1, \frac{1}{2} \rangle$ (d) $\mathbf{u} = \langle 2, -4 \rangle$
        $\mathbf{v} = \langle 2, 3 \rangle$ $\mathbf{v} = \langle 5, 5 \rangle$

6. A skydiver is falling at a constant downward velocity of 120 miles per hour. In the figure, vector $\mathbf{u}$ represents the skydiver's velocity. A steady breeze pushes the skydiver to the east at 40 miles per hour. Vector $\mathbf{v}$ represents the wind velocity.

   (a) Write the vectors $\mathbf{u}$ and $\mathbf{v}$ in component form.

   (b) Let $\mathbf{s} = \mathbf{u} + \mathbf{v}$. Use the figure to sketch $\mathbf{s}$. To print an enlarged copy of the graph, go to the website, www.mathgraphs.com.

   (c) Find the magnitude of $\mathbf{s}$. What information does the magnitude give you about the skydiver's fall?

   (d) If there were no wind, the skydiver would fall in a path perpendicular to the ground. At what angle to the ground is the path of the skydiver when the skydiver is affected by the 40 mile per hour wind from due west?

   (e) The skydiver is blown to the west at 30 miles per hour. Draw a new figure that gives a visual representation of the problem and find the skydiver's new velocity.

7. Write the vector **w** in terms of **u** and **v**, given that the terminal point of **w** bisects the line segment (see figure).

8. Prove that if **u** is orthogonal to **v** and **w**, then **u** is orthogonal to

$$c\mathbf{v} + d\mathbf{w}$$

for any scalars c and d (see figure).

9. Two forces of the same magnitude $\mathbf{F}_1$ and $\mathbf{F}_2$ act at angles θ_1 and θ_2, respectively. Use a diagram to compare the work done by $\mathbf{F}_1$ with the work done by $\mathbf{F}_2$ in moving along the vector PQ if

(a) $\theta_1 = -\theta_2$

(b) $\theta_1 = 60°$ and $\theta_2 = 30°$.

10. Four basic forces are in action during flight: weight, lift, thrust, and drag. To fly through the air, an object must overcome its own *weight*. To do this, it must create an upward force called *lift*. To generate lift, a forward motion called *thrust* is needed. The thrust must be great enough to overcome air resistance, which is called *drag*.

For a commercial jet aircraft, a quick climb is important to maximize efficiency, because the performance of an aircraft at high altitudes is enhanced. In addition, it is necessary to clear obstacles such as buildings and mountains and reduce noise in residential areas. In the diagram, the angle θ is called the climb angle. The velocity of the plane can be represented by a vector **v** with a vertical component $\|\mathbf{v}\| \sin \theta$ (called climb speed) and a horizontal component $\|\mathbf{v}\| \cos \theta$, where $\|\mathbf{v}\|$ is the speed of the plane.

When taking off, a pilot must decide how much of the thrust to apply to each component. The more the thrust is applied to the horizontal component, the faster the airplane will gain speed. The more the thrust is applied to the vertical component, the quicker the airplane will climb.

FIGURE FOR 10

(a) Complete the table for an airplane that has a speed of $\|\mathbf{v}\| = 100$ miles per hour.

θ	0.5°	1.0°	1.5°	2.0°	2.5°	3.0°
$\|\mathbf{v}\| \sin \theta$						
$\|\mathbf{v}\| \cos \theta$						

(b) Does an airplane's speed equal the sum of the vertical and horizontal components of its velocity? If not, how could you find the speed of an airplane whose velocity components were known?

(c) Use the result of part (b) to find the speed of an airplane with the given velocity components.

(i) $\|\mathbf{v}\| \sin \theta = 5.235$ miles per hour
$\|\mathbf{v}\| \cos \theta = 149.909$ miles per hour

(ii) $\|\mathbf{v}\| \sin \theta = 10.463$ miles per hour
$\|\mathbf{v}\| \cos \theta = 149.634$ miles per hour

Complex Numbers

Concepts of complex numbers can be used to create beautiful pictures called fractals.

Gregory Sams/SPL PhotoResearchers, Inc.

SELECTED APPLICATIONS

Concepts of complex numbers have many real-life applications. The applications listed below represent a small sample of the applications in this chapter.

4.1 Complex Numbers

The Imaginary Unit i

Some quadratic equations have no real solutions. For instance, the quadratic equation $x^2 + 1 = 0$ has no real solution because there is no real number x that can be squared to produce -1. To overcome this deficiency, mathematicians created an expanded system of numbers using the **imaginary unit i,** defined as

$$i = \sqrt{-1} \qquad \text{Imaginary unit}$$

where $i^2 = -1$. By adding real numbers to real multiples of this imaginary unit, the set of **complex numbers** is obtained. Each complex number can be written in the **standard form $a + bi$.** For instance, the standard form of the complex number $-5 + \sqrt{-9}$ is $-5 + 3i$ because

$$-5 + \sqrt{-9} = -5 + \sqrt{3^2(-1)} = -5 + 3\sqrt{-1} = -5 + 3i.$$

In the standard form $a + bi$, the real number a is called the **real part** of the **complex number $a + bi$,** and the number bi (where b is a real number) is called the **imaginary part** of the complex number.

Definition of a Complex Number

If a and b are real numbers, the number $a + bi$ is a **complex number,** and it is said to be written in **standard form.** If $b = 0$, the number $a + bi = a$ is a real number. If $b \neq 0$, the number $a + bi$ is called an **imaginary number.** A number of the form bi, where $b \neq 0$, is called a **pure imaginary number.**

The set of real numbers is a subset of the set of complex numbers, as shown in Figure 4.1. This is true because every real number a can be written as a complex number using $b = 0$. That is, for every real number a, you can write $a = a + 0i$.

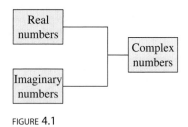

FIGURE 4.1

Equality of Complex Numbers

Two complex numbers $a + bi$ and $c + di$, written in standard form, are equal to each other

$$a + bi = c + di \qquad \text{Equality of two complex numbers}$$

if and only if $a = c$ and $b = d$.

Operations with Complex Numbers

To add (or subtract) two complex numbers, you add (or subtract) the real and imaginary parts of the numbers separately.

For each operation on complex numbers, you can show the parallel operation on polynomials.

Addition and Subtraction of Complex Numbers

If $a + bi$ and $c + di$ are two complex numbers written in standard form, their sum and difference are defined as follows.

Sum: $(a + bi) + (c + di) = (a + c) + (b + d)i$

Difference: $(a + bi) - (c + di) = (a - c) + (b - d)i$

The **additive identity** in the complex number system is zero (the same as in the real number system). Furthermore, the **additive inverse** of the complex number $a + bi$ is

$$-(a + bi) = -a - bi.$$ Additive inverse

So, you have

$$(a + bi) + (-a - bi) = 0 + 0i = 0.$$

Example 1   Adding and Subtracting Complex Numbers

a. $(4 + 7i) + (1 - 6i) = 4 + 7i + 1 - 6i$ Remove parentheses.

$= (4 + 1) + (7i - 6i)$ Group like terms.

$= 5 + i$ Write in standard form.

b. $(1 + 2i) - (4 + 2i) = 1 + 2i - 4 - 2i$ Remove parentheses.

$= (1 - 4) + (2i - 2i)$ Group like terms.

$= -3 + 0$ Simplify.

$= -3$ Write in standard form.

c. $3i - (-2 + 3i) - (2 + 5i) = 3i + 2 - 3i - 2 - 5i$

$= (2 - 2) + (3i - 3i - 5i)$

$= 0 - 5i$

$= -5i$

d. $(3 + 2i) + (4 - i) - (7 + i) = 3 + 2i + 4 - i - 7 - i$

$= (3 + 4 - 7) + (2i - i - i)$

$= 0 + 0i$

$= 0$

✓*CHECKPOINT*   Now try Exercise 17.

Note in Examples 1(b) and 1(d) that the sum of two complex numbers can be a real number.

Exploration

Complete the following.

$i^1 = i$	$i^7 = \blacksquare$
$i^2 = -1$	$i^8 = \blacksquare$
$i^3 = -i$	$i^9 = \blacksquare$
$i^4 = 1$	$i^{10} = \blacksquare$
$i^5 = \blacksquare$	$i^{11} = \blacksquare$
$i^6 = \blacksquare$	$i^{12} = \blacksquare$

What pattern do you see? Write a brief description of how you would find i raised to any positive integer power.

Many of the properties of real numbers are valid for complex numbers as well. Here are some examples.

Associative Properties of Addition and Multiplication

Commutative Properties of Addition and Multiplication

Distributive Property of Multiplication Over Addition

Notice below how these properties are used when two complex numbers are multiplied.

$$(a + bi)(c + di) = a(c + di) + bi(c + di) \qquad \text{Distributive Property}$$
$$= ac + (ad)i + (bc)i + (bd)i^2 \qquad \text{Distributive Property}$$
$$= ac + (ad)i + (bc)i + (bd)(-1) \qquad i^2 = -1$$
$$= ac - bd + (ad)i + (bc)i \qquad \text{Commutative Property}$$
$$= (ac - bd) + (ad + bc)i \qquad \text{Associative Property}$$

Rather than trying to memorize this multiplication rule, you should simply remember how the Distributive Property is used to multiply two complex numbers.

Example 2   Multiplying Complex Numbers

a.
$$4(-2 + 3i) = 4(-2) + 4(3i) \qquad \text{Distributive Property}$$
$$= -8 + 12i \qquad \text{Simplify.}$$

b.
$$(2 - i)(4 + 3i) = 2(4 + 3i) - i(4 + 3i) \qquad \text{Distributive Property}$$
$$= 8 + 6i - 4i - 3i^2 \qquad \text{Distributive Property}$$
$$= 8 + 6i - 4i - 3(-1) \qquad i^2 = -1$$
$$= (8 + 3) + (6i - 4i) \qquad \text{Group like terms.}$$
$$= 11 + 2i \qquad \text{Write in standard form.}$$

c.
$$(3 + 2i)(3 - 2i) = 3(3 - 2i) + 2i(3 - 2i) \qquad \text{Distributive Property}$$
$$= 9 - 6i + 6i - 4i^2 \qquad \text{Distributive Property}$$
$$= 9 - 6i + 6i - 4(-1) \qquad i^2 = -1$$
$$= 9 + 4 \qquad \text{Simplify.}$$
$$= 13 \qquad \text{Write in standard form.}$$

d.
$$(3 + 2i)^2 = (3 + 2i)(3 + 2i) \qquad \text{Square of a binomial}$$
$$= 3(3 + 2i) + 2i(3 + 2i) \qquad \text{Distributive Property}$$
$$= 9 + 6i + 6i + 4i^2 \qquad \text{Distributive Property}$$
$$= 9 + 6i + 6i + 4(-1) \qquad i^2 = -1$$
$$= 9 + 12i - 4 \qquad \text{Simplify.}$$
$$= 5 + 12i \qquad \text{Write in standard form.}$$

✓**CHECKPOINT**   Now try Exercise 27.

STUDY TIP

The procedure described above is similar to multiplying two polynomials and combining like terms, as in the FOIL Method. For instance, you can use the FOIL Method to multiply the two complex numbers from Example 2(b).

$$\overset{\text{F}\quad\text{O}\quad\text{I}\quad\text{L}}{(2 - i)(4 + 3i) = 8 + 6i - 4i - 3i^2}$$

Complex Conjugates

Notice in Example 2(c) that the product of two complex numbers can be a real number. This occurs with pairs of complex numbers of the form $a + bi$ and $a - bi$, called **complex conjugates.**

A comparison with the method of rationalizing denominators (Section P.2) may be helpful.

$$(a + bi)(a - bi) = a^2 - abi + abi - b^2i^2$$
$$= a^2 - b^2(-1)$$
$$= a^2 + b^2$$

Example 3   Multiplying Conjugates

Multiply each complex number by its complex conjugate.

a. $1 + i$ **b.** $4 - 3i$

Solution

a. The complex conjugate of $1 + i$ is $1 - i$.
$$(1 + i)(1 - i) = 1^2 - i^2 = 1 - (-1) = 2$$

b. The complex conjugate of $4 - 3i$ is $4 + 3i$.
$$(4 - 3i)(4 + 3i) = 4^2 - (3i)^2 = 16 - 9i^2 = 16 - 9(-1) = 25$$

✓*CHECKPOINT*   Now try Exercise 37.

To write the quotient of $a + bi$ and $c + di$ in standard form, where c and d are not both zero, multiply the numerator and denominator by the complex conjugate of the *denominator* to obtain

$$\frac{a + bi}{c + di} = \frac{a + bi}{c + di}\left(\frac{c - di}{c - di}\right)$$

$$= \frac{(ac + bd) + (bc - ad)i}{c^2 + d^2}. \qquad \text{Standard form}$$

STUDY TIP

Note that when you multiply the numerator and denominator of a quotient of complex numbers by

$$\frac{c - di}{c - di}$$

you are actually multiplying the quotient by a form of 1. You are not changing the original expression, you are only creating an expression that is equivalent to the original expression.

Example 4   Writing a Quotient of Complex Numbers in Standard Form

$$\frac{2 + 3i}{4 - 2i} = \frac{2 + 3i}{4 - 2i}\left(\frac{4 + 2i}{4 + 2i}\right) \qquad \text{Multiply numerator and denominator by complex conjugate of denominator.}$$

$$= \frac{8 + 4i + 12i + 6i^2}{16 - 4i^2} \qquad \text{Expand.}$$

$$= \frac{8 - 6 + 16i}{16 + 4} \qquad i^2 = -1$$

$$= \frac{2 + 16i}{20} \qquad \text{Simplify.}$$

$$= \frac{1}{10} + \frac{4}{5}i \qquad \text{Write in standard form.}$$

✓*CHECKPOINT*   Now try Exercise 49.

Complex Solutions of Quadratic Equations

When using the Quadratic Formula to solve a quadratic equation, you often obtain a result such as $\sqrt{-3}$, which you know is not a real number. By factoring out $i = \sqrt{-1}$, you can write this number in standard form.

$$\sqrt{-3} = \sqrt{3(-1)} = \sqrt{3}\sqrt{-1} = \sqrt{3}i$$

The number $\sqrt{3}i$ is called the *principal square root* of -3.

STUDY TIP

The definition of principal square root uses the rule

$$\sqrt{ab} = \sqrt{a}\sqrt{b}$$

for $a > 0$ and $b < 0$. This rule is not valid if *both* a and b are negative. For example,

$$\sqrt{-5}\sqrt{-5} = \sqrt{5(-1)}\sqrt{5(-1)}$$
$$= \sqrt{5}i\sqrt{5}i$$
$$= \sqrt{25}i^2$$
$$= 5i^2 = -5$$

whereas

$$\sqrt{(-5)(-5)} = \sqrt{25} = 5.$$

To avoid problems with square roots of negative numbers, be sure to convert complex numbers to standard form *before* multiplying.

> **Principal Square Root of a Negative Number**
>
> If a is a positive number, the **principal square root** of the negative number $-a$ is defined as
>
> $$\sqrt{-a} = \sqrt{a}i.$$

Example 5 **Writing Complex Numbers in Standard Form**

a. $\sqrt{-3}\sqrt{-12} = \sqrt{3}i\sqrt{12}i = \sqrt{36}i^2 = 6(-1) = -6$

b. $\sqrt{-48} - \sqrt{-27} = \sqrt{48}i - \sqrt{27}i = 4\sqrt{3}i - 3\sqrt{3}i = \sqrt{3}i$

c. $\left(-1 + \sqrt{-3}\right)^2 = \left(-1 + \sqrt{3}i\right)^2$
$$= (-1)^2 - 2\sqrt{3}i + \left(\sqrt{3}\right)^2(i^2)$$
$$= 1 - 2\sqrt{3}i + 3(-1)$$
$$= -2 - 2\sqrt{3}i$$

✓**CHECKPOINT** Now try Exercise 59.

Example 6 **Complex Solutions of a Quadratic Equation**

Solve (a) $x^2 + 4 = 0$ and (b) $3x^2 - 2x + 5 = 0$.

Solution

a. $x^2 + 4 = 0$ Write original equation.

       $x^2 = -4$ Subtract 4 from each side.

       $x = \pm 2i$ Extract square roots.

b. $3x^2 - 2x + 5 = 0$ Write original equation.

$$x = \frac{-(-2) \pm \sqrt{(-2)^2 - 4(3)(5)}}{2(3)}$$ Quadratic Formula

$$= \frac{2 \pm \sqrt{-56}}{6}$$ Simplify.

$$= \frac{2 \pm 2\sqrt{14}i}{6}$$ Write $\sqrt{-56}$ in standard form.

$$= \frac{1}{3} \pm \frac{\sqrt{14}}{3}i$$ Write in standard form.

✓**CHECKPOINT** Now try Exercise 65.

Activities

1. Find the product and write the result in standard form.

$$\left(4 - \sqrt{-9}\right)\left(2 + \sqrt{-9}\right)$$

Answer: $17 + 6i$

2. Write $\dfrac{3 + i}{i}$ in standard form.

Answer: $1 - 3i$

3. Use the Quadratic Formula to solve $2x^2 - 5x + 7 = 0$.

Answer: $x = \dfrac{5}{4} \pm \dfrac{\sqrt{31}}{4}i$

4.1 | Exercises

VOCABULARY CHECK:

1. Match the type of complex number with its definition.

    (a) Real Number   (i) $a + bi, \quad a \neq 0, \quad b \neq 0$

    (b) Imaginary number   (ii) $a + bi, \quad a = 0, \quad b \neq 0$

    (c) Pure imaginary number   (iii) $a + bi, \quad b = 0$

In Exercises 2–4, fill in the blanks.

2. The imaginary unit i is defined as $i = $ _____, where $i^2 = $ _____.

3. If a is a positive number, the _____ _____ root of the negative number $-a$ is defined as $\sqrt{-a} = \sqrt{a}\,i$.

4. The numbers $a + bi$ and $a - bi$ are called _____ _____, and their product is a real number $a^2 + b^2$.

PREREQUISITE SKILLS REVIEW: Practice and review algebra skills needed for this section at **www.Eduspace.com.**

In Exercises 1–4, find real numbers a and b such that the equation is true.

1. $a + bi = -10 + 6i$
2. $a + bi = 13 + 4i$
3. $(a - 1) + (b + 3)i = 5 + 8i$
4. $(a + 6) + 2bi = 6 - 5i$

In Exercises 5–16, write the complex number in standard form.

5. $4 + \sqrt{-9}$
6. $3 + \sqrt{-16}$
7. $2 - \sqrt{-27}$
8. $1 + \sqrt{-8}$
9. $\sqrt{-75}$
10. $\sqrt{-4}$
11. 8
12. 45
13. $-6i + i^2$
14. $-4i^2 + 2i$
15. $\sqrt{-0.09}$
16. $\sqrt{-0.0004}$

In Exercises 17–26, perform the addition or subtraction and write the result in standard form.

17. $(5 + i) + (6 - 2i)$
18. $(13 - 2i) + (-5 + 6i)$
19. $(8 - i) - (4 - i)$
20. $(3 + 2i) - (6 + 13i)$
21. $\left(-2 + \sqrt{-8}\right) + \left(5 - \sqrt{-50}\right)$
22. $\left(8 + \sqrt{-18}\right) - \left(4 + 3\sqrt{2}i\right)$
23. $13i - (14 - 7i)$
24. $22 + (-5 + 8i) + 10i$
25. $-\left(\frac{3}{2} + \frac{5}{2}i\right) + \left(\frac{5}{3} + \frac{11}{3}i\right)$
26. $(1.6 + 3.2i) + (-5.8 + 4.3i)$

In Exercises 27–36, perform the operation and write the result in standard form.

27. $(1 + i)(3 - 2i)$
28. $(6 - 2i)(2 - 3i)$
29. $6i(5 - 2i)$
30. $-8i(9 + 4i)$
31. $\left(\sqrt{14} + \sqrt{10}i\right)\left(\sqrt{14} - \sqrt{10}i\right)$
32. $\left(\sqrt{3} + \sqrt{15}i\right)\left(\sqrt{3} - \sqrt{15}i\right)$
33. $(4 + 5i)^2$
34. $(2 - 3i)^2$
35. $(2 + 3i)^2 + (2 - 3i)^2$
36. $(1 - 2i)^2 - (1 + 2i)^2$

In Exercises 37–44, write the complex conjugate of the complex number. Then multiply the number by its complex conjugate.

37. $6 + 3i$
38. $7 - 12i$
39. $-1 - \sqrt{5}i$
40. $-3 + \sqrt{2}i$
41. $\sqrt{-20}$
42. $\sqrt{-15}$
43. $\sqrt{8}$
44. $1 + \sqrt{8}$

In Exercises 45–54, write the quotient in standard form.

45. $\dfrac{5}{i}$
46. $-\dfrac{14}{2i}$
47. $\dfrac{2}{4 - 5i}$
48. $\dfrac{5}{1 - i}$
49. $\dfrac{3 + i}{3 - i}$
50. $\dfrac{6 - 7i}{1 - 2i}$
51. $\dfrac{6 - 5i}{i}$
52. $\dfrac{8 + 16i}{2i}$
53. $\dfrac{3i}{(4 - 5i)^2}$
54. $\dfrac{5i}{(2 + 3i)^2}$

In Exercises 55–58, perform the operation and write the result in standard form.

55. $\dfrac{2}{1 + i} - \dfrac{3}{1 - i}$
56. $\dfrac{2i}{2 + i} + \dfrac{5}{2 - i}$
57. $\dfrac{i}{3 - 2i} + \dfrac{2i}{3 + 8i}$
58. $\dfrac{1 + i}{i} - \dfrac{3}{4 - i}$

In Exercises 59–64, write the complex number in standard form.

59. $\sqrt{-6} \cdot \sqrt{-2}$ **60.** $\sqrt{-5} \cdot \sqrt{-10}$

61. $\left(\sqrt{-10}\right)^2$ **62.** $\left(\sqrt{-75}\right)^2$

63. $\left(3 + \sqrt{-5}\right)\left(7 - \sqrt{-10}\right)$ **64.** $\left(2 - \sqrt{-6}\right)^2$

In Exercises 65–74, use the Quadratic Formula to solve the quadratic equation.

65. $x^2 - 2x + 2 = 0$ **66.** $x^2 + 6x + 10 = 0$

67. $4x^2 + 16x + 17 = 0$ **68.** $9x^2 - 6x + 37 = 0$

69. $4x^2 + 16x + 15 = 0$ **70.** $16t^2 - 4t + 3 = 0$

71. $\frac{3}{2}x^2 - 6x + 9 = 0$ **72.** $\frac{7}{8}x^2 - \frac{3}{4}x + \frac{5}{16} = 0$

73. $1.4x^2 - 2x - 10 = 0$ **74.** $4.5x^2 - 3x + 12 = 0$

In Exercises 75–82, simplify the complex number and write it in standard form.

75. $-6i^3 + i^2$ **76.** $4i^2 - 2i^3$

77. $-5i^5$ **78.** $(-i)^3$

79. $\left(\sqrt{-75}\right)^3$ **80.** $\left(\sqrt{-2}\right)^6$

81. $\dfrac{1}{i^3}$ **82.** $\dfrac{1}{(2i)^3}$

Model It

83. *Impedance*   The opposition to current in an electrical circuit is called its impedance. The impedance z in a parallel circuit with two pathways satisfies the equation

$$\frac{1}{z} = \frac{1}{z_1} + \frac{1}{z_2}$$

where z_1 is the impedance (in ohms) of pathway 1 and z_2 is the impedance of pathway 2.

(a) The impedance of each pathway in a parallel circuit is found by adding the impedances of all components in the pathway. Use the table to find z_1 and z_2.

(b) Find the impedance z.

	Resistor	Inductor	Capacitor
Symbol	$-\!\!\bigwedge\!\!-$ $a\Omega$	$-\!\!\mathrm{mmm}\!\!-$ $b\Omega$	$-\!\!\mid\mid\!\!-$ $c\Omega$
Impedance	a	bi	$-ci$

---

84. Cube each complex number.

    (a) 2 (b) $-1 + \sqrt{3}\,i$ (c) $-1 - \sqrt{3}\,i$

85. Raise each complex number to the fourth power.

    (a) 2 (b) -2 (c) $2i$ (d) $-2i$

86. Write each of the powers of i as i, $-i$, 1, or -1.

    (a) i^{40} (b) i^{25} (c) i^{50} (d) i^{67}

Synthesis

True or False?   **In Exercises 87–89, determine whether the statement is true or false. Justify your answer.**

87. There is no complex number that is equal to its complex conjugate.

88. $-i\sqrt{6}$ is a solution of $x^4 - x^2 + 14 = 56$.

89. $i^{44} + i^{150} - i^{74} - i^{109} + i^{61} = -1$

90. *Error Analysis*   Describe the error.

$$\sqrt{-6}\sqrt{-6} = \sqrt{(-6)(-6)} = \sqrt{36} = 6$$

91. *Proof*   Prove that the complex conjugate of the product of two complex numbers $a_1 + b_1 i$ and $a_2 + b_2 i$ is the product of their complex conjugates.

92. *Proof*   Prove that the complex conjugate of the sum of two complex numbers $a_1 + b_1 i$ and $a_2 + b_2 i$ is the sum of their complex conjugates.

Skills Review

In Exercises 93–96, perform the operation and write the result in standard form.

93. $(4 + 3x) + (8 - 6x - x^2)$

94. $(x^3 - 3x^2) - (6 - 2x - 4x^2)$

95. $\left(3x - \frac{1}{2}\right)(x + 4)$ **96.** $(2x - 5)^2$

In Exercises 97–100, solve the equation and check your solution.

97. $-x - 12 = 19$ **98.** $8 - 3x = -34$

99. $4(5x - 6) - 3(6x + 1) = 0$

100. $5[x - (3x + 11)] = 20x - 15$

101. *Volume of an Oblate Spheroid*

    Solve for a: $V = \frac{4}{3}\pi a^2 b$

102. *Newton's Law of Universal Gravitation*

    Solve for r: $F = \alpha \dfrac{m_1 m_2}{r^2}$

103. *Mixture Problem*   A five-liter container contains a mixture with a concentration of 50%. How much of this mixture must be withdrawn and replaced by 100% concentrate to bring the mixture up to 60% concentration?

4.2   Complex Solutions of Equations

What you should learn

- Determine the numbers of solutions of polynomial equations.
- Find solutions of polynomial equations.
- Find zeros of polynomial functions and find polynomial functions given the zeros of the functions.

Why you should learn it

Finding zeros of polynomial functions is an important part of solving real-life problems. For instance, in Exercise 79 on page 346, the zeros of a polynomial function can help you analyze the profit function for a microwave oven.

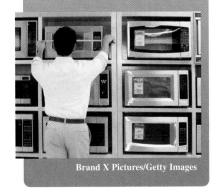

Brand X Pictures/Getty Images

The Number of Solutions of a Polynomial Equation

The Fundamental Theorem of Algebra implies that a polynomial equation of degree n has precisely n solutions in the complex number system. These solutions can be real or complex and may be repeated. The Fundamental Theorem of Algebra and the Linear Factorization Theorem are listed below for your review. For a proof of the Linear Factorization Theorem, see Proofs in Mathematics on page 364.

The Fundamental Theorem of Algebra

If $f(x)$ is a polynomial of degree n, where $n > 0$, then f has at least one zero in the complex number system.

Note that finding zeros of a polynomial function f is equivalent to finding solutions to the polynomial equation $f(x) = 0$.

Linear Factorization Theorem

If $f(x)$ is a polynomial of degree n, where $n > 0$, then f has precisely n linear factors $f(x) = a_n(x - c_1)(x - c_2) \cdots (x - c_n)$, where $c_1, c_2, \ldots, c_n$ are complex numbers.

Example 1   Solutions of Polynomial Equations

a. The first-degree equation $x - 2 = 0$ has exactly *one* solution: $x = 2$.

b. The second-degree equation

$$x^2 - 6x + 9 = 0 \qquad \text{Second-degree equation}$$
$$(x - 3)(x - 3) = 0 \qquad \text{Factor.}$$

has exactly *two* solutions: $x = 3$ and $x = 3$. (This is called a *repeated solution*.)

c. The third-degree equation

$$x^3 + 4x = 0 \qquad \text{Third-degree equation}$$
$$x(x - 2i)(x + 2i) = 0 \qquad \text{Factor.}$$

has exactly *three* solutions: $x = 0$, $x = 2i$, and $x = -2i$.

d. The fourth-degree equation

$$x^4 - 1 = 0 \qquad \text{Fourth-degree equation}$$
$$(x - 1)(x + 1)(x - i)(x + i) = 0 \qquad \text{Factor.}$$

has exactly *four* solutions: $x = 1$, $x = -1$, $x = i$, and $x = -i$.

✔CHECKPOINT   Now try Exercise 1.

You can use a graph to check the number of *real* solutions of an equation. As shown in Figure 4.2, the graph of $f(x) = x^4 - 1$ has two x-intercepts, which implies that the equation has two real solutions.

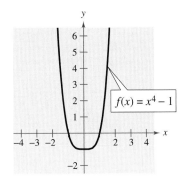

$f(x) = x^4 - 1$

FIGURE 4.2

Every second-degree equation, $ax^2 + bx + c = 0$, has precisely two solutions given by the Quadratic Formula.

$$x = \frac{-b \pm \sqrt{b^2 - 4ac}}{2a}$$

The expression inside the radical, $b^2 - 4ac$, is called the **discriminant,** and can be used to determine whether the solutions are real, repeated, or complex.

1. If $b^2 - 4ac < 0$, the equation has two complex solutions.
2. If $b^2 - 4ac = 0$, the equation has one repeated real solution.
3. If $b^2 - 4ac > 0$, the equation has two distinct real solutions.

Example 2   Using the Discriminant

Use the discriminant to find the number of real solutions of each equation.

a. $4x^2 - 20x + 25 = 0$ **b.** $13x^2 + 7x + 2 = 0$ **c.** $5x^2 - 8x = 0$

Solution

a. For this equation, $a = 4$, $b = -20$, and $c = 25$. So, the discriminant is

$$b^2 - 4ac = (-20)^2 - 4(4)(25) = 400 - 400 = 0.$$

Because the discriminant is zero, there is one repeated real solution.

b. For this equation, $a = 13$, $b = 7$, and $c = 2$. So, the discriminant is

$$b^2 - 4ac = 7^2 - 4(13)(2) = 49 - 104 = -55.$$

Because the discriminant is negative, there are two complex solutions.

c. For this equation, $a = 5$, $b = -8$, and $c = 0$. So, the discriminant is

$$b^2 - 4ac = (-8)^2 - 4(5)(0) = 64 - 0 = 64.$$

Because the discriminant is positive, there are two distinct real solutions.

✓CHECKPOINT   Now try Exercise 5.

Figure 4.3 shows the graphs of the functions corresponding to the equations in Example 2. Notice that with one repeated solution, the graph *touches* the x-axis at its x-intercept. With two complex solutions, the graph has no x-intercepts. With two real solutions, the graph *crosses* the x-axis at its x-intercepts.

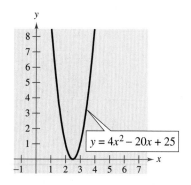

(a) **Repeated real solution**

FIGURE 4.3

(b) **No real solution**

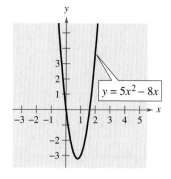

(c) **Two distinct real solutions**

Finding Solutions of Polynomial Equations

Example 3   Solving a Quadratic Equation

Solve $x^2 + 2x + 2 = 0$. Write complex solutions in standard form.

Solution

Using $a = 1$, $b = 2$, and $c = 2$, you can apply the Quadratic Formula as follows.

$$x = \frac{-b \pm \sqrt{b^2 - 4ac}}{2a} \qquad \text{Quadratic Formula}$$

$$= \frac{-2 \pm \sqrt{2^2 - 4(1)(2)}}{2(1)} \qquad \text{Substitute 1 for } a, 2 \text{ for } b, \text{ and 2 for } c.$$

$$= \frac{-2 \pm \sqrt{-4}}{2} \qquad \text{Simplify.}$$

$$= \frac{-2 \pm 2i}{2} \qquad \text{Simplify.}$$

$$= -1 \pm i \qquad \text{Write in standard form.}$$

✓CHECKPOINT   Now try Exercise 19.

In Example 3, the two complex solutions are **conjugates.** That is, they are of the form $a \pm bi$. This is not a coincidence, as indicated by the following theorem.

Complex Solutions Occur in Conjugate Pairs

If $a + bi$, $b \neq 0$, is a solution of a polynomial equation with real coefficients, the conjugate $a - bi$ is also a solution of the equation.

Be sure you see that this result is true only if the polynomial has *real* coefficients. For instance, the result applies to the equation $x^2 + 1 = 0$, but not to the equation $x - i = 0$.

Example 4   Solving a Polynomial Equation

Solve $x^4 - x^2 - 20 = 0$.

Solution

$$x^4 - x^2 - 20 = 0 \qquad \text{Write original equation.}$$

$$(x^2 - 5)(x^2 + 4) = 0 \qquad \text{Partially factor.}$$

$$\left(x + \sqrt{5}\right)\left(x - \sqrt{5}\right)(x + 2i)(x - 2i) = 0 \qquad \text{Factor completely.}$$

Setting each factor equal to zero yields the solutions $x = -\sqrt{5}$, $x = \sqrt{5}$, $x = -2i$, and $x = 2i$.

✓CHECKPOINT   Now try Exercise 47.

Finding Zeros of Polynomial Functions

The problem of finding the *zeros* of a polynomial function is essentially the same problem as finding the solutions of a polynomial equation. For instance, the zeros of the polynomial function

$$f(x) = 3x^2 - 4x + 5$$

are simply the solutions of the polynomial equation

$$3x^2 - 4x + 5 = 0.$$

Example 5   Finding the Zeros of a Polynomial Function

Find all the zeros of

$$f(x) = x^4 - 3x^3 + 6x^2 + 2x - 60$$

given that $1 + 3i$ is a zero of f.

Algebraic Solution

Because complex zeros occur in conjugate pairs, you know that $1 - 3i$ is also a zero of f. This means that both

$$[x - (1 + 3i)] \quad \text{and} \quad [x - (1 - 3i)]$$

are factors of f. Multiplying these two factors produces

$$[x - (1 + 3i)][x - (1 - 3i)] = [(x - 1) - 3i][(x - 1) + 3i]$$
$$= (x - 1)^2 - 9i^2$$
$$= x^2 - 2x + 10.$$

Using long division, you can divide $x^2 - 2x + 10$ into f to obtain the following.

$$
\begin{array}{r}
x^2 - x - 6 \\
x^2 - 2x + 10 \overline{\smash{\big)}\ x^4 - 3x^3 + 6x^2 + 2x - 60} \\
\underline{x^4 - 2x^3 + 10x^2} \\
-x^3 - 4x^2 + 2x \\
\underline{-x^3 + 2x^2 - 10x} \\
-6x^2 + 12x - 60 \\
\underline{-6x^2 + 12x - 60} \\
0
\end{array}
$$

So, you have

$$f(x) = (x^2 - 2x + 10)(x^2 - x - 6)$$
$$= (x^2 - 2x + 10)(x - 3)(x + 2)$$

and you can conclude that the zeros of f are $x = 1 + 3i$, $x = 1 - 3i$, $x = 3$, and $x = -2$.

✔*CHECKPOINT*   Now try Exercise 49.

Graphical Solution

Complex zeros always occur in conjugate pairs, so you know that $1 - 3i$ is also a zero of f. Because the polynomial is a fourth-degree polynomial, you know that there are at most two other zeros of the function. Use a graphing utility to graph

$$y = x^4 - 3x^3 + 6x^2 + 2x - 60$$

as shown in Figure 4.4.

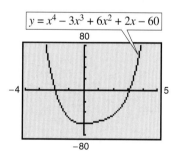

FIGURE **4.4**

You can see that -2 and 3 appear to be x-intercepts of the graph of the function. Use the *zero* or *root* feature or the *zoom* and *trace* features of the graphing utility to confim that $x = -2$ and $x = 3$ are x-intercepts of the graph. So, you can conclude that the zeros of f are $x = 1 + 3i$, $x = 1 - 3i$, $x = 3$, and $x = -2$.

Reinforce to students that a graphing utility is helpful in determining real zeros, which in turn are useful in finding complex zeros. Try this technique with the function $f(x) = x^5 + x^3 + 2x^2 - 12x + 8$, which has zeros $x = 1$, $x = -2$, $x = 2i$, and $x = -2i$. (Note that $x = 1$ is a repeated zero.)

Example 6   **Finding a Polynomial with Given Zeros**

Find a fourth-degree polynomial function with real coefficients that has -1, -1, and $3i$ as zeros.

Solution

Because $3i$ is a zero *and* the polynomial is stated to have real coefficients, you know that the conjugate $-3i$ must also be a zero. So, from the Linear Factorization Theorem, $f(x)$ can be written as

$$f(x) = a(x + 1)(x + 1)(x - 3i)(x + 3i).$$

For simplicity, let $a = 1$ to obtain

$$f(x) = (x^2 + 2x + 1)(x^2 + 9)$$

$$= x^4 + 2x^3 + 10x^2 + 18x + 9.$$

✓CHECKPOINT   Now try Exercise 61.

Example 7   **Finding a Polynomial with Given Zeros**

Find a cubic polynomial function f with real coefficients that has 2 and $1 - i$ as zeros, such that $f(1) = 3$.

Solution

Because $1 - i$ is a zero of f, so is $1 + i$. So,

$$f(x) = a(x - 2)[x - (1 - i)][x - (1 + i)]$$

$$= a(x - 2)[(x - 1) + i][(x - 1) - i]$$

$$= a(x - 2)[(x - 1)^2 - i^2]$$

$$= a(x - 2)(x^2 - 2x + 2)$$

$$= a(x^3 - 4x^2 + 6x - 4).$$

To find the value of a, use the fact that $f(1) = 3$ and obtain

$$f(1) = a[1^3 - 4(1)^2 + 6(1) - 4]$$

$$3 = -a$$

$$-3 = a.$$

So, $a = -3$ and it follows that

$$f(x) = -3(x^3 - 4x^2 + 6x - 4)$$

$$= -3x^3 + 12x^2 - 18x + 12.$$

✓CHECKPOINT   Now try Exercise 65.

Activities

1. Write $f(x) = x^4 - 16$ as a product of linear factors.

   Answer:

   $f(x) = (x - 2)(x + 2)(x - 2i)(x + 2i)$

2. Find a third-degree polynomial function with real coefficients that has 2, $3 + i$, and $3 - i$ as zeros.

   Answer: $f(x) = x^3 - 8x^2 + 22x - 20$

3. Use the zero $x = 4i$ to find all the zeros of

   $f(x) = x^4 + 4x^3 + 11x^2 + 64x - 80.$

   Answer: $-5, 1, 4i, -4i$

WRITING ABOUT **M**ATHEMATICS

Solutions, Zeros, and Intercepts Write a paragraph explaining the relationships among the solutions of a polynomial equation, the zeros of a polynomial function, and the x-intercepts of the graph of a polynomial function. Include examples in your paragraph.

4.2 Exercises

VOCABULARY CHECK: Fill in the blanks.

1. The theorem that states, if $f(x)$ is a polynomial of degree n $(n > 0)$, then f has at least one zero in the complex number system, is called the _____ Theorem of _____.

2. The theorem that states, if $f(x)$ is a polynomial of degree $n(n > 0)$, then f has exactly n linear factors of the form $f(x) = a_n(x - c_1)(x - c_2) \cdots (x - c_n)$, where $c_1, c_2, \ldots, c_n$ are complex numbers, is called the _____ _____ Theorem.

3. Two complex solutions of a polynomial equation with real coefficients are called _____.

4. The expression inside the radical of the Quadratic Formula, $b^2 - 4ac$, is called the _____ and is used to determine types of solutions of a quadratic equatio

PREREQUISITE SKILLS REVIEW: Practice and review algebra skills needed for this section at **www.Eduspace.com.**

In Exercises 1–4, determine the number of solutions of the equation in the complex number system.

1. $2x^3 + 3x + 1 = 0$
2. $x^6 + 4x^2 + 12 = 0$
3. $50 - 2x^4 = 0$
4. $14 - x + 4x^2 - 7x^5 = 0$

In Exercises 5–12, use the discriminant to determine the number of real solutions of the quadratic equation.

5. $2x^2 - 5x + 5 = 0$
6. $2x^2 - x - 1 = 0$
7. $\frac{1}{5}x^2 + \frac{6}{5}x - 8 = 0$
8. $\frac{1}{3}x^2 - 5x + 25 = 0$
9. $2x^2 - x - 15 = 0$
10. $-2x^2 + 11x - 2 = 0$
11. $x^2 + 2x + 10 = 0$
12. $x^2 - 4x + 53 = 0$

In Exercises 13–26, solve the equation. Write complex solutions in standard form.

13. $x^2 - 5 = 0$
14. $3x^2 - 1 = 0$
15. $(x + 5)^2 - 6 = 0$
16. $16 - (x - 1)^2 = 0$
17. $x^2 - 8x + 16 = 0$
18. $4x^2 + 4x + 1 = 0$
19. $x^2 + 2x + 5 = 0$
20. $54 + 16x - x^2 = 0$
21. $4x^2 - 4x + 5 = 0$
22. $4x^2 - 4x + 21 = 0$
23. $230 + 20x - 0.5x^2 = 0$
24. $125 - 30x + 0.4x^2 = 0$
25. $8 + (x + 3)^2 = 0$
26. $6 - (x - 1)^2 = 0$

Graphical and Analytical Analysis **In Exercises 27–30, (a) use a graphing utility to graph the function, (b) find all the zeros of the function, and (c) describe the relationship between the number of real zeros and the number of x-intercepts of the graph.**

27. $f(x) = x^3 - 4x^2 + x - 4$
28. $f(x) = x^3 - 4x^2 - 4x + 16$
29. $f(x) = x^4 + 4x^2 + 4$
30. $f(x) = x^4 - 3x^2 - 4$

In Exercises 31–48, find all the zeros of the function and write the polynomial as a product of linear factors.

31. $f(x) = x^2 + 25$
32. $f(x) = x^2 - x + 56$
33. $h(x) = x^2 - 4x + 1$
34. $g(x) = x^2 + 10x + 23$
35. $f(x) = x^4 - 81$
36. $f(y) = y^4 - 625$
37. $f(z) = z^2 - 2z + 2$
38. $h(x) = x^2 - 6x - 10$
39. $g(x) = x^3 + 3x^2 - 3x - 9$
40. $f(x) = x^3 - 8x^2 - 12x + 96$
41. $h(x) = x^3 - 4x^2 + 16x - 64$
42. $h(x) = x^3 + 5x^2 + 2x + 10$
43. $f(x) = 2x^3 - x^2 + 36x - 18$
44. $g(x) = 4x^3 + 3x^2 + 96x + 72$
45. $g(x) = x^4 - 4x^3 + 36x^2 - 144x$
46. $h(x) = x^4 + x^3 + 100x^2 + 100x$
47. $f(x) = x^4 + 10x^2 + 9$
48. $f(x) = x^4 + 29x^2 + 100$

In Exercises 49–56, use the given zero to find all the zeros of the function.

Function	Zero
49. $f(x) = 2x^3 + 3x^2 + 50x + 75$	$5i$
50. $f(x) = x^3 + x^2 + 9x + 9$	$3i$
51. $f(x) = 2x^4 - x^3 + 7x^2 - 4x - 4$	$2i$
52. $g(x) = x^3 - 7x^2 - x + 87$	$5 + 2i$
53. $g(x) = 4x^3 + 23x^2 + 34x - 10$	$-3 + i$
54. $h(x) = 3x^3 - 4x^2 + 8x + 8$	$1 - \sqrt{3}i$
55. $f(x) = x^4 + 3x^3 - 5x^2 - 21x + 22$	$-3 + \sqrt{2}i$
56. $f(x) = x^3 + 4x^2 + 14x + 20$	$-1 - 3i$

In Exercises 57–62, find a polynomial function with real coefficients that has the given zeros. (There are many correct answers.)

57. $1, 5i, -5i$

58. $4, 3i, -3i$

59. $6, -5 + 2i, -5 - 2i$

60. $2, 4 + i, 4 - i$

61. $\frac{2}{3}, -1, 3 + \sqrt{2}i$

62. $-5, -5, 1 + \sqrt{3}i$

In Exercises 63–68, find a cubic polynomial function f with real coefficients that has the given zeros and the given function value.

	Zeros	Function Value
63.	$1, 2i$	$f(-1) = 10$
64.	$2, i$	$f(-1) = 6$
65.	$-1, 2 + i$	$f(2) = -9$
66.	$-2, 1 + 2i$	$f(2) = 10$
67.	$\frac{1}{2}, 1 + \sqrt{3}i$	$f(1) = -3$
68.	$\frac{3}{2}, 2 + \sqrt{2}i$	$f(1) = -6$

In Exercises 69–74, find a cubic polynomial function f with real coefficients that has the given complex zeros and x-intercept. (There are many correct answers.)

	Complex Zeros	x-Intercept
69.	$x = 4 \pm 2i$	$(-2, 0)$
70.	$x = 3 \pm i$	$(1, 0)$
71.	$x = 2 \pm \sqrt{6}i$	$(-1, 0)$
72.	$x = 2 \pm \sqrt{5}i$	$(2, 0)$
73.	$x = -2 \pm \sqrt{7}i$	$(3, 0)$
74.	$x = -3 \pm \sqrt{2}i$	$(-2, 0)$

75. Find the fourth-degree polynomial function f with real coefficients that has the zeros $x = \pm \sqrt{5}i$ and the x-intercepts shown in the graph.

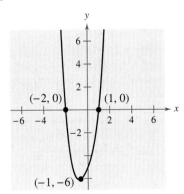

76. Find the fourth-degree polynomial function f with real coefficients that has the zeros $x = \pm \sqrt{2}i$ and the x-intercepts shown in the graph.

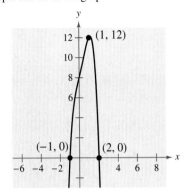

77. *Height of a Ball*   A ball is kicked upward from ground level with an initial velocity of 48 feet per second. The height h (in feet) of the ball is given by $h(t) = -16t^2 + 48t,\ \ 0 \le t \le 3$, where t is the time (in seconds).

(a) Complete the table to find the heights h of the ball for the given times t.

t	0	0.5	1	1.5	2	2.5	3
H							

(b) From the table in part (a), does it appear that the ball reaches a height of 64 feet?

(c) Determine algebraically if the ball reaches a height of 64 feet.

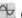 (d) Use a graphing utility to graph the function. Determine graphically if the ball reaches a height of 64 feet.

(e) Compare your results from parts (b), (c), and (d).

78. *Height of a Baseball*   A baseball is thrown upward from a height of 5 feet with an initial velocity of 79 feet per second. The height h (in feet) of the baseball is given by $h = -16t^2 + 79t + 5,\ \ 0 \le t \le 5$, where t is the time (in seconds).

(a) Complete the table to find the heights h of the baseball for the given times t.

t	0	1	2	3	4	5
H						

(b) From the table in part (a), does it appear that the baseball reaches a height of 110 feet?

(c) Determine algebraically if the baseball reaches a height of 110 feet.

(d) Use a graphing utility to graph the function. Determine graphically if the baseball reaches a height of 110 feet.

 (e) Compare your results from parts (b), (c), and (d).

79. *Profit* The demand equation for a microwave oven is given by $p = 140 - 0.0001x$, where p is the unit price (in dollars) of the microwave oven and x is the number of units sold. The cost equation for the microwave oven is $C = 80x + 150,000$, where C is the total cost (in dollars) and x is the number of units produced. The total profit P obtained by producing and selling x units is $P = xp - C$. You are working in the marketing department of the company and have been asked to determine the following.

(a) The profit function

(b) The profit when 250,000 units are sold

(c) The unit price when 250,000 units are sold

(d) If possible, the unit price that will yield a profit of 10 million dollars.

Model It

80. *Data Analysis: Sales* The sales S (in billions of dollars) for Winn-Dixie Stores, Inc. for selected years from 1994 to 2004 are shown in the table. (Source: Winn-Dixie Stores, Inc.)

Year	Sales, S
1994	11.1
1996	13.0
1998	13.6
2000	13.7
2002	12.3
2004	10.6

(a) Use the *regression* feature of a graphing utility to find a quadratic model for the data. Let t represent the year, with $t = 4$ corresponding to 1994.

(b) Use a graphing utility to graph the model you found in part (a).

(c) Use your graph from part (b) to determine the year in which sales reached $14 billion. Is this possible?

(d) Determine algebraically the year in which sales reached $14 billion. Is this possible? Explain.

True or False? In Exercises 81 and 82, decide whether the statement is true or false. Justify your answer.

81. It is possible for a third-degree polynomial function with integer coefficients to have no real zeros.

82. If $x = -i$ is a zero of the function given by

$$f(x) = x^3 + ix^2 + ix - 1$$

then $x = i$ must also be a zero of f.

Think About It In Exercises 83–88, determine (if possible) the zeros of the function g if the function f has zeros at $x = r_1, x = r_2,$ and $x = r_3$.

83. $g(x) = -f(x)$ **84.** $g(x) = 3f(x)$

85. $g(x) = f(x - 5)$ **86.** $g(x) = f(2x)$

87. $g(x) = 3 + f(x)$ **88.** $g(x) = f(-x)$

89. Find a quadratic function f (with integer coefficients) that has $\pm \sqrt{b}i$ as zeros. Assume that b is a positive integer.

90. Find a quadratic function f (with integer coefficients) that has $a \pm bi$ as zeros. Assume that b is a positive integer and a is an integer not equal to zero.

Skills Review

In Exercises 91–94, perform the operation and simplify.

91. $(-3 + 6i) - (8 - 3i)$ **92.** $(12 - 5i) + 16i$

93. $(6 - 2i)(1 + 7i)$ **94.** $(9 - 5i)(9 + 5i)$

In Exercises 95–100, use the graph of f to sketch the graph of g. To print an enlarged copy of the graph, go to the website *www.mathgraphs.com*.

95. $g(x) = f(x - 2)$

96. $g(x) = f(x) - 2$

97. $g(x) = 2f(x)$

98. $g(x) = f(-x)$

99. $g(x) = f(2x)$

100. $g(x) = f\left(\frac{1}{2}x\right)$

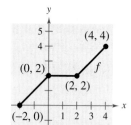

In Exercises 101–104, find the angle θ between the vectors.

101. $\mathbf{u} = \langle -6, 1 \rangle, \mathbf{v} = \langle 0, 3 \rangle$

102. $\mathbf{u} = \langle 4, -2 \rangle, \mathbf{v} = \langle 1, 4 \rangle$

103. $\mathbf{u} = \langle 5, 4 \rangle, \mathbf{v} = \langle 3, -1 \rangle$

104. $\mathbf{u} = \langle 8, 0 \rangle, \mathbf{v} = \langle 2, 2 \rangle$

105. *Work* Determine the work done by a crane lifting a 5700-pound minivan 10 feet.

106. *Work* A force of 60 pounds in the direction of 25° above the horizontal is required to pull a couch across a floor. The couch is pulled 10 feet. Determine the work done in pulling the couch.

---

107. Make a Decision To work an extended application analyzing Head Start enrollment in the United States from 1985 to 2004, visit this text's website at *college.hmco.com*. (Data Source: U.S. Department of Health and Human Services)

4.3   Trigonometric Form of a Complex Number

What you should learn

- Plot complex numbers in the complex plane and find absolute values of complex numbers.
- Write the trigonometric forms of complex numbers.
- Multiply and divide complex numbers written in trigonometric form.

Why you should learn it

You can perform the operations of multiplication and division on complex numbers by learning to write complex numbers in trigonometric form. For instance, in Exercises 59–66 on page 353, you can multiply and divide complex numbers in trigonometric form and standard form.

The Complex Plane

Just as real numbers can be represented by points on the real number line, you can represent a complex number

$$z = a + bi$$

as the point (a, b) in a coordinate plane (the **complex plane**). The horizontal axis is called the **real axis** and the vertical axis is called the **imaginary axis,** as shown in Figure 4.5.

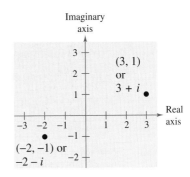

FIGURE **4.5**

The **absolute value** of the complex number $a + bi$ is defined as the distance between the origin $(0, 0)$ and the point (a, b).

> ### Definition of the Absolute Value of a Complex Number
> The **absolute value** of the complex number $z = a + bi$ is
> $$|a + bi| = \sqrt{a^2 + b^2}.$$

If the complex number $a + bi$ is a real number (that is, if $b = 0$), then this definition agrees with that given for the absolute value of a real number

$$|a + 0i| = \sqrt{a^2 + 0^2} = |a|.$$

Example 1   Finding the Absolute Value of a Complex Number

Plot $z = -2 + 5i$ and find its absolute value.

Solution

The number is plotted in Figure 4.6. It has an absolute value of

$$|z| = \sqrt{(-2)^2 + 5^2}$$
$$= \sqrt{29}.$$

✔CHECKPOINT   Now try Exercise 3.

FIGURE **4.6**

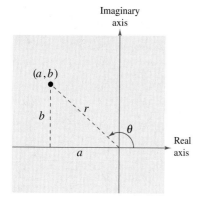

FIGURE **4.7**

Trigonometric Form of a Complex Number

In Section 4.1, you learned how to add, subtract, multiply, and divide complex numbers. To work effectively with *powers* and *roots* of complex numbers, it is helpful to write complex numbers in trigonometric form. In Figure 4.7, consider the nonzero complex number $a + bi$. By letting θ be the angle from the positive real axis (measured counterclockwise) to the line segment connecting the origin and the point (a, b), you can write

$$a = r \cos \theta \qquad \text{and} \qquad b = r \sin \theta$$

where $r = \sqrt{a^2 + b^2}$. Consequently, you have

$$a + bi = (r \cos \theta) + (r \sin \theta)i$$

from which you can obtain the **trigonometric form of a complex number.**

> **Trigonometric Form of a Complex Number**
>
> The **trigonometric form** of the complex number $z = a + bi$ is
>
> $$z = r(\cos \theta + i \sin \theta)$$
>
> where $a = r \cos \theta$, $b = r \sin \theta$, $r = \sqrt{a^2 + b^2}$, and $\tan \theta = b/a$. The number r is the **modulus** of z, and θ is called an **argument** of z.

The trigonometric form of a complex number is also called the *polar form*. Because there are infinitely many choices for θ, the trigonometric form of a complex number is not unique. Normally, θ is restricted to the interval $0 \le \theta < 2\pi$, although on occasion it is convenient to use $\theta < 0$.

Example 2   Writing a Complex Number in Trigonometric Form

Write the complex number $z = -2 - 2\sqrt{3}i$ in trigonometric form.

Solution

The absolute value of z is

$$r = \left| -2 - 2\sqrt{3}i \right| = \sqrt{(-2)^2 + \left(-2\sqrt{3}\right)^2} = \sqrt{16} = 4$$

and the reference angle θ' is given by

$$\tan \theta' = \frac{b}{a} = \frac{-2\sqrt{3}}{-2} = \sqrt{3}.$$

Because $\tan(\pi/3) = \sqrt{3}$ and because $z = -2 - 2\sqrt{3}i$ lies in Quadrant III, you choose θ to be $\theta = \pi + \pi/3 = 4\pi/3$. So, the trigonometric form is

$$z = r(\cos \theta + i \sin \theta)$$

$$= 4\left(\cos \frac{4\pi}{3} + i \sin \frac{4\pi}{3} \right).$$

See Figure 4.8.

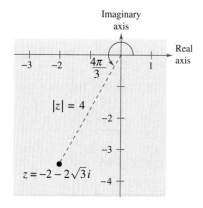

FIGURE **4.8**

✓**CHECKPOINT**   Now try Exercise 13.

Reasons for using the trigonometric forms of complex numbers are to multiply efficiently, raise to powers, and find roots of complex numbers.

Example 3   **Writing a Complex Number in Trigonometric Form**

Write the complex number in trigonometric form.

$$z = 6 + 2i$$

Solution

The absolute value of z is

$$r = |6 + 2i|$$
$$= \sqrt{6^2 + 2^2}$$
$$= \sqrt{40}$$
$$= 2\sqrt{10}$$

and the angle θ is

$$\tan \theta = \frac{b}{a} = \frac{2}{6} = \frac{1}{3}.$$

Because $z = 6 + 2i$ is in Quadrant I, you can conclude that

$$\theta = \arctan \frac{1}{3} \approx 0.32175 \text{ radian} \approx 18.4°.$$

So, the trigonometric form of z is

$$z = r(\cos \theta + i \sin \theta)$$
$$= 2\sqrt{10}\left[\cos\left(\arctan \frac{1}{3}\right) + i \sin\left(\arctan \frac{1}{3}\right)\right]$$
$$\approx 2\sqrt{10}(\cos 18.4° + i \sin 18.4°).$$

This result is illustrated graphically in Figure 4.9.

✔CHECKPOINT   Now try Exercise 19.

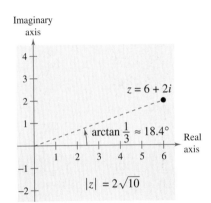

Imaginary axis

$z = 6 + 2i$

$\arctan \frac{1}{3} \approx 18.4°$

Real axis

$|z| = 2\sqrt{10}$

FIGURE 4.9

Example 4   **Writing a Complex Number in Standard Form**

Write the complex number in standard form $a + bi$.

$$z = \sqrt{8}\left[\cos\left(-\frac{\pi}{3}\right) + i \sin\left(-\frac{\pi}{3}\right)\right]$$

Solution

Because $\cos(-\pi/3) = \frac{1}{2}$ and $\sin(-\pi/3) = -\sqrt{3}/2$, you can write

$$z = \sqrt{8}\left[\cos\left(-\frac{\pi}{3}\right) + i \sin\left(-\frac{\pi}{3}\right)\right]$$
$$= 2\sqrt{2}\left(\frac{1}{2} - \frac{\sqrt{3}}{2}i\right)$$
$$= \sqrt{2} - \sqrt{6}i.$$

✔CHECKPOINT   Now try Exercise 35.

Technology

A graphing utility can be used to convert a complex number in trigonometric (or polar) form to standard form. For specific keystrokes, see the user's manual for your graphing utility.

Multiplication and Division of Complex Numbers

The trigonometric form adapts nicely to multiplication and division of complex numbers. Suppose you are given two complex numbers

$$z_1 = r_1(\cos \theta_1 + i \sin \theta_1) \quad \text{and} \quad z_2 = r_2(\cos \theta_2 + i \sin \theta_2).$$

The product of z_1 and z_2 is given by

$$z_1 z_2 = r_1 r_2(\cos \theta_1 + i \sin \theta_1)(\cos \theta_2 + i \sin \theta_2)$$
$$= r_1 r_2[(\cos \theta_1 \cos \theta_2 - \sin \theta_1 \sin \theta_2) + i(\sin \theta_1 \cos \theta_2 + \cos \theta_1 \sin \theta_2)].$$

Using the sum and difference formulas for cosine and sine, you can rewrite this equation as

$$z_1 z_2 = r_1 r_2[\cos(\theta_1 + \theta_2) + i \sin(\theta_1 + \theta_2)].$$

This establishes the first part of the following rule. The second part is left for you to verify (see Exercise 73).

Product and Quotient of Two Complex Numbers

Let $z_1 = r_1(\cos \theta_1 + i \sin \theta_1)$ and $z_2 = r_2(\cos \theta_2 + i \sin \theta_2)$ be complex numbers.

$$z_1 z_2 = r_1 r_2[\cos(\theta_1 + \theta_2) + i \sin(\theta_1 + \theta_2)] \qquad \text{Product}$$

$$\frac{z_1}{z_2} = \frac{r_1}{r_2}[\cos(\theta_1 - \theta_2) + i \sin(\theta_1 - \theta_2)], \quad z_2 \neq 0 \qquad \text{Quotient}$$

Note that this rule says that to *multiply* two complex numbers you multiply moduli and add arguments, whereas to *divide* two complex numbers you divide moduli and subtract arguments.

Example 5   Dividing Complex Numbers

Find the quotient z_1/z_2 of the complex numbers.

$$z_1 = 24(\cos 300° + i \sin 300°) \qquad z_2 = 8(\cos 75° + i \sin 75°)$$

Solution

$$\frac{z_1}{z_2} = \frac{24(\cos 300° + i \sin 300°)}{8(\cos 75° + i \sin 75°)}$$

$$= \frac{24}{8}[\cos(300° - 75°) + i \sin(300° - 75°)] \qquad \text{Divide moduli and subtract arguments.}$$

$$= 3(\cos 225° + i \sin 225°)$$

$$= 3\left[\left(-\frac{\sqrt{2}}{2}\right) + i\left(-\frac{\sqrt{2}}{2}\right)\right]$$

$$= -\frac{3\sqrt{2}}{2} - \frac{3\sqrt{2}}{2}i$$

✓CHECKPOINT   Now try Exercise 53.

| Example 6 | **Multiplying Complex Numbers** |

Find the product z_1z_2 of the complex numbers.

$$z_1 = 2\left(\cos\frac{2\pi}{3} + i\sin\frac{2\pi}{3}\right) \qquad z_2 = 8\left(\cos\frac{11\pi}{6} + i\sin\frac{11\pi}{6}\right)$$

Solution

$$z_1z_2 = 2\left(\cos\frac{2\pi}{3} + i\sin\frac{2\pi}{3}\right) \cdot 8\left(\cos\frac{11\pi}{6} + i\sin\frac{11\pi}{6}\right)$$

$$= 16\left[\cos\left(\frac{2\pi}{3} + \frac{11\pi}{6}\right) + i\sin\left(\frac{2\pi}{3} + \frac{11\pi}{6}\right)\right] \qquad \text{Multiply moduli and add arguments.}$$

$$= 16\left(\cos\frac{5\pi}{2} + i\sin\frac{5\pi}{2}\right)$$

$$= 16\left(\cos\frac{\pi}{2} + i\sin\frac{\pi}{2}\right)$$

$$= 16[0 + i(1)]$$

$$= 16i$$

✓*CHECKPOINT*   Now try Exercise 47.

You can check the result in Example 6 by first converting the complex numbers to the standard forms $z_1 = -1 + \sqrt{3}i$ and $z_2 = 4\sqrt{3} - 4i$ and then multiplying algebraically, as in Section 4.1.

$$z_1z_2 = \left(-1 + \sqrt{3}i\right)\left(4\sqrt{3} - 4i\right)$$

$$= -4\sqrt{3} + 4i + 12i + 4\sqrt{3}$$

$$= 16i$$

Activities

1. Find the absolute value and the trigonometric form of the complex number $z = 4 - 4i$.

    Answer: $r = 4\sqrt{2}$

    $$z = 4\sqrt{2}\left(\cos\frac{7\pi}{4} + i\sin\frac{7\pi}{4}\right)$$

2. Multiply, using the trigonometric form, then rewrite your answer in standard form.

    $$\left[3\left(\cos\frac{\pi}{3} + i\sin\frac{\pi}{3}\right)\right] \cdot$$

    $$\left[\frac{2}{3}\left(\cos\frac{5\pi}{3} + i\sin\frac{5\pi}{3}\right)\right]$$

    Answer: $2(\cos 2\pi + i\sin 2\pi) = 2$

3. Represent the complex number $-4 + 2i$ graphically.

    Answer:

𝒲RITING ABOUT 𝓜ATHEMATICS

Multiplying Complex Numbers Graphically Discuss how you can graphically approximate the product of the complex numbers. Then, approximate the values of the products and check your answers analytically.

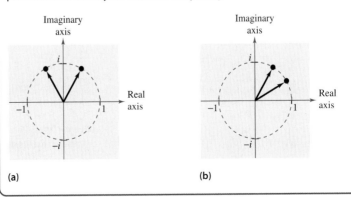

(a)   (b)

4.3 | Exercises

VOCABULARY CHECK: Fill in the blanks.

1. The _____ _____ of a complex number $a + bi$ is the distance between the origin $(0, 0)$ and the point (a, b).

2. The _____ _____ of a complex number $z = a + bi$ is given by $z = r(\cos\theta + i\sin\theta)$, where r is the _____ of z and θ is the _____ of z.

3. Let $z_1 = r_1(\cos\theta_1 + i\sin\theta_1)$ and $z_2 = r_2(\cos\theta_2 + i\sin\theta_2)$ be complex numbers, then the product $z_1 z_2 =$ _____ and the quotient $z_1/z_2 =$ _____ $(z_2 \neq 0)$.

PREREQUISITE SKILLS REVIEW: Practice and review algebra skills needed for this section at **www.Eduspace.com.**

In Exercises 1–6, plot the complex number and find its absolute value.

1. $-7i$

2. -7

3. $-4 + 4i$

4. $5 - 12i$

5. $6 - 7i$

6. $-8 + 3i$

In Exercises 7–10, write the complex number in trigonometric form.

7.

8.

9.

10.

In Exercises 11–30, represent the complex number graphically, and find the trigonometric form of the number.

11. $3 - 3i$

12. $2 + 2i$

13. $\sqrt{3} + i$

14. $4 - 4\sqrt{3}i$

15. $-2(1 + \sqrt{3}i)$

16. $\frac{5}{2}(\sqrt{3} - i)$

17. $-5i$

18. $4i$

19. $-7 + 4i$

20. $3 - i$

21. 7

22. 4

23. $3 + \sqrt{3}i$

24. $2\sqrt{2} - i$

25. $-3 - i$

26. $1 + 3i$

27. $5 + 2i$

28. $8 + 3i$

29. $-8 - 5\sqrt{3}i$

30. $-9 - 2\sqrt{10}i$

In Exercises 31–40, represent the complex number graphically, and find the standard form of the number.

31. $3(\cos 120° + i\sin 120°)$

32. $5(\cos 135° + i\sin 135°)$

33. $\frac{3}{2}(\cos 300° + i\sin 300°)$

34. $\frac{1}{4}(\cos 225° + i\sin 225°)$

35. $3.75\left(\cos\dfrac{3\pi}{4} + i\sin\dfrac{3\pi}{4}\right)$

36. $6\left(\cos\dfrac{5\pi}{12} + i\sin\dfrac{5\pi}{12}\right)$

37. $8\left(\cos\dfrac{\pi}{2} + i\sin\dfrac{\pi}{2}\right)$

38. $7(\cos 0 + i\sin 0)$

39. $3[\cos(18° \ 45') + i\sin(18° \ 45')]$

40. $6[\cos(230° \ 30') + i\sin(230° \ 30')]$

In Exercises 41–44, use a graphing utility to represent the complex number in standard form.

41. $5\left(\cos\dfrac{\pi}{9} + i\sin\dfrac{\pi}{9}\right)$

42. $10\left(\cos\dfrac{2\pi}{5} + i\sin\dfrac{2\pi}{5}\right)$

43. $3(\cos 165.5° + i\sin 165.5°)$

44. $9(\cos 58° + i\sin 58°)$

In Exercises 45 and 46, represent the powers z, z^2, z^3, and z^4 graphically. Describe the pattern.

45. $z = \dfrac{\sqrt{2}}{2}(1 + i)$

46. $z = \dfrac{1}{2}(1 + \sqrt{3}i)$

In Exercises 47–58, perform the operation and leave the result in trigonometric form.

47. $\left[2\left(\cos\dfrac{\pi}{4} + i\sin\dfrac{\pi}{4}\right)\right]\left[6\left(\cos\dfrac{\pi}{12} + i\sin\dfrac{\pi}{12}\right)\right]$

48. $\left[\dfrac{3}{4}\left(\cos\dfrac{\pi}{3} + i\sin\dfrac{\pi}{3}\right)\right]\left[4\left(\cos\dfrac{3\pi}{4} + i\sin\dfrac{3\pi}{4}\right)\right]$

49. $\left[\frac{5}{3}(\cos 140° + i\sin 140°)\right]\left[\frac{2}{3}(\cos 60° + i\sin 60°)\right]$

50. $[0.5(\cos 100° + i\sin 100°)] \cdot$
$[0.8(\cos 300° + i\sin 300°)]$

51. $[0.45(\cos 310° + i\sin 310°)] \cdot$
$[0.60(\cos 200° + i\sin 200°)]$

52. $(\cos 5° + i\sin 5°)(\cos 20° + i\sin 20°)$

53. $\dfrac{\cos 50° + i\sin 50°}{\cos 20° + i\sin 20°}$

54. $\dfrac{2(\cos 120° + i\sin 120°)}{4(\cos 40° + i\sin 40°)}$

55. $\dfrac{\cos(5\pi/3) + i\sin(5\pi/3)}{\cos\pi + i\sin\pi}$

56. $\dfrac{5(\cos 4.3 + i\sin 4.3)}{4(\cos 2.1 + i\sin 2.1)}$

57. $\dfrac{12(\cos 52° + i\sin 52°)}{3(\cos 110° + i\sin 110°)}$

58. $\dfrac{6(\cos 40° + i\sin 40°)}{7(\cos 100° + i\sin 100°)}$

In Exercises 59–66, (a) write the trigonometric forms of the complex numbers, (b) perform the indicated operation using the trigonometric forms, and (c) perform the indicated operation using the standard forms, and check your result with that of part (b).

59. $(2 + 2i)(1 - i)$

60. $\left(\sqrt{3} + i\right)(1 + i)$

61. $-2i(1 + i)$

62. $4\left(1 - \sqrt{3}i\right)$

63. $\dfrac{3 + 4i}{1 - \sqrt{3}i}$

64. $\dfrac{1 + \sqrt{3}i}{6 - 3i}$

65. $\dfrac{5}{2 + 3i}$

66. $\dfrac{4i}{-4 + 2i}$

In Exercises 67–70, sketch the graph of all complex numbers z satisfying the given condition.

67. $|z| = 2$

68. $|z| = 3$

69. $\theta = \dfrac{\pi}{6}$

70. $\theta = \dfrac{5\pi}{4}$

Synthesis

True or False?   In Exercises 71 and 72, determine whether the statement is true or false. Justify your answer.

71. Although the square of the complex number bi is given by $(bi)^2 = -b^2$, the absolute value of the complex number $z = a + bi$ is defined as
$$|a + bi| = \sqrt{a^2 + b^2}.$$

72. The product of two complex numbers
$$z_1 = r_1(\cos\theta_1 + i\sin\theta_1)$$
and
$$z_2 = r_2(\cos\theta_2 + i\sin\theta_2).$$
is zero only when $r_1 = 0$ and/or $r_2 = 0$.

73. Given two complex numbers $z_1 = r_1(\cos\theta_1 + i\sin\theta_1)$ and $z_2 = r_2(\cos\theta_2 + i\sin\theta_2)$, $z_2 \neq 0$, show that
$$\frac{z_1}{z_2} = \frac{r_1}{r_2}\left[\cos(\theta_1 - \theta_2) + i\sin(\theta_1 - \theta_2)\right].$$

74. Show that $\bar{z} = r[\cos(-\theta) + i\sin(-\theta)]$ is the complex conjugate of $z = r(\cos\theta + i\sin\theta)$.

75. Use the trigonometric forms of z and $\bar{z}$ in Exercise 74 to find (a) $z\bar{z}$ and (b) $z/\bar{z}$, $\bar{z} \neq 0$.

76. Show that the negative of $z = r(\cos\theta + i\sin\theta)$ is $-z = r[\cos(\theta + \pi) + i\sin(\theta + \pi)]$.

Skills Review

In Exercises 77–82, solve the right triangle shown in the figure. Round your answers to two decimal places.

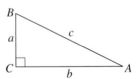

77. $A = 22°, \quad a = 8$

78. $B = 66°, \quad a = 33.5$

79. $A = 30°, \quad b = 112.6$

80. $B = 6°, \quad b = 211.2$

81. $A = 42°\, 15', \quad c = 11.2$

82. $B = 81°\, 30', \quad c = 6.8$

Harmonic Motion   In Exercises 83–86, for the simple harmonic motion described by the trigonometric function, find the maximum displacement and the least positive value of t for which $d = 0$.

83. $d = 16\cos\dfrac{\pi}{4}t$

84. $d = \dfrac{1}{8}\cos 12\pi t$

85. $d = \dfrac{1}{16}\sin\dfrac{5}{4}\pi t$

86. $d = \dfrac{1}{12}\sin 60\pi t$

4.4 DeMoivre's Theorem

Powers of Complex Numbers

The trigonometric form of a complex number is used to raise a complex number to a power. To accomplish this, consider repeated use of the multiplication rule.

$$z = r(\cos \theta + i \sin \theta)$$

$$z^2 = r(\cos \theta + i \sin \theta)r(\cos \theta + i \sin \theta) = r^2(\cos 2\theta + i \sin 2\theta)$$

$$z^3 = r^2(\cos 2\theta + i \sin 2\theta)r(\cos \theta + i \sin \theta) = r^3(\cos 3\theta + i \sin 3\theta)$$

$$z^4 = r^4(\cos 4\theta + i \sin 4\theta)$$

$$z^5 = r^5(\cos 5\theta + i \sin 5\theta)$$

$$\vdots$$

This pattern leads to DeMoivre's Theorem, which is named after the French mathematician Abraham DeMoivre (1667–1754).

DeMoivre's Theorem

If $z = r(\cos \theta + i \sin \theta)$ is a complex number and n is a positive integer, then

$$z^n = [r(\cos \theta + i \sin \theta)]^n = r^n(\cos n\theta + i \sin n\theta).$$

Example 1 **Finding a Power of a Complex Number**

Use DeMoivre's Theorem to find $\left(-1 + \sqrt{3}i\right)^{12}$.

Solution

First convert the complex number to trigonometric form using

$$r = \sqrt{(-1)^2 + \left(\sqrt{3}\right)^2} = 2 \quad \text{and} \quad \theta = \arctan \frac{\sqrt{3}}{-1} = \frac{2\pi}{3}.$$

So, the trigonometric form is

$$-1 + \sqrt{3}i = 2\left(\cos \frac{2\pi}{3} + i \sin \frac{2\pi}{3}\right).$$

Then, by DeMoivre's Theorem, you have

$$(-1 + \sqrt{3}i)^{12} = \left[2\left(\cos \frac{2\pi}{3} + i \sin \frac{2\pi}{3}\right)\right]^{12}$$

$$= 2^{12}\left[\cos(12)\frac{2\pi}{3} + i \sin(12)\frac{2\pi}{3}\right]$$

$$= 4096(\cos 8\pi + i \sin 8\pi)$$

$$= 4096(1 + 0) = 4096.$$

✓**CHECKPOINT** Now try Exercise 1.

Historical Note
Abraham DeMoivre
(1667–1754) is remembered for
his work in probability theory
and DeMoivre's Theorem. His
book *The Doctrine of Chances*
(published in 1718) includes
the theory of recurring series
and the theory of partial
fractions.

Roots of Complex Numbers

Recall that a consequence of the Fundamental Theorem of Algebra is that a polynomial equation of degree n has n solutions in the complex number system. So, the equation $x^6 = 1$ has six solutions, and in this particular case you can find the six solutions by factoring and using the Quadratic Formula.

$$x^6 - 1 = (x^3 - 1)(x^3 + 1)$$
$$= (x - 1)(x^2 + x + 1)(x + 1)(x^2 - x + 1) = 0$$

Consequently, the solutions are

$$x = \pm 1, \qquad x = \frac{-1 \pm \sqrt{3}\, i}{2}, \qquad \text{and} \qquad x = \frac{1 \pm \sqrt{3}\, i}{2}.$$

Each of these numbers is a sixth root of 1. In general, the **nth root of a complex number** is defined as follows.

Definition of an nth Root of a Complex Number

The complex number $u = a + bi$ is an **nth root** of the complex number z if

$$z = u^n = (a + bi)^n.$$

To find a formula for an nth root of a complex number, let u be an nth root of z, where

$$u = s(\cos \beta + i \sin \beta)$$

and

$$z = r(\cos \theta + i \sin \theta).$$

By DeMoivre's Theorem and the fact that $u^n = z$, you have

$$s^n (\cos n\beta + i \sin n\beta) = r(\cos \theta + i \sin \theta).$$

Taking the absolute value of each side of this equation, it follows that $s^n = r$. Substituting back into the previous equation and dividing by r, you get

$$\cos n\beta + i \sin n\beta = \cos \theta + i \sin \theta.$$

So, it follows that

$$\cos n\beta = \cos \theta$$

and

$$\sin n\beta = \sin \theta.$$

Because both sine and cosine have a period of 2π, these last two equations have solutions if and only if the angles differ by a multiple of 2π. Consequently, there must exist an integer k such that

$$n\beta = \theta + 2\pi k$$

$$\beta = \frac{\theta + 2\pi k}{n}.$$

By substituting this value of β into the trigonometric form of u, you get the result stated on the following page.

Exploration

The nth roots of a complex number are useful for solving some polynomial equations. For instance, explain how you can use DeMoivre's Theorem to solve the polynomial equation

$$x^4 + 16 = 0.$$

[*Hint:* Write -16 as $16(\cos \pi + i \sin \pi)$.]

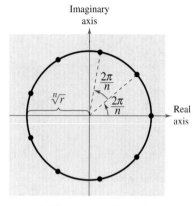

FIGURE 4.10

> ### Finding *n*th Roots of a Complex Number
>
> For a positive integer n, the complex number $z = r(\cos \theta + i \sin \theta)$ has exactly n distinct *n*th roots given by
>
> $$\sqrt[n]{r}\left(\cos \frac{\theta + 2\pi k}{n} + i \sin \frac{\theta + 2\pi k}{n}\right)$$
>
> where $k = 0, 1, 2, \ldots, n - 1$.

When k exceeds $n - 1$, the roots begin to repeat. For instance, if $k = n$, the angle

$$\frac{\theta + 2\pi n}{n} = \frac{\theta}{n} + 2\pi$$

is coterminal with θ/n, which is also obtained when $k = 0$.

The formula for the *n*th roots of a complex number z has a nice geometric interpretation, as shown in Figure 4.10. Note that because the *n*th roots of z all have the same magnitude $\sqrt[n]{r}$, they all lie on a circle of radius $\sqrt[n]{r}$ with center at the origin. Furthermore, because successive *n*th roots have arguments that differ by $2\pi/n$, the n roots are equally spaced around the circle.

You have already found the sixth roots of 1 by factoring and by using the Quadratic Formula. Example 2 shows how you can solve the same problem with the formula for *n*th roots.

Example 2   Finding the *n*th Roots of a Real Number

Find all the sixth roots of 1.

Solution

First write 1 in the trigonometric form $1 = 1(\cos 0 + i \sin 0)$. Then, by the *n*th root formula, with $n = 6$ and $r = 1$, the roots have the form

$$\sqrt[6]{1}\left(\cos \frac{0 + 2\pi k}{6} + i \sin \frac{0 + 2\pi k}{6}\right) = \cos \frac{\pi k}{3} + i \sin \frac{\pi k}{3}.$$

So, for $k = 0, 1, 2, 3, 4,$ and 5, the sixth roots are as follows. (See Figure 4.11.)

$$\cos 0 + i \sin 0 = 1$$

$$\cos \frac{\pi}{3} + i \sin \frac{\pi}{3} = \frac{1}{2} + \frac{\sqrt{3}}{2} i \qquad \text{Increment by } \frac{2\pi}{n} = \frac{2\pi}{6} = \frac{\pi}{3}$$

$$\cos \frac{2\pi}{3} + i \sin \frac{2\pi}{3} = -\frac{1}{2} + \frac{\sqrt{3}}{2} i$$

$$\cos \pi + i \sin \pi = -1$$

$$\cos \frac{4\pi}{3} + i \sin \frac{4\pi}{3} = -\frac{1}{2} - \frac{\sqrt{3}}{2} i$$

$$\cos \frac{5\pi}{3} + i \sin \frac{5\pi}{3} = \frac{1}{2} - \frac{\sqrt{3}}{2} i$$

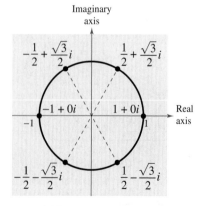

FIGURE 4.11

✓ CHECKPOINT   Now try Exercise 37.

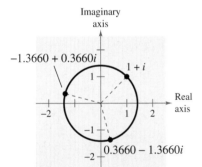

FIGURE **4.12**

STUDY TIP

Note in Example 3 that the absolute value of z is

$r = |-2 + 2i|$

$\;= \sqrt{(-2)^2 + 2^2}$

$\;= \sqrt{8}$

and the angle θ is given by

$\tan \theta = \dfrac{b}{a} = \dfrac{2}{-2} = -1.$

In Figure 4.11, notice that the roots obtained in Example 2 all have a magnitude of 1 and are equally spaced around the unit circle. Also notice that the complex roots occur in conjugate pairs, as discussed in Section 4.2. The n distinct nth roots of 1 are called the **nth roots of unity.**

Example 3   **Finding the nth Roots of a Complex Number**

Find the three cube roots of $z = -2 + 2i$.

Solution

Because z lies in Quadrant II, the trigonometric form of z is

$z = -2 + 2i$

$\;= \sqrt{8}\,(\cos 135° + i \sin 135°). \qquad \theta = \arctan(2/-2) = 135°$

By the formula for nth roots, the cube roots have the form

$$\sqrt[6]{8}\left(\cos \frac{135° + 360°k}{3} + i \sin \frac{135° + 360°k}{3}\right).$$

Finally, for $k = 0$, 1, and 2, you obtain the roots

$$\sqrt[6]{8}\left[\left(\cos \frac{135° + 360°(0)}{3} + i \sin \frac{135° + 360°(0)}{3}\right)\right] = \sqrt{2}(\cos 45° + i \sin 45°)$$

$$= 1 + i$$

$$\sqrt[6]{8}\left[\left(\cos \frac{135° + 360°(1)}{3} + i \sin \frac{135° + 360°(1)}{3}\right)\right] = \sqrt{2}(\cos 165° + i \sin 165°)$$

$$\approx -1.3660 + 0.3660i$$

$$\sqrt[6]{8}\left[\left(\cos \frac{135° + 360°(2)}{3} + i \sin \frac{135° + 360°(2)}{3}\right)\right] = \sqrt{2}(\cos 285° + i \sin 285°)$$

$$\approx 0.3660 - 1.3660i.$$

See Figure 4.12

✓CHECKPOINT   Now try Exercise 43.

$\mathcal{W}$RITING ABOUT $\mathcal{M}$ATHEMATICS

A Famous Mathematical Formula   The famous formula

$e^{a + bi} = e^a(\cos b + i \sin b)$

is called Euler's Formula, after the Swiss mathematician Leonhard Euler (1707–1783). Although the interpretation of this formula is beyond the scope of this text, we decided to include it because it gives rise to one of the most wonderful equations in mathematics.

$e^{\pi i} + 1 = 0$

This elegant equation relates the five most famous numbers in mathematics—0, 1, π, e, and i—in a single equation (e is called the natural base and is discussed in Section 5.1). Show how Euler's Formula can be used to derive this equation.

4.4 Exercises

VOCABULARY CHECK: Fill in the blanks.

1. The theorem states that if $z = r(\cos \theta + i \sin \theta)$ is a complex number and n is a positive integer, then $z^n = r^n(\cos n\theta + i \sin n\theta)$, is called _____ Theorem.

2. The complex number $u = a + bi$ is an _____ _____ of the complex number z if $z = u^n = (a + bi)^n$.

3. The n distinct nth roots of 1 are called the nth roots of _____.

PREREQUISITE SKILLS REVIEW: Practice and review algebra skills needed for this section at **www.Eduspace.com**.

In Exercises 1–24, use DeMoivre's Theorem to find the indicated power of the complex number. Write the result in standard form.

1. $(1 + i)^5$

2. $(2 + 2i)^6$

3. $(-1 + i)^{10}$

4. $(3 - 2i)^8$

5. $2(\sqrt{3} + i)^7$

6. $4(1 - \sqrt{3}i)^3$

7. $[5(\cos 20° + i \sin 20°)]^3$

8. $[3(\cos 150° + i \sin 150°)]^4$

9. $\left(\cos \dfrac{\pi}{4} + i \sin \dfrac{\pi}{4}\right)^{12}$

10. $\left[2\left(\cos \dfrac{\pi}{2} + i \sin \dfrac{\pi}{2}\right)\right]^8$

11. $[5(\cos 3.2 + i \sin 3.2)]^4$

12. $(\cos 0 + i \sin 0)^{20}$

13. $(3 - 2i)^5$

14. $(2 + 5i)^6$

15. $(\sqrt{5} - 4i)^3$

16. $(\sqrt{3} + 2i)^4$

17. $[3(\cos 15° + i \sin 15°)]^4$

18. $[2(\cos 10° + i \sin 10°)]^8$

19. $[5(\cos 95° + i \sin 95°)]^3$

20. $[4(\cos 110° + i \sin 110°)]^4$

21. $\left[2\left(\cos \dfrac{\pi}{10} + i \sin \dfrac{\pi}{10}\right)\right]^5$

22. $\left[2\left(\cos \dfrac{\pi}{8} + i \sin \dfrac{\pi}{8}\right)\right]^6$

23. $\left[3\left(\cos \dfrac{2\pi}{3} + i \sin \dfrac{2\pi}{3}\right)\right]^3$

24. $\left[3\left(\cos \dfrac{\pi}{12} + i \sin \dfrac{\pi}{12}\right)\right]^5$

In Exercises 25–44, (a) use the theorem on page 356 to find the indicated roots of the complex number, (b) represent each of the roots graphically, and (c) write each of the roots in standard form.

25. Square roots of $5(\cos 120° + i \sin 120°)$

26. Square roots of $16(\cos 60° + i \sin 60°)$

27. Cube roots of $8\left(\cos \dfrac{2\pi}{3} + i \sin \dfrac{2\pi}{3}\right)$

28. Cube roots of $64\left(\cos \dfrac{\pi}{3} + i \sin \dfrac{\pi}{3}\right)$

29. Fifth roots of $243\left(\cos \dfrac{\pi}{6} + i \sin \dfrac{\pi}{6}\right)$

30. Fifth roots of $32\left(\cos \dfrac{5\pi}{6} + i \sin \dfrac{5\pi}{6}\right)$

31. Square roots of $-25i$

32. Square roots of $-36i$

33. Fourth roots of $81i$

34. Fourth roots of $625i$

35. Cube roots of $-\frac{125}{2}(1 + \sqrt{3}i)$

36. Cube roots of $-4\sqrt{2}(1 - i)$

37. Fourth roots of 16

38. Fourth roots of i

39. Fifth roots of 1

40. Cube roots of 1000

41. Cube roots of -125

42. Fourth roots of -4

43. Fifth roots of $128(-1 + i)$

44. Sixth roots of $64i$

In Exercises 45–60, use the theorem on page 356 to find all the solutions of the equation and represent the solutions graphically.

45. $x^4 + i = 0$

46. $x^3 - i = 0$

47. $x^6 + 1 = 0$

48. $x^3 + 1 = 0$

49. $x^5 + 243 = 0$

50. $x^3 + 125 = 0$

51. $x^5 - 32 = 0$

52. $x^3 - 27 = 0$

53. $x^4 + 16i = 0$

54. $x^3 + 27i = 0$

55. $x^4 - 16i = 0$

56. $x^6 + 64i = 0$

57. $x^3 - (1 - i) = 0$

58. $x^5 - (1 - i) = 0$

59. $x^6 + (1 + i) = 0$

60. $x^4 + (1 + i) = 0$

Synthesis

True or False?   **In Exercises 61 and 62, determine whether the statement is true or false. Justify your answer.**

61. Geometrically, the nth roots of any complex number z are all equally spaced around the unit circle centered at the origin.

62. By DeMoivre's Theorem,
$$\left(4 + \sqrt{6}i\right)^8 = \cos(32) + i\sin(8\sqrt{6}).$$

63. Show that $-\frac{1}{2}\left(1 + \sqrt{3}i\right)$ is a sixth root of 1.

64. Show that $2^{-1/4}(1 - i)$ is a fourth root of -2.

Graphical Reasoning   **In Exercises 65 and 66, use the graph of the roots of a complex number.**

(a) Write each of the roots in trigonometric form.

(b) Identify the complex number whose roots are given.

(c) Use a graphing utility to verify the results of part (b).

65.

66.

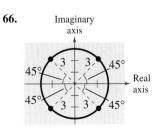

Skills Review

In Exercises 67–70, find the slope and the y-intercept (if possible) of the equation of the line. Then sketch the line.

67. $x - 4y = -1$

68. $7x + 6y = -8$

69. $x + 5 = 0$

70. $y - 9 = 0$

In Exercises 71–74, determine whether the function has an inverse function. If it does, find its inverse function.

71. $f(x) = 5x - 1$

72. $g(x) = \dfrac{2}{x}$

73. $h(x) = \sqrt{4x + 3}$

74. $f(x) = (x - 3)^2$

In Exercises 75–84, use the figure and trigonometric identities to find the exact value of the trigonometric function.

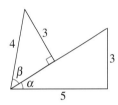

75. $\cos(\alpha + \beta)$

76. $\sin(\alpha + \beta)$

77. $\sin(\alpha - \beta)$

78. $\cos(\beta - \alpha)$

79. $\tan(\alpha + \beta)$

80. $\tan(\beta - \alpha)$

81. $\tan 2\alpha$

82. $\sin 2\beta$

83. $\sin\dfrac{\beta}{2}$

84. $\cos\dfrac{\alpha}{2}$

In Exercises 85–88, find a unit vector in the direction of the given vector.

85. $\mathbf{u} = \langle 10, 0 \rangle$

86. $\mathbf{v} = \langle -3, 7 \rangle$

87. $\mathbf{v} = 12\mathbf{i} - 5\mathbf{j}$

88. $\mathbf{w} = 8\mathbf{j}$

In Exercises 89–96, use the dot product to find the magnitude of u.

89. $\mathbf{u} = \langle -3, 4 \rangle$

90. $\mathbf{u} = \langle -5, 7 \rangle$

91. $\mathbf{u} = \langle -9, 40 \rangle$

92. $\mathbf{u} = \langle -5, -12 \rangle$

93. $\mathbf{u} = 22\mathbf{i} + 3\mathbf{j}$

94. $\mathbf{u} = 16\mathbf{i} + 4\mathbf{j}$

95. $\mathbf{u} = 13\mathbf{i} + 6\mathbf{j}$

96. $\mathbf{u} = 24\mathbf{i} + 16\mathbf{j}$

4 Chapter Summary

What did you learn?

4 Review Exercises

4.1 In Exercises 1–4, write the complex number in standard form.

1. $6 + \sqrt{-4}$

2. $3 - \sqrt{-25}$

3. $i^2 + 3i$

4. $-5i + i^2$

In Exercises 5–10, perform the operation and write the result in standard form.

5. $(7 + 5i) + (-4 + 2i)$

6. $\left(\dfrac{\sqrt{2}}{2} - \dfrac{\sqrt{2}}{2}i\right) - \left(\dfrac{\sqrt{2}}{2} + \dfrac{\sqrt{2}}{2}i\right)$

7. $5i(13 - 8i)$

8. $(1 + 6i)(5 - 2i)$

9. $(10 - 8i)(2 - 3i)$

10. $i(6 + i)(3 - 2i)$

In Exercises 11 and 12, write the quotient in standard form.

11. $\dfrac{6 + i}{4 - i}$

12. $\dfrac{3 + 2i}{5 + i}$

In Exercises 13 and 14, perform the operation and write the result in standard form.

13. $\dfrac{4}{2 - 3i} + \dfrac{2}{1 + i}$

14. $\dfrac{1}{2 + i} - \dfrac{5}{1 + 4i}$

In Exercises 15–18, find all solutions of the equation.

15. $3x^2 + 1 = 0$

16. $2 + 8x^2 = 0$

17. $x^2 - 2x + 10 = 0$

18. $6x^2 + 3x + 27 = 0$

4.2 In Exercises 19–22, determine the number of solutions of the equation in the complex number system.

19. $x^5 - 2x^4 + 3x^2 - 5 = 0$

20. $-2x^6 + 7x^3 + x^2 + 4x - 19 = 0$

21. $\frac{1}{2}x^4 + \frac{2}{3}x^3 - x^2 + \frac{3}{10} = 0$

22. $\frac{3}{4}x^3 + \frac{1}{2}x^2 + \frac{3}{2}x + 2 = 0$

In Exercises 23–26, use the discriminant to determine the number of real solutions of the quadratic equation.

23. $6x^2 + x - 2 = 0$

24. $9x^2 - 12x + 4 = 0$

25. $0.13x^2 - 0.45x + 0.65 = 0$

26. $4x^2 + \frac{4}{3}x + \frac{1}{9} = 0$

In Exercises 27–34, solve the equation. Write complex solutions in standard form.

27. $x^2 - 2x = 0$

28. $6x - x^2 = 0$

29. $x^2 - 3x + 5 = 0$

30. $x^2 - 4x + 9 = 0$

31. $x^2 + 8x + 10 = 0$

32. $3 + 4x - x^2 = 0$

33. $2x^2 + 3x + 6 = 0$

34. $4x^2 - x + 10 = 0$

35. *Profit* The demand equation for a DVD player is $p = 140 - 0.0001x$, where p is the unit price (in dollars) of the DVD player and x is the number of units produced and sold. The cost equation for the DVD player is $C = 75x + 100{,}000$, where C is the total cost (in dollars) and x is the number of units produced. The total profit obtained by producing and selling x units is

$$P = xp - C.$$

You work in the marketing department of the company that produces this DVD player and are asked to determine a price p that would yield a profit of 9 million dollars. Is this possible? Explain.

36. *Consumer Awareness* The average prices p (in dollars) for a personal computer from 1997 to 2002 can be modeled by

$$p = 16.52t^2 - 436.0t + 3704, \quad 7 \le t \le 12$$

where t represents the year, with $t = 7$ corresponding to 1997. According to this model, will the average price of a personal computer drop to \$800? Explain your reasoning. (Source: IDC; Consumer Electronics Association)

In Exercises 37–42, find all the zeros of the function and write the polynomial as a product of linear factors.

37. $r(x) = 2x^2 + 2x + 3$

38. $s(x) = 2x^2 + 5x + 4$

39. $f(x) = 2x^3 - 3x^2 + 50x - 75$

40. $f(x) = 4x^3 - x^2 + 128x - 32$

41. $f(x) = 4x^4 + 3x^2 - 10$

42. $f(x) = 5x^4 + 126x^2 + 25$

In Exercises 43–50, use the given zero to find all the zeros of the function. Write the polynomial as a product of linear factors.

	Function	*Zero*
43.	$f(x) = x^3 + 3x^2 - 24x + 28$	2
44.	$f(x) = 10x^3 + 21x^2 - x - 6$	-2
45.	$f(x) = x^3 + 3x^2 - 5x + 25$	-5
46.	$g(x) = x^3 - 8x^2 + 29x - 52$	4
47.	$h(x) = 2x^3 - 19x^2 + 58x + 34$	$5 + 3i$
48.	$f(x) = 5x^3 - 4x^2 + 20x - 16$	$2i$
49.	$f(x) = x^4 + 5x^3 + 2x^2 - 50x - 84$	$-3 + \sqrt{5}i$
50.	$g(x) = x^4 - 6x^3 + 18x^2 - 26x + 21$	$2 + \sqrt{3}i$

In Exercises 51–58, find a polynomial function with real coefficients that has the given zeros. (There are many correct answers.)

51. $1, 1, \frac{1}{4}, -\frac{2}{3}$

52. $-2, 2, 3, 3$

53. $3, 2 - \sqrt{3}, 2 + \sqrt{3}$

54. $5, 1 - \sqrt{2}, 1 + \sqrt{2}$

55. $\frac{2}{3}, 4, \sqrt{3}i, -\sqrt{3}i$

56. $2, -3, 1 - 2i, 1 + 2i$

57. $-\sqrt{2}i, \sqrt{2}i, -5i, 5i$

58. $-2i, 2i, -4i, 4i$

In Exercises 59 and 60, find a cubic polynomial function f with real coefficients that has the given zeros and the given function value.

Zeros	Function Value
59. $5, 1 - i$	$f(1) = -8$
60. $2, 4 + i$	$f(3) = 4$

4.3 In Exercises 61–64, plot the complex number and find its absolute value.

61. $8i$

62. $-6i$

63. $5 + 3i$

64. $-10 - 4i$

In Exercises 65–68, write the complex number in trigonometric form.

65. $5 - 5i$

66. $5 + 12i$

67. $-3\sqrt{3} + 3i$

68. -9

In Exercises 69 and 70, (a) write the two complex numbers in trigonometric form, and (b) use the trigonometric form to find $z_1 z_2$ and z_1/z_2, $z_2 \neq 0$.

69. $z_1 = 2\sqrt{3} - 2i, \quad z_2 = -10i$

70. $z_1 = -3(1 + i), \quad z_2 = 2(\sqrt{3} + i)$

4.4 In Exercises 71–74, use DeMoivre's Theorem to find the indicated power of the complex number. Write the result in standard form.

71. $\left[5\left(\cos \dfrac{\pi}{12} + i \sin \dfrac{\pi}{12} \right) \right]^4$

72. $\left[2\left(\cos \dfrac{4\pi}{15} + i \sin \dfrac{4\pi}{15} \right) \right]^5$

73. $(2 + 3i)^6$

74. $(1 - i)^8$

In Exercises 75 and 76, (a) use the theorem on page 356 to find the indicated roots of the complex number, (b) represent each of the roots graphically, and (c) write each of the roots in standard form.

75. Sixth roots of $-729i$

76. Fourth roots of 256

In Exercises 77–80, use the theorem on page 356 to find all solutions of the equation and represent the solutions graphically.

77. $x^4 + 81 = 0$

78. $x^5 - 32 = 0$

79. $x^3 + 8i = 0$

80. $(x^3 - 1)(x^2 + 1) = 0$

Synthesis

True or False? In Exercises 81–83, determine whether the statement is true or false. Justify your answer.

81. $\sqrt{-18}\sqrt{-2} = \sqrt{(-18)(-2)}$

82. The equation $325x^2 - 717x + 398 = 0$ has no solution.

83. A fourth-degree polynomial with real coefficients can have $-5, 128i, 4i$, and 5 as its zeros.

84. Write quadratic equations that have (a) two distinct real solutions, (b) two complex solutions, and (c) no real solution.

Graphical Reasoning In Exercises 85 and 86, use the graph of the roots of a complex number.

(a) Write each of the roots in trigonometric form.

(b) Identify the complex number whose roots are given.

(c) Use a graphing utility to verify the results of part (b).

85.

86.

87. The figure shows z_1 and z_2. Describe $z_1 z_2$ and z_1/z_2.

FIGURE FOR 87   FIGURE FOR 88

88. One of the fourth roots of a complex number z is shown in the figure.

(a) How many roots are not shown?

(b) Describe the other roots.

4 | Chapter Test

Take this test as you would take a test in class. When you are finished, check your work against the answers given in the back of the book.

1. Write the complex number $-3 + \sqrt{-81}$ in standard form.

In Exercises 2–4, perform the operations and write the result in standard form.

2. $10i - \left(3 + \sqrt{-25}\right)$ **3.** $(2 + 6i)^2$ **4.** $\left(2 + \sqrt{3}i\right)\left(2 - \sqrt{3}i\right)$

5. Write the quotient in standard form: $\dfrac{5}{2 + i}$.

6. Use the Quadratic Formula to solve the equation $2x^2 - 2x + 3 = 0$.

In Exercises 7 and 8, determine the number of solutions of the equation in the complex number system.

7. $x^5 + x^3 - x + 1 = 0$ **8.** $x^4 - 3x^3 + 2x^2 - 4x - 5 = 0$

In Exercises 9 and 10, find all the zeros of the function.

9. $f(x) = x^3 - 6x^2 + 5x - 30$ **10.** $f(x) = x^4 - 2x^2 - 24$

In Exercises 11 and 12, use the given zero(s) to find all the zeros of the function. Write the polynomial as a product of linear factors.

Function	Zero(s)
11. $h(x) = x^4 - 2x^2 - 8$	$-2, 2$
12. $g(v) = 2v^3 - 11v^2 + 22v - 15$	$\frac{3}{2}$

In Exercises 13 and 14, find a polynomial function with real coefficients that has the given zeros. (There are many correct answers.)

13. $0, 3, 3 + i, 3 - i$ **14.** $1 + \sqrt{6}i, 1 - \sqrt{6}i, 3, 3$

15. Is it possible for a polynomial function with integer coefficients to have exactly one complex zero? Explain.

16. Write the complex number $z = 5 - 5i$ in trigonometric form.

17. Write the complex number $z = 6(\cos 120° + i \sin 120°)$ in standard form.

In Exercises 18 and 19, use DeMoivre's Theorem to find the indicated power of the complex number. Write the result in standard form.

18. $\left[3\left(\cos \dfrac{7\pi}{6} + i \sin \dfrac{7\pi}{6}\right)\right]^8$ **19.** $(3 - 3i)^6$

20. Find the fourth roots of $256\left(1 + \sqrt{3}i\right)$.

21. Find all solutions of the equation $x^3 - 27i = 0$ and represent the solutions graphically.

22. A projectile is fired upward from ground level with an initial velocity of 88 feet per second. The height h (in feet) of the projectile is given by

$$h = -16t^2 + 88t, \quad 0 \le t \le 5.5$$

where t is the time (in seconds). You are told that the projectile reaches a height of 125 feet. Is ths possible? Explain.

Proofs in Mathematics

The Linear Factorization Theorem is closely related to the Fundamental Theorem of Algebra. The Fundamental Theorem of Algebra has a long and interesting history. In the early work with polynomial equations, The Fundamental Theorem of Algebra was thought to have been not true, because imaginary solutions were not considered. In fact, in the very early work by mathematicians such as Abu al-Khwarizmi (c. 800 A.D.), negative solutions were also not considered.

Once imaginary numbers were accepted, several mathematicians attempted to give a general proof of the Fundamental Theorem of Algebra. These mathematicians included Gottfried von Leibniz (1702), Jean D'Alembert (1746), Leonhard Euler (1749), Joseph-Louis Lagrange (1772), and Pierre Simon Laplace (1795). The mathematician usually credited with the first correct proof of the Fundamental Theorem of Algebra is Carl Friedrich Gauss, who published the proof in his doctoral thesis in 1799.

Linear Factorization Theorem *(p. 339)*

If $f(x)$ is a polynomial of degree n, where $n > 0$, then f has precisely n linear factors

$$f(x) = a_n(x - c_1)(x - c_2) \cdots (x - c_n)$$

where $c_1, c_2, \ldots, c_n$ are complex numbers.

Proof

Using the Fundamental Theorem of Algebra, you know that f must have at least one zero, c_1. Consequently, $(x - c_1)$ is a factor of $f(x)$, and you have

$$f(x) = (x - c_1)f_1(x).$$

If the degree of $f_1(x)$ is greater than zero, you again apply the Fundamental Theorem to conclude that f_1 must have a zero c_2, which implies that

$$f(x) = (x - c_1)(x - c_2)f_2(x).$$

It is clear that the degree of $f_1(x)$ is $n - 1$, that the degree of $f_2(x)$ is $n - 2$, and that you can repeatedly apply the Fundamental Theorem n times until you obtain

$$f(x) = a_n(x - c_1)(x - c_2) \cdots (x - c_n)$$

where a_n is the leading coefficient of the polynomial $f(x)$.

This collection of thought-provoking and challenging exercises further explores and expands upon concepts learned in this chapter.

1. (a) The complex numbers

$$z = 2, z = \frac{-2 + 2\sqrt{3}i}{2}, \quad \text{and} \quad z = \frac{-2 - 2\sqrt{3}i}{2}$$

are represented graphically (see figure). Evaluate the expression z^3 for each complex number. What do you observe?

(b) The complex numbers

$$z = 3, z = \frac{-3 + 3\sqrt{3}i}{2}, \quad \text{and} \quad z = \frac{-3 - 3\sqrt{3}i}{2}$$

are represented graphically (see figure). Evaluate the expression z^3 for each complex number. What do you observe?

(c) Use your results from parts (a) and (b) to generalize your findings.

2. The multiplicative inverse of z is a complex number z_m such that $z \cdot z_m = 1$. Find the multiplicative inverse of each complex number.

(a) $z = 1 + i$

(b) $z = 3 - i$

(c) $z = -2 + 8i$

3. Show that the product of a complex number $a + bi$ and its conjugate is a real number.

4. Let

$$z = a + bi, \bar{z} = a - bi, w = c + di, \text{ and } \bar{w} = c - di.$$

Prove each statement.

(a) $\overline{z + w} = \bar{z} + \bar{w}$

(b) $\overline{z - w} = \bar{z} - \bar{w}$

(c) $\overline{zw} = \bar{z} \cdot \bar{w}$

(d) $\overline{z/w} = \bar{z}/\bar{w}$

(e) $(\bar{z})^2 = \overline{z^2}$

(f) $\bar{\bar{z}} = z$

(g) $\bar{z} = z$ if z is real.

5. Find the values of k such that the equation

$$x^2 - 2kx + k = 0$$

has (a) two real solutions and (b) two complex solutions.

 6. Use a graphing utility to graph the function given by

$$f(x) = x^4 - 4x^2 + k$$

for different values of k. Find values of k such that the zeros of f satisfy the specified characteristics. (Some parts do not have unique answers.)

(a) Four real zeros

(b) Two real zeros and two complex zeros

(c) Four complex zeros

7. Will the answers to Exercise 6 change for the function g?

(a) $g(x) = f(x - 2)$

(b) $g(x) = f(2x)$

8. A third-degree polynomial function f has real zeros $-2, \frac{1}{2}$, and 3, and its leading coefficient is negative.

(a) Write an equation for f.

(b) Sketch the graph of f.

(c) How many different polynomial functions are possible for f?

9. The graph of one of the following functions is shown below. Identify the function shown in the graph. Explain why each of the others is not the correct function. Use a graphing utility to verify your result.

(a) $f(x) = x^2(x + 2)(x - 3.5)$

(b) $g(x) = (x + 2)(x - 3.5)$

(c) $h(x) = (x + 2)(x - 3.5)(x^2 + 1)$

(d) $k(x) = (x + 1)(x + 2)(x - 3.5)$

10. Use the information in the table to answer each question.

Interval	Value of $f(x)$
$(-\infty, -2)$	Positive
$(-2, 1)$	Negative
$(1, 4)$	Negative
$(4, \infty)$	Positive

(a) What are the three real zeros of the polynomial function f?

(b) What can be said about the behavior of the graph of f at $x = 1$?

(c) What is the least possible degree of f? Explain. Can the degree of f ever be odd? Explain.

(d) Is the leading coefficient of f positive or negative? Explain.

(e) Write an equation for f.

(f) Sketch a graph of the function you wrote in part (e).

11. A **fractal** is a geometric figure that consists of a pattern that is repeated infinitely on a smaller and smaller scale. The most famous fractal is called the **Mandelbrot Set,** named after the Polish-born mathematician Benoit Mandelbrot. To draw the Mandelbrot Set, consider the following sequence of numbers.

$$c, c^2 + c, (c^2 + c)^2 + c, [(c^2 + c)^2 + c]^2 + c, \ldots$$

The behavior of this sequence depends on the value of the complex number c. If the sequence is bounded (the absolute value of each number in the sequence, $|a + bi| = \sqrt{a^2 + b^2}$, is less than some fixed number N), the complex number c is in the Mandelbrot Set, and if the sequence is unbounded (the absolute value of the terms of the sequence become infinitely large), the complex number c is not in the Mandelbrot Set. Determine whether the complex number c is in the Mandelbrot Set.

(a) $c = i$ (b) $c = 1 + i$ (c) $c = -2$

12. (a) Complete the table.

Function	Zeros	Sum of zeros	Product of zeros
$f_1(x) = x^2 - 5x + 6$			
$f_2(x) = x^3 - 7x + 6$			
$f_3(x) = x^4 + 2x^3 + x^2 + 8x - 12$			
$f_4(x) = x^5 - 3x^4 - 9x^3 + 25x^2 - 6x$			

(b) Use the table to make a conjecture relating the sum of the zeros of a polynomial function to the coefficients of the polynomial function.

(c) Use the table to make a conjecture relating the product of the zeros of a polynomial function to the coefficients of the polynomial function.

13. Use the Quadratic Formula and, if necessary, DeMoivre's Theorem to solve each equation with complex coefficients.

(a) $x^2 - (4 + 2i)x + 2 + 4i = 0$

(b) $x^2 - (3 + 2i)x + 5 + i = 0$

(c) $2x^2 + (5 - 8i)x - 13 - i = 0$

(d) $3x^2 - (11 + 14i)x + 1 - 9i = 0$

14. Show that the solutions to

$$|z - 1| \cdot |\bar{z} - 1| = 1$$

are the points (x, y) in the complex plane such that $(x - 1)^2 + y^2 = 1$. Identify the graph of the solution set. $\bar{z}$ is the conjugate of z. (*Hint:* Let $z = x + yi$.)

15. Let $z = a + bi$ and $\bar{z} = a - bi$. Show that the equation

$$z^2 - \bar{z}^2 = 0$$

has only real solutions, whereas the equation

$$z^2 + \bar{z}^2 = 0$$

has complex solutions.

Exponential and Logarithmic Functions

5

Carbon dating is a method used to determine the ages of archeological artifacts up to 50,000 years old. For example, archeologists are using carbon dating to determine the ages of the great pyramids of Egypt.

© Sylvain Grandadam/Getty Images

SELECTED APPLICATIONS

Exponential and logarithmic functions have many real-life applications. The applications listed below represent a small sample of the applications in this chapter.

- Computer Virus,
  Exercise 65, page 377

- Data Analysis: Meteorology,
  Exercise 70, page 378

- Sound Intensity,
  Exercise 90, page 388

- Galloping Speeds of Animals,
  Exercise 85, page 394

- Average Heights,
  Exercise 115, page 405

- Carbon Dating,
  Exercise 41, page 416

- IQ Scores,
  Exercise 47, page 416

- Forensics,
  Exercise 63, page 418

- Compound Interest,
  Exercise 135, page 423

5.1   Exponential Functions and Their Graphs

What you should learn

- Recognize and evaluate exponential functions with base a.
- Graph exponential functions and use the One-to-One Property.
- Recognize, evaluate, and graph exponential functions with base e.
- Use exponential functions to model and solve real-life problems.

Why you should learn it

Exponential functions can be used to model and solve real-life problems. For instance, in Exercise 70 on page 378, an exponential function is used to model the atmospheric pressure at different altitudes.

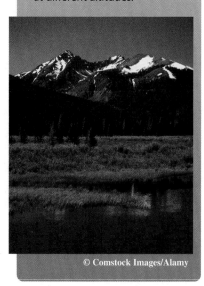
© Comstock Images/Alamy

The *HM mathSpace*® CD-ROM and *Eduspace*® for this text contain additional resources related to the concepts discussed in this chapter.

Exponential Functions

So far, this text has dealt mainly with **algebraic functions,** which include polynomial functions and rational functions. In this chapter, you will study two types of nonalgebraic functions—*exponential functions* and *logarithmic functions.* These functions are examples of **transcendental functions.**

Definition of Exponential Function

The **exponential function** f with base a is denoted by

$$f(x) = a^x$$

where $a > 0$, $a \neq 1$, and x is any real number.

The base $a = 1$ is excluded because it yields $f(x) = 1^x = 1$. This is a constant function, not an exponential function.

You have evaluated a^x for integer and rational values of x. For example, you know that $4^3 = 64$ and $4^{1/2} = 2$. However, to evaluate 4^x for any real number x, you need to interpret forms with *irrational* exponents. For the purposes of this text, it is sufficient to think of

$$a^{\sqrt{2}} \quad (\text{where } \sqrt{2} \approx 1.41421356)$$

as the number that has the successively closer approximations

$$a^{1.4}, a^{1.41}, a^{1.414}, a^{1.4142}, a^{1.41421}, \ldots$$

Example 1   **Evaluating Exponential Functions**

Use a calculator to evaluate each function at the indicated value of x.

Function	Value
a. $f(x) = 2^x$	$x = -3.1$
b. $f(x) = 2^{-x}$	$x = \pi$
c. $f(x) = 0.6^x$	$x = \frac{3}{2}$

Solution

Function Value	Graphing Calculator Keystrokes	Display
a. $f(-3.1) = 2^{-3.1}$	2 [^] [(−)] 3.1 [ENTER]	0.1166291
b. $f(\pi) = 2^{-\pi}$	2 [^] [(−)] π [ENTER]	0.1133147
c. $f\left(\frac{3}{2}\right) = (0.6)^{3/2}$	.6 [^] [(] 3 [÷] 2 [)] [ENTER]	0.4647580

✓CHECKPOINT   Now try Exercise 1.

When evaluating exponential functions with a calculator, remember to enclose fractional exponents in parentheses. Because the calculator follows the order of operations, parentheses are crucial in order to obtain the correct result.

Graphs of Exponential Functions

The graphs of all exponential functions have similar characteristics, as shown in Examples 2, 3, and 5.

Example 2   Graphs of $y = a^x$

In the same coordinate plane, sketch the graph of each function.

a. $f(x) = 2^x$ **b.** $g(x) = 4^x$

Solution

The table below lists some values for each function, and Figure 5.1 shows the graphs of the two functions. Note that both graphs are increasing. Moreover, the graph of $g(x) = 4^x$ is increasing more rapidly than the graph of $f(x) = 2^x$.

x	-3	-2	-1	0	1	2
2^x	$\frac{1}{8}$	$\frac{1}{4}$	$\frac{1}{2}$	1	2	4
4^x	$\frac{1}{64}$	$\frac{1}{16}$	$\frac{1}{4}$	1	4	16

✓CHECKPOINT   Now try Exercise 11.

The table in Example 2 was evaluated by hand. You could, of course, use a graphing utility to construct tables with even more values.

Example 3   Graphs of $y = a^{-x}$

In the same coordinate plane, sketch the graph of each function.

a. $F(x) = 2^{-x}$ **b.** $G(x) = 4^{-x}$

Solution

The table below lists some values for each function, and Figure 5.2 shows the graphs of the two functions. Note that both graphs are decreasing. Moreover, the graph of $G(x) = 4^{-x}$ is decreasing more rapidly than the graph of $F(x) = 2^{-x}$.

x	-2	-1	0	1	2	3
2^{-x}	4	2	1	$\frac{1}{2}$	$\frac{1}{4}$	$\frac{1}{8}$
4^{-x}	16	4	1	$\frac{1}{4}$	$\frac{1}{16}$	$\frac{1}{64}$

✓CHECKPOINT   Now try Exercise 13.

In Example 3, note that by using one of the properties of exponents, the functions $F(x) = 2^{-x}$ and $G(x) = 4^{-x}$ can be rewritten with positive exponents.

$$F(x) = 2^{-x} = \frac{1}{2^x} = \left(\frac{1}{2}\right)^x \quad \text{and} \quad G(x) = 4^{-x} = \frac{1}{4^x} = \left(\frac{1}{4}\right)^x$$

FIGURE 5.1

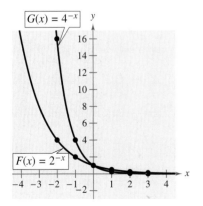

FIGURE 5.2

Comparing the functions in Examples 2 and 3, observe that

$$F(x) = 2^{-x} = f(-x) \quad \text{and} \quad G(x) = 4^{-x} = g(-x).$$

Consequently, the graph of F is a reflection (in the y-axis) of the graph of f. The graphs of G and g have the same relationship. The graphs in Figures 5.1 and 5.2 are typical of the exponential functions $y = a^x$ and $y = a^{-x}$. They have one y-intercept and one horizontal asymptote (the x-axis), and they are continuous. The basic characteristics of these exponential functions are summarized in Figures 5.3 and 5.4.

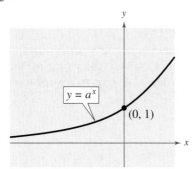

FIGURE 5.3

Graph of $y = a^x$, $a > 1$
- Domain: $(-\infty, \infty)$
- Range: $(0, \infty)$
- Intercept: $(0, 1)$
- Increasing
- x-axis is a horizontal asymptote ($a^x \rightarrow 0$ as $x \rightarrow -\infty$)
- Continuous

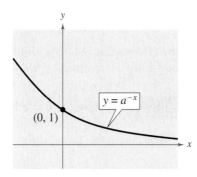

FIGURE 5.4

Graph of $y = a^{-x}$, $a > 1$
- Domain: $(-\infty, \infty)$
- Range: $(0, \infty)$
- Intercept: $(0, 1)$
- Decreasing
- x-axis is a horizontal asymptote ($a^{-x} \rightarrow 0$ as $x \rightarrow \infty$)
- Continuous

From Figures 5.3 and 5.4, you can see that the graph of an exponential function is always increasing or always decreasing. As a result, the graphs pass the Horizontal Line Test, and therefore the functions are one-to-one functions. You can use the following **One-to-One Property** to solve simple exponential equations.

For $a > 0$ and $a \neq 1$, $a^x = a^y$ if and only if $x = y$.   One to One Property

Example 4   **Using the One-to-One Property**

a.
$9 = 3^{x+1}$	Original equation
$3^2 = 3^{x+1}$	$9 = 3^2$
$2 = x + 1$	One-to-One Property
$1 = x$	Solve for x.

b. $\left(\frac{1}{2}\right)^x = 8 \Longrightarrow 2^{-x} = 2^3 \Longrightarrow x = -3$

✓CHECKPOINT   Now try Exercise 45.

In the following example, notice how the graph of $y = a^x$ can be used to sketch the graphs of functions of the form $f(x) = b \pm a^{x+c}$.

Example 5   Transformations of Graphs of Exponential Functions

Each of the following graphs is a transformation of the graph of $f(x) = 3^x$.

a. Because $g(x) = 3^{x+1} = f(x+1)$, the graph of g can be obtained by shifting the graph of f one unit to the *left*, as shown in Figure 5.5.

b. Because $h(x) = 3^x - 2 = f(x) - 2$, the graph of h can be obtained by shifting the graph of f *downward* two units, as shown in Figure 5.6.

c. Because $k(x) = -3^x = -f(x)$, the graph of k can be obtained by *reflecting* the graph of f in the x-axis, as shown in Figure 5.7.

d. Because $j(x) = 3^{-x} = f(-x)$, the graph of j can be obtained by *reflecting* the graph of f in the y-axis, as shown in Figure 5.8.

FIGURE 5.5   *Horizontal shift*

FIGURE 5.6   *Vertical shift*

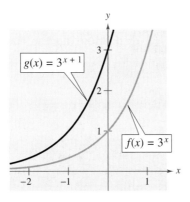

FIGURE 5.7   *Reflection in x-axis*

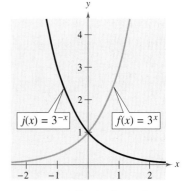

FIGURE 5.8   *Reflection in y-axis*

✓CHECKPOINT   Now try Exercise 17.

Notice that the transformations in Figures 5.5, 5.7, and 5.8 keep the x-axis as a horizontal asymptote, but the transformation in Figure 5.6 yields a new horizontal asymptote of $y = -2$. Also, be sure to note how the y-intercept is affected by each transformation.

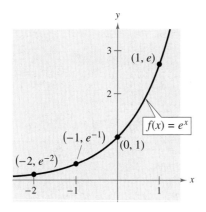

FIGURE **5.9**

The Natural Base *e*

In many applications, the most convenient choice for a base is the irrational number

$$e \approx 2.718281828 \ldots .$$

This number is called the **natural base.** The function given by $f(x) = e^x$ is called the **natural exponential function.** Its graph is shown in Figure 5.9. Be sure you see that for the exponential function $f(x) = e^x$, e is the constant $2.718281828 \ldots$, whereas x is the variable.

> *Exploration*
>
> Use a graphing utility to graph $y_1 = (1 + 1/x)^x$ and $y_2 = e$ in the same viewing window. Using the *trace* feature, explain what happens to the graph of y_1 as x increases.

Example 6   Evaluating the Natural Exponential Function

Use a calculator to evaluate the function given by $f(x) = e^x$ at each indicated value of x.

a. $x = -2$ **b.** $x = -1$ **c.** $x = 0.25$ **d.** $x = -0.3$

Solution

Function Value	Graphing Calculator Keystrokes	Display
a. $f(-2) = e^{-2}$	e^x $(-)$ 2 ENTER	0.1353353
b. $f(-1) = e^{-1}$	e^x $(-)$ 1 ENTER	0.3678794
c. $f(0.25) = e^{0.25}$	e^x 0.25 ENTER	1.2840254
d. $f(-0.3) = e^{-0.3}$	e^x $(-)$ 0.3 ENTER	0.7408182

 CHECKPOINT   Now try Exercise 27.

FIGURE **5.10**

FIGURE **5.11**

Example 7   Graphing Natural Exponential Functions

Sketch the graph of each natural exponential function.

a. $f(x) = 2e^{0.24x}$ **b.** $g(x) = \frac{1}{2}e^{-0.58x}$

Solution

To sketch these two graphs, you can use a graphing utility to construct a table of values, as shown below. After constructing the table, plot the points and connect them with smooth curves, as shown in Figures 5.10 and 5.11. Note that the graph in Figure 5.10 is increasing, whereas the graph in Figure 5.11 is decreasing.

x	-3	-2	-1	0	1	2	3
$f(x)$	0.974	1.238	1.573	2.000	2.542	3.232	4.109
$g(x)$	2.849	1.595	0.893	0.500	0.280	0.157	0.088

CHECKPOINT   Now try Exercise 35.

m	$\left(1 + \dfrac{1}{m}\right)^m$
1	2
10	2.59374246
100	2.704813829
1,000	2.716923932
10,000	2.718145927
100,000	2.718268237
1,000,000	2.718280469
10,000,000	2.718281693
$\downarrow$	$\downarrow$
∞	e

Applications

One of the most familiar examples of exponential growth is that of an investment earning *continuously compounded interest*. Using exponential functions, you can *develop* a formula for interest compounded n times per year and show how it leads to continuous compounding.

Suppose a principal P is invested at an annual interest rate r, compounded once a year. If the interest is added to the principal at the end of the year, the new balance P_1 is

$$P_1 = P + Pr$$
$$= P(1 + r).$$

This pattern of multiplying the previous principal by $1 + r$ is then repeated each successive year, as shown below.

Year	Balance After Each Compounding
0	$P = P$
1	$P_1 = P(1 + r)$
2	$P_2 = P_1(1 + r) = P(1 + r)(1 + r) = P(1 + r)^2$
3	$P_3 = P_2(1 + r) = P(1 + r)^2(1 + r) = P(1 + r)^3$
$\vdots$	$\vdots$
t	$P_t = P(1 + r)^t$

To accommodate more frequent (quarterly, monthly, or daily) compounding of interest, let n be the number of compoundings per year and let t be the number of years. Then the rate per compounding is r/n and the account balance after t years is

$$A = P\left(1 + \frac{r}{n}\right)^{nt}. \qquad \text{Amount (balance) with } n \text{ compoundings per year}$$

If you let the number of compoundings n increase without bound, the process approaches what is called **continuous compounding.** In the formula for n compoundings per year, let $m = n/r$. This produces

$$A = P\left(1 + \frac{r}{n}\right)^{nt} \qquad \text{Amount with } n \text{ compoundings per year}$$

$$= P\left(1 + \frac{r}{mr}\right)^{mrt} \qquad \text{Substitute } mr \text{ for } n.$$

$$= P\left(1 + \frac{1}{m}\right)^{mrt} \qquad \text{Simplify.}$$

$$= P\left[\left(1 + \frac{1}{m}\right)^m\right]^{rt}. \qquad \text{Property of exponents}$$

As m increases without bound, the table at the left shows that $[1 + (1/m)]^m \to e$ as $m \to \infty$. From this, you can conclude that the formula for continuous compounding is

$$A = Pe^{rt}. \qquad \text{Substitute } e \text{ for } (1 + 1/m)^m.$$

Activities

1. Sketch the graphs of the functions $f(x) = e^x$ and $g(x) = 1 + e^x$ on the same coordinate system.

2. Determine the balance A at the end of 20 years if $1500 is invested at 6.5% interest and the interest is compounded (a) quarterly and (b) continuously.
   Answer: (a) $5446.73   (b) $5503.95

3. The number of fruit flies in an experimental population after t hours is given by $Q(t) = 20e^{0.03t}$, $t \geq 0$.
   a. Find the initial number of fruit flies in the population.
   b. How large is the population of fruit flies after 72 hours?
   Answer: (a) 20 flies   (b) 173 flies

Group Activity

The sequence $3, 6, 9, 12, 15, \ldots$ is given by $f(n) = 3n$ and is an example of linear growth. The sequence $3, 9, 27, 81, 243, \ldots$ is given by $f(n) = 3^n$ and is an example of exponential growth. Explain the difference between these two types of growth. For each of the following sequences, indicate whether the sequence represents linear growth or exponential growth, and find a linear or exponential function that represents the sequence. Give several other examples of linear and exponential growth.

a. $\frac{1}{2}, \frac{1}{4}, \frac{1}{8}, \frac{1}{16}, \frac{1}{32}, \ldots$

b. $4, 8, 12, 16, 20, \ldots$

c. $\frac{2}{3}, \frac{4}{3}, 2, \frac{8}{3}, \frac{10}{3}, 4, \ldots$

d. $5, 25, 125, 625, \ldots$

> ### Formulas for Compound Interest
>
> After t years, the balance A in an account with principal P and annual interest rate r (in decimal form) is given by the following formulas.
>
> **1.** For n compoundings per year: $A = P\left(1 + \dfrac{r}{n}\right)^{nt}$
>
> **2.** For continuous compounding: $A = Pe^{rt}$

Example 8   Compound Interest

A total of $12,000 is invested at an annual interest rate of 9%. Find the balance after 5 years if it is compounded

a. quarterly.

b. monthly.

c. continuously.

Solution

a. For quarterly compounding, you have $n = 4$. So, in 5 years at 9%, the balance is

$$A = P\left(1 + \frac{r}{n}\right)^{nt}$$ Formula for compound interest

$$= 12{,}000\left(1 + \frac{0.09}{4}\right)^{4(5)}$$ Substitute for P, r, n, and t.

$$\approx \$18{,}726.11.$$ Use a calculator.

b. For monthly compounding, you have $n = 12$. So, in 5 years at 9%, the balance is

$$A = P\left(1 + \frac{r}{n}\right)^{nt}$$ Formula for compound interest

$$= 12{,}000\left(1 + \frac{0.09}{12}\right)^{12(5)}$$ Substitute for P, r, n, and t.

$$\approx \$18{,}788.17.$$ Use a calculator.

c. For continuous compounding, the balance is

$$A = Pe^{rt}$$ Formula for continuous compounding

$$= 12{,}000e^{0.09(5)}$$ Substitute for P, r, and t.

$$\approx \$18{,}819.75.$$ Use a calculator.

✔**CHECKPOINT**   Now try Exercise 53.

In Example 8, note that continuous compounding yields more than quarterly or monthly compounding. This is typical of the two types of compounding. That is, for a given principal, interest rate, and time, continuous compounding will always yield a larger balance than compounding n times a year.

Example 9   Radioactive Decay

In 1986, a nuclear reactor accident occurred in Chernobyl in what was then the Soviet Union. The explosion spread highly toxic radioactive chemicals, such as plutonium, over hundreds of square miles, and the government evacuated the city and the surrounding area. To see why the city is now uninhabited, consider the model

$$P = 10\left(\frac{1}{2}\right)^{t/24,100}$$

which represents the amount of plutonium P that remains (from an initial amount of 10 pounds) after t years. Sketch the graph of this function over the interval from $t = 0$ to $t = 100,000$, where $t = 0$ represents 1986. How much of the 10 pounds will remain in the year 2010? How much of the 10 pounds will remain after 100,000 years?

Solution

The graph of this function is shown in Figure 5.12. Note from this graph that plutonium has a *half-life* of about 24,100 years. That is, after 24,100 years, *half* of the original amount will remain. After another 24,100 years, one-quarter of the original amount will remain, and so on. In the year 2010 ($t = 24$), there will still be

$$P = 10\left(\frac{1}{2}\right)^{24/24,100} \approx 10\left(\frac{1}{2}\right)^{0.0009959} \approx 9.993 \text{ pounds}$$

of plutonium remaining. After 100,000 years, there will still be

$$P = 10\left(\frac{1}{2}\right)^{100,000/24,100} \approx 10\left(\frac{1}{2}\right)^{4.1494} \approx 0.564 \text{ pound}$$

of plutonium remaining.

✓**CHECKPOINT**   Now try Exercise 67.

Radioactive Decay

$$P = 10\left(\frac{1}{2}\right)^{t/24,100}$$

(24,100, 5)

(100,000, 0.564)

Plutonium (in pounds)

Years of decay

FIGURE **5.12**

✍RITING ABOUT ℳATHEMATICS

Identifying Exponential Functions Which of the following functions generated the two tables below? Discuss how you were able to decide. What do these functions have in common? Are any of them the same? If so, explain why.

a. $f_1(x) = 2^{(x+3)}$ **b.** $f_2(x) = 8\left(\frac{1}{2}\right)^x$ **c.** $f_3(x) = \left(\frac{1}{2}\right)^{(x-3)}$

d. $f_4(x) = \left(\frac{1}{2}\right)^x + 7$ **e.** $f_5(x) = 7 + 2^x$ **f.** $f_6(x) = (8)2^x$

x	-1	0	1	2	3
$g(x)$	7.5	8	9	11	15

x	-2	-1	0	1	2
$h(x)$	32	16	8	4	2

Create two different exponential functions of the forms $y = a(b)^x$ and $y = c^x + d$ with y-intercepts of $(0, -3)$.

| **5.1** | **Exercises** | The *HM mathSpace®* CD-ROM and *Eduspace®* for this text contain step-by-step solutions to all odd-numbered exercises. They also provide Tutorial Exercises for additional help. |

VOCABULARY CHECK: Fill in the blanks.

1. Polynomials and rational functions are examples of _____ functions.

2. Exponential and logarithmic functions are examples of nonalgebraic functions, also called _____ functions.

3. The exponential function given by $f(x) = e^x$ is called the _____ _____ function, and the base e is called the _____ base.

4. To find the amount A in an account after t years with principal P and an annual interest rate r compounded n times per year, you can use the formula _____.

5. To find the amount A in an account after t years with principal P and an annual interest rate r compounded continuously, you can use the formula _____.

PREREQUISITE SKILLS REVIEW: Practice and review algebra skills needed for this section at **www.Eduspace.com**.

In Exercises 1–6, evaluate the function at the indicated value of *x*. Round your result to three decimal places.

Function	Value
1. $f(x) = 3.4^x$	$x = 5.6$
2. $f(x) = 2.3^x$	$x = \frac{3}{2}$
3. $f(x) = 5^x$	$x = -\pi$
4. $f(x) = \left(\frac{2}{3}\right)^{5x}$	$x = \frac{3}{10}$
5. $g(x) = 5000(2^x)$	$x = -1.5$
6. $f(x) = 200(1.2)^{12x}$	$x = 24$

In Exercises 7–10, match the exponential function with its graph. [The graphs are labeled (a), (b), (c), and (d).]

(a)

(b)

(c)

(d)
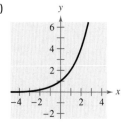

7. $f(x) = 2^x$

8. $f(x) = 2^x + 1$

9. $f(x) = 2^{-x}$

10. $f(x) = 2^{x-2}$

 In Exercises 11–16, use a graphing utility to construct a table of values for the function. Then sketch the graph of the function.

11. $f(x) = \left(\frac{1}{2}\right)^x$ **12.** $f(x) = \left(\frac{1}{2}\right)^{-x}$

13. $f(x) = 6^{-x}$ **14.** $f(x) = 6^x$

15. $f(x) = 2^{x-1}$ **16.** $f(x) = 4^{x-3} + 3$

In Exercises 17–22, use the graph of *f* to describe the transformation that yields the graph of *g*.

17. $f(x) = 3^x$, $g(x) = 3^{x-4}$

18. $f(x) = 4^x$, $g(x) = 4^x + 1$

19. $f(x) = -2^x$, $g(x) = 5 - 2^x$

20. $f(x) = 10^x$, $g(x) = 10^{-x+3}$

21. $f(x) = \left(\frac{7}{2}\right)^x$, $g(x) = -\left(\frac{7}{2}\right)^{-x+6}$

22. $f(x) = 0.3^x$, $g(x) = -0.3^x + 5$

 In Exercises 23–26, use a graphing utility to graph the exponential function.

23. $y = 2^{-x^2}$ **24.** $y = 3^{-|x|}$

25. $y = 3^{x-2} + 1$ **26.** $y = 4^{x+1} - 2$

In Exercises 27–32, evaluate the function at the indicated value of *x*. Round your result to three decimal places.

Function	Value
27. $h(x) = e^{-x}$	$x = \frac{3}{4}$
28. $f(x) = e^x$	$x = 3.2$
29. $f(x) = 2e^{-5x}$	$x = 10$
30. $f(x) = 1.5e^{x/2}$	$x = 240$
31. $f(x) = 5000e^{0.06x}$	$x = 6$
32. $f(x) = 250e^{0.05x}$	$x = 20$

In Exercises 33–38, use a graphing utility to construct a table of values for the function. Then sketch the graph of the function.

33. $f(x) = e^x$

34. $f(x) = e^{-x}$

35. $f(x) = 3e^{x+4}$

36. $f(x) = 2e^{-0.5x}$

37. $f(x) = 2e^{x-2} + 4$

38. $f(x) = 2 + e^{x-5}$

 In Exercises 39–44, use a graphing utility to graph the exponential function.

39. $y = 1.08^{-5x}$

40. $y = 1.08^{5x}$

41. $s(t) = 2e^{0.12t}$

42. $s(t) = 3e^{-0.2t}$

43. $g(x) = 1 + e^{-x}$

44. $h(x) = e^{x-2}$

In Exercise 45–52, use the One-to-One Property to solve the equation for x.

45. $3^{x+1} = 27$

46. $2^{x-3} = 16$

47. $2^{x-2} = \dfrac{1}{32}$

48. $\left(\dfrac{1}{5}\right)^{x+1} = 125$

49. $e^{3x+2} = e^3$

50. $e^{2x-1} = e^4$

51. $e^{x^2-3} = e^{2x}$

52. $e^{x^2+6} = e^{5x}$

Compound Interest   **In Exercises 53–56, complete the table to determine the balance A for P dollars invested at rate r for t years and compounded n times per year.**

n	1	2	4	12	365	Continuous
A						

53. $P = \$2500$, $r = 2.5\%$, $t = 10$ years

54. $P = \$1000$, $r = 4\%$, $t = 10$ years

55. $P = \$2500$, $r = 3\%$, $t = 20$ years

56. $P = \$1000$, $r = 6\%$, $t = 40$ years

Compound Interest   **In Exercises 57–60, complete the table to determine the balance A for $12,000 invested at rate r for t years, compounded continuously.**

t	10	20	30	40	50
A					

57. $r = 4\%$

58. $r = 6\%$

59. $r = 6.5\%$

60. $r = 3.5\%$

61. *Trust Fund*   On the day of a child's birth, a deposit of $25,000 is made in a trust fund that pays 8.75% interest, compounded continuously. Determine the balance in this account on the child's 25th birthday.

62. *Trust Fund*   A deposit of $5000 is made in a trust fund that pays 7.5% interest, compounded continuously. It is specified that the balance will be given to the college from which the donor graduated after the money has earned interest for 50 years. How much will the college receive?

63. *Inflation*   If the annual rate of inflation averages 4% over the next 10 years, the approximate costs C of goods or services during any year in that decade will be modeled by $C(t) = P(1.04)^t$, where t is the time in years and P is the present cost. The price of an oil change for your car is presently $23.95. Estimate the price 10 years from now.

64. *Demand*   The demand equation for a product is given by

$$p = 5000\left(1 - \frac{4}{4 + e^{-0.002x}}\right)$$

where p is the price and x is the number of units.

 (a) Use a graphing utility to graph the demand function for $x > 0$ and $p > 0$.

(b) Find the price p for a demand of $x = 500$ units.

 (c) Use the graph in part (a) to approximate the greatest price that will still yield a demand of at least 600 units.

65. *Computer Virus*   The number V of computers infected by a computer virus increases according to the model $V(t) = 100e^{4.6052t}$, where t is the time in hours. Find (a) $V(1)$, (b) $V(1.5)$, and (c) $V(2)$.

66. *Population*   The population P (in millions) of Russia from 1996 to 2004 can be approximated by the model $P = 152.26e^{-0.0039t}$, where t represents the year, with $t = 6$ corresponding to 1996. (Source: Census Bureau, International Data Base)

(a) According to the model, is the population of Russia increasing or decreasing? Explain.

(b) Find the population of Russia in 1998 and 2000.

(c) Use the model to predict the population of Russia in 2010.

67. *Radioactive Decay*   Let Q represent a mass of radioactive radium (^{226}Ra) (in grams), whose half-life is 1599 years. The quantity of radium present after t years is $Q = 25\left(\frac{1}{2}\right)^{t/1599}$.

(a) Determine the initial quantity (when $t = 0$).

(b) Determine the quantity present after 1000 years.

 (c) Use a graphing utility to graph the function over the interval $t = 0$ to $t = 5000$.

68. *Radioactive Decay*   Let Q represent a mass of carbon 14 (^{14}C) (in grams), whose half-life is 5715 years. The quantity of carbon 14 present after t years is $Q = 10\left(\frac{1}{2}\right)^{t/5715}$.

(a) Determine the initial quantity (when $t = 0$).

(b) Determine the quantity present after 2000 years.

(c) Sketch the graph of this function over the interval $t = 0$ to $t = 10,000$.

Model It

69. Data Analysis: Biology To estimate the amount of defoliation caused by the gypsy moth during a given year, a forester counts the number x of egg masses on $\frac{1}{40}$ of an acre (circle of radius 18.6 feet) in the fall. The percent of defoliation y the next spring is shown in the table. (Source: USDA, Forest Service)

Egg masses, x	Percent of defoliation, y
0	12
25	44
50	81
75	96
100	99

A model for the data is given by

$$y = \frac{100}{1 + 7e^{-0.069x}}.$$

 (a) Use a graphing utility to create a scatter plot of the data and graph the model in the same viewing window.

(b) Create a table that compares the model with the sample data.

(c) Estimate the percent of defoliation if 36 egg masses are counted on $\frac{1}{40}$ acre.

 (d) You observe that $\frac{2}{3}$ of a forest is defoliated the following spring. Use the graph in part (a) to estimate the number of egg masses per $\frac{1}{40}$ acre.

70. Data Analysis: Meteorology A meteorologist measures the atmospheric pressure P (in pascals) at altitude h (in kilometers). The data are shown in the table.

Altitude, h	Pressure, P
0	101,293
5	54,735
10	23,294
15	12,157
20	5,069

A model for the data is given by $P = 107,428e^{-0.150h}$.

(a) Sketch a scatter plot of the data and graph the model on the same set of axes.

(b) Estimate the atmospheric pressure at a height of 8 kilometers.

Synthesis

True or False? In Exercises 71 and 72, determine whether the statement is true or false. Justify your answer.

71. The line $y = -2$ is an asymptote for the graph of $f(x) = 10^x - 2$.

72. $e = \dfrac{271,801}{99,990}$.

Think About It In Exercises 73–76, use properties of exponents to determine which functions (if any) are the same.

73. $f(x) = 3^{x-2}$
$g(x) = 3^x - 9$
$h(x) = \frac{1}{9}(3^x)$

74. $f(x) = 4^x + 12$
$g(x) = 2^{2x+6}$
$h(x) = 64(4^x)$

75. $f(x) = 16(4^{-x})$
$g(x) = \left(\frac{1}{4}\right)^{x-2}$
$h(x) = 16(2^{-2x})$

76. $f(x) = e^{-x} + 3$
$g(x) = e^{3-x}$
$h(x) = -e^{x-3}$

77. Graph the functions given by $y = 3^x$ and $y = 4^x$ and use the graphs to solve each inequality.

(a) $4^x < 3^x$ 　　(b) $4^x > 3^x$

 78. Use a graphing utility to graph each function. Use the graph to find where the function is increasing and decreasing, and approximate any relative maximum or minimum values.

(a) $f(x) = x^2 e^{-x}$ 　　(b) $g(x) = x2^{3-x}$

 79. Graphical Analysis Use a graphing utility to graph

$$f(x) = \left(1 + \frac{0.5}{x}\right)^x \quad \text{and} \quad g(x) = e^{0.5}$$

in the same viewing window. What is the relationship between f and g as x increases and decreases without bound?

80. Think About It Which functions are exponential?

(a) $3x$ 　(b) $3x^2$ 　(c) 3^x 　(d) 2^{-x}

Skills Review

In Exercises 81 and 82, solve for y.

81. $x^2 + y^2 = 25$ 　　　　**82.** $x - |y| = 2$

In Exercises 83 and 84, sketch the graph of the function.

83. $f(x) = \dfrac{2}{9 + x}$ 　　　　**84.** $f(x) = \sqrt{7 - x}$

85. Make a Decision To work an extended application analyzing the population per square mile of the United States, visit this text's website at *college.hmco.com*. (Data Source: U.S. Census Bureau)

5.2 Logarithmic Functions and Their Graphs

What you should learn

- Recognize and evaluate logarithmic functions with base a.
- Graph logarithmic functions.
- Recognize, evaluate, and graph natural logarithmic functions.
- Use logarithmic functions to model and solve real-life problems.

Why you should learn it

Logarithmic functions are often used to model scientific observations. For instance, in Exercise 89 on page 388, a logarithmic function is used to model human memory.

© Ariel Skelley/Corbis

Logarithmic Functions

In Section P.10, you studied the concept of an inverse function. There, you learned that if a function is one-to-one—that is, if the function has the property that no horizontal line intersects the graph of the function more than once—the function must have an inverse function. By looking back at the graphs of the exponential functions introduced in Section 5.1, you will see that every function of the form $f(x) = a^x$ passes the Horizontal Line Test and therefore must have an inverse function. This inverse function is called the **logarithmic function with base a.**

Definition of Logarithmic Function with Base a

For $x > 0$, $a > 0$, and $a \neq 1$,

$$y = \log_a x \text{ if and only if } x = a^y.$$

The function given by

$$f(x) = \log_a x \qquad \text{Read as "log base } a \text{ of } x."$$

is called the **logarithmic function with base a.**

The equations

$$y = \log_a x \qquad \text{and} \qquad x = a^y$$

are equivalent. The first equation is in logarithmic form and the second is in exponential form. For example, the logarithmic equation $2 = \log_3 9$ can be rewritten in exponential form as $9 = 3^2$. The exponential equation $5^3 = 125$ can be rewritten in logarithmic form as $\log_5 125 = 3$.

When evaluating logarithms, remember that *a logarithm is an exponent.* This means that $\log_a x$ is the exponent to which a must be raised to obtain x. For instance, $\log_2 8 = 3$ because 2 must be raised to the third power to get 8.

Example 1   **Evaluating Logarithms**

Use the definition of logarithmic function to evaluate each logarithm at the indicated value of x.

a. $f(x) = \log_2 x, \quad x = 32$ **b.** $f(x) = \log_3 x, \quad x = 1$

c. $f(x) = \log_4 x, \quad x = 2$ **d.** $f(x) = \log_{10} x, \quad x = \frac{1}{100}$

Solution

a. $f(32) = \log_2 32 = 5$ because $2^5 = 32$.

b. $f(1) = \log_3 1 = 0$ because $3^0 = 1$.

c. $f(2) = \log_4 2 = \frac{1}{2}$ because $4^{1/2} = \sqrt{4} = 2$.

d. $f\left(\frac{1}{100}\right) = \log_{10} \frac{1}{100} = -2$ because $10^{-2} = \frac{1}{10^2} = \frac{1}{100}$.

✓CHECKPOINT   Now try Exercise 17.

STUDY TIP

Remember that a logarithm is an exponent. So, to evaluate the logarithmic expression $\log_a x$, you need to ask the question, "To what power must a be raised to obtain x?"

Complete the table for
$f(x) = 10^x$.

x	-2	-1	0	1	2
$f(x)$					

Complete the table for
$f(x) = \log x$.

x	$\frac{1}{100}$	$\frac{1}{10}$	1	10	100
$f(x)$					

Compare the two tables. What is the relationship between $f(x) = 10^x$ and $f(x) = \log x$?

The logarithmic function with base 10 is called the **common logarithmic function.** It is denoted by $\log_{10}$ or simply by log. On most calculators, this function is denoted by $\boxed{\text{LOG}}$. Example 2 shows how to use a calculator to evaluate common logarithmic functions. You will learn how to use a calculator to calculate logarithms to any base in the next section.

Example 2 **Evaluating Common Logarithms on a Calculator**

Use a calculator to evaluate the function given by $f(x) = \log x$ at each value of x.

a. $x = 10$ **b.** $x = \frac{1}{3}$ **c.** $x = 2.5$ **d.** $x = -2$

Solution

Function Value	Graphing Calculator Keystrokes	Display
a. $f(10) = \log 10$	$\boxed{\text{LOG}}$ 10 $\boxed{\text{ENTER}}$	1
b. $f\left(\frac{1}{3}\right) = \log \frac{1}{3}$	$\boxed{\text{LOG}}$ $\boxed{(}$ 1 $\boxed{\div}$ 3 $\boxed{)}$ $\boxed{\text{ENTER}}$	-0.4771213
c. $f(2.5) = \log 2.5$	$\boxed{\text{LOG}}$ 2.5 $\boxed{\text{ENTER}}$	0.3979400
d. $f(-2) = \log(-2)$	$\boxed{\text{LOG}}$ $\boxed{(-)}$ 2 $\boxed{\text{ENTER}}$	ERROR

Note that the calculator displays an error message (or a complex number) when you try to evaluate $\log(-2)$. The reason for this is that there is no real number power to which 10 can be raised to obtain -2.

✓**CHECKPOINT** Now try Exercise 23.

The following properties follow directly from the definition of the logarithmic function with base a.

The logarithmic function can be one of the most difficult concepts for students to understand. Remind students that a logarithm is an exponent. Converting back and forth from logarithmic form to exponential form supports this concept.

> **Properties of Logarithms**
>
> **1.** $\log_a 1 = 0$ because $a^0 = 1$.
>
> **2.** $\log_a a = 1$ because $a^1 = a$.
>
> **3.** $\log_a a^x = x$ and $a^{\log_a x} = x$ Inverse Properties
>
> **4.** If $\log_a x = \log_a y$, then $x = y$.   One-to-One Property

Example 3 **Using Properties of Logarithms**

a. Simplify: $\log_4 1$ **b.** Simplify: $\log_{\sqrt{7}} \sqrt{7}$ **c.** Simplify: $6^{\log_6 20}$

Solution

a. Using Property 1, it follows that $\log_4 1 = 0$.

b. Using Property 2, you can conclude that $\log_{\sqrt{7}} \sqrt{7} = 1$.

c. Using the Inverse Property (Property 3), it follows that $6^{\log_6 20} = 20$.

✓**CHECKPOINT** Now try Exercise 27.

You can use the One-to-One Property (Property 4) to solve simple logarithmic equations, as shown in Example 4.

Example 4   **Using the One-to-One Property**

a. $\log_3 x = \log_3 12$ Original equation

         $x = 12$ One-to-One Property

b. $\log(2x + 1) = \log x \implies 2x + 1 = x \implies x = -1$

c. $\log_4(x^2 - 6) = \log_4 10 \implies x^2 - 6 = 10 \implies x^2 = 16 \implies x = \pm 4$

✓CHECKPOINT   Now try Exercise 79.

Graphs of Logarithmic Functions

To sketch the graph of $y = \log_a x$, you can use the fact that the graphs of inverse functions are reflections of each other in the line $y = x$.

Example 5   **Graphs of Exponential and Logarithmic Functions**

In the same coordinate plane, sketch the graph of each function.

a. $f(x) = 2^x$ **b.** $g(x) = \log_2 x$

Solution

a. For $f(x) = 2^x$, construct a table of values. By plotting these points and connecting them with a smooth curve, you obtain the graph shown in Figure 5.13.

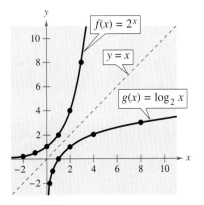

FIGURE 5.13

x	-2	-1	0	1	2	3
$f(x) = 2^x$	$\frac{1}{4}$	$\frac{1}{2}$	1	2	4	8

b. Because $g(x) = \log_2 x$ is the inverse function of $f(x) = 2^x$, the graph of g is obtained by plotting the points $(f(x), x)$ and connecting them with a smooth curve. The graph of g is a reflection of the graph of f in the line $y = x$, as shown in Figure 5.13.

✓CHECKPOINT   Now try Exercise 31.

Example 6   **Sketching the Graph of a Logarithmic Function**

Sketch the graph of the common logarithmic function $f(x) = \log x$. Identify the vertical asymptote.

Solution

Begin by constructing a table of values. Note that some of the values can be obtained without a calculator by using the Inverse Property of Logarithms. Others require a calculator. Next, plot the points and connect them with a smooth curve, as shown in Figure 5.14. The vertical asymptote is $x = 0$ (y-axis).

FIGURE 5.14

x	Without calculator				With calculator		
	$\frac{1}{100}$	$\frac{1}{10}$	1	10	2	5	8
$f(x) = \log x$	-2	-1	0	1	0.301	0.699	0.903

✓CHECKPOINT   Now try Exercise 37.

The nature of the graph in Figure 5.14 is typical of functions of the form $f(x) = \log_a x, a > 1$. They have one x-intercept and one vertical asymptote. Notice how slowly the graph rises for $x > 1$. The basic characteristics of logarithmic graphs are summarized in Figure 5.15.

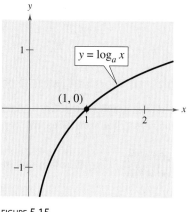

FIGURE 5.15

Graph of $y = \log_a x, a > 1$
- Domain: $(0, \infty)$
- Range: $(-\infty, \infty)$
- x-intercept: $(1, 0)$
- Increasing
- One-to-one, therefore has an inverse function
- y-axis is a vertical asymptote $(\log_a x \to -\infty$ as $x \to 0^+)$.
- Continuous
- Reflection of graph of $y = a^x$ about the line $y = x$

The basic characteristics of the graph of $f(x) = a^x$ are shown below to illustrate the inverse relation between $f(x) = a^x$ and $g(x) = \log_a x$.

- Domain: $(-\infty, \infty)$ • Range: $(0, \infty)$
- y-intercept: $(0,1)$ • x-axis is a horizontal asymptote $(a^x \to 0$ as $x \to -\infty)$.

In the next example, the graph of $y = \log_a x$ is used to sketch the graphs of functions of the form $f(x) = b \pm \log_a(x + c)$. Notice how a horizontal shift of the graph results in a horizontal shift of the vertical asymptote.

STUDY TIP

You can use your understanding of transformations to identify vertical asymptotes of logarithmic functions. For instance, in Example 7(a) the graph of $g(x) = f(x - 1)$ shifts the graph of $f(x)$ one unit to the right. So, the vertical asymptote of $g(x)$ is $x = 1$, one unit to the right of the vertical asymptote of the graph of $f(x)$.

Example 7   Shifting Graphs of Logarithmic Functions

The graph of each of the functions is similar to the graph of $f(x) = \log x$.

a. Because $g(x) = \log(x - 1) = f(x - 1)$, the graph of g can be obtained by shifting the graph of f one unit to the right, as shown in Figure 5.16.

b. Because $h(x) = 2 + \log x = 2 + f(x)$, the graph of h can be obtained by shifting the graph of f two units upward, as shown in Figure 5.17.

FIGURE 5.16

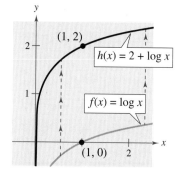

FIGURE 5.17

✓CHECKPOINT   Now try Exercise 39.

The Natural Logarithmic Function

By looking back at the graph of the natural exponential function introduced in Section 5.1 on page 372, you will see that $f(x) = e^x$ is one-to-one and so has an inverse function. This inverse function is called the **natural logarithmic function** and is denoted by the special symbol $\ln x$, read as "the natural log of x" or "el en of x." Note that the natural logarithm is written without a base. The base is understood to be e.

Reflection of graph of $f(x) = e^x$ about the line $y = x$

FIGURE 5.18

> ### The Natural Logarithmic Function
>
> The function defined by
>
> $$f(x) = \log_e x = \ln x, \quad x > 0$$
>
> is called the **natural logarithmic function.**

The definition above implies that the natural logarithmic function and the natural exponential function are inverse functions of each other. So, every logarithmic equation can be written in an equivalent exponential form and every exponential equation can be written in logarithmic form. That is, $y = \ln x$ and $x = e^y$ are equivalent equations.

Because the functions given by $f(x) = e^x$ and $g(x) = \ln x$ are inverse functions of each other, their graphs are reflections of each other in the line $y = x$. This reflective property is illustrated in Figure 5.18.

On most calculators, the natural logarithm is denoted by [LN], as illustrated in Example 8.

Example 8   Evaluating the Natural Logarithmic Function

Use a calculator to evaluate the function given by $f(x) = \ln x$ for each value of x.

a. $x = 2$ **b.** $x = 0.3$ **c.** $x = -1$ **d.** $x = 1 + \sqrt{2}$

Solution

Function Value	Graphing Calculator Keystrokes	Display
a. $f(2) = \ln 2$	[LN] 2 [ENTER]	0.6931472
b. $f(0.3) = \ln 0.3$	[LN] .3 [ENTER]	−1.2039728
c. $f(-1) = \ln(-1)$	[LN] [(−)] 1 [ENTER]	ERROR
d. $f(1 + \sqrt{2}) = \ln(1 + \sqrt{2})$	[LN] [(] 1 [+] [√] 2 [)] [ENTER]	0.8813736

✓CHECKPOINT   Now try Exercise 61.

In Example 8, be sure you see that $\ln(-1)$ gives an error message on most calculators. (Some calculators may display a complex number.) This occurs because the domain of $\ln x$ is the set of positive real numbers (see Figure 5.18). So, $\ln(-1)$ is undefined.

The four properties of logarithms listed on page 380 are also valid for natural logarithms.

STUDY TIP

Notice that as with every other logarithmic function, the domain of the natural logarithmic function is the set of *positive real numbers*—be sure you see that $\ln x$ is not defined for zero or for negative numbers.

Properties of Natural Logarithms

1. $\ln 1 = 0$ because $e^0 = 1$.

2. $\ln e = 1$ because $e^1 = e$.

3. $\ln e^x = x$ and $e^{\ln x} = x$ Inverse Properties

4. If $\ln x = \ln y$, then $x = y$.   One-to-One Property

Example 9   **Using Properties of Natural Logarithms**

Use the properties of natural logarithms to simplify each expression.

a. $\ln \dfrac{1}{e}$ **b.** $e^{\ln 5}$ **c.** $\dfrac{\ln 1}{3}$ **d.** $2 \ln e$

Solution

a. $\ln \dfrac{1}{e} = \ln e^{-1} = -1$ Inverse Property   **b.** $e^{\ln 5} = 5$ Inverse Property

c. $\dfrac{\ln 1}{3} = \dfrac{0}{3} = 0$ Property 1   **d.** $2 \ln e = 2(1) = 2$ Property 2

✔CHECKPOINT   Now try Exercise 65.

Example 10   **Finding the Domains of Logarithmic Functions**

Find the domain of each function.

a. $f(x) = \ln(x - 2)$ **b.** $g(x) = \ln(2 - x)$ **c.** $h(x) = \ln x^2$

Solution

a. Because $\ln(x - 2)$ is defined only if $x - 2 > 0$, it follows that the domain of f is $(2, \infty)$. The graph of f is shown in Figure 5.19.

b. Because $\ln(2 - x)$ is defined only if $2 - x > 0$, it follows that the domain of g is $(-\infty, 2)$. The graph of g is shown in Figure 5.20.

c. Because $\ln x^2$ is defined only if $x^2 > 0$, it follows that the domain of h is all real numbers except $x = 0$. The graph of h is shown in Figure 5.21.

FIGURE 5.19

FIGURE 5.20

FIGURE 5.21

✔CHECKPOINT   Now try Exercise 69.

Memory Model

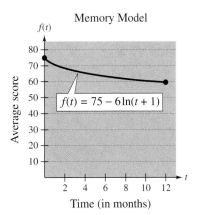

$f(t) = 75 - 6\ln(t + 1)$

FIGURE 5.22

Application

Example 11   Human Memory Model

Students participating in a psychology experiment attended several lectures on a subject and were given an exam. Every month for a year after the exam, the students were retested to see how much of the material they remembered. The average scores for the group are given by the *human memory model*

$$f(t) = 75 - 6 \ln(t + 1), \quad 0 \le t \le 12$$

where t is the time in months. The graph of f is shown in Figure 5.22.

a. What was the average score on the original ($t = 0$) exam?
b. What was the average score at the end of $t = 2$ months?
c. What was the average score at the end of $t = 6$ months?

Solution

a. The original average score was

$$f(0) = 75 - 6 \ln(0 + 1) \qquad \text{Substitute 0 for } t.$$
$$= 75 - 6 \ln 1 \qquad \text{Simplify.}$$
$$= 75 - 6(0) \qquad \text{Property of natural logarithms}$$
$$= 75. \qquad \text{Solution}$$

b. After 2 months, the average score was

$$f(2) = 75 - 6 \ln(2 + 1) \qquad \text{Substitute 2 for } t.$$
$$= 75 - 6 \ln 3 \qquad \text{Simplify.}$$
$$\approx 75 - 6(1.0986) \qquad \text{Use a calculator.}$$
$$\approx 68.4. \qquad \text{Solution}$$

c. After 6 months, the average score was

$$f(6) = 75 - 6 \ln(6 + 1) \qquad \text{Substitute 6 for } t.$$
$$= 75 - 6 \ln 7 \qquad \text{Simplify.}$$
$$\approx 75 - 6(1.9459) \qquad \text{Use a calculator.}$$
$$\approx 63.3. \qquad \text{Solution}$$

✓**CHECKPOINT**   Now try Exercise 89.

Alternative Writing About Mathematics

Use a graphing utility to graph $f(x) = \ln x$. How will the graphs of $h(x) = \ln x + 5$, $j(x) = \ln(x - 3)$, and $l(x) = \ln x - 4$ differ from the graph of f?

How will the basic graph of f be affected when a constant c is introduced: $g(x) = c \ln x$? Use a graphing utility to graph g with several different positive values of c, and summarize the effect of c.

WRITING ABOUT **M**ATHEMATICS

Analyzing a Human Memory Model Use a graphing utility to determine the time in months when the average score in Example 11 was 60. Explain your method of solving the problem. Describe another way that you can use a graphing utility to determine the answer.

5.2 Exercises

VOCABULARY CHECK: Fill in the blanks.

1. The inverse function of the exponential function given by $f(x) = a^x$ is called the _____ function with base a.

2. The common logarithmic function has base _____ .

3. The logarithmic function given by $f(x) = \ln x$ is called the _____ logarithmic function and has base _____.

4. The Inverse Property of logarithms and exponentials states that $\log_a a^x = x$ and _____.

5. The One-to-One Property of natural logarithms states that if $\ln x = \ln y$, then _____.

PREREQUISITE SKILLS REVIEW: Practice and review algebra skills needed for this section at **www.Eduspace.com.**

In Exercises 1–8, write the logarithmic equation in exponential form. For example, the exponential form of $\log_5 25 = 2$ is $5^2 = 25$.

1. $\log_4 64 = 3$ **2.** $\log_3 81 = 4$

3. $\log_7 \frac{1}{49} = -2$ **4.** $\log \frac{1}{1000} = -3$

5. $\log_{32} 4 = \frac{2}{5}$ **6.** $\log_{16} 8 = \frac{3}{4}$

7. $\log_{36} 6 = \frac{1}{2}$ **8.** $\log_8 4 = \frac{2}{3}$

In Exercises 9–16, write the exponential equation in logarithmic form. For example, the logarithmic form of $2^3 = 8$ is $\log_2 8 = 3$.

9. $5^3 = 125$ **10.** $8^2 = 64$

11. $81^{1/4} = 3$ **12.** $9^{3/2} = 27$

13. $6^{-2} = \frac{1}{36}$ **14.** $4^{-3} = \frac{1}{64}$

15. $7^0 = 1$ **16.** $10^{-3} = 0.001$

In Exercises 17–22, evaluate the function at the indicated value of x without using a calculator.

Function	Value
17. $f(x) = \log_2 x$	$x = 16$
18. $f(x) = \log_{16} x$	$x = 4$
19. $f(x) = \log_7 x$	$x = 1$
20. $f(x) = \log x$	$x = 10$
21. $g(x) = \log_a x$	$x = a^2$
22. $g(x) = \log_b x$	$x = b^{-3}$

 In Exercises 23–26, use a calculator to evaluate $f(x) = \log x$ at the indicated value of x. Round your result to three decimal places.

23. $x = \frac{4}{5}$ **24.** $x = \frac{1}{500}$

25. $x = 12.5$ **26.** $x = 75.25$

In Exercises 27–30, use the properties of logarithms to simplify the expression.

27. $\log_3 3^4$ **28.** $\log_{1.5} 1$

29. $\log_\pi \pi$ **30.** $9^{\log_9 15}$

In Exercises 31–38, find the domain, x-intercept, and vertical asymptote of the logarithmic function and sketch its graph.

31. $f(x) = \log_4 x$ **32.** $g(x) = \log_6 x$

33. $y = -\log_3 x + 2$ **34.** $h(x) = \log_4(x - 3)$

35. $f(x) = -\log_6(x + 2)$ **36.** $y = \log_5(x - 1) + 4$

37. $y = \log\left(\dfrac{x}{5}\right)$ **38.** $y = \log(-x)$

In Exercises 39–44, use the graph of $g(x) = \log_3 x$ to match the given function with its graph. Then describe the relationship between the graphs of f and g. [The graphs are labeled (a), (b), (c), (d), (e), and (f).]

(a)

(b)

(c)

(d)

(e)

(f)

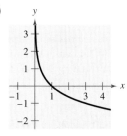

39. $f(x) = \log_3 x + 2$

40. $f(x) = -\log_3 x$

41. $f(x) = -\log_3(x + 2)$

42. $f(x) = \log_3(x - 1)$

43. $f(x) = \log_3(1 - x)$

44. $f(x) = -\log_3(-x)$

In Exercises 45–52, write the logarithmic equation in exponential form.

45. $\ln \frac{1}{2} = -0.693 \ldots$

46. $\ln \frac{2}{5} = -0.916 \ldots$

47. $\ln 4 = 1.386 \ldots$

48. $\ln 10 = 2.302 \ldots$

49. $\ln 250 = 5.521 \ldots$

50. $\ln 679 = 6.520 \ldots$

51. $\ln 1 = 0$

52. $\ln e = 1$

In Exercises 53–60, write the exponential equation in logarithmic form.

53. $e^3 = 20.0855 \ldots$

54. $e^2 = 7.3890 \ldots$

55. $e^{1/2} = 1.6487 \ldots$

56. $e^{1/3} = 1.3956 \ldots$

57. $e^{-0.5} = 0.6065 \ldots$

58. $e^{-4.1} = 0.0165 \ldots$

59. $e^x = 4$

60. $e^{2x} = 3$

 In Exercises 61–64, use a calculator to evaluate the function at the indicated value of x. Round your result to three decimal places.

Function	Value
61. $f(x) = \ln x$	$x = 18.42$
62. $f(x) = 3 \ln x$	$x = 0.32$
63. $g(x) = 2 \ln x$	$x = 0.75$
64. $g(x) = -\ln x$	$x = \frac{1}{2}$

In Exercises 65–68, evaluate $g(x) = \ln x$ at the indicated value of x without using a calculator.

65. $x = e^3$

66. $x = e^{-2}$

67. $x = e^{-2/3}$

68. $x = e^{-5/2}$

In Exercises 69–72, find the domain, x-intercept, and vertical asymptote of the logarithmic function and sketch its graph.

69. $f(x) = \ln(x - 1)$

70. $h(x) = \ln(x + 1)$

71. $g(x) = \ln(-x)$

72. $f(x) = \ln(3 - x)$

 In Exercises 73–78, use a graphing utility to graph the function. Be sure to use an appropriate viewing window.

73. $f(x) = \log(x + 1)$

74. $f(x) = \log(x - 1)$

75. $f(x) = \ln(x - 1)$

76. $f(x) = \ln(x + 2)$

77. $f(x) = \ln x + 2$

78. $f(x) = 3 \ln x - 1$

In Exercises 79–86, use the One-to-One Property to solve the equation for x.

79. $\log_2(x + 1) = \log_2 4$

80. $\log_2(x - 3) = \log_2 9$

81. $\log(2x + 1) = \log 15$

82. $\log(5x + 3) = \log 12$

83. $\ln(x + 2) = \ln 6$

84. $\ln(x - 4) = \ln 2$

85. $\ln(x^2 - 2) = \ln 23$

86. $\ln(x^2 - x) = \ln 6$

Model It

87. Monthly Payment   The model

$$t = 12.542 \ln\left(\frac{x}{x - 1000}\right), \quad x > 1000$$

approximates the length of a home mortgage of $150,000 at 8% in terms of the monthly payment. In the model, t is the length of the mortgage in years and x is the monthly payment in dollars (see figure).

(a) Use the model to approximate the lengths of a $150,000 mortgage at 8% when the monthly payment is $1100.65 and when the monthly payment is $1254.68.

(b) Approximate the total amounts paid over the term of the mortgage with a monthly payment of $1100.65 and with a monthly payment of $1254.68.

(c) Approximate the total interest charges for a monthly payment of $1100.65 and for a monthly payment of $1254.68.

(d) What is the vertical asymptote for the model? Interpret its meaning in the context of the problem.

88. *Compound Interest* A principal P, invested at $9\frac{1}{2}\%$ and compounded continuously, increases to an amount K times the original principal after t years, where t is given by $t = (\ln K)/0.095$.

(a) Complete the table and interpret your results.

K	1	2	4	6	8	10	12
t							

(b) Sketch a graph of the function.

89. *Human Memory Model* Students in a mathematics class were given an exam and then retested monthly with an equivalent exam. The average scores for the class are given by the human memory model $f(t) = 80 - 17 \log(t + 1)$, $0 \le t \le 12$ where t is the time in months.

 (a) Use a graphing utility to graph the model over the specified domain.

(b) What was the average score on the original exam ($t = 0$)?

(c) What was the average score after 4 months?

(d) What was the average score after 10 months?

90. *Sound Intensity* The relationship between the number of decibels β and the intensity of a sound I in watts per square meter is

$$\beta = 10 \log\left(\frac{I}{10^{-12}}\right).$$

(a) Determine the number of decibels of a sound with an intensity of 1 watt per square meter.

(b) Determine the number of decibels of a sound with an intensity of 10^{-2} watt per square meter.

(c) The intensity of the sound in part (a) is 100 times as great as that in part (b). Is the number of decibels 100 times as great? Explain.

Synthesis

True or False? In Exercises 91 and 92, determine whether the statement is true or false. Justify your answer.

91. You can determine the graph of $f(x) = \log_6 x$ by graphing $g(x) = 6^x$ and reflecting it about the x-axis.

92. The graph of $f(x) = \log_3 x$ contains the point $(27, 3)$.

In Exercises 93–96, sketch the graph of f and g and describe the relationship between the graphs of f and g. What is the relationship between the functions f and g?

93. $f(x) = 3^x, \quad g(x) = \log_3 x$

94. $f(x) = 5^x, \quad g(x) = \log_5 x$

95. $f(x) = e^x, \quad g(x) = \ln x$

96. $f(x) = 10^x, \quad g(x) = \log x$

 97. *Graphical Analysis* Use a graphing utility to graph f and g in the same viewing window and determine which is increasing at the greater rate as x approaches $+\infty$. What can you conclude about the rate of growth of the natural logarithmic function?

(a) $f(x) = \ln x, \quad g(x) = \sqrt{x}$

(b) $f(x) = \ln x, \quad g(x) = \sqrt[4]{x}$

98. (a) Complete the table for the function given by

$$f(x) = \frac{\ln x}{x}.$$

x	1	5	10	10^2	10^4	10^6
f(x)						

(b) Use the table in part (a) to determine what value $f(x)$ approaches as x increases without bound.

(c) Use a graphing utility to confirm the result of part (b).

99. *Think About It* The table of values was obtained by evaluating a function. Determine which of the statements may be true and which must be false.

x	1	2	8
y	0	1	3

(a) y is an exponential function of x.

(b) y is a logarithmic function of x.

(c) x is an exponential function of y.

(d) y is a linear function of x.

100. *Writing* Explain why $\log_a x$ is defined only for $0 < a < 1$ and $a > 1$.

 In Exercises 101 and 102, (a) use a graphing utility to graph the function, (b) use the graph to determine the intervals in which the function is increasing and decreasing, and (c) approximate any relative maximum or minimum values of the function.

101. $f(x) = |\ln x|$

102. $h(x) = \ln(x^2 + 1)$

Skills Review

In Exercises 103–108, evaluate the function for $f(x) = 3x + 2$ and $g(x) = x^3 - 1$.

103. $(f + g)(2)$

104. $(f - g)(-1)$

105. $(fg)(6)$

106. $\left(\dfrac{f}{g}\right)(0)$

107. $(f \circ g)(7)$

108. $(g \circ f)(-3)$

What you should learn

- Use the change-of-base formula to rewrite and evaluate logarithmic expressions.
- Use properties of logarithms to evaluate or rewrite logarithmic expressions.
- Use properties of logarithms to expand or condense logarithmic expressions.
- Use logarithmic functions to model and solve real-life problems.

Why you should learn it

Logarithmic functions can be used to model and solve real-life problems. For instance, in Exercises 81–83 on page 394, a logarithmic function is used to model the relationship between the number of decibels and the intensity of a sound.

AP Photo/Stephen Chernin

Change of Base

Most calculators have only two types of log keys, one for common logarithms (base 10) and one for natural logarithms (base e). Although common logs and natural logs are the most frequently used, you may occasionally need to evaluate logarithms to other bases. To do this, you can use the following **change-of-base formula.**

Change-of-Base Formula

Let a, b, and x be positive real numbers such that $a \neq 1$ and $b \neq 1$. Then $\log_a x$ can be converted to a different base as follows.

Base b	Base 10	Base e
$\log_a x = \dfrac{\log_b x}{\log_b a}$	$\log_a x = \dfrac{\log x}{\log a}$	$\log_a x = \dfrac{\ln x}{\ln a}$

One way to look at the change-of-base formula is that logarithms to base a are simply *constant multiples* of logarithms to base b. The constant multiplier is $1/(\log_b a)$.

Example 1   **Changing Bases Using Common Logarithms**

a. $\log_4 25 = \dfrac{\log 25}{\log 4}$ $\log_a x = \dfrac{\log x}{\log a}$

$\approx \dfrac{1.39794}{0.60206}$ Use a calculator.

≈ 2.3219 Simplify.

b. $\log_2 12 = \dfrac{\log 12}{\log 2} \approx \dfrac{1.07918}{0.30103} \approx 3.5850$

✓**CHECKPOINT**   Now try Exercise 1(a).

Example 2   **Changing Bases Using Natural Logarithms**

a. $\log_4 25 = \dfrac{\ln 25}{\ln 4}$ $\log_a x = \dfrac{\ln x}{\ln a}$

$\approx \dfrac{3.21888}{1.38629}$ Use a calculator.

≈ 2.3219 Simplify.

b. $\log_2 12 = \dfrac{\ln 12}{\ln 2} \approx \dfrac{2.48491}{0.69315} \approx 3.5850$

✓**CHECKPOINT**   Now try Exercise 1(b).

Encourage your students to know these properties well. They will be used for solving logarithmic and exponential equations, as well as in calculus.

STUDY TIP

There is no general property that can be used to rewrite $\log_a(u \pm v)$. Specifically, $\log_a(u + v)$ is *not* equal to $\log_a u + \log_a v$.

Remind your students to note the domain when applying properties of logarithms to a logarithmic function. For example, the domain of $f(x) = \ln x^2$ is all real numbers $x \neq 0$, whereas the domain of $g(x) = 2 \ln x$ is all real numbers $x > 0$.

The Granger Collection

Historical Note

John Napier, a Scottish mathematician, developed logarithms as a way to simplify some of the tedious calculations of his day. Beginning in 1594, Napier worked about 20 years on the invention of logarithms. Napier was only partially successful in his quest to simplify tedious calculations. Nonetheless, the development of logarithms was a step forward and received immediate recognition.

Properties of Logarithms

You know from the preceding section that the logarithmic function with base a is the *inverse function* of the exponential function with base a. So, it makes sense that the properties of exponents should have corresponding properties involving logarithms. For instance, the exponential property $a^0 = 1$ has the corresponding logarithmic property $\log_a 1 = 0$.

Properties of Logarithms

Let a be a positive number such that $a \neq 1$, and let n be a real number. If u and v are positive real numbers, the following properties are true.

	Logarithm with Base a	*Natural Logarithm*
1. Product Property:	$\log_a(uv) = \log_a u + \log_a v$	$\ln(uv) = \ln u + \ln v$
2. Quotient Property:	$\log_a \dfrac{u}{v} = \log_a u - \log_a v$	$\ln \dfrac{u}{v} = \ln u - \ln v$
3. Power Property:	$\log_a u^n = n \log_a u$	$\ln u^n = n \ln u$

For proofs of the properties listed above, see Proofs in Mathematics on page 426.

Example 3 Using Properties of Logarithms

Write each logarithm in terms of $\ln 2$ and $\ln 3$.

a. $\ln 6$ **b.** $\ln \dfrac{2}{27}$

Solution

a. $\ln 6 = \ln(2 \cdot 3)$ Rewrite 6 as $2 \cdot 3$.

     $= \ln 2 + \ln 3$ Product Property

b. $\ln \dfrac{2}{27} = \ln 2 - \ln 27$ Quotient Property

         $= \ln 2 - \ln 3^3$ Rewrite 27 as 3^3.

         $= \ln 2 - 3 \ln 3$ Power Property

✔CHECKPOINT   Now try Exercise 17.

Example 4 Using Properties of Logarithms

Find the exact value of each expression without using a calculator.

a. $\log_5 \sqrt[3]{5}$ **b.** $\ln e^6 - \ln e^2$

Solution

a. $\log_5 \sqrt[3]{5} = \log_5 5^{1/3} = \frac{1}{3} \log_5 5 = \frac{1}{3}(1) = \frac{1}{3}$

b. $\ln e^6 - \ln e^2 = \ln \dfrac{e^6}{e^2} = \ln e^4 = 4 \ln e = 4(1) = 4$

✔CHECKPOINT   Now try Exercise 23.

Rewriting Logarithmic Expressions

The properties of logarithms are useful for rewriting logarithmic expressions in forms that simplify the operations of algebra. This is true because these properties convert complicated products, quotients, and exponential forms into simpler sums, differences, and products, respectively.

A common error made in expanding logarithmic expressions is to rewrite log ax^n as n log ax instead of as log $a + n$ log x.

Example 5   **Expanding Logarithmic Expressions**

Expand each logarithmic expression.

a. $\log_4 5x^3y$ **b.** $\ln \dfrac{\sqrt{3x-5}}{7}$

Solution

a. $\log_4 5x^3y = \log_4 5 + \log_4 x^3 + \log_4 y$ Product Property

$\qquad\qquad = \log_4 5 + 3\log_4 x + \log_4 y$ Power Property

b. $\ln \dfrac{\sqrt{3x-5}}{7} = \ln \dfrac{(3x-5)^{1/2}}{7}$ Rewrite using rational exponent.

$\qquad\qquad = \ln(3x-5)^{1/2} - \ln 7$ Quotient Property

$\qquad\qquad = \dfrac{1}{2}\ln(3x-5) - \ln 7$ Power Property

✓CHECKPOINT   Now try Exercise 47.

In Example 5, the properties of logarithms were used to *expand* logarithmic expressions. In Example 6, this procedure is reversed and the properties of logarithms are used to *condense* logarithmic expressions.

Exploration

Use a graphing utility to graph the functions given by

$y_1 = \ln x - \ln(x - 3)$

and

$y_2 = \ln \dfrac{x}{x-3}$

in the same viewing window. Does the graphing utility show the functions with the same domain? If so, should it? Explain your reasoning.

A common error made in condensing logarithmic expressions is to rewrite

$\log x - \log y$ as $\dfrac{\log x}{\log y}$ instead of as $\log \dfrac{x}{y}$.

Example 6   **Condensing Logarithmic Expressions**

Condense each logarithmic expression.

a. $\frac{1}{2}\log x + 3\log(x+1)$ **b.** $2\ln(x+2) - \ln x$
c. $\frac{1}{3}[\log_2 x + \log_2(x+1)]$

Solution

a. $\frac{1}{2}\log x + 3\log(x+1) = \log x^{1/2} + \log(x+1)^3$ Power Property

$\qquad\qquad = \log\left[\sqrt{x}(x+1)^3\right]$ Product Property

b. $2\ln(x+2) - \ln x = \ln(x+2)^2 - \ln x$ Power Property

$\qquad\qquad = \ln \dfrac{(x+2)^2}{x}$ Quotient Property

c. $\frac{1}{3}[\log_2 x + \log_2(x+1)] = \frac{1}{3}\{\log_2[x(x+1)]\}$ Product Property

$\qquad\qquad = \log_2[x(x+1)]^{1/3}$ Power Property

$\qquad\qquad = \log_2 \sqrt[3]{x(x+1)}$ Rewrite with a radical.

✓CHECKPOINT   Now try Exercise 69.

Application

One method of determining how the x- and y-values for a set of nonlinear data are related is to take the natural logarithm of each of the x- and y-values. If the points are graphed and fall on a line, then you can determine that the x- and y-values are related by the equation

$$\ln y = m \ln x$$

where m is the slope of the line.

Example 7 Finding a Mathematical Model

The table shows the mean distance x and the period (the time it takes a planet to orbit the sun) y for each of the six planets that are closest to the sun. In the table, the mean distance is given in terms of astronomical units (where Earth's mean distance is defined as 1.0), and the period is given in years. Find an equation that relates y and x.

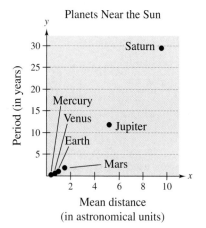

Planets Near the Sun

FIGURE 5.23

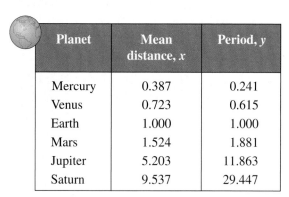

Planet	Mean distance, x	Period, y
Mercury	0.387	0.241
Venus	0.723	0.615
Earth	1.000	1.000
Mars	1.524	1.881
Jupiter	5.203	11.863
Saturn	9.537	29.447

Solution

The points in the table above are plotted in Figure 5.23. From this figure it is not clear how to find an equation that relates y and x. To solve this problem, take the natural logarithm of each of the x- and y-values in the table. This produces the following results.

Planet	Mercury	Venus	Earth	Mars	Jupiter	Saturn
$\ln x$	-0.949	-0.324	0.000	0.421	1.649	2.255
$\ln y$	-1.423	-0.486	0.000	0.632	2.473	3.383

Now, by plotting the points in the second table, you can see that all six of the points appear to lie in a line (see Figure 5.24). Choose any two points to determine the slope of the line. Using the two points $(0.421, 0.632)$ and $(0, 0)$, you can determine that the slope of the line is

$$m = \frac{0.632 - 0}{0.421 - 0} \approx 1.5 = \frac{3}{2}.$$

By the point-slope form, the equation of the line is $Y = \frac{3}{2}X$, where $Y = \ln y$ and $X = \ln x$. You can therefore conclude that $\ln y = \frac{3}{2}\ln x$.

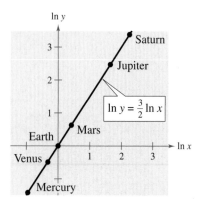

FIGURE 5.24

✓CHECKPOINT Now try Exercise 85.

5.3 | Exercises

VOCABULARY CHECK:

In Exercises 1 and 2, fill in the blanks.

1. To evaluate a logarithm to any base, you can use the _____ formula.
2. The change-of-base formula for base e is given by $\log_a x =$ _____ .

In Exercises 3–5, match the property of logarithms with its name.

3. $\log_a(uv) = \log_a u + \log_a v$ (a) Power Property
4. $\ln u^n = n \ln u$ (b) Quotient Property
5. $\log_a \dfrac{u}{v} = \log_a u - \log_a v$ (c) Product Property

PREREQUISITE SKILLS REVIEW: Practice and review algebra skills needed for this section at **www.Eduspace.com.**

In Exercises 1–8, rewrite the logarithm as a ratio of (a) common logarithms and (b) natural logarithms.

1. $\log_5 x$
2. $\log_3 x$
3. $\log_{1/5} x$
4. $\log_{1/3} x$
5. $\log_x \frac{3}{10}$
6. $\log_x \frac{3}{4}$
7. $\log_{2.6} x$
8. $\log_{7.1} x$

In Exercises 9–16, evaluate the logarithm using the change-of-base formula. Round your result to three decimal places.

9. $\log_3 7$
10. $\log_7 4$
11. $\log_{1/2} 4$
12. $\log_{1/4} 5$
13. $\log_9 0.4$
14. $\log_{20} 0.125$
15. $\log_{15} 1250$
16. $\log_3 0.015$

In Exercises 17–22, use the properties of logarithms to rewrite and simplify the logarithmic expression.

17. $\log_4 8$
18. $\log_2(4^2 \cdot 3^4)$
19. $\log_5 \frac{1}{250}$
20. $\log \frac{9}{300}$
21. $\ln(5e^6)$
22. $\ln \frac{6}{e^2}$

In Exercises 23–38, find the exact value of the logarithmic expression without using a calculator. (If this is not possible, state the reason.)

23. $\log_3 9$
24. $\log_5 \frac{1}{125}$
25. $\log_2 \sqrt[4]{8}$
26. $\log_6 \sqrt[3]{6}$
27. $\log_4 16^{1.2}$
28. $\log_3 81^{-0.2}$
29. $\log_3(-9)$
30. $\log_2(-16)$
31. $\ln e^{4.5}$

32. $3 \ln e^4$
33. $\ln \dfrac{1}{\sqrt{e}}$
34. $\ln \sqrt[4]{e^3}$
35. $\ln e^2 + \ln e^5$
36. $2 \ln e^6 - \ln e^5$
37. $\log_5 75 - \log_5 3$
38. $\log_4 2 + \log_4 32$

In Exercises 39–60, use the properties of logarithms to expand the expression as a sum, difference, and/or constant multiple of logarithms. (Assume all variables are positive.)

39. $\log_4 5x$
40. $\log_3 10z$
41. $\log_8 x^4$
42. $\log_{10} \frac{y}{2}$
43. $\log_5 \frac{5}{x}$
44. $\log_6 \frac{1}{z^3}$
45. $\ln \sqrt{z}$
46. $\ln \sqrt[3]{t}$
47. $\ln xyz^2$
48. $\log 4x^2 y$
49. $\ln z(z-1)^2, \ z > 1$
50. $\ln\left(\dfrac{x^2 - 1}{x^3}\right), \ x > 1$
51. $\log_2 \dfrac{\sqrt{a-1}}{9}, \ a > 1$
52. $\ln \dfrac{6}{\sqrt{x^2 + 1}}$
53. $\ln \sqrt[3]{\dfrac{x}{y}}$
54. $\ln \sqrt{\dfrac{x^2}{y^3}}$
55. $\ln \dfrac{x^4 \sqrt{y}}{z^5}$
56. $\log_2 \dfrac{\sqrt{x}\, y^4}{z^4}$
57. $\log_5 \dfrac{x^2}{y^2 z^3}$
58. $\log_{10} \dfrac{xy^4}{z^5}$
59. $\ln \sqrt[4]{x^3(x^2 + 3)}$
60. $\ln \sqrt{x^2(x + 2)}$

In Exercises 61–78, condense the expression to the logarithm of a single quantity.

61. $\ln x + \ln 3$

62. $\ln y + \ln t$

63. $\log_4 z - \log_4 y$

64. $\log_5 8 - \log_5 t$

65. $2 \log_2(x + 4)$

66. $\frac{2}{3} \log_7(z - 2)$

67. $\frac{1}{4} \log_3 5x$

68. $-4 \log_6 2x$

69. $\ln x - 3 \ln(x + 1)$

70. $2 \ln 8 + 5 \ln (z - 4)$

71. $\log x - 2 \log y + 3 \log z$

72. $3 \log_3 x + 4 \log_3 y - 4 \log_3 z$

73. $\ln x - 4[\ln(x + 2) + \ln(x - 2)]$

74. $4[\ln z + \ln(z + 5)] - 2 \ln(z - 5)$

75. $\frac{1}{3}[2 \ln(x + 3) + \ln x - \ln(x^2 - 1)]$

76. $2[3 \ln x - \ln(x + 1) - \ln(x - 1)]$

77. $\frac{1}{3}[\log_8 y + 2 \log_8(y + 4)] - \log_8(y - 1)$

78. $\frac{1}{2}[\log_4(x + 1) + 2 \log_4(x - 1)] + 6 \log_4 x$

In Exercises 79 and 80, compare the logarithmic quantities. If two are equal, explain why.

79. $\dfrac{\log_2 32}{\log_2 4}, \quad \log_2 \dfrac{32}{4}, \quad \log_2 32 - \log_2 4$

80. $\log_7 \sqrt{70}, \quad \log_7 35, \quad \frac{1}{2} + \log_7 \sqrt{10}$

Sound Intensity **In Exercises 81–83, use the following information. The relationship between the number of decibels β and the intensity of a sound I in watts per square meter is given by**

$$\beta = 10 \log\left(\frac{I}{10^{-12}}\right).$$

81. Use the properties of logarithms to write the formula in simpler form, and determine the number of decibels of a sound with an intensity of 10^{-6} watt per square meter.

82. Find the difference in loudness between an average office with an intensity of 1.26×10^{-7} watt per square meter and a broadcast studio with an intensity of 3.16×10^{-5} watt per square meter.

83. You and your roommate are playing your stereos at the same time and at the same intensity. How much louder is the music when both stereos are playing compared with just one stereo playing?

Model It

84. *Human Memory Model*   Students participating in a psychology experiment attended several lectures and were given an exam. Every month for a year after the exam, the students were retested to see how much of the material they remembered. The average scores for the group can be modeled by the human memory model

$$f(t) = 90 - 15 \log(t + 1), \quad 0 \le t \le 12$$

where t is the time in months.

(a) Use the properties of logarithms to write the function in another form.

(b) What was the average score on the original exam $(t = 0)$?

(c) What was the average score after 4 months?

(d) What was the average score after 12 months?

 (e) Use a graphing utility to graph the function over the specified domain.

(f) Use the graph in part (e) to determine when the average score will decrease to 75.

(g) Verify your answer to part (f) numerically.

---

85. *Galloping Speeds of Animals*   Four-legged animals run with two different types of motion: trotting and galloping. An animal that is trotting has at least one foot on the ground at all times, whereas an animal that is galloping has all four feet off the ground at some point in its stride. The number of strides per minute at which an animal breaks from a trot to a gallop depends on the weight of the animal. Use the table to find a logarithmic equation that relates an animal's weight x (in pounds) and its lowest galloping speed y (in strides per minute).

Weight, x	Galloping Speed, y
25	191.5
35	182.7
50	173.8
75	164.2
500	125.9
1000	114.2

86. Comparing Models   A cup of water at an initial temperature of 78° C is placed in a room at a constant temperature of 21° C. The temperature of the water is measured every 5 minutes during a half-hour period. The results are recorded as ordered pairs of the form (t, T), where t is the time (in minutes) and T is the temperature (in degrees Celsius).

$(0, 78.0°), (5, 66.0°), (10, 57.5°), (15, 51.2°), (20, 46.3°),$
$(25, 42.4°), (30, 39.6°)$

(a) The graph of the model for the data should be asymptotic with the graph of the temperature of the room. Subtract the room temperature from each of the temperatures in the ordered pairs. Use a graphing utility to plot the data points (t, T) and $(t, T - 21)$.

(b) An exponential model for the data $(t, T - 21)$ is given by

$$T - 21 = 54.4(0.964)^t.$$

Solve for T and graph the model. Compare the result with the plot of the original data.

(c) Take the natural logarithms of the revised temperatures. Use a graphing utility to plot the points $(t, \ln(T - 21))$ and observe that the points appear to be linear. Use the *regression* feature of the graphing utility to fit a line to these data. This resulting line has the form

$$\ln(T - 21) = at + b.$$

Use the properties of the logarithms to solve for T. Verify that the result is equivalent to the model in part (b).

(d) Fit a rational model to the data. Take the reciprocals of the y-coordinates of the revised data points to generate the points

$$\left(t, \frac{1}{T - 21}\right).$$

Use a graphing utility to graph these points and observe that they appear to be linear. Use the *regression* feature of a graphing utility to fit a line to these data. The resulting line has the form

$$\frac{1}{T - 21} = at + b.$$

Solve for T, and use a graphing utility to graph the rational function and the original data points.

(e) Write a short paragraph explaining why the transformations of the data were necessary to obtain each model. Why did taking the logarithms of the temperatures lead to a linear scatter plot? Why did taking the reciprocals of the temperature lead to a linear scatter plot?

Synthesis

True or False?   In Exercises 87–92, determine whether the statement is true or false given that $f(x) = \ln x$. Justify your answer.

87. $f(0) = 0$

88. $f(ax) = f(a) + f(x), \quad a > 0, x > 0$

89. $f(x - 2) = f(x) - f(2), \quad x > 2$

90. $\sqrt{f(x)} = \frac{1}{2}f(x)$

91. If $f(u) = 2f(v)$, then $v = u^2$.

92. If $f(x) < 0$, then $0 < x < 1$.

93. Proof   Prove that $\log_b \dfrac{u}{v} = \log_b u - \log_b v$.

94. Proof   Prove that $\log_b u^n = n \log_b u$.

 In Exercises 95–100, use the change-of-base formula to rewrite the logarithm as a ratio of logarithms. Then use a graphing utility to graph both functions in the same viewing window to verify that the functions are equivalent.

95. $f(x) = \log_2 x$

96. $f(x) = \log_4 x$

97. $f(x) = \log_{1/2} x$

98. $f(x) = \log_{1/4} x$

99. $f(x) = \log_{11.8} x$

100. $f(x) = \log_{12.4} x$

101. Think About It   Consider the functions below.

$$f(x) = \ln \frac{x}{2}, \quad g(x) = \frac{\ln x}{\ln 2}, \quad h(x) = \ln x - \ln 2$$

Which two functions should have identical graphs? Verify your answer by sketching the graphs of all three functions on the same set of coordinate axes.

102. Exploration   For how many integers between 1 and 20 can the natural logarithms be approximated given that $\ln 2 \approx 0.6931$, $\ln 3 \approx 1.0986$, and $\ln 5 \approx 1.6094$? Approximate these logarithms (do not use a calculator).

Skills Review

In Exercises 103–106, simplify the expression.

103. $\dfrac{24xy^{-2}}{16x^{-3}y}$

104. $\left(\dfrac{2x^2}{3y}\right)^{-3}$

105. $(18x^3y^4)^{-3}(18x^3y^4)^3$

106. $xy(x^{-1} + y^{-1})^{-1}$

In Exercises 107–110, solve the equation.

107. $3x^2 + 2x - 1 = 0$

108. $4x^2 - 5x + 1 = 0$

109. $\dfrac{2}{3x + 1} = \dfrac{x}{4}$

110. $\dfrac{5}{x - 1} = \dfrac{2x}{3}$

5.4 Exponential and Logarithmic Equations

What you should learn

- Solve simple exponential and logarithmic equations.
- Solve more complicated exponential equations.
- Solve more complicated logarithmic equations.
- Use exponential and logarithmic equations to model and solve real-life problems.

Why you should learn it

Exponential and logarithmic equations are used to model and solve life science applications. For instance, in Exercise 112, on page 405, a logarithmic function is used to model the number of trees per acre given the average diameter of the trees.

© James Marshall/Corbis

Introduction

So far in this chapter, you have studied the definitions, graphs, and properties of exponential and logarithmic functions. In this section, you will study procedures for *solving equations* involving these exponential and logarithmic functions.

There are two basic strategies for solving exponential or logarithmic equations. The first is based on the One-to-One Properties and was used to solve simple exponential and logarithmic equations in Sections 5.1 and 5.2. The second is based on the Inverse Properties. For $a > 0$ and $a \neq 1$, the following properties are true for all x and y for which $\log_a x$ and $\log_a y$ are defined.

One-to-One Properties

$a^x = a^y$ if and only if $x = y$.

$\log_a x = \log_a y$ if and only if $x = y$.

Inverse Properties

$a^{\log_a x} = x$

$\log_a a^x = x$

Example 1   **Solving Simple Equations**

Original Equation	Rewritten Equation	Solution	Property
a. $2^x = 32$	$2^x = 2^5$	$x = 5$	One-to-One
b. $\ln x - \ln 3 = 0$	$\ln x = \ln 3$	$x = 3$	One-to-One
c. $\left(\frac{1}{3}\right)^x = 9$	$3^{-x} = 3^2$	$x = -2$	One-to-One
d. $e^x = 7$	$\ln e^x = \ln 7$	$x = \ln 7$	Inverse
e. $\ln x = -3$	$e^{\ln x} = e^{-3}$	$x = e^{-3}$	Inverse
f. $\log x = -1$	$10^{\log x} = 10^{-1}$	$x = 10^{-1} = \frac{1}{10}$	Inverse

✔CHECKPOINT   Now try Exercise 13.

The strategies used in Example 1 are summarized as follows.

Strategies for Solving Exponential and Logarithmic Equations

1. Rewrite the original equation in a form that allows the use of the One-to-One Properties of exponential or logarithmic functions.

2. Rewrite an *exponential* equation in logarithmic form and apply the Inverse Property of logarithmic functions.

3. Rewrite a *logarithmic* equation in exponential form and apply the Inverse Property of exponential functions.

Solving Exponential Equations

Example 2   **Solving Exponential Equations**

Solve each equation and approximate the result to three decimal places if necessary.

a. $e^{-x^2} = e^{-3x-4}$ **b.** $3(2^x) = 42$

Solution

a.

$e^{-x^2} = e^{-3x-4}$	Write original equation.
$-x^2 = -3x - 4$	One-to-One Property
$x^2 - 3x - 4 = 0$	Write in general form.
$(x + 1)(x - 4) = 0$	Factor.
$(x + 1) = 0 \Longrightarrow x = -1$	Set 1st factor equal to 0.
$(x - 4) = 0 \Longrightarrow x = 4$	Set 2nd factor equal to 0.

The solutions are $x = -1$ and $x = 4$. Check these in the original equation.

b.

$3(2^x) = 42$	Write original equation.
$2^x = 14$	Divide each side by 3.
$\log_2 2^x = \log_2 14$	Take log (base 2) of each side.
$x = \log_2 14$	Inverse Property
$x = \dfrac{\ln 14}{\ln 2} \approx 3.807$	Change-of-base formula

The solution is $x = \log_2 14 \approx 3.807$. Check this in the original equation.

✓*CHECKPOINT*   Now try Exercise 25.

In Example 2(b), the exact solution is $x = \log_2 14$ and the approximate solution is $x \approx 3.807$. An exact answer is preferred when the solution is an intermediate step in a larger problem. For a final answer, an approximate solution is easier to comprehend.

Example 3   **Solving an Exponential Equation**

Solve $e^x + 5 = 60$ and approximate the result to three decimal places.

Solution

$e^x + 5 = 60$	Write original equation.
$e^x = 55$	Subtract 5 from each side.
$\ln e^x = \ln 55$	Take natural log of each side.
$x = \ln 55 \approx 4.007$	Inverse Property

The solution is $x = \ln 55 \approx 4.007$. Check this in the original equation.

✓*CHECKPOINT*   Now try Exercise 51.

STUDY TIP

Remember that the natural logarithmic function has a base of e.

STUDY TIP

Remember that to evaluate a logarithm such as $\log_3 7.5$, you need to use the change-of-base formula.

$$\log_3 7.5 = \frac{\ln 7.5}{\ln 3} \approx 1.834$$

| Example 4 | **Solving an Exponential Equation** |

Solve $2(3^{2t-5}) - 4 = 11$ and approximate the result to three decimal places.

Solution

$2(3^{2t-5}) - 4 = 11$	Write original equation.
$2(3^{2t-5}) = 15$	Add 4 to each side.
$3^{2t-5} = \dfrac{15}{2}$	Divide each side by 2.
$\log_3 3^{2t-5} = \log_3 \dfrac{15}{2}$	Take log (base 3) of each side.
$2t - 5 = \log_3 \dfrac{15}{2}$	Inverse Property
$2t = 5 + \log_3 7.5$	Add 5 to each side.
$t = \dfrac{5}{2} + \dfrac{1}{2} \log_3 7.5$	Divide each side by 2.
$t \approx 3.417$	Use a calculator.

The solution is $t = \frac{5}{2} + \frac{1}{2} \log_3 7.5 \approx 3.417$. Check this in the original equation.

✓**CHECKPOINT** Now try Exercise 53.

When an equation involves two or more exponential expressions, you can still use a procedure similar to that demonstrated in Examples 2, 3, and 4. However, the algebra is a bit more complicated.

| Example 5 | **Solving an Exponential Equation of Quadratic Type** |

Solve $e^{2x} - 3e^x + 2 = 0$.

Algebraic Solution

$e^{2x} - 3e^x + 2 = 0$	Write original equation.
$(e^x)^2 - 3e^x + 2 = 0$	Write in quadratic form.
$(e^x - 2)(e^x - 1) = 0$	Factor.
$e^x - 2 = 0$	Set 1st factor equal to 0.
$x = \ln 2$	Solution
$e^x - 1 = 0$	Set 2nd factor equal to 0.
$x = 0$	Solution

The solutions are $x = \ln 2 \approx 0.693$ and $x = 0$. Check these in the original equation.

✓**CHECKPOINT** Now try Exercise 67.

Graphical Solution

Use a graphing utility to graph $y = e^{2x} - 3e^x + 2$. Use the *zero* or *root* feature or the *zoom* and *trace* features of the graphing utility to approximate the values of x for which $y = 0$. In Figure 5.25, you can see that the zeros occur at $x = 0$ and at $x \approx 0.693$. So, the solutions are $x = 0$ and $x \approx 0.693$.

FIGURE 5.25

Solving Logarithmic Equations

To solve a logarithmic equation, you can write it in exponential form.

$\ln x = 3$	Logarithmic form
$e^{\ln x} = e^3$	Exponentiate each side.
$x = e^3$	Exponential form

This procedure is called *exponentiating* each side of an equation.

Example 6   Solving Logarithmic Equations

a. $\ln x = 2$ Original equation

$e^{\ln x} = e^2$ Exponentiate each side.

$x = e^2$ Inverse Property

b. $\log_3(5x - 1) = \log_3(x + 7)$ Original equation

$5x - 1 = x + 7$ One-to-One Property

$4x = 8$ Add $-x$ and 1 to each side.

$x = 2$ Divide each side by 4.

c. $\log_6(3x + 14) - \log_6 5 = \log_6 2x$ Original equation

$\log_6\left(\dfrac{3x + 14}{5}\right) = \log_6 2x$ Quotient Property of Logarithms

$\dfrac{3x + 14}{5} = 2x$ One-to-One Property

$3x + 14 = 10x$ Cross multiply.

$-7x = -14$ Isolate x.

$x = 2$ Divide each side by -7.

 CHECKPOINT   Now try Exercise 77.

Example 7   Solving a Logarithmic Equation

Solve $5 + 2 \ln x = 4$ and approximate the result to three decimal places.

Solution

$5 + 2 \ln x = 4$	Write original equation.
$2 \ln x = -1$	Subtract 5 from each side.
$\ln x = -\dfrac{1}{2}$	Divide each side by 2.
$e^{\ln x} = e^{-1/2}$	Exponentiate each side.
$x = e^{-1/2}$	Inverse Property
$x \approx 0.607$	Use a calculator.

 CHECKPOINT   Now try Exercise 85.

Activities

1. Solve for x: $7^x = 3$.

    Answer: $x = \dfrac{\ln 3}{\ln 7} \approx 0.5646$

2. Solve for x:
   $\log(x + 4) + \log(x + 1) = 1$.
   Answer: $x = 1$ ($x = -6$ is not in the domain.)

Example 8   **Solving a Logarithmic Equation**

Solve $2 \log_5 3x = 4$.

Solution

$2 \log_5 3x = 4$	Write original equation.
$\log_5 3x = 2$	Divide each side by 2.
$5^{\log_5 3x} = 5^2$	Exponentiate each side (base 5).
$3x = 25$	Inverse Property
$x = \dfrac{25}{3}$	Divide each side by 3.

The solution is $x = \frac{25}{3}$. Check this in the original equation.

✓**CHECKPOINT**   Now try Exercise 87.

> **STUDY TIP**
>
> Notice in Example 9 that the logarithmic part of the equation is condensed into a single logarithm before exponentiating each side of the equation.

Because the domain of a logarithmic function generally does not include all real numbers, you should be sure to check for extraneous solutions of logarithmic equations.

Example 9   **Checking for Extraneous Solutions**

Solve $\log 5x + \log(x - 1) = 2$.

Algebraic Solution

$\log 5x + \log(x - 1) = 2$	Write original equation.
$\log[5x(x - 1)] = 2$	Product Property of Logarithms
$10^{\log(5x^2 - 5x)} = 10^2$	Exponentiate each side (base 10).
$5x^2 - 5x = 100$	Inverse Property
$x^2 - x - 20 = 0$	Write in general form.
$(x - 5)(x + 4) = 0$	Factor.
$x - 5 = 0$	Set 1st factor equal to 0.
$x = 5$	Solution
$x + 4 = 0$	Set 2nd factor equal to 0.
$x = -4$	Solution

The solutions appear to be $x = 5$ and $x = -4$. However, when you check these in the original equation, you can see that $x = 5$ is the only solution.

✓**CHECKPOINT**   Now try Exercise 99.

Graphical Solution

Use a graphing utility to graph $y_1 = \log 5x + \log(x - 1)$ and $y_2 = 2$ in the same viewing window. From the graph shown in Figure 5.26, it appears that the graphs intersect at one point. Use the *intersect* feature or the *zoom* and *trace* features to determine that the graphs intersect at approximately $(5, 2)$. So, the solution is $x = 5$. Verify that 5 is an exact solution algebraically.

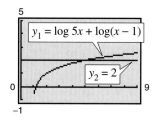

FIGURE 5.26

In Example 9, the domain of $\log 5x$ is $x > 0$ and the domain of $\log(x - 1)$ is $x > 1$, so the domain of the original equation is $x > 1$. Because the domain is all real numbers greater than 1, the solution $x = -4$ is extraneous. The graph in Figure 5.26 verifies this concept.

Applications

Example 10   Doubling an Investment

You have deposited $500 in an account that pays 6.75% interest, compounded continuously. How long will it take your money to double?

Solution

Using the formula for continuous compounding, you can find that the balance in the account is

$$A = Pe^{rt}$$

$$A = 500e^{0.0675t}.$$

To find the time required for the balance to double, let $A = 1000$ and solve the resulting equation for t.

$500e^{0.0675t} = 1000$	Let $A = 1000$.
$e^{0.0675t} = 2$	Divide each side by 500.
$\ln e^{0.0675t} = \ln 2$	Take natural log of each side.
$0.0675t = \ln 2$	Inverse Property
$t = \dfrac{\ln 2}{0.0675}$	Divide each side by 0.0675.
$t \approx 10.27$	Use a calculator.

The balance in the account will double after approximately 10.27 years. This result is demonstrated graphically in Figure 5.27.

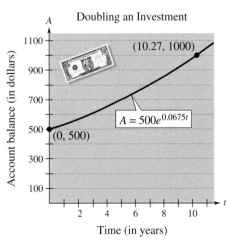

FIGURE 5.27

✓CHECKPOINT   Now try Exercise 107.

In Example 10, an approximate answer of 10.27 years is given. Within the context of the problem, the exact solution, $(\ln 2)/0.0675$ years, does not make sense as an answer.

Endangered Animal Species

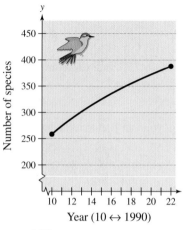

FIGURE 5.28

Example 11 **Endangered Animals**

The number y of endangered animal species in the United States from 1990 to 2002 can be modeled by

$$y = -119 + 164 \ln t, \quad 10 \leq t \leq 22$$

where t represents the year, with $t = 10$ corresponding to 1990 (see Figure 5.28). During which year did the number of endangered animal species reach 357? (Source: U.S. Fish and Wildlife Service)

Solution

$-119 + 164 \ln t = y$	Write original equation.
$-119 + 164 \ln t = 357$	Substitute 357 for y.
$164 \ln t = 476$	Add 119 to each side.
$\ln t = \dfrac{476}{164}$	Divide each side by 164.
$e^{\ln t} \approx e^{476/164}$	Exponentiate each side.
$t \approx e^{476/164}$	Inverse Property
$t \approx 18$	Use a calculator.

The solution is $t \approx 18$. Because $t = 10$ represents 1990, it follows that the number of endangered animals reached 357 in 1998.

✔CHECKPOINT   Now try Exercise 113.

𝒲RITING ABOUT 𝓜ATHEMATICS

Comparing Mathematical Models The table shows the U.S. Postal Service rates y for sending an express mail package for selected years from 1985 through 2002, where $x = 5$ represents 1985. (Source: U.S. Postal Service)

Year, x	Rate, y
5	10.75
8	12.00
11	13.95
15	15.00
19	15.75
21	16.00
22	17.85

a. Create a scatter plot of the data. Find a linear model for the data, and add its graph to your scatter plot. According to this model, when will the rate for sending an express mail package reach $19.00?

b. Create a new table showing values for $\ln x$ and $\ln y$ and create a scatter plot of these transformed data. Use the method illustrated in Example 7 in Section 5.3 to find a model for the transformed data, and add its graph to your scatter plot. According to this model, when will the rate for sending an express mail package reach $19.00?

c. Solve the model in part (b) for y, and add its graph to your scatter plot in part (a). Which model better fits the original data? Which model will better predict future rates? Explain.

5.4   Exercises

VOCABULARY CHECK: Fill in the blanks.

1. To _____ an equation in x means to find all values of x for which the equation is true.

2. To solve exponential and logarithmic equations, you can use the following One-to-One and Inverse Properties.

   (a) $a^x = a^y$ if and only if _____.

   (b) $\log_a x = \log_a y$ if and only if _____.

   (c) $a^{\log_a x} =$ _____

   (d) $\log_a a^x =$ _____

3. An _____ solution does not satisfy the original equation.

PREREQUISITE SKILLS REVIEW: Practice and review algebra skills needed for this section at **www.Eduspace.com.**

In Exercises 1–8, determine whether each x-value is a solution (or an approximate solution) of the equation.

1. $4^{2x-7} = 64$
   (a) $x = 5$
   (b) $x = 2$

2. $2^{3x+1} = 32$
   (a) $x = -1$
   (b) $x = 2$

3. $3e^{x+2} = 75$
   (a) $x = -2 + e^{25}$
   (b) $x = -2 + \ln 25$
   (c) $x \approx 1.219$

4. $2e^{5x+2} = 12$
   (a) $x = \frac{1}{5}(-2 + \ln 6)$
   (b) $x = \dfrac{\ln 6}{5 \ln 2}$
   (c) $x \approx -0.0416$

5. $\log_4(3x) = 3$
   (a) $x \approx 21.333$
   (b) $x = -4$
   (c) $x = \frac{64}{3}$

6. $\log_2(x + 3) = 10$
   (a) $x = 1021$
   (b) $x = 17$
   (c) $x = 10^2 - 3$

7. $\ln(2x + 3) = 5.8$
   (a) $x = \frac{1}{2}(-3 + \ln 5.8)$
   (b) $x = \frac{1}{2}(-3 + e^{5.8})$
   (c) $x \approx 163.650$

8. $\ln(x - 1) = 3.8$
   (a) $x = 1 + e^{3.8}$
   (b) $x \approx 45.701$
   (c) $x = 1 + \ln 3.8$

In Exercises 9–20, solve for x.

9. $4^x = 16$
10. $3^x = 243$
11. $\left(\frac{1}{2}\right)^x = 32$
12. $\left(\frac{1}{4}\right)^x = 64$
13. $\ln x - \ln 2 = 0$
14. $\ln x - \ln 5 = 0$
15. $e^x = 2$
16. $e^x = 4$
17. $\ln x = -1$
18. $\ln x = -7$
19. $\log_4 x = 3$
20. $\log_5 x = -3$

In Exercises 21–24, approximate the point of intersection of the graphs of f and g. Then solve the equation $f(x) = g(x)$ algebraically to verify your approximation.

21. $f(x) = 2^x$
    $g(x) = 8$

22. $f(x) = 27^x$
    $g(x) = 9$

23. $f(x) = \log_3 x$
    $g(x) = 2$

24. $f(x) = \ln(x - 4)$
    $g(x) = 0$

In Exercises 25–66, solve the exponential equation algebraically. Approximate the result to three decimal places.

25. $e^x = e^{x^2-2}$

26. $e^{2x} = e^{x^2-8}$

27. $e^{x^2-3} = e^{x-2}$

28. $e^{-x^2} = e^{x^2-2x}$

29. $4(3^x) = 20$

30. $2(5^x) = 32$

31. $2e^x = 10$

32. $4e^x = 91$

33. $e^x - 9 = 19$

34. $6^x + 10 = 47$

35. $3^{2x} = 80$

36. $6^{5x} = 3000$

37. $5^{-t/2} = 0.20$

38. $4^{-3t} = 0.10$

39. $3^{x-1} = 27$

40. $2^{x-3} = 32$

41. $2^{3-x} = 565$

42. $8^{-2-x} = 431$

43. $8(10^{3x}) = 12$

44. $5(10^{x-6}) = 7$

45. $3(5^{x-1}) = 21$

46. $8(3^{6-x}) = 40$

47. $e^{3x} = 12$

48. $e^{2x} = 50$

49. $500e^{-x} = 300$

50. $1000e^{-4x} = 75$

51. $7 - 2e^x = 5$

52. $-14 + 3e^x = 11$

53. $6(2^{3x-1}) - 7 = 9$

54. $8(4^{6-2x}) + 13 = 41$

55. $e^{2x} - 4e^x - 5 = 0$

56. $e^{2x} - 5e^x + 6 = 0$

57. $e^{2x} - 3e^x - 4 = 0$

58. $e^{2x} + 9e^x + 36 = 0$

59. $\dfrac{500}{100 - e^{x/2}} = 20$

60. $\dfrac{400}{1 + e^{-x}} = 350$

61. $\dfrac{3000}{2 + e^{2x}} = 2$

62. $\dfrac{119}{e^{6x} - 14} = 7$

63. $\left(1 + \dfrac{0.065}{365}\right)^{365t} = 4$

64. $\left(4 - \dfrac{2.471}{40}\right)^{9t} = 21$

65. $\left(1 + \dfrac{0.10}{12}\right)^{12t} = 2$

66. $\left(16 - \dfrac{0.878}{26}\right)^{3t} = 30$

 In Exercises 67–74, use a graphing utility to graph and solve the equation. Approximate the result to three decimal places. Verify your result algebraically.

67. $6e^{1-x} = 25$

68. $-4e^{-x-1} + 15 = 0$

69. $3e^{3x/2} = 962$

70. $8e^{-2x/3} = 11$

71. $e^{0.09t} = 3$

72. $-e^{1.8x} + 7 = 0$

73. $e^{0.125t} - 8 = 0$

74. $e^{2.724x} = 29$

In Exercises 75–102, solve the logarithmic equation algebraically. Approximate the result to three decimal places.

75. $\ln x = -3$

76. $\ln x = 2$

77. $\ln 2x = 2.4$

78. $\ln 4x = 1$

79. $\log x = 6$

80. $\log 3z = 2$

81. $3\ln 5x = 10$

82. $2 \ln x = 7$

83. $\ln \sqrt{x + 2} = 1$

84. $\ln \sqrt{x - 8} = 5$

85. $7 + 3 \ln x = 5$

86. $2 - 6 \ln x = 10$

87. $6 \log_3(0.5x) = 11$

88. $5 \log_{10}(x - 2) = 11$

89. $\ln x - \ln(x + 1) = 2$

90. $\ln x + \ln(x + 1) = 1$

91. $\ln x + \ln(x - 2) = 1$

92. $\ln x + \ln(x + 3) = 1$

93. $\ln(x + 5) = \ln(x - 1) - \ln(x + 1)$

94. $\ln(x + 1) - \ln(x - 2) = \ln x$

95. $\log_2(2x - 3) = \log_2(x + 4)$

96. $\log(x - 6) = \log(2x + 1)$

97. $\log(x + 4) - \log x = \log(x + 2)$

98. $\log_2 x + \log_2(x + 2) = \log_2(x + 6)$

99. $\log_4 x - \log_4(x - 1) = \frac{1}{2}$

100. $\log_3 x + \log_3(x - 8) = 2$

101. $\log 8x - \log\left(1 + \sqrt{x}\right) = 2$

102. $\log 4x - \log\left(12 + \sqrt{x}\right) = 2$

 In Exercises 103–106, use a graphing utility to graph and solve the equation. Approximate the result to three decimal places. Verify your result algebraically.

103. $7 = 2^x$

104. $500 = 1500e^{-x/2}$

105. $3 - \ln x = 0$

106. $10 - 4 \ln(x - 2) = 0$

Compound Interest   In Exercises 107 and 108, $2500 is invested in an account at interest rate r, compounded continuously. Find the time required for the amount to (a) double and (b) triple.

107. $r = 0.085$

108. $r = 0.12$

109. *Demand*   The demand equation for a microwave oven is given by

$$p = 500 - 0.5(e^{0.004x}).$$

Find the demand x for a price of (a) $p = \$350$ and (b) $p = \$300$.

110. *Demand*   The demand equation for a hand-held electronic organizer is

$$p = 5000\left(1 - \dfrac{4}{4 + e^{-0.002x}}\right).$$

Find the demand x for a price of (a) $p = \$600$ and (b) $p = \$400$.

111. *Forest Yield*   The yield V (in millions of cubic feet per acre) for a forest at age t years is given by

$$V = 6.7e^{-48.1/t}.$$

 (a) Use a graphing utility to graph the function.

(b) Determine the horizontal asymptote of the function. Interpret its meaning in the context of the problem.

(c) Find the time necessary to obtain a yield of 1.3 million cubic feet.

112. Trees per Acre   The number N of trees of a given species per acre is approximated by the model $N = 68(10^{-0.04x})$, $5 \le x \le 40$ where x is the average diameter of the trees (in inches) 3 feet above the ground. Use the model to approximate the average diameter of the trees in a test plot when $N = 21$.

113. Medicine   The number y of hospitals in the United States from 1995 to 2002 can be modeled by

$$y = 7312 - 630.0 \ln t, \quad 5 \le t \le 12$$

where t represents the year, with $t = 5$ corresponding to 1995. During which year did the number of hospitals reach 5800?   (Source: Health Forum)

114. Sports   The number y of daily fee golf facilities in the United States from 1995 to 2003 can be modeled by $y = 4381 + 1883.6 \ln t$, $5 \le t \le 13$ where t represents the year, with $t = 5$ corresponding to 1995. During which year did the number of daily fee golf facilities reach 9000?   (Source: National Golf Foundation)

115. Average Heights   The percent m of American males between the ages of 18 and 24 who are no more than x inches tall is modeled by

$$m(x) = \frac{100}{1 + e^{-0.6114(x - 69.71)}}$$

and the percent f of American females between the ages of 18 and 24 who are no more than x inches tall is modeled by

$$f(x) = \frac{100}{1 + e^{-0.66607(x - 64.51)}}.$$

(Source: U.S. National Center for Health Statistics)

(a) Use the graph to determine any horizontal asymptotes of the graphs of the functions. Interpret the meaning in the context of the problem.

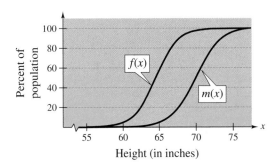

(b) What is the average height of each sex?

116. Learning Curve   In a group project in learning theory, a mathematical model for the proportion P of correct responses after n trials was found to be

$$P = \frac{0.83}{1 + e^{-0.2n}}.$$

 (a) Use a graphing utility to graph the function.

 (b) Use the graph to determine any horizontal asymptotes of the graph of the function. Interpret the meaning of the upper asymptote in the context of this problem.

(c) After how many trials will 60% of the responses be correct?

Model It

117. Automobiles   Automobiles are designed with crumple zones that help protect their occupants in crashes. The crumple zones allow the occupants to move short distances when the automobiles come to abrupt stops. The greater the distance moved, the fewer g's the crash victims experience. (One g is equal to the acceleration due to gravity. For very short periods of time, humans have withstood as much as 40 g's.) In crash tests with vehicles moving at 90 kilometers per hour, analysts measured the numbers of g's experienced during deceleration by crash dummies that were permitted to move x meters during impact. The data are shown in the table.

x	g's
0.2	158
0.4	80
0.6	53
0.8	40
1.0	32

A model for the data is given by

$$y = -3.00 + 11.88 \ln x + \frac{36.94}{x}$$

where y is the number of g's.

(a) Complete the table using the model.

x	0.2	0.4	0.6	0.8	1.0
y					

 (b) Use a graphing utility to graph the data points and the model in the same viewing window. How do they compare?

(c) Use the model to estimate the distance traveled during impact if the passenger deceleration must not exceed 30 g's.

(d) Do you think it is practical to lower the number of g's experienced during impact to fewer than 23? Explain your reasoning.

118. *Data Analysis* An object at a temperature of 160°C was removed from a furnace and placed in a room at 20°C. The temperature T of the object was measured each hour h and recorded in the table. A model for the data is given by $T = 20[1 + 7(2^{-h})]$. The graph of this model is shown in the figure.

Hour, h	Temperature, T
0	160°
1	90°
2	56°
3	38°
4	29°
5	24°

(a) Use the graph to identify the horizontal asymptote of the model and interpret the asymptote in the context of the problem.

(b) Use the model to approximate the time when the temperature of the object was 100°C.

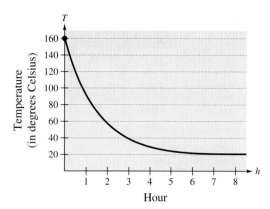

Hour

Synthesis

True or False? **In Exercises 119–122, rewrite each verbal statement as an equation. Then decide whether the statement is true or false. Justify your answer.**

119. The logarithm of the product of two numbers is equal to the sum of the logarithms of the numbers.

120. The logarithm of the sum of two numbers is equal to the product of the logarithms of the numbers.

121. The logarithm of the difference of two numbers is equal to the difference of the logarithms of the numbers.

122. The logarithm of the quotient of two numbers is equal to the difference of the logarithms of the numbers.

123. *Think About It* Is it possible for a logarithmic equation to have more than one extraneous solution? Explain.

124. *Finance* You are investing P dollars at an annual interest rate of r, compounded continuously, for t years. Which of the following would result in the highest value of the investment? Explain your reasoning.

(a) Double the amount you invest.

(b) Double your interest rate.

(c) Double the number of years.

125. *Think About It* Are the times required for the investments in Exercises 107 and 108 to quadruple twice as long as the times for them to double? Give a reason for your answer and verify your answer algebraically.

126. *Writing* Write two or three sentences stating the general guidelines that you follow when solving (a) exponential equations and (b) logarithmic equations.

Skills Review

In Exercises 127–130, simplify the expression.

127. $\sqrt{48x^2y^5}$

128. $\sqrt{32} - 2\sqrt{25}$

129. $\sqrt[3]{25} \cdot \sqrt[3]{15}$

130. $\dfrac{3}{\sqrt{10} - 2}$

In Exercises 131–134, sketch a graph of the function.

131. $f(x) = |x| + 9$

132. $f(x) = |x + 2| - 8$

133. $g(x) = \begin{cases} 2x, & x < 0 \\ -x^2 + 4, & x \geq 0 \end{cases}$

134. $g(x) = \begin{cases} x - 3, & x \leq -1 \\ x^2 + 1, & x > -1 \end{cases}$

In Exercises 135–138, evaluate the logarithm using the change-of-base formula. Approximate your result to three decimal places.

135. $\log_6 9$

136. $\log_3 4$

137. $\log_{3/4} 5$

138. $\log_8 22$

5.5 Exponential and Logarithmic Models

What you should learn

- Recognize the five most common types of models involving exponential and logarithmic functions.
- Use exponential growth and decay functions to model and solve real-life problems.
- Use Gaussian functions to model and solve real-life problems.
- Use logistic growth functions to model and solve real-life problems.
- Use logarithmic functions to model and solve real-life problems.

Why you should learn it

Exponential growth and decay models are often used to model the population of a country. For instance, in Exercise 36 on page 415, you will use exponential growth and decay models to compare the populations of several countries.

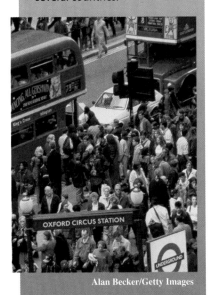

Alan Becker/Getty Images

Introduction

The five most common types of mathematical models involving exponential functions and logarithmic functions are as follows.

1. **Exponential growth model:** $y = ae^{bx}, \quad b > 0$
2. **Exponential decay model:** $y = ae^{-bx}, \quad b > 0$
3. **Gaussian model:** $y = ae^{-(x-b)^2/c}$
4. **Logistic growth model:** $y = \dfrac{a}{1 + be^{-rx}}$
5. **Logarithmic models:** $y = a + b \ln x, \quad y = a + b \log x$

The basic shapes of the graphs of these functions are shown in Figure 5.29.

EXPONENTIAL GROWTH MODEL

EXPONENTIAL DECAY MODEL

GAUSSIAN MODEL

LOGISTIC GROWTH MODEL

NATURAL LOGARITHMIC MODEL

COMMON LOGARITHMIC MODEL

FIGURE 5.29

You can often gain quite a bit of insight into a situation modeled by an exponential or logarithmic function by identifying and interpreting the function's asymptotes. Use the graphs in Figure 5.29 to identify the asymptotes of the graph of each function.

This section shows students real-world applications of logarithmic and exponential functions.

Exponential Growth and Decay

Example 1 **Digital Television**

Estimates of the numbers (in millions) of U.S. households with digital television from 2003 through 2007 are shown in the table. The scatter plot of the data is shown in Figure 5.30. (Source: eMarketer)

Year	Households
2003	44.2
2004	49.0
2005	55.5
2006	62.5
2007	70.3

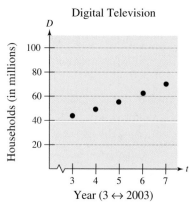

Digital Television

FIGURE 5.30

An exponential growth model that approximates these data is given by

$$D = 30.92e^{0.1171t}, \quad 3 \le t \le 7$$

where D is the number of households (in millions) and $t = 3$ represents 2003. Compare the values given by the model with the estimates shown in the table. According to this model, when will the number of U.S. households with digital television reach 100 million?

Solution

The following table compares the two sets of figures. The graph of the model and the original data are shown in Figure 5.31.

Year	2003	2004	2005	2006	2007
Households	44.2	49.0	55.5	62.5	70.3
Model	43.9	49.4	55.5	62.4	70.2

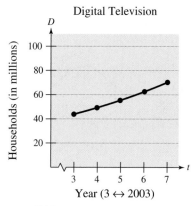

Digital Television

FIGURE 5.31

To find when the number of U.S. households with digital television will reach 100 million, let $D = 100$ in the model and solve for t.

$30.92e^{0.1171t} = D$	Write original model.
$30.92e^{0.1171t} = 100$	Let $D = 100$.
$e^{0.1171t} \approx 3.2342$	Divide each side by 30.92.
$\ln e^{0.1171t} \approx \ln 3.2342$	Take natural log of each side.
$0.1171t \approx 1.1738$	Inverse Property
$t \approx 10.0$	Divide each side by 0.1171.

According to the model, the number of U.S. households with digital television will reach 100 million in 2010.

✓CHECKPOINT Now try Exercise 35.

Technology

Some graphing utilities have an *exponential regression* feature that can be used to find exponential models that represent data. If you have such a graphing utility, try using it to find an exponential model for the data given in Example 1. How does your model compare with the model given in Example 1?

In Example 1, you were given the exponential growth model. But suppose this model were not given; how could you find such a model? One technique for doing this is demonstrated in Example 2.

Example 2 Modeling Population Growth

In a research experiment, a population of fruit flies is increasing according to the law of exponential growth. After 2 days there are 100 flies, and after 4 days there are 300 flies. How many flies will there be after 5 days?

Solution

Let y be the number of flies at time t. From the given information, you know that $y = 100$ when $t = 2$ and $y = 300$ when $t = 4$. Substituting this information into the model $y = ae^{bt}$ produces

$$100 = ae^{2b} \quad \text{and} \quad 300 = ae^{4b}.$$

To solve for b, solve for a in the first equation.

$$100 = ae^{2b} \quad \Longrightarrow \quad a = \frac{100}{e^{2b}} \qquad \text{Solve for } a \text{ in the first equation.}$$

Then substitute the result into the second equation.

$$300 = ae^{4b} \qquad\qquad \text{Write second equation.}$$

$$300 = \left(\frac{100}{e^{2b}}\right)e^{4b} \qquad \text{Substitute } 100/e^{2b} \text{ for } a.$$

$$\frac{300}{100} = e^{2b} \qquad\qquad \text{Divide each side by 100.}$$

$$\ln 3 = 2b \qquad\qquad \text{Take natural log of each side.}$$

$$\frac{1}{2}\ln 3 = b \qquad\qquad \text{Solve for } b.$$

Using $b = \frac{1}{2}\ln 3$ and the equation you found for a, you can determine that

$$a = \frac{100}{e^{2[(1/2)\ln 3]}} \qquad \text{Substitute } \tfrac{1}{2}\ln 3 \text{ for } b.$$

$$= \frac{100}{e^{\ln 3}} \qquad\qquad \text{Simplify.}$$

$$= \frac{100}{3} \qquad\qquad \text{Inverse Property}$$

$$\approx 33.33. \qquad\qquad \text{Simplify.}$$

So, with $a \approx 33.33$ and $b = \frac{1}{2}\ln 3 \approx 0.5493$, the exponential growth model is

$$y = 33.33e^{0.5493t}$$

as shown in Figure 5.32. This implies that, after 5 days, the population will be

$$y = 33.33e^{0.5493(5)} \approx 520 \text{ flies.}$$

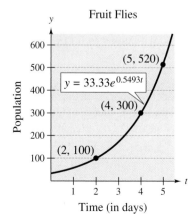

FIGURE 5.32

✓CHECKPOINT   Now try Exercise 37.

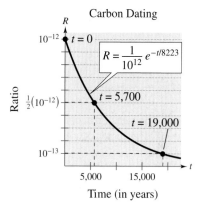

Carbon Dating

FIGURE 5.33

In living organic material, the ratio of the number of radioactive carbon isotopes (carbon 14) to the number of nonradioactive carbon isotopes (carbon 12) is about 1 to 10^{12}. When organic material dies, its carbon 12 content remains fixed, whereas its radioactive carbon 14 begins to decay with a half-life of about 5700 years. To estimate the age of dead organic material, scientists use the following formula, which denotes the ratio of carbon 14 to carbon 12 present at any time t (in years).

$$R = \frac{1}{10^{12}} e^{-t/8223}$$ Carbon dating model

The graph of R is shown in Figure 5.33. Note that R decreases as t increases.

Example 3   Carbon Dating

Estimate the age of a newly discovered fossil in which the ratio of carbon 14 to carbon 12 is

$$R = \frac{1}{10^{13}}.$$

Solution

In the carbon dating model, substitute the given value of R to obtain the following.

$$\frac{1}{10^{12}} e^{-t/8223} = R$$ Write original model.

$$\frac{e^{-t/8223}}{10^{12}} = \frac{1}{10^{13}}$$ Let $R = \frac{1}{10^{13}}$.

$$e^{-t/8223} = \frac{1}{10}$$ Multiply each side by 10^{12}.

$$\ln e^{-t/8223} = \ln \frac{1}{10}$$ Take natural log of each side.

$$-\frac{t}{8223} \approx -2.3026$$ Inverse Property

$$t \approx 18{,}934$$ Multiply each side by -8223.

So, to the nearest thousand years, the age of the fossil is about 19,000 years.

✓CHECKPOINT   Now try Exercise 41.

The value of b in the exponential decay model $y = ae^{-bt}$ determines the *decay* of radioactive isotopes. For instance, to find how much of an initial 10 grams of ^{226}Ra isotope with a half-life of 1599 years is left after 500 years, substitute this information into the model $y = ae^{-bt}$.

$$\frac{1}{2}(10) = 10e^{-b(1599)} \quad\Longrightarrow\quad \ln\frac{1}{2} = -1599b \quad\Longrightarrow\quad b = -\frac{\ln\frac{1}{2}}{1599}$$

Using the value of b found above and $a = 10$, the amount left is

$$y = 10e^{-[-\ln(1/2)/1599](500)} \approx 8.05 \text{ grams.}$$

Gaussian Models

As mentioned at the beginning of this section, Gaussian models are of the form

$$y = ae^{-(x-b)^2/c}.$$

This type of model is commonly used in probability and statistics to represent populations that are **normally distributed.** The graph of a Gaussian model is called a **bell-shaped curve.** Try graphing the normal distribution with a graphing utility. Can you see why it is called a bell-shaped curve?

For *standard* normal distributions, the model takes the form

$$y = \frac{1}{\sqrt{2\pi}}e^{-x^2/2}.$$

The **average value** for a population can be found from the bell-shaped curve by observing where the maximum y-value of the function occurs. The x-value corresponding to the maximum y-value of the function represents the average value of the independent variable—in this case, x.

Example 4   **SAT Scores**

In 2004, the Scholastic Aptitude Test (SAT) math scores for college-bound seniors roughly followed the normal distribution given by

$$y = 0.0035e^{-(x-518)^2/25,992}, \quad 200 \leq x \leq 800$$

where x is the SAT score for mathematics. Sketch the graph of this function. From the graph, estimate the average SAT score.   (Source: College Board)

Solution

The graph of the function is shown in Figure 5.34. On this bell-shaped curve, the maximum value of the curve represents the average score. From the graph, you can estimate that the average mathematics score for college-bound seniors in 2004 was 518.

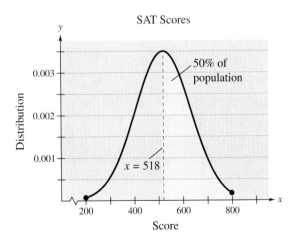

FIGURE 5.34

✓CHECKPOINT   Now try Exercise 47.

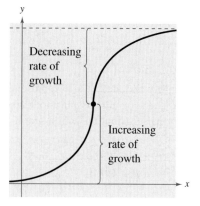

FIGURE 5.35

Logistic Growth Models

Some populations initially have rapid growth, followed by a declining rate of growth, as indicated by the graph in Figure 5.35. One model for describing this type of growth pattern is the **logistic curve** given by the function

$$y = \frac{a}{1 + be^{-rx}}$$

where y is the population size and x is the time. An example is a bacteria culture that is initially allowed to grow under ideal conditions, and then under less favorable conditions that inhibit growth. A logistic growth curve is also called a **sigmoidal curve.**

Example 5   Spread of a Virus

On a college campus of 5000 students, one student returns from vacation with a contagious and long-lasting flu virus. The spread of the virus is modeled by

$$y = \frac{5000}{1 + 4999e^{-0.8t}}, \quad t \geq 0$$

where y is the total number of students infected after t days. The college will cancel classes when 40% or more of the students are infected.

a. How many students are infected after 5 days?

b. After how many days will the college cancel classes?

Solution

a. After 5 days, the number of students infected is

$$y = \frac{5000}{1 + 4999e^{-0.8(5)}} = \frac{5000}{1 + 4999e^{-4}} \approx 54.$$

b. Classes are canceled when the number infected is $(0.40)(5000) = 2000$.

$$2000 = \frac{5000}{1 + 4999e^{-0.8t}}$$

$$1 + 4999e^{-0.8t} = 2.5$$

$$e^{-0.8t} = \frac{1.5}{4999}$$

$$\ln e^{-0.8t} = \ln \frac{1.5}{4999}$$

$$-0.8t = \ln \frac{1.5}{4999}$$

$$t = -\frac{1}{0.8} \ln \frac{1.5}{4999}$$

$$t \approx 10.1$$

So, after about 10 days, at least 40% of the students will be infected, and the college will cancel classes. The graph of the function is shown in Figure 5.36.

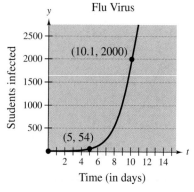

FIGURE 5.36

✓CHECKPOINT   Now try Exercise 49.

BAY ISMOYO/AFP/Getty Images

On December 26, 2004, an earthquake of magnitude 9.0 struck northern Sumatra and many other Asian countries. This earthquake caused a deadly tsunami and was the fourth largest earthquake in the world since 1900.

Alternative Writing About Mathematics

Use your school's library, the Internet, or some other reference source to find an application that fits one of the five models discussed in this section. After you have collected data for the model, plot the corresponding points and find an equation that describes the points you have plotted.

Logarithmic Models

Example 6   **Magnitudes of Earthquakes**

On the Richter scale, the magnitude R of an earthquake of intensity I is given by

$$R = \log \frac{I}{I_0}$$

where $I_0 = 1$ is the minimum intensity used for comparison. Find the intensities per unit of area for each earthquake. (Intensity is a measure of the wave energy of an earthquake.)

a. Northern Sumatra in 2004: $R = 9.0$

b. Southeastern Alaska in 2004: $R = 6.8$

Solution

a. Because $I_0 = 1$ and $R = 9.0$, you have

$$9.0 = \log \frac{I}{1}$$ Substitute 1 for I_0 and 9.0 for R.

$$10^{9.0} = 10^{\log I}$$ Exponentiate each side.

$$I = 10^{9.0} \approx 100,000,000.$$ Inverse Property

b. For $R = 6.8$, you have

$$6.8 = \log \frac{I}{1}$$ Substitute 1 for I_0 and 6.8 for R.

$$10^{6.8} = 10^{\log I}$$ Exponentiate each side.

$$I = 10^{6.8} \approx 6,310,000.$$ Inverse Property

Note that an increase of 2.2 units on the Richter scale (from 6.8 to 9.0) represents an increase in intensity by a factor of

$$\frac{1,000,000,000}{6,310,000} \approx 158.$$

In other words, the intensity of the earthquake in Sumatra was about 158 times greater than that of the earthquake in Alaska.

✓CHECKPOINT   Now try Exercise 51.

t	Year	Population, P
1	1910	92.23
2	1920	106.02
3	1930	123.20
4	1940	132.16
5	1950	151.33
6	1960	179.32
7	1970	203.30
8	1980	226.54
9	1990	248.72
10	2000	281.42

𝓦RITING ABOUT **𝓜**ATHEMATICS

Comparing Population Models The populations P (in millions) of the United States for the census years from 1910 to 2000 are shown in the table at the left. Least squares regression analysis gives the best quadratic model for these data as $P = 1.0328t^2 + 9.607t + 81.82$, and the best exponential model for these data as $P = 82.677e^{0.124t}$. Which model better fits the data? Describe how you reached your conclusion. (Source: U.S. Census Bureau)

5.5 Exercises

VOCABULARY CHECK: Fill in the blanks.

1. An exponential growth model has the form _____ and an exponential decay model has the form _____.

2. A logarithmic model has the form _____ or _____.

3. Gaussian models are commonly used in probability and statistics to represent populations that are _____ _____.

4. The graph of a Gaussian model is _____ shaped, where the _____ _____ is the maximum y-value of the graph.

5. A logistic curve is also called a _____ curve.

PREREQUISITE SKILLS REVIEW: Practice and review algebra skills needed for this section at **www.Eduspace.com.**

In Exercises 1–6, match the function with its graph. [The graphs are labeled (a), (b), (c), (d), (e), and (f).]

(a)

(b)

(c)

(d)

(e)

(f)
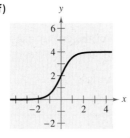

1. $y = 2e^{x/4}$

2. $y = 6e^{-x/4}$

3. $y = 6 + \log(x + 2)$

4. $y = 3e^{-(x-2)^2/5}$

5. $y = \ln(x + 1)$

6. $y = \dfrac{4}{1 + e^{-2x}}$

Compound Interest In Exercises 7–14, complete the table for a savings account in which interest is compounded continuously.

	Initial Investment	Annual % Rate	Time to Double	Amount After 10 Years
7.	$1000	3.5%		
8.	$750	$10\frac{1}{2}$%		
9.	$750		$7\frac{3}{4}$ yr	
10.	$10,000		12 yr	
11.	$500			$1505.00
12.	$600			$19,205.00
13.		4.5%		$10,000.00
14.		2%		$2000.00

Compound Interest In Exercises 15 and 16, determine the principal P that must be invested at rate r, compounded monthly, so that $500,000 will be available for retirement in t years.

15. $r = 7\frac{1}{2}\%, t = 20$

16. $r = 12\%, t = 40$

Compound Interest In Exercises 17 and 18, determine the time necessary for $1000 to double if it is invested at interest rate r compounded (a) annually, (b) monthly, (c) daily, and (d) continuously.

17. $r = 11\%$

18. $r = 10\frac{1}{2}\%$

19. *Compound Interest* Complete the table for the time t necessary for P dollars to triple if interest is compounded continuously at rate r.

r	2%	4%	6%	8%	10%	12%
t						

 20. *Modeling Data* Draw a scatter plot of the data in Exercise 19. Use the *regression* feature of a graphing utility to find a model for the data.

21. Compound Interest Complete the table for the time t necessary for P dollars to triple if interest is compounded annually at rate r.

r	2%	4%	6%	8%	10%	12%
t						

 22. Modeling Data Draw a scatter plot of the data in Exercise 21. Use the *regression* feature of a graphing utility to find a model for the data.

23. Comparing Models If $1 is invested in an account over a 10-year period, the amount in the account, where t represents the time in years, is given by $A = 1 + 0.075[\![t]\!]$ or $A = e^{0.07t}$ depending on whether the account pays simple interest at $7\frac{1}{2}\%$ or continuous compound interest at 7%. Graph each function on the same set of axes. Which grows at a higher rate? (Remember that $[\![t]\!]$ is the greatest integer function discussed in Section P.7.)

24. Comparing Models If $1 is invested in an account over a 10-year period, the amount in the account, where t represents the time in years, is given by

$$A = 1 + 0.06[\![t]\!] \quad \text{or} \quad A = \left(1 + \frac{0.055}{365}\right)^{[\![365t]\!]}$$

depending on whether the account pays simple interest at 6% or compound interest at $5\frac{1}{2}\%$ compounded daily. Use a graphing utility to graph each function in the same viewing window. Which grows at a higher rate?

Radioactive Decay In Exercises 25–30, complete the table for the radioactive isotope.

Isotope	Half-life (years)	Initial Quantity	Amount After 1000 Years
25. ^{226}Ra	1599	10 g	
26. ^{226}Ra	1599		1.5 g
27. ^{14}C	5715		2 g
28. ^{14}C	5715	3 g	
29. ^{239}Pu	24,100		2.1 g
30. ^{239}Pu	24,100		0.4 g

In Exercises 31–34, find the exponential model $y = ae^{bx}$ that fits the points shown in the graph or table.

31.

32.

33.

x	0	4
y	5	1

34.

x	0	3
y	1	$\frac{1}{4}$

35. Population The population P (in thousands) of Pittsburgh, Pennsylvania from 2000 through 2003 can be modeled by $P = 2430e^{-0.0029t}$, where t represents the year, with $t = 0$ corresponding to 2000. (Source: U.S. Census Bureau)

(a) According to the model, was the population of Pittsburgh increasing or decreasing from 2000 to 2003? Explain your reasoning.

(b) What were the populations of Pittsburgh in 2000 and 2003?

(c) According to the model, when will the population be approximately 2.3 million?

Model It

36. Population The table shows the populations (in millions) of five countries in 2000 and the projected populations (in millions) for the year 2010. (Source: U.S. Census Bureau)

Country	2000	2010
Bulgaria	7.8	7.1
Canada	31.3	34.3
China	1268.9	1347.6
United Kingdom	59.5	61.2
United States	282.3	309.2

(a) Find the exponential growth or decay model $y = ae^{bt}$ or $y = ae^{-bt}$ for the population of each country by letting $t = 0$ correspond to 2000. Use the model to predict the population of each country in 2030.

(b) You can see that the populations of the United States and the United Kingdom are growing at different rates. What constant in the equation $y = ae^{bt}$ is determined by these different growth rates? Discuss the relationship between the different growth rates and the magnitude of the constant.

(c) You can see that the population of China is increasing while the population of Bulgaria is decreasing. What constant in the equation $y = ae^{bt}$ reflects this difference? Explain.

37. Website Growth The number y of hits a new search-engine website receives each month can be modeled by

$$y = 4080e^{kt}$$

where t represents the number of months the website has been operating. In the website's third month, there were 10,000 hits. Find the value of k, and use this result to predict the number of hits the website will receive after 24 months.

38. Value of a Painting The value V (in millions of dollars) of a famous painting can be modeled by

$$V = 10e^{kt}$$

where t represents the year, with $t = 0$ corresponding to 1990. In 2004, the same painting was sold for \$65 million. Find the value of k, and use this result to predict the value of the painting in 2010.

39. Bacteria Growth The number N of bacteria in a culture is modeled by

$$N = 100e^{kt}$$

where t is the time in hours. If $N = 300$ when $t = 5$, estimate the time required for the population to double in size.

40. Bacteria Growth The number N of bacteria in a culture is modeled by

$$N = 250e^{kt}$$

where t is the time in hours. If $N = 280$ when $t = 10$, estimate the time required for the population to double in size.

41. Carbon Dating

(a) The ratio of carbon 14 to carbon 12 in a piece of wood discovered in a cave is $R = 1/8^{14}$. Estimate the age of the piece of wood.

(b) The ratio of carbon 14 to carbon 12 in a piece of paper buried in a tomb is $R = 1/13^{11}$. Estimate the age of the piece of paper.

42. Radioactive Decay Carbon 14 dating assumes that the carbon dioxide on Earth today has the same radioactive content as it did centuries ago. If this is true, the amount of ^{14}C absorbed by a tree that grew several centuries ago should be the same as the amount of ^{14}C absorbed by a tree growing today. A piece of ancient charcoal contains only 15% as much radioactive carbon as a piece of modern charcoal. How long ago was the tree burned to make the ancient charcoal if the half-life of ^{14}C is 5715 years?

43. Depreciation A 2005 Jeep Wrangler that costs \$30,788 new has a book value of \$18,000 after 2 years.

(a) Find the linear model $V = mt + b$.

(b) Find the exponential model $V = ae^{kt}$.

(c) Use a graphing utility to graph the two models in the same viewing window. Which model depreciates faster in the first 2 years?

(d) Find the book values of the vehicle after 1 year and after 3 years using each model.

(e) Explain the advantages and disadvantages of using each model to a buyer and a seller.

44. Depreciation A Dell Inspiron 8600 laptop computer that costs \$1150 new has a book value of \$550 after 2 years.

(a) Find the linear model $V = mt + b$.

(b) Find the exponential model $V = ae^{kt}$.

(c) Use a graphing utility to graph the two models in the same viewing window. Which model depreciates faster in the first 2 years?

(d) Find the book values of the computer after 1 year and after 3 years using each model.

(e) Explain the advantages and disadvantages to a buyer and a seller of using each model.

45. Sales The sales S (in thousands of units) of a new CD burner after it has been on the market for t years are modeled by

$$S(t) = 100(1 - e^{kt}).$$

Fifteen thousand units of the new product were sold the first year.

(a) Complete the model by solving for k.

(b) Sketch the graph of the model.

(c) Use the model to estimate the number of units sold after 5 years.

46. Learning Curve The management at a plastics factory has found that the maximum number of units a worker can produce in a day is 30. The learning curve for the number N of units produced per day after a new employee has worked t days is modeled by

$$N = 30(1 - e^{kt}).$$

After 20 days on the job, a new employee produces 19 units.

(a) Find the learning curve for this employee (first, find the value of k).

(b) How many days should pass before this employee is producing 25 units per day?

47. IQ Scores The IQ scores from a sample of a class of returning adult students at a small northeastern college roughly follow the normal distribution

$$y = 0.0266e^{-(x - 100)^2/450}, \quad 70 \le x \le 115$$

where x is the IQ score.

(a) Use a graphing utility to graph the function.

(b) From the graph in part (a), estimate the average IQ score of an adult student.

48. Education   The time (in hours per week) a student utilizes a math-tutoring center roughly follows the normal distribution

$$y = 0.7979e^{-(x-5.4)^2/0.5}, \quad 4 \le x \le 7$$

where x is the number of hours.

(a) Use a graphing utility to graph the function.

(b) From the graph in part (a), estimate the average number of hours per week a student uses the tutor center.

49. Population Growth   A conservation organization releases 100 animals of an endangered species into a game preserve. The organization believes that the preserve has a carrying capacity of 1000 animals and that the growth of the pack will be modeled by the logistic curve

$$p(t) = \frac{1000}{1 + 9e^{-0.1656t}}$$

where t is measured in years (see figure).

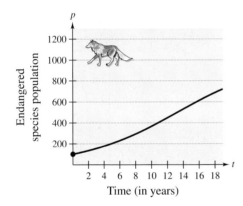

Time (in years)

(a) Estimate the population after 5 years.

(b) After how many years will the population be 500?

(c) Use a graphing utility to graph the function. Use the graph to determine the horizontal asymptotes, and interpret the meaning of the larger p-value in the context of the problem.

50. Sales   After discontinuing all advertising for a tool kit in 2000, the manufacturer noted that sales began to drop according to the model

$$S = \frac{500,000}{1 + 0.6e^{kt}}$$

where S represents the number of units sold and $t = 0$ represents 2000. In 2004, the company sold 300,000 units.

(a) Complete the model by solving for k.

(b) Estimate sales in 2008.

Geology   **In Exercises 51 and 52, use the Richter scale**

$$R = \log \frac{I}{I_0}$$

for measuring the magnitudes of earthquakes.

51. Find the intensity I of an earthquake measuring R on the Richter scale (let $I_0 = 1$).

(a) Central Alaska in 2002, $R = 7.9$

(b) Hokkaido, Japan in 2003, $R = 8.3$

(c) Illinois in 2004, $R = 4.2$

52. Find the magnitude R of each earthquake of intensity I (let $I_0 = 1$).

(a) $I = 80,500,000$ (b) $I = 48,275,000$

(c) $I = 251,200$

Intensity of Sound   **In Exercises 53–56, use the following information for determining sound intensity. The level of sound β, in decibels, with an intensity of I, is given by**

$$\beta = 10 \log \frac{I}{I_0}$$

where I_0 is an intensity of 10^{-12} watt per square meter, corresponding roughly to the faintest sound that can be heard by the human ear. In Exercises 53 and 54, find the level of sound β.

53. (a) $I = 10^{-10}$ watt per m² (quiet room)

(b) $I = 10^{-5}$ watt per m² (busy street corner)

(c) $I = 10^{-8}$ watt per m² (quiet radio)

(d) $I = 10^{0}$ watt per m² (threshold of pain)

54. (a) $I = 10^{-11}$ watt per m² (rustle of leaves)

(b) $I = 10^{2}$ watt per m² (jet at 30 meters)

(c) $I = 10^{-4}$ watt per m² (door slamming)

(d) $I = 10^{-2}$ watt per m² (siren at 30 meters)

55. Due to the installation of noise suppression materials, the noise level in an auditorium was reduced from 93 to 80 decibels. Find the percent decrease in the intensity level of the noise as a result of the installation of these materials.

56. Due to the installation of a muffler, the noise level of an engine was reduced from 88 to 72 decibels. Find the percent decrease in the intensity level of the noise as a result of the installation of the muffler.

pH Levels   **In Exercises 57–62, use the acidity model given by pH $= -\log[H^+]$, where acidity (pH) is a measure of the hydrogen ion concentration $[H^+]$ (measured in moles of hydrogen per liter) of a solution.**

57. Find the pH if $[H^+] = 2.3 \times 10^{-5}$.

58. Find the pH if $[H^+] = 11.3 \times 10^{-6}$.

59. Compute $[H^+]$ for a solution in which pH $= 5.8$.

60. Compute $[H^+]$ for a solution in which pH $= 3.2$.

61. Apple juice has a pH of 2.9 and drinking water has a pH of 8.0. The hydrogen ion concentration of the apple juice is how many times the concentration of drinking water?

62. The pH of a solution is decreased by one unit. The hydrogen ion concentration is increased by what factor?

63. *Forensics* At 8:30 A.M., a coroner was called to the home of a person who had died during the night. In order to estimate the time of death, the coroner took the person's temperature twice. At 9:00 A.M. the temperature was 85.7°F, and at 11:00 a.m. the temperature was 82.8°F. From these two temperatures, the coroner was able to determine that the time elapsed since death and the body temperature were related by the formula

$$t = -10 \ln \frac{T - 70}{98.6 - 70}$$

where t is the time in hours elapsed since the person died and T is the temperature (in degrees Fahrenheit) of the person's body. Assume that the person had a normal body temperature of 98.6°F at death, and that the room temperature was a constant 70°F. (This formula is derived from a general cooling principle called *Newton's Law of Cooling*.) Use the formula to estimate the time of death of the person.

64. *Home Mortgage* A $120,000 home mortgage for 35 years at $7\frac{1}{2}\%$ has a monthly payment of $809.39. Part of the monthly payment is paid toward the interest charge on the unpaid balance, and the remainder of the payment is used to reduce the principal. The amount that is paid toward the interest is

$$u = M - \left(M - \frac{Pr}{12}\right)\left(1 + \frac{r}{12}\right)^{12t}$$

and the amount that is paid toward the reduction of the principal is

$$v = \left(M - \frac{Pr}{12}\right)\left(1 + \frac{r}{12}\right)^{12t}.$$

In these formulas, P is the size of the mortgage, r is the interest rate, M is the monthly payment, and t is the time in years.

(a) Use a graphing utility to graph each function in the same viewing window. (The viewing window should show all 35 years of mortgage payments.)

(b) In the early years of the mortgage, is the larger part of the monthly payment paid toward the interest or the principal? Approximate the time when the monthly payment is evenly divided between interest and principal reduction.

(c) Repeat parts (a) and (b) for a repayment period of 20 years ($M = \$966.71$). What can you conclude?

65. *Home Mortgage* The total interest u paid on a home mortgage of P dollars at interest rate r for t years is

$$u = P\left[\frac{rt}{1 - \left(\dfrac{1}{1 + r/12}\right)^{12t}} - 1\right].$$

Consider a $120,000 home mortgage at $7\frac{1}{2}\%$.

(a) Use a graphing utility to graph the total interest function.

(b) Approximate the length of the mortgage for which the total interest paid is the same as the size of the mortgage. Is it possible that some people are paying twice as much in interest charges as the size of the mortgage?

66. *Data Analysis* The table shows the time t (in seconds) required to attain a speed of s miles per hour from a standing start for a car.

Speed, s	Time, t
30	3.4
40	5.0
50	7.0
60	9.3
70	12.0
80	15.8
90	20.0

Two models for these data are as follows.

$$t_1 = 40.757 + 0.556s - 15.817 \ln s$$

$$t_2 = 1.2259 + 0.0023s^2$$

(a) Use the *regression* feature of a graphing utility to find a linear model t_3 and an exponential model t_4 for the data.

(b) Use a graphing utility to graph the data and each model in the same viewing window.

(c) Create a table comparing the data with estimates obtained from each model.

(d) Use the results of part (c) to find the sum of the absolute values of the differences between the data and the estimated values given by each model. Based on the four sums, which model do you think better fits the data? Explain.

Synthesis

True or False?   In Exercises 67–70, determine whether the statement is true or false. Justify your answer.

67. The domain of a logistic growth function cannot be the set of real numbers.

68. A logistic growth function will always have an x-intercept.

69. The graph of

$$f(x) = \frac{4}{1 + 6e^{-2x}} + 5$$

is the graph of

$$g(x) = \frac{4}{1 + 6e^{-2x}}$$

shifted to the right five units.

70. The graph of a Gaussian model will never have an x-intercept.

71. Identify each model as linear, logarithmic, exponential, logistic, or none of the above. Explain your reasoning.

(a)

(b)

(c)

(d)

(e)

(f)

72. *Writing*   Use your school's library, the Internet, or some other reference source to write a paper describing John Napier's work with logarithms.

Skills Review

In Exercises 73–78, (a) plot the points, (b) find the distance between the points, (c) find the midpoint of the line segment joining the points, and (d) find the slope of the line passing through the points.

73. $(-1, 2), (0, 5)$

74. $(4, -3), (-6, 1)$

75. $(3, 3), (14, -2)$

76. $(7, 0), (10, 4)$

77. $\left(\frac{1}{2}, -\frac{1}{4}\right), \left(\frac{3}{4}, 0\right)$

78. $\left(\frac{7}{3}, \frac{1}{6}\right), \left(-\frac{2}{3}, -\frac{1}{3}\right)$

In Exercises 79–88, sketch the graph of the equation.

79. $y = 10 - 3x$

80. $y = -4x - 1$

81. $y = -2x^2 - 3$

82. $y = 2x^2 - 7x - 30$

83. $3x^2 - 4y = 0$

84. $-x^2 - 8y = 0$

85. $y = \dfrac{4}{1 - 3x}$

86. $y = \dfrac{x^2}{-x - 2}$

87. $x^2 + (y - 8)^2 = 25$

88. $(x - 4)^2 + (y + 7) = 4$

In Exercises 89–92, graph the exponential function.

89. $f(x) = 2^{x-1} + 5$

90. $f(x) = -2^{-x-1} - 1$

91. $f(x) = 3^x - 4$

92. $f(x) = -3^x + 4$

93. **Make a Decision**   To work an extended application analyzing the net sales for Kohl's Corporation from 1992 to 2004, visit this text's website at *college.hmco.com*. *(Data Source: Kohl's Illinois, Inc.)*

5 Chapter Summary

What did you learn?

Section 5.1	Review Exercises
☐ Recognize and evaluate exponential functions with base *a* (p. 368).	1–6
☐ Graph exponential functions and use the One-to-One Property (p. 369).	7–26
☐ Recognize, evaluate, and graph exponential functions with base *e* (p. 372).	27–34
☐ Use exponential functions to model and solve real-life problems (p. 373).	35–40

Section 5.2	
☐ Recognize and evaluate logarithmic functions with base *a* (p. 379).	41–52
☐ Graph logarithmic functions (p. 381).	53–58
☐ Recognize, evaluate, and graph natural logarithmic functions (p. 383).	59–68
☐ Use logarithmic functions to model and solve real-life problems (p. 385).	69, 70

Section 5.3	
☐ Use the change-of-base formula to rewrite and evaluate logarithmic expressions (p. 389).	71–74
☐ Use properties of logarithms to evaluate or rewrite logarithmic expressions (p. 390).	75–78
☐ Use properties of logarithms to expand or condense logarithmic expressions (p. 391).	79–94
☐ Use logarithmic functions to model and solve real-life problems (p. 392).	95, 96

Section 5.4	
☐ Solve simple exponential and logarithmic equations (p. 396).	97–104
☐ Solve more complicated exponential equations (p. 397).	105–118
☐ Solve more complicated logarithmic equations (p. 399).	119–134
☐ Use exponential and logarithmic equations to model and solve real-life problems (p. 401).	135, 136

Section 5.5	
☐ Recognize the five most common types of models involving exponential and logarithmic functions (p. 407).	137–142
☐ Use exponential growth and decay functions to model and solve real-life problems (p. 408).	143–148
☐ Use Gaussian functions to model and solve real-life problems (p. 411).	149
☐ Use logistic growth functions to model and solve real-life problems (p. 412).	150
☐ Use logarithmic functions to model and solve real-life problems (p. 413).	151, 152

| **5** | **Review Exercises** |

In Exercises 1–6, evaluate the function at the indicated value of *x*. Round your result to three decimal places.

Function	*Value*
1. $f(x) = 6.1^x$	$x = 2.4$
2. $f(x) = 30^x$	$x = \sqrt{3}$
3. $f(x) = 2^{-0.5x}$	$x = \pi$
4. $f(x) = 1278^{x/5}$	$x = 1$
5. $f(x) = 7(0.2^x)$	$x = -\sqrt{11}$
6. $f(x) = -14(5^x)$	$x = -0.8$

In Exercises 7–10, match the function with its graph. [The graphs are labeled (a), (b), (c), and (d).]

(a)

(b)

(c)

(d)

7. $f(x) = 4^x$

8. $f(x) = 4^{-x}$

9. $f(x) = -4^x$

10. $f(x) = 4^x + 1$

In Exercises 11–14, use the graph of *f* to describe the transformation that yields the graph of *g*.

11. $f(x) = 5^x$, $\quad g(x) = 5^{x-1}$

12. $f(x) = 4^x$, $\quad g(x) = 4^x - 3$

13. $f(x) = \left(\frac{1}{2}\right)^x$, $\quad g(x) = -\left(\frac{1}{2}\right)^{x+2}$

14. $f(x) = \left(\frac{2}{3}\right)^x$, $\quad g(x) = 8 - \left(\frac{2}{3}\right)^x$

 In Exercises 15–22, use a graphing utility to construct a table of values for the function. Then sketch the graph of the function.

15. $f(x) = 4^{-x} + 4$

16. $f(x) = -4^x - 3$

17. $f(x) = -2.65^{x+1}$

18. $f(x) = 2.65^{x-1}$

19. $f(x) = 5^{x-2} + 4$

20. $f(x) = 2^{x-6} - 5$

21. $f(x) = \left(\frac{1}{2}\right)^{-x} + 3$

22. $f(x) = \left(\frac{1}{8}\right)^{x+2} - 5$

In Exercises 23–26, use the One-to-One Property to solve the equation for *x*.

23. $3^{x+2} = \dfrac{1}{9}$

24. $\left(\dfrac{1}{3}\right)^{x-2} = 81$

25. $e^{5x-7} = e^{15}$

26. $e^{8-2x} = e^{-3}$

In Exercises 27–30, evaluate the function given by $f(x) = e^x$ at the indicated value of *x*. Round your result to three decimal places.

27. $x = 8$

28. $x = \frac{5}{8}$

29. $x = -1.7$

30. $x = 0.278$

 In Exercises 31–34, use a graphing utility to construct a table of values for the function. Then sketch the graph of the function.

31. $h(x) = e^{-x/2}$

32. $h(x) = 2 - e^{-x/2}$

33. $f(x) = e^{x+2}$

34. $s(t) = 4e^{-2/t}, \quad t > 0$

Compound Interest In Exercises 35 and 36, complete the table to determine the balance *A* for *P* dollars invested at rate *r* for *t* years and compounded *n* times per year.

n	1	2	4	12	365	Continuous
A						

35. $P = \$3500$, $r = 6.5\%$, $t = 10$ years

36. $P = \$2000$, $r = 5\%$, $t = 30$ years

37. ***Waiting Times*** The average time between incoming calls at a switchboard is 3 minutes. The probability *F* of waiting less than *t* minutes until the next incoming call is approximated by the model $F(t) = 1 - e^{-t/3}$. A call has just come in. Find the probability that the next call will be within

(a) $\frac{1}{2}$ minute. (b) 2 minutes. (c) 5 minutes.

38. ***Depreciation*** After *t* years, the value *V* of a car that originally cost $14,000 is given by $V(t) = 14{,}000\left(\frac{3}{4}\right)^t$.

 (a) Use a graphing utility to graph the function.

(b) Find the value of the car 2 years after it was purchased.

(c) According to the model, when does the car depreciate most rapidly? Is this realistic? Explain.

39. Trust Fund On the day a person is born, a deposit of $50,000 is made in a trust fund that pays 8.75% interest, compounded continuously.

(a) Find the balance on the person's 35th birthday.

(b) How much longer would the person have to wait for the balance in the trust fund to double?

40. Radioactive Decay Let Q represent a mass of plutonium 241 (^{241}Pu) (in grams), whose half-life is 14.4 years. The quantity of plutonium 241 present after t years is given by $Q = 100\left(\frac{1}{2}\right)^{t/14.4}$.

(a) Determine the initial quantity (when $t = 0$).

(b) Determine the quantity present after 10 years.

(c) Sketch the graph of this function over the interval $t = 0$ to $t = 100$.

5.2 **In Exercises 41–44, write the exponential equation in logarithmic form.**

41. $4^3 = 64$ **42.** $25^{3/2} = 125$

43. $e^{0.8} = 2.2255\ldots$ **44.** $e^0 = 1$

In Exercises 45–48, evaluate the function at the indicated value of x without using a calculator.

Function	Value
45. $f(x) = \log x$	$x = 1000$
46. $g(x) = \log_9 x$	$x = 3$
47. $g(x) = \log_2 x$	$x = \frac{1}{8}$
48. $f(x) = \log_4 x$	$x = \frac{1}{4}$

In Exercises 49–52, use the One-to-One Property to solve the equation for x.

49. $\log_4 (x + 7) = \log_4 14$ **50.** $\log_8 (3x - 10) = \log_8 5$

51. $\ln(x + 9) = \ln 4$ **52.** $\ln(2x - 1) = \ln 11$

In Exercises 53–58, find the domain, x-intercept, and vertical asymptote of the logarithmic function and sketch its graph.

53. $g(x) = \log_7 x$ **54.** $g(x) = \log_5 x$

55. $f(x) = \log\left(\frac{x}{3}\right)$ **56.** $f(x) = 6 + \log x$

57. $f(x) = 4 - \log(x + 5)$ **58.** $f(x) = \log(x - 3) + 1$

In Exercises 59–64, use a calculator to evaluate the function given by $f(x) = \ln x$ at the indicated value of x. Round your result to three decimal places if necessary.

59. $x = 22.6$ **60.** $x = 0.98$

61. $x = e^{-12}$ **62.** $x = e^7$

63. $x = \sqrt{7} + 5$ **64.** $x = \dfrac{\sqrt{3}}{8}$

In Exercises 65–68, find the domain, x-intercept, and vertical asymptote of the logarithmic function and sketch its graph.

65. $f(x) = \ln x + 3$ **66.** $f(x) = \ln(x - 3)$

67. $h(x) = \ln(x^2)$ **68.** $f(x) = \frac{1}{4}\ln x$

69. Antler Spread The antler spread a (in inches) and shoulder height h (in inches) of an adult male American elk are related by the model $h = 116 \log(a + 40) - 176$. Approximate the shoulder height of a male American elk with an antler spread of 55 inches.

70. Snow Removal The number of miles s of roads cleared of snow is approximated by the model

$$s = 25 - \frac{13 \ln(h/12)}{\ln 3}, \quad 2 \le h \le 15$$

where h is the depth of the snow in inches. Use this model to find s when $h = 10$ inches.

5.3 **In Exercises 71–74, evaluate the logarithm using the change-of-base formula. Do each exercise twice, once with common logarithms and once with natural logarithms. Round your the results to three decimal places.**

71. $\log_4 9$ **72.** $\log_{12} 200$

73. $\log_{1/2} 5$ **74.** $\log_3 0.28$

In Exercises 75–78, use the properties of logarithms to rewrite and simplify the logarithmic expression.

75. $\log 18$ **76.** $\log_2 \left(\frac{1}{12}\right)$

77. $\ln 20$ **78.** $\ln(3e^{-4})$

In Exercises 79–86, use the properties of logarithms to expand the expression as a sum, difference, and/or constant multiple of logarithms. (Assume all variables are positive.)

79. $\log_5 5x^2$ **80.** $\log 7x^4$

81. $\log_3 \dfrac{6}{\sqrt[3]{x}}$ **82.** $\log_7 \dfrac{\sqrt{x}}{4}$

83. $\ln x^2 y^2 z$ **84.** $\ln 3xy^2$

85. $\ln\left(\dfrac{x + 3}{xy}\right)$ **86.** $\ln\left(\dfrac{y - 1}{4}\right)^2, \quad y > 1$

In Exercises 87–94, condense the expression to the logarithm of a single quantity.

87. $\log_2 5 + \log_2 x$ **88.** $\log_6 y - 2 \log_6 z$

89. $\ln x - \frac{1}{4} \ln y$ **90.** $3 \ln x + 2 \ln(x + 1)$

91. $\frac{1}{3} \log_8(x + 4) + 7 \log_8 y$ **92.** $-2 \log x - 5 \log(x + 6)$

93. $\frac{1}{2} \ln(2x - 1) - 2 \ln(x + 1)$

94. $5 \ln(x - 2) - \ln(x + 2) - 3 \ln x$

95. *Climb Rate*　The time t (in minutes) for a small plane to climb to an altitude of h feet is modeled by

$$t = 50 \log \frac{18{,}000}{18{,}000 - h}$$

where 18,000 feet is the plane's absolute ceiling.

(a) Determine the domain of the function in the context of the problem.

 (b) Use a graphing utility to graph the function and identify any asymptotes.

(c) As the plane approaches its absolute ceiling, what can be said about the time required to increase its altitude?

(d) Find the time for the plane to climb to an altitude of 4000 feet.

96. *Human Memory Model*　Students in a learning theory study were given an exam and then retested monthly for 6 months with an equivalent exam. The data obtained in the study are given as the ordered pairs (t, s), where t is the time in months after the initial exam and s is the average score for the class. Use these data to find a logarithmic equation that relates t and s.

$(1, 84.2), (2, 78.4), (3, 72.1),$

$(4, 68.5), (5, 67.1), (6, 65.3)$

5.4　In Exercises 97–104, solve for x.

97. $8^x = 512$　　　　**98.** $6^x = \frac{1}{216}$

99. $e^x = 3$　　　　**100.** $e^x = 6$

101. $\log_4 x = 2$　　　　**102.** $\log_6 x = -1$

103. $\ln x = 4$　　　　**104.** $\ln x = -3$

In Exercises 105–114, solve the exponential equation algebraically. Approximate your result to three decimal places.

105. $e^x = 12$　　　　**106.** $e^{3x} = 25$

107. $e^{4x} = e^{x^2+3}$　　　　**108.** $14e^{3x+2} = 560$

109. $2^x + 13 = 35$　　　　**110.** $6^x - 28 = -8$

111. $-4(5^x) = -68$　　　　**112.** $2(12^x) = 190$

113. $e^{2x} - 7e^x + 10 = 0$　　　　**114.** $e^{2x} - 6e^x + 8 = 0$

 In Exercises 115–118, use a graphing utility to graph and solve the equation. Approximate the result to three decimal places.

115. $2^{0.6x} - 3x = 0$　　　　**116.** $4^{-0.2x} + x = 0$

117. $25e^{-0.3x} = 12$　　　　**118.** $4e^{1.2x} = 9$

In Exercises 119–130, solve the logarithmic equation algebraically. Approximate the result to three decimal places.

119. $\ln 3x = 8.2$　　　　**120.** $\ln 5x = 7.2$

121. $2 \ln 4x = 15$　　　　**122.** $4 \ln 3x = 15$

123. $\ln x - \ln 3 = 2$　　　　**124.** $\ln \sqrt{x + 8} = 3$

125. $\ln \sqrt{x + 1} = 2$　　　　**126.** $\ln x - \ln 5 = 4$

127. $\log_8 (x - 1) = \log_8 (x - 2) - \log_8 (x + 2)$

128. $\log_6 (x + 2) - \log_6 x = \log_6 (x + 5)$

129. $\log(1 - x) = -1$

130. $\log(-x - 4) = 2$

 In Exercises 131–134, use a graphing utility to graph and solve the equation. Approximate the result to three decimal places.

131. $2 \ln(x + 3) + 3x = 8$　　**132.** $6 \log(x^2 + 1) - x = 0$

133. $4 \ln(x + 5) - x = 10$　　**134.** $x - 2 \log(x + 4) = 0$

135. *Compound Interest*　You deposit $7550 in an account that pays 7.25% interest, compounded continuously. How long will it take for the money to triple?

136. *Meteorology*　The speed of the wind S (in miles per hour) near the center of a tornado and the distance d (in miles) the tornado travels are related by the model $S = 93 \log d + 65$. On March 18, 1925, a large tornado struck portions of Missouri, Illinois, and Indiana with a wind speed at the center of about 283 miles per hour. Approximate the distance traveled by this tornado.

5.5　In Exercises 137–142, match the function with its graph. [The graphs are labeled (a), (b), (c), (d), (e), and (f).]

(a)

(b)

(c)

(d)

(e)

(f)
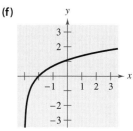

137. $y = 3e^{-2x/3}$

138. $y = 4e^{2x/3}$

139. $y = \ln(x + 3)$

140. $y = 7 - \log(x + 3)$

141. $y = 2e^{-(x+4)^2/3}$

142. $y = \dfrac{6}{1 + 2e^{-2x}}$

In Exercises 143 and 144, find the exponential model $y = ae^{bx}$ that passes through the points.

143. $(0, 2), (4, 3)$

144. $\left(0, \frac{1}{2}\right), (5, 5)$

145. Population   The population P of South Carolina (in thousands) from 1990 through 2003 can be modeled by $P = 3499e^{0.0135t}$, where t represents the year, with $t = 0$ corresponding to 1990. According to this model, when will the population reach 4.5 million? (Source: U.S. Census Bureau)

146. Radioactive Decay   The half-life of radioactive uranium II (^{234}U) is about 250,000 years. What percent of a present amount of radioactive uranium II will remain after 5000 years?

147. Compound Interest   A deposit of $10,000 is made in a savings account for which the interest is compounded continuously. The balance will double in 5 years.

(a) What is the annual interest rate for this account?

(b) Find the balance after 1 year.

148. Wildlife Population   A species of bat is in danger of becoming extinct. Five years ago, the total population of the species was 2000. Two years ago, the total population of the species was 1400. What was the total population of the species one year ago?

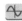 **149. Test Scores**   The test scores for a biology test follow a normal distribution modeled by

$$y = 0.0499e^{-(x-71)^2/128}, \quad 40 \le x \le 100$$

where x is the test score.

(a) Use a graphing utility to graph the equation.

(b) From the graph in part (a), estimate the average test score.

150. Typing Speed   In a typing class, the average number N of words per minute typed after t weeks of lessons was found to be

$$N = \frac{157}{1 + 5.4e^{-0.12t}}.$$

Find the time necessary to type (a) 50 words per minute and (b) 75 words per minute.

151. Sound Intensity   The relationship between the number of decibels β and the intensity of a sound I in watts per square centimeter is

$$\beta = 10 \log\left(\frac{I}{10^{-16}}\right).$$

Determine the intensity of a sound in watts per square centimeter if the decibel level is 125.

152. Geology   On the Richter scale, the magnitude R of an earthquake of intensity I is given by

$$R = \log \frac{I}{I_0}$$

where $I_0 = 1$ is the minimum intensity used for comparison. Find the intensity per unit of area for each value of R.

(a) $R = 8.4$ (b) $R = 6.85$ (c) $R = 9.1$

Synthesis

True or False?   **In Exercises 153 and 154, determine whether the equation is true or false. Justify your answer.**

153. $\log_b b^{2x} = 2x$

154. $\ln(x + y) = \ln x + \ln y$

155. The graphs of $y = e^{kt}$ are shown where $k = a, b, c,$ and d. Which of the four values are negative? Which are positive? Explain your reasoning.

(a)   (b)

(c)   (d)

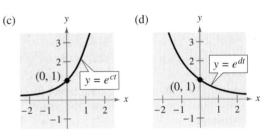

5 Chapter Test

Take this test as you would take a test in class. When you are finished, check your work against the answers given in the back of the book.

In Exercises 1–4, evaluate the expression. Approximate your result to three decimal places.

1. $12.4^{2.79}$ **2.** $4^{3\pi/2}$ **3.** $e^{-7/10}$ **4.** $e^{3.1}$

In Exercises 5–7, construct a table of values. Then sketch the graph of the function.

5. $f(x) = 10^{-x}$ **6.** $f(x) = -6^{x-2}$ **7.** $f(x) = 1 - e^{2x}$

8. Evaluate (a) $\log_7 7^{-0.89}$ and (b) $4.6 \ln e^2$.

In Exercises 9–11, construct a table of values. Then sketch the graph of the function. Identify any asymptotes.

9. $f(x) = -\log x - 6$ **10.** $f(x) = \ln(x - 4)$ **11.** $f(x) = 1 + \ln(x + 6)$

In Exercises 12–14, evaluate the logarithm using the change-of-base formula. Round your result to three decimal places.

12. $\log_7 44$ **13.** $\log_{2/5} 0.9$ **14.** $\log_{24} 68$

In Exercises 15–17, use the properties of logarithms to expand the expression as a sum, difference, and/or constant multiple of logarithms.

15. $\log_2 3a^4$ **16.** $\ln \dfrac{5\sqrt{x}}{6}$ **17.** $\log \dfrac{7x^2}{yz^3}$

In Exercises 18–20, condense the expression to the logarithm of a single quantity.

18. $\log_3 13 + \log_3 y$ **19.** $4 \ln x - 4 \ln y$

20. $2 \ln x + \ln(x - 5) - 3 \ln y$

Exponential Growth

(9, 11,277)

(0, 2745)

FIGURE FOR **27**

In Exercises 21–26, solve the equation algebraically. Approximate your result to three decimal places.

21. $5^x = \dfrac{1}{25}$ **22.** $3e^{-5x} = 132$

23. $\dfrac{1025}{8 + e^{4x}} = 5$ **24.** $\ln x = \dfrac{1}{2}$

25. $18 + 4 \ln x = 7$ **26.** $\log x - \log(8 - 5x) = 2$

27. Find an exponential growth model for the graph shown in the figure.

28. The half-life of radioactive actinium (^{227}Ac) is 21.77 years. What percent of a present amount of radioactive actinium will remain after 19 years?

29. A model that can be used for predicting the height H (in centimeters) of a child based on his or her age is $H = 70.228 + 5.104x + 9.222 \ln x$, $\frac{1}{4} \le x \le 6$, where x is the age of the child in years. (Source: Snapshots of Applications in Mathematics)

  (a) Construct a table of values. Then sketch the graph of the model.

  (b) Use the graph from part (a) to estimate the height of a four-year-old child. Then calculate the actual height using the model.

Proofs in Mathematics

Each of the following three properties of logarithms can be proved by using properties of exponential functions.

Properties of Logarithms *(p. 390)*

Let a be a positive number such that $a \neq 1$, and let n be a real number. If u and v are positive real numbers, the following properties are true.

	Logarithm with Base a	*Natural Logarithm*
1. Product Property:	$\log_a(uv) = \log_a u + \log_a v$	$\ln(uv) = \ln u + \ln v$
2. Quotient Property:	$\log_a \dfrac{u}{v} = \log_a u - \log_a v$	$\ln \dfrac{u}{v} = \ln u - \ln v$
3. Power Property:	$\log_a u^n = n \log_a u$	$\ln u^n = n \ln u$

Proof

Let

$$x = \log_a u \quad \text{and} \quad y = \log_a v.$$

The corresponding exponential forms of these two equations are

$$a^x = u \quad \text{and} \quad a^y = v.$$

To prove the Product Property, multiply u and v to obtain

$$uv = a^x a^y = a^{x+y}.$$

The corresponding logarithmic form of $uv = a^{x+y}$ is $\log_a(uv) = x + y$. So,

$$\log_a(uv) = \log_a u + \log_a v.$$

To prove the Quotient Property, divide u by v to obtain

$$\frac{u}{v} = \frac{a^x}{a^y} = a^{x-y}.$$

The corresponding logarithmic form of $u/v = a^{x-y}$ is $\log_a(u/v) = x - y$. So,

$$\log_a \frac{u}{v} = \log_a u - \log_a v.$$

To prove the Power Property, substitute a^x for u in the expression $\log_a u^n$, as follows.

$$
\begin{aligned}
\log_a u^n &= \log_a (a^x)^n && \text{Substitute } a^x \text{ for } u. \\
&= \log_a a^{nx} && \text{Property of exponents} \\
&= nx && \text{Inverse Property of Logarithms} \\
&= n \log_a u && \text{Substitute } \log_a u \text{ for } x.
\end{aligned}
$$

So, $\log_a u^n = n \log_a u$.

This collection of thought-provoking and challenging exercises further explores and expands upon concepts learned in this chapter.

1. Graph the exponential function given by $y = a^x$ for $a = 0.5$, 1.2, and 2.0. Which of these curves intersects the line $y = x$? Determine all positive numbers a for which the curve $y = a^x$ intersects the line $y = x$.

 2. Use a graphing utility to graph $y_1 = e^x$ and each of the functions $y_2 = x^2$, $y_3 = x^3$, $y_4 = \sqrt{x}$, and $y_5 = |x|$. Which function increases at the greatest rate as x approaches $+\infty$?

 3. Use the result of Exercise 2 to make a conjecture about the rate of growth of $y_1 = e^x$ and $y = x^n$, where n is a natural number and x approaches $+\infty$.

 4. Use the results of Exercises 2 and 3 to describe what is implied when it is stated that a quantity is growing exponentially.

5. Given the exponential function

$$f(x) = a^x$$

show that

(a) $f(u + v) = f(u) \cdot f(v)$.

(b) $f(2x) = [f(x)]^2$.

6. Given that

$$f(x) = \frac{e^x + e^{-x}}{2} \text{ and } g(x) = \frac{e^x - e^{-x}}{2}$$

show that

$$[f(x)]^2 - [g(x)]^2 = 1.$$

 7. Use a graphing utility to compare the graph of the function given by $y = e^x$ with the graph of each given function. [$n!$ (read "n factorial") is defined as $n! = 1 \cdot 2 \cdot 3 \cdots (n - 1) \cdot n$.]

(a) $y_1 = 1 + \dfrac{x}{1!}$

(b) $y_2 = 1 + \dfrac{x}{1!} + \dfrac{x^2}{2!}$

(c) $y_3 = 1 + \dfrac{x}{1!} + \dfrac{x^2}{2!} + \dfrac{x^3}{3!}$

8. Identify the pattern of successive polynomials given in Exercise 7. Extend the pattern one more term and compare the graph of the resulting polynomial function with the graph of $y = e^x$. What do you think this pattern implies?

9. Graph the function given by

$$f(x) = e^x - e^{-x}.$$

From the graph, the function appears to be one-to-one. Assuming that the function has an inverse function, find $f^{-1}(x)$.

10. Find a pattern for $f^{-1}(x)$ if

$$f(x) = \frac{a^x + 1}{a^x - 1}$$

where $a > 0$, $a \neq 1$.

11. By observation, identify the equation that corresponds to the graph. Explain your reasoning.

(a) $y = 6e^{-x^2/2}$

(b) $y = \dfrac{6}{1 + e^{-x/2}}$

(c) $y = 6(1 - e^{-x^2/2})$

12. You have two options for investing $500. The first earns 7% compounded annually and the second earns 7% simple interest. The figure shows the growth of each investment over a 30-year period.

(a) Identify which graph represents each type of investment. Explain your reasoning.

(b) Verify your answer in part (a) by finding the equations that model the investment growth and graphing the models.

(c) Which option would you choose? Explain your reasoning.

13. Two different samples of radioactive isotopes are decaying. The isotopes have initial amounts of c_1 and c_2, as well as half-lives of k_1 and k_2, respectively. Find the time required for the samples to decay to equal amounts.

14. A lab culture initially contains 500 bacteria. Two hours later, the number of bacteria has decreased to 200. Find the exponential decay model of the form

$$B = B_0 a^{kt}$$

that can be used to approximate the number of bacteria after t hours.

 15. The table shows the colonial population estimates of the American colonies from 1700 to 1780. (Source: U.S. Census Bureau)

Year	Population
1700	250,900
1710	331,700
1720	466,200
1730	629,400
1740	905,600
1750	1,170,800
1760	1,593,600
1770	2,148,100
1780	2,780,400

In each of the following, let y represent the population in the year t, with $t = 0$ corresponding to 1700.

(a) Use the *regression* feature of a graphing utility to find an exponential model for the data.

(b) Use the *regression* feature of the graphing utility to find a quadratic model for the data.

(c) Use the graphing utility to plot the data and the models from parts (a) and (b) in the same viewing window.

(d) Which model is a better fit for the data? Would you use this model to predict the population of the United States in 2010? Explain your reasoning.

16. Show that $\dfrac{\log_a x}{\log_{a/b} x} = 1 + \log_a \dfrac{1}{b}$.

17. Solve $(\ln x)^2 = \ln x^2$.

 18. Use a graphing utility to compare the graph of the function given by $y = \ln x$ with the graph of each given function.

(a) $y_1 = x - 1$

(b) $y_2 = (x - 1) - \frac{1}{2}(x - 1)^2$

(c) $y_3 = (x - 1) - \frac{1}{2}(x - 1)^2 + \frac{1}{3}(x - 1)^3$

19. Identify the pattern of successive polynomials given in Exercise 18. Extend the pattern one more term and compare the graph of the resulting polynomial function with the graph of $y = \ln x$. What do you think the pattern implies?

20. Using

$$y = ab^x \qquad \text{and} \qquad y = ax^b$$

take the natural logarithm of each side of each equation. What are the slope and y-intercept of the line relating x and $\ln y$ for $y = ab^x$? What are the slope and y-intercept of the line relating $\ln x$ and $\ln y$ for $y = ax^b$?

In Exercises 21 and 22, use the model

$$y = 80.4 - 11 \ln x, \quad 100 \le x \le 1500$$

which approximates the minimum required ventilation rate in terms of the air space per child in a public school classroom. In the model, x is the air space per child in cubic feet and y is the ventilation rate per child in cubic feet per minute.

 21. Use a graphing utility to graph the model and approximate the required ventilation rate if there is 300 cubic feet of air space per child.

22. A classroom is designed for 30 students. The air conditioning system in the room has the capacity of moving 450 cubic feet of air per minute.

(a) Determine the ventilation rate per child, assuming that the room is filled to capacity.

(b) Estimate the air space required per child.

(c) Determine the minimum number of square feet of floor space required for the room if the ceiling height is 30 feet.

In Exercises 23–26, (a) use a graphing utility to create a scatter plot of the data, (b) decide whether the data could best be modeled by a linear model, an exponential model, or a logarithmic model, (c) explain why you chose the model you did in part (b), (d) use the *regression* feature of a graphing utility to find the model you chose in part (b) for the data and graph the model with the scatter plot, and (e) determine how well the model you chose fits the data.

23. (1, 2.0), (1.5, 3.5), (2, 4.0), (4, 5.8), (6, 7.0), (8, 7.8)

24. (1, 4.4), (1.5, 4.7), (2, 5.5), (4, 9.9), (6, 18.1), (8, 33.0)

25. (1, 7.5), (1.5, 7.0), (2, 6.8), (4, 5.0), (6, 3.5), (8, 2.0)

26. (1, 5.0), (1.5, 6.0), (2, 6.4), (4, 7.8), (6, 8.6), (8, 9.0)

Topics in Analytic Geometry

6

The nine planets move about the sun in elliptical orbits. You can use the techniques presented in this chapter to determine the distances between the planets and the center of the sun.

Kauko Helavuo/Getty Images

SELECTED APPLICATIONS

Analytic geometry concepts have many real-life applications. The applications listed below represent a small sample of the applications in this chapter.

- Inclined Plane,
  Exercise 56, page 436

- Revenue,
  Exercise 59, page 443

- Architecture,
  Exercise 57, page 453

- Satellite Orbit,
  Exercise 60, page 454

- LORAN,
  Exercise 42, page 463

- Running Path,
  Exercise 44, page 464

- Projectile Motion,
  Exercises 57 and 58, page 479

- Planetary Motion,
  Exercises 51–56, page 500

- Locating an Explosion,
  Exercise 40, page 504

6.1   Lines

What you should learn

- Find the inclination of a line.
- Find the angle between two lines.
- Find the distance between a point and a line.

Why you should learn it

The inclination of a line can be used to measure heights indirectly. For instance, in Exercise 56 on page 436, the inclination of a line can be used to determine the change in elevation from the base to the top of the Johnstown Inclined Plane.

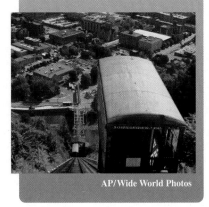

AP/Wide World Photos

The *HM mathSpace®* CD-ROM and *Eduspace®* for this text contain additional resources related to the concepts discussed in this chapter.

Inclination of a Line

In Section P.4, you learned that the graph of the linear equation

$$y = mx + b$$

is a nonvertical line with slope m and y-intercept $(0, b)$. There, the slope of a line was described as the rate of change in y with respect to x. In this section, you will look at the slope of a line in terms of the angle of inclination of the line.

Every nonhorizontal line must intersect the x-axis. The angle formed by such an intersection determines the **inclination** of the line, as specified in the following definition.

Definition of Inclination

The **inclination** of a nonhorizontal line is the positive angle θ (less than π) measured counterclockwise from the x-axis to the line. (See Figure 6.1.)

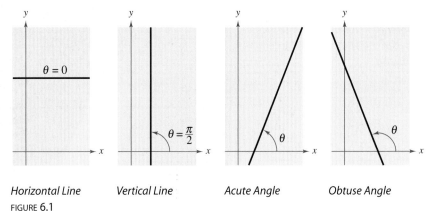

Horizontal Line   Vertical Line   Acute Angle   Obtuse Angle
FIGURE 6.1

The inclination of a line is related to its slope in the following manner.

Inclination and Slope

If a nonvertical line has inclination θ and slope m, then

$$m = \tan \theta.$$

For a proof of the relation between inclination and slope, see Proofs in Mathematics on page 510.

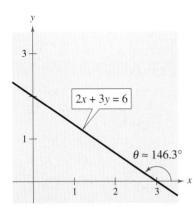

FIGURE **6.2**

Example 1   **Finding the Inclination of a Line**

Find the inclination of the line $2x + 3y = 6$.

Solution

The slope of this line is $m = -\frac{2}{3}$. So, its inclination is determined from the equation

$$\tan \theta = -\frac{2}{3}.$$

From Figure 6.2, it follows that $\frac{\pi}{2} < \theta < \pi$. This means that

$$\theta = \pi + \arctan\left(-\frac{2}{3}\right)$$

$$\approx \pi + (-0.588)$$

$$= \pi - 0.588$$

$$\approx 2.554.$$

The angle of inclination is about 2.554 radians or about 146.3°.

✓CHECKPOINT   Now try Exercise 19.

The Angle Between Two Lines

Two distinct lines in a plane are either parallel or intersecting. If they intersect and are nonperpendicular, their intersection forms two pairs of opposite angles. One pair is acute and the other pair is obtuse. The smaller of these angles is called the **angle between the two lines.** As shown in Figure 6.3, you can use the inclinations of the two lines to find the angle between the two lines. If two lines have inclinations θ_1 and θ_2, where $\theta_1 < \theta_2$ and $\theta_2 - \theta_1 < \pi/2$, the angle between the two lines is

$$\theta = \theta_2 - \theta_1.$$

You can use the formula for the tangent of the difference of two angles

$$\tan \theta = \tan(\theta_2 - \theta_1)$$

$$= \frac{\tan \theta_2 - \tan \theta_1}{1 + \tan \theta_1 \tan \theta_2}$$

to obtain the formula for the angle between two lines.

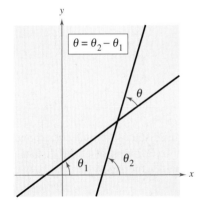

FIGURE **6.3**

Angle Between Two Lines

If two nonperpendicular lines have slopes m_1 and m_2, the angle between the two lines is

$$\tan \theta = \left| \frac{m_2 - m_1}{1 + m_1 m_2} \right|.$$

Example 2 Finding the Angle Between Two Lines

Find the angle between the two lines.

Line 1: $2x - y - 4 = 0$ *Line 2:* $3x + 4y - 12 = 0$

Solution

The two lines have slopes of $m_1 = 2$ and $m_2 = -\frac{3}{4}$, respectively. So, the tangent of the angle between the two lines is

$$\tan \theta = \left| \frac{m_2 - m_1}{1 + m_1 m_2} \right| = \left| \frac{(-3/4) - 2}{1 + (2)(-3/4)} \right| = \left| \frac{-11/4}{-2/4} \right| = \frac{11}{2}.$$

Finally, you can conclude that the angle is

$$\theta = \arctan \frac{11}{2} \approx 1.391 \text{ radians} \approx 79.70°$$

as shown in Figure 6.4.

CHECKPOINT Now try Exercise 27.

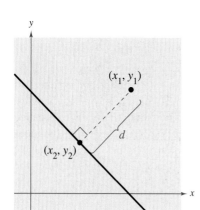

FIGURE 6.4

The Distance Between a Point and a Line

Finding the distance between a line and a point not on the line is an application of perpendicular lines. This distance is defined as the length of the perpendicular line segment joining the point and the line, as shown in Figure 6.5.

> **Distance Between a Point and a Line**
>
> The distance between the point (x_1, y_1) and the line $Ax + By + C = 0$ is
>
> $$d = \frac{|Ax_1 + By_1 + C|}{\sqrt{A^2 + B^2}}.$$

FIGURE 6.5

Remember that the values of A, B, and C in this distance formula correspond to the general equation of a line, $Ax + By + C = 0$. For a proof of the distance between a point and a line, see Proofs in Mathematics on page 510.

Example 3 Finding the Distance Between a Point and a Line

Find the distance between the point $(4, 1)$ and the line $y = 2x + 1$.

Solution

The general form of the equation is

$$-2x + y - 1 = 0.$$

So, the distance between the point and the line is

$$d = \frac{|-2(4) + 1(1) + (-1)|}{\sqrt{(-2)^2 + 1^2}} = \frac{8}{\sqrt{5}} \approx 3.58 \text{ units.}$$

The line and the point are shown in Figure 6.6.

CHECKPOINT Now try Exercise 39.

FIGURE 6.6

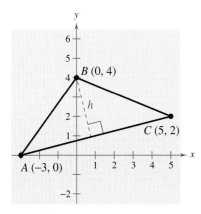

FIGURE 6.7

| Example 4 | **An Application of Two Distance Formulas** |

Figure 6.7 shows a triangle with vertices $A(-3, 0)$, $B(0, 4)$, and $C(5, 2)$.

a. Find the altitude h from vertex B to side AC.

b. Find the area of the triangle.

Solution

a. To find the altitude, use the formula for the distance between line AC and the point $(0, 4)$. The equation of line AC is obtained as follows.

$$\text{Slope:} \quad m = \frac{2 - 0}{5 - (-3)} = \frac{2}{8} = \frac{1}{4}$$

$$\text{Equation:} \qquad y - 0 = \frac{1}{4}(x + 3) \qquad \text{Point-slope form}$$

$$4y = x + 3 \qquad \text{Multiply each side by 4.}$$

$$x - 4y + 3 = 0 \qquad \text{General form}$$

So, the distance between this line and the point $(0, 4)$ is

$$\text{Altitude} = h = \frac{\left|1(0) + (-4)(4) + 3\right|}{\sqrt{1^2 + (-4)^2}} = \frac{13}{\sqrt{17}} \text{ units.}$$

b. Using the formula for the distance between two points, you can find the length of the base AC to be

$$b = \sqrt{[5 - (-3)]^2 + (2 - 0)^2} \qquad \text{Distance Formula}$$

$$= \sqrt{8^2 + 2^2} \qquad \text{Simplify.}$$

$$= \sqrt{68} \qquad \text{Simplify.}$$

$$= 2\sqrt{17} \text{ units.} \qquad \text{Simplify.}$$

Finally, the area of the triangle in Figure 6.7 is

$$A = \frac{1}{2}bh \qquad \text{Formula for the area of a triangle}$$

$$= \frac{1}{2}(2\sqrt{17})\left(\frac{13}{\sqrt{17}}\right) \qquad \text{Substitute for } b \text{ and } h.$$

$$= 13 \text{ square units.} \qquad \text{Simplify.}$$

✓**CHECKPOINT**   Now try Exercise 45.

WRITING ABOUT **M**ATHEMATICS

Inclination and the Angle Between Two Lines Discuss why the inclination of a line can be an angle that is larger than $\pi/2$, but the angle between two lines cannot be larger than $\pi/2$. Decide whether the following statement is true or false: "The inclination of a line is the angle between the line and the x-axis." Explain.

6.1 Exercises

The *HM mathSpace®* CD-ROM and *Eduspace®* for this text contain step-by-step solutions to all odd-numbered exercises. They also provide Tutorial Exercises for additional help.

VOCABULARY CHECK: Fill in the blanks.

1. The _____ of a nonhorizontal line is the positive angle θ (less than π) measured counterclockwise from the *x*-axis to the line.

2. If a nonvertical line has inclination θ and slope m, then $m = $ _____ .

3. If two nonperpendicular lines have slopes m_1 and m_2, the angle between the two lines is $\tan\theta = $ _____ .

4. The distance between the point (x_1, y_1) and the line $Ax + By + C = 0$ is given by $d = $ _____ .

PREREQUISITE SKILLS REVIEW: Practice and review algebra skills needed for this section at **www.Eduspace.com.**

In Exercises 1–8, find the slope of the line with inclination θ.

1.

$\theta = \dfrac{\pi}{6}$

2.

$\theta = \dfrac{\pi}{4}$

3.

$\theta = \dfrac{3\pi}{4}$

4.

$\theta = \dfrac{2\pi}{3}$

5. $\theta = \dfrac{\pi}{3}$ radians

6. $\theta = \dfrac{5\pi}{6}$ radians

7. $\theta = 1.27$ radians

8. $\theta = 2.88$ radians

In Exercises 9–14, find the inclination θ (in radians and degrees) of the line with a slope of m.

9. $m = -1$

10. $m = -2$

11. $m = 1$

12. $m = 2$

13. $m = \frac{3}{4}$

14. $m = -\frac{5}{2}$

In Exercises 15–18, find the inclination θ (in radians and degrees) of the line passing through the points.

15. $(6, 1), (10, 8)$

16. $(12, 8), (-4, -3)$

17. $(-2, 20), (10, 0)$

18. $(0, 100), (50, 0)$

In Exercises 19–22, find the inclination θ (in radians and degrees) of the line.

19. $6x - 2y + 8 = 0$

20. $4x + 5y - 9 = 0$

21. $5x + 3y = 0$

22. $x - y - 10 = 0$

In Exercises 23–32, find the angle θ (in radians and degrees) between the lines.

23. $3x + y = 3$
    $x - y = 2$

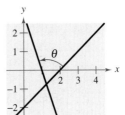

24. $x + 3y = 2$
    $x - 2y = -3$

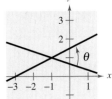

25. $x - y = 0$
    $3x - 2y = -1$

26. $2x - y = 2$
    $4x + 3y = 24$

27. $x - 2y = 7$
    $6x + 2y = 5$

28. $5x + 2y = 16$
    $3x - 5y = -1$

29. $x + 2y = 8$
    $x - 2y = 2$

30. $3x - 5y = 3$
    $3x + 5y = 12$

31. $0.05x - 0.03y = 0.21$
$0.07x + 0.02y = 0.16$

32. $0.02x - 0.05y = -0.19$
$0.03x + 0.04y = 0.52$

Angle Measurement   In Exercises 33–36, find the slope of each side of the triangle and use the slopes to find the measures of the interior angles.

33.

34.

35.

36.

In Exercises 37–44, find the distance between the point and the line.

Point	Line
37. $(0, 0)$	$4x + 3y = 0$
38. $(0, 0)$	$2x - y = 4$
39. $(2, 3)$	$4x + 3y = 10$
40. $(-2, 1)$	$x - y = 2$
41. $(6, 2)$	$x + 1 = 0$
42. $(10, 8)$	$y - 4 = 0$
43. $(0, 8)$	$6x - y = 0$
44. $(4, 2)$	$x - y = 20$

In Exercises 45–48, the points represent the vertices of a triangle. (a) Draw triangle *ABC* in the coordinate plane, (b) find the altitude from vertex *B* of the triangle to side *AC*, and (c) find the area of the triangle.

45. $A = (0, 0)$, $B = (1, 4)$, $C = (4, 0)$

46. $A = (0, 0)$, $B = (4, 5)$, $C = (5, -2)$

47. $A = \left(-\frac{1}{2}, \frac{1}{2}\right)$, $B = (2, 3)$, $C = \left(\frac{5}{2}, 0\right)$

48. $A = (-4, -5)$, $B = (3, 10)$, $C = (6, 12)$

In Exercises 49 and 50, find the distance between the parallel lines.

49. $x + y = 1$
$x + y = 5$

50. $3x - 4y = 1$
$3x - 4y = 10$

51. *Road Grade*   A straight road rises with an inclination of 0.10 radian from the horizontal (see figure). Find the slope of the road and the change in elevation over a two-mile stretch of the road.

52. *Road Grade*   A straight road rises with an inclination of 0.20 radian from the horizontal. Find the slope of the road and the change in elevation over a one-mile stretch of the road.

53. *Pitch of a Roof*   A roof has a rise of 3 feet for every horizontal change of 5 feet (see figure). Find the inclination of the roof.

54. Conveyor Design   A moving conveyor is built so that it rises 1 meter for each 3 meters of horizontal travel.

(a) Draw a diagram that gives a visual representation of the problem.

(b) Find the inclination of the conveyor.

(c) The conveyor runs between two floors in a factory. The distance between the floors is 5 meters. Find the length of the conveyor.

55. Truss   Find the angles α and β shown in the drawing of the roof truss.

Model It

56. Inclined Plane   The Johnstown Inclined Plane in Johnstown, Pennsylvania is an inclined railway that was designed to carry people to the hilltop community of Westmont. It also proved useful in carrying people and vehicles to safety during severe floods. The railway is 896.5 feet long with a 70.9% uphill grade (see figure).

(a) Find the inclination θ of the railway.

(b) Find the change in elevation from the base to the top of the railway.

(c) Using the origin of a rectangular coordinate system as the base of the inclined plane, find the equation of the line that models the railway track.

(d) Sketch a graph of the equation you found in part (c).

Synthesis

True or False?   **In Exercises 57 and 58, determine whether the statement is true or false. Justify your answer.**

57. A line that has an inclination greater than $\pi/2$ radians has a negative slope.

58. To find the angle between two lines whose angles of inclination θ_1 and θ_2 are known, substitute θ_1 and θ_2 for m_1 and m_2, respectively, in the formula for the angle between two lines.

59. Exploration   Consider a line with slope m and y-intercept $(0, 4)$.

(a) Write the distance d between the origin and the line as a function of m.

(b) Graph the function in part (a).

(c) Find the slope that yields the maximum distance between the origin and the line.

(d) Find the asymptote of the graph in part (b) and interpret its meaning in the context of the problem.

60. Exploration   Consider a line with slope m and y-intercept $(0, 4)$.

(a) Write the distance d between the point $(3, 1)$ and the line as a function of m.

(b) Graph the function in part (a).

(c) Find the slope that yields the maximum distance between the point and the line.

(d) Is it possible for the distance to be 0? If so, what is the slope of the line that yields a distance of 0?

(e) Find the asymptote of the graph in part (b) and interpret its meaning in the context of the problem.

Skills Review

In Exercises 61–66, find all x-intercepts and y-intercepts of the graph of the quadratic function.

61. $f(x) = (x - 7)^2$

62. $f(x) = (x + 9)^2$

63. $f(x) = (x - 5)^2 - 5$

64. $f(x) = (x + 11)^2 + 12$

65. $f(x) = x^2 - 7x - 1$

66. $f(x) = x^2 + 9x - 22$

In Exercises 67–72, write the quadratic function in standard form by completing the square. Identify the vertex of the function.

67. $f(x) = 3x^2 + 2x - 16$

68. $f(x) = 2x^2 - x - 21$

69. $f(x) = 5x^2 + 34x - 7$

70. $f(x) = -x^2 - 8x - 15$

71. $f(x) = 6x^2 - x - 12$

72. $f(x) = -8x^2 - 34x - 21$

In Exercises 73–76, graph the quadratic function.

73. $f(x) = (x - 4)^2 + 3$

74. $f(x) = 6 - (x + 1)^2$

75. $g(x) = 2x^2 - 3x + 1$

76. $g(x) = -x^2 + 6x - 8$

6.2   Introduction to Conics: Parabolas

Conics

Conic sections were discovered during the classical Greek period, 600 to 300 B.C. The early Greeks were concerned largely with the geometric properties of conics. It was not until the 17th century that the broad applicability of conics became apparent and played a prominent role in the early development of calculus.

A **conic section** (or simply **conic**) is the intersection of a plane and a double-napped cone. Notice in Figure 6.8 that in the formation of the four basic conics, the intersecting plane does not pass through the vertex of the cone. When the plane does pass through the vertex, the resulting figure is a **degenerate conic,** as shown in Figure 6.9.

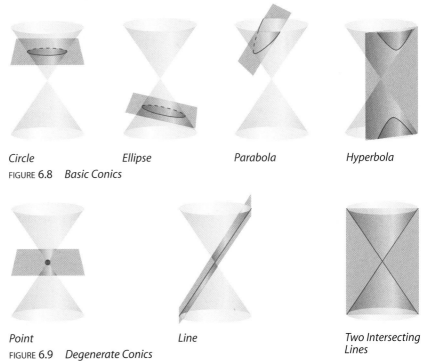

Circle   Ellipse   Parabola   Hyperbola

FIGURE **6.8**   *Basic Conics*

Point   Line   Two Intersecting Lines

FIGURE **6.9**   *Degenerate Conics*

There are several ways to approach the study of conics. You could begin by defining conics in terms of the intersections of planes and cones, as the Greeks did, or you could define them algebraically, in terms of the general second-degree equation

$$Ax^2 + Bxy + Cy^2 + Dx + Ey + F = 0.$$

However, you will study a third approach, in which each of the conics is defined as a **locus** (collection) of points satisfying a geometric property. For example, in Section P.3, you learned that a circle is defined as the collection of all points (x, y) that are equidistant from a fixed point (h, k). This leads to the standard form of the equation of a circle

$$(x - h)^2 + (y - k)^2 = r^2. \qquad \text{Equation of circle}$$

Parabolas

In Section P.3, you learned that the graph of the quadratic function

$$f(x) = ax^2 + bx + c$$

is a parabola that opens upward or downward. The following definition of a parabola is more general in the sense that it is independent of the orientation of the parabola.

Definition of Parabola

A **parabola** is the set of all points (x, y) in a plane that are equidistant from a fixed line (**directrix**) and a fixed point (**focus**) not on the line.

The midpoint between the focus and the directrix is called the **vertex,** and the line passing through the focus and the vertex is called the **axis** of the parabola. Note in Figure 6.10 that a parabola is symmetric with respect to its axis. Using the definition of a parabola, you can derive the following **standard form** of the equation of a parabola whose directrix is parallel to the x-axis or to the y-axis.

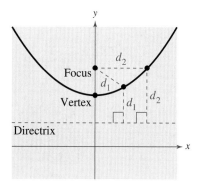

FIGURE 6.10   *Parabola*

Standard Equation of a Parabola

The **standard form of the equation of a parabola** with vertex at (h, k) is as follows.

$$(x - h)^2 = 4p(y - k), \ p \neq 0 \qquad \text{Vertical axis, directrix: } y = k - p$$

$$(y - k)^2 = 4p(x - h), \ p \neq 0 \qquad \text{Horizontal axis, directrix: } x = h - p$$

The focus lies on the axis p units (*directed distance*) from the vertex. If the vertex is at the origin $(0, 0)$, the equation takes one of the following forms.

$$x^2 = 4py \qquad\qquad \text{Vertical axis}$$

$$y^2 = 4px \qquad\qquad \text{Horizontal axis}$$

See Figure 6.11.

For a proof of the standard form of the equation of a parabola, see Proofs in Mathematics on page 511.

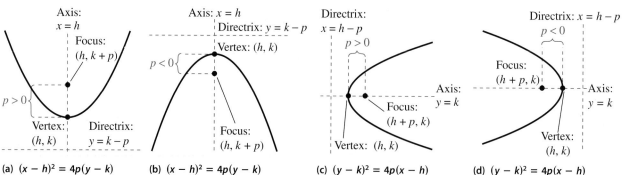

(a) $(x - h)^2 = 4p(y - k)$
Vertical axis: $p > 0$

(b) $(x - h)^2 = 4p(y - k)$
Vertical axis: $p < 0$

(c) $(y - k)^2 = 4p(x - h)$
Horizontal axis: $p > 0$

(d) $(y - k)^2 = 4p(x - h)$
Horizontal axis: $p < 0$

FIGURE 6.11

STUDY TIP

You may want to review the technique of completing the square found in Section P.2, which will be used to rewrite each of the conics in standard form.

Example 1   Vertex at the Origin

Find the standard equation of the parabola with vertex at the origin and focus $(2, 0)$.

Solution

The axis of the parabola is horizontal, passing through $(0, 0)$ and $(2, 0)$, as shown in Figure 6.12.

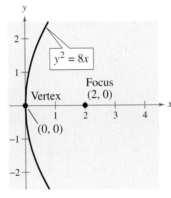

FIGURE 6.12

So, the standard form is $y^2 = 4px$, where $h = 0$, $k = 0$, and $p = 2$. So, the equation is $y^2 = 8x$.

✓CHECKPOINT   Now try Exercise 33.

Example 2   Finding the Focus of a Parabola

Find the focus of the parabola given by $y = -\frac{1}{2}x^2 - x + \frac{1}{2}$.

Solution

To find the focus, convert to standard form by completing the square.

$y = -\frac{1}{2}x^2 - x + \frac{1}{2}$	Write original equation.
$-2y = x^2 + 2x - 1$	Multiply each side by -2.
$1 - 2y = x^2 + 2x$	Add 1 to each side.
$1 + 1 - 2y = x^2 + 2x + 1$	Complete the square.
$2 - 2y = x^2 + 2x + 1$	Combine like terms.
$-2(y - 1) = (x + 1)^2$	Standard form

Comparing this equation with

$$(x - h)^2 = 4p(y - k)$$

you can conclude that $h = -1$, $k = 1$, and $p = -\frac{1}{2}$. Because p is negative, the parabola opens downward, as shown in Figure 6.13. So, the focus of the parabola is $(h, k + p) = \left(-1, \frac{1}{2}\right)$.

✓CHECKPOINT   Now try Exercise 21.

FIGURE 6.13

FIGURE 6.14

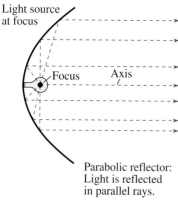

Light source at focus

Focus   Axis

Parabolic reflector: Light is reflected in parallel rays.

FIGURE 6.15

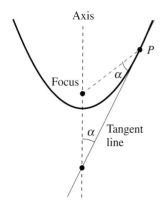

Axis

P

Focus α

α Tangent line

FIGURE 6.16

Example 3   Finding the Standard Equation of a Parabola

Find the standard form of the equation of the parabola with vertex $(2, 1)$ and focus $(2, 4)$.

Solution

Because the axis of the parabola is vertical, passing through $(2, 1)$ and $(2, 4)$, consider the equation

$$(x - h)^2 = 4p(y - k)$$

where $h = 2$, $k = 1$, and $p = 4 - 1 = 3$. So, the standard form is

$$(x - 2)^2 = 12(y - 1).$$

You can obtain the more common quadratic form as follows.

$$(x - 2)^2 = 12(y - 1)$$ Write original equation.

$$x^2 - 4x + 4 = 12y - 12$$ Multiply.

$$x^2 - 4x + 16 = 12y$$ Add 12 to each side.

$$\frac{1}{12}(x^2 - 4x + 16) = y$$ Divide each side by 12.

The graph of this parabola is shown in Figure 6.14.

✓CHECKPOINT   Now try Exercise 45.

Application

A line segment that passes through the focus of a parabola and has endpoints on the parabola is called a **focal chord.** The specific focal chord perpendicular to the axis of the parabola is called the **latus rectum.**

Parabolas occur in a wide variety of applications. For instance, a parabolic reflector can be formed by revolving a parabola around its axis. The resulting surface has the property that all incoming rays parallel to the axis are reflected through the focus of the parabola. This is the principle behind the construction of the parabolic mirrors used in reflecting telescopes. Conversely, the light rays emanating from the focus of a parabolic reflector used in a flashlight are all parallel to one another, as shown in Figure 6.15.

A line is **tangent** to a parabola at a point on the parabola if the line intersects, but does not cross, the parabola at the point. Tangent lines to parabolas have special properties related to the use of parabolas in constructing reflective surfaces.

Reflective Property of a Parabola

The tangent line to a parabola at a point P makes equal angles with the following two lines (see Figure 6.16).

1. The line passing through P and the focus

2. The axis of the parabola

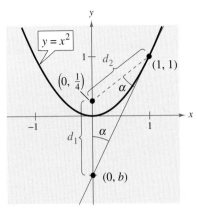

FIGURE 6.17

Technology

Use a graphing utility to confirm the result of Example 4. By graphing

$$y_1 = x^2 \quad \text{and} \quad y_2 = 2x - 1$$

in the same viewing window, you should be able to see that the line touches the parabola at the point $(1, 1)$.

| Example 4 | **Finding the Tangent Line at a Point on a Parabola** |

Find the equation of the tangent line to the parabola given by $y = x^2$ at the point $(1, 1)$.

Solution

For this parabola, $p = \frac{1}{4}$ and the focus is $\left(0, \frac{1}{4}\right)$, as shown in Figure 6.17. You can find the y-intercept $(0, b)$ of the tangent line by equating the lengths of the two sides of the isosceles triangle shown in Figure 6.17:

$$d_1 = \frac{1}{4} - b$$

and

$$d_2 = \sqrt{(1 - 0)^2 + \left[1 - \left(\frac{1}{4}\right)\right]^2} = \frac{5}{4}.$$

Note that $d_1 = \frac{1}{4} - b$ rather than $b - \frac{1}{4}$. The order of subtraction for the distance is important because the distance must be positive. Setting $d_1 = d_2$ produces

$$\frac{1}{4} - b = \frac{5}{4}$$

$$b = -1.$$

So, the slope of the tangent line is

$$m = \frac{1 - (-1)}{1 - 0} = 2$$

and the equation of the tangent line in slope-intercept form is

$$y = 2x - 1.$$

✓CHECKPOINT   Now try Exercise 55.

Activities

1. Find the vertex, focus, and directrix of the parabola $x^2 - 6x - 4y + 5 = 0$.
   Answer: Vertex $(3, -1)$; Focus $(3, 0)$;
   Directrix $y = -2$

2. Find the standard form of the equation of the parabola with vertex $(4, 0)$ and directrix $x = 5$.
   Answer: $y^2 = -4(x - 4)$

3. Find an equation of the tangent line to the parabola $y = 2x^2$ at the point $(1, 2)$.
   Answer: $y = 4x - 2$

𝒲RITING ABOUT 𝑀ATHEMATICS

Television Antenna Dishes Cross sections of television antenna dishes are parabolic in shape. Use the figure shown to write a paragraph explaining why these dishes are parabolic.

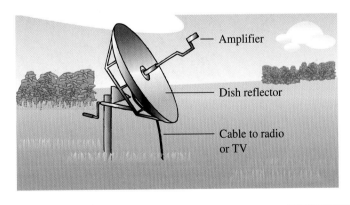

6.2 | Exercises

VOCABULARY CHECK: Fill in the blanks.

1. A _____ is the intersection of a plane and a double-napped cone.

2. A collection of points satisfying a geometric property can also be referred to as a _____ of points.

3. A _____ is defined as the set of all points (x, y) in a plane that are equidistant from a fixed line, called the _____, and a fixed point, called the _____, not on the line.

4. The line that passes through the focus and vertex of a parabola is called the _____ of the parabola.

5. The _____ of a parabola is the midpoint between the focus and the directrix.

6. A line segment that passes through the focus of a parabola and has endpoints on the parabola is called a _____ _____ .

7. A line is _____ to a parabola at a point on the parabola if the line intersects, but does not cross, the parabola at the point.

PREREQUISITE SKILLS REVIEW: Practice and review algebra skills needed for this section at **www.Eduspace.com.**

In Exercises 1–4, describe in words how a plane could intersect with the double-napped cone shown to form the conic section.

1. Circle
2. Ellipse
3. Parabola
4. Hyperbola

In Exercises 5–10, match the equation with its graph. [The graphs are labeled (a), (b), (c), (d), (e), and (f).]

(a)

(b)

(c)

(d)

(e)

(f)
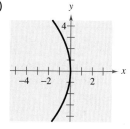

5. $y^2 = -4x$
6. $x^2 = 2y$
7. $x^2 = -8y$
8. $y^2 = -12x$
9. $(y - 1)^2 = 4(x - 3)$
10. $(x + 3)^2 = -2(y - 1)$

In Exercises 11–24, find the vertex, focus, and directrix of the parabola and sketch its graph.

11. $y = \frac{1}{2}x^2$
12. $y = -2x^2$
13. $y^2 = -6x$
14. $y^2 = 3x$
15. $x^2 + 6y = 0$
16. $x + y^2 = 0$
17. $(x - 1)^2 + 8(y + 2) = 0$
18. $(x + 5) + (y - 1)^2 = 0$
19. $\left(x + \frac{3}{2}\right)^2 = 4(y - 2)$
20. $\left(x + \frac{1}{2}\right)^2 = 4(y - 1)$
21. $y = \frac{1}{4}(x^2 - 2x + 5)$
22. $x = \frac{1}{4}(y^2 + 2y + 33)$
23. $y^2 + 6y + 8x + 25 = 0$
24. $y^2 - 4y - 4x = 0$

 In Exercises 25–28, find the vertex, focus, and directrix of the parabola. Use a graphing utility to graph the parabola.

25. $x^2 + 4x + 6y - 2 = 0$
26. $x^2 - 2x + 8y + 9 = 0$
27. $y^2 + x + y = 0$
28. $y^2 - 4x - 4 = 0$

In Exercises 29–40, find the standard form of the equation of the parabola with the given characteristic(s) and vertex at the origin.

29.

30.

31. Focus: $\left(0, -\frac{3}{2}\right)$

32. Focus: $\left(\frac{5}{2}, 0\right)$

33. Focus: $(-2, 0)$

34. Focus: $(0, -2)$

35. Directrix: $y = -1$

36. Directrix: $y = 3$

37. Directrix: $x = 2$

38. Directrix: $x = -3$

39. Horizontal axis and passes through the point $(4, 6)$

40. Vertical axis and passes through the point $(-3, -3)$

In Exercises 41–50, find the standard form of the equation of the parabola with the given characteristics.

41.

42.

43.

44.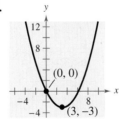

45. Vertex: $(5, 2)$; focus: $(3, 2)$

46. Vertex: $(-1, 2)$; focus: $(-1, 0)$

47. Vertex: $(0, 4)$; directrix: $y = 2$

48. Vertex: $(-2, 1)$; directrix: $x = 1$

49. Focus: $(2, 2)$; directrix: $x = -2$

50. Focus: $(0, 0)$; directrix: $y = 8$

In Exercises 51 and 52, change the equation of the parabola so that its graph matches the description.

51. $(y - 3)^2 = 6(x + 1)$; upper half of parabola

52. $(y + 1)^2 = 2(x - 4)$; lower half of parabola

In Exercises 53 and 54, the equations of a parabola and a tangent line to the parabola are given. Use a graphing utility to graph both equations in the same viewing window. Determine the coordinates of the point of tangency.

Parabola	*Tangent Line*
53. $y^2 - 8x = 0$	$x - y + 2 = 0$
54. $x^2 + 12y = 0$	$x + y - 3 = 0$

In Exercises 55–58, find an equation of the tangent line to the parabola at the given point, and find the *x*-intercept of the line.

55. $x^2 = 2y$, $(4, 8)$

56. $x^2 = 2y$, $\left(-3, \frac{9}{2}\right)$

57. $y = -2x^2$, $(-1, -2)$

58. $y = -2x^2$, $(2, -8)$

59. *Revenue*   The revenue R (in dollars) generated by the sale of x units of a patio furniture set is given by

$$(x - 106)^2 = -\frac{4}{5}(R - 14{,}045).$$

Use a graphing utility to graph the function and approximate the number of sales that will maximize revenue.

60. *Revenue*   The revenue R (in dollars) generated by the sale of x units of a digital camera is given by

$$(x - 135)^2 = -\frac{5}{7}(R - 25{,}515).$$

Use a graphing utility to graph the function and approximate the number of sales that will maximize revenue.

61. *Satellite Antenna*   The receiver in a parabolic television dish antenna is 4.5 feet from the vertex and is located at the focus (see figure). Write an equation for a cross section of the reflector. (Assume that the dish is directed upward and the vertex is at the origin.)

Model It

62. Suspension Bridge Each cable of the Golden Gate Bridge is suspended (in the shape of a parabola) between two towers that are 1280 meters apart. The top of each tower is 152 meters above the roadway. The cables touch the roadway midway between the towers.

(a) Draw a sketch of the bridge. Locate the origin of a rectangular coordinate system at the center of the roadway. Label the coordinates of the known points.

(b) Write an equation that models the cables.

(c) Complete the table by finding the height y of the suspension cables over the roadway at a distance of x meters from the center of the bridge.

Distance, x	Height, y
0	
250	
400	
500	
1000	

63. Road Design Roads are often designed with parabolic surfaces to allow rain to drain off. A particular road that is 32 feet wide is 0.4 foot higher in the center than it is on the sides (see figure).

32 ft   0.4 ft

Not drawn to scale

Cross section of road surface

(a) Find an equation of the parabola that models the road surface. (Assume that the origin is at the center of the road.)

(b) How far from the center of the road is the road surface 0.1 foot lower than in the middle?

64. Highway Design Highway engineers design a parabolic curve for an entrance ramp from a straight street to an interstate highway (see figure). Find an equation of the parabola.

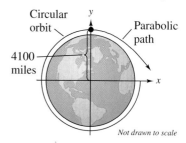

FIGURE FOR 64

65. Satellite Orbit A satellite in a 100-mile-high circular orbit around Earth has a velocity of approximately 17,500 miles per hour. If this velocity is multiplied by $\sqrt{2}$, the satellite will have the minimum velocity necessary to escape Earth's gravity and it will follow a parabolic path with the center of Earth as the focus (see figure).

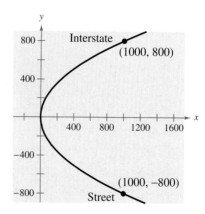

Not drawn to scale

(a) Find the escape velocity of the satellite.

(b) Find an equation of the parabolic path of the satellite (assume that the radius of Earth is 4000 miles).

66. Path of a Softball The path of a softball is modeled by

$$-12.5(y - 7.125) = (x - 6.25)^2$$

where the coordinates x and y are measured in feet, with $x = 0$ corresponding to the position from which the ball was thrown.

(a) Use a graphing utility to graph the trajectory of the softball.

(b) Use the *trace* feature of the graphing utility to approximate the highest point and the range of the trajectory.

Projectile Motion In Exercises 67 and 68, consider the path of a projectile projected horizontally with a velocity of *v* feet per second at a height of *s* feet, where the model for the path is

$$x^2 = -\frac{v^2}{16}(y - s).$$

In this model (in which air resistance is disregarded), *y* is the height (in feet) of the projectile and *x* is the horizontal distance (in feet) the projectile travels.

67. A ball is thrown from the top of a 75-foot tower with a velocity of 32 feet per second.

    (a) Find the equation of the parabolic path.

    (b) How far does the ball travel horizontally before striking the ground?

68. A cargo plane is flying at an altitude of 30,000 feet and a speed of 540 miles per hour. A supply crate is dropped from the plane. How many *feet* will the crate travel horizontally before it hits the ground?

Synthesis

True or False?   **In Exercises 69 and 70, determine whether the statement is true or false. Justify your answer.**

69. It is possible for a parabola to intersect its directrix.

70. If the vertex and focus of a parabola are on a horizontal line, then the directrix of the parabola is vertical.

 71. *Exploration*   Consider the parabola $x^2 = 4py$.

    (a) Use a graphing utility to graph the parabola for $p = 1$, $p = 2$, $p = 3$, and $p = 4$. Describe the effect on the graph when p increases.

    (b) Locate the focus for each parabola in part (a).

    (c) For each parabola in part (a), find the length of the chord passing through the focus and parallel to the directrix (see figure). How can the length of this chord be determined directly from the standard form of the equation of the parabola?

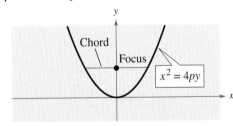

    (d) Explain how the result of part (c) can be used as a sketching aid when graphing parabolas.

72. *Geometry*   The area of the shaded region in the figure is

$$A = \frac{8}{3}p^{1/2}b^{3/2}.$$

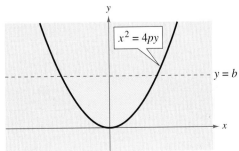

    (a) Find the area when $p = 2$ and $b = 4$.

    (b) Give a geometric explanation of why the area approaches 0 as p approaches 0.

73. *Exploration*   Let (x_1, y_1) be the coordinates of a point on the parabola $x^2 = 4py$. The equation of the line tangent to the parabola at the point is

$$y - y_1 = \frac{x_1}{2p}(x - x_1).$$

What is the slope of the tangent line?

74. *Writing*   In your own words, state the reflective property of a parabola.

Skills Review

75. Find a polynomial with real coefficients that has the zeros $3, 2 + i,$ and $2 - i$.

76. Find all the zeros of

$$f(x) = 2x^3 - 3x^2 + 50x - 75$$

if one of the zeros is $x = \frac{3}{2}$.

77. Find all the zeros of the function

$$g(x) = 6x^4 + 7x^3 - 29x^2 - 28x + 20$$

if two of the zeros are $x = \pm 2$.

 78. Use a graphing utility to graph the function given by

$$h(x) = 2x^4 + x^3 - 19x^2 - 9x + 9.$$

Use the graph to approximate the zeros of h.

In Exercises 79–86, use the information to solve the triangle. Round your answers to two decimal places.

79. $A = 35°, a = 10, b = 7$

80. $B = 54°, b = 18, c = 11$

81. $A = 40°, B = 51°, c = 3$

82. $B = 26°, C = 104°, a = 19$

83. $a = 7, b = 10, c = 16$

84. $a = 58, b = 28, c = 75$

85. $A = 65°, b = 5, c = 12$

86. $B = 71°, a = 21, c = 29$

6.3 Ellipses

Harvard College Observatory/
SPL/Photo Researchers, Inc.

Introduction

The second type of conic is called an **ellipse,** and is defined as follows.

Definition of Ellipse

An **ellipse** is the set of all points (x, y) in a plane, the sum of whose distances from two distinct fixed points **(foci)** is constant. See Figure 6.18.

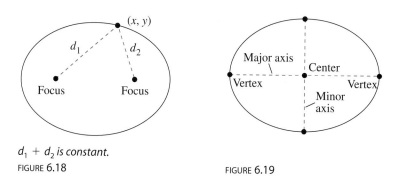

$d_1 + d_2$ is constant.
FIGURE 6.18

FIGURE 6.19

The line through the foci intersects the ellipse at two points called **vertices.** The chord joining the vertices is the **major axis,** and its midpoint is the **center** of the ellipse. The chord perpendicular to the major axis at the center is the **minor axis** of the ellipse. See Figure 6.19.

You can visualize the definition of an ellipse by imagining two thumbtacks placed at the foci, as shown in Figure 6.20. If the ends of a fixed length of string are fastened to the thumbtacks and the string is *drawn taut* with a pencil, the path traced by the pencil will be an ellipse.

FIGURE 6.20

To derive the standard form of the equation of an ellipse, consider the ellipse in Figure 6.21 with the following points: center, (h, k); vertices, $(h \pm a, k)$; foci, $(h \pm c, k)$. Note that the center is the midpoint of the segment joining the foci.

$$2\sqrt{b^2 + c^2} = 2a$$
$$b^2 + c^2 = a^2$$

FIGURE 6.21

When discussing ellipses, you might also choose to discuss the latera recta as background for Exercises 62–66.

The sum of the distances from any point on the ellipse to the two foci is constant. Using a vertex point, this constant sum is

$$(a + c) + (a - c) = 2a \qquad \text{Length of major axis}$$

or simply the length of the major axis. Now, if you let (x, y) be *any* point on the ellipse, the sum of the distances between (x, y) and the two foci must also be $2a$. That is,

$$\sqrt{[x - (h - c)]^2 + (y - k)^2} + \sqrt{[x - (h + c)]^2 + (y - k)^2} = 2a.$$

Finally, in Figure 6.21, you can see that $b^2 = a^2 - c^2$, which implies that the equation of the ellipse is

$$b^2(x - h)^2 + a^2(y - k)^2 = a^2b^2$$

$$\frac{(x - h)^2}{a^2} + \frac{(y - k)^2}{b^2} = 1.$$

You would obtain a similar equation in the derivation by starting with a vertical major axis. Both results are summarized as follows.

STUDY TIP

Consider the equation of the ellipse

$$\frac{(x - h)^2}{a^2} + \frac{(y - k)^2}{b^2} = 1.$$

If you let $a = b$, then the equation can be rewritten as

$$(x - h)^2 + (y - k)^2 = a^2$$

which is the standard form of the equation of a circle with radius $r = a$ (see Section P.3). Geometrically, when $a = b$ for an ellipse, the major and minor axes are of equal length, and so the graph is a circle.

Standard Equation of an Ellipse

The **standard form of the equation of an ellipse,** with center (h, k) and major and minor axes of lengths $2a$ and $2b$, respectively, where $0 < b < a$, is

$$\frac{(x - h)^2}{a^2} + \frac{(y - k)^2}{b^2} = 1 \qquad \text{Major axis is horizontal.}$$

$$\frac{(x - h)^2}{b^2} + \frac{(y - k)^2}{a^2} = 1. \qquad \text{Major axis is vertical.}$$

The foci lie on the major axis, c units from the center, with $c^2 = a^2 - b^2$. If the center is at the origin $(0, 0)$, the equation takes one of the following forms.

$$\frac{x^2}{a^2} + \frac{y^2}{b^2} = 1 \quad \begin{array}{l}\text{Major axis is}\\\text{horizontal.}\end{array} \qquad \frac{x^2}{b^2} + \frac{y^2}{a^2} = 1 \quad \begin{array}{l}\text{Major axis is}\\\text{vertical.}\end{array}$$

Figure 6.22 shows both the horizontal and vertical orientations for an ellipse.

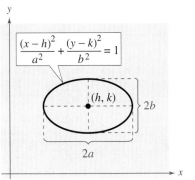

Major axis is horizontal.
FIGURE 6.22

Major axis is vertical.

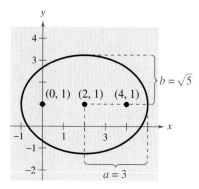

FIGURE **6.23**

Remind your students that completing the square must be performed twice to write the equation of the ellipse in standard form in Example 2.

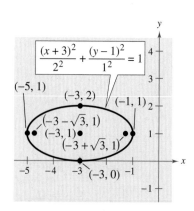

FIGURE **6.24**

Example 1   **Finding the Standard Equation of an Ellipse**

Find the standard form of the equation of the ellipse having foci at $(0, 1)$ and $(4, 1)$ and a major axis of length 6, as shown in Figure 6.23.

Solution

Because the foci occur at $(0, 1)$ and $(4, 1)$, the center of the ellipse is $(2, 1)$ and the distance from the center to one of the foci is $c = 2$. Because $2a = 6$, you know that $a = 3$. Now, from $c^2 = a^2 - b^2$, you have

$$b = \sqrt{a^2 - c^2} = \sqrt{3^2 - 2^2} = \sqrt{5}.$$

Because the major axis is horizontal, the standard equation is

$$\frac{(x - 2)^2}{3^2} + \frac{(y - 1)^2}{(\sqrt{5})^2} = 1.$$

This equation simplifies to

$$\frac{(x - 2)^2}{9} + \frac{(y - 1)^2}{5} = 1.$$

✔**CHECKPOINT**   Now try Exercise 49.

Example 2   **Sketching an Ellipse**

Sketch the ellipse given by $x^2 + 4y^2 + 6x - 8y + 9 = 0$.

Solution

Begin by writing the original equation in standard form. In the fourth step, note that 9 and 4 are added to *both* sides of the equation when completing the squares.

$$x^2 + 4y^2 + 6x - 8y + 9 = 0 \qquad \text{Write original equation.}$$

$$\left(x^2 + 6x + \right) + \left(4y^2 - 8y + \right) = -9 \qquad \text{Group terms.}$$

$$\left(x^2 + 6x + \right) + 4\left(y^2 - 2y + \right) = -9 \qquad \text{Factor 4 out of } y\text{-terms.}$$

$$(x^2 + 6x + 9) + 4(y^2 - 2y + 1) = -9 + 9 + 4(1)$$

$$(x + 3)^2 + 4(y - 1)^2 = 4 \qquad \text{Write in completed square form.}$$

$$\frac{(x + 3)^2}{4} + \frac{(y - 1)^2}{1} = 1 \qquad \text{Divide each side by 4.}$$

$$\frac{(x + 3)^2}{2^2} + \frac{(y - 1)^2}{1^2} = 1 \qquad \text{Write in standard form.}$$

From this standard form, it follows that the center is $(h, k) = (-3, 1)$. Because the denominator of the x-term is $a^2 = 2^2$, the endpoints of the major axis lie two units to the right and left of the center. Similarly, because the denominator of the y-term is $b^2 = 1^2$, the endpoints of the minor axis lie one unit up and down from the center. Now, from $c^2 = a^2 - b^2$, you have $c = \sqrt{2^2 - 1^2} = \sqrt{3}$. So, the foci of the ellipse are $\left(-3 - \sqrt{3}, 1\right)$ and $\left(-3 + \sqrt{3}, 1\right)$. The ellipse is shown in Figure 6.24.

✔**CHECKPOINT**   Now try Exercise 25.

Example 3 **Analyzing an Ellipse**

Find the center, vertices, and foci of the ellipse $4x^2 + y^2 - 8x + 4y - 8 = 0$.

Solution

By completing the square, you can write the original equation in standard form.

$$4x^2 + y^2 - 8x + 4y - 8 = 0 \quad \text{Write original equation.}$$

$$\left(4x^2 - 8x + \boxed{}\right) + \left(y^2 + 4y + \boxed{}\right) = 8 \quad \text{Group terms.}$$

$$4\left(x^2 - 2x + \boxed{}\right) + \left(y^2 + 4y + \boxed{}\right) = 8 \quad \text{Factor 4 out of } x\text{-terms.}$$

$$4(x^2 - 2x + 1) + (y^2 + 4y + 4) = 8 + 4(1) + 4$$

$$4(x - 1)^2 + (y + 2)^2 = 16 \quad \text{Write in completed square form.}$$

$$\frac{(x - 1)^2}{4} + \frac{(y + 2)^2}{16} = 1 \quad \text{Divide each side by 16.}$$

$$\frac{(x - 1)^2}{2^2} + \frac{(y + 2)^2}{4^2} = 1 \quad \text{Write in standard form.}$$

The major axis is vertical, where $h = 1$, $k = -2$, $a = 4$, $b = 2$, and

$$c = \sqrt{a^2 - b^2} = \sqrt{16 - 4} = \sqrt{12} = 2\sqrt{3}.$$

So, you have the following.

Center: $(1, -2)$ Vertices: $(1, -6)$ Foci: $\left(1, -2 - 2\sqrt{3}\right)$

                          $(1, 2)$ $\left(1, -2 + 2\sqrt{3}\right)$

The graph of the ellipse is shown in Figure 6.25.

✓CHECKPOINT   Now try Exercise 29.

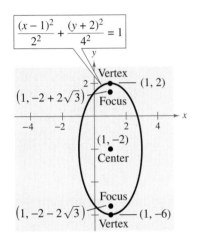

$$\frac{(x - 1)^2}{2^2} + \frac{(y + 2)^2}{4^2} = 1$$

FIGURE 6.25

Technology

You can use a graphing utility to graph an ellipse by graphing the upper and lower portions in the same viewing window. For instance, to graph the ellipse in Example 3, first solve for y to get

$$y_1 = -2 + 4\sqrt{1 - \frac{(x - 1)^2}{4}} \quad \text{and} \quad y_2 = -2 - 4\sqrt{1 - \frac{(x - 1)^2}{4}}.$$

Use a viewing window in which $-6 \leq x \leq 9$ and $-7 \leq y \leq 3$. You should obtain the graph shown below.

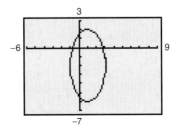

Application

Ellipses have many practical and aesthetic uses. For instance, machine gears, supporting arches, and acoustic designs often involve elliptical shapes. The orbits of satellites and planets are also ellipses. Example 4 investigates the elliptical orbit of the moon about Earth.

Example 4 An Application Involving an Elliptical Orbit

The moon travels about Earth in an elliptical orbit with Earth at one focus, as shown in Figure 6.26. The major and minor axes of the orbit have lengths of 768,800 kilometers and 767,640 kilometers, respectively. Find the greatest and smallest distances (the *apogee* and *perigee*), respectively from Earth's center to the moon's center.

Solution

Because $2a = 768,800$ and $2b = 767,640$, you have

$$a = 384,400 \text{ and } b = 383,820$$

which implies that

$$c = \sqrt{a^2 - b^2}$$
$$= \sqrt{384,400^2 - 383,820^2}$$
$$\approx 21,108.$$

So, the greatest distance between the center of Earth and the center of the moon is

$$a + c \approx 384,400 + 21,108 = 405,508 \text{ kilometers}$$

and the smallest distance is

$$a - c \approx 384,400 - 21,108 = 363,292 \text{ kilometers}.$$

✓**CHECKPOINT**   Now try Exercise 59.

Eccentricity

One of the reasons it was difficult for early astronomers to detect that the orbits of the planets are ellipses is that the foci of the planetary orbits are relatively close to their centers, and so the orbits are nearly circular. To measure the ovalness of an ellipse, you can use the concept of **eccentricity.**

> ### Definition of Eccentricity
> The **eccentricity** e of an ellipse is given by the ratio
>
> $$e = \frac{c}{a}.$$

Note that $0 < e < 1$ for *every* ellipse.

767,640 km

Moon

Earth

768,800 km

Perigee   Apogee

FIGURE **6.26**

STUDY TIP

Note in Example 4 and Figure 6.26 that Earth *is not* the center of the moon's orbit.

Ask students to make a conjecture about the eccentricity of a circle before going further.

Activities

1. Find the center, foci, vertices, and eccentricity of the ellipse

$$\frac{(x-2)^2}{25} + \frac{(y+1)^2}{9} = 1.$$

Answer: Center $(2, -1)$;
Foci $(-2, -1), (6, -1)$;
Vertices $(-3, -1), (7, -1)$; $e = \frac{4}{5}$

2. Rewrite the equation of the ellipse in standard form:

$$9x^2 + 5y^2 + 36x - 30y + 36 = 0.$$

Answer: $\dfrac{(x+2)^2}{\left(\sqrt{5}\right)^2} + \dfrac{(y-3)^2}{3^2} = 1$

To see how this ratio is used to describe the shape of an ellipse, note that because the foci of an ellipse are located along the major axis between the vertices and the center, it follows that

$$0 < c < a.$$

For an ellipse that is nearly circular, the foci are close to the center and the ratio c/a is small, as shown in Figure 6.27. On the other hand, for an elongated ellipse, the foci are close to the vertices, and the ratio c/a is close to 1, as shown in Figure 6.28.

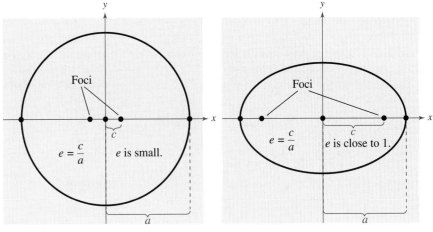

FIGURE 6.27   FIGURE 6.28

The orbit of the moon has an eccentricity of $e \approx 0.0549$, and the eccentricities of the nine planetary orbits are as follows.

Mercury: $e \approx 0.2056$	Saturn: $e \approx 0.0542$
Venus: $e \approx 0.0068$	Uranus: $e \approx 0.0472$
Earth: $e \approx 0.0167$	Neptune: $e \approx 0.0086$
Mars: $e \approx 0.0934$	Pluto: $e \approx 0.2488$
Jupiter: $e \approx 0.0484$	

The time it takes Saturn to orbit the sun is equal to 29.4 Earth years.

---

*W*RITING ABOUT *M*ATHEMATICS

Ellipses and Circles

a. Show that the equation of an ellipse can be written as

$$\frac{(x-h)^2}{a^2} + \frac{(y-k)^2}{a^2(1-e^2)} = 1.$$

b. For the equation in part (a), let $a = 4$, $h = 1$, and $k = 2$, and use a graphing utility to graph the ellipse for $e = 0.95$, $e = 0.75$, $e = 0.5$, $e = 0.25$, and $e = 0.1$. Discuss the changes in the shape of the ellipse as e approaches 0.

c. Make a conjecture about the shape of the graph in part (b) when $e = 0$. What is the equation of this ellipse? What is another name for an ellipse with an eccentricity of 0?

> # 6.3 | Exercises

VOCABULARY CHECK: Fill in the blanks.

1. An _____ is the set of all points (x, y) in a plane, the sum of whose distances from two distinct fixed points, called _____, is constant.

2. The chord joining the vertices of an ellipse is called the _____ _____, and its midpoint is the _____ of the ellipse.

3. The chord perpendicular to the major axis at the center of the ellipse is called the _____ _____ of the ellipse.

4. The concept of _____ is used to measure the ovalness of an ellipse.

PREREQUISITE SKILLS REVIEW: Practice and review algebra skills needed for this section at **www.Eduspace.com**.

In Exercises 1–6, match the equation with its graph. [The graphs are labeled (a), (b), (c), (d), (e), and (f).]

(a)

(b)

(c)

(d)

(e)

(f)

1. $\dfrac{x^2}{4} + \dfrac{y^2}{9} = 1$

2. $\dfrac{x^2}{9} + \dfrac{y^2}{4} = 1$

3. $\dfrac{x^2}{4} + \dfrac{y^2}{25} = 1$

4. $\dfrac{x^2}{4} + \dfrac{y^2}{4} = 1$

5. $\dfrac{(x - 2)^2}{16} + (y + 1)^2 = 1$

6. $\dfrac{(x + 2)^2}{9} + \dfrac{(y + 2)^2}{4} = 1$

In Exercises 7–30, identify the conic as a circle or an ellipse. Then find the center, radius, vertices, foci, and eccentricity of the conic (if applicable), and sketch its graph.

7. $\dfrac{x^2}{25} + \dfrac{y^2}{16} = 1$

8. $\dfrac{x^2}{81} + \dfrac{y^2}{144} = 1$

9. $\dfrac{x^2}{25} + \dfrac{y^2}{25} = 1$

10. $\dfrac{x^2}{9} + \dfrac{y^2}{9} = 1$

11. $\dfrac{x^2}{5} + \dfrac{y^2}{9} = 1$

12. $\dfrac{x^2}{64} + \dfrac{y^2}{28} = 1$

13. $\dfrac{(x + 3)^2}{16} + \dfrac{(y - 5)^2}{25} = 1$

14. $\dfrac{(x - 4)^2}{12} + \dfrac{(y + 3)^2}{16} = 1$

15. $\dfrac{x^2}{4/9} + \dfrac{(y + 1)^2}{4/9} = 1$

16. $\dfrac{(x + 5)^2}{9/4} + (y - 1)^2 = 1$

17. $(x + 2)^2 + \dfrac{(y + 4)^2}{1/4} = 1$

18. $\dfrac{(x - 3)^2}{25/4} + \dfrac{(y - 1)^2}{25/4} = 1$

19. $9x^2 + 4y^2 + 36x - 24y + 36 = 0$

20. $9x^2 + 4y^2 - 54x + 40y + 37 = 0$

21. $x^2 + y^2 - 2x + 4y - 31 = 0$

22. $x^2 + 5y^2 - 8x - 30y - 39 = 0$

23. $3x^2 + y^2 + 18x - 2y - 8 = 0$

24. $6x^2 + 2y^2 + 18x - 10y + 2 = 0$

25. $x^2 + 4y^2 - 6x + 20y - 2 = 0$

26. $x^2 + y^2 - 4x + 6y - 3 = 0$

27. $9x^2 + 9y^2 + 18x - 18y + 14 = 0$

28. $16x^2 + 25y^2 - 32x + 50y + 16 = 0$

29. $9x^2 + 25y^2 - 36x - 50y + 60 = 0$

30. $16x^2 + 16y^2 - 64x + 32y + 55 = 0$

In Exercises 31–34, use a graphing utility to graph the ellipse. Find the center, foci, and vertices. (Recall that it may be necessary to solve the equation for *y* and obtain two equations.)

31. $5x^2 + 3y^2 = 15$

32. $3x^2 + 4y^2 = 12$

33. $12x^2 + 20y^2 - 12x + 40y - 37 = 0$

34. $36x^2 + 9y^2 + 48x - 36y - 72 = 0$

In Exercises 35–42, find the standard form of the equation of the ellipse with the given characteristics and center at the origin.

35.   **36.**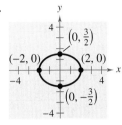

37. Vertices: $(\pm 6, 0)$; foci: $(\pm 2, 0)$

38. Vertices: $(0, \pm 8)$; foci: $(0, \pm 4)$

39. Foci: $(\pm 5, 0)$; major axis of length 12

40. Foci: $(\pm 2, 0)$; major axis of length 8

41. Vertices: $(0, \pm 5)$; passes through the point $(4, 2)$

42. Major axis vertical; passes through the points $(0, 4)$ and $(2, 0)$

In Exercises 43–56, find the standard form of the equation of the ellipse with the given characteristics.

43.   **44.**

45.   **46.**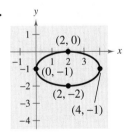

47. Vertices: $(0, 4), (4, 4)$; minor axis of length 2

48. Foci: $(0, 0), (4, 0)$; major axis of length 8

49. Foci: $(0, 0), (0, 8)$; major axis of length 16

50. Center: $(2, -1)$; vertex: $\left(2, \frac{1}{2}\right)$; minor axis of length 2

51. Center: $(0, 4)$; $a = 2c$; vertices: $(-4, 4), (4, 4)$

52. Center: $(3, 2)$; $a = 3c$; foci: $(1, 2), (5, 2)$

53. Vertices: $(0, 2), (4, 2)$; endpoints of the minor axis: $(2, 3), (2, 1)$

54. Vertices: $(5, 0), (5, 12)$; endpoints of the minor axis: $(1, 6), (9, 6)$

55. Find an equation of the ellipse with vertices $(\pm 5, 0)$ and eccentricity $e = \frac{3}{5}$.

56. Find an equation of the ellipse with vertices $(0, \pm 8)$ and eccentricity $e = \frac{1}{2}$.

57. *Architecture* A semielliptical arch over a tunnel for a one-way road through a mountain has a major axis of 50 feet and a height at the center of 10 feet.

(a) Draw a rectangular coordinate system on a sketch of the tunnel with the center of the road entering the tunnel at the origin. Identify the coordinates of the known points.

(b) Find an equation of the semielliptical arch over the tunnel.

(c) You are driving a moving truck that has a width of 8 feet and a height of 9 feet. Will the moving truck clear the opening of the arch?

58. *Architecture* A fireplace arch is to be constructed in the shape of a semiellipse. The opening is to have a height of 2 feet at the center and a width of 6 feet along the base (see figure). The contractor draws the outline of the ellipse using tacks as described at the beginning of this section. Give the required positions of the tacks and the length of the string.

Model It

59. *Comet Orbit* Halley's comet has an elliptical orbit, with the sun at one focus. The eccentricity of the orbit is approximately 0.967. The length of the major axis of the orbit is approximately 35.88 astronomical units. (An astronomical unit is about 93 million miles.)

(a) Find an equation of the orbit. Place the center of the orbit at the origin, and place the major axis on the x-axis.

(b) Use a graphing utility to graph the equation of the orbit.

(c) Find the greatest (aphelion) and smallest (perihelion) distances from the sun's center to the comet's center.

60. *Satellite Orbit*   The first artificial satellite to orbit Earth was Sputnik I (launched by the former Soviet Union in 1957). Its highest point above Earth's surface was 947 kilometers, and its lowest point was 228 kilometers (see figure). The center of Earth was the focus of the elliptical orbit, and the radius of Earth is 6378 kilometers. Find the eccentricity of the orbit.

228 km — 947 km

61. *Motion of a Pendulum*   The relation between the velocity y (in radians per second) of a pendulum and its angular displacement θ from the vertical can be modeled by a semiellipse. A 12-centimeter pendulum crests $(y = 0)$ when the angular displacement is -0.2 radian and 0.2 radian. When the pendulum is at equilibrium $(\theta = 0)$, the velocity is -1.6 radians per second.

(a) Find an equation that models the motion of the pendulum. Place the center at the origin.

(b) Graph the equation from part (a).

(c) Which half of the ellipse models the motion of the pendulum?

62. *Geometry*   A line segment through a focus of an ellipse with endpoints on the ellipse and perpendicular to the major axis is called a **latus rectum** of the ellipse. Therefore, an ellipse has two latera recta. Knowing the length of the latera recta is helpful in sketching an ellipse because it yields other points on the curve (see figure). Show that the length of each latus rectum is $2b^2/a$.

Latera recta

In Exercises 63–66, sketch the graph of the ellipse, using latera recta (see Exercise 62).

63. $\dfrac{x^2}{9} + \dfrac{y^2}{16} = 1$ **64.** $\dfrac{x^2}{4} + \dfrac{y^2}{1} = 1$

65. $5x^2 + 3y^2 = 15$ **66.** $9x^2 + 4y^2 = 36$

Synthesis

True or False?   In Exercises 67 and 68, determine whether the statement is true or false. Justify your answer.

67. The graph of $x^2 + 4y^4 - 4 = 0$ is an ellipse.

68. It is easier to distinguish the graph of an ellipse from the graph of a circle if the eccentricity of the ellipse is large (close to 1).

69. *Exploration*   Consider the ellipse

$$\frac{x^2}{a^2} + \frac{y^2}{b^2} = 1, \quad a + b = 20.$$

(a) The area of the ellipse is given by $A = \pi ab$. Write the area of the ellipse as a function of a.

(b) Find the equation of an ellipse with an area of 264 square centimeters.

(c) Complete the table using your equation from part (a), and make a conjecture about the shape of the ellipse with maximum area.

a	8	9	10	11	12	13
A						

 (d) Use a graphing utility to graph the area function and use the graph to support your conjecture in part (c).

70. *Think About It*   At the beginning of this section it was noted that an ellipse can be drawn using two thumbtacks, a string of fixed length (greater than the distance between the two tacks), and a pencil. If the ends of the string are fastened at the tacks and the string is drawn taut with a pencil, the path traced by the pencil is an ellipse.

(a) What is the length of the string in terms of a?

(b) Explain why the path is an ellipse.

Skills Review

In Exercises 71–74, evaluate the logarithm using the change-of-base formula. Round your result to three decimal places.

71. $\log_4 7.1$ **72.** $\log_{15} 100$

73. $\log_{1/2} 22$ **74.** $\log_{2/3} 6$

In Exercises 75–78, use the properties of logarithms to rewrite and simplify the logarithmic expression.

75. $\log_3 135$ **76.** $\log \dfrac{1}{150}$

77. $\ln(9e^4)$ **78.** $\ln \dfrac{4}{e^{1.5}}$

6.4 Hyperbolas

What you should learn

- Write equations of hyperbolas in standard form.
- Find asymptotes of and graph hyperbolas.
- Use properties of hyperbolas to solve real-life problems.
- Classify conics from their general equations.

Why you should learn it

Hyperbolas can be used to model and solve many types of real-life problems. For instance, in Exercise 42 on page 463, hyperbolas are used in long distance radio navigation for aircraft and ships.

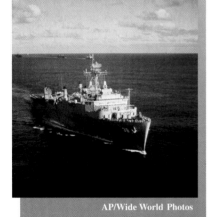

AP/Wide World Photos

Introduction

The third type of conic is called a **hyperbola.** The definition of a hyperbola is similar to that of an ellipse. The difference is that for an ellipse the *sum* of the distances between the foci and a point on the ellipse is fixed, whereas for a hyperbola the *difference* of the distances between the foci and a point on the hyperbola is fixed.

Definition of Hyperbola

A **hyperbola** is the set of all points (x, y) in a plane, the difference of whose distances from two distinct fixed points (**foci**) is a positive constant. See Figure 6.29.

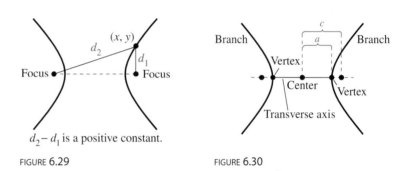

$d_2 - d_1$ is a positive constant.

FIGURE 6.29

FIGURE 6.30

The graph of a hyperbola has two disconnected **branches.** The line through the two foci intersects the hyperbola at its two **vertices.** The line segment connecting the vertices is the **transverse axis,** and the midpoint of the transverse axis is the **center** of the hyperbola. See Figure 6.30. The development of the standard form of the equation of a hyperbola is similar to that of an ellipse. Note in the definition below that a, b, and c are related differently for hyperbolas than for ellipses.

Standard Equation of a Hyperbola

The **standard form of the equation of a hyperbola** with center (h, k) is

$$\frac{(x - h)^2}{a^2} - \frac{(y - k)^2}{b^2} = 1 \qquad \text{Transverse axis is horizontal.}$$

$$\frac{(y - k)^2}{a^2} - \frac{(x - h)^2}{b^2} = 1. \qquad \text{Transverse axis is vertical.}$$

The vertices are a units from the center, and the foci are c units from the center. Moreover, $c^2 = a^2 + b^2$. If the center of the hyperbola is at the origin $(0, 0)$, the equation takes one of the following forms.

$$\frac{x^2}{a^2} - \frac{y^2}{b^2} = 1 \quad \text{Transverse axis is horizontal.} \qquad \frac{y^2}{a^2} - \frac{x^2}{b^2} = 1 \quad \text{Transverse axis is vertical.}$$

Figure 6.31 shows both the horizontal and vertical orientations for a hyperbola.

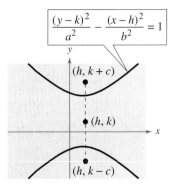

Transverse axis is horizontal.

Transverse axis is vertical.

FIGURE **6.31**

Example 1 Finding the Standard Equation of a Hyperbola

Find the standard form of the equation of the hyperbola with foci $(-1, 2)$ and $(5, 2)$ and vertices $(0, 2)$ and $(4, 2)$.

Solution

By the Midpoint Formula, the center of the hyperbola occurs at the point $(2, 2)$. Furthermore, $c = 5 - 2 = 3$ and $a = 4 - 2 = 2$, and it follows that

$$b = \sqrt{c^2 - a^2} = \sqrt{3^2 - 2^2} = \sqrt{9 - 4} = \sqrt{5}.$$

So, the hyperbola has a horizontal transverse axis and the standard form of the equation is

$$\frac{(x - 2)^2}{2^2} - \frac{(y - 2)^2}{\left(\sqrt{5}\right)^2} = 1. \qquad \text{See Figure 6.32.}$$

This equation simplifies to

$$\frac{(x - 2)^2}{4} - \frac{(y - 2)^2}{5} = 1.$$

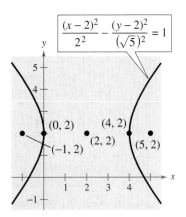

FIGURE **6.32**

✓CHECKPOINT Now try Exercise 27.

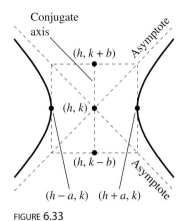

Conjugate axis

$(h, k + b)$

Asymptote

(h, k)

$(h, k - b)$

Asymptote

$(h - a, k)$ $(h + a, k)$

FIGURE 6.33

Asymptotes of a Hyperbola

Each hyperbola has two **asymptotes** that intersect at the center of the hyperbola, as shown in Figure 6.33. The asymptotes pass through the vertices of a rectangle of dimensions $2a$ by $2b$, with its center at (h, k). The line segment of length $2b$ joining $(h, k + b)$ and $(h, k - b)$ [or $(h + b, k)$ and $(h - b, k)$] is the **conjugate axis** of the hyperbola.

Asymptotes of a Hyperbola

The equations of the asymptotes of a hyperbola are

$$y = k \pm \frac{b}{a}(x - h)$$ Transverse axis is horizontal.

$$y = k \pm \frac{a}{b}(x - h).$$ Transverse axis is vertical.

Example 2   Using Asymptotes to Sketch a Hyperbola

Sketch the hyperbola whose equation is $4x^2 - y^2 = 16$.

Solution

Divide each side of the original equation by 16, and rewrite the equation in standard form.

$$\frac{x^2}{2^2} - \frac{y^2}{4^2} = 1$$ Write in standard form.

From this, you can conclude that $a = 2$, $b = 4$, and the transverse axis is horizontal. So, the vertices occur at $(-2, 0)$ and $(2, 0)$, and the endpoints of the conjugate axis occur at $(0, -4)$ and $(0, 4)$. Using these four points, you are able to sketch the rectangle shown in Figure 6.34. Now, from $c^2 = a^2 + b^2$, you have $c = \sqrt{2^2 + 4^2} = \sqrt{20} = 2\sqrt{5}$. So, the foci of the hyperbola are $\left(-2\sqrt{5}, 0\right)$ and $\left(2\sqrt{5}, 0\right)$. Finally, by drawing the asymptotes through the corners of this rectangle, you can complete the sketch shown in Figure 6.35. Note that the asymptotes are $y = 2x$ and $y = -2x$.

FIGURE 6.34

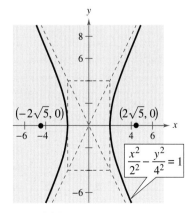

FIGURE 6.35

✔CHECKPOINT   Now try Exercise 7.

Example 3   **Finding the Asymptotes of a Hyperbola**

Sketch the hyperbola given by $4x^2 - 3y^2 + 8x + 16 = 0$ and find the equations of its asymptotes and the foci.

Solution

$$4x^2 - 3y^2 + 8x + 16 = 0 \qquad \text{Write original equation.}$$

$$(4x^2 + 8x) - 3y^2 = -16 \qquad \text{Group terms.}$$

$$4(x^2 + 2x) - 3y^2 = -16 \qquad \text{Factor 4 from } x\text{-terms.}$$

$$4(x^2 + 2x + 1) - 3y^2 = -16 + 4 \qquad \text{Add 4 to each side.}$$

$$4(x + 1)^2 - 3y^2 = -12 \qquad \text{Write in completed square form.}$$

$$-\frac{(x + 1)^2}{3} + \frac{y^2}{4} = 1 \qquad \text{Divide each side by } -12.$$

$$\frac{y^2}{2^2} - \frac{(x + 1)^2}{\left(\sqrt{3}\right)^2} = 1 \qquad \text{Write in standard form.}$$

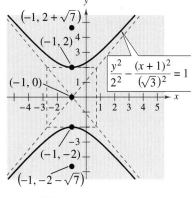

From this equation you can conclude that the hyperbola has a vertical transverse axis, centered at $(-1, 0)$, has vertices $(-1, 2)$ and $(-1, -2)$, and has a conjugate axis with endpoints $\left(-1 - \sqrt{3}, 0\right)$ and $\left(-1 + \sqrt{3}, 0\right)$. To sketch the hyperbola, draw a rectangle through these four points. The asymptotes are the lines passing through the corners of the rectangle. Using $a = 2$ and $b = \sqrt{3}$, you can conclude that the equations of the asymptotes are

$$y = \frac{2}{\sqrt{3}}(x + 1) \qquad \text{and} \qquad y = -\frac{2}{\sqrt{3}}(x + 1).$$

Finally, you can determine the foci by using the equation $c^2 = a^2 + b^2$. So, you have $c = \sqrt{2^2 + \left(\sqrt{3}\right)^2} = \sqrt{7}$, and the foci are $\left(-1, -2 - \sqrt{7}\right)$ and $\left(-1, -2 + \sqrt{7}\right)$. The hyperbola is shown in Figure 6.36.

FIGURE 6.36

✓**CHECKPOINT**   Now try Exercise 13.

Technology

You can use a graphing utility to graph a hyperbola by graphing the upper and lower portions in the same viewing window. For instance, to graph the hyperbola in Example 3, first solve for y to get

$$y_1 = 2\sqrt{1 + \frac{(x + 1)^2}{3}} \qquad \text{and} \qquad y_2 = -2\sqrt{1 + \frac{(x + 1)^2}{3}}.$$

Use a viewing window in which $-9 \le x \le 9$ and $-6 \le y \le 6$. You should obtain the graph shown below. Notice that the graphing utility does not draw the asymptotes. However, if you trace along the branches, you will see that the values of the hyperbola approach the asymptotes.

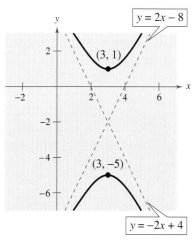

FIGURE **6.37**

Example 4   **Using Asymptotes to Find the Standard Equation**

Find the standard form of the equation of the hyperbola having vertices $(3, -5)$ and $(3, 1)$ and having asymptotes

$$y = 2x - 8 \qquad \text{and} \qquad y = -2x + 4$$

as shown in Figure 6.37.

Solution

By the Midpoint Formula, the center of the hyperbola is $(3, -2)$. Furthermore, the hyperbola has a vertical transverse axis with $a = 3$. From the original equations, you can determine the slopes of the asymptotes to be

$$m_1 = 2 = \frac{a}{b} \qquad \text{and} \qquad m_2 = -2 = -\frac{a}{b}$$

and, because $a = 3$ you can conclude

$$2 = \frac{a}{b} \implies 2 = \frac{3}{b} \implies b = \frac{3}{2}.$$

So, the standard form of the equation is

$$\frac{(y + 2)^2}{3^2} - \frac{(x - 3)^2}{\left(\frac{3}{2}\right)^2} = 1.$$

✓CHECKPOINT   Now try Exercise 35.

As with ellipses, the *eccentricity* of a hyperbola is

$$e = \frac{c}{a} \qquad \text{Eccentricity}$$

and because $c > a$, it follows that $e > 1$. If the eccentricity is large, the branches of the hyperbola are nearly flat, as shown in Figure 6.38. If the eccentricity is close to 1, the branches of the hyperbola are more narrow, as shown in Figure 6.39.

Additional Example

Find the standard form of the equation of the hyperbola having vertices $(-4, 2)$ and $(1, 2)$ and having foci $(-7, 2)$ and $(4, 2)$.

Answer: $\dfrac{\left(x + \frac{3}{2}\right)^2}{\left(\frac{5}{2}\right)^2} - \dfrac{(y - 2)^2}{(\sqrt{24})^2} = 1$

FIGURE **6.38**

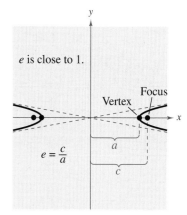

FIGURE **6.39**

Applications

The following application was developed during World War II. It shows how the properties of hyperbolas can be used in radar and other detection systems.

Example 5 **An Application Involving Hyperbolas**

Two microphones, 1 mile apart, record an explosion. Microphone A receives the sound 2 seconds before microphone B. Where did the explosion occur? (Assume sound travels at 1100 feet per second.)

Solution

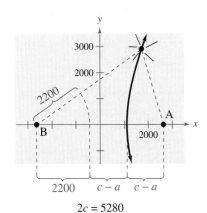

$$2c = 5280$$
$$2200 + 2(c - a) = 5280$$

FIGURE **6.40**

Assuming sound travels at 1100 feet per second, you know that the explosion took place 2200 feet farther from B than from A, as shown in Figure 6.40. The locus of all points that are 2200 feet closer to A than to B is one branch of the hyperbola

$$\frac{x^2}{a^2} - \frac{y^2}{b^2} = 1$$

where

$$c = \frac{5280}{2} = 2640$$

and

$$a = \frac{2200}{2} = 1100.$$

So, $b^2 = c^2 - a^2 = 2640^2 - 1100^2 = 5{,}759{,}600$, and you can conclude that the explosion occurred somewhere on the right branch of the hyperbola

$$\frac{x^2}{1{,}210{,}000} - \frac{y^2}{5{,}759{,}600} = 1.$$

✓**CHECKPOINT**   Now try Exercise 41.

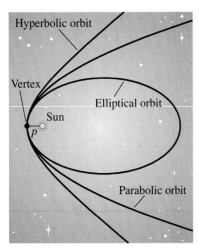

FIGURE **6.41**

Another interesting application of conic sections involves the orbits of comets in our solar system. Of the 610 comets identified prior to 1970, 245 have elliptical orbits, 295 have parabolic orbits, and 70 have hyperbolic orbits. The center of the sun is a focus of each of these orbits, and each orbit has a vertex at the point where the comet is closest to the sun, as shown in Figure 6.41. Undoubtedly, there have been many comets with parabolic or hyperbolic orbits that were not identified. We only get to see such comets *once*. Comets with elliptical orbits, such as Halley's comet, are the only ones that remain in our solar system.

If p is the distance between the vertex and the focus (in meters), and v is the velocity of the comet at the vertex in (meters per second), then the type of orbit is determined as follows.

1. Ellipse: $v < \sqrt{2GM/p}$
2. Parabola: $v = \sqrt{2GM/p}$
3. Hyperbola: $v > \sqrt{2GM/p}$

In each of these relations, $M = 1.989 \times 10^{30}$ kilograms (the mass of the sun) and $G \approx 6.67 \times 10^{-11}$ cubic meter per kilogram-second squared (the universal gravitational constant).

General Equations of Conics

Classifying a Conic from Its General Equation

The graph of $Ax^2 + Cy^2 + Dx + Ey + F = 0$ is one of the following.

1. *Circle:* $A = C$

2. *Parabola:* $AC = 0$ $A = 0$ or $C = 0$, but not both.

3. *Ellipse:* $AC > 0$ A and C have like signs.

4. *Hyperbola:* $AC < 0$ A and C have unlike signs.

The test above is valid *if* the graph is a conic. The test does not apply to equations such as $x^2 + y^2 = -1$, whose graph is not a conic.

Example 6   Classifying Conics from General Equations

Classify the graph of each equation.

a. $4x^2 - 9x + y - 5 = 0$
b. $4x^2 - y^2 + 8x - 6y + 4 = 0$
c. $2x^2 + 4y^2 - 4x + 12y = 0$
d. $2x^2 + 2y^2 - 8x + 12y + 2 = 0$

Solution

a. For the equation $4x^2 - 9x + y - 5 = 0$, you have
$$AC = 4(0) = 0. \quad \text{Parabola}$$
So, the graph is a parabola.

b. For the equation $4x^2 - y^2 + 8x - 6y + 4 = 0$, you have
$$AC = 4(-1) < 0. \quad \text{Hyperbola}$$
So, the graph is a hyperbola.

c. For the equation $2x^2 + 4y^2 - 4x + 12y = 0$, you have
$$AC = 2(4) > 0. \quad \text{Ellipse}$$
So, the graph is an ellipse.

d. For the equation $2x^2 + 2y^2 - 8x + 12y + 2 = 0$, you have
$$A = C = 2. \quad \text{Circle}$$
So, the graph is a circle.

✓CHECKPOINT   Now try Exercise 49.

*W*RITING ABOUT *M*ATHEMATICS

Sketching Conics Sketch each of the conics described in Example 6. Write a paragraph describing the procedures that allow you to sketch the conics efficiently.

> ## 6.4 | Exercises

VOCABULARY CHECK: Fill in the blanks.

1. A _____ is the set of all points (x, y) in a plane, the difference of whose distances from two distinct fixed points, called _____, is a positive constant.

2. The graph of a hyperbola has two disconnected parts called _____.

3. The line segment connecting the vertices of a hyperbola is called the _____ _____, and the midpoint of the line segment is the _____ of the hyperbola.

4. Each hyperbola has two _____ that intersect at the center of the hyperbola.

5. The general form of the equation of a conic is given by _____.

PREREQUISITE SKILLS REVIEW: Practice and review algebra skills needed for this section at **www.Eduspace.com.**

In Exercises 1–4, match the equation with its graph. [The graphs are labeled (a), (b), (c), and (d).]

(a)

(b)

(c)

(d)

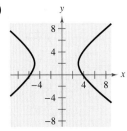

1. $\dfrac{y^2}{9} - \dfrac{x^2}{25} = 1$

2. $\dfrac{y^2}{25} - \dfrac{x^2}{9} = 1$

3. $\dfrac{(x - 1)^2}{16} - \dfrac{y^2}{4} = 1$

4. $\dfrac{(x + 1)^2}{16} - \dfrac{(y - 2)^2}{9} = 1$

In Exercises 5–16, find the center, vertices, foci, and the equations of the asymptotes of the hyperbola, and sketch its graph using the asymptotes as an aid.

5. $x^2 - y^2 = 1$

6. $\dfrac{x^2}{9} - \dfrac{y^2}{25} = 1$

7. $\dfrac{y^2}{25} - \dfrac{x^2}{81} = 1$

8. $\dfrac{x^2}{36} - \dfrac{y^2}{4} = 1$

9. $\dfrac{(x - 1)^2}{4} - \dfrac{(y + 2)^2}{1} = 1$

10. $\dfrac{(x + 3)^2}{144} - \dfrac{(y - 2)^2}{25} = 1$

11. $\dfrac{(y + 6)^2}{1/9} - \dfrac{(x - 2)^2}{1/4} = 1$

12. $\dfrac{(y - 1)^2}{1/4} - \dfrac{(x + 3)^2}{1/16} = 1$

13. $9x^2 - y^2 - 36x - 6y + 18 = 0$

14. $x^2 - 9y^2 + 36y - 72 = 0$

15. $x^2 - 9y^2 + 2x - 54y - 80 = 0$

16. $16y^2 - x^2 + 2x + 64y + 63 = 0$

 In Exercises 17–20, find the center, vertices, foci, and the equations of the asymptotes of the hyperbola. Use a graphing utility to graph the hyperbola and its asymptotes.

17. $2x^2 - 3y^2 = 6$

18. $6y^2 - 3x^2 = 18$

19. $9y^2 - x^2 + 2x + 54y + 62 = 0$

20. $9x^2 - y^2 + 54x + 10y + 55 = 0$

In Exercises 21–26, find the standard form of the equation of the hyperbola with the given characteristics and center at the origin.

21. Vertices: $(0, \pm 2)$; foci: $(0, \pm 4)$

22. Vertices: $(\pm 4, 0)$; foci: $(\pm 6, 0)$

23. Vertices: $(\pm 1, 0)$; asymptotes: $y = \pm 5x$

24. Vertices: $(0, \pm 3)$; asymptotes: $y = \pm 3x$

25. Foci: $(0, \pm 8)$; asymptotes: $y = \pm 4x$

26. Foci: $(\pm 10, 0)$; asymptotes: $y = \pm \frac{3}{4}x$

In Exercises 27–38, find the standard form of the equation of the hyperbola with the given characteristics.

27. Vertices: $(2, 0), (6, 0)$; foci: $(0, 0), (8, 0)$

28. Vertices: $(2, 3), (2, -3)$; foci: $(2, 6), (2, -6)$

29. Vertices: $(4, 1), (4, 9)$; foci: $(4, 0), (4, 10)$

30. Vertices: $(-2, 1), (2, 1)$; foci: $(-3, 1), (3, 1)$

31. Vertices: $(2, 3), (2, -3)$;

passes through the point $(0, 5)$

32. Vertices: $(-2, 1), (2, 1)$;

passes through the point $(5, 4)$

33. Vertices: $(0, 4), (0, 0)$;

passes through the point $\left(\sqrt{5}, -1\right)$

34. Vertices: $(1, 2), (1, -2)$;

passes through the point $\left(0, \sqrt{5}\right)$

35. Vertices: $(1, 2), (3, 2)$;

asymptotes: $y = x$, $y = 4 - x$

36. Vertices: $(3, 0), (3, 6)$;

asymptotes: $y = 6 - x$, $y = x$

37. Vertices: $(0, 2), (6, 2)$;

asymptotes: $y = \frac{2}{3}x$, $y = 4 - \frac{2}{3}x$

38. Vertices: $(3, 0), (3, 4)$;

asymptotes: $y = \frac{2}{3}x$, $y = 4 - \frac{2}{3}x$

39. *Art* A sculpture has a hyperbolic cross section (see figure).

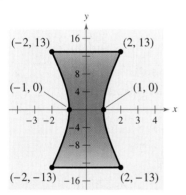

(a) Write an equation that models the curved sides of the sculpture.

(b) Each unit in the coordinate plane represents 1 foot. Find the width of the sculpture at a height of 5 feet.

40. *Sound Location* You and a friend live 4 miles apart (on the same "east-west" street) and are talking on the phone. You hear a clap of thunder from lightning in a storm, and 18 seconds later your friend hears the thunder. Find an equation that gives the possible places where the lightning could have occurred. (Assume that the coordinate system is measured in feet and that sound travels at 1100 feet per second.)

41. *Sound Location* Three listening stations located at $(3300, 0)$, $(3300, 1100)$, and $(-3300, 0)$ monitor an explosion. The last two stations detect the explosion 1 second and 4 seconds after the first, respectively. Determine the coordinates of the explosion. (Assume that the coordinate system is measured in feet and that sound travels at 100 feet per second.)

Model It

42. *LORAN* Long distance radio navigation for aircraft and ships uses synchronized pulses transmitted by widely separated transmitting stations. These pulses travel at the speed of light (186,000 miles per second). The difference in the times of arrival of these pulses at an aircraft or ship is constant on a hyperbola having the transmitting stations as foci. Assume that two stations, 300 miles apart, are positioned on the rectangular coordinate system at points with coordinates $(-150, 0)$ and $(150, 0)$, and that a ship is traveling on a hyperbolic path with coordinates $(x, 75)$ (see figure).

(a) Find the x-coordinate of the position of the ship if the time difference between the pulses from the transmitting stations is 1000 microseconds (0.001 second).

(b) Determine the distance between the ship and station 1 when the ship reaches the shore.

(c) The ship wants to enter a bay located between the two stations. The bay is 30 miles from station 1. What should the time difference be between the pulses?

(d) The ship is 60 miles offshore when the time difference in part (c) is obtained. What is the position of the ship?

43. ***Hyperbolic Mirror*** A hyperbolic mirror (used in some telescopes) has the property that a light ray directed at a focus will be reflected to the other focus. The focus of a hyperbolic mirror (see figure) has coordinates (24, 0). Find the vertex of the mirror if the mount at the top edge of the mirror has coordinates (24, 24).

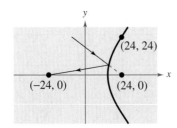

44. ***Running Path*** Let $(0, 0)$ represent a water fountain located in a city park. Each day you run through the park along a path given by

$$x^2 + y^2 - 200x - 52,500 = 0$$

where x and y are measured in meters.

(a) What type of conic is your path? Explain your reasoning.

(b) Write the equation of the path in standard form. Sketch a graph of the equation.

(c) After you run, you walk to the water fountain. If you stop running at $(-100, 150)$, how far must you walk for a drink of water?

In Exercises 45–60, classify the graph of the equation as a circle, a parabola, an ellipse, or a hyperbola.

45. $x^2 + y^2 - 6x + 4y + 9 = 0$

46. $x^2 + 4y^2 - 6x + 16y + 21 = 0$

47. $4x^2 - y^2 - 4x - 3 = 0$

48. $y^2 - 6y - 4x + 21 = 0$

49. $y^2 - 4x^2 + 4x - 2y - 4 = 0$

50. $x^2 + y^2 - 4x + 6y - 3 = 0$

51. $x^2 - 4x - 8y + 2 = 0$

52. $4x^2 + y^2 - 8x + 3 = 0$

53. $4x^2 + 3y^2 + 8x - 24y + 51 = 0$

54. $4y^2 - 2x^2 - 4y - 8x - 15 = 0$

55. $25x^2 - 10x - 200y - 119 = 0$

56. $4y^2 + 4x^2 - 24x + 35 = 0$

57. $4x^2 + 16y^2 - 4x - 32y + 1 = 0$

58. $2y^2 + 2x + 2y + 1 = 0$

59. $100x^2 + 100y^2 - 100x + 400y + 409 = 0$

60. $4x^2 - y^2 + 4x + 2y - 1 = 0$

Synthesis

True or False? **In Exercises 61 and 62, determine whether the statement is true or false. Justify your answer.**

61. In the standard form of the equation of a hyperbola, the larger the ratio of b to a, the larger the eccentricity of the hyperbola.

62. In the standard form of the equation of a hyperbola, the trivial solution of two intersecting lines occurs when $b = 0$.

63. Consider a hyperbola centered at the origin with a horizontal transverse axis. Use the definition of a hyperbola to derive its standard form.

64. ***Writing*** Explain how the central rectangle of a hyperbola can be used to sketch its asymptotes.

65. ***Think About It*** Change the equation of the hyperbola so that its graph is the bottom half of the hyperbola.

$$9x^2 - 54x - 4y^2 + 8y + 41 = 0$$

66. ***Exploration*** A circle and a parabola can have 0, 1, 2, 3, or 4 points of intersection. Sketch the circle given by $x^2 + y^2 = 4$. Discuss how this circle could intersect a parabola with an equation of the form $y = x^2 + C$. Then find the values of C for each of the five cases described below. Use a graphing utility to verify your results.

(a) No points of intersection

(b) One point of intersection

(c) Two points of intersection

(d) Three points of intersection

(e) Four points of intersection

Skills Review

In Exercises 67–72, factor the polynomial completely.

67. $x^3 - 16x$

68. $x^2 + 14x + 49$

69. $2x^3 - 24x^2 + 72x$

70. $6x^3 - 11x^2 - 10x$

71. $16x^3 + 54$

72. $4 - x + 4x^2 - x^3$

In Exercises 73–76, sketch a graph of the function. Include two full periods.

73. $y = 2 \cos x + 1$

74. $y = \sin \pi x$

75. $y = \tan 2x$

76. $y = -\frac{1}{2} \sec x$

6.5   Rotation of Conics

What you should learn

• Rotate the coordinate axes to eliminate the *xy*-term in equations of conics.
• Use the discriminant to classify conics.

Why you should learn it

As illustrated in Exercises 7–18 on page 471, rotation of the coordinate axes can help you identify the graph of a general second-degree equation.

Rotation

In the preceding section, you learned that the equation of a conic with axes parallel to one of the coordinate axes has a standard form that can be written in the general form

$$Ax^2 + Cy^2 + Dx + Ey + F = 0.$$ Horizontal or vertical axis

In this section, you will study the equations of conics whose axes are rotated so that they are not parallel to either the *x*-axis or the *y*-axis. The general equation for such conics contains an *xy*-term.

$$Ax^2 + Bxy + Cy^2 + Dx + Ey + F = 0$$ Equation in *xy*-plane

To eliminate this *xy*-term, you can use a procedure called **rotation of axes.** The objective is to rotate the *x*- and *y*-axes until they are parallel to the axes of the conic. The rotated axes are denoted as the *x′*-axis and the *y′*-axis, as shown in Figure 6.42.

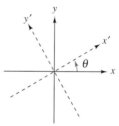

FIGURE **6.42**

After the rotation, the equation of the conic in the new *x′y′*-plane will have the form

$$A'(x')^2 + C'(y')^2 + D'x' + E'y' + F' = 0.$$ Equation in *x′y′*-plane

Because this equation has no *xy*-term, you can obtain a standard form by completing the square. The following theorem identifies how much to rotate the axes to eliminate the *xy*-term and also the equations for determining the new coefficients *A′*, *C′*, *D′*, *E′*, and *F′*.

Rotation of Axes to Eliminate an *xy*-Term

The general second-degree equation $Ax^2 + Bxy + Cy^2 + Dx + Ey + F = 0$ can be rewritten as

$$A'(x')^2 + C'(y')^2 + D'x' + E'y' + F' = 0$$

by rotating the coordinate axes through an angle θ, where

$$\cot 2\theta = \frac{A - C}{B}.$$

The coefficients of the new equation are obtained by making the substitutions $x = x'\cos\theta - y'\sin\theta$ and $y = x'\sin\theta + y'\cos\theta$.

Example 1 **Rotation of Axes for a Hyperbola**

Write the equation $xy - 1 = 0$ in standard form.

Solution

Because $A = 0$, $B = 1$, and $C = 0$, you have

$$\cot 2\theta = \frac{A - C}{B} = 0 \quad \Longrightarrow \quad 2\theta = \frac{\pi}{2} \quad \Longrightarrow \quad \theta = \frac{\pi}{4}$$

which implies that

$$x = x' \cos \frac{\pi}{4} - y' \sin \frac{\pi}{4}$$

$$= x'\left(\frac{1}{\sqrt{2}}\right) - y'\left(\frac{1}{\sqrt{2}}\right)$$

$$= \frac{x' - y'}{\sqrt{2}}$$

and

$$y = x' \sin \frac{\pi}{4} + y' \cos \frac{\pi}{4}$$

$$= x'\left(\frac{1}{\sqrt{2}}\right) + y'\left(\frac{1}{\sqrt{2}}\right)$$

$$= \frac{x' + y'}{\sqrt{2}}.$$

The equation in the $x'y'$-system is obtained by substituting these expressions in the equation $xy - 1 = 0$.

$$\left(\frac{x' - y'}{\sqrt{2}}\right)\left(\frac{x' + y'}{\sqrt{2}}\right) - 1 = 0$$

$$\frac{(x')^2 - (y')^2}{2} - 1 = 0$$

$$\frac{(x')^2}{(\sqrt{2})^2} - \frac{(y')^2}{(\sqrt{2})^2} = 1 \qquad \text{Write in standard form.}$$

In the $x'y'$-system, this is a hyperbola centered at the origin with vertices at $(\pm \sqrt{2}, 0)$, as shown in Figure 6.43. To find the coordinates of the vertices in the xy-system, substitute the coordinates $(\pm \sqrt{2}, 0)$ in the equations

$$x = \frac{x' - y'}{\sqrt{2}} \qquad \text{and} \qquad y = \frac{x' + y'}{\sqrt{2}}.$$

This substitution yields the vertices $(1, 1)$ and $(-1, -1)$ in the xy-system. Note also that the asymptotes of the hyperbola have equations $y' = \pm x'$, which correspond to the original x- and y-axes.

✓CHECKPOINT   Now try Exercise 7.

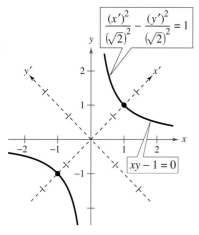

Vertices:

In $x'y'$-system: $(\sqrt{2}, 0), (-\sqrt{2}, 0)$

In xy-system: $(1, 1), (-1, -1)$

FIGURE 6.43

Example 2   **Rotation of Axes for an Ellipse**

Sketch the graph of $7x^2 - 6\sqrt{3}xy + 13y^2 - 16 = 0$.

Solution

Because $A = 7$, $B = -6\sqrt{3}$, and $C = 13$, you have

$$\cot 2\theta = \frac{A - C}{B} = \frac{7 - 13}{-6\sqrt{3}} = \frac{1}{\sqrt{3}}$$

which implies that $\theta = \pi/6$. The equation in the $x'y'$-system is obtained by making the substitutions

$$x = x' \cos \frac{\pi}{6} - y' \sin \frac{\pi}{6}$$

$$= x'\left(\frac{\sqrt{3}}{2}\right) - y'\left(\frac{1}{2}\right)$$

$$= \frac{\sqrt{3}x' - y'}{2}$$

and

$$y = x' \sin \frac{\pi}{6} + y' \cos \frac{\pi}{6}$$

$$= x'\left(\frac{1}{2}\right) + y'\left(\frac{\sqrt{3}}{2}\right)$$

$$= \frac{x' + \sqrt{3}y'}{2}$$

in the original equation. So, you have

$$7x^2 - 6\sqrt{3}xy + 13y^2 - 16 = 0$$

$$7\left(\frac{\sqrt{3}x' - y'}{2}\right)^2 - 6\sqrt{3}\left(\frac{\sqrt{3}x' - y'}{2}\right)\left(\frac{x' + \sqrt{3}y'}{2}\right)$$

$$+ 13\left(\frac{x' + \sqrt{3}y'}{2}\right)^2 - 16 = 0$$

which simplifies to

$$4(x')^2 + 16(y')^2 - 16 = 0$$

$$4(x')^2 + 16(y')^2 = 16$$

$$\frac{(x')^2}{4} + \frac{(y')^2}{1} = 1$$

$$\frac{(x')^2}{2^2} + \frac{(y')^2}{1^2} = 1. \qquad \text{Write in standard form.}$$

This is the equation of an ellipse centered at the origin with vertices $(\pm 2, 0)$ in the $x'y'$-system, as shown in Figure 6.44.

✓CHECKPOINT   Now try Exercise 13.

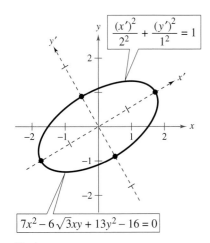

$$\frac{(x')^2}{2^2} + \frac{(y')^2}{1^2} = 1$$

$$7x^2 - 6\sqrt{3}xy + 13y^2 - 16 = 0$$

Vertices:
In $x'y'$-system: $(\pm 2, 0), (0, \pm 1)$
In xy-system: $\left(\sqrt{3}, 1\right), \left(-\sqrt{3}, -1\right),$
$\left(\frac{1}{2}, -\frac{\sqrt{3}}{2}\right), \left(-\frac{1}{2}, \frac{\sqrt{3}}{2}\right)$

FIGURE **6.44**

| Example 3 | **Rotation of Axes for a Parabola** |

Sketch the graph of $x^2 - 4xy + 4y^2 + 5\sqrt{5}y + 1 = 0$.

Solution

Because $A = 1$, $B = -4$, and $C = 4$, you have

$$\cot 2\theta = \frac{A - C}{B} = \frac{1 - 4}{-4} = \frac{3}{4}.$$

Using this information, draw a right triangle as shown in Figure 6.45. From the figure, you can see that $\cos 2\theta = \frac{3}{5}$. To find the values of $\sin \theta$ and $\cos \theta$, you can use the half-angle formulas in the forms

$$\sin \theta = \sqrt{\frac{1 - \cos 2\theta}{2}} \quad \text{and} \quad \cos \theta = \sqrt{\frac{1 + \cos 2\theta}{2}}.$$

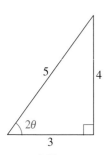

FIGURE **6.45**

So,

$$\sin \theta = \sqrt{\frac{1 - \cos 2\theta}{2}} = \sqrt{\frac{1 - \frac{3}{5}}{2}} = \sqrt{\frac{1}{5}} = \frac{1}{\sqrt{5}}$$

$$\cos \theta = \sqrt{\frac{1 + \cos 2\theta}{2}} = \sqrt{\frac{1 + \frac{3}{5}}{2}} = \sqrt{\frac{4}{5}} = \frac{2}{\sqrt{5}}.$$

Consequently, you use the substitutions

$$x = x' \cos \theta - y' \sin \theta$$

$$= x'\left(\frac{2}{\sqrt{5}}\right) - y'\left(\frac{1}{\sqrt{5}}\right) = \frac{2x' - y'}{\sqrt{5}}$$

$$y = x' \sin \theta + y' \cos \theta$$

$$= x'\left(\frac{1}{\sqrt{5}}\right) + y'\left(\frac{2}{\sqrt{5}}\right) = \frac{x' + 2y'}{\sqrt{5}}.$$

Substituting these expressions in the original equation, you have

$$x^2 - 4xy + 4y^2 + 5\sqrt{5}y + 1 = 0$$

$$\left(\frac{2x' - y'}{\sqrt{5}}\right)^2 - 4\left(\frac{2x' - y'}{\sqrt{5}}\right)\left(\frac{x' + 2y'}{\sqrt{5}}\right) + 4\left(\frac{x' + 2y'}{\sqrt{5}}\right)^2 + 5\sqrt{5}\left(\frac{x' + 2y'}{\sqrt{5}}\right) + 1 = 0$$

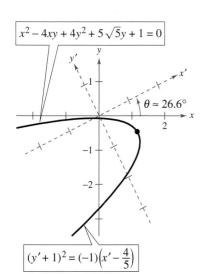

$x^2 - 4xy + 4y^2 + 5\sqrt{5}y + 1 = 0$

$(y' + 1)^2 = (-1)\left(x' - \frac{4}{5}\right)$

which simplifies as follows.

$$5(y')^2 + 5x' + 10y' + 1 = 0$$

$$5[(y')^2 + 2y'] = -5x' - 1 \qquad \text{Group terms.}$$

$$5(y' + 1)^2 = -5x' + 4 \qquad \text{Write in completed square form.}$$

$$(y' + 1)^2 = (-1)\left(x' - \frac{4}{5}\right) \qquad \text{Write in standard form.}$$

The graph of this equation is a parabola with vertex $\left(\frac{4}{5}, -1\right)$. Its axis is parallel to the x'-axis in the $x'y'$-system, and because $\sin \theta = 1/\sqrt{5}$, $\theta \approx 26.6°$, as shown in Figure 6.46.

Vertex:

In $x'y'$-system: $\left(\frac{4}{5}, -1\right)$

In xy-system: $\left(\frac{13}{5\sqrt{5}}, -\frac{6}{5\sqrt{5}}\right)$

FIGURE **6.46**

✓**CHECKPOINT**   Now try Exercise 17.

Invariants Under Rotation

In the rotation of axes theorem listed at the beginning of this section, note that the constant term is the same in both equations, $F' = F$. Such quantities are **invariant under rotation.** The next theorem lists some other rotation invariants.

Rotation Invariants

The rotation of the coordinate axes through an angle θ that transforms the equation $Ax^2 + Bxy + Cy^2 + Dx + Ey + F = 0$ into the form

$$A'(x')^2 + C'(y')^2 + D'x' + E'y' + F' = 0$$

has the following rotation invariants.

1. $F = F'$

2. $A + C = A' + C'$

3. $B^2 - 4AC = (B')^2 - 4A'C'$

STUDY TIP

If there is an xy-term in the equation of a conic, you should realize then that the conic is rotated. Before rotating the axes, you should use the discriminant to classify the conic.

You can use the results of this theorem to classify the graph of a second-degree equation *with* an xy-term in much the same way you do for a second-degree equation *without* an xy-term. Note that because $B' = 0$, the invariant $B^2 - 4AC$ reduces to

$$B^2 - 4AC = -4A'C'. \qquad \text{Discriminant}$$

This quantity is called the **discriminant** of the equation

$$Ax^2 + Bxy + Cy^2 + Dx + Ey + F = 0.$$

Now, from the classification procedure given in Section 6.4, you know that the sign of $A'C'$ determines the type of graph for the equation

$$A'(x')^2 + C'(y')^2 + D'x' + E'y' + F' = 0.$$

Consequently, the sign of $B^2 - 4AC$ will determine the type of graph for the original equation, as given in the following classification.

Activities

1. Find the angle of rotation necessary to eliminate the xy-term in the equation $3x^2 + 2xy - y^2 + x - 1 = 0$.
   Answer: $\theta \approx 13.28°$

2. Rotate the axes to eliminate the xy-term in the equation $x^2 + 2xy + y^2 - \sqrt{2}y = 0$.
   Answer: $2(x')^2 - x' - y' = 0$, $\theta = \dfrac{\pi}{4}$

3. Use the discriminant to classify each equation.
   a. $x^2 - 4xy + y^2 + 2x - 3y + 1 = 0$
   b. $4x^2 - xy + 4y^2 + 2x - 3y + 1 = 0$
   c. $x^2 - xy + 4y^2 + 2x - 3y + 1 = 0$
   d. $x^2 - 4xy + 4y^2 + 2x - 3y + 1 = 0$
   Answer: (a) Hyperbola, (b) Ellipse or circle, (c) Ellipse or circle, (d) Parabola

---

Classification of Conics by the Discriminant

The graph of the equation $Ax^2 + Bxy + Cy^2 + Dx + Ey + F = 0$ is, except in degenerate cases, determined by its discriminant as follows.

1. *Ellipse or circle:* $B^2 - 4AC < 0$

2. *Parabola:* $B^2 - 4AC = 0$

3. *Hyperbola:* $B^2 - 4AC > 0$

For example, in the general equation

$$3x^2 + 7xy + 5y^2 - 6x - 7y + 15 = 0$$

you have $A = 3$, $B = 7$, and $C = 5$. So the discriminant is

$$B^2 - 4AC = 7^2 - 4(3)(5) = 49 - 60 = -11.$$

Because $-11 < 0$, the graph of the equation is an ellipse or a circle.

Example 4 Rotation and Graphing Utilities

For each equation, classify the graph of the equation, use the Quadratic Formula to solve for y, and then use a graphing utility to graph the equation.

a. $2x^2 - 3xy + 2y^2 - 2x = 0$ **b.** $x^2 - 6xy + 9y^2 - 2y + 1 = 0$
c. $3x^2 + 8xy + 4y^2 - 7 = 0$

Solution

a. Because $B^2 - 4AC = 9 - 16 < 0$, the graph is a circle or an ellipse. Solve for y as follows.

$$2x^2 - 3xy + 2y^2 - 2x = 0 \qquad \text{Write original equation.}$$

$$2y^2 - 3xy + (2x^2 - 2x) = 0 \qquad \text{Quadratic form } ay^2 + by + c = 0$$

$$y = \frac{-(-3x) \pm \sqrt{(-3x)^2 - 4(2)(2x^2 - 2x)}}{2(2)}$$

$$y = \frac{3x \pm \sqrt{x(16 - 7x)}}{4}$$

Graph both of the equations to obtain the ellipse shown in Figure 6.47.

$$y_1 = \frac{3x + \sqrt{x(16 - 7x)}}{4} \qquad \text{Top half of ellipse}$$

$$y_2 = \frac{3x - \sqrt{x(16 - 7x)}}{4} \qquad \text{Bottom half of ellipse}$$

b. Because $B^2 - 4AC = 36 - 36 = 0$, the graph is a parabola.

$$x^2 - 6xy + 9y^2 - 2y + 1 = 0 \qquad \text{Write original equation.}$$

$$9y^2 - (6x + 2)y + (x^2 + 1) = 0 \qquad \text{Quadratic form } ay^2 + by + c = 0$$

$$y = \frac{(6x + 2) \pm \sqrt{(6x + 2)^2 - 4(9)(x^2 + 1)}}{2(9)}$$

Graphing both of the equations to obtain the parabola shown in Figure 6.48.

c. Because $B^2 - 4AC = 64 - 48 > 0$, the graph is a hyperbola.

$$3x^2 + 8xy + 4y^2 - 7 = 0 \qquad \text{Write original equation.}$$

$$4y^2 + 8xy + (3x^2 - 7) = 0 \qquad \text{Quadratic form } ay^2 + by + c = 0$$

$$y = \frac{-8x \pm \sqrt{(8x)^2 - 4(4)(3x^2 - 7)}}{2(4)}$$

The graphs of these two equations yield the hyperbola shown in Figure 6.49.

✔**CHECKPOINT**   Now try Exercise 33.

FIGURE **6.47**

FIGURE **6.48**

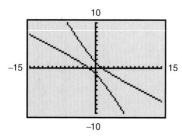

FIGURE **6.49**

---

*W*RITING ABOUT *M*ATHEMATICS

Classifying a Graph as a Hyperbola The graph of $f(x) = 1/x$ is a hyperbola. Use the techniques in this section to verify this, and justify each step. Compare your results with those of another student.

6.5 | Exercises

VOCABULARY CHECK: Fill in the blanks.

1. The procedure used to eliminate the xy-term in a general second-degree equation is called _____ of _____.

2. After rotating the coordinate axes through an angle θ, the general second-degree equation in the new $x'y'$-plane will have the form _____.

3. Quantities that are equal in both the original equation of a conic and the equation of the rotated conic are _____ _____ _____.

4. The quantity $B^2 - 4AC$ is called the _____ of the equation $Ax^2 + Bxy + Cy^2 + Dx + Ey + F = 0$.

PREREQUISITE SKILLS REVIEW: Practice and review algebra skills needed for this section at **www.Eduspace.com.**

In Exercises 1–6, the $x'y'$-coordinate system has been rotated θ degrees from the xy-coordinate system. The coordinates of a point in the xy-coordinate system are given. Find the coordinates of the point in the rotated coordinate system.

1. $\theta = 90°$, $(0, 3)$
2. $\theta = 45°$, $(3, 3)$
3. $\theta = 30°$, $(1, 3)$
4. $\theta = 60°$, $(3, 1)$
5. $\theta = 45°$, $(2, 1)$
6. $\theta = 30°$, $(2, 4)$

In Exercises 7–18, rotate the axes to eliminate the xy-term in the equation. Then write the equation in standard form. Sketch the graph of the resulting equation, showing both sets of axes.

7. $xy + 1 = 0$
8. $xy - 2 = 0$
9. $x^2 - 2xy + y^2 - 1 = 0$
10. $xy + x - 2y + 3 = 0$
11. $xy - 2y - 4x = 0$
12. $2x^2 - 3xy - 2y^2 + 10 = 0$
13. $5x^2 - 6xy + 5y^2 - 12 = 0$
14. $13x^2 + 6\sqrt{3}xy + 7y^2 - 16 = 0$
15. $3x^2 - 2\sqrt{3}xy + y^2 + 2x + 2\sqrt{3}y = 0$
16. $16x^2 - 24xy + 9y^2 - 60x - 80y + 100 = 0$
17. $9x^2 + 24xy + 16y^2 + 90x - 130y = 0$
18. $9x^2 + 24xy + 16y^2 + 80x - 60y = 0$

In Exercises 19–26, use a graphing utility to graph the conic. Determine the angle θ through which the axes are rotated. Explain how you used the graphing utility to obtain the graph.

19. $x^2 + 2xy + y^2 = 20$
20. $x^2 - 4xy + 2y^2 = 6$
21. $17x^2 + 32xy - 7y^2 = 75$

22. $40x^2 + 36xy + 25y^2 = 52$
23. $32x^2 + 48xy + 8y^2 = 50$
24. $24x^2 + 18xy + 12y^2 = 34$
25. $4x^2 - 12xy + 9y^2 + \left(4\sqrt{13} - 12\right)x - \left(6\sqrt{13} + 8\right)y = 91$
26. $6x^2 - 4xy + 8y^2 + \left(5\sqrt{5} - 10\right)x - \left(7\sqrt{5} + 5\right)y = 80$

In Exercises 27–32, match the graph with its equation. [The graphs are labeled (a), (b), (c), (d), (e), and (f).]

(a)

(b)

(c)

(d)

(e)

(f)

27. $xy + 2 = 0$

28. $x^2 + 2xy + y^2 = 0$

29. $-2x^2 + 3xy + 2y^2 + 3 = 0$

30. $x^2 - xy + 3y^2 - 5 = 0$

31. $3x^2 + 2xy + y^2 - 10 = 0$

32. $x^2 - 4xy + 4y^2 + 10x - 30 = 0$

In Exercises 33–40, (a) use the discriminant to classify the graph, (b) use the Quadratic Formula to solve for y, and (c) use a graphing utility to graph the equation.

33. $16x^2 - 8xy + y^2 - 10x + 5y = 0$

34. $x^2 - 4xy - 2y^2 - 6 = 0$

35. $12x^2 - 6xy + 7y^2 - 45 = 0$

36. $2x^2 + 4xy + 5y^2 + 3x - 4y - 20 = 0$

37. $x^2 - 6xy - 5y^2 + 4x - 22 = 0$

38. $36x^2 - 60xy + 25y^2 + 9y = 0$

39. $x^2 + 4xy + 4y^2 - 5x - y - 3 = 0$

40. $x^2 + xy + 4y^2 + x + y - 4 = 0$

In Exercises 41–44, sketch (if possible) the graph of the degenerate conic.

41. $y^2 - 9x^2 = 0$

42. $x^2 + y^2 - 2x + 6y + 10 = 0$

43. $x^2 + 2xy + y^2 - 1 = 0$

44. $x^2 - 10xy + y^2 = 0$

In Exercises 45–58, find any points of intersection of the graphs algebraically and then verify using a graphing utility.

45. $-x^2 + y^2 + 4x - 6y + 4 = 0$
$\quad\quad x^2 + y^2 - 4x - 6y + 12 = 0$

46. $-x^2 - y^2 - 8x + 20y - 7 = 0$
$\quad\quad x^2 + 9y^2 + 8x + 4y + 7 = 0$

47. $-4x^2 - y^2 - 16x + 24y - 16 = 0$
$\quad\quad 4x^2 + y^2 + 40x - 24y + 208 = 0$

48. $\quad\quad x^2 - 4y^2 - 20x - 64y - 172 = 0$
$\quad 16x^2 + 4y^2 - 320x + 64y + 1600 = 0$

49. $x^2 - y^2 - 12x + 16y - 64 = 0$
$\quad\quad x^2 + y^2 - 12x - 16y + 64 = 0$

50. $x^2 + 4y^2 - 2x - 8y + 1 = 0$
$\quad\quad -x^2 + 2x - 4y - 1 = 0$

51. $-16x^2 - y^2 + 24y - 80 = 0$
$\quad\quad 16x^2 + 25y^2 - 400 = 0$

52. $16x^2 - y^2 + 16y - 128 = 0$
$\quad\quad y^2 - 48x - 16y - 32 = 0$

53. $x^2 + y^2 - 4 = 0$
$\quad\quad 3x - y^2 = 0$

54. $4x^2 + 9y^2 - 36y = 0$
$\quad\quad x^2 + 9y - 27 = 0$

55. $x^2 + 2y^2 - 4x + 6y - 5 = 0$
$\quad\quad -x + y - 4 = 0$

56. $x^2 + 2y^2 - 4x + 6y - 5 = 0$
$\quad\quad x^2 - 4x - y + 4 = 0$

57. $xy + x - 2y + 3 = 0$
$\quad\quad x^2 + 4y^2 - 9 = 0$

58. $5x^2 - 2xy + 5y^2 - 12 = 0$
$\quad\quad x + y - 1 = 0$

Synthesis

True or False? **In Exercises 59 and 60, determine whether the statement is true or false. Justify your answer.**

59. The graph of the equation

$$x^2 + xy + ky^2 + 6x + 10 = 0$$

where k is any constant less than $\frac{1}{4}$, is a hyperbola.

60. After a rotation of axes is used to eliminate the xy-term from an equation of the form

$$Ax^2 + Bxy + Cy^2 + Dx + Ey + F = 0$$

the coefficients of the x^2- and y^2-terms remain A and C, respectively.

61. Show that the equation

$$x^2 + y^2 = r^2$$

is invariant under rotation of axes.

62. Find the lengths of the major and minor axes of the ellipse graphed in Exercise 14.

Skills Review

In Exercises 63–70, graph the function.

63. $f(x) = |x + 3|$

64. $f(x) = |x - 4| + 1$

65. $g(x) = \sqrt{4 - x^2}$

66. $g(x) = \sqrt{3x - 2}$

67. $h(t) = -(t - 2)^3 + 3$

68. $h(t) = \frac{1}{2}(t + 4)^3$

69. $f(t) = [\![t - 5]\!] + 1$

70. $f(t) = -2[\![t]\!] + 3$

In Exercises 71–74, find the area of the triangle.

71. $C = 110°, a = 8, b = 12$

72. $B = 70°, a = 25, c = 16$

73. $a = 11, b = 18, c = 10$

74. $a = 23, b = 35, c = 27$

6.6   Parametric Equations

What you should learn

- Evaluate sets of parametric equations for given values of the parameter.
- Sketch curves that are represented by sets of parametric equations.
- Rewrite sets of parametric equations as single rectangular equations by eliminating the parameter.
- Find sets of parametric equations for graphs.

Why you should learn it

Parametric equations are useful for modeling the path of an object. For instance, in Exercise 59 on page 479, you will use a set of parametric equations to model the path of a baseball.

Jed Jacobsohn/Getty Images

Plane Curves

Up to this point you have been representing a graph by a single equation involving the *two* variables x and y. In this section, you will study situations in which it is useful to introduce a *third* variable to represent a curve in the plane.

To see the usefulness of this procedure, consider the path followed by an object that is propelled into the air at an angle of 45°. If the initial velocity of the object is 48 feet per second, it can be shown that the object follows the parabolic path

$$y = -\frac{x^2}{72} + x \qquad \text{Rectangular equation}$$

as shown in Figure 6.50. However, this equation does not tell the whole story. Although it does tell you *where* the object has been, it doesn't tell you *when* the object was at a given point (x, y) on the path. To determine this time, you can introduce a third variable t, called a **parameter.** It is possible to write both x and y as functions of t to obtain the **parametric equations**

$$x = 24\sqrt{2}\,t \qquad \text{Parametric equation for } x$$

$$y = -16t^2 + 24\sqrt{2}\,t. \qquad \text{Parametric equation for } y$$

From this set of equations you can determine that at time $t = 0$, the object is at the point $(0, 0)$. Similarly, at time $t = 1$, the object is at the point $\left(24\sqrt{2},\ 24\sqrt{2} - 16\right)$, and so on, as shown in Figure 6.50.

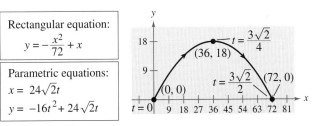

Curvilinear Motion: Two Variables for Position, One Variable for Time
FIGURE 6.50

For this particular motion problem, x and y are continuous functions of t, and the resulting path is a **plane curve.** (Recall that a *continuous function* is one whose graph can be traced without lifting the pencil from the paper.)

Definition of Plane Curve

If f and g are continuous functions of t on an interval I, the set of ordered pairs $(f(t), g(t))$ is a **plane curve** C. The equations

$$x = f(t) \qquad \text{and} \qquad y = g(t)$$

are **parametric equations** for C, and t is the **parameter.**

Sketching a Plane Curve

When sketching a curve represented by a pair of parametric equations, you still plot points in the xy-plane. Each set of coordinates (x, y) is determined from a value chosen for the parameter t. Plotting the resulting points in the order of *increasing* values of t traces the curve in a specific direction. This is called the **orientation** of the curve.

Example 1   Sketching a Curve

Sketch the curve given by the parametric equations

$$x = t^2 - 4 \quad \text{and} \quad y = \frac{t}{2}, \quad -2 \le t \le 3.$$

Solution

Using values of t in the interval, the parametric equations yield the points (x, y) shown in the table.

t	x	y
-2	0	-1
-1	-3	$-1/2$
0	-4	0
1	-3	$1/2$
2	0	1
3	5	$3/2$

FIGURE 6.51

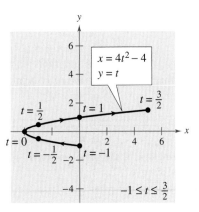

FIGURE 6.52

By plotting these points in the order of increasing t, you obtain the curve C shown in Figure 6.51. Note that the arrows on the curve indicate its orientation as t increases from -2 to 3. So, if a particle were moving on this curve, it would start at $(0, -1)$ and then move along the curve to the point $\left(5, \frac{3}{2}\right)$.

✔CHECKPOINT   Now try Exercises 1(a) and (b).

Note that the graph shown in Figure 6.51 does not define y as a function of x. This points out one benefit of parametric equations—they can be used to represent graphs that are more general than graphs of functions.

It often happens that two different sets of parametric equations have the same graph. For example, the set of parametric equations

$$x = 4t^2 - 4 \quad \text{and} \quad y = t, \quad -1 \le t \le \frac{3}{2}$$

has the same graph as the set given in Example 1. However, by comparing the values of t in Figures 6.51 and 6.52, you see that this second graph is traced out more *rapidly* (considering t as time) than the first graph. So, in applications, different parametric representations can be used to represent various *speeds* at which objects travel along a given path.

Eliminating the Parameter

Example 1 uses simple point plotting to sketch the curve. This tedious process can sometimes be simplified by finding a rectangular equation (in x and y) that has the same graph. This process is called **eliminating the parameter.**

Parametric equations	⟹	Solve for t in one equation.	⟹	Substitute in other equation.	⟹	Rectangular equation
$x = t^2 - 4$ $y = t/2$		$t = 2y$		$x = (2y)^2 - 4$		$x = 4y^2 - 4$

Now you can recognize that the equation $x = 4y^2 - 4$ represents a parabola with a horizontal axis and vertex $(-4, 0)$.

When converting equations from parametric to rectangular form, you may need to alter the domain of the rectangular equation so that its graph matches the graph of the parametric equations. Such a situation is demonstrated in Example 2.

Emphasize that converting equations from parametric to rectangular form is primarily an aid in graphing.

Exploration

Most graphing utilities have a *parametric* mode. If yours does, enter the parametric equations from Example 2. Over what values should you let t vary to obtain the graph shown in Figure 6.53?

Example 2   Eliminating the Parameter

Sketch the curve represented by the equations

$$x = \frac{1}{\sqrt{t + 1}} \quad \text{and} \quad y = \frac{t}{t + 1}$$

by eliminating the parameter and adjusting the domain of the resulting rectangular equation.

Solution

Solving for t in the equation for x produces

$$x = \frac{1}{\sqrt{t + 1}} \quad \Longrightarrow \quad x^2 = \frac{1}{t + 1}$$

which implies that

$$t = \frac{1 - x^2}{x^2}.$$

Now, substituting in the equation for y, you obtain the rectangular equation

$$y = \frac{t}{t + 1} = \frac{\dfrac{(1 - x^2)}{x^2}}{\left[\dfrac{(1 - x^2)}{x^2}\right] + 1} = \frac{\dfrac{1 - x^2}{x^2}}{\dfrac{1 - x^2}{x^2} + 1} \cdot \frac{x^2}{x^2} = 1 - x^2.$$

From this rectangular equation, you can recognize that the curve is a parabola that opens downward and has its vertex at $(0, 1)$. Also, this rectangular equation is defined for all values of x, but from the parametric equation for x you can see that the curve is defined only when $t > -1$. This implies that you should restrict the domain of x to positive values, as shown in Figure 6.53.

✓CHECKPOINT   Now try Exercise 1(c).

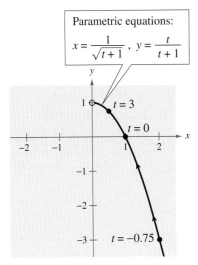

Parametric equations:

$x = \dfrac{1}{\sqrt{t + 1}}, \ y = \dfrac{t}{t + 1}$

FIGURE 6.53

STUDY TIP

To eliminate the parameter in equations involving trigonometric functions, try using the identities

$$\sin^2 \theta + \cos^2 \theta = 1$$

$$\sec^2 \theta - \tan^2 \theta = 1$$

as shown in Example 3.

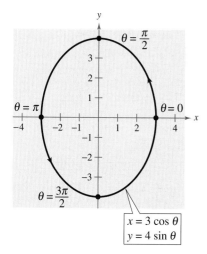

FIGURE 6.54

It is not necessary for the parameter in a set of parametric equations to represent time. The next example uses an *angle* as the parameter.

Example 3   Eliminating an Angle Parameter

Sketch the curve represented by

$$x = 3 \cos \theta \qquad \text{and} \qquad y = 4 \sin \theta, \quad 0 \le \theta \le 2\pi$$

by eliminating the parameter.

Solution

Begin by solving for $\cos \theta$ and $\sin \theta$ in the equations.

$$\cos \theta = \frac{x}{3} \qquad \text{and} \qquad \sin \theta = \frac{y}{4} \qquad \text{Solve for } \cos \theta \text{ and } \sin \theta.$$

Use the identity $\sin^2 \theta + \cos^2 \theta = 1$ to form an equation involving only x and y.

$$\cos^2 \theta + \sin^2 \theta = 1 \qquad \qquad \text{Pythagorean identity}$$

$$\left(\frac{x}{3}\right)^2 + \left(\frac{y}{4}\right)^2 = 1 \qquad \qquad \text{Substitute } \frac{x}{3} \text{ for } \cos \theta \text{ and } \frac{y}{4} \text{ for } \sin \theta.$$

$$\frac{x^2}{9} + \frac{y^2}{16} = 1 \qquad \qquad \text{Rectangular equation}$$

From this rectangular equation, you can see that the graph is an ellipse centered at $(0, 0)$, with vertices $(0, 4)$ and $(0, -4)$ and minor axis of length $2b = 6$, as shown in Figure 6.54. Note that the elliptic curve is traced out *counterclockwise* as θ varies from 0 to 2π.

✓*CHECKPOINT*   Now try Exercise 13.

In Examples 2 and 3, it is important to realize that eliminating the parameter is primarily an *aid to curve sketching*. If the parametric equations represent the path of a moving object, the graph alone is not sufficient to describe the object's motion. You still need the parametric equations to tell you the *position*, *direction*, and *speed* at a given time.

Finding Parametric Equations for a Graph

You have been studying techniques for sketching the graph represented by a set of parametric equations. Now consider the *reverse* problem—that is, how can you find a set of parametric equations for a given graph or a given physical description? From the discussion following Example 1, you know that such a representation is not unique. That is, the equations

$$x = 4t^2 - 4 \qquad \text{and} \qquad y = t, \; -1 \le t \le \frac{3}{2}$$

produced the same graph as the equations

$$x = t^2 - 4 \qquad \text{and} \qquad y = \frac{t}{2}, \; -2 \le t \le 3.$$

This is further demonstrated in Example 4.

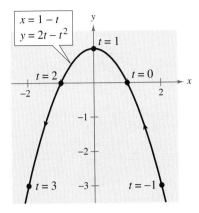

$x = 1 - t$
$y = 2t - t^2$

FIGURE 6.55

Example 4   Finding Parametric Equations for a Graph

Find a set of parametric equations to represent the graph of $y = 1 - x^2$, using the following parameters.

a. $t = x$ **b.** $t = 1 - x$

Solution

a. Letting $t = x$, you obtain the parametric equations

$$x = t \quad \text{and} \quad y = 1 - x^2 = 1 - t^2.$$

b. Letting $t = 1 - x$, you obtain the parametric equations

$$x = 1 - t \quad \text{and} \quad y = 1 - x^2 = 1 - (1 - t)^2 = 2t - t^2.$$

In Figure 6.55, note how the resulting curve is oriented by the increasing values of t. For part (a), the curve would have the opposite orientation.

✓CHECKPOINT   Now try Exercise 37.

Point out that a single rectangular equation can have many different parametric representations. To reinforce this, demonstrate along with parts (a) and (b) of Example 4 the parametric equation representations of the graph of $y = 1 - x^2$ using the parameters $t = 2x$ and $t = 2 - 3x$. A graphing utility can be a helpful tool in demonstrating that each of these representations yields the same graph.

Example 5   Parametric Equations for a Cycloid

Describe the **cycloid** traced out by a point P on the circumference of a circle of radius a as the circle rolls along a straight line in a plane.

Solution

As the parameter, let θ be the measure of the circle's rotation, and let the point $P = (x, y)$ begin at the origin. When $\theta = 0$, P is at the origin; when $\theta = \pi$, P is at a maximum point $(\pi a, 2a)$; and when $\theta = 2\pi$, P is back on the x-axis at $(2\pi a, 0)$. From Figure 6.56, you can see that $\angle APC = 180° - \theta$. So, you have

$$\sin \theta = \sin(180° - \theta) = \sin(\angle APC) = \frac{AC}{a} = \frac{BD}{a}$$

$$\cos \theta = -\cos(180° - \theta) = -\cos(\angle APC) = \frac{AP}{-a}$$

which implies that $AP = -a \cos \theta$ and $BD = a \sin \theta$. Because the circle rolls along the x-axis, you know that $OD = \overset{\frown}{PD} = a\theta$. Furthermore, because $BA = DC = a$, you have

$$x = OD - BD = a\theta - a \sin \theta \quad \text{and} \quad y = BA + AP = a - a \cos \theta.$$

So, the parametric equations are $x = a(\theta - \sin \theta)$ and $y = a(1 - \cos \theta)$.

STUDY TIP

STUDY TIP

In Example 5, $\overset{\frown}{PD}$ represents the arc of the circle between points P and D.

Technology

Use a graphing utility in *parametric* mode to obtain a graph similar to Figure 6.56 by graphing the following equations.

$$X_{1T} = T - \sin T$$

$$Y_{1T} = 1 - \cos T$$

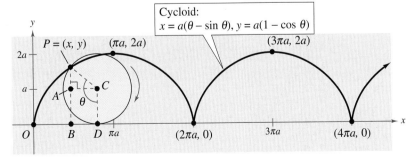

Cycloid:
$x = a(\theta - \sin \theta), y = a(1 - \cos \theta)$

FIGURE 6.56

✓CHECKPOINT   Now try Exercise 63.

6.6 | Exercises

VOCABULARY CHECK: Fill in the blanks.

1. If f and g are continuous functions of t on an interval I, the set of ordered pairs $(f(t), g(t))$ is a _____ _____ C. The equations $x = f(t)$ and $y = g(t)$ are _____ equations for C, and t is the _____.

2. The _____ of a curve is the direction in which the curve is traced out for increasing values of the parameter.

3. The process of converting a set of parametric equations to a corresponding rectangular equation is called _____ the _____.

PREREQUISITE SKILLS REVIEW: Practice and review algebra skills needed for this section at **www.Eduspace.com.**

1. Consider the parametric equations $x = \sqrt{t}$ and $y = 3 - t$.

(a) Create a table of x- and y-values using $t = 0, 1, 2, 3$, and 4.

(b) Plot the points (x, y) generated in part (a), and sketch a graph of the parametric equations.

(c) Find the rectangular equation by eliminating the parameter. Sketch its graph. How do the graphs differ?

2. Consider the parametric equations $x = 4 \cos^2 \theta$ and $y = 2 \sin \theta$.

(a) Create a table of x- and y-values using $\theta = -\pi/2$, $-\pi/4, 0, \pi/4$, and $\pi/2$.

(b) Plot the points (x, y) generated in part (a), and sketch a graph of the parametric equations.

(c) Find the rectangular equation by eliminating the parameter. Sketch its graph. How do the graphs differ?

In Exercises 3–22, (a) sketch the curve represented by the parametric equations (indicate the orientation of the curve) and (b) eliminate the parameter and write the corresponding rectangular equation whose graph represents the curve. Adjust the domain of the resulting rectangular equation if necessary.

3. $x = 3t - 3$
$y = 2t + 1$

4. $x = 3 - 2t$
$y = 2 + 3t$

5. $x = \frac{1}{4}t$
$y = t^2$

6. $x = t$
$y = t^3$

7. $x = t + 2$
$y = t^2$

8. $x = \sqrt{t}$
$y = 1 - t$

9. $x = t + 1$
$y = \dfrac{t}{t + 1}$

10. $x = t - 1$
$y = \dfrac{t}{t - 1}$

11. $x = 2(t + 1)$
$y = |t - 2|$

12. $x = |t - 1|$
$y = t + 2$

13. $x = 3 \cos \theta$
$y = 3 \sin \theta$

14. $x = 2 \cos \theta$
$y = 3 \sin \theta$

15. $x = 4 \sin 2\theta$
$y = 2 \cos 2\theta$

16. $x = \cos \theta$
$y = 2 \sin 2\theta$

17. $x = 4 + 2 \cos \theta$
$y = -1 + \sin \theta$

18. $x = 4 + 2 \cos \theta$
$y = 2 + 3 \sin \theta$

19. $x = e^{-t}$
$y = e^{3t}$

20. $x = e^{2t}$
$y = e^t$

21. $x = t^3$
$y = 3 \ln t$

22. $x = \ln 2t$
$y = 2t^2$

In Exercises 23 and 24, determine how the plane curves differ from each other.

23. (a) $x = t$
$y = 2t + 1$

(b) $x = \cos \theta$
$y = 2 \cos \theta + 1$

(c) $x = e^{-t}$
$y = 2e^{-t} + 1$

(d) $x = e^t$
$y = 2e^t + 1$

24. (a) $x = t$
$y = t^2 - 1$

(b) $x = t^2$
$y = t^4 - 1$

(c) $x = \sin t$
$y = \sin^2 t - 1$

(d) $x = e^t$
$y = e^{2t} - 1$

In Exercises 25–28, eliminate the parameter and obtain the standard form of the rectangular equation.

25. Line through (x_1, y_1) and (x_2, y_2):
$x = x_1 + t(x_2 - x_1), \ y = y_1 + t(y_2 - y_1)$

26. Circle: $x = h + r \cos \theta, \ y = k + r \sin \theta$

27. Ellipse: $x = h + a \cos \theta, \ y = k + b \sin \theta$

28. Hyperbola: $x = h + a \sec \theta, \ y = k + b \tan \theta$

In Exercises 29–36, use the results of Exercises 25–28 to find a set of parametric equations for the line or conic.

29. Line: passes through $(0, 0)$ and $(6, -3)$

30. Line: passes through $(2, 3)$ and $(6, -3)$

31. Circle: center: $(3, 2)$; radius: 4

32. Circle: center: $(-3, 2)$; radius: 5

33. Ellipse: vertices: $(\pm 4, 0)$; foci: $(\pm 3, 0)$

34. Ellipse: vertices: $(4, 7)$, $(4, -3)$;

  foci: $(4, 5)$, $(4, -1)$

35. Hyperbola: vertices: $(\pm 4, 0)$; foci: $(\pm 5, 0)$

36. Hyperbola: vertices: $(\pm 2, 0)$; foci: $(\pm 4, 0)$

In Exercises 37–44, find a set of parametric equations for the rectangular equation using (a) $t = x$ and (b) $t = 2 - x$.

37. $y = 3x - 2$ **38.** $x = 3y - 2$

39. $y = x^2$ **40.** $y = x^3$

41. $y = x^2 + 1$ **42.** $y = 2 - x$

43. $y = \dfrac{1}{x}$ **44.** $y = \dfrac{1}{2x}$

 In Exercises 45–52, use a graphing utility to graph the curve represented by the parametric equations.

45. Cycloid: $x = 4(\theta - \sin \theta)$, $y = 4(1 - \cos \theta)$

46. Cycloid: $x = \theta + \sin \theta$, $y = 1 - \cos \theta$

47. Prolate cycloid: $x = \theta - \frac{3}{2} \sin \theta$, $y = 1 - \frac{3}{2} \cos \theta$

48. Prolate cycloid: $x = 2\theta - 4 \sin \theta$, $y = 2 - 4 \cos \theta$

49. Hypocycloid: $x = 3 \cos^3 \theta$, $y = 3 \sin^3 \theta$

50. Curtate cycloid: $x = 8\theta - 4 \sin \theta$, $y = 8 - 4 \cos \theta$

51. Witch of Agnesi: $x = 2 \cot \theta$, $y = 2 \sin^2 \theta$

52. Folium of Descartes: $x = \dfrac{3t}{1 + t^3}$, $y = \dfrac{3t^2}{1 + t^3}$

In Exercises 53–56, match the parametric equations with the correct graph and describe the domain and range. [The graphs are labeled (a), (b), (c), and (d).]

(a)

(b)

(c)

(d)

53. Lissajous curve: $x = 2 \cos \theta$, $y = \sin 2\theta$

54. Evolute of ellipse: $x = 4 \cos^3 \theta$, $y = 6 \sin^3 \theta$

55. Involute of circle: $x = \frac{1}{2}(\cos \theta + \theta \sin \theta)$

        $y = \frac{1}{2}(\sin \theta - \theta \cos \theta)$

56. Serpentine curve: $x = \frac{1}{2} \cot \theta$, $y = 4 \sin \theta \cos \theta$

Projectile Motion   **A projectile is launched at a height of h feet above the ground at an angle of θ with the horizontal. The initial velocity is v_0 feet per second and the path of the projectile is modeled by the parametric equations**

$$x = (v_0 \cos \theta)t \quad \text{and} \quad y = h + (v_0 \sin \theta)t - 16t^2.$$

In Exercises 57 and 58, use a graphing utility to graph the paths of a projectile launched from ground level at each value of θ and v_0. For each case, use the graph to approximate the maximum height and the range of the projectile.

57. (a) $\theta = 60°$, $v_0 = 88$ feet per second

  (b) $\theta = 60°$, $v_0 = 132$ feet per second

  (c) $\theta = 45°$, $v_0 = 88$ feet per second

  (d) $\theta = 45°$, $v_0 = 132$ feet per second

58. (a) $\theta = 15°$, $v_0 = 60$ feet per second

  (b) $\theta = 15°$, $v_0 = 100$ feet per second

  (c) $\theta = 30°$, $v_0 = 60$ feet per second

  (d) $\theta = 30°$, $v_0 = 100$ feet per second

Model It

59. *Sports*   The center field fence in Yankee Stadium is 7 feet high and 408 feet from home plate. A baseball is hit at a point 3 feet above the ground. It leaves the bat at an angle of θ degrees with the horizontal at a speed of 100 miles per hour (see figure).

Not drawn to scale

 (a) Write a set of parametric equations that model the path of the baseball.

 (b) Use a graphing utility to graph the path of the baseball when $\theta = 15°$. Is the hit a home run?

 (c) Use a graphing utility to graph the path of the baseball when $\theta = 23°$. Is the hit a home run?

 (d) Find the minimum angle required for the hit to be a home run.

60. *Sports* An archer releases an arrow from a bow at a point 5 feet above the ground. The arrow leaves the bow at an angle of $10°$ with the horizontal and at an initial speed of 240 feet per second.

(a) Write a set of parametric equations that model the path of the arrow.

(b) Assuming the ground is level, find the distance the arrow travels before it hits the ground. (Ignore air resistance.)

 (c) Use a graphing utility to graph the path of the arrow and approximate its maximum height.

(d) Find the total time the arrow is in the air.

61. *Projectile Motion* Eliminate the parameter t from the parametric equations

$$x = (v_0 \cos \theta)t \quad \text{and} \quad y = h + (v_0 \sin \theta)t - 16t^2$$

for the motion of a projectile to show that the rectangular equation is

$$y = -\frac{16 \sec^2 \theta}{v_0^2}x^2 + (\tan \theta)x + h.$$

62. *Path of a Projectile* The path of a projectile is given by the rectangular equation

$$y = 7 + x - 0.02x^2.$$

(a) Use the result of Exercise 61 to find h, v_0, and θ. Find the parametric equations of the path.

 (b) Use a graphing utility to graph the rectangular equation for the path of the projectile. Confirm your answer in part (a) by sketching the curve represented by the parametric equations.

 (c) Use a graphing utility to approximate the maximum height of the projectile and its range.

63. *Curtate Cycloid* A wheel of radius a units rolls along a straight line without slipping. The curve traced by a point P that is b units from the center $(b < a)$ is called a **curtate cycloid** (see figure). Use the angle θ shown in the figure to find a set of parametric equations for the curve.

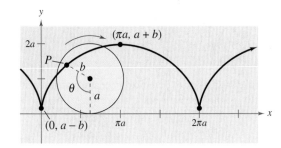

64. *Epicycloid* A circle of radius one unit rolls around the outside of a circle of radius two units without slipping. The curve traced by a point on the circumference of the smaller circle is called an **epicycloid** (see figure). Use the angle θ shown in the figure to find a set of parametric equations for the curve.

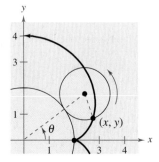

Synthesis

True or False? **In Exercises 65 and 66, determine whether the statement is true or false. Justify your answer.**

65. The two sets of parametric equations $x = t$, $y = t^2 + 1$ and $x = 3t$, $y = 9t^2 + 1$ have the same rectangular equation.

66. The graph of the parametric equations $x = t^2$ and $y = t^2$ is the line $y = x$.

67. *Writing* Write a short paragraph explaining why parametric equations are useful.

68. *Writing* Explain the process of sketching a plane curve given by parametric equations. What is meant by the orientation of the curve?

Skills Review

In Exercises 69–72, find the reference angle θ', and sketch θ and θ' in standard position.

69. $\theta = 105°$

70. $\theta = 230°$

71. $\theta = -\dfrac{2\pi}{3}$

72. $\theta = \dfrac{5\pi}{6}$

6.7 Polar Coordinates

What you should learn

- Plot points on the polar coordinate system.
- Convert points from rectangular to polar form and vice versa.
- Convert equations from rectangular to polar form and vice versa.

Why you should learn it

Polar coordinates offer a different mathematical perspective on graphing. For instance, in Exercises 1–8 on page 485, you are asked to find multiple representations of polar coordinates.

Introduction

So far, you have been representing graphs of equations as collections of points (x, y) on the rectangular coordinate system, where x and y represent the directed distances from the coordinate axes to the point (x, y). In this section, you will study a different system called the **polar coordinate system.**

To form the polar coordinate system in the plane, fix a point O, called the **pole** (or **origin**), and construct from O an initial ray called the **polar axis,** as shown in Figure 6.57. Then each point P in the plane can be assigned **polar coordinates** (r, θ) as follows.

1. $r = $ *directed distance* from O to P
2. $\theta = $ *directed angle*, counterclockwise from polar axis to segment $\overline{OP}$

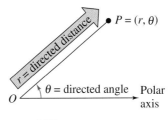

FIGURE 6.57

Example 1 **Plotting Points on the Polar Coordinate System**

a. The point $(r, \theta) = (2, \pi/3)$ lies two units from the pole on the terminal side of the angle $\theta = \pi/3$, as shown in Figure 6.58.

b. The point $(r, \theta) = (3, -\pi/6)$ lies three units from the pole on the terminal side of the angle $\theta = -\pi/6$, as shown in Figure 6.59.

c. The point $(r, \theta) = (3, 11\pi/6)$ coincides with the point $(3, -\pi/6)$, as shown in Figure 6.60.

FIGURE 6.58

FIGURE 6.59

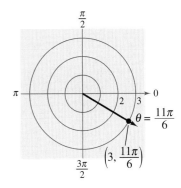

FIGURE 6.60

CHECKPOINT   Now try Exercise 1.

In rectangular coordinates, each point (x, y) has a unique representation. This is not true for polar coordinates. For instance, the coordinates (r, θ) and $(r, \theta + 2\pi)$ represent the same point, as illustrated in Example 1. Another way to obtain multiple representations of a point is to use negative values for r. Because r is a *directed distance*, the coordinates (r, θ) and $(-r, \theta + \pi)$ represent the same point. In general, the point (r, θ) can be represented as

$$(r, \theta) = (r, \theta \pm 2n\pi) \quad \text{or} \quad (r, \theta) = (-r, \theta \pm (2n + 1)\pi)$$

where n is any integer. Moreover, the pole is represented by $(0, \theta)$, where θ is any angle.

Example 2 **Multiple Representations of Points**

Plot the point $(3, -3\pi/4)$ and find three additional polar representations of this point, using $-2\pi < \theta < 2\pi$.

Solution

The point is shown in Figure 6.61. Three other representations are as follows.

$$\left(3, -\frac{3\pi}{4} + 2\pi\right) = \left(3, \frac{5\pi}{4}\right) \qquad \text{Add } 2\pi \text{ to } \theta.$$

$$\left(-3, -\frac{3\pi}{4} - \pi\right) = \left(-3, -\frac{7\pi}{4}\right) \qquad \text{Replace } r \text{ by } -r; \text{ subtract } \pi \text{ from } \theta.$$

$$\left(-3, -\frac{3\pi}{4} + \pi\right) = \left(-3, \frac{\pi}{4}\right) \qquad \text{Replace } r \text{ by } -r; \text{ add } \pi \text{ to } \theta.$$

✓CHECKPOINT   Now try Exercise 3.

Coordinate Conversion

To establish the relationship between polar and rectangular coordinates, let the polar axis coincide with the positive x-axis and the pole with the origin, as shown in Figure 6.62. Because (x, y) lies on a circle of radius r, it follows that $r^2 = x^2 + y^2$. Moreover, for $r > 0$, the definitions of the trigonometric functions imply that

$$\tan \theta = \frac{y}{x}, \qquad \cos \theta = \frac{x}{r}, \qquad \text{and} \qquad \sin \theta = \frac{y}{r}.$$

If $r < 0$, you can show that the same relationships hold.

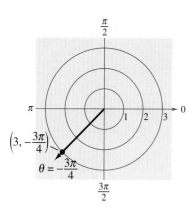

$$\left(3, -\tfrac{3\pi}{4}\right) = \left(3, \tfrac{5\pi}{4}\right) = \left(-3, -\tfrac{7\pi}{4}\right) = \left(-3, \tfrac{\pi}{4}\right) = \cdots$$

FIGURE 6.61

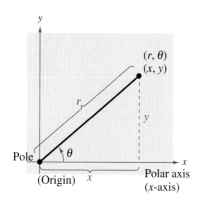

FIGURE 6.62

Coordinate Conversion

The polar coordinates (r, θ) are related to the rectangular coordinates (x, y) as follows.

Polar-to-Rectangular

$$x = r \cos \theta$$
$$y = r \sin \theta$$

Rectangular-to-Polar

$$\tan \theta = \frac{y}{x}$$
$$r^2 = x^2 + y^2$$

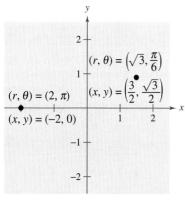

FIGURE 6.63

Activities

1. Find three additional polar representations of the point $\left(-2, \frac{\pi}{3}\right)$.

   Answer: $\left(2, \frac{4\pi}{3}\right), \left(2, -\frac{2\pi}{3}\right)$, and $\left(-2, -\frac{5\pi}{3}\right)$

2. Convert the point $(-4, 2)$ from rectangular to polar form.

   Answer: $\left(-2\sqrt{5}, -0.4636\right)$ or $\left(2\sqrt{5}, 2.6779\right)$

FIGURE 6.64

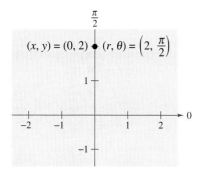

FIGURE 6.65

| **Example 3** | **Polar-to-Rectangular Conversion** |

Convert each point to rectangular coordinates.

a. $(2, \pi)$ **b.** $\left(\sqrt{3}, \frac{\pi}{6}\right)$

Solution

a. For the point $(r, \theta) = (2, \pi)$, you have the following.

$$x = r \cos \theta = 2 \cos \pi = -2$$
$$y = r \sin \theta = 2 \sin \pi = 0$$

The rectangular coordinates are $(x, y) = (-2, 0)$. (See Figure 6.63.)

b. For the point $(r, \theta) = \left(\sqrt{3}, \frac{\pi}{6}\right)$, you have the following.

$$x = \sqrt{3} \cos \frac{\pi}{6} = \sqrt{3}\left(\frac{\sqrt{3}}{2}\right) = \frac{3}{2}$$
$$y = \sqrt{3} \sin \frac{\pi}{6} = \sqrt{3}\left(\frac{1}{2}\right) = \frac{\sqrt{3}}{2}$$

The rectangular coordinates are $(x, y) = \left(\frac{3}{2}, \frac{\sqrt{3}}{2}\right)$. (See Figure 6.63.)

✓**CHECKPOINT**   Now try Exercise 13.

| **Example 4** | **Rectangular-to-Polar Conversion** |

Convert each point to polar coordinates.

a. $(-1, 1)$ **b.** $(0, 2)$

Solution

a. For the second-quadrant point $(x, y) = (-1, 1)$, you have

$$\tan \theta = \frac{y}{x} = -1$$
$$\theta = \frac{3\pi}{4}.$$

Because θ lies in the same quadrant as (x, y), use positive r.

$$r = \sqrt{x^2 + y^2} = \sqrt{(-1)^2 + (1)^2} = \sqrt{2}$$

So, *one* set of polar coordinates is $(r, \theta) = \left(\sqrt{2}, 3\pi/4\right)$, as shown in Figure 6.64.

b. Because the point $(x, y) = (0, 2)$ lies on the positive y-axis, choose

$$\theta = \frac{\pi}{2} \quad \text{and} \quad r = 2.$$

This implies that *one* set of polar coordinates is $(r, \theta) = (2, \pi/2)$, as shown in Figure 6.65.

✓**CHECKPOINT**   Now try Exercise 19.

Multiple representations of points and equations in the polar system can cause confusion. You may want to discuss several examples.

Activities

1. Convert the polar equation $r = 3 \cos \theta$ to rectangular form.

   Answer: $x^2 + y^2 - 3x = 0$

2. Convert the rectangular equation $x = 4$ to polar form.

   Answer: $r = 4 \sec \theta$

FIGURE **6.66**

FIGURE **6.67**

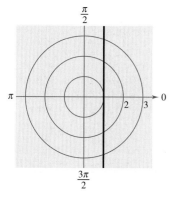

FIGURE **6.68**

Equation Conversion

By comparing Examples 3 and 4, you can see that point conversion from the polar to the rectangular system is straightforward, whereas point conversion from the rectangular to the polar system is more involved. For equations, the opposite is true. To convert a rectangular equation to polar form, you simply replace x by $r \cos \theta$ and y by $r \sin \theta$. For instance, the rectangular equation $y = x^2$ can be written in polar form as follows.

$$y = x^2 \qquad \text{Rectangular equation}$$
$$r \sin \theta = (r \cos \theta)^2 \qquad \text{Polar equation}$$
$$r = \sec \theta \tan \theta \qquad \text{Simplest form}$$

On the other hand, converting a polar equation to rectangular form requires considerable ingenuity.

Example 5 demonstrates several polar-to-rectangular conversions that enable you to sketch the graphs of some polar equations.

Example 5   **Converting Polar Equations to Rectangular Form**

Describe the graph of each polar equation and find the corresponding rectangular equation.

a. $r = 2$ **b.** $\theta = \dfrac{\pi}{3}$ **c.** $r = \sec \theta$

Solution

a. The graph of the polar equation $r = 2$ consists of all points that are two units from the pole. In other words, this graph is a circle centered at the origin with a radius of 2, as shown in Figure 6.66. You can confirm this by converting to rectangular form, using the relationship $r^2 = x^2 + y^2$.

$$\underbrace{r = 2}_{\text{Polar equation}} \quad \Longrightarrow \quad r^2 = 2^2 \quad \Longrightarrow \quad \underbrace{x^2 + y^2 = 2^2}_{\text{Rectangular equation}}$$

b. The graph of the polar equation $\theta = \pi/3$ consists of all points on the line that makes an angle of $\pi/3$ with the positive polar axis, as shown in Figure 6.67. To convert to rectangular form, make use of the relationship $\tan \theta = y/x$.

$$\underbrace{\theta = \frac{\pi}{3}}_{\text{Polar equation}} \quad \Longrightarrow \quad \tan \theta = \sqrt{3} \quad \Longrightarrow \quad \underbrace{y = \sqrt{3}x}_{\text{Rectangular equation}}$$

c. The graph of the polar equation $r = \sec \theta$ is not evident by simple inspection, so convert to rectangular form by using the relationship $r \cos \theta = x$.

$$\underbrace{r = \sec \theta}_{\text{Polar equation}} \quad \Longrightarrow \quad r \cos \theta = 1 \quad \Longrightarrow \quad \underbrace{x = 1}_{\text{Rectangular equation}}$$

Now you see that the graph is a vertical line, as shown in Figure 6.68.

✓*CHECKPOINT*   Now try Exercise 65.

6.7 | Exercises

VOCABULARY CHECK: Fill in the blanks.

1. The origin of the polar coordinate system is called the _____.

2. For the point (r, θ), r is the _____ _____ from O to P and θ is the _____ _____ counterclockwise from the polar axis to the line segment $\overline{OP}$.

3. To plot the point (r, θ), use the _____ coordinate system.

4. The polar coordinates (r, θ) are related to the rectangular coordinates (x, y) as follows:

   $x =$ _____ $\tan \theta =$ _____

   $y =$ _____ $r^2 =$ _____

PREREQUISITE SKILLS REVIEW: Practice and review algebra skills needed for this section at **www.Eduspace.com.**

In Exercises 1–8, plot the point given in polar coordinates and find two additional polar representations of the point, using $-2\pi < \theta < 2\pi$.

1. $\left(4, -\dfrac{\pi}{3}\right)$

2. $\left(-1, -\dfrac{3\pi}{4}\right)$

3. $\left(0, -\dfrac{7\pi}{6}\right)$

4. $\left(16, \dfrac{5\pi}{2}\right)$

5. $\left(\sqrt{2}, 2.36\right)$

6. $(-3, -1.57)$

7. $\left(2\sqrt{2}, 4.71\right)$

8. $(-5, -2.36)$

In Exercises 9–16, a point in polar coordinates is given. Convert the point to rectangular coordinates.

9. $\left(3, \dfrac{\pi}{2}\right)$

10. $\left(3, \dfrac{3\pi}{2}\right)$

11. $\left(-1, \dfrac{5\pi}{4}\right)$

12. $(0, -\pi)$

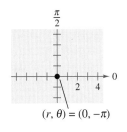

13. $\left(2, \dfrac{3\pi}{4}\right)$

14. $\left(-2, \dfrac{7\pi}{6}\right)$

15. $(-2.5, 1.1)$

16. $(8.25, 3.5)$

In Exercises 17–26, a point in rectangular coordinates is given. Convert the point to polar coordinates.

17. $(1, 1)$

18. $(-3, -3)$

19. $(-6, 0)$

20. $(0, -5)$

21. $(-3, 4)$

22. $(3, -1)$

23. $\left(-\sqrt{3}, -\sqrt{3}\right)$

24. $\left(\sqrt{3}, -1\right)$

25. $(6, 9)$

26. $(5, 12)$

 In Exercises 27–32, use a graphing utility to find one set of polar coordinates for the point given in rectangular coordinates.

27. $(3, -2)$

28. $(-5, 2)$

29. $\left(\sqrt{3}, 2\right)$

30. $\left(3, \sqrt{2}, 3\sqrt{2}\right)$

31. $\left(\dfrac{5}{2}, \dfrac{4}{3}\right)$

32. $\left(\dfrac{7}{4}, \dfrac{3}{2}\right)$

In Exercises 33–48, convert the rectangular equation to polar form. Assume $a > 0$.

33. $x^2 + y^2 = 9$

34. $x^2 + y^2 = 16$

35. $y = 4$

36. $y = x$

37. $x = 10$

38. $x = 4a$

39. $3x - y + 2 = 0$

40. $3x + 5y - 2 = 0$

41. $xy = 16$

42. $2xy = 1$

43. $y^2 - 8x - 16 = 0$

44. $(x^2 + y^2)^2 = 9(x^2 - y^2)$

45. $x^2 + y^2 = a^2$

46. $x^2 + y^2 = 9a^2$

47. $x^2 + y^2 - 2ax = 0$

48. $x^2 + y^2 - 2ay = 0$

In Exercises 49–64, convert the polar equation to rectangular form.

49. $r = 4 \sin \theta$

50. $r = 2 \cos \theta$

51. $\theta = \dfrac{2\pi}{3}$

52. $\theta = \dfrac{5\pi}{3}$

53. $r = 4$

54. $r = 10$

55. $r = 4 \csc \theta$

56. $r = -3 \sec \theta$

57. $r^2 = \cos \theta$

58. $r^2 = \sin 2\theta$

59. $r = 2 \sin 3\theta$

60. $r = 3 \cos 2\theta$

61. $r = \dfrac{2}{1 + \sin \theta}$

62. $r = \dfrac{1}{1 - \cos \theta}$

63. $r = \dfrac{6}{2 - 3 \sin \theta}$

64. $r = \dfrac{6}{2 \cos \theta - 3 \sin \theta}$

In Exercises 65–70, describe the graph of the polar equation and find the corresponding rectangular equation. Sketch its graph.

65. $r = 6$

66. $r = 8$

67. $\theta = \dfrac{\pi}{6}$

68. $\theta = \dfrac{3\pi}{4}$

69. $r = 3 \sec \theta$

70. $r = 2 \csc \theta$

Synthesis

True or False? **In Exercises 71 and 72, determine whether the statement is true or false. Justify your answer.**

71. If $\theta_1 = \theta_2 + 2\pi n$ for some integer n, then (r, θ_1) and (r, θ_2) represent the same point on the polar coordinate system.

72. If $|r_1| = |r_2|$, then (r_1, θ) and (r_2, θ) represent the same point on the polar coordinate system.

73. Convert the polar equation $r = 2(h \cos \theta + k \sin \theta)$ to rectangular form and verify that it is the equation of a circle. Find the radius of the circle and the rectangular coordinates of the center of the circle.

74. Convert the polar equation $r = \cos \theta + 3 \sin \theta$ to rectangular form and identify the graph.

75. *Think About It*

(a) Show that the distance between the points (r_1, θ_1) and (r_2, θ_2) is $\sqrt{r_1^2 + r_2^2 - 2r_1 r_2 \cos(\theta_1 - \theta_2)}$.

(b) Describe the positions of the points relative to each other for $\theta_1 = \theta_2$. Simplify the Distance Formula for this case. Is the simplification what you expected? Explain.

(c) Simplify the Distance Formula for $\theta_1 - \theta_2 = 90°$. Is the simplification what you expected? Explain.

(d) Choose two points on the polar coordinate system and find the distance between them. Then choose different polar representations of the same two points and apply the Distance Formula again. Discuss the result.

76. *Exploration*

(a) Set the window format of your graphing utility on rectangular coordinates and locate the cursor at any position off the coordinate axes. Move the cursor horizontally and observe any changes in the displayed coordinates of the points. Explain the changes in the coordinates. Now repeat the process moving the cursor vertically.

(b) Set the window format of your graphing utility on polar coordinates and locate the cursor at any position off the coordinate axes. Move the cursor horizontally and observe any changes in the displayed coordinates of the points. Explain the changes in the coordinates. Now repeat the process moving the cursor vertically.

(c) Explain why the results of parts (a) and (b) are not the same.

Skills Review

In Exercises 77–80, use the properties of logarithms to expand the expression as a sum, difference, and/or constant multiple of logarithms. (Assume all variables are positive.)

77. $\log_6 \dfrac{x^2 z}{3y}$

78. $\log_4 \dfrac{\sqrt{2x}}{y}$

79. $\ln x(x + 4)^2$

80. $\ln 5x^2(x^2 + 1)$

In Exercises 81–84, condense the expression to the logarithm of a single quantity.

81. $\log_7 x - \log_7 3y$

82. $\log_5 a + 8 \log_5(x + 1)$

83. $\dfrac{1}{2} \ln x + \ln(x - 2)$

84. $\ln 6 + \ln y - \ln(x - 3)$

6.8 Graphs of Polar Equations

Introduction

In previous chapters, you spent a lot of time learning how to sketch graphs on rectangular coordinate systems. You began with the basic point-plotting method, which was then enhanced by sketching aids such as symmetry, intercepts, asymptotes, periods, and shifts. This section approaches curve sketching on the polar coordinate system similarly, beginning with a demonstration of point plotting.

Example 1   Graphing a Polar Equation by Point Plotting

Sketch the graph of the polar equation $r = 4 \sin \theta$.

Solution

The sine function is periodic, so you can get a full range of r-values by considering values of θ in the interval $0 \le \theta \le 2\pi$, as shown in the following table.

θ	0	$\dfrac{\pi}{6}$	$\dfrac{\pi}{3}$	$\dfrac{\pi}{2}$	$\dfrac{2\pi}{3}$	$\dfrac{5\pi}{6}$	π	$\dfrac{7\pi}{6}$	$\dfrac{3\pi}{2}$	$\dfrac{11\pi}{6}$	2π
r	0	2	$2\sqrt{3}$	4	$2\sqrt{3}$	2	0	-2	-4	-2	0

If you plot these points as shown in Figure 6.69, it appears that the graph is a circle of radius 2 whose center is at the point $(x, y) = (0, 2)$.

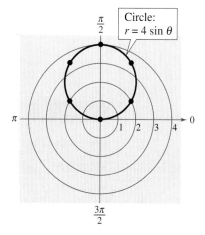

FIGURE 6.69

✓CHECKPOINT   Now try Exercise 21.

You can confirm the graph in Figure 6.69 by converting the polar equation to rectangular form and then sketching the graph of the rectangular equation. You can also use a graphing utility set to *polar* mode and graph the polar equation or set the graphing utility to *parametric* mode and graph a parametric representation.

Symmetry

In Figure 6.69, note that as θ increases from 0 to 2π the graph is traced out twice. Moreover, note that the graph is *symmetric with respect to the line $\theta = \pi/2$*. Had you known about this symmetry and retracing ahead of time, you could have used fewer points.

Symmetry with respect to the line $\theta = \pi/2$ is one of three important types of symmetry to consider in polar curve sketching. (See Figure 6.70.)

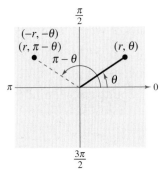

Symmetry with Respect to the

Line $\theta = \dfrac{\pi}{2}$

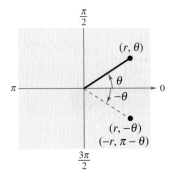

Symmetry with Respect to the
Polar Axis

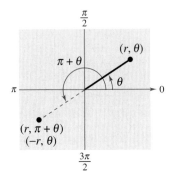

Symmetry with Respect to the
Pole

FIGURE 6.70

Tests for Symmetry in Polar Coordinates

The graph of a polar equation is symmetric with respect to the following if the given substitution yields an equivalent equation.

1. *The line $\theta = \pi/2$:* Replace (r, θ) by $(r, \pi - \theta)$ or $(-r, -\theta)$.

2. *The polar axis:* Replace (r, θ) by $(r, -\theta)$ or $(-r, \pi - \theta)$.

3. *The pole:* Replace (r, θ) by $(r, \pi + \theta)$ or $(-r, \theta)$.

Example 2 Using Symmetry to Sketch a Polar Graph

Use symmetry to sketch the graph of $r = 3 + 2\cos\theta$.

Solution

Replacing (r, θ) by $(r, -\theta)$ produces $r = 3 + 2\cos(-\theta) = 3 + 2\cos\theta$. So, you can conclude that the curve is symmetric with respect to the polar axis. Plotting the points in the table and using polar axis symmetry, you obtain the graph shown in Figure 6.71. This graph is called a **limaçon.**

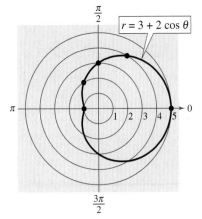

$r = 3 + 2\cos\theta$

FIGURE 6.71

θ	0	$\dfrac{\pi}{3}$	$\dfrac{\pi}{2}$	$\dfrac{2\pi}{3}$	π
r	5	4	3	2	1

✔CHECKPOINT   Now try Exercise 27.

The three tests for symmetry in polar coordinates listed on page 488 are sufficient to guarantee symmetry, but they are not necessary. For instance, Figure 6.72 shows the graph of $r = \theta + 2\pi$ to be symmetric with respect to the line $\theta = \pi/2$, and yet the tests on page 488 fail to indicate symmetry because neither of the following replacements yields an equivalent equation.

Original Equation	Replacement	New Equation
$r = \theta + 2\pi$	(r, θ) by $(-r, -\theta)$	$-r = -\theta + 2\pi$
$r = \theta + 2\pi$	(r, θ) by $(r, \pi - \theta)$	$r = -\theta + 3\pi$

The equations discussed in Examples 1 and 2 are of the form

$$r = 4 \sin \theta = f(\sin \theta) \quad \text{and} \quad r = 3 + 2 \cos \theta = g(\cos \theta).$$

The graph of the first equation is symmetric with respect to the line $\theta = \pi/2$, and the graph of the second equation is symmetric with respect to the polar axis. This observation can be generalized to yield the following tests.

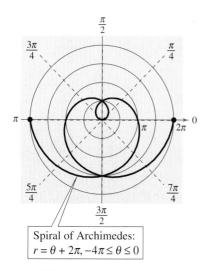

Spiral of Archimedes:
$r = \theta + 2\pi, -4\pi \le \theta \le 0$

FIGURE 6.72

Quick Tests for Symmetry in Polar Coordinates

1. The graph of $r = f(\sin \theta)$ is symmetric with respect to the line $\theta = \dfrac{\pi}{2}$.

2. The graph of $r = g(\cos \theta)$ is symmetric with respect to the polar axis.

Zeros and Maximum r-Values

Two additional aids to graphing of polar equations involve knowing the θ-values for which $|r|$ is maximum and knowing the θ-values for which $r = 0$. For instance, in Example 1, the maximum value of $|r|$ for $r = 4 \sin \theta$ is $|r| = 4$, and this occurs when $\theta = \pi/2$, as shown in Figure 6.69. Moreover, $r = 0$ when $\theta = 0$.

Example 3   Sketching a Polar Graph

Sketch the graph of $r = 1 - 2 \cos \theta$.

Solution

From the equation $r = 1 - 2 \cos \theta$, you can obtain the following.

Symmetry: With respect to the polar axis

Maximum value of $|r|$: $r = 3$ when $\theta = \pi$

Zero of r: $r = 0$ when $\theta = \pi/3$

The table shows several θ-values in the interval $[0, \pi]$. By plotting the corresponding points, you can sketch the graph shown in Figure 6.73.

θ	0	$\dfrac{\pi}{6}$	$\dfrac{\pi}{3}$	$\dfrac{\pi}{2}$	$\dfrac{2\pi}{3}$	$\dfrac{5\pi}{6}$	π
r	-1	-0.73	0	1	2	2.73	3

Note how the negative r-values determine the *inner loop* of the graph in Figure 6.73. This graph, like the one in Figure 6.71, is a limaçon.

✓**CHECKPOINT**   Now try Exercise 29.

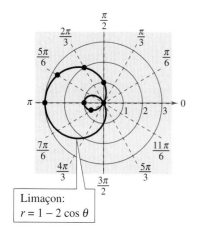

Limaçon:
$r = 1 - 2 \cos \theta$

FIGURE 6.73

Some curves reach their zeros and maximum r-values at more than one point, as shown in Example 4.

Example 4 Sketching a Polar Graph

Sketch the graph of $r = 2 \cos 3\theta$.

Solution

Symmetry:	With respect to the polar axis
Maximum value of $|r|$:	$|r| = 2$ when $3\theta = 0, \pi, 2\pi, 3\pi$ or $\theta = 0, \pi/3, 2\pi/3, \pi$
Zeros of r:	$r = 0$ when $3\theta = \pi/2, 3\pi/2, 5\pi/2$ or $\theta = \pi/6, \pi/2, 5\pi/6$

θ	0	$\dfrac{\pi}{12}$	$\dfrac{\pi}{6}$	$\dfrac{\pi}{4}$	$\dfrac{\pi}{3}$	$\dfrac{5\pi}{12}$	$\dfrac{\pi}{2}$
r	2	$\sqrt{2}$	0	$-\sqrt{2}$	-2	$-\sqrt{2}$	0

By plotting these points and using the specified symmetry, zeros, and maximum values, you can obtain the graph shown in Figure 6.74. This graph is called a **rose curve,** and each of the loops on the graph is called a *petal* of the rose curve. Note how the entire curve is generated as θ increases from 0 to π.

$0 \le \theta \le \dfrac{\pi}{6}$

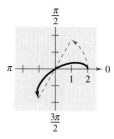

$0 \le \theta \le \dfrac{\pi}{3}$

$0 \le \theta \le \dfrac{\pi}{2}$

$0 \le \theta \le \dfrac{2\pi}{3}$

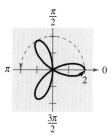

$0 \le \theta \le \dfrac{5\pi}{6}$

$0 \le \theta \le \pi$

FIGURE **6.74**

✓CHECKPOINT Now try Exercise 33.

Special Polar Graphs

Several important types of graphs have equations that are simpler in polar form than in rectangular form. For example, the circle

$$r = 4 \sin \theta$$

in Example 1 has the more complicated rectangular equation

$$x^2 + (y - 2)^2 = 4.$$

Several other types of graphs that have simple polar equations are shown below.

Limaçons

$r = a \pm b \cos \theta$

$r = a \pm b \sin \theta$

$(a > 0, b > 0)$

  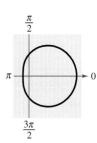

$\dfrac{a}{b} < 1$	$\dfrac{a}{b} = 1$	$1 < \dfrac{a}{b} < 2$	$\dfrac{a}{b} \geq 2$
Limaçon with inner loop	Cardioid (heart-shaped)	Dimpled limaçon	Convex limaçon

Rose Curves

n petals if n is odd,
$2n$ petals if n is even
$(n \geq 2)$

$r = a \cos n\theta$	$r = a \cos n\theta$	$r = a \sin n\theta$	$r = a \sin n\theta$
Rose curve	Rose curve	Rose curve	Rose curve

Circles and Lemniscates

  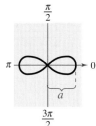

The quick tests for symmetry presented in this section can be especially useful in graphing these special polar graphs.

$r = a \cos \theta$	$r = a \sin \theta$	$r^2 = a^2 \sin 2\theta$	$r^2 = a^2 \cos 2\theta$
Circle	Circle	Lemniscate	Lemniscate

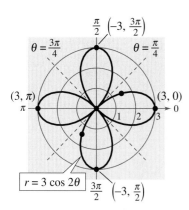

$r = 3 \cos 2\theta$

FIGURE **6.75**

Activities

1. Test $r^2 = 3 \sin \theta$ for symmetry.

   Answer: Symmetric with respect to the pole

2. Find the maximum value of $|r|$ and any zeros of r for the polar equation $r = 2 + 2 \cos \theta$.

   Answer: Maximum value of $|r| = 4$ when $\theta = 0$, and $r = 0$ when $\theta = \pi$

3. Identify the shape of the graph of the polar equation $r = 2 \sin 3\theta$.

   Answer: Rose curve with 3 petals

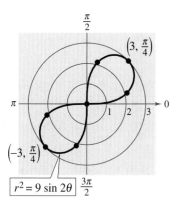

$r^2 = 9 \sin 2\theta$

FIGURE **6.76**

| Example 5 | **Sketching a Rose Curve** |

Sketch the graph of $r = 3 \cos 2\theta$.

Solution

Type of curve:	Rose curve with $2n = 4$ petals
Symmetry:	With respect to polar axis, the line $\theta = \pi/2$, and the pole
Maximum value of $|r|$:	$|r| = 3$ when $\theta = 0, \pi/2, \pi, 3\pi/2$
Zeros of r:	$r = 0$ when $\theta = \pi/4, 3\pi/4$

Using this information together with the additional points shown in the following table, you obtain the graph shown in Figure 6.75.

θ	0	$\dfrac{\pi}{6}$	$\dfrac{\pi}{4}$	$\dfrac{\pi}{3}$
r	3	$\dfrac{3}{2}$	0	$-\dfrac{3}{2}$

✓CHECKPOINT   Now try Exercise 35.

| Example 6 | **Sketching a Lemniscate** |

Sketch the graph of $r^2 = 9 \sin 2\theta$.

Solution

Type of curve:	Lemniscate
Symmetry:	With respect to the pole
Maximum value of $|r|$:	$|r| = 3$ when $\theta = \dfrac{\pi}{4}$
Zeros of r:	$r = 0$ when $\theta = 0, \dfrac{\pi}{2}$

If $\sin 2\theta < 0$, this equation has no solution points. So, you restrict the values of θ to those for which $\sin 2\theta \geq 0$.

$$0 \leq \theta \leq \frac{\pi}{2} \quad \text{or} \quad \pi \leq \theta \leq \frac{3\pi}{2}$$

Moreover, using symmetry, you need to consider only the first of these two intervals. By finding a few additional points (see table below), you can obtain the graph shown in Figure 6.76.

θ	0	$\dfrac{\pi}{12}$	$\dfrac{\pi}{4}$	$\dfrac{5\pi}{12}$	$\dfrac{\pi}{2}$
$r = \pm 3\sqrt{\sin 2\theta}$	0	$\dfrac{\pm 3}{\sqrt{2}}$	± 3	$\dfrac{\pm 3}{\sqrt{2}}$	0

✓CHECKPOINT   Now try Exercise 39.

6.8 | Exercises

VOCABULARY CHECK: Fill in the blanks.

1. The graph of $r = f(\sin \theta)$ is symmetric with respect to the line _____.
2. The graph of $r = g(\cos \theta)$ is symmetric with respect to the _____ _____.
3. The equation $r = 2 + \cos \theta$ represents a _____ _____.
4. The equation $r = 2 \cos \theta$ represents a _____.
5. The equation $r^2 = 4 \sin 2\theta$ represents a _____.
6. The equation $r = 1 + \sin \theta$ represents a _____.

PREREQUISITE SKILLS REVIEW: Practice and review algebra skills needed for this section at **www.Eduspace.com**.

In Exercises 1–6, identify the type of polar graph.

1.

$r = 3 \cos 2\theta$

2.

$r = 5 - 5 \sin \theta$

3.

$r = 3(1 - 2 \cos \theta)$

4.

$r^2 = 16 \cos 2\theta$

5.

$r = 6 \sin 2\theta$

6.

$r = 3 \cos \theta$

In Exercises 7–12, test for symmetry with respect to $\theta = \pi/2$, the polar axis, and the pole.

7. $r = 5 + 4 \cos \theta$
8. $r = 16 \cos 3\theta$
9. $r = \dfrac{2}{1 + \sin \theta}$
10. $r = \dfrac{3}{2 + \cos \theta}$
11. $r^2 = 16 \cos 2\theta$
12. $r^2 = 36 \sin 2\theta$

In Exercises 13–16, find the maximum value of $|r|$ and any zeros of r.

13. $r = 10(1 - \sin \theta)$
14. $r = 6 + 12 \cos \theta$
15. $r = 4 \cos 3\theta$
16. $r = 3 \sin 2\theta$

In Exercises 17–40, sketch the graph of the polar equation using symmetry, zeros, maximum r-values, and any other additional points.

17. $r = 5$
18. $r = 2$
19. $r = \dfrac{\pi}{6}$
20. $r = -\dfrac{3\pi}{4}$
21. $r = 3 \sin \theta$
22. $r = 4 \cos \theta$
23. $r = 3(1 - \cos \theta)$
24. $r = 4(1 - \sin \theta)$
25. $r = 4(1 + \sin \theta)$
26. $r = 2(1 + \cos \theta)$
27. $r = 3 + 6 \sin \theta$
28. $r = 4 - 3 \sin \theta$
29. $r = 1 - 2 \sin \theta$
30. $r = 1 - 2 \cos \theta$
31. $r = 3 - 4 \cos \theta$
32. $r = 4 + 3 \cos \theta$
33. $r = 5 \sin 2\theta$
34. $r = 3 \cos 2\theta$
35. $r = 2 \sec \theta$
36. $r = 5 \csc \theta$
37. $r = \dfrac{3}{\sin \theta - 2 \cos \theta}$
38. $r = \dfrac{6}{2 \sin \theta - 3 \cos \theta}$
39. $r^2 = 9 \cos 2\theta$
40. $r^2 = 4 \sin \theta$

In Exercises 41–46, use a graphing utility to graph the polar equation. Describe your viewing window.

41. $r = 8 \cos \theta$
42. $r = \cos 2\theta$
43. $r = 3(2 - \sin \theta)$
44. $r = 2 \cos(3\theta - 2)$
45. $r = 8 \sin \theta \cos^2 \theta$
46. $r = 2 \csc \theta + 5$

In Exercises 47–52, use a graphing utility to graph the polar equation. Find an interval for θ for which the graph is traced *only once*.

47. $r = 3 - 4 \cos \theta$
48. $r = 5 + 4 \cos \theta$

49. $r = 2 \cos\left(\dfrac{3\theta}{2}\right)$

50. $r = 3 \sin\left(\dfrac{5\theta}{2}\right)$

51. $r^2 = 9 \sin 2\theta$

52. $r^2 = \dfrac{1}{\theta}$

In Exercises 53–56, use a graphing utility to graph the polar equation and show that the indicated line is an asymptote of the graph.

Name of Graph	*Polar Equation*	*Asymptote*
53. Conchoid	$r = 2 - \sec \theta$	$x = -1$
54. Conchoid	$r = 2 + \csc \theta$	$y = 1$
55. Hyperbolic spiral	$r = \dfrac{3}{\theta}$	$y = 3$
56. Strophoid	$r = 2 \cos 2\theta \sec \theta$	$x = -2$

Synthesis

True or False? **In Exercises 57 and 58, determine whether the statement is true or false. Justify your answer.**

57. In the polar coordinate system, if a graph that has symmetry with respect to the polar axis were folded on the line $\theta = 0$, the portion of the graph above the polar axis would coincide with the portion of the graph below the polar axis.

58. In the polar coordinate system, if a graph that has symmetry with respect to the pole were folded on the line $\theta = 3\pi/4$, the portion of the graph on one side of the fold would coincide with the portion of the graph on the other side of the fold.

59. ***Exploration*** Sketch the graph of $r = 6 \cos \theta$ over each interval. Describe the part of the graph obtained in each case.

(a) $0 \le \theta \le \dfrac{\pi}{2}$ (b) $\dfrac{\pi}{2} \le \theta \le \pi$

(c) $-\dfrac{\pi}{2} \le \theta \le \dfrac{\pi}{2}$ (d) $\dfrac{\pi}{4} \le \theta \le \dfrac{3\pi}{4}$

60. ***Graphical Reasoning*** Use a graphing utility to graph the polar equation $r = 6[1 + \cos(\theta - \phi)]$ for (a) $\phi = 0$, (b) $\phi = \pi/4$, and (c) $\phi = \pi/2$. Use the graphs to describe the effect of the angle ϕ. Write the equation as a function of $\sin \theta$ for part (c).

61. The graph of $r = f(\theta)$ is rotated about the pole through an angle ϕ. Show that the equation of the rotated graph is $r = f(\theta - \phi)$.

62. Consider the graph of $r = f(\sin \theta)$.

(a) Show that if the graph is rotated counterclockwise $\pi/2$ radians about the pole, the equation of the rotated graph is $r = f(-\cos \theta)$.

(b) Show that if the graph is rotated counterclockwise π radians about the pole, the equation of the rotated graph is $r = f(-\sin \theta)$.

(c) Show that if the graph is rotated counterclockwise $3\pi/2$ radians about the pole, the equation of the rotated graph is $r = f(\cos \theta)$.

In Exercises 63–66, use the results of Exercises 61 and 62.

63. Write an equation for the limaçon $r = 2 - \sin \theta$ after it has been rotated through the given angle.

(a) $\dfrac{\pi}{4}$ (b) $\dfrac{\pi}{2}$ (c) π (d) $\dfrac{3\pi}{2}$

64. Write an equation for the rose curve $r = 2 \sin 2\theta$ after it has been rotated through the given angle.

(a) $\dfrac{\pi}{6}$ (b) $\dfrac{\pi}{2}$ (c) $\dfrac{2\pi}{3}$ (d) π

65. Sketch the graph of each equation.

(a) $r = 1 - \sin \theta$ (b) $r = 1 - \sin\left(\theta - \dfrac{\pi}{4}\right)$

66. Sketch the graph of each equation.

(a) $r = 3 \sec \theta$ (b) $r = 3 \sec\left(\theta - \dfrac{\pi}{4}\right)$

(c) $r = 3 \sec\left(\theta + \dfrac{\pi}{3}\right)$ (d) $r = 3 \sec\left(\theta - \dfrac{\pi}{2}\right)$

67. ***Exploration*** Use a graphing utility to graph and identify $r = 2 + k \sin \theta$ for $k = 0, 1, 2,$ and 3.

68. ***Exploration*** Consider the equation $r = 3 \sin k\theta$.

(a) Use a graphing utility to graph the equation for $k = 1.5$. Find the interval for θ over which the graph is traced only once.

(b) Use a graphing utility to graph the equation for $k = 2.5$. Find the interval for θ over which the graph is traced only once.

(c) Is it possible to find an interval for θ over which the graph is traced only once for any rational number k? Explain.

Skills Review

In Exercises 69 and 70, find the standard form of the equation of the ellipse with the given characteristics. Then sketch the ellipse.

69. Vertices: $(-4, 2), (2, 2)$; minor axis of length 4

70. Foci: $(3, 2), (3, -4)$; major axis of length 8

6.9   Polar Equations of Conics

What you should learn

- Define conics in terms of eccentricity.
- Write and graph equations of conics in polar form.
- Use equations of conics in polar form to model real-life problems.

Why you should learn it

The orbits of planets and satellites can be modeled with polar equations. For instance, in Exercise 58 on page 500, a polar equation is used to model the orbit of a satellite.

Digital Image © 1996 Corbis;
Original image courtesy of NASA/Corbis

Alternative Definition of Conic

In Sections 6.3 and 6.4, you learned that the rectangular equations of ellipses and hyperbolas take simple forms when the origin lies at their *centers*. As it happens, there are many important applications of conics in which it is more convenient to use one of the *foci* as the origin. In this section, you will learn that polar equations of conics take simple forms if one of the foci lies at the pole.

To begin, consider the following alternative definition of conic that uses the concept of eccentricity.

> **Alternative Definition of Conic**
>
> The locus of a point in the plane that moves so that its distance from a fixed point (focus) is in a constant ratio to its distance from a fixed line (directrix) is a **conic.** The constant ratio is the **eccentricity** of the conic and is denoted by e. Moreover, the conic is an **ellipse** if $e < 1$, a **parabola** if $e = 1$, and a **hyperbola** if $e > 1$. (See Figure 6.77.)

In Figure 6.77, note that for each type of conic, the focus is at the pole.

  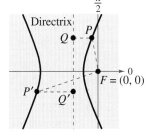

Ellipse: $0 < e < 1$
$\dfrac{PF}{PQ} < 1$

Parabola: $e = 1$
$\dfrac{PF}{PQ} = 1$

Hyperbola $e > 1$
$\dfrac{PF}{PQ} = \dfrac{P'F}{P'Q'} > 1$

FIGURE **6.77**

Polar Equations of Conics

The benefit of locating a focus of a conic at the pole is that the equation of the conic takes on a simpler form. For a proof of the polar equations of conics, see Proofs in Mathematics on page 512.

> **Polar Equations of Conics**
>
> The graph of a polar equation of the form
>
> **1.** $r = \dfrac{ep}{1 \pm e \cos \theta}$ or   **2.** $r = \dfrac{ep}{1 \pm e \sin \theta}$
>
> is a conic, where $e > 0$ is the eccentricity and $|p|$ is the distance between the focus (pole) and the directrix.

Equations of the form

$$r = \frac{ep}{1 \pm e \cos \theta} = g(\cos \theta) \qquad \text{Vertical directrix}$$

correspond to conics with a vertical directrix and symmetry with respect to the polar axis. Equations of the form

$$r = \frac{ep}{1 \pm e \sin \theta} = g(\sin \theta) \qquad \text{Horizontal directrix}$$

correspond to conics with a horizontal directrix and symmetry with respect to the line $\theta = \pi/2$. Moreover, the converse is also true—that is, any conic with a focus at the pole and having a horizontal or vertical directrix can be represented by one of the given equations.

Example 1   Identifying a Conic from Its Equation

Identify the type of conic represented by the equation $r = \dfrac{15}{3 - 2 \cos \theta}$.

Algebraic Solution

To identify the type of conic, rewrite the equation in the form $r = (ep)/(1 \pm e \cos \theta)$.

$$r = \frac{15}{3 - 2 \cos \theta} \qquad \text{Write original equation.}$$

$$= \frac{5}{1 - (2/3) \cos \theta} \qquad \begin{array}{l}\text{Divide numerator and} \\ \text{denominator by 3.}\end{array}$$

Because $e = \frac{2}{3} < 1$, you can conclude that the graph is an ellipse.

✔CHECKPOINT   Now try Exercise 11.

Graphical Solution

You can start sketching the graph by plotting points from $\theta = 0$ to $\theta = \pi$. Because the equation is of the form $r = g(\cos \theta)$, the graph of r is symmetric with respect to the polar axis. So, you can complete the sketch, as shown in Figure 6.78. From this, you can conclude that the graph is an ellipse.

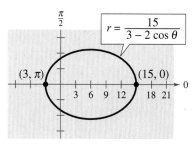

FIGURE **6.78**

For the ellipse in Figure 6.78, the major axis is horizontal and the vertices lie at $(15, 0)$ and $(3, \pi)$. So, the length of the *major* axis is $2a = 18$. To find the length of the *minor* axis, you can use the equations $e = c/a$ and $b^2 = a^2 - c^2$ to conclude that

$$b^2 = a^2 - c^2$$

$$= a^2 - (ea)^2$$

$$= a^2(1 - e^2). \qquad \text{Ellipse}$$

Because $e = \frac{2}{3}$, you have $b^2 = 9^2\left[1 - \left(\frac{2}{3}\right)^2\right] = 45$, which implies that $b = \sqrt{45} = 3\sqrt{5}$. So, the length of the minor axis is $2b = 6\sqrt{5}$. A similar analysis for hyperbolas yields

$$b^2 = c^2 - a^2$$

$$= (ea)^2 - a^2$$

$$= a^2(e^2 - 1). \qquad \text{Hyperbola}$$

Additional Example
Identify the conic and sketch its graph.
$$r = \frac{4}{2 + 2 \sin \theta}$$
Solution
Parabola

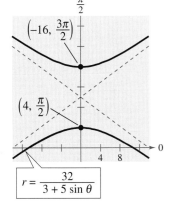

FIGURE **6.79**

Technology

Use a graphing utility set in *polar* mode to verify the four orientations shown at the right. Remember that *e* must be positive, but *p* can be positive or negative.

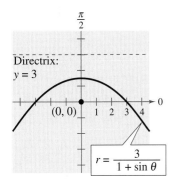

FIGURE **6.80**

Example 2   **Sketching a Conic from Its Polar Equation**

Identify the conic $r = \dfrac{32}{3 + 5 \sin \theta}$ and sketch its graph.

Solution

Dividing the numerator and denominator by 3, you have

$$r = \frac{32/3}{1 + (5/3) \sin \theta}.$$

Because $e = \frac{5}{3} > 1$, the graph is a hyperbola. The transverse axis of the hyperbola lies on the line $\theta = \pi/2$, and the vertices occur at $(4, \pi/2)$ and $(-16, 3\pi/2)$. Because the length of the transverse axis is 12, you can see that $a = 6$. To find b, write

$$b^2 = a^2(e^2 - 1) = 6^2 \left[\left(\frac{5}{3} \right)^2 - 1 \right] = 64.$$

So, $b = 8$. Finally, you can use a and b to determine that the asymptotes of the hyperbola are $y = 10 \pm \frac{3}{4}x$. The graph is shown in Figure 6.79.

✔CHECKPOINT   Now try Exercise 19.

In the next example, you are asked to find a polar equation of a specified conic. To do this, let p be the distance between the pole and the directrix.

1. *Horizontal directrix above the pole:* $r = \dfrac{ep}{1 + e \sin \theta}$

2. *Horizontal directrix below the pole:* $r = \dfrac{ep}{1 - e \sin \theta}$

3. *Vertical directrix to the right of the pole:* $r = \dfrac{ep}{1 + e \cos \theta}$

4. *Vertical directrix to the left of the pole:* $r = \dfrac{ep}{1 - e \cos \theta}$

Example 3   **Finding the Polar Equation of a Conic**

Find the polar equation of the parabola whose focus is the pole and whose directrix is the line $y = 3$.

Solution

From Figure 6.80, you can see that the directrix is horizontal and above the pole, so you can choose an equation of the form

$$r = \frac{ep}{1 + e \sin \theta}.$$

Moreover, because the eccentricity of a parabola is $e = 1$ and the distance between the pole and the directrix is $p = 3$, you have the equation

$$r = \frac{3}{1 + \sin \theta}.$$

✔CHECKPOINT   Now try Exercise 33.

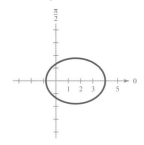
Applications

Kepler's Laws (listed below), named after the German astronomer Johannes Kepler (1571–1630), can be used to describe the orbits of the planets about the sun.

1. Each planet moves in an elliptical orbit with the sun at one focus.

2. A ray from the sun to the planet sweeps out equal areas of the ellipse in equal times.

3. The square of the period (the time it takes for a planet to orbit the sun) is proportional to the cube of the mean distance between the planet and the sun.

Although Kepler simply stated these laws on the basis of observation, they were later validated by Isaac Newton (1642–1727). In fact, Newton was able to show that each law can be deduced from a set of universal laws of motion and gravitation that govern the movement of all heavenly bodies, including comets and satellites. This is illustrated in the next example, which involves the comet named after the English mathematician and physicist Edmund Halley (1656–1742).

If you use Earth as a reference with a period of 1 year and a distance of 1 astronomical unit (an *astronomical unit* is defined as the mean distance between Earth and the sun, or about 93 million miles), the proportionality constant in Kepler's third law is 1. For example, because Mars has a mean distance to the sun of $d = 1.524$ astronomical units, its period P is given by $d^3 = P^2$. So, the period of Mars is $P \approx 1.88$ years.

Example 4 Halley's Comet

Halley's comet has an elliptical orbit with an eccentricity of $e \approx 0.967$. The length of the major axis of the orbit is approximately 35.88 astronomical units. Find a polar equation for the orbit. How close does Halley's comet come to the sun?

Solution

Using a vertical axis, as shown in Figure 6.81, choose an equation of the form $r = ep/(1 + e\sin\theta)$. Because the vertices of the ellipse occur when $\theta = \pi/2$ and $\theta = 3\pi/2$, you can determine the length of the major axis to be the sum of the r-values of the vertices. That is,

$$2a = \frac{0.967p}{1 + 0.967} + \frac{0.967p}{1 - 0.967} \approx 29.79p \approx 35.88.$$

So, $p \approx 1.204$ and $ep \approx (0.967)(1.204) \approx 1.164$. Using this value of ep in the equation, you have

$$r = \frac{1.164}{1 + 0.967\sin\theta}$$

where r is measured in astronomical units. To find the closest point to the sun (the focus), substitute $\theta = \pi/2$ in this equation to obtain

$$r = \frac{1.164}{1 + 0.967\sin(\pi/2)} \approx 0.59 \text{ astronomical unit} \approx 55,000,000 \text{ miles.}$$

✔CHECKPOINT   Now try Exercise 57.

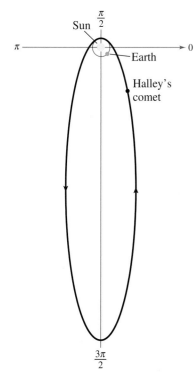

FIGURE **6.81**

| **Exercises**

VOCABULARY CHECK:

In Exercises 1–3, fill in the blanks.

1. The locus of a point in the plane that moves so that its distance from a fixed point (focus) is in a constant ratio to its distance from a fixed line (directrix) is a _____.

2. The constant ratio is the _____ of the conic and is denoted by _____.

3. An equation of the form $r = \dfrac{ep}{1 + e \cos \theta}$ has a _____ directrix to the _____ of the pole.

4. Match the conic with its eccentricity.

   (a) $e < 1$ (b) $e = 1$ (c) $e > 1$

   (i) parabola   (ii) hyperbola   (iii) ellipse

PREREQUISITE SKILLS REVIEW: Practice and review algebra skills needed for this section at **www.Eduspace.com.**

In Exercises 1–4, write the polar equation of the conic for $e = 1$, $e = 0.5$, **and** $e = 1.5$. **Identify the conic for each equation. Verify your answers with a graphing utility.**

1. $r = \dfrac{4e}{1 + e \cos \theta}$

2. $r = \dfrac{4e}{1 - e \cos \theta}$

3. $r = \dfrac{4e}{1 - e \sin \theta}$

4. $r = \dfrac{4e}{1 + e \sin \theta}$

5. $r = \dfrac{2}{1 + \cos \theta}$

6. $r = \dfrac{3}{2 - \cos \theta}$

7. $r = \dfrac{3}{1 + 2 \sin \theta}$

8. $r = \dfrac{2}{1 - \sin \theta}$

9. $r = \dfrac{4}{2 + \cos \theta}$

10. $r = \dfrac{4}{1 - 3 \sin \theta}$

In Exercises 5–10, match the polar equation with its graph. [The graphs are labeled (a), (b), (c), (d), (e), and (f).]

(a)

(b)

(c)

(d)

(e)

(f)
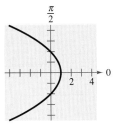

In Exercises 11–24, identify the conic and sketch its graph.

11. $r = \dfrac{2}{1 - \cos \theta}$

12. $r = \dfrac{3}{1 + \sin \theta}$

13. $r = \dfrac{5}{1 + \sin \theta}$

14. $r = \dfrac{6}{1 + \cos \theta}$

15. $r = \dfrac{2}{2 - \cos \theta}$

16. $r = \dfrac{3}{3 + \sin \theta}$

17. $r = \dfrac{6}{2 + \sin \theta}$

18. $r = \dfrac{9}{3 - 2 \cos \theta}$

19. $r = \dfrac{3}{2 + 4 \sin \theta}$

20. $r = \dfrac{5}{-1 + 2 \cos \theta}$

21. $r = \dfrac{3}{2 - 6 \cos \theta}$

22. $r = \dfrac{3}{2 + 6 \sin \theta}$

23. $r = \dfrac{4}{2 - \cos \theta}$

24. $r = \dfrac{2}{2 + 3 \sin \theta}$

 In Exercises 25–28, use a graphing utility to graph the polar equation. Identify the graph.

25. $r = \dfrac{-1}{1 - \sin \theta}$

26. $r = \dfrac{-5}{2 + 4 \sin \theta}$

27. $r = \dfrac{3}{-4 + 2 \cos \theta}$

28. $r = \dfrac{4}{1 - 2 \cos \theta}$

In Exercises 29–32, use a graphing utility to graph the rotated conic.

29. $r = \dfrac{2}{1 - \cos(\theta - \pi/4)}$ (See Exercise 11.)

30. $r = \dfrac{3}{3 + \sin(\theta - \pi/3)}$ (See Exercise 16.)

31. $r = \dfrac{6}{2 + \sin(\theta + \pi/6)}$ (See Exercise 17.)

32. $r = \dfrac{5}{-1 + 2\cos(\theta + 2\pi/3)}$ (See Exercise 20.)

In Exercises 33–48, find a polar equation of the conic with its focus at the pole.

	Conic	Eccentricity	Directrix
33.	Parabola	$e = 1$	$x = -1$
34.	Parabola	$e = 1$	$y = -2$
35.	Ellipse	$e = \frac{1}{2}$	$y = 1$
36.	Ellipse	$e = \frac{3}{4}$	$y = -3$
37.	Hyperbola	$e = 2$	$x = 1$
38.	Hyperbola	$e = \frac{3}{2}$	$x = -1$

	Conic	Vertex or Vertices
39.	Parabola	$(1, -\pi/2)$
40.	Parabola	$(6, 0)$
41.	Parabola	$(5, \pi)$
42.	Parabola	$(10, \pi/2)$
43.	Ellipse	$(2, 0), (10, \pi)$
44.	Ellipse	$(2, \pi/2), (4, 3\pi/2)$
45.	Ellipse	$(20, 0), (4, \pi)$
46.	Hyperbola	$(2, 0), (8, 0)$
47.	Hyperbola	$(1, 3\pi/2), (9, 3\pi/2)$
48.	Hyperbola	$(4, \pi/2), (1, \pi/2)$

49. *Planetary Motion* The planets travel in elliptical orbits with the sun at one focus. Assume that the focus is at the pole, the major axis lies on the polar axis, and the length of the major axis is $2a$ (see figure). Show that the polar equation of the orbit is $r = a(1 - e^2)/(1 - e\cos\theta)$ where e is the eccentricity.

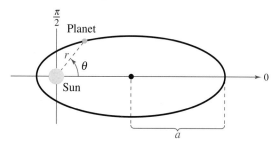

50. *Planetary Motion* Use the result of Exercise 49 to show that the minimum distance (*perihelion distance*) from the sun to the planet is $r = a(1 - e)$ and the maximum distance (*aphelion distance*) is $r = a(1 + e)$.

Planetary Motion **In Exercises 51–56, use the results of Exercises 49 and 50 to find the polar equation of the planet's orbit and the perihelion and aphelion distances.**

51. Earth $\quad a = 95.956 \times 10^6$ miles, $e = 0.0167$

52. Saturn $\quad a = 1.427 \times 10^9$ kilometers, $e = 0.0542$

53. Venus $\quad a = 108.209 \times 10^6$ kilometers, $e = 0.0068$

54. Mercury $\quad a = 35.98 \times 10^6$ miles, $e = 0.2056$

55. Mars $\quad a = 141.63 \times 10^6$ miles, $e = 0.0934$

56. Jupiter $\quad a = 778.41 \times 10^6$ kilometers, $e = 0.0484$

57. *Astronomy* The comet Encke has an elliptical orbit with an eccentricity of $e \approx 0.847$. The length of the major axis of the orbit is approximately 4.42 astronomical units. Find a polar equation for the orbit. How close does the comet come to the sun?

Model It

58. *Satellite Tracking* A satellite in a 100-mile-high circular orbit around Earth has a velocity of approximately 17,500 miles per hour. If this velocity is multiplied by $\sqrt{2}$, the satellite will have the minimum velocity necessary to escape Earth's gravity and it will follow a parabolic path with the center of Earth as the focus (see figure).

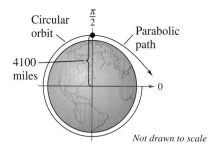

Not drawn to scale

(a) Find a polar equation of the parabolic path of the satellite (assume the radius of Earth is 4000 miles).

 (b) Use a graphing utility to graph the equation you found in part (a).

(c) Find the distance between the surface of the Earth and the satellite when $\theta = 30°$.

(d) Find the distance between the surface of Earth and the satellite when $\theta = 60°$.

Synthesis

True or False?   **In Exercises 59–61, determine whether the statement is true or false. Justify your answer.**

59. For a given value of $e > 1$ over the interval $\theta = 0$ to $\theta = 2\pi$, the graph of

$$r = \frac{ex}{1 - e \cos \theta}$$

is the same as the graph of

$$r = \frac{e(-x)}{1 + e \cos \theta}.$$

60. The graph of

$$r = \frac{4}{-3 - 3 \sin \theta}$$

has a horizontal directrix above the pole.

61. The conic represented by the following equation is an ellipse.

$$r^2 = \frac{16}{9 - 4 \cos\left(\theta + \frac{\pi}{4}\right)}$$

62. ***Writing***   In your own words, define the term *eccentricity* and explain how it can be used to classify conics.

63. Show that the polar equation of the ellipse

$$\frac{x^2}{a^2} + \frac{y^2}{b^2} = 1 \quad \text{is} \quad r^2 = \frac{b^2}{1 - e^2 \cos^2 \theta}.$$

64. Show that the polar equation of the hyperbola

$$\frac{x^2}{a^2} - \frac{y^2}{b^2} = 1 \quad \text{is} \quad r^2 = \frac{-b^2}{1 - e^2 \cos^2 \theta}.$$

In Exercises 65–70, use the results of Exercises 63 and 64 to write the polar form of the equation of the conic.

65. $\dfrac{x^2}{169} + \dfrac{y^2}{144} = 1$

66. $\dfrac{x^2}{25} + \dfrac{y^2}{16} = 1$

67. $\dfrac{x^2}{9} - \dfrac{y^2}{16} = 1$

68. $\dfrac{x^2}{36} - \dfrac{y^2}{4} = 1$

69. Hyperbola   One focus: $(5, \pi/2)$

Vertices: $(4, \pi/2), (4, -\pi/2)$

70. Ellipse   One focus: $(4, 0)$

Vertices: $(5, 0), (5, \pi)$

71. ***Exploration***   Consider the polar equation

$$r = \frac{4}{1 - 0.4 \cos \theta}.$$

(a) Identify the conic without graphing the equation.

(b) Without graphing the following polar equations, describe how each differs from the given polar equation.

$$r_1 = \frac{4}{1 + 0.4 \cos \theta}, \quad r_2 = \frac{4}{1 - 0.4 \sin \theta}$$

(c) Use a graphing utility to verify your results in part (b).

72. ***Exploration***   The equation

$$r = \frac{ep}{1 \pm e \sin \theta}$$

is the equation of an ellipse with $e < 1$. What happens to the lengths of both the major axis and the minor axis when the value of e remains fixed and the value of p changes? Use an example to explain your reasoning.

Skills Review

In Exercises 73–78, solve the trigonometric equation.

73. $4\sqrt{3} \tan \theta - 3 = 1$

74. $6 \cos x - 2 = 1$

75. $12 \sin^2 \theta = 9$

76. $9 \csc^2 x - 10 = 2$

77. $2 \cot x = 5 \cos \dfrac{\pi}{2}$

78. $\sqrt{2} \sec \theta = 2 \csc \dfrac{\pi}{4}$

In Exercises 79–82, find the exact value of the trigonometric function given that u and v are in Quadrant IV and $\sin u = -\frac{3}{5}$ and $\cos v = 1/\sqrt{2}$.

79. $\cos(u + v)$

80. $\sin(u + v)$

81. $\cos(u - v)$

82. $\sin(u - v)$

In Exercises 83 and 84, find the exact values of $\sin 2u$, $\cos 2u$, and $\tan 2u$ using the double-angle formulas.

83. $\sin u = \dfrac{4}{5}, \dfrac{\pi}{2} < u < \pi$

84. $\tan u = -\sqrt{3}, \dfrac{3\pi}{2} < u < 2\pi$

6 Chapter Summary

What did you learn?

6   Review Exercises

6.1 In Exercises 1–4, find the inclination θ (in radians and degrees) of the line with the given characteristics.

1. Passes through the points $(-1, 2)$ and $(2, 5)$
2. Passes through the points $(3, 4)$ and $(-2, 7)$
3. Equation: $y = 2x + 4$
4. Equation: $6x - 7y - 5 = 0$

In Exercises 5–8, find the angle θ (in radians and degrees) between the lines.

5. $4x + y = 2$
     $-5x + y = -1$

6. $-5x + 3y = 3$
     $-2x + 3y = 1$

7. $2x - 7y = 8$
     $0.4x + y = 0$

8. $0.02x + 0.07y = 0.18$
     $0.09x - 0.04y = 0.17$

In Exercises 9 and 10, find the distance between the point and the line.

Point	Line
9. $(1, 2)$	$x - y - 3 = 0$
10. $(0, 4)$	$x + 2y - 2 = 0$

6.2 In Exercises 11 and 12, state what type of conic is formed by the intersection of the plane and the double-napped cone.

11.

12.

In Exercises 13–16, find the standard form of the equation of the parabola with the given characteristics. Then graph the parabola.

13. Vertex: $(0, 0)$
    Focus: $(4, 0)$
14. Vertex: $(2, 0)$
    Focus: $(0, 0)$
15. Vertex: $(0, 2)$
    Directrix: $x = -3$
16. Vertex: $(2, 2)$
    Directrix: $y = 0$

In Exercises 17 and 18, find an equation of the tangent line to the parabola at the given point, and find the x-intercept of the line.

17. $x^2 = -2y$, $(2, -2)$
18. $x^2 = -2y$, $(-4, -8)$

19. *Architecture* A parabolic archway is 12 meters high at the vertex. At a height of 10 meters, the width of the archway is 8 meters (see figure). How wide is the archway at ground level?

FIGURE FOR 19   FIGURE FOR 20

20. *Flashlight* The light bulb in a flashlight is at the focus of its parabolic reflector, 1.5 centimeters from the vertex of the reflector (see figure). Write an equation of a cross section of the flashlight's reflector with its focus on the positive x-axis and its vertex at the origin.

6.3 In Exercises 21–24, find the standard form of the equation of the ellipse with the given characteristics. Then graph the ellipse.

21. Vertices: $(-3, 0), (7, 0)$; foci: $(0, 0), (4, 0)$
22. Vertices: $(2, 0), (2, 4)$; foci: $(2, 1), (2, 3)$
23. Vertices: $(0, 1), (4, 1)$; endpoints of the minor axis: $(2, 0), (2, 2)$
24. Vertices: $(-4, -1), (-4, 11)$; endpoints of the minor axis: $(-6, 5), (-2, 5)$

25. *Architecture* A semielliptical archway is to be formed over the entrance to an estate. The arch is to be set on pillars that are 10 feet apart and is to have a height (atop the pillars) of 4 feet. Where should the foci be placed in order to sketch the arch?

26. *Wading Pool* You are building a wading pool that is in the shape of an ellipse. Your plans give an equation for the elliptical shape of the pool measured in feet as

$$\frac{x^2}{324} + \frac{y^2}{196} = 1.$$

Find the longest distance across the pool, the shortest distance, and the distance between the foci.

In Exercises 27–30, find the center, vertices, foci, and eccentricity of the ellipse.

27. $\dfrac{(x + 2)^2}{81} + \dfrac{(y - 1)^2}{100} = 1$

28. $\dfrac{(x - 5)^2}{1} + \dfrac{(y + 3)^2}{36} = 1$

29. $16x^2 + 9y^2 - 32x + 72y + 16 = 0$

30. $4x^2 + 25y^2 + 16x - 150y + 141 = 0$

6.4 In Exercises 31–34, find the standard form of the equation of the hyperbola with the given characteristics.

31. Vertices: $(0, \pm 1)$; foci: $(0, \pm 3)$

32. Vertices: $(2, 2), (-2, 2)$; foci: $(4, 2), (-4, 2)$

33. Foci: $(0, 0), (8, 0)$; asymptotes: $y = \pm 2(x - 4)$

34. Foci: $(3, \pm 2)$; asymptotes: $y = \pm 2(x - 3)$

In Exercises 35–38, find the center, vertices, foci, and the equations of the asymptotes of the hyperbola, and sketch its graph using the asymptotes as an aid.

35. $\dfrac{(x - 3)^2}{16} - \dfrac{(y + 5)^2}{4} = 1$

36. $\dfrac{(y - 1)^2}{4} - x^2 = 1$

37. $9x^2 - 16y^2 - 18x - 32y - 151 = 0$

38. $-4x^2 + 25y^2 - 8x + 150y + 121 = 0$

39. *LORAN* Radio transmitting station A is located 200 miles east of transmitting station B. A ship is in an area to the north and 40 miles west of station A. Synchronized radio pulses transmitted at 186,000 miles per second by the two stations are received 0.0005 second sooner from station A than from station B. How far north is the ship?

40. *Locating an Explosion* Two of your friends live 4 miles apart and on the same "east-west" street, and you live halfway between them. You are having a three-way phone conversation when you hear an explosion. Six seconds later, your friend to the east hears the explosion, and your friend to the west hears it 8 seconds after you do. Find equations of two hyperbolas that would locate the explosion. (Assume that the coordinate system is measured in feet and that sound travels at 1100 feet per second.)

In Exercises 41–44, classify the graph of the equation as a circle, a parabola, an ellipse, or a hyperbola.

41. $5x^2 - 2y^2 + 10x - 4y + 17 = 0$

42. $-4y^2 + 5x + 3y + 7 = 0$

43. $3x^2 + 2y^2 - 12x + 12y + 29 = 0$

44. $4x^2 + 4y^2 - 4x + 8y - 11 = 0$

6.5 In Exercises 45–48, rotate the axes to eliminate the *xy*-term in the equation. Then write the equation in standard form. Sketch the graph of the resulting equation, showing both sets of axes.

45. $xy - 4 = 0$

46. $x^2 - 10xy + y^2 + 1 = 0$

47. $5x^2 - 2xy + 5y^2 - 12 = 0$

48. $4x^2 + 8xy + 4y^2 + 7\sqrt{2}x + 9\sqrt{2}y = 0$

In Exercises 49–52, (a) use the discriminant to classify the graph, (b) use the Quadratic Formula to solve for *y*, and (c) use a graphing utility to graph the equation.

49. $16x^2 - 24xy + 9y^2 - 30x - 40y = 0$

50. $13x^2 - 8xy + 7y^2 - 45 = 0$

51. $x^2 + y^2 + 2xy + 2\sqrt{2}x - 2\sqrt{2}y + 2 = 0$

52. $x^2 - 10xy + y^2 + 1 = 0$

6.6 In Exercises 53 and 54, complete the table for each set of parametric equations. Plot the points (x, y) and sketch a graph of the parametric equations.

53. $x = 3t - 2$ and $y = 7 - 4t$

t	-3	-2	-1	0	1	2	3
x							
y							

54. $x = \dfrac{1}{5}t$ and $y = \dfrac{4}{t - 1}$

t	-1	0	2	3	4	5
x						
y						

In Exercises 55–60, (a) sketch the curve represented by the parametric equations (indicate the orientation of the curve) and (b) eliminate the parameter and write the corresponding rectangular equation whose graph represents the curve. Adjust the domain of the resulting rectangular equation, if necessary. (c) Verify your result with a graphing utility.

55. $x = 2t$
$y = 4t$

56. $x = 1 + 4t$
$y = 2 - 3t$

57. $x = t^2$
$y = \sqrt{t}$

58. $x = t + 4$
$y = t^2$

59. $x = 6 \cos \theta$
$y = 6 \sin \theta$

60. $x = 3 + 3 \cos \theta$
$y = 2 + 5 \sin \theta$

61. Find a parametric representation of the circle with center $(5, 4)$ and radius 6.

62. Find a parametric representation of the ellipse with center $(-3, 4)$, major axis horizontal and eight units in length, and minor axis six units in length.

63. Find a parametric representation of the hyperbola with vertices $(0, \pm 4)$ and foci $(0, \pm 5)$.

64. *Involute of a Circle* The *involute* of a circle is described by the endpoint P of a string that is held taut as it is unwound from a spool (see figure). The spool does not rotate. Show that a parametric representation of the involute of a circle is

$x = r(\cos \theta + \theta \sin \theta)$

$y = r(\sin \theta - \theta \cos \theta)$.

6.7 In Exercises 65–68, plot the point given in polar coordinates and find two additional polar representations of the point, using $-2\pi < \theta < 2\pi$.

65. $\left(2, \dfrac{\pi}{4}\right)$

66. $\left(-5, -\dfrac{\pi}{3}\right)$

67. $(-7, 4.19)$

68. $\left(\sqrt{3}, 2.62\right)$

In Exercises 69–72, a point in polar coordinates is given. Convert the point to rectangular coordinates.

69. $\left(-1, \dfrac{\pi}{3}\right)$

70. $\left(2, \dfrac{5\pi}{4}\right)$

71. $\left(3, \dfrac{3\pi}{4}\right)$

72. $\left(0, \dfrac{\pi}{2}\right)$

In Exercises 73–76, a point in rectangular coordinates is given. Convert the point to polar coordinates.

73. $(0, 2)$

74. $\left(-\sqrt{5}, \sqrt{5}\right)$

75. $(4, 6)$

76. $(3, -4)$

In Exercises 77–82, convert the rectangular equation to polar form.

77. $x^2 + y^2 = 49$

78. $x^2 + y^2 = 20$

79. $x^2 + y^2 - 6y = 0$

80. $x^2 + y^2 - 4x = 0$

81. $xy = 5$

82. $xy = -2$

In Exercises 83–88, convert the polar equation to rectangular form.

83. $r = 5$

84. $r = 12$

85. $r = 3 \cos \theta$

86. $r = 8 \sin \theta$

87. $r^2 = \sin \theta$

88. $r^2 = \cos 2\theta$

6.8 In Exercises 89–98, determine the symmetry of r, the maximum value of $|r|$, and any zeros of r. Then sketch the graph of the polar equation (plot additional points if necessary).

89. $r = 4$

90. $r = 11$

91. $r = 4 \sin 2\theta$

92. $r = \cos 5\theta$

93. $r = -2(1 + \cos \theta)$

94. $r = 3 - 4 \cos \theta$

95. $r = 2 + 6 \sin \theta$

96. $r = 5 - 5 \cos \theta$

97. $r = -3 \cos 2\theta$

98. $r = \cos 2\theta$

In Exercises 99–102, identify the type of polar graph and use a graphing utility to graph the equation.

99. $r = 3(2 - \cos \theta)$

100. $r = 3(1 - 2 \cos \theta)$

101. $r = 4 \cos 3\theta$

102. $r^2 = 9 \cos 2\theta$

6.9 In Exercises 103–106, identify the conic and sketch its graph.

103. $r = \dfrac{1}{1 + 2 \sin \theta}$

104. $r = \dfrac{2}{1 + \sin \theta}$

105. $r = \dfrac{4}{5 - 3 \cos \theta}$

106. $r = \dfrac{16}{4 + 5 \cos \theta}$

In Exercises 107–110, find a polar equation of the conic with its focus at the pole.

107. Parabola Vertex: $(2, \pi)$

108. Parabola Vertex: $(2, \pi/2)$

109. Ellipse Vertices: $(5, 0), (1, \pi)$

110. Hyperbola Vertices: $(1, 0), (7, 0)$

111. *Explorer 18* On November 26, 1963, the United States launched Explorer 18. Its low and high points above the surface of Earth were 119 miles and 122,800 miles, respectively (see figure). The center of Earth was at one focus of the orbit. Find the polar equation of the orbit and find the distance between the surface of Earth (assume Earth has a radius of 4000 miles) and the satellite when $\theta = \pi/3$.

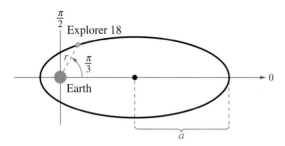

112. *Asteroid* An asteroid takes a parabolic path with Earth as its focus. It is about 6,000,000 miles from Earth at its closest approach. Write the polar equation of the path of the asteroid with its vertex at $\theta = \pi/2$. Find the distance between the asteroid and Earth when $\theta = -\pi/3$.

Synthesis

True or False? **In Exercises 113–116, determine whether the statement is true or false. Justify your answer.**

113. When $B = 0$ in an equation of the form

$$Ax^2 + Bxy + Cy^2 + Dx + Ey + F = 0$$

the graph of the equation can be a parabola only if $C = 0$ also.

114. The graph of $\frac{1}{4}x^2 - y^4 = 1$ is a hyperbola.

115. Only one set of parametric equations can represent the line $y = 3 - 2x$.

116. There is a unique polar coordinate representation of each point in the plane.

117. Consider an ellipse with the major axis horizontal and 10 units in length. The number b in the standard form of the equation of the ellipse must be less than what real number? Explain the change in the shape of the ellipse as b approaches this number.

118. The graph of the parametric equations $x = 2 \sec t$ and $y = 3 \tan t$ is shown in the figure. How would the graph change for the equations $x = 2 \sec(-t)$ and $y = 3 \tan(-t)$?

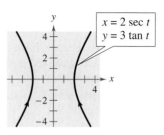

FIGURE FOR **118**

119. A moving object is modeled by the parametric equations $x = 4 \cos t$ and $y = 3 \sin t$, where t is time (see figure). How would the path change for the following?

(a) $x = 4 \cos 2t, \quad y = 3 \sin 2t$

(b) $x = 5 \cos t, \quad y = 3 \sin t$

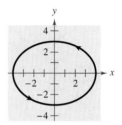

120. Identify the type of symmetry each of the following polar points has with the point in the figure.

(a) $\left(-4, \dfrac{\pi}{6}\right)$ (b) $\left(4, -\dfrac{\pi}{6}\right)$ (c) $\left(-4, -\dfrac{\pi}{6}\right)$

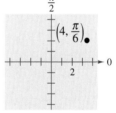

121. What is the relationship between the graphs of the rectangular and polar equations?

(a) $x^2 + y^2 = 25, \quad r = 5$ (b) $x - y = 0, \quad \theta = \dfrac{\pi}{4}$

122. *Geometry* The area of the ellipse in the figure is twice the area of the circle. What is the length of the major axis? (*Hint:* The area of an ellipse is $A = \pi ab$.)

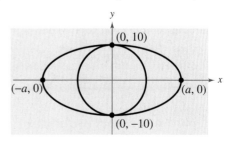

6	**Chapter Test**

Take this test as you would take a test in class. When you are finished, check your work against the answers given in the back of the book.

1. Find the inclination of the line $2x - 7y + 3 = 0$.

2. Find the angle between the lines $3x + 2y - 4 = 0$ and $4x - y + 6 = 0$.

3. Find the distance between the point $(7, 5)$ and the line $y = 5 - x$.

In Exercises 4–7, classify the conic and write the equation in standard form. Identify the center, vertices, foci, and asymptotes (if applicable). Then sketch the graph of the conic.

4. $y^2 - 4x + 4 = 0$

5. $x^2 - 4y^2 - 4x = 0$

6. $9x^2 + 16y^2 + 54x - 32y - 47 = 0$

7. $2x^2 + 2y^2 - 8x - 4y + 9 = 0$

8. Find the standard form of the equation of the parabola with vertex $(3, -2)$, with a vertical axis, and passing through the point $(0, 4)$.

9. Find the standard form of the equation of the hyperbola with foci $(0, 0)$ and $(0, 4)$ and asymptotes $y = \pm\frac{1}{2}x + 2$.

10. (a) Determine the number of degrees the axis must be rotated to eliminate the xy-term of the conic $x^2 + 6xy + y^2 - 6 = 0$.

    (b) Graph the conic from part (a) and use a graphing utility to confirm your result.

11. Sketch the curve represented by the parametric equations $x = 2 + 3\cos\theta$ and $y = 2\sin\theta$. Eliminate the parameter and write the corresponding rectangular equation.

12. Find a set of parametric equations of the line passing through the points $(2, -3)$ and $(6, 4)$. (There are many correct answers.)

13. Convert the polar coordinate $\left(-2, \dfrac{5\pi}{6}\right)$ to rectangular form.

14. Convert the rectangular coordinate $(2, -2)$ to polar form and find two additional polar representations of this point.

15. Convert the rectangular equation $x^2 + y^2 - 4y = 0$ to polar form.

In Exercises 16–19, sketch the graph of the polar equation. Identify the type of graph.

16. $r = \dfrac{4}{1 + \cos\theta}$

17. $r = \dfrac{4}{2 + \cos\theta}$

18. $r = 2 + 3\sin\theta$

19. $r = 3\sin 2\theta$

20. Find a polar equation of the ellipse with focus at the pole, eccentricity $e = \frac{1}{4}$, and directrix $y = 4$.

21. A straight road rises with an inclination of 0.15 radian from the horizontal. Find the slope of the road and the change in elevation over a one-mile stretch of the road.

22. A baseball is hit at a point 3 feet above the ground toward the left field fence. The fence is 10 feet high and 375 feet from home plate. The path of the baseball can be modeled by the parametric equations $x = (115\cos\theta)t$ and $y = 3 + (115\sin\theta)t - 16t^2$. Will the baseball go over the fence if it is hit at an angle of $\theta = 30°$? Will the baseball go over the fence if $\theta = 35°$?

6 | Cumulative Test for Chapters 4–6

Take this test to review material from earlier chapters. When you are finished, check your work against the answers given in the back of the book.

1. Write the complex number $3 - \sqrt{-25}$ in standard form.

In Exercises 2–4, perform the operations and write the result in standard form.

2. $6i - \left(2 + \sqrt{-81}\right)$ **3.** $(2i - 3)^2$ **4.** $\left(\sqrt{3} + i\right)\left(\sqrt{3} - i\right)$

5. Write the quotient in standard form: $\dfrac{4i}{1 + 2i}$.

In Exercises 6 and 7, find all the zeros of the function.

6. $f(x) = x^3 + 2x^2 + 4x + 8$ **7.** $f(x) = x^4 + 4x^3 - 21x^2$

8. Find a polynomial with real coefficients that has -5, -2, and $2 + \sqrt{3}i$ as its zeros.

9. Write the complex number $z = -2 + 2i$ in trigonometric form.

10. Find the product $[4(\cos 30° + i \sin 30°)][6(\cos 120° + i \sin 120°)]$. Write the result in standard form.

In Exercises 11 and 12, use DeMoivre's Theorem to find the indicated power of the complex number. Write the result in standard form.

11. $\left[2\left(\cos \dfrac{2\pi}{3} + i \sin \dfrac{2\pi}{3}\right)\right]^4$ **12.** $\left(-\sqrt{3} - i\right)^6$

13. Find the three cube roots of 1.

14. Write all the solutions of the equation $x^4 - 81i = 0$.

In Exercises 15 and 16, use the graph of f to describe the transformation that yields the graph of g.

15. $f(x) = \left(\tfrac{2}{5}\right)^x$, $g(x) = -\left(\tfrac{2}{5}\right)^{-x+3}$

16. $f(x) = 2.2^x$, $g(x) = -2.2^x + 4$

In Exercises 17–20, use a calculator to evaluate each expression. Round your result to three decimal places.

17. $\log_{10} 98$ **18.** $\log_{10}\left(\tfrac{6}{7}\right)$ **19.** $\ln \sqrt{31}$ **20.** $\ln\left(\sqrt{40} - 5\right)$

In Exercises 21–23, evaluate the logarithm using the change-of-base formula. Round your answer to three decimal places.

21. $\log_7 1.8$ **22.** $\log_3 0.149$ **23.** $\log_{1/2} 17$

24. Use the properties of logarithms to expand $\ln\left(\dfrac{x^2 - 16}{x^4}\right)$, where $x > 4$.

25. Write $2 \ln x - \tfrac{1}{2}\ln(x + 5)$ as a logarithm of a single quantity.

In Exercises 26–29, solve the equation algebraically. Round the result to three decimal places.

26. $6e^{2x} = 72$

27. $4^{x-5} + 21 = 30$

28. $\log_2 x + \log_2 5 = 6$

29. $\ln 4x - \ln 2 = 8$

30. Use a graphing utility to graph $f(x) = \dfrac{1000}{1 + 4e^{-0.2x}}$ and determine the horizontal asymptotes.

31. The number of bacteria N in a culture is given by the model $N = 175e^{kt}$, where t is the time in hours. If $N = 420$ when $t = 8$, estimate the time required for the population to double in size.

32. The population P of Texas (in thousands) from 1990 through 2003 can be modeled by $P = 16{,}989e^{0.0207t}$, where t represents the year, with $t = 0$ corresponding to 1990. According to this model, when will the population reach 24 million? (Source: U.S. Census Bureau)

33. Find the angle between the lines $2x + y - 3 = 0$ and $x - 3y + 6 = 0$.

34. Find the distance between the point $(6, -3)$ and the line $y = 2x - 4$.

In Exercises 35–38, classify the conic and write the equation in standard form. Identify the center, vertices, foci, and asymptotes (if any). Then sketch the graph.

35. $9x^2 + 4y^2 - 36x + 8y + 4 = 0$

36. $4x^2 - y^2 - 4 = 0$

37. $x^2 + y^2 + 2x - 6y - 12 = 0$

38. $y^2 + 2x + 2 = 0$

39. Find an equation in rectangular coordinates of the circle with center $(2, -4)$ and passing through the point $(0, 4)$.

40. Find an equation in rectangular coordinates of the hyperbola with foci $(0, 0)$ and $(0, 6)$ and asymptotes $y = \pm\dfrac{2\sqrt{5}}{5}x + 3$.

41. (a) Determine the number of degrees the axes must be rotated to eliminate the xy-term of the conic $x^2 + xy + y^2 + 2x - 3y - 30 = 0$.

(b) Graph the conic and use a graphing utility to confirm your result.

42. Sketch the curve represented by the parametric equations $x = 3 + 4\cos\theta$ and $y = \sin\theta$. Eliminate the parameter and write the corresponding rectangular equation.

43. Find a set of parametric equations of the line passing through the points $(3, -2)$ and $(-3, 4)$. (The answer is not unique.)

44. Plot the point $(-2, -3\pi/4)$ and find three additional polar representations for $-2\pi < \theta < 2\pi$.

45. Convert the rectangular equation $x^2 + y^2 - 6y = 0$ to polar form.

46. Convert the polar equation $r = \dfrac{2}{4 - 5\cos\theta}$ to rectangular form.

In Exercises 47 and 48, sketch the graph of the polar equation.

47. $r = \dfrac{3}{2 + \cos\theta}$

48. $r = \dfrac{4}{1 + \sin\theta}$

49. Match each polar equation with its graph at the left.

(a) $r = 2 + 3\sin\theta$ (b) $r = 3\sin\theta$ (c) $r = 3\sin 2\theta$

(i)

(ii)

(iii)

FIGURE FOR 49

Inclination and Slope   *(p. 430)*

If a nonvertical line has inclination θ and slope m, then $m = \tan \theta$.

Proof

If $m = 0$, the line is horizontal and $\theta = 0$. So, the result is true for horizontal lines because $m = 0 = \tan 0$.

If the line has a positive slope, it will intersect the x-axis. Label this point $(x_1, 0)$, as shown in the figure. If (x_2, y_2) is a second point on the line, the slope is

$$m = \frac{y_2 - 0}{x_2 - x_1} = \frac{y_2}{x_2 - x_1} = \tan \theta.$$

The case in which the line has a negative slope can be proved in a similar manner.

Distance Between a Point and a Line   *(p. 432)*

The distance between the point (x_1, y_1) and the line $Ax + By + C = 0$ is

$$d = \frac{|Ax_1 + By_1 + C|}{\sqrt{A^2 + B^2}}.$$

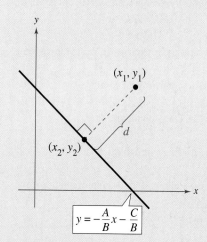

Proof

For simplicity's sake, assume that the given line is neither horizontal nor vertical (see figure). By writing the equation $Ax + By + C = 0$ in slope-intercept form

$$y = -\frac{A}{B}x - \frac{C}{B}$$

you can see that the line has a slope of $m = -A/B$. So, the slope of the line passing through (x_1, y_1) and perpendicular to the given line is B/A, and its equation is $y - y_1 = (B/A)(x - x_1)$. These two lines intersect at the point (x_2, y_2), where

$$x_2 = \frac{B(Bx_1 - Ay_1) - AC}{A^2 + B^2} \quad \text{and} \quad y_2 = \frac{A(-Bx_1 + Ay_1) - BC}{A^2 + B^2}.$$

Finally, the distance between (x_1, y_1) and (x_2, y_2) is

$$d = \sqrt{(x_2 - x_1)^2 + (y_2 - y_1)^2}$$

$$= \sqrt{\left(\frac{B^2x_1 - ABy_1 - AC}{A^2 + B^2} - x_1\right)^2 + \left(\frac{-ABx_1 + A^2y_1 - BC}{A^2 + B^2} - y_1\right)^2}$$

$$= \sqrt{\frac{A^2(Ax_1 + By_1 + C)^2 + B^2(Ax_1 + By_1 + C)^2}{(A^2 + B^2)^2}}$$

$$= \frac{|Ax_1 + By_1 + C|}{\sqrt{A^2 + B^2}}.$$

Parabolic Paths

There are many natural occurrences of parabolas in real life. For instance, the famous astronomer Galileo discovered in the 17th century that an object that is projected upward and obliquely to the pull of gravity travels in a parabolic path. Examples of this are the center of gravity of a jumping dolphin and the path of water molecules in a drinking fountain.

Parabola with vertical axis

Parabola with horizontal axis

Standard Equation of a Parabola   *(p. 438)*

The standard form of the equation of a parabola with vertex at (h, k) is as follows.

$$(x - h)^2 = 4p(y - k), \quad p \neq 0 \qquad \text{Vertical axis, directrix: } y = k - p$$

$$(y - k)^2 = 4p(x - h), \quad p \neq 0 \qquad \text{Horizontal axis, directrix: } x = h - p$$

The focus lies on the axis p units (*directed distance*) from the vertex. If the vertex is at the origin $(0, 0)$, the equation takes one of the following forms.

$$x^2 = 4py \qquad\qquad \text{Vertical axis}$$

$$y^2 = 4px \qquad\qquad \text{Horizontal axis}$$

Proof

For the case in which the directrix is parallel to the x-axis and the focus lies above the vertex, as shown in the top figure, if (x, y) is any point on the parabola, then, by definition, it is equidistant from the focus $(h, k + p)$ and the directrix $y = k - p$. So, you have

$$\sqrt{(x - h)^2 + [y - (k + p)]^2} = y - (k - p)$$

$$(x - h)^2 + [y - (k + p)]^2 = [y - (k - p)]^2$$

$$(x - h)^2 + y^2 - 2y(k + p) + (k + p)^2 = y^2 - 2y(k - p) + (k - p)^2$$

$$(x - h)^2 + y^2 - 2ky - 2py + k^2 + 2pk + p^2 = y^2 - 2ky + 2py + k^2 - 2pk + p^2$$

$$(x - h)^2 - 2py + 2pk = 2py - 2pk$$

$$(x - h)^2 = 4p(y - k).$$

For the case in which the directrix is parallel to the y-axis and the focus lies to the right of the vertex, as shown in the bottom figure, if (x, y) is any point on the parabola, then, by definition, it is equidistant from the focus $(h + p, k)$ and the directrix $x = h - p$. So, you have

$$\sqrt{[x - (h + p)]^2 + (y - k)^2} = x - (h - p)$$

$$[x - (h + p)]^2 + (y - k)^2 = [x - (h - p)]^2$$

$$x^2 - 2x(h + p) + (h + p)^2 + (y - k)^2 = x^2 - 2x(h - p) + (h - p)^2$$

$$x^2 - 2hx - 2px + h^2 + 2ph + p^2 + (y - k)^2 = x^2 - 2hx + 2px + h^2 - 2ph + p^2$$

$$-2px + 2ph + (y - k)^2 = 2px - 2ph$$

$$(y - k)^2 = 4p(x - h).$$

Note that if a parabola is centered at the origin, then the two equations above would simplify to $x^2 = 4py$ and $y^2 = 4px$, respectively.

Polar Equations of Conics *(p. 495)*

The graph of a polar equation of the form

1. $r = \dfrac{ep}{1 \pm e \cos \theta}$

or

2. $r = \dfrac{ep}{1 \pm e \sin \theta}$

is a conic, where $e > 0$ is the eccentricity and $|p|$ is the distance between the focus (pole) and the directrix.

Proof

A proof for $r = ep/(1 + e \cos \theta)$ with $p > 0$ is shown here. The proofs of the other cases are similar. In the figure, consider a vertical directrix, p units to the right of the focus $F = (0, 0)$. If $P = (r, \theta)$ is a point on the graph of

$$r = \frac{ep}{1 + e \cos \theta}$$

the distance between P and the directrix is

$$
\begin{aligned}
PQ &= |p - x| \\
&= |p - r \cos \theta| \\
&= \left| p - \left(\frac{ep}{1 + e \cos \theta} \right) \cos \theta \right| \\
&= \left| p \left(1 - \frac{e \cos \theta}{1 + e \cos \theta} \right) \right| \\
&= \left| \frac{p}{1 + e \cos \theta} \right| \\
&= \left| \frac{r}{e} \right|.
\end{aligned}
$$

Moreover, because the distance between P and the pole is simply $PF = |r|$, the ratio of PF to PQ is

$$
\begin{aligned}
\frac{PF}{PQ} &= \frac{|r|}{\left| \dfrac{r}{e} \right|} \\
&= |e| \\
&= e
\end{aligned}
$$

and, by definition, the graph of the equation must be a conic.

P.S. Problem Solving

This collection of thought-provoking and challenging exercises further explores and expands upon concepts learned in this chapter.

1. Several mountain climbers are located in a mountain pass between two peaks. The angles of elevation to the two peaks are 0.84 radian and 1.10 radians. A range finder shows that the distances to the peaks are 3250 feet and 6700 feet, respectively (see figure).

(a) Find the angle between the two lines of sight to the peaks.

(b) Approximate the amount of vertical climb that is necessary to reach the summit of each peak.

2. Statuary Hall is an elliptical room in the United States Capitol in Washington D.C. The room is also called the Whispering Gallery because a person standing at one focus of the room can hear even a whisper spoken by a person standing at the other focus. This occurs because any sound that is emitted from one focus of an ellipse will reflect off the side of the ellipse to the other focus. Statuary Hall is 46 feet wide and 97 feet long.

(a) Find an equation that models the shape of the room.

(b) How far apart are the two foci?

(c) What is the area of the floor of the room? (The area of an ellipse is $A = \pi ab$.)

3. Find the equation(s) of all parabolas that have the x-axis as the axis of symmetry and focus at the origin.

4. Find the area of the square inscribed in the ellipse below.

$$\frac{x^2}{a^2} + \frac{y^2}{b^2} = 1$$

5. A tour boat travels between two islands that are 12 miles apart (see figure). For a trip between the islands, there is enough fuel for a 20-mile trip.

Not drawn to scale

(a) Explain why the region in which the boat can travel is bounded by an ellipse.

(b) Let $(0, 0)$ represent the center of the ellipse. Find the coordinates of each island.

(c) The boat travels from one island, straight past the other island to the vertex of the ellipse, and back to the second island. How many miles does the boat travel? Use your answer to find the coordinates of the vertex.

(d) Use the results from parts (b) and (c) to write an equation for the ellipse that bounds the region in which the boat can travel.

6. Find an equation of the hyperbola such that for any point on the hyperbola, the difference between its distances from the points $(2, 2)$ and $(10, 2)$ is 6.

7. Prove that the graph of the equation

$$Ax^2 + Cy^2 + Dx + Ey + F = 0$$

is one of the following (except in degenerate cases).

Conic	Condition
(a) Circle	$A = C$
(b) Parabola	$A = 0$ or $C = 0$ (but not both)
(c) Ellipse	$AC > 0$
(d) Hyperbola	$AC < 0$

8. The following sets of parametric equations model projectile motion.

$x = (v_0 \cos \theta)t$ $\qquad$ $x = (v_0 \cos \theta)t$

$y = (v_0 \sin \theta)t$ $\qquad$ $y = h + (v_0 \sin \theta)t - 16t^2$

(a) Under what circumstances would you use each model?

(b) Eliminate the parameter for each set of equations.

(c) In which case is the path of the moving object not affected by a change in the velocity v? Explain.

9. As t increases, the ellipse given by the parametric equations

$$x = \cos t \text{ and } y = 2 \sin t$$

is traced out *counterclockwise*. Find a parametric representation for which the same ellipse is traced out *clockwise*.

 10. A **hypocycloid** has the parametric equations

$$x = (a - b) \cos t + b \cos\left(\frac{a - b}{b}t\right)$$

and

$$y = (a - b) \sin t - b \sin\left(\frac{a - b}{b}t\right).$$

Use a graphing utility to graph the hypocycloid for each value of a and b. Describe each graph.

(a) $a = 2, b = 1$ (b) $a = 3, b = 1$

(c) $a = 4, b = 1$ (d) $a = 10, b = 1$

(e) $a = 3, b = 2$ (f) $a = 4, b = 3$

11. The curve given by the parametric equations

$$x = \frac{1 - t^2}{1 + t^2}$$

and

$$y = \frac{t(1 - t^2)}{1 + t^2}$$

is called a **strophoid.**

(a) Find a rectangular equation of the strophoid.

(b) Find a polar equation of the strophoid.

 (c) Use a graphing utility to graph the strophoid.

 12. The rose curves described in this chapter are of the form

$$r = a \cos n\theta \quad \text{or} \quad r = a \sin n\theta$$

where n is a positive integer that is greater than or equal to 2. Use a graphing utility to graph $r = a \cos n\theta$ and $r = a \sin n\theta$ for some noninteger values of n. Describe the graphs.

13. What conic section is represented by the polar equation

$$r = a \sin \theta + b \cos \theta?$$

14. The graph of the polar equation

$$r = e^{\cos\theta} - 2 \cos 4\theta + \sin^5\left(\frac{\theta}{12}\right)$$

is called the *butterfly curve*, as shown in the figure.

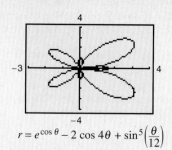

$$r = e^{\cos\theta} - 2 \cos 4\theta + \sin^5\left(\frac{\theta}{12}\right)$$

FIGURE FOR **14**

(a) The graph above was produced using $0 \le \theta \le 2\pi$. Does this show the entire graph? Explain your reasoning.

(b) Approximate the maximum r-value of the graph. Does this value change if you use $0 \le \theta \le 4\pi$ instead of $0 \le \theta \le 2\pi$? Explain.

 15. Use a graphing utility to graph the polar equation

$$r = \cos 5\theta + n \cos \theta$$

for $0 \le \theta \le \pi$ for the integers $n = -5$ to $n = 5$. As you graph these equations, you should see the graph change shape from a heart to a bell. Write a short paragraph explaining what values of n produce the heart portion of the curve and what values of n produce the bell portion of the curve.

16. The planets travel in elliptical orbits with the sun at one focus. The polar equation of the orbit of a planet with one focus at the pole and major axis of length $2a$ is

$$r = \frac{(1 - e^2)a}{1 - e \cos \theta}$$

where e is the eccentricity. The minimum distance (perihelion) from the sun to a planet is $r = a(1 - e)$ and the maximum distance (aphelion) is $r = a(1 + e)$. The length of the major axis for the planet Neptune is $a = 9.000 \times 10^9$ kilometers and the eccentricity is $e = 0.0086$. The length of the major axis for the planet Pluto is $a = 10.813 \times 10^9$ kilometers and the eccentricity is $e = 0.2488$.

(a) Find the polar equation of the orbit of each planet.

(b) Find the perihelion and aphelion distances for each planet.

 (c) Use a graphing utility to graph the polar equation of each planet's orbit in the same viewing window.

(d) Do the orbits of the two planets intersect? Will the two planets ever collide? Why or why not?

(e) Is Pluto ever closer to the sun than Neptune? Why is Pluto called the ninth planet and Neptune the eighth planet?

Answers to All Exercises and Tests

Chapter P

Section P.1   *(page 9)*

(page 9)

Vocabulary Check   *(page 9)*

1. rational   **2.** irrational   **3.** absolute value
4. composite   **5.** prime   **6.** variables; constants
7. terms   **8.** coefficient   **9.** Zero-Factor Property

---

1. (a) $5, 1, 2$ (b) $0, 5, 1, 2$
   (c) $-9, 5, 0, 1, -4, 2, -11$
   (d) $-\frac{7}{2}, \frac{2}{3}, -9, 5, 0, 1, -4, 2, -11$ (e) $\sqrt{2}$
2. (a) $12, 5$ (b) $12, 5, 0$ (c) $-7, 12, 5, 0, -3$
   (d) $-7, -\frac{7}{3}, 0, 3.12, \frac{5}{4}, -3, 12, 5$ (e) $\sqrt{5}$
3. (a) 1 (b) 1 (c) $-13, 1, -6$
   (d) $2.01, -13, 1, -6, 0.666\ldots$
   (e) $0.010110111\ldots$
4. (a) 4 (b) 4 (c) $-75, 4$
   (d) $-75, 4, -4.63, 0.7575$
   (e) $2.3030030003\ldots, \sqrt{10}$
5. (a) $\frac{6}{3}, 8$ (b) $\frac{6}{3}, 8$ (c) $\frac{6}{3}, -1, 8, -22$
   (d) $-\frac{1}{3}, \frac{6}{3}, -7.5, -1, 8, -22$ (e) $-\pi, \frac{1}{2}\sqrt{2}$
6. (a) $25, \sqrt{9}, 7, 13$ (b) $25, \sqrt{9}, 7, 13$
   (c) $25, -17, \sqrt{9}, 7, 13$
   (d) $-\frac{12}{5}, \sqrt{9}, 3.12, 25, -17, -11.1, 13$ (e) $\frac{\pi}{2}$
7. 0.625 **8.** $0.\overline{3}$ **9.** $0.\overline{123}$ **10.** $0.\overline{54}$
11. $-1 < 2.5$ **12.** $-6 < -2.5$
13. $-4 > -8$

14.

   $-3.5 < 1$
15. $\frac{3}{2} < 7$

16.

   $1 < \frac{16}{3}$
17. $\frac{5}{6} > \frac{2}{3}$

18.

   $-\frac{8}{7} < -\frac{3}{7}$
19. (a) $x \le 5$ denotes the set of all real numbers less than or equal to 5.

(b)   **(c)** Unbounded
20. (a) $x \ge -2$ denotes the set of all real numbers greater than or equal to -2.
   (b)   **(c)** Unbounded
21. (a) $x < 0$ denotes the set of all real numbers less than 0.
   (b)   **(c)** Unbounded
22. (a) $x > 3$ denotes the set of all real numbers greater than 3.
   (b)   **(c)** Unbounded
23. (a) $[4, \infty)$ denotes the set of all real numbers greater than or equal to 4.
   (b)   **(c)** Unbounded
24. (a) $(-\infty, 2)$ denotes the set of all real numbers less than 2.
   (b)   **(c)** Unbounded
25. (a) $-2 < x < 2$ denotes the set of all real numbers greater than -2 and less than 2.
   (b)   **(c)** Bounded
26. (a) $0 \le x \le 5$ denotes the set of all real numbers greater than or equal to zero and less than or equal to 5.
   (b)   **(c)** Bounded
27. (a) $-1 \le x < 0$ denotes the set of all real numbers greater than or equal to -1 and less than 0.
   (b)   **(c)** Bounded
28. (a) $0 < x \le 6$ denotes the set of positive real numbers less than or equal to 6.
   (b)   **(c)** Bounded
29. (a) $[-2, 5)$ denotes the set of all real numbers greater than or equal to -2 and less than 5.
   (b)   **(c)** Bounded
30. (a) $(-1, 2]$ denotes all real numbers greater than -1 and less than or equal to 2.
   (b)   **(c)** Bounded
31. $-2 < x \le 4$ **32.** $-6 \le y < 0$ **33.** $y \ge 0$
34. $y \le 25$ **35.** $10 \le t \le 22$ **36.** $-3 \le k < 5$
37. $W > 65$ **38.** $2.5\% \le r \le 5\%$ **39.** 10 **40.** 0
41. 5 **42.** 3 **43.** -1 **44.** -6 **45.** -1
46. -9 **47.** -1 **48.** 1 **49.** $|-3| > -|-3|$
50. $|-4| = |4|$ **51.** $-5 = -|5|$
52. $-|-6| < |-6|$ **53.** $-|-2| = -|2|$

54. $-(-2) > -2$ **55.** 51   **56.** 51   **57.** $\frac{5}{2}$
58. $\frac{5}{2}$ **59.** $\frac{128}{75}$ **60.** 14.99
61. $|\$113{,}356 - \$112{,}700| = \$656 > \$500$
$0.05(\$112{,}700) = \$5635$

Because the actual expenses differ from the budget by more than $500, there is failure to meet the "budget variance test."

62. $|\$9772 - \$9400| = \$372 < \$500$
$0.05(\$9400) = \$470$

Because the difference between the actual expenses and the budget is less than $500 and less than 5% of the budgeted amount, there is compliance with the "budget variance test."

63. $|\$37{,}335 - \$37{,}640| = \$305 < \$500$
$0.05(\$37{,}640) = \$1882$

Because the difference between the actual expenses and the budget is less than $500 and less than 5% of the budgeted amount, there is compliance with the "budget variance test."

64. $|\$2613 - \$2575| = \$38 < \$500$
$0.05(\$2575) = \$128.75$

Because the difference between the actual expenses and the budget is less than $500 and less than 5% of the budgeted amount, there is compliance with the "budget variance test."

65. (a)

Year	Expenditures (in billions)	Surplus or deficit (in billions)
1960	$92.2	$0.3 (s)
1970	$195.6	$2.8 (d)
1980	$590.9	$73.8 (d)
1990	$1253.2	$221.2 (d)
2000	$1788.8	$236.4 (s)

(b)

66.

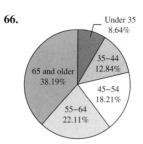

67. $|x - 5| \le 3$ **68.** $|x + 10| \ge 6$ **69.** $|y| \ge 6$
70. $|y - a| \le 2$ **71.** $|326 - 351| = 25$ miles
72. $|48° - 82°| = 34°$
73. $7x$ and 4 are the terms; 7 is the coefficient.
74. $6x^3$ and $-5x$ are the terms; 6 and -5 are the coefficients.
75. $\sqrt{3}x^2$, $-8x$, and -11 are the terms; $\sqrt{3}$ and -8 are the coefficients.
76. $3\sqrt{3}x^2$ and 1 are the terms; $3\sqrt{3}$ is the coefficient.
77. $4x^3$, $x/2$, and -5 are the terms; 4 and $\frac{1}{2}$ are the coefficients.
78. $3x^4$ and $-x^2/4$ are the terms; 3 and $-\frac{1}{4}$ are the coefficients.
79. (a) -10 (b) -6 **80.** (a) 30   (b) -12
81. (a) 14   (b) 2   **82.** (a) -10 (b) 0
83. (a) Division by 0 is undefined.   (b) 0
84. (a) $\frac{1}{2}$ (b) Division by 0 is undefined.
85. Commutative Property of Addition
86. Multiplicative Inverse Property
87. Multiplicative Inverse Property
88. Additive Inverse Property
89. Distributive Property
90. Additive Identity Property
91. Multiplicative Identity Property
92. Distributive Property
93. Associative Property of Addition
94. Associative and Commutative Properties of Multiplication
95. Distributive Property
96. $\frac{1}{7}(7 \cdot 12) = \left(\frac{1}{7} \cdot 7\right)12$ Associative Property of Multiplication
$= 1 \cdot 12$ Multiplicative Inverse Property
$= 12$ Multiplicative Identity Property
97. $\frac{1}{2}$ **98.** $\frac{2}{7}$ **99.** $\frac{3}{8}$ **100.** $\frac{59}{66}$ **101.** 48
102. -3 **103.** $\dfrac{5x}{12}$ **104.** $\dfrac{5x}{27}$
105. (a)

n	1	0.5	0.01	0.0001	0.000001
$5/n$	5	10	500	50,000	5,000,000

(b) The value of $5/n$ approaches infinity as n approaches 0.

106. (a)

n	1	10	100	10,000	100,000
$5/n$	5	0.5	0.05	0.0005	0.00005

(b) The value of $5/n$ approaches 0 as n increases without bound.

107. False. If $a < b$, then $\dfrac{1}{a} > \dfrac{1}{b}$, where $a \neq b \neq 0$.

108. False. The denominators cannot be added when adding fractions.

109. (a) No. If one variable is negative and the other is positive, the expressions are unequal.

(b) $|u + v| \leq |u| + |v|$

The expressions are equal when u and v have the same sign. If u and v differ in sign, $|u + v|$ is less than $|u| + |v|$.

110. Yes. y is nonnegative if $y \geq 0$. y is positive if $y > 0$.

111. The only even prime number is 2, because its only factors are itself and 1.

112.

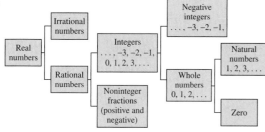

113. (a) Negative   (b) Negative

114. (a) Positive   (b) Positive

115. Yes. $|a| = -a$ if $a < 0$.

Section P.2   *(page 22)*

Vocabulary Check   *(page 22)*

1. equation   **2.** solve   **3.** identities; conditional

4. $ax + b = 0$ **5.** extraneous

6. quadratic equation

7. factoring; extracting square roots; completing the square; Quadratic Formula

---

1. Identity   **2.** Conditional equation

3. Conditional equation   **4.** Identity   **5.** Identity

6. Identity   **7.** Identity   **8.** Identity

9. Conditional equation   **10.** Conditional equation

11. 4   **12.** -12 **13.** -9 **14.** 3   **15.** 5

16. -5 **17.** 9   **18.** $-\frac{5}{8}$ **19.** No solution

20. All real numbers x **21.** -4 **22.** -5

23. $-\frac{6}{5}$ **24.** 6   **25.** 9   **26.** 50

27. No solution. The x-terms sum to zero.

28. No solution. The x-terms sum to zero.   **29.** 10

30. $\frac{1}{2}$ **31.** 4   **32.** 0   **33.** 3   **34.** $\frac{9}{7}$ **35.** 0

36. $\frac{5}{3}$ **37.** No solution. The variable is divided out.

38. $\frac{11}{6}$ **39.** No solution. The solution is extraneous.

40. $-\frac{13}{3}$ **41.** 2   **42.** $\frac{7}{4}$

43. No solution. The solution is extraneous.

44. No solution. The solution is extraneous.

45. 0   **46.** $-\frac{1}{2}$ **47.** All real numbers x

48. No solution. The x-terms sum to zero.

49. $2x^2 + 8x - 3 = 0$ **50.** $x^2 - 16x = 0$

51. $x^2 - 6x + 6 = 0$ **52.** $-3x^2 - 42x - 134 = 0$

53. $3x^2 - 90x - 10 = 0$ **54.** $4x^2 - 2x + 1 = 0$

55. $0, -\frac{1}{2}$ **56.** $-\frac{1}{3}, \frac{1}{3}$ **57.** $4, -2$ **58.** $9, 1$

59. -5 **60.** $-\frac{3}{2}$ **61.** $3, -\frac{1}{2}$ **62.** $-\frac{3}{2}, 11$

63. $2, -6$ **64.** $2, 6$ **65.** $-\frac{20}{3}, -4$ **66.** $-8, 16$

67. $-a$ **68.** $-a - b, -a + b$ **69.** ± 7 **70.** ± 13

71. $\pm\sqrt{11}$ **72.** $\pm 4\sqrt{2}$ **73.** $\pm 3\sqrt{3}$ **74.** ± 2

75. $8, 16$ **76.** $-18, -8$ **77.** $-2 \pm \sqrt{14}$

78. $5 \pm \sqrt{30}$ **79.** $\dfrac{1 \pm 3\sqrt{2}}{2}$ **80.** $-\dfrac{7}{4} \pm \dfrac{\sqrt{11}}{2}$

81. 2   **82.** $-\frac{9}{2}$ **83.** $4, -8$ **84.** $3, -1$

85. $\sqrt{11} - 6, -\sqrt{11} - 6$ **86.** $-4 \pm \sqrt{2}$

87. $1 \pm \dfrac{\sqrt{6}}{3}$ **88.** $\dfrac{2}{3} \pm \sqrt{2}$ **89.** $2 \pm 2\sqrt{3}$

90. No solution   **91.** $\dfrac{-5 \pm \sqrt{89}}{4}$ **92.** $\dfrac{11}{2}, -\dfrac{9}{2}$

93. $\frac{1}{2}, -1$ **94.** $1, -\frac{1}{2}$ **95.** $\frac{1}{4}, -\frac{3}{4}$ **96.** $\frac{3}{5}, \frac{1}{5}$

97. $1 \pm \sqrt{3}$ **98.** $5 \pm \sqrt{3}$ **99.** $-7 \pm \sqrt{5}$

100. $-3 \pm \sqrt{13}$ **101.** $-4 \pm 2\sqrt{5}$ **102.** $\dfrac{1}{2} \pm \dfrac{\sqrt{5}}{2}$

103. $\dfrac{2}{3} \pm \dfrac{\sqrt{7}}{3}$ **104.** $\dfrac{5}{4} \pm \dfrac{\sqrt{3}}{4}$ **105.** $-\dfrac{4}{3}$

106. $-\dfrac{1}{3} \pm \dfrac{\sqrt{11}}{6}$ **107.** $-\dfrac{1}{2} \pm \sqrt{2}$ **108.** $\dfrac{5}{4} \pm \dfrac{\sqrt{5}}{2}$

109. $\dfrac{2}{7}$ **110.** $-\dfrac{3}{2} \pm \dfrac{\sqrt{13}}{2}$ **111.** $2 \pm \dfrac{\sqrt{6}}{2}$

112. $-\dfrac{8}{5} \pm \dfrac{\sqrt{3}}{5}$ **113.** $6 \pm \sqrt{11}$ **114.** $-7 \pm \sqrt{13}$

115. $-\dfrac{3}{8} \pm \dfrac{\sqrt{265}}{8}$ **116.** $\dfrac{686 \pm 196\sqrt{6}}{25}$

117. $0.976, -0.643$ **118.** $1.400, -0.150$

119. $1.355, -14.071$ **120.** $2.137, 18.063$

121. $1.687, -0.488$ **122.** $0.672, -0.968$

123. $-0.290, -2.200$ **124.** $-2.995, 2.971$

125. $1 \pm \sqrt{2}$ **126.** $0, -3$ **127.** $6, -12$ **128.** 7

129. $\frac{1}{2} \pm \sqrt{3}$ **130.** $-\frac{3}{2} \pm \sqrt{3}$ **131.** $-\frac{1}{2}$

132. $-\dfrac{b}{a}, \dfrac{b}{a}$ **133.** $\dfrac{3}{4} \pm \dfrac{\sqrt{97}}{4}$ **134.** ± 1

135. $0, \pm\dfrac{3\sqrt{2}}{2}$ **136.** $0, \pm\dfrac{5}{2}$ **137.** ± 3 **138.** ± 2

139. -6 **140.** $\frac{8}{3}$ **141.** $-3, 0$ **142.** $0, \frac{4}{3}$

143. $3, 1, -1$ **144.** -2 **145.** ± 1 **146.** ± 2

147. $\pm\sqrt{3}, \pm 1$ **148.** ± 2 **149.** $\pm\frac{1}{2}, \pm 4$

150. $\pm\dfrac{\sqrt{7}}{6}$ **151.** $1, -2$ **152.** $-\sqrt[3]{2}$ **153.** 50

154. $\frac{9}{16}$ **155.** 26   **156.** -4 **157.** -16

158. $\frac{124}{3}$ **159.** $2, -5$ **160.** $3, -2$ **161.** 0

162. No solution   **163.** 9   **164.** 1   **165.** $-3 \pm 16\sqrt{2}$

166. $-29, 25$ **167.** $\pm\sqrt{14}$ **168.** $\dfrac{1 \pm 5\sqrt{5}}{2}$ **169.** 1

170. $0, 1, \frac{3}{5}$ **171.** $2, -\frac{3}{2}$ **172.** $2, -12$

173. $\dfrac{-3 \pm \sqrt{21}}{6}$ **174.** $-3, 1$ **175.** $4, -5$

176. $\dfrac{3}{4}, -1$ **177.** $\dfrac{1 \pm \sqrt{31}}{3}$ **178.** ± 1 **179.** $3, -2$

180. $\frac{5}{3}, -3$ **181.** $\sqrt{3}, -3$ **182.** $-6, -3, 3$

183. $3, \dfrac{-1 - \sqrt{17}}{2}$ **184.** $10, -1$

185. (a) 61.2 inches

(b) Yes. The estimated height of a male with a 19-inch femur is 69.4 inches.

(c)

Height, x	Female femur length	Male femur length
60	15.48	14.79
70	19.80	19.28
80	24.12	23.77
90	28.44	28.26
100	32.76	32.75
110	37.08	37.24

100 inches

(d) $x \approx 100.59$; There would not be a problem because it is not likely for either a male or a female to be 100 inches tall (which is 8 feet 4 inches tall).

186. 23,437.5 miles

187. $y = -0.25t + 8$; after about 28 hours

188. (a)

(b) $w(w + 14) = 1632$
(c) $w = 34$ feet
$l = 48$ feet

189. 6 inches $\times$ 6 inches $\times$ 2 inches

190. $\dfrac{5\sqrt{2}}{2} \approx 3.54$ centimeters

191. $\dfrac{20\sqrt{3}}{3} \approx 11.55$ inches

192. ≈ 550 miles per hour and 600 miles per hour

193. (a) 1998   (b) During 2007

194. 26,250 passengers   **195.** 500 units

196. 250,000 units

197. False. $x(3 - x) = 10$
$3x - x^2 = 10$

The equation cannot be written in the form $ax + b = 0$.

198. False. The product must equal zero for the Zero-Factor Property to be used.

199. False. See Example 14 on page A55.
200. False. $|x| = 0$ has only one solution to check, 0.
201. Equivalent equations have the same solution set, and one is derived from the other by steps for generating equivalent equations.
$2x = 5, \; 2x + 3 = 8$
202. Remove symbols of grouping, combine like terms, reduce fractions.

Add (or subtract) the same quantity to (from) each side of the equation.

Multiply (or divide) each side of the equation by the same nonzero quantity.

Interchange the two sides of the equation.
203. Yes. The student should have subtracted $15x$ from both sides to make the right side of the equation equal to zero. Factoring out an x shows that there are two solutions, $x = 0$ and $x = 6$.
204. (a) and (b) $x = -5, -\frac{10}{3}$
(c) The method used in part (a) reduces the number of algebraic steps.
205. $x^2 - 3x - 18 = 0$ **206.** $x^2 + 15x + 44 = 0$
207. $x^2 - 22x + 112 = 0$ **208.** $30x^2 + 7x - 2 = 0$
209. $x^2 - 2x - 1 = 0$ **210.** $x^2 + 6x + 4 = 0$
211. $a = 9, b = 9$
212. Isolate the absolute value by subtracting x from both sides of the equation. The expression inside the absolute value can be positive or negative, so two separate equations must be solved.
213. (a) $x = 0, -\dfrac{b}{a}$ (b) $x = 0, 1$

Section P.3   *(page 36)*

Vocabulary Check   *(page 36)*

1. Cartesian   2. Distance Formula
3. Midpoint Formula   4. solution or solution point
5. graph   6. intercepts   7. y-axis
8. circle; (h, k); r

1. A: $(2, 6)$, B: $(-6, -2)$, C: $(4, -4)$, D: $(-3, 2)$
2. A: $\left(\frac{3}{2}, -4\right)$, B: $(0, -2)$, C: $\left(-3, \frac{5}{2}\right)$, D: $(-6, 0)$
3.   **4.**

5.

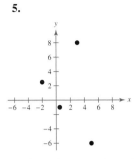

6.

7. $(-3, 4)$ **8.** $(4, -8)$ **9.** Quadrant IV

10. Quadrant III   **11.** Quadrant II   **12.** Quadrant I

13. Quadrant III   **14.** Quadrant III

15. Quadrant I or III   **16.** Quadrant II or IV

17.   **18.**

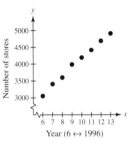

19. 8   **20.** 7   **21.** 5   **22.** 10

23. (a) 4, 3, 5   (b) $4^2 + 3^2 = 5^2$

24. (a) 5, 12, 13   (b) $5^2 + 12^2 = 13^2$

25. (a)   (b) 10
        (c) $(5, 4)$

26. (a)   (b) 13
        (c) $\left(\frac{7}{2}, 6\right)$

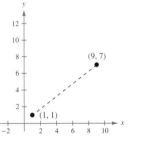

27. (a)   (b) 17
        (c) $\left(0, \frac{5}{2}\right)$

28. (a)   (b) 15
        (c) $\left(-\frac{5}{2}, 2\right)$

29. (a)   (b) $2\sqrt{10}$
        (c) $(2, 3)$

30. (a)   (b) $8\sqrt{2}$
        (c) $(6, 6)$

31. (a)   (b) $\dfrac{\sqrt{82}}{3}$
        (c) $\left(-1, \frac{7}{6}\right)$

CHAPTER P

32. (a)

(b) $\dfrac{\sqrt{2}}{6}$

(c) $\left(-\frac{1}{4}, -\frac{5}{12}\right)$

33. (a)

(b) $\sqrt{110.97}$

(c) $(1.25, 3.6)$

34. (a)

(b) $\sqrt{556.52}$

(c) $(-5.6, 8.6)$

35. $\left(\sqrt{5}\right)^2 + \left(\sqrt{45}\right)^2 = \left(\sqrt{50}\right)^2$

36. Distances between the points: $\sqrt{29}, \sqrt{58}, \sqrt{29}$

37. $\approx 250\%$ **38.** (a) $\approx 25\%$ (b) $\approx 147\%$

39. $2\sqrt{505} \approx 45$ yards   **40.** $30\sqrt{41} \approx 192$ kilometers

41. (1998, \$7652.25 million); (2000, \$8701.5 million); (2002, \$9750.75 million)

42. (1998, \$19,384.5 million); (2000, \$20,223 million); (2002, \$21,061.5 million)

43. (a) Yes   (b) Yes   **44.** (a) Yes   (b) No

45. (a) No   (b) Yes   **46.** (a) Yes   (b) No

47.

x	-1	0	1	2	$\frac{5}{2}$
y	7	5	3	1	0
(x, y)	$(-1, 7)$	$(0, 5)$	$(1, 3)$	$(2, 1)$	$\left(\frac{5}{2}, 0\right)$

48.

x	-1	0	1	2	3
y	4	0	-2	-2	0
(x, y)	$(-1, 4)$	$(0, 0)$	$(1, -2)$	$(2, -2)$	$(3, 0)$

49. x-intercepts: $(\pm 2, 0)$
    y-intercept: $(0, 16)$

50. x-intercept: $(-3, 0)$
    y-intercept: $(0, 9)$

51. x-intercept: $\left(\frac{6}{5}, 0\right)$
    y-intercept: $(0, -6)$

52. x-intercept: $\left(\frac{8}{3}, 0\right)$
    y-intercept: $(0, 8)$

53. x-intercept: $(-4, 0)$
    y-intercepts: $(0, 2)$

54. x-intercept: $\left(\frac{1}{2}, 0\right)$
    y-intercepts: $(0, 1)$

55. x-intercept: $\left(\frac{7}{3}, 0\right)$
    y-intercept: $(0, 7)$

56. x-intercept: $(-10, 0)$
    y-intercept: $(0, -10)$

57. x-intercepts: $(0, 0), (2, 0)$
    y-intercept: $(0, 0)$

58. x-intercepts: $\left(\pm\sqrt{5}, 0\right)$
    y-intercept: $(0, -25)$

59. x-intercept: $(6, 0)$
    y-intercepts: $\left(0, \pm\sqrt{6}\right)$

60. x-intercept: $(-1, 0)$
    y-intercepts: $(0, \pm 1)$

61.

62.

63.

64.

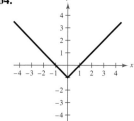

65. y-axis symmetry   **66.** x-axis symmetry

67. Origin symmetry   **68.** y-axis symmetry

69. Origin symmetry   **70.** y-axis symmetry

71. x-axis symmetry   **72.** Origin symmetry

73.

74.

83.

84.

75.

76.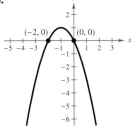

85. $(x - 2)^2 + (y + 1)^2 = 16$

86. $(x + 7)^2 + (y + 4)^2 = 49$

87. $(x + 1)^2 + (y - 2)^2 = 5$

88. $(x - 3)^2 + (y + 2)^2 = 25$

89. $(x - 3)^2 + (y - 4)^2 = 25$ **90.** $x^2 + y^2 = 17$

91. Center: $(0, 0)$; Radius: 5   **92.** Center: $(0, 0)$; Radius: 4

77.

78.

93. Center: $(1, -3)$; Radius: 3   **94.** Center: $(0, 1)$; Radius: 1

79.

80.

95. Center: $\left(\frac{1}{2}, \frac{1}{2}\right)$; Radius: $\frac{3}{2}$ **96.** Center: $(2, -3)$; Radius: $\frac{4}{3}$

81.

82.

CHAPTER P

97.   **98.**

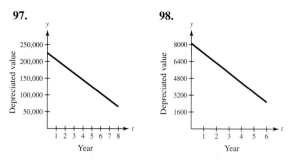

99. (a)

x	5	10	20	30	40	50
y	430.43	107.33	26.56	11.60	6.36	3.94

x	60	70	80	90	100
y	2.62	1.83	1.31	0.96	0.71

(b)

When $x = 85.5$, the resistance is 1.1 ohms.

(c) Answers will vary.

(d) As the diameter of the copper wire increases, the resistance decreases.

100. (a) and (b)

(c) 66.0 years   (d) 2005: 77.0 years; 2010: 77.1 years

(e) Answers will vary.

101. False. The Midpoint Formula would be used 15 times.

102. True. Two sides of the triangle have lengths of $\sqrt{149}$, and the third side has a length of $\sqrt{18}$.

103. False. A graph is symmetric with respect to the x-axis if, whenever (x, y) is on the graph $(x, -y)$ is also on the graph.

104. True. It is possible for a graph to have no intercepts, one intercept, or several intercepts, as shown in Figure P.21.

105. Point on x-axis: $y = 0$; Point on y-axis: $x = 0$

106. No. It depends on the magnitudes of the quantities measured.

107. Use the Midpoint Formula to prove that the diagonals of the parallelogram bisect each other.

$$\left(\frac{b + a}{2}, \frac{c + 0}{2}\right) = \left(\frac{a + b}{2}, \frac{c}{2}\right)$$
$$\left(\frac{a + b + 0}{2}, \frac{c + 0}{2}\right) = \left(\frac{a + b}{2}, \frac{c}{2}\right)$$

108. The viewing window is incorrect. Change the viewing window. Answers will vary.

109. (a) $a = 1, b = 0$ (b) $a = 0, b = 1$

110.

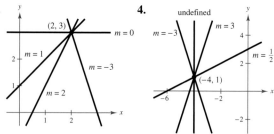

(a) The point is reflected through the y-axis.

(b) The point is reflected through the x-axis.

(c) The point is reflected through the origin.

Section P.4   *(page 49)*

Vocabulary Check   *(page 49)*

1. linear   **2.** slope   **3.** parallel

4. perpendicular   **5.** rate or rate of change

6. linear extrapolation

7. a. iii   b. i   c. v   d. ii   e. iv

1. (a) L_2 (b) L_3 (c) L_1

2. (a) L_2 (b) L_1 (c) L_3

3.   **4.**

5. $\frac{3}{2}$ **6.** $\frac{5}{2}$ **7.** -4 **8.** -1

9. $m = 5$;
y-intercept: $(0, 3)$

10. $m = 1$;
y-intercept: $(0, -10)$

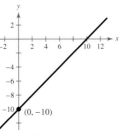

17. $m = 0$;
y-intercept: $(0, 3)$

18. $m = 0$;
y-intercept: $(0, -4)$

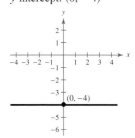

11. $m = -\frac{1}{2}$;
y-intercept: $(0, 4)$

12. $m = -\frac{3}{2}$;
y-intercept: $(0, 6)$

19. m is undefined.
There is no y-intercept.

20. m is undefined.
There is no y-intercept.

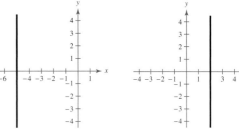

13. m is undefined.
There is no y-intercept.

14. $m = 0$;
y-intercept: $\left(0, -\frac{5}{3}\right)$

21.

$m = 2$

22.

$m = -4$

23.

m is undefined.

24.

$m = -\frac{5}{2}$

15. $m = -\frac{7}{6}$;
y-intercept: $(0, 5)$

16. $m = -\frac{2}{3}$;
y-intercept: $(0, 3)$

25.

$m = -\frac{1}{7}$

26.

$m = -\frac{8}{3}$

27.

$m = 0.15$

28.

$m = 1.425$

29. Answers will vary. Sample answer:
$(0, 1), (3, 1), (-1, 1)$

30. Answers will vary. Sample answer:
$(-4, 0), (-4, 3), (-4, 5)$

31. Answers will vary. Sample answer:
$(6, -5), (7, -4), (8, -3)$

32. Answers will vary. Sample answer:
$(0, 4), (9, -5), (11, -7)$

33. Answers will vary. Sample answer:
$(-8, 0), (-8, 2), (-8, 3)$

34. Answers will vary. Sample answer:
$(-4, -1), (-2, -1), (0, -1)$

35. Answers will vary. Sample answer:
$(-4, 6), (-3, 8), (-2, 10)$

36. Answers will vary. Sample answer:
$(-2, -5), (1, -11), (3, -15)$

37. Answers will vary. Sample answer:
$(9, -1), (11, 0), (13, 1)$

38. Answers will vary. Sample answer:
$(-3, -5), (1, -7), (5, -9)$

39. $y = 3x - 2$

40. $y = -x + 10$

41. $y = -2x$

42. $y = 4x$

43. $y = -\frac{1}{3}x + \frac{4}{3}$

44. $y = \frac{3}{4}x - \frac{7}{2}$

45. $x = 6$

46. $x = -10$

47. $y = \frac{5}{2}$

48. $y = \frac{3}{2}$

49. $y = 5x + 27.3$

50. $y = -2.5x - 2.75$

51. $y = -\frac{3}{5}x + 2$

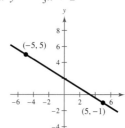

52. $y = \frac{7}{8}x - \frac{1}{2}$

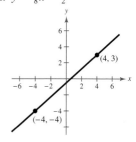

53. $x = -8$ **54.** $y = 4$

55. $y = -\frac{1}{2}x + \frac{3}{2}$ **56.** $y = -\frac{1}{3}x + \frac{4}{3}$

57. $y = -\frac{6}{5}x - \frac{18}{25}$ **58.** $y = -\frac{3}{25}x + \frac{159}{100}$

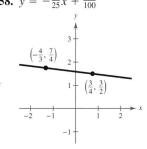

59. $y = 0.4x + 0.2$ **60.** $y = -0.3x - 1.8$

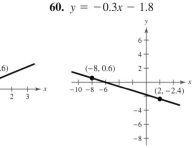

61. $y = -1$ **62.** $y = -2$

63. $x = \frac{7}{3}$ **64.** $x = 1.5$

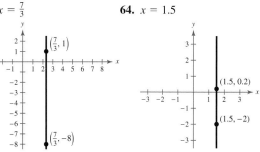

65. Perpendicular   **66.** Neither parallel nor perpendicular
67. Parallel   **68.** Perpendicular
69. (a) $y = 2x - 3$ (b) $y = -\frac{1}{2}x + 2$
70. (a) $y = -x - 1$ (b) $y = x + 5$
71. (a) $y = -\frac{3}{4}x + \frac{3}{8}$ (b) $y = \frac{4}{3}x + \frac{127}{72}$
72. (a) $y = -\frac{5}{3}x + \frac{53}{24}$ (b) $y = \frac{3}{5}x + \frac{9}{40}$
73. (a) $y = 0$ (b) $x = -1$
74. (a) $y = -2$ (b) $x = 4$
75. (a) $x = 2$ (b) $y = 5$
76. (a) $x = -5$ (b) $y = 1$
77. (a) $y = x + 4.3$ (b) $y = -x + 9.3$
78. (a) $y = -3x - 13.1$ (b) $y = \frac{1}{3}x - 0.1$
79. $3x + 2y - 6 = 0$ **80.** $4x - 3y + 12 = 0$
81. $12x + 3y + 2 = 0$ **82.** $3x - y - 2 = 0$
83. $x + y - 3 = 0$ **84.** $x + y - 1 = 0$
85. Line (b) is perpendicular to line (c).

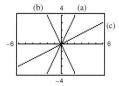

86. Line (a) is parallel to line (c). Line (b) is perpendicular to line (a) and line (c).

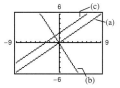

87. Line (a) is parallel to line (b).
Line (c) is perpendicular to line (a) and line (b).

88. Line (a) is parallel to line (b). Line (c) is perpendicular to line (a) and line (b).

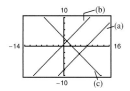

89. $3x - 2y - 1 = 0$ **90.** $5x + 13y + 2 = 0$

91. $80x + 12y + 139 = 0$ **92.** $128x + 168y + 39 = 0$

93. (a) Sales increasing 135 units per year
   (b) No change in sales
   (c) Sales decreasing 40 units per year

94. (a) Revenues increasing \$400 per day.
   (b) Revenues increasing \$100 per day.
   (c) No change in revenues.

95. (a) Salary increased greatest from 1990 to 1992; Least from 1992 to 1994
   (b) Slope of line from 1990 to 2002 is about 2351.83
   (c) Salary increased an average of \$2351.83 over the 12 years between 1990 and 2002

96. (a) Net profit showed largest increase in years 2002–2003; Least increase in years 2000–2001
   (b) Slope of line connecting 1994 and 2003 is about 9.18
   (c) Annual profits increased a factor of about 9.18 over those years

97. 12 feet

98. (a) and (b)

Horizontal measurements

   (c) $y = -\frac{1}{12}x$
   (d) For every 12 horizontal measurements, the vertical measurement decreases by 1.
   (e) "8.3% grade"

99. $V(t) = 3165 - 125t$ **100.** $V(t) = 4.5t + 133.5$

101. b; The slope is -20, which represents the decrease in the amount of the loan each week. The y-intercept is $(0, 200)$ which represents the original amount of the loan.

102. c; The slope is 2, which represents the hourly wage per unit produced. The y-intercept is $(0, 8.50)$ which represents the initial hourly wage.

103. a; The slope is 0.32, which represents the increase in travel cost for each mile driven. The y-intercept is $(0, 30)$ which represents the amount per day for food.

104. d; The slope is -100, which represents the decrease in the value of the word processor each year. The y-intercept is $(0, 750)$ which represents the initial purchase price of the computer.

105. $y = 0.4825t - 2.2325$; $y(18) \approx \$6.45$; $y(20) \approx \$7.42$

106. $y(t) = -749.5t + 10,821.5$; $y(18) = -2669.5$; $y(20) = -4168.5$; Answers will vary.

107. $V = -175t + 875$ **108.** $V = 25,000 - 2300t$

109. (a) $y(t) = 179.5t + 40,571$
   (b) $y(8) = 42,007$; $y(10) = 42,366$ (c) $m = 179.5$

110. (a) Average annual change from 1990 to 2003 was 934.
   (b) 1994; 40,267; 1998: 44,003; 2002: 47,739
   (c) $y(t) = 934t + 36,531$; $m = 934$
   (d) Answers will vary.

111. $S = 0.85L$ **112.** $W = 0.75x + 11.50$

113. (a) $C = 16.75t + 36,500$ (b) $R = 27t$
   (c) $P = 10.25t - 36,500$ (d) $t \approx 3561$ hours

114. (a) $x = -\frac{1}{15}p + \frac{266}{3}$ (b) 45   (c) 49

115. (a)   (b) $y = 8x + 50$

   (c)   (d) $m = 8$, 8 meters

116. $W = 0.07S + 2500$ **117.** $C = 0.38x + 120$

118. Answers will vary. Sample answer: $y = 327.3t - 1854$

119. (a) and (b)

Cellular phone subscribers (in millions)

Year (0 ↔ 1990)

   (c) Answers will vary. Sample answer:
      $y = 11.72x - 14.1$
   (d) Answers will vary. Sample answer: The y-intercept indicates that initially there were -14.1 million subscribers which doesn't make sense in the context of this problem. Each year, the number of cellular phone subscribers increases by 11.72 million.
   (e) The model is accurate.
   (f) Answers will vary. Sample answer: 196.9 million

120. (a) and (b)

Average quiz score

(c) $y = 4x + 19$ (d) 87

(e) Vertical shift four units upward

121. False. The slope with the greatest magnitude corresponds to the steepest line.

122. False. The slope of the first line is $\frac{2}{7}$ and the slope of the second line is $-\frac{11}{7}$.

123. Find the distance between each two points and use the Pythagorean Theorem.

124. The slope of a vertical line is undefined because division by zero is undefined.

125. No. The slope cannot be determined without knowing the scale on the y-axis. The slopes could be the same.

126. The line with a slope of -4 is steeper. The slope with the greatest magnitude corresponds to the steepest line.

127. V-intercept: initial cost; Slope: annual depreciation

128. No. The slopes of two perpendicular lines have opposite signs (assume that neither line is vertical or horizontal).

129. Answers will vary.

Section P.5   *(page 63)*

Vocabulary Check   *(page 63)*

1. domain; range; function

2. verbally; numerically; graphically; algebraically

3. independent; dependent   **4.** piecewise-defined

5. implied domain   **6.** difference quotient

1. Yes   **2.** No   **3.** No   **4.** Yes

5. Yes, each input value has exactly one output value.

6. No, the input values of 0 and 1 each have two different output values.

7. No, the input values of 7 and 10 each have two different output values.

8. Yes, it does not matter that each input value has the same output value.

9. (a) Function

   (b) Not a function, because the element 1 in A corresponds to two elements, -2 and 1, in B.

   (c) Function

   (d) Not a function, because not every element in A is matched with an element in B.

10. (a) Not a function, because the element c in A corresponds to two elements, 2 and 3, in B.

   (b) Function

   (c) Not a function from A to B. (It is instead a function from B to A.)

   (d) Function

11. Each is a function. For each year there corresponds one and only one circulation.

12. 11 million   **13.** Not a function   **14.** Not a function

15. Function   **16.** Not a function   **17.** Function

18. Not a function   **19.** Not a function   **20.** Function

21. Function   **22.** Not a function   **23.** Not a function

24. Function   **25.** (a) -1 (b) -9 (c) $2x - 5$

26. (a) 7   (b) 0   (c) $1 - 3s$

27. (a) 36π (b) $\frac{9}{2}\pi$ (c) $\frac{32}{3}\pi r^3$

28. (a) 0   (b) -0.75 (c) $x^2 + 2x$

29. (a) 1   (b) 2.5   (c) $3 - 2|x|$

30. (a) 2   (b) 5   (c) $\sqrt{x} + 2$

31. (a) $-\dfrac{1}{9}$ (b) Undefined   (c) $\dfrac{1}{y^2 + 6y}$

32. (a) $\dfrac{11}{4}$ (b) Undefined   (c) $\dfrac{2x^2 + 3}{x^2}$

33. (a) 1   (b) -1 (c) $\dfrac{|x - 1|}{x - 1}$

34. (a) 6   (b) 6   (c) $x^2 + 4$

35. (a) -1 (b) 2   (c) 6

36. (a) 6   (b) 3   (c) 10

37. (a) -7 (b) 4   (c) 9

38. (a) 19   (b) 17   (c) 0

39.

x	-2	-1	0	1	2
$f(x)$	1	-2	-3	-2	1

40.

x	3	4	5	6	7
$g(x)$	0	1	$\sqrt{2}$	$\sqrt{3}$	2

41.

t	-5	-4	-3	-2	-1
$h(t)$	1	$\frac{1}{2}$	0	$\frac{1}{2}$	1

42.

s	0	1	$\frac{3}{2}$	$\frac{5}{2}$	4
$f(s)$	-1	-1	-1	1	1

43.

x	-2	-1	0	1	2
$f(x)$	5	$\frac{9}{2}$	4	1	0

44.

x	1	2	3	4	5
$f(x)$	8	5	0	1	2

45. 5   **46.** $-\frac{1}{5}$ **47.** $\frac{4}{3}$ **48.** $\pm 2\sqrt{3}$ **49.** ± 3

50. 3, 5   **51.** $0, \pm 1$ **52.** $1, \pm 2$ **53.** $2, -1$

54. $0, \pm 2$ **55.** 3, 0   **56.** 4

57. All real numbers   **58.** All real numbers

59. All real numbers t except $t = 0$

60. All real numbers y except $y = -5$

61. All real numbers y such that $y \geq 0$

62. All real numbers

63. All real numbers x such that $-1 \leq x \leq 1$

64. All real numbers x such that $x \leq -3$ and $x \geq 0$

65. All real numbers x except $x = 0, -2$

66. All real numbers x except $x = 0, 2$

67. All real numbers s such that $s \geq 1$ except $s = 4$

68. All real numbers x such that $x > -6$

69. All real numbers x such that $x > 0$

70. All real numbers x such that $x < -3, x > 3$

71. $\{(-2, 4), (-1, 1), (0, 0), (1, 1), (2, 4)\}$

72. $\{(-2, 1), (-1, -2), (0, -3), (1, -2), (2, 1)\}$

73. $\{(-2, 4), (-1, 3), (0, 2), (1, 3), (2, 4)\}$

74. $\{(-2, 1), (-1, 0), (0, 1), (1, 2), (2, 3)\}$

75. $g(x) = cx^2; c = -2$ **76.** $f(x) = cx; c = \frac{1}{4}$

77. $r(x) = \dfrac{c}{x}; c = 32$ **78.** $h(x) = c\sqrt{|x|}; c = 3$

79. $3 + h, h \neq 0$ **80.** $-(5 + h), \; h \neq 0$

81. $3x^2 + 3xh + h^2 + 3, h \neq 0$

82. $8x + 4h - 2, h \neq 0$ **83.** $-\dfrac{x + 3}{9x^2}, x \neq 3$

84. $\dfrac{1}{t - 2}, t \neq 1$ **85.** $\dfrac{\sqrt{5x} - 5}{x - 5}$ **86.** $\dfrac{x^{2/3} - 4}{x - 8}$

87. $A = \dfrac{P^2}{16}$ **88.** $A = \dfrac{C^2}{4\pi}$

89. (a) The maximum volume is 1024 cubic centimeters.

  (b)

    Yes, V is a function of x.

  (c) $V = x(24 - 2x)^2, 0 < x < 12$

90. (a) The maximum profit is \$3375.

  (b)

    Yes, P is a function of x.

  (c) $P = 45x - 0.15x^2, x > 100$

91. $A = \dfrac{x^2}{2(x - 2)}, x > 2$

92. $A = 2xy = 2x\sqrt{36 - x^2}, \; 0 < x < 6$

93. Yes, the ball will be at a height of 6 feet.

94. 1991: \$42 billion   **95.** 1990: \$27,300

    1992: \$47 billion   1991: \$28,052

    1993: \$52 billion   1992: \$29,168

    1994: \$57 billion   1993: \$30,648

    1995: \$62 billion   1994: \$32,492

    1996: \$67 billion   1995: \$34,700

    1997: \$72 billion   1996: \$37,272

    1998: \$85.6 billion   1997: \$40,208

    1999: \$104.3 billion   1998: \$41,300

    2000: \$123 billion   1999: \$43,800

    2001: \$141.7 billion   2000: \$46,300

    2002: \$160.4 billion   2001: \$48,800

                            2002: \$51,300

96. (a) $V = x^2y = x^2(108 - 4x)$

         $= 108x^2 - 4x^3, 0 < x < 27$

  (b)

  (c) 18 inches $\times$ 18 inches $\times$ 36 inches

97. (a) $C = 12.30x + 98{,}000$

  (b) $R = 17.98x$

  (c) $P = 5.68x - 98{,}000$

98. (a) $C = 6000 + 0.95x$ (b) $\overline{C} = \dfrac{C}{x} = \dfrac{6000}{x} + 0.95$

99. (a) $R = \dfrac{240n - n^2}{20}, n \geq 80$

  (b)

n	90	100	110	120	130	140	150
$R(n)$	\$675	\$700	\$715	\$720	\$715	\$700	\$675

The revenue is maximum when 120 people take the trip.

100. (a)

y	5	10	20
$F(y)$	26,474.08	149,760.00	847,170.49

y	30	40
$F(y)$	2,334,527.36	4,792,320.00

The deeper the water, the greater the force.

(b) 21 feet   (c) 21.37 feet

101. (a)

(b) $h = \sqrt{d^2 - 3000^2}, d \geq 3000$

102. (a) 1.428. Approximately 1.428 species of fish became threatened and/or endangered each year between 1996 and 2003.

(b) $N = \begin{cases} 2x + 104, & 6 \leq x \leq 7 \\ 2x + 103, & 8 \leq x \leq 11 \\ 126, & 12 \leq x \leq 13 \end{cases}$

(c)

x	6	7	8	9	10	11	12	13
N	116	118	119	121	123	125	126	126

(d) The algebraic model is a good fit to the actual data.

(e) $N = 1.5x + 107$; The model found in part (b) is a better model than the model given by the graphing utility.

103. False. The range is $[-1, \infty)$.

104. True. Each x-value corresponds to one y-value.

105. The domain is the set of inputs of the function, and the range is the set of outputs.

106. Domain of $f(x)$: all real numbers x such that $x \geq -2$

Domain of $g(x)$: all real numbers x

107. (a) Yes. The amount you pay in sales tax will increase as the price of the item purchased increases.

(b) No. The length of time that you study will not necessarily determine how well you do on an exam.

108. (a) No, not necessarily. It depends on how you manage your money.

(b) Yes. The speed of the baseball will increase as the height from which it was dropped increases.

Section P.6   *(page 76)*

Vocabulary Check   *(page 76)*

1. ordered pairs   **2.** Vertical Line Test
3. zeros   **4.** decreasing
5. maximum   **6.** average rate of change; secant
7. odd   **8.** even

1. Domain: $(-\infty, -1], [1, \infty)$ **2.** Domain: $(-\infty, \infty)$
Range: $[0, \infty)$ Range: $[0, \infty)$
3. Domain: $[-4, 4]$ **4.** Domain: $(-\infty, 1), (1, \infty)$
Range: $[0, 4]$ Range: $-1, 1$
5. (a) 0   (b) -1 (c) 0   (d) -2
6. (a) 4   (b) 4   (c) 2   (d) 0
7. (a) -3 (b) 0   (c) 1   (d) -3
8. (a) 0   (b) 1   (c) 2   (d) 3
9. Function   **10.** Function   **11.** Not a function
12. Not a function   **13.** Function   **14.** Not a function
15. $-\frac{5}{2}, 6$ **16.** $-8, \frac{2}{3}$ **17.** 0   **18.** 2, 7
19. $0, \pm\sqrt{2}$ **20.** $\pm 3, 4$ **21.** $\pm\frac{1}{2}, 6$ **22.** $0, \pm\frac{5}{3}$
23. $\frac{1}{2}$ **24.** $-\frac{2}{3}$

25.

$-\frac{5}{3}$

26.

$0, 7$

27.

$-\frac{11}{2}$

28.

26

29.

$\frac{1}{3}$

30.

$\pm\frac{3\sqrt{2}}{2}$

31. Increasing on $(-\infty, \infty)$
32. Increasing on $(2, \infty)$; Decreasing on $(-\infty, 2)$
33. Increasing on $(-\infty, 0)$ and $(2, \infty)$
Decreasing on $(0, 2)$
34. Increasing on $(1, \infty)$; Decreasing on $(-\infty, -1)$
35. Increasing on $(-\infty, 0)$ and $(2, \infty)$; Constant on $(0, 2)$

CHAPTER P

36. Increasing on $(-\infty, -1)$, $(0, \infty)$; Decreasing on $(-1, 0)$

37. Increasing on $(1, \infty)$; Decreasing on $(-\infty, -1)$
Constant on $(-1, 1)$

38. Increasing on $(-\infty, -2)$, $(0, \infty)$
Decreasing on $(-2, -1)$, $(-1, 0)$

39.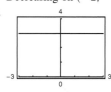
Constant on $(-\infty, \infty)$

40.
Increasing on $(-\infty, \infty)$

41.
Decreasing on $(-\infty, 0)$
Increasing on $(0, \infty)$

42.
Increasing on $(0, \infty)$
Decreasing on $(-\infty, 0)$

43.
Increasing on $(-\infty, 0)$
Decreasing on $(0, \infty)$

44.
Increasing on $(-1, 0)$, $(1, \infty)$
Decreasing on $(-\infty, -1)$, $(0, 1)$

45.
Decreasing on $(-\infty, 1)$

46.
Increasing on $(-2, \infty)$
Decreasing on $(-3, -2)$

47.
Increasing on $(0, \infty)$

48.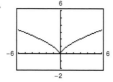
Decreasing on $(-\infty, 0)$
Increasing on $(0, \infty)$
Relative minimum: $(1, -9)$

49.

50.
Relative minimum: $\left(\frac{1}{3}, -\frac{16}{3}\right)$

51.
Relative maximum: $(1.5, 0.25)$

52.
Relative maximum: $(2.25, 10.125)$

53.
Relative maximum: $(-1.79, 8.21)$
Relative minimum: $(1.12, -4.06)$

54.
Relative maximum: $(-0.15, 1.08)$
Relative minimum: $(2.15, -5.08)$

55.
$(-\infty, 4]$

56.
$\left[-\frac{1}{2}, \infty\right)$

57.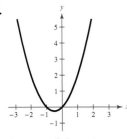

$(-\infty, -1], [0, \infty)$

58.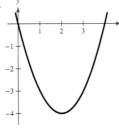

$(-\infty, 0], [4, \infty)$

59.

$[1, \infty)$

60.

$[-2, \infty)$

61.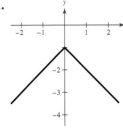

$f(x) < 0$ for all x

62.

$(-\infty, \infty)$

63. The average rate of change from $x_1 = 0$ to $x_2 = 3$ is -2.
64. The average rate of change from $x_1 = 0$ to $x_2 = 3$ is 3.
65. The average rate of change from $x_1 = 1$ to $x_2 = 5$ is 18.
66. The average rate of change from $x_1 = 1$ to $x_2 = 5$ is 4.
67. The average rate of change from $x_1 = 1$ to $x_2 = 3$ is 0.
68. The average rate of change from $x_1 = 1$ to $x_2 = 6$ is 0.
69. The average rate of change from $x_1 = 3$ to $x_2 = 11$ is $-\frac{1}{4}$.
70. The average rate of change from $x_1 = 3$ to $x_2 = 8$ is $-\frac{1}{5}$.
71. Even; y-axis symmetry
72. Neither even nor odd; no symmetry
73. Odd; origin symmetry   **74.** Odd; origin symmetry
75. Neither even nor odd; no symmetry
76. Even; y-axis symmetry
77. $h = -x^2 + 4x - 3$ **78.** $h = 3 - 4x + x^2$
79. $h = 2x - x^2$ **80.** $h = 2 - \sqrt[3]{x}$
81. $L = \frac{1}{2}y^2$ **82.** $L = 2 - \sqrt[3]{2y}$ **83.** $L = 4 - y^2$
84. $L = \dfrac{2}{y}$
85. (a)   (b) 30 watts

86. (a)

(b) The model fits the data very well.
(c) The temperature was increasing from 6 A.M. $(x = 0)$ to noon $(x = 6)$, and again from 2 A.M. $(x = 20)$ to 6 A.M. $(x = 24)$. The temperature was decreasing from noon to 2 A.M.
(d) The maximum temperature was 63.93°F and the minimum temperature was 33.98°F.
(e) Answers will vary.
87. (a) Ten thousands   (b) Ten millions   (c) Percents
88. (a) $A = 64 - 2x^2$, $0 \le x \le 4$
(b) $32 \le A \le 64$

(c) Square with sides of $4\sqrt{2}$ meters
89. (a)

(b) The average rate of change from 2002 to 2007 is 408.56. The estimated revenue is increasing each year at a fast pace.

90. (a)

(b) The average rate of change of the model from 1992 to 2002 is 14.73. This shows that the number of students enrolling in college increased at a fairly steady rate.
(c) The five-year time period when the rate of change was the least was from 1992 to 1997 with a 6.05 average rate, and the greatest from 1997 to 2002 with a 23.41 average rate.
91. (a) $s = -16t^2 + 64t + 6$
(b)

(c) Average rate of change $= 16$
(d) The slope of the secant line is positive.
(e) Secant line: $16t + 6$

CHAPTER P

(f)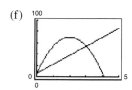

92. (a) $s = -16t^2 + 72t + 6.5$

(b)

(c) Average rate of change $= 8$
(d) The slope of the secant line is positive.
(e) Secant line $= 8t + 6.5$
(f)

93. (a) $s = -16t^2 + 120t$

(b)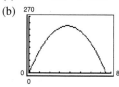

(c) Average rate of change $= -8$
(d) The slope of the secant line is negative.
(e) Secant line: $-8t + 240$
(f)

94. (a) $s = -16t^2 + 96t$

(b)

(c) Average rate of change $= -16$
(d) The slope of the secant line is negative.
(e) Secant line $= -16t + 160$
(f)

95. (a) $s = -16t^2 + 120$

(b)

(c) Average rate of change $= -32$
(d) The slope of the secant line is negative.
(e) Secant line: $-32t + 120$
(f)

96. (a) $s = -16t^2 + 80$

(b)

(c) Average rate of change $= -48$
(d) The slope of the secant line is negative.
(e) Secant line $= -48t + 112$
(f)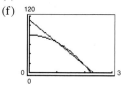

97. False. The function $f(x) = \sqrt{x^2 + 1}$ has a domain of all real numbers.

98. False. An odd function is symmetric with respect to the origin, so its domain must include negative values.

99. (a) Even. The graph is a reflection in the x-axis.
(b) Even. The graph is a reflection in the y-axis.
(c) Even. The graph is a vertical translation of f.
(d) Neither. The graph is a horizontal translation of f.

100. Yes. For each value of y there corresponds one and only one value of x.

101. (a) $\left(\frac{3}{2}, 4\right)$ (b) $\left(\frac{3}{2}, -4\right)$

102. (a) $\left(\frac{5}{3}, -7\right)$ (b) $\left(\frac{5}{3}, 7\right)$

103. (a) $(-4, 9)$ (b) $(-4, -9)$

104. (a) $(-5, -1)$ (b) $(-5, 1)$

105. (a) (b)

(c) (d)

(e) (f)

All the graphs pass through the origin. The graphs of the odd powers of x are symmetric with respect to the origin, and the graphs of the even powers are symmetric with respect to the y-axis. As the powers increase, the graphs become flatter in the interval $-1 < x < 1$.

106. Both graphs will pass through the origin. $y = x^7$ will be symmetric with respect to the origin, and $y = x^8$ will be symmetric with respect to the y-axis.

Section P.7 *(page 86)*

1.
(a) $f(x) = -2x + 6$
(b)

2.
(a) $f(x) = \frac{5}{2}x - \frac{1}{2}$
(b)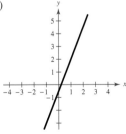

3.
(a) $f(x) = -3x + 11$
(b)

4.
(a) $f(x) = 5x - 6$
(b)

5.
(a) $f(x) = -1$
(b)

6.
(a) $f(x) = -\frac{1}{2}x + 7$
(b)

7.
(a) $f(x) = \frac{6}{7}x - \frac{45}{7}$
(b)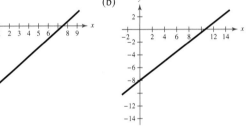

8.
(a) $f(x) = \frac{3}{4}x - 8$
(b)

9.

10.

11.

12.

13.

14.

15.

16.

17.

18.

19.

20.

21.

22.

23.

24.

25.

26.

27.

28.

29. (a) 2   (b) 2   (c) −4   (d) 3
30. (a) −6   (b) 0   (c) 18   (d) 6
31. (a) 1   (b) 3   (c) 7   (d) −19
32. (a) 7   (b) −1   (c) 31   (d) 11
33. (a) 6   (b) −11   (c) 6   (d) −22
34. (a) 8   (b) 2   (c) 6   (d) 13
35. (a) −10   (b) −4   (c) −1   (d) 41
36. (a) −22   (b) −85   (c) 6   (d) −29

37.

38.

39.

40.

41.

42.

43.

44.

45.

46.

47.

48.

61. (a)

Time (in minutes)

(b) $5.64

49.

50.
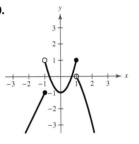

62. (a) C_2 is the appropriate model, because the cost does not increase until after the next minute of conversation has started.

(b)

Time (in minutes)

$7.89

51. (a)

(b) Domain: $(-\infty, \infty)$;
Range: $[0, 2)$

(c) Sawtooth pattern

52. (a)

(b) Domain: $(-\infty, \infty)$;
Range: $[0, 2)$

(c) Sawtooth pattern

53. (a) $f(x) = |x|$ (b) $g(x) = |x + 2| - 1$

54. (a) $f(x) = \sqrt{x}$ (b) $g(x) = 1 + \sqrt{x + 2}$

55. (a) $f(x) = x^3$ (b) $g(x) = (x - 1)^3 - 2$

56. (a) $f(x) = \dfrac{1}{x}$ (b) $g(x) = \dfrac{1}{x} - 2$

57. (a) $f(x) = 2$ (b) $g(x) = 2$

58. (a) $f(x) = x^2$ (b) $g(x) = 1 - (x + 2)^2$

59. (a) $f(x) = x$ (b) $g(x) = x - 2$

60. (a) $f(x) = [\![x]\!]$ (b) $g(x) = [\![x - 1]\!]$

63. (a)

Weight (in pounds)

(b) $50.25

64. (a) $C = 9.8 + 2.5[\![x]\!]$

(b)

Weight (in pounds)

65. (a) $W(30) = 360$; $W(40) = 480$;
$W(45) = 570$; $W(50) = 660$

(b) $W(h) = \begin{cases} 12h, & 0 < h \le 45 \\ 18(h - 45) + 540, & h > 45 \end{cases}$

66. $f(t) = \begin{cases} t, & 0 \le t \le 2 \\ 2t - 2, & 2 < t \le 8 \\ \frac{1}{2}t + 10, & 8 < t \le 9 \end{cases}$

Total accumulation = 14.5 inches

67. (a) $f(x) = \begin{cases} 0.505x^2 - 1.47x + 6.3, & 1 \le x \le 6 \\ -1.97x + 26.3, & 6 < x \le 12 \end{cases}$

Answers will vary. Sample answer: The domain is determined by inspection of a graph of the data with the two models.

(b)

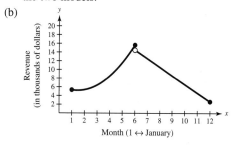

(c) $f(5) = 11.575, f(11) = 4.63$; These values represent the revenue for the months of May and November, respectively.

(d) These values are quite close to the actual data values.

68.

Interval	Input Pipe	Drainpipe 1	Drainpipe 2
$[0, 5]$	Open	Closed	Closed
$[5, 10]$	Open	Open	Closed
$[10, 20]$	Closed	Closed	Closed
$[20, 30]$	Closed	Closed	Open
$[30, 40]$	Open	Open	Open
$[40, 45]$	Open	Closed	Open
$[45, 50]$	Open	Open	Open
$[50, 60]$	Open	Open	Closed

69. False. A piecewise-defined function is a function that is defined by two or more equations over a specified domain. That domain may or may not include x- and y-intercepts.

70. True. The solution sets are the same.

71. $f(x) = \begin{cases} -\frac{4}{3}x + 6, & 0 \le x \le 3 \\ -\frac{2}{5}x + \frac{16}{5}, & 3 < x \le 8 \end{cases}$

72. $f(x) = \begin{cases} x^2, & x \le 2 \\ 7 - x, & x > 2 \end{cases}$

Section P.8 *(page 94)*

Vocabulary Check *(page 94)*

1. rigid **2.** $-f(x); f(-x)$ **3.** nonrigid
4. horizontal shrink; horizontal stretch
5. vertical stretch; vertical shrink
6. (a) iv (b) ii (c) iii (d) i

1. (a) (b)

(c)

2. (a) (b)

(c)

3. (a)

(b)

(c)

4. (a)

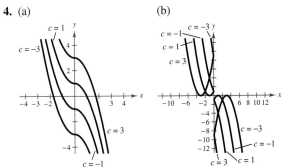

(b)

5. (a)

(b)

(c)

(d)

(e)

(f)

(g)

6. (a)

(b)

(c)

(d)

(e)

(f)

(g)

8. (a)

(b)

7. (a)

(b)

(c)

(d)

(c)

(d)

(e)

(f)

(e)

(f)

(g)

(g)

9. (a) $y = x^2 - 1$ (b) $y = 1 - (x + 1)^2$
(c) $y = -(x - 2)^2 + 6$ (d) $y = (x - 5)^2 - 3$

10. (a) $y = 1 - x^3$ (b) $y = (x - 1)^3 + 1$
(c) $y = -(x + 3)^3 - 1$ (d) $y = (x - 10)^3 - 4$

11. (a) $y = |x| + 5$ (b) $y = -|x + 3|$
(c) $y = |x - 2| - 4$ (d) $y = -|x - 6| - 1$

12. (a) $y = \sqrt{x} - 3$ (b) $y = \sqrt{x + 1} - 7$
(c) $y = -\sqrt{x - 5} + 5$ (d) $y = -\sqrt{-x + 3} - 4$

13. Horizontal shift of $y = x^3$; $y = (x - 2)^3$

14. Vertical shrink of $y = x$; $y = \frac{1}{2}x$

15. Reflection in the x-axis of $y = x^2$; $y = -x^2$

16. Vertical shift of $y = [\![x]\!]$; $y = [\![x]\!] + 4$

17. Reflection in the x-axis and vertical shift of $y = \sqrt{x}$; $y = 1 - \sqrt{x}$

18. Horizontal shift of $y = |x|$; $y = |x + 2|$

19. (a) $f(x) = x^2$

(b) Reflection in the x-axis, and vertical shift 12 units upward, of $f(x) = x^2$

(c)

(d) $g(x) = 12 - f(x)$

20. (a) $f(x) = x^2$

(b) Horizontal shift eight units to the right, of $f(x) = x^2$

(c)

(d) $g(x) = f(x - 8)$

21. (a) $f(x) = x^3$

(b) Vertical shift seven units upward, of $f(x) = x^3$

(c)

(d) $g(x) = f(x) + 7$

22. (a) $f(x) = x^3$

(b) Reflection in the x-axis, and a vertical shift of one unit downward, of $f(x) = x^3$

(c)

(d) $g(x) = -f(x) - 1$

23. (a) $f(x) = x^2$

(b) Vertical shrink of two-thirds, and vertical shift four units upward, of $f(x) = x^2$

(c)

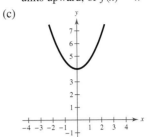

(d) $g(x) = \frac{2}{3}f(x) + 4$

24. (a) $f(x) = x^2$

(b) Vertical stretch of two, and horizontal shift seven units to the right, of $f(x) = x^2$

(c)

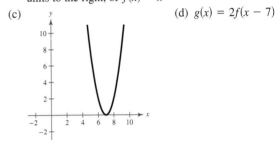

(d) $g(x) = 2f(x - 7)$

25. (a) $f(x) = x^2$

(b) Reflection in the x-axis, horizontal shift five units to the left, and vertical shift two units upward, of $f(x) = x^2$

(c)

(d) $g(x) = 2 - f(x + 5)$

26. (a) $f(x) = x^2$

(b) Reflection in the x-axis, horizontal shift 10 units to the left, and vertical shift five units upward, of $f(x) = x^2$

(c)

(d) $g(x) = -f(x + 10) + 5$

27. (a) $f(x) = \sqrt{x}$

(b) Horizontal shrink of $\frac{1}{3}$, of $f(x) = \sqrt{x}$

(c) 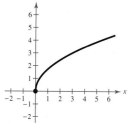 (d) $g(x) = f(3x)$

28. (a) $f(x) = \sqrt{x}$

(b) Horizontal stretch of four, of $f(x) = \sqrt{x}$

(c) 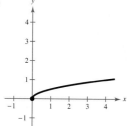 (d) $g(x) = f\left(\frac{1}{4}x\right)$

29. (a) $f(x) = x^3$

(b) Vertical shift two units upward, and horizontal shift one unit to the right, of $f(x) = x^3$

(c) (d) $g(x) = f(x-1) + 2$

30. (a) $f(x) = x^3$

(b) Vertical shift 10 units downward, and horizontal shift three units to the left, of $f(x) = x^3$

(c)

(d) $g(x) = f(x+3) - 10$

31. (a) $f(x) = |x|$

(b) Reflection in the x-axis, and vertical shift two units downward, of $f(x) = |x|$

(c) 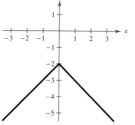 (d) $g(x) = -f(x) - 2$

32. (a) $f(x) = |x|$

(b) Reflection in the x-axis, horizontal shift five units to the left, and vertical shift six units upward, of $f(x) = |x|$

(c) 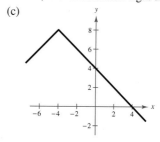 (d) $g(x) = 6 - f(x+5)$

33. (a) $f(x) = |x|$

(b) Reflection in the x-axis, horizontal shift four units to the left, and vertical shift eight units upward, of $f(x) = |x|$

(c)

(d) $g(x) = -f(x+4) + 8$

34. (a) $f(x) = |x|$

(b) Reflection in the y-axis, horizontal shift three units to the right, and vertical shift nine units upward, of $f(x) = |x|$

(c)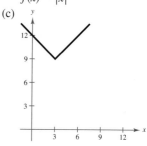

(d) $g(x) = f(-(x-3)) + 9$

35. (a) $f(x) = [\![x]\!]$
   (b) Reflection in the x-axis, and vertical shift three units upward, of $f(x) = [\![x]\!]$
   (c)
   (d) $g(x) = 3 - f(x)$

36. (a) $f(x) = \|x\|$
   (b) Horizontal shift five units to the left, and vertical stretch of two, of $f(x) = \|x\|$
   (c)
   (d) $g(x) = 2f(x + 5)$

37. (a) $f(x) = \sqrt{x}$
   (b) Horizontal shift of nine units to the right, of $f(x) = \sqrt{x}$
   (c)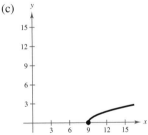
   (d) $g(x) = f(x - 9)$

38. (a) $f(x) = \sqrt{x}$
   (b) Horizontal shift of four units to the left, and vertical shift eight units upward, of $f(x) = \sqrt{x}$
   (c)
   (d) $g(x) = f(x + 4) + 8$

39. (a) $f(x) = \sqrt{x}$
   (b) Reflection in the y-axis, horizontal shift of seven units to the right, and vertical shift two units downward, of $f(x) = \sqrt{x}$
   (c)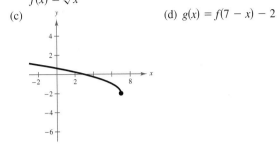
   (d) $g(x) = f(7 - x) - 2$

40. (a) $f(x) = \sqrt{x}$
   (b) Reflection in the x-axis, horizontal shift of one unit to the left, and vertical shift six units downward, of $f(x) = \sqrt{x}$
   (c)
   (d) $g(x) = -f(x + 1) - 6$

41. (a) $f(x) = \sqrt{x}$
   (b) Horizontal stretch, and vertical shift four units downward, of $f(x) = \sqrt{x}$
   (c)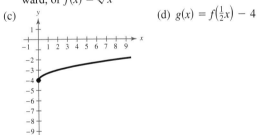
   (d) $g(x) = f\left(\frac{1}{2}x\right) - 4$

42. (a) $f(x) = \sqrt{x}$
   (b) Horizontal shrink of $\frac{1}{3}$, and vertical shift one unit upward, of $f(x) = \sqrt{x}$
   (c)
   (d) $g(x) = f(3x) + 1$

CHAPTER P

43. $f(x) = (x - 2)^2 - 8$ **44.** $f(x) = -(x + 3)^2 - 7$
45. $f(x) = (x - 13)^3$ **46.** $f(x) = (-x + 6)^3 - 6$
47. $f(x) = -|x| - 10$ **48.** $f(x) = |x + 1| - 7$
49. $f(x) = -\sqrt{-x + 6}$ **50.** $f(x) = -\sqrt{-x} + 9$
51. (a) $y = -3x^2$ (b) $y = 4x^2 + 3$
52. (a) $y = \frac{1}{4}x^3$ (b) $y = -2x^3$
53. (a) $y = -\frac{1}{2}|x|$ (b) $y = 3|x| - 3$
54. (a) $y = 8\sqrt{x}$ (b) $y = -\frac{1}{4}\sqrt{x}$
55. Vertical stretch of $y = x^3$; $y = 2x^3$
56. Vertical stretch of $y = |x|$; $y = 6|x|$
57. Reflection in the x-axis and vertical shrink of $y = x^2$;
$y = -\frac{1}{2}x^2$
58. Horizontal stretch of $y = [\![x]\!]$; $y = [\![\frac{1}{2}x]\!]$
59. Reflection in the y-axis and vertical shrink of $y = \sqrt{x}$;
$y = \frac{1}{2}\sqrt{-x}$
60. Reflection in the x-axis, vertical stretch, and vertical shift
two units downward, of $y = |x|$;
$y = -2|x| - 2$
61. $y = -(x - 2)^3 + 2$ **62.** $y = |x + 4| - 2$
63. $y = -\sqrt{x} - 3$ **64.** $y = (x - 2)^2 + 4$
65. (a)   (b)

(c)   (d)

(e)   (f)

66. (a)   (b)

(c)   (d)

(e)   (f)

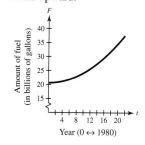

67. (a) Horizontal stretch of 0.035 and a vertical shift of 20.6
units upward.

(b) 0.77-billion-gallon increase in fuel usage by trucks
each year
(c) $f(t) = 20.6 + 0.035(t + 10)^2$. The graph is shifted
10 units to the left.
(d) 52.1 billion gallons. Yes.

68. (a) Horizontal shift of 20.396 units to the left and vertical shrink of 0.0054

Year $(0 \leftrightarrow 1990)$

(b) $f(t) = 0.0054(t + 30.396)^2$. The graph is shifted 10 units to the left.

69. True. $|-x| = |x|$

70. False. The point $(-2, -67)$ will lie on the transformation.

71. (a) $g(t) = \frac{3}{4}f(t)$ (b) $g(t) = f(t) + 10{,}000$
   (c) $g(t) = f(t - 2)$

72. If you consider the x-axis to be a mirror, the graph of $y = -f(x)$ is the mirror image of the graph of $y = f(x)$.

73. $(-2, 0), (-1, 1), (0, 2)$ **74.** Answers will vary.

Section P.9   *(page 104)*

Vocabulary Check   *(page 104)*

1. addition; subtraction; multiplication; division
2. composition   **3.** $g(x)$ **4.** inner; outer

1.

2.

3.

4.

5. (a) $2x$ (b) 4 (c) $x^2 - 4$
   (d) $\dfrac{x + 2}{x - 2}$; all real numbers x except $x = 2$

6. (a) $x - 3$ (b) $3x - 7$ (c) $-2x^2 + 9x - 10$
   (d) $\dfrac{2x - 5}{2 - x}$; all real numbers x except $x = 2$

7. (a) $x^2 + 4x - 5$ (b) $x^2 - 4x + 5$ (c) $4x^3 - 5x^2$
   (d) $\dfrac{x^2}{4x - 5}$; all real numbers x except $x = \dfrac{5}{4}$

8. (a) $2x - 1$ (b) $2x - 9$ (c) $8x - 20$
   (d) $\frac{1}{2}x - \frac{5}{4}$; all real numbers x

9. (a) $x^2 + 6 + \sqrt{1 - x}$ (b) $x^2 + 6 - \sqrt{1 - x}$
   (c) $(x^2 + 6)\sqrt{1 - x}$
   (d) $\dfrac{(x^2 + 6)\sqrt{1 - x}}{1 - x}$; all real numbers x such that $x < 1$

10. (a) $\sqrt{x^2 - 4} + \dfrac{x^2}{x^2 + 1}$ (b) $\sqrt{x^2 - 4} - \dfrac{x^2}{x^2 + 1}$
   (c) $\dfrac{x^2\sqrt{x^2 - 4}}{x^2 + 1}$
   (d) $\dfrac{(x^2 + 1)\sqrt{x^2 - 4}}{x^2}$; all real numbers x such that $|x| \geq 2$

11. (a) $\dfrac{x + 1}{x^2}$ (b) $\dfrac{x - 1}{x^2}$ (c) $\dfrac{1}{x^3}$
   (d) x; all real numbers x except $x = 0$

12. (a) $\dfrac{x^4 + x^3 + x}{x + 1}$ (b) $\dfrac{-x^4 - x^3 + x}{x + 1}$ (c) $\dfrac{x^4}{x + 1}$
   (d) $\dfrac{1}{x^2(x + 1)}$; all real numbers x except $x = 0, -1$

13. 3 **14.** 7 **15.** 5 **16.** -1
17. $9t^2 - 3t + 5$ **18.** $t^2 - 3t - 1$ **19.** 74
20. -370 **21.** 26 **22.** $-\frac{1}{4}$ **23.** $\frac{3}{5}$ **24.** 43

25.

26.

27.

28.

29.

30.

   $f(x), g(x)$ $g(x), f(x)$

31. (a) $(x - 1)^2$ (b) $x^2 - 1$ (c) x^4
32. (a) $20 - 3x$ (b) $-3x$ (c) $9x + 20$

33. (a) x (b) x (c) $\sqrt[3]{\sqrt[3]{x-1}-1}$

34. (a) $\dfrac{1}{x^3}$ (b) $\dfrac{1}{x^3}$ (c) x^9

35. (a) $\sqrt{x^2+4}$ (b) $x+4$
Domains of f and $g \circ f$: $x \geq -4$
Domains of g and $f \circ g$: all real numbers

36. (a) $\sqrt[3]{x^3-4}$ (b) $x-4$
Domains of f, g, $f \circ g$, and $g \circ f$: all real numbers

37. (a) $x+1$ (b) $\sqrt{x^2+1}$
Domains of f and $g \circ f$: all real numbers
Domains of g and $f \circ g$: all real numbers x such that $x \geq 0$

38. (a) x^4 (b) x^4
Domains of f, g, $f \circ g$, and $g \circ f$: all real numbers

39. (a) $|x+6|$ (b) $|x|+6$
Domains of f, g, $f \circ g$, and $g \circ f$: all real numbers

40. (a) $|-1-x|$ (b) $3-|x-4|$
Domains of f, g, $f \circ g$, and $g \circ f$: all real numbers

41. (a) $\dfrac{1}{x+3}$ (b) $\dfrac{1}{x}+3$
Domains of f and $g \circ f$: all real numbers x except $x=0$
Domains of g: all real numbers
Domains of $f \circ g$: all real numbers x except $x=-3$

42. (a) $\dfrac{3}{x^2+2x}$ (b) $\dfrac{x^2+2}{x^2-1}$
Domains of f and $g \circ f$: all real numbers x except $x=\pm 1$
Domains of g: all real numbers
Domains of $f \circ g$: all real numbers x except $x=0,-2$

43. (a) 3   (b) 0   **44.** (a) -1 (b) 0
45. (a) 0   (b) 4   **46.** (a) 2   (b) 2
47. $f(x)=x^2$, $g(x)=2x+1$
48. $f(x)=x^3$, $g(x)=1-x$
49. $f(x)=\sqrt[3]{x}$, $g(x)=x^2-4$
50. $f(x)=\sqrt{x}$, $g(x)=9-x$

51. $f(x)=\dfrac{1}{x}$, $g(x)=x+2$

52. $f(x)=\dfrac{4}{x^2}$, $g(x)=5x+2$

53. $f(x)=\dfrac{x+3}{4+x}$, $g(x)=-x^2$

54. $f(x)=\dfrac{27x+6\sqrt[3]{x}}{10-27x}$, $g(x)=x^3$

55. $T=\frac{3}{4}x+\frac{1}{15}x^2$

56. (a) $R_3=734-7.22t-0.8t^2$
(b)

57. (a) $c(t)=\dfrac{p(t)+b(t)-d(t)}{p(t)} \times 100$
(b) $c(5)$ is the population change in the year 2005.

58. (a) $p(t)=d(t)+c(t)$
(b) $p(5)=$ Number of dogs and cats in the year 2005.
(c) The function $h(t)$ represents the number of dogs and cats per capita.

59. (a) $(A+N)(t)=5.31t^2-102.0t+1338$
$(A+N)(4)=1014.96$
$(A+N)(8)=861.84$
$(A+N)(12)=878.64$
(b) $(A-N)(t)=1.41t^2-17.6t+132$
$(A-N)(4)=84.16$
$(A-N)(8)=81.44$
$(A-N)(12)=123.84$

60. (a) $h(t)=\dfrac{25.95t^2-231.2t+3356}{3.02t+252.0}$, which represents the millions of dollars spent on exercise equipment compared with the millions of people in the United States.
(b) $h(7)=11.02$; $h(10)=12.90$; $h(12)=14.98$

61. (a) $y_1=10.20t+92.7$
$y_2=3.357t^2-26.46t+379.5$
$y_3=-0.465t^2+9.71t+7.4$
(b) $y_1+y_2+y_3=2.892t^2-6.55t+479.6$; this amount represents the amount spent on health care in the United States.
(c)

(d) In 2008, \$1298.708 billion is estimated to be spent on health services and supplies, and in 2010, \$1505.4 billion is estimated.

62. (a) For each time t there corresponds one and only one temperature T.
(b) $60°$, $72°$
(c) All the temperature changes occur 1 hour later.
(d) The temperature is decreased by 1 degree.
(e) $T(t)=\begin{cases} 60, & 0 \leq t \leq 6 \\ 12t-12, & 6 < t < 7 \\ 72, & 7 \leq t < 20 \\ -12t+312, & 20 < t < 21 \\ 60, & 21 \leq t \leq 24 \end{cases}$

63. (a) $r(x) = \dfrac{x}{2}$ (b) $A(r) = \pi r^2$

(c) $(A \circ r)(x) = \pi \left(\dfrac{x}{2}\right)^2$; $(A \circ r)(x)$ represents the area of the circular base of the tank on the square foundation with side length x.

64. $(A \circ r)(t) = 0.36\pi t^2$; $(A \circ r)(t)$ represents the area of the circle at time t.

65. (a) $N(T(t)) = 30(3t^2 + 2t + 20)$ This represents the number of bacteria in the food as a function of time.

(b) $t = 2.846$ hours

66. (a) $(C \circ x)(t) = 3000t + 750$; $(C \circ x)(t)$ represents the cost of t hours of production.

(b) $t = 4.75$ hours

67. $g(f(x))$ represents 3 percent of an amount over \$500,000.

68. (a) $R(p) = p - 2000$ (b) $S(p) = 0.9p$

(c) $(R \circ S)(p) = 0.9p - 2000$; $(R \circ S)(p)$ represents the dealership discount after the factory discount.
$(S \circ R)(p) = 0.9(p - 2000)$; $(S \circ R)(p)$ represents the dealership discount after the factory discount.

(d) $(R \circ S)(p) = \$16{,}450$
$(S \circ R)(p) = \$16{,}650$
\$16,450 is the lower cost because 10% of the price of the car is larger than \$2000.

69. False. $(f \circ g)(x) = 6x + 1$ and $(g \circ f)(x) = 6x + 6$

70. True. The range of g must be a subset of the domain of f for $(f \circ g)(x)$ to be defined.

71 and 72. Proofs will vary.

Section P.10   (page 114)

Vocabulary Check   (page 114)

1. inverse; f-inverse   **2.** range; domain
3. $y = x$ **4.** one-to-one   **5.** horizontal

1. $f^{-1}(x) = \tfrac{1}{6}x$ **2.** $f^{-1}(x) = 3x$
3. $f^{-1}(x) = x - 9$ **4.** $f^{-1}(x) = x + 4$
5. $f^{-1}(x) = \dfrac{x - 1}{3}$
6. $f^{-1}(x) = 5x + 1$ **7.** $f^{-1}(x) = x^3$
8. $f^{-1}(x) = \sqrt[5]{x}$
9. c   **10.** b   **11.** a   **12.** d
13. (a) $f(g(x)) = f\left(\dfrac{x}{2}\right) = 2\left(\dfrac{x}{2}\right) = x$

$g(f(x)) = g(2x) = \dfrac{(2x)}{2} = x$

(b)
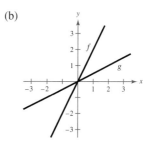

14. (a) $f(g(x)) = f(x + 5) = (x + 5) - 5 = x$
$g(f(x)) = g(x - 5) = (x - 5) + 5 = x$

(b)

15. (a) $f(g(x)) = f\left(\dfrac{x - 1}{7}\right) = 7\left(\dfrac{x - 1}{7}\right) + 1 = x$

$g(f(x)) = g(7x + 1) = \dfrac{(7x + 1) - 1}{7} = x$

(b)

16. (a) $f(g(x)) = f\left(\dfrac{3 - x}{4}\right) = 3 - 4\left(\dfrac{3 - x}{4}\right) = x$

$g(f(x)) = g(3 - 4x) = \dfrac{3 - (3 - 4x)}{4} = x$

(b)

CHAPTER P

17. (a) $f(g(x)) = f(\sqrt[3]{8x}) = \dfrac{(\sqrt[3]{8x})^3}{8} = x$

$g(f(x)) = g\left(\dfrac{x^3}{8}\right) = \sqrt[3]{8\left(\dfrac{x^3}{8}\right)} = x$

(b)

18. (a) $f(g(x)) = f\left(\dfrac{1}{x}\right) = \dfrac{1}{1/x} = x$

$g(f(x)) = g\left(\dfrac{1}{x}\right) = \dfrac{1}{1/x} = x$

(b)

19. (a) $f(g(x)) = f(x^2 + 4),\ x \ge 0$

$\quad = \sqrt{(x^2 + 4) - 4} = x$

$g(f(x)) = g(\sqrt{x - 4})$

$\quad = (\sqrt{x - 4})^2 + 4 = x$

(b)

20. (a) $f(g(x)) = f(\sqrt[3]{1 - x})$

$\quad = 1 - (\sqrt[3]{1 - x})^3 = x$

$g(f(x)) = g(1 - x^3)$

$\quad = \sqrt[3]{1 - (1 - x^3)} = x$

(b)

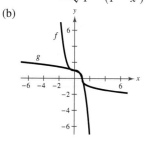

21. (a) $f(g(x)) = f(\sqrt{9 - x}),\ x \le 9$

$\quad = 9 - (\sqrt{9 - x})^2 = x$

$g(f(x)) = g(9 - x^2),\ x \ge 0$

$\quad = \sqrt{9 - (9 - x^2)} = x$

(b)

22. (a) $f(g(x)) = f\left(\dfrac{1 - x}{x}\right),\quad 0 < x \le 1$

$\quad = \dfrac{1}{1 + (1 - x)/x} = \dfrac{x}{x + 1 - x} = x$

$g(f(x)) = g\left(\dfrac{1}{1 + x}\right),\quad x \ge 0$

$\quad = \dfrac{1 - 1/(1 + x)}{1/(1 + x)} = \dfrac{1 + x - 1}{1} = x$

(b)

23. (a) $f(g(x)) = f\left(-\dfrac{5x + 1}{x - 1}\right) = \dfrac{-\left(\dfrac{5x + 1}{x - 1}\right) - 1}{-\left(\dfrac{5x + 1}{x - 1}\right) + 5}$

$\quad = \dfrac{-5x - 1 - x + 1}{-5x - 1 + 5x - 5} = x$

$g(f(x)) = g\left(\dfrac{x - 1}{x + 5}\right) = \dfrac{-5\left(\dfrac{x - 1}{x + 5}\right) - 1}{\dfrac{x - 1}{x + 5} - 1}$

$\quad = \dfrac{-5x + 5 - x - 5}{x - 1 - x - 5} = x$

(b)

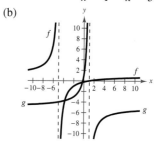

24. (a) $f(g(x)) = f\left(\dfrac{2x + 3}{x - 1}\right) = \dfrac{\left(\dfrac{2x + 3}{x - 1}\right) + 3}{\left(\dfrac{2x + 3}{x - 1}\right) - 2}$

$= \dfrac{2x + 3 + 3x - 3}{2x + 3 - 2x + 2} = x$

(b) $g(f(x)) = g\left(\dfrac{x + 3}{x - 2}\right) = \dfrac{2\left(\dfrac{x + 3}{x - 2}\right) + 3}{\left(\dfrac{x + 3}{x - 2}\right) - 1}$

$= \dfrac{2x + 6 + 3x - 6}{x + 3 - x + 2} = x$

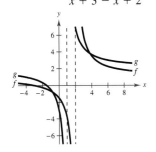

25. No   **26.** Yes

27.

x	-2	0	2	4	6	8
$f^{-1}(x)$	-2	-1	0	1	2	3

28.

x	-10	-7	-4	-1	2	5
$f^{-1}(x)$	-3	-2	-1	0	1	2

29. Yes   **30.** No   **31.** No   **32.** Yes

33.

The function has an inverse.

34.

The function does not have an inverse.

35.

The function does not have an inverse.

36.

The function has an inverse.

37.

The function does not have an inverse.

38.

The function does not have an inverse.

39. (a) $f^{-1}(x) = \dfrac{x + 3}{2}$

(b)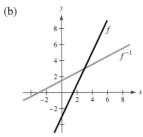

(c) The graph of f^{-1} is the reflection of the graph of f in the line $y = x$.

(d) The domains and ranges of f and f^{-1} are all real numbers.

40. (a) $f^{-1}(x) = \dfrac{x - 1}{3}$

(b)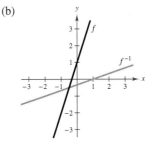

(c) The graph of f^{-1} is the reflection of the graph of f in the line $y = x$.

(d) The domains and ranges of f and f^{-1} are all real numbers.

41. (a) $f^{-1}(x) = \sqrt[5]{x + 2}$

(b)

(c) The graph of f^{-1} is the reflection of the graph of f in the line $y = x$.

(d) The domains and ranges of f and f^{-1} are all real numbers.

42. (a) $f^{-1}(x) = \sqrt[3]{x - 1}$

(b)

(c) The graph of f^{-1} is the reflection of the graph of f in the line $y = x$.

(d) The domains and ranges of f and f^{-1} are all real numbers.

43. (a) $f^{-1}(x) = x^2$, $x \geq 0$

(b)

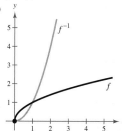

(c) The graph of f^{-1} is the reflection of the graph of f in the line $y = x$.

(d) The domains and ranges of f and f^{-1} are all real numbers x such that $x \geq 0$.

44. (a) $f^{-1}(x) = \sqrt{x}$

(b)

(c) The graph of f^{-1} is the reflection of the graph of f in the line $y = x$.

(d) The domains and ranges of f and f^{-1} are all real numbers such that $x \geq 0$.

45. (a) $f^{-1}(x) = \sqrt{4 - x^2}$, $0 \leq x \leq 2$

(b)

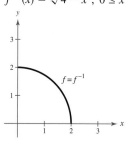

(c) The graph of f^{-1} is the same as the graph of f.

(d) The domains and ranges of f and f^{-1} are all real numbers x such that $0 \leq x \leq 2$.

46. (a) $f^{-1}(x) = -\sqrt{x + 2}$

(b)

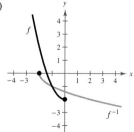

(c) The graph of f^{-1} is the reflection of the graph of f in the line $y = x$.

(d) The domain of f and the range of f^{-1} are all real numbers x such that $x \leq 0$. The domain of f^{-1} and the range of f are all real numbers x such that $x \geq -2$.

47. (a) $f^{-1}(x) = \dfrac{4}{x}$

(b)

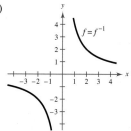

(c) The graph of f^{-1} is the same as the graph of f.

(d) The domains and ranges of f and f^{-1} are all real numbers x except $x = 0$.

48. (a) $f^{-1}(x) = -\dfrac{2}{x}$

(b)

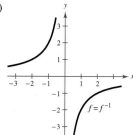

(c) The graph of f^{-1} is the same as the graph of f.

(d) The domains and ranges of f and f^{-1} are all real numbers x except $x = 0$.

49. (a) $f^{-1}(x) = \dfrac{2x + 1}{x - 1}$

(b)

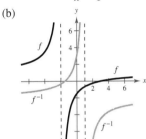

(c) The graph of f^{-1} is the reflection of the graph of f in the line $y = x$.

(d) The domain of f and the range of f^{-1} are all real numbers x except $x = 2$. The domain of f^{-1} and the range of f are all real numbers x except $x = 1$.

50. (a) $f^{-1}(x) = \dfrac{-2x - 3}{x - 1}$

(b)

(c) The graph of f^{-1} is the reflection of the graph of f in the line $y = x$.

(d) The domain of f and the range of f^{-1} are all real numbers x except $x = -2$. The domain of f^{-1} and the range of f are all real numbers x except $x = 1$.

51. (a) $f^{-1}(x) = x^3 + 1$

(b)

(c) The graph of f^{-1} is the reflection of the graph of f in the line $y = x$.

(d) The domains and ranges of f and f^{-1} are all real numbers.

52. (a) $f^{-1}(x) = x^{5/3}$

(b)

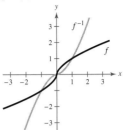

(c) The graph of f^{-1} is the reflection of the graph of f in the line $y = x$.

(d) The domains and ranges of f and f^{-1} are all real numbers.

53. (a) $f^{-1}(x) = \dfrac{5x - 4}{6 - 4x}$

(b)

(c) The graph of f^{-1} is the reflection of the graph of f in the line $y = x$.

(d) The domain of f and the range of f^{-1} are all real numbers x except $x = -\frac{5}{4}$. The domain of f^{-1} and the range of f are all real numbers x except $x = \frac{3}{2}$.

54. (a) $f^{-1}(x) = \dfrac{-6x - 4}{2x - 8}$

(b)

(c) The graph of f^{-1} is the reflection of the graph of f in the line $y = x$.

(d) The domain of f and the range of f^{-1} are all real numbers x except $x = -3$. The domain of f^{-1} and the range of f are all real numbers x except $x = -4$.

55. No inverse   **56.** No inverse   **57.** $g^{-1}(x) = 8x$

58. $f^{-1}(x) = \dfrac{x - 5}{3}$ **59.** No inverse

60. $f^{-1}(x) = \dfrac{5x - 4}{3}$ **61.** $f^{-1}(x) = \sqrt{x} - 3$

62. No inverse **63.** No inverse **64.** No inverse
65. No inverse **66.** $f^{-1}(x) = 2 - x, \quad x \geq 0$

67. $f^{-1}(x) = \dfrac{x^2 - 3}{2}, \quad x \geq 0$

68. $f^{-1}(x) = x^2 + 2, \quad x \geq 0$ **69.** 32 **70.** 0
71. 600 **72.** $\sqrt[9]{-4}$ **73.** $2\sqrt[3]{x + 3}$ **74.** $2\sqrt[3]{x} + 3$

75. $\dfrac{x + 1}{2}$ **76.** $\dfrac{x - 3}{2}$ **77.** $\dfrac{x + 1}{2}$ **78.** $\dfrac{x - 3}{2}$

79. (a) $f^{-1}(108{,}209) = 11$
  (b) f^{-1} represents the year for a given number of house-holds in the United States.
  (c) $y = 1578.68t + 90{,}183.63$
  (d) $f^{-1} = \dfrac{t - 90{,}183.63}{1578.68}$ (e) $f^{-1}(117{,}022) = 17$
  (f) $f^{-1}(108{,}209) = 11.418$; the results are similar.
80. (a) Yes
  (b) f^{-1} represents the time in years for a given sales level.
  (c) $f^{-1}(1825) = 10$
  (d) No, it wouldn't exist because $f(12) = 2794$.
81. (a) Yes
  (b) f^{-1} yields the year for a given number of miles traveled by motor vehicles.
  (c) $f^{-1}(2632) = 8$
  (d) No. $f(t)$ would not pass the Horizontal Line Test.
82. (a) $y = \dfrac{x - 8}{0.75}$
  (b) $y =$ number of units produced; $x =$ hourly wage
  (c) 19 units
83. (a) $y = \sqrt{\dfrac{x - 245.50}{0.03}}, \quad 245.5 < x < 545.5$
  $x =$ degrees Fahrenheit; $y = \%$ load
  (b) (c) $0 < x < 92.11$
84. (a) $y = \dfrac{80 - x}{0.35}$
  $x =$ total cost; $y =$ number of pounds of the less expensive ground beef
  (b) $62.5 \leq x \leq 80$ (c) 20 pounds
85. False. $f(x) = x^2$ has no inverse.
86. True. If $f(x) = x - 6$ and $f^{-1}(x) = x + 6$, then the y-intercept of f is $(0, -6)$ and the x-intercept of f^{-1} is $(-6, 0)$.
87 and 88. Proofs will vary.

89.

x	1	3	4	6
y	1	2	6	7

x	1	2	6	7
$f^{-1}(x)$	1	3	4	6

90.

x	-2	-1	1	3
y	-5	-2	2	3

x	-5	-2	2	3
$f^{-1}(x)$	-2	-1	1	3

91.

x	-2	-1	3	4
y	6	0	-2	-3

x	-3	-2	0	6
$f^{-1}(x)$	4	3	-1	-2

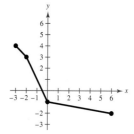

92.

x	-4	-2	0	3
y	3	4	0	-1

The graph of f does not pass the Horizontal Line Test, so $f^{-1}(x)$ does not exist.

93. $k = \frac{1}{4}$ **94.** $k = -\frac{1}{2}$

Review Exercises *(page 120)*

1. (a) 11   (b) 11   (c) 11, -14
   (d) 11, -14, $-\frac{8}{9}$, $\frac{5}{2}$, 0.4   (e) $\sqrt{6}$

2. (a) None   (b) 0   (c) -22, 0
   (d) -22, $-\frac{10}{3}$, 0, 5.2, $\frac{3}{7}$ (e) $\sqrt{15}$

3. (a) $0.8\overline{3}$ (b) 0.875

$\frac{5}{6} < \frac{7}{8}$

4. (a) 0.36   (b) $0.\overline{714285}$

$\frac{9}{25} < \frac{5}{7}$

5. The set consists of all real numbers less than or equal to 7.

6. The set consists of all real numbers greater than 1.

7. 155   **8.** 106   **9.** $|x - 7| \geq 4$
10. $|x - 25| \leq 10$ **11.** (a) -7 (b) -19
12. (a) 21   (b) -3
13. Associative Property of Addition
14. Distributive Property   **15.** Additive Identity Property
16. Multiplicative Inverse Property
17. -11 **18.** -1 **19.** $\frac{1}{12}$ **20.** 2   **21.** -144
22. -100 **23.** Identity   **24.** Conditional equation
25. 20   **26.** $-\frac{9}{2}$ **27.** $-\frac{1}{2}$ **28.** $-\frac{17}{3}$ **29.** -30
30. 18   **31.** 9   **32.** $\frac{11}{3}$ **33.** $-\frac{5}{2}$, 3   **34.** $-\frac{7}{2}$, 4
35. $\pm\sqrt{2}$ **36.** $\pm\frac{5}{4}$ **37.** $-4 \pm 3\sqrt{2}$
38. $8 \pm \sqrt{15}$ **39.** $6 \pm \sqrt{6}$ **40.** $-3 \pm 2\sqrt{3}$
41. $-\frac{5}{4} \pm \frac{\sqrt{241}}{4}$ **42.** $\frac{1}{2} \pm \frac{\sqrt{249}}{6}$ **43.** 0, $\frac{12}{5}$
44. 0, $\frac{3}{2}$ **45.** $\pm\sqrt{2}, \pm\sqrt{3}$ **46.** 0, -3, $\pm\frac{2}{3}$ **47.** 5
48. 66   **49.** No solution   **50.** $\dfrac{38 + 5\sqrt{3}}{24}$
51. -124, 126   **52.** 79   **53.** -5, 15   **54.** 2, -5
55. 1, 3   **56.** 2, 3

57.

58.

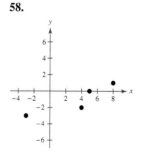

59. Quadrant IV   **60.** Quadrant I or II
61. (a)

(b) 5
(c) $\left(-1, \dfrac{13}{2}\right)$

62. (a)

(b) $3\sqrt{13}$
(c) $\left(1, \frac{3}{2}\right)$

63. (a)

(b) 9.9
(c) $(2.8, 4.1)$

64. (a)

(b) $\sqrt{14.4}$
(c) $(-1.8, -0.6)$

65. \$656.45 million
66. (a)

(b) $80°F$

CHAPTER P

67.

x	−2	−1	0	1	2
y	−11	−8	−5	−2	1

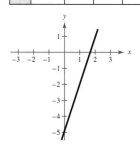

68.

x	−4	−2	0	2	4
y	4	3	2	1	0

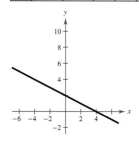

69.

x	−1	0	1	2	3	4
y	4	0	−2	−2	0	4

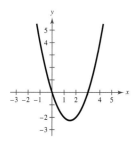

70.

x	−2	−1	0	1	2	3
y	1	−6	−9	−8	−3	6

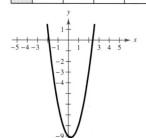

71. x-intercepts: $(1, 0)$, $(5, 0)$
y-intercept: $(0, 5)$

72. x-intercepts: $(2, 0)$, $(−4, 0)$
y-intercept: $(0, −2)$

73. No symmetry   **74.** No symmetry

          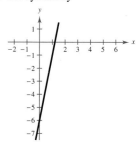

75. y-axis symmetry   **76.** y-axis symmetry

          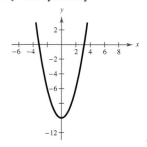

77. No symmetry   **78.** No symmetry

          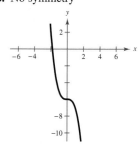

79. No symmetry   **80.** y-axis symmetry

          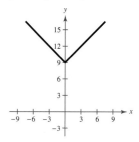

81. Center: $(0, 0)$;
Radius: 3

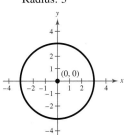

82. Center: $(0, 0)$;
Radius: 2

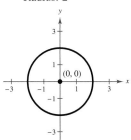

91. $m = 3$
y-intercept: 13

92. $m = -10$
y-intercept: 9

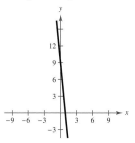

83. Center: $\left(\frac{1}{2}, -1\right)$;
Radius: 6

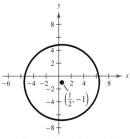

84. Center: $\left(-4, \frac{3}{2}\right)$;
Radius: 10

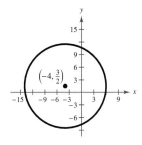

93. $m = -\frac{5}{2}$
y-intercept: -1

94. $m = \frac{5}{6}$
y-intercept: 5

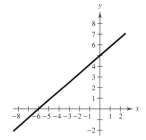

85. $(x - 2)^2 + (y + 3)^2 = 13$
86. $(x - 1)^2 + \left(y + \frac{13}{2}\right)^2 = \frac{85}{4}$
87. $m = -2$
y-intercept: -7

88. $m = 4$
y-intercept: -3

95.

$m = -\frac{1}{2}$

96.

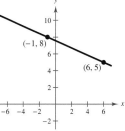

$m = -\frac{3}{7}$

89. $m = 0$
y-intercept: 6

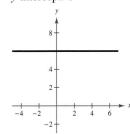

90. $m = $ undefined
y-intercept: none

97.

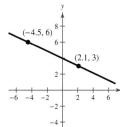

$m = -\frac{5}{11}$

99. $y = \frac{3}{2}x - 5$

98.

$m = 0$

100. $y = 6$

101. $y = -\frac{1}{2}x + 2$ **102.** $x = -8$

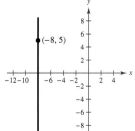

103. $x = 0$ **104.** $y = \frac{3}{2}x + 2$ **105.** $y = -\frac{4}{3}x + \frac{8}{3}$

106. $y = -\frac{1}{5}x + \frac{1}{5}$

107. (a) $y = \frac{5}{4}x - \frac{23}{4}$ (b) $y = -\frac{4}{5}x + \frac{2}{5}$

108. (a) $y = -\frac{2}{3}x - \frac{7}{3}$ (b) $y = \frac{3}{2}x + 15$

109. \$210,000

110. (a) $V = 3.75t + 66.25$

(b)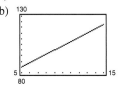

(c) \$103.75

111. (a) Not a function, because 20 in the domain corresponds to two values in the range and because 10 in A is not matched with any element in B.

(b) A function, because each input value has exactly one output value

(c) A function, because each input value has exactly one output value

(d) Not a function, because 30 in A is not matched with any element in B

112. (a) Not a function, because u in the domain corresponds to two values in the range.

(b) A function, because each input value has exactly one output value.

(c) A function, because each input value has exactly one output value.

(d) Not a function, because w in the domain corresponds to two values in the range and because u in A is not matched with any element in B.

113. No   **114.** Yes   **115.** Yes   **116.** No

117. (a) -3 (b) -1 (c) 2 (d) 6

118. (a) 2 (b) $\dfrac{2}{13}$ (c) $\dfrac{4}{t^2 + 1}$ (d) 4

119. Domain: All real numbers x such that $-5 \le x \le 5$

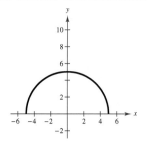

120. Domain: all real numbers x

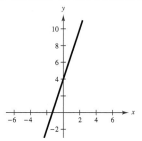

121. Domain: all real numbers s except $s = 3$

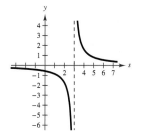

122. Domain: all real numbers x such that $x \le -8$ and $x \ge 0$

123. Domain: all real numbers x except $x = 3, -2$

124. Domain: all real numbers

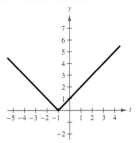

125. (a) 16 feet per second   (b) 1.5 seconds
   (c) −16 feet per second
126. (a) $C(x) = 5.35x + 16,000$
   (b) $P(x) = R - C = 2.85x - 16,000$
127. $4x + 2h + 3, \ h \neq 0$
128. $3x^2 + 3xh + h^2 - 10x - 5h + 1, \ h \neq 0$
129. Function   **130.** Function   **131.** Not a function
132. Not a function   **133.** $\frac{7}{3}, 3$ **134.** $-1, \frac{1}{5}$
135. $-\frac{3}{8}$ **136.** $1, \pm 5$
137. Increasing on $(0, \infty)$
   Decreasing on $(-\infty, -1)$
   Constant on $(-1, 0)$
138. Increasing on $(-2, 0)$ and $(2, \infty)$
   Decreasing on $(-\infty, -2)$ and $(0, 2)$
139.

140.

141.

142.

143. 4   **144.** 28

145. $\dfrac{1 - \sqrt{2}}{2}$ **146.** -0.2 **147.** Neither even nor odd
148. Even   **149.** Odd   **150.** Even

151. $f(x) = -3x$

152. $f(x) = -\frac{3}{4}x - 5$

153. $f(x) = \frac{5}{3}x + \frac{10}{3}$

154. $f(x) = \frac{7}{8}x + \frac{217}{80}$

155.

156.

157.

158.

159.

160.

CHAPTER P

161.

162.

170. (a) $f(x) = |x|$
(b) Horizontal shift of three units to the left and vertical shift of five units downward
(c) 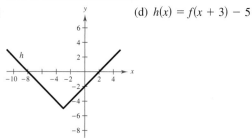   (d) $h(x) = f(x + 3) - 5$

163.

164.
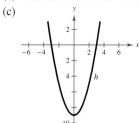

165. $y = x^3$ **166.** $y = \sqrt{x}$

167. (a) $f(x) = x^2$
(b) Vertical shift of nine units downward
(c)

(d) $h(x) = f(x) - 9$

168. (a) $f(x) = x^3$
(b) Horizontal shift of two units to the right and vertical shift of two units upward
(c) 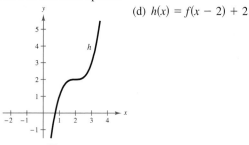   (d) $h(x) = f(x - 2) + 2$

169. (a) $f(x) = \sqrt{x}$
(b) Horizontal shift of seven units to the right
(c)   (d) $h(x) = f(x - 7)$

171. (a) $f(x) = x^2$
(b) Reflection in the x-axis, horizontal shift of three units to the left, and vertical shift of one unit upward
(c)

(d) $h(x) = -f(x + 3) + 1$

172. (a) $f(x) = x^3$
(b) Reflection in the x-axis, horizontal shift of five units to the right, and vertical shift of five units downward
(c)

(d) $h(x) = -f(x - 5) - 5$

173. (a) $f(x) = [\![x]\!]$
(b) Reflection in the x-axis and vertical shift of six units upward
(c)

(d) $h(x) = -f(x) + 6$

174. (a) $f(x) = \sqrt{x}$
(b) Reflection in the x-axis, horizontal shift of one unit to the left, and vertical shift of nine units upward
(c)
(d) $h(x) = -f(x + 1) + 9$

175. (a) $f(x) = |x|$
(b) Reflections in the x-axis and the y-axis, horizontal shift of four units to the right, and vertical shift of six units upward
(c)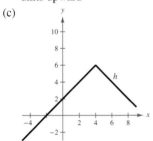
(d) $h(x) = -f(-x + 4) + 6$

176. (a) $f(x) = x^2$
(b) Reflection in the x-axis, horizontal shift of one unit to the left, and vertical shift of three units downward
(c)
(d) $h(x) = -f(x + 1) - 3$

177. (a) $f(x) = [\![x]\!]$
(b) Horizontal shift of nine units to the right and vertical stretch
(c)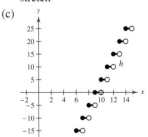
(d) $h(x) = 5 f(x - 9)$

178. (a) $f(x) = x^3$
(b) Reflection in the x-axis and vertical shrink
(c)
(d) $h(x) = -\frac{1}{3}f(x)$

179. (a) $f(x) = \sqrt{x}$
(b) Reflection in the x-axis, vertical stretch, and horizontal shift of four units to the right
(c)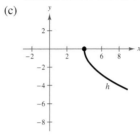
(d) $h(x) = -2 f(x - 4)$

180. (a) $f(x) = |x|$
(b) Vertical shrink and vertical shift of one unit downward
(c)
(d) $h(x) = \frac{1}{2}f(x) - 1$

181. (a) $x^2 + 2x + 2$ (b) $x^2 - 2x + 4$
(c) $2x^3 - x^2 + 6x - 3$
(d) $\dfrac{x^2 + 3}{2x - 1}$; all real numbers x except $x = \dfrac{1}{2}$

182. (a) $x^2 - 4 + \sqrt{3 - x}$ (b) $x^2 - 4 - \sqrt{3 - x}$
(c) $(x^2 - 4)\sqrt{3 - x}$
(d) $\dfrac{x^2 - 4}{\sqrt{3 - x}}$; all real numbers x such that $x < 3$

183. (a) $x - \frac{8}{3}$ (b) $x - 8$
Domains of $f, g, f \circ g$, and $g \circ f$: all real numbers

184. (a) $x + 3$ (b) $\sqrt[3]{x^3 + 3}$
Domains of $f, g, f \circ g$, and $g \circ f$: all real numbers

185. $f(x) = x^3, g(x) = 6x - 5$

186. $f(x) = \sqrt[3]{x}, g(x) = x + 2$

187. (a) $(v + d)(t) = -36.04t^2 + 804.6t - 1112$

(b)

(c) $(v + d)(10) = 3330$

188. (a) $N(T(t)) = 100t^2 + 275$

The composition function $N(T(t))$ represents the number of bacteria in the food as a function of time.

(b) $t = 2.18$ hours

189. $\quad f^{-1}(x) = x + 7$

$f(f^{-1}(x)) = x + 7 - 7 = x$

$f^{-1}(f(x)) = x - 7 + 7 = x$

190. $f^{-1}(x) = x - 5$

$f(f^{-1}(x)) = x - 5 + 5 = x$

$f^{-1}(f(x)) = x + 5 - 5 = x$

191. The function has an inverse.

192. The function does not have an inverse.

193. The function has an inverse.

194. The function does not have an inverse.

195. The function has an inverse.

196. The function has an inverse.

197. (a) $f^{-1}(x) = 2x + 6$

(b)

(c) The graph of f^{-1} is the reflection of the graph of f in the line $y = x$.

(d) Both f and f^{-1} have domains and ranges that are all real numbers.

198. (a) $f^{-1}(x) = \dfrac{x + 7}{5}$

(b)

(c) The graph of f^{-1} is the reflection of the graph of f in the line $y = x$.

(d) Both f and f^{-1} have domains and ranges that are all real numbers.

199. (a) $f^{-1}(x) = x^2 - 1, \ x \geq 0$

(b)

(c) The graph of f^{-1} is the reflection of the graph of f in the line $y = x$.

(d) The graph of f has a domain of all real numbers x such that $x \geq -1$ and a range of $[0, \infty)$. The graph of f^{-1} has a domain of all real numbers x such that $x \geq 0$ and a range of $[-1, \infty)$.

200. (a) $f^{-1}(x) = \sqrt[3]{x - 2}$

(b)

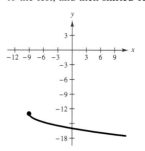

(c) The graph of f^{-1} is the reflection of the graph of f in the line $y = x$.

(d) Both f and f^{-1} have domains and ranges that are all real numbers.

201. $x \geq 4; \ f^{-1}(x) = \sqrt{\dfrac{x}{2} + 4}$

202. $x \geq 2; \ f^{-1}(x) = x + 2, \ x \geq 0$

203. False. The graph is reflected in the x-axis, shifted 9 units to the left, and then shifted 13 units downward.

204. True. If $f(x) = x^3$ and $g(x) = \sqrt[3]{x}$, then the domain of g is all real numbers, which is equal to the range of f and vice versa.

205. Some solutions to certain types of equations may be extraneous solutions, which do not satisfy the original equations. So, checking is crucial.

206. A function from a set A to a set B is a relation that assigns to each element x in the set A exactly one element y in the set B.

207. The Vertical Line Test is used to determine if the graph of y is a function of x. The Horizontal Line Test is used to determine if a function has an inverse function.

Chapter Test *(page 125)*

1. $-\frac{10}{3} > -|-4|$ **2.** 9.15

3. Additive Identity Property   **4.** $\frac{128}{11}$ **5.** $-4, 5$

6. No solution   **7.** $\pm\sqrt{2}, \pm\sqrt{3}i$ **8.** 4   **9.** $-2, \frac{8}{3}$

10.

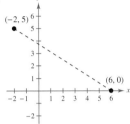

Midpoint: $\left(2, \frac{5}{2}\right)$; Distance: $\sqrt{89}$

11. No symmetry   **12.** No symmetry

     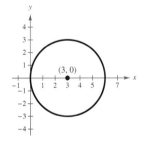

13. Origin symmetry   **14.** Center: $(3, 0)$; Radius: 3

15. (a) $4x - 7y + 44 = 0$ (b) $7x + 4y - 53 = 0$

16. (a) -9 (b) 1   (c) $|x - 4| - 15$

17. (a)

(b) All real numbers x

(c) Increasing on $(-0.31, 0), (0.31, \infty)$
  Decreasing on $(-\infty, -0.31), (0, 0.31)$

(d) Even

18. (a)

(b) All real numbers x such that $x \le 3$

(c) Increasing on $(-\infty, 2)$
  Decreasing on $(2, 3)$

(d) Neither even nor odd

19. (a)

     (b) All real numbers x

(c) Increasing on $(-5, \infty)$
  Decreasing on $(-\infty, -5)$

(d) Neither even nor odd

20. (a) $f(x) = [\![x]\!]$ (b) Reflection in the x-axis

(c)

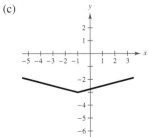

21. (a) $f(x) = \sqrt{x}$

(b) Reflection in the x-axis, vertical shift of eight units upward, and horizontal shift of five units to the left

(c)

22. (a) $f(x) = |x|$

(b) Vertical shrink, vertical shift of three units downward, and horizontal shift of one unit to the left

(c)

CHAPTER P

23. (a) $2x^2 - 4x - 2$ (b) $4x^2 + 4x - 12$
  (c) $-3x^4 - 12x^3 + 22x^2 + 28x - 35$
  (d) $\dfrac{3x^2 - 7}{-x^2 - 4x + 5}$, $x \neq 1, -5$
  (e) $3x^4 + 24x^3 + 18x^2 - 120x + 68$
  (f) $-9x^4 + 30x^2 - 16$

24. (a) $\dfrac{1 + 2x^{3/2}}{x}$, $x > 0$ (b) $\dfrac{1 - 2x^{3/2}}{x}$, $x > 0$
  (c) $\dfrac{2\sqrt{x}}{x}$, $x > 0$ (d) $\dfrac{1}{2x^{3/2}}$, $x > 0$
  (e) $\dfrac{\sqrt{x}}{2x}$, $x > 0$ (f) $\dfrac{2\sqrt{x}}{x}$, $x > 0$

25. $f^{-1}(x) = \sqrt[3]{x - 8}$ **26.** No inverse
27. $f^{-1}(x) = \left(\frac{8}{3}x\right)^{2/3}$, $x \geq 0$

Problem Solving *(page 127)*

1. (a) $W_1 = 2000 + 0.07S$ (b) $W_2 = 2300 + 0.05S$
  (c)

  Both jobs pay the same monthly salary if sales equal $15,000.
  (d) No. Job 1 would pay $3400 and job 2 would pay $3300.
2. Mapping numbers onto letters is not a function since each number corresponds to three letters.
  Mapping letters onto numbers is a function since every letter is assigned exactly one number.
3. (a) The function will be even.
  (b) The function will be odd.
  (c) The function will be neither even nor odd.
4. $f(x) = x$ $g(x) = -x$

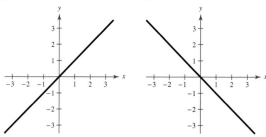

  Both graphs are already symmetric with respect to the line $y = x$.
  General formula: $y = -x + c, c \geq 0$
5. $f(x) = a_{2n}x^{2n} + a_{2n-2}x^{2n-2} + \cdots + a_2 x^2 + a_0$
  $f(-x) = a_{2n}(-x)^{2n} + a_{2n-2}(-x)^{2n-2}$
    $+ \cdots + a_2(-x)^2 + a_0$
    $= f(x)$

6. $(6, 8)$
  $f(x) = \begin{cases} \frac{12}{7}x - \frac{16}{7}, & 2.5 \leq x \leq 6 \\ -\frac{12}{7}x + \frac{128}{7}, & 6 < x \leq 9.5 \end{cases}$

7. (a) $81\frac{2}{3}$ hours (b) $25\frac{5}{7}$ miles per hour
  (c) $y = \dfrac{-180}{7}x + 3400$
  Domain: $0 \leq x \leq \dfrac{1190}{9}$
  Range: $0 \leq y \leq 3400$
  (d)

8. (a) 1 (b) 1.5 (c) 1.75 (d) 1.875 (e) 1.9375
  (f) Yes, 2.
  (g) a. $y = x - 1$
    b. $y = 1.5x - 1.5$
    c. $y = 1.75x - 1.75$
    d. $y = 1.875x - 1.875$
    e. $y = 1.9375x - 1.9375$
  (h) $y = 2x - 2$
9. (a) $(f \circ g)(x) = 4x + 24$
  (b) $(f \circ g)^{-1}(x) = \frac{1}{4}x - 6$
  (c) $f^{-1}(x) = \frac{1}{4}x$; $g^{-1}(x) = x - 6$
  (d) $(g^{-1} \circ f^{-1})(x) = \frac{1}{4}x - 6$
  (e) $(f \circ g)(x) = 8x^3 + 1$; $(f \circ g)^{-1}(x) = \frac{1}{2}\sqrt[3]{x - 1}$;
    $f^{-1}(x) = \sqrt[3]{x - 1}$; $g^{-1}(x) = \frac{1}{2}x$;
    $(g^{-1} \circ f^{-1})(x) = \frac{1}{2}\sqrt[3]{x - 1}$
  (f) Answers will vary.
  (g) $(f \circ g)^{-1}(x) = (g^{-1} \circ f^{-1})(x)$
10. (a) $T = \frac{1}{2}\sqrt{4 + x^2} + \frac{1}{4}\sqrt{x^2 - 6x + 10}$
  (b) $0 \leq x \leq 3$
  (c)
  (d) $x = 1$
  (e) The distance $x = 1$ yields a time of 1.68 hours.

11. (a)

(b)

(c)

(d)

(e)

(f)

(c)

(d)

(e)

(f)

(g)

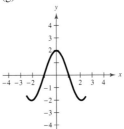

12. (a) Domain: all real numbers x except $x = 1$
Range: all real numbers y except $y = 0$

(b) $f(f(x)) = \dfrac{x - 1}{x}$
Domain: all real numbers x except $x = 0, 1$

(c) $f(f(f(x))) = x$
The graph is not a line because there are holes at $x = 0$ and $x = 1$.

13. Proof

14. (a)

(b)

15. (a)

x	-4	-2	0	4
$f(f^{-1}(x))$	-4	-2	0	4

(b)

x	-3	-2	0	1
$(f + f^{-1})(x)$	5	1	-3	-5

(c)

x	-3	-2	0	1
$(f \circ f^{-1})(x)$	4	0	2	6

(d)

x	-4	-3	0	4
$|f^{-1}(x)|$	2	1	1	3

CHAPTER P

Chapter 1

Section 1.1   *(page 138)*

Vocabulary Check   *(page 138)*

1. Trigonometry   **2.** angle   **3.** coterminal

4. radian   **5.** acute; obtuse

6. complementary; supplementary   **7.** degree

8. linear   **9.** angular   **10.** $A = \frac{1}{2}r^2\theta$

---

1. 2 radians   **2.** 5.5 radians   **3.** -3 radians

4. -4 radians   **5.** 1 radian   **6.** 6.5 radians

7. (a) Quadrant I   (b) Quadrant III

8. (a) Quadrant III   (b) Quadrant III

9. (a) Quadrant IV   (b) Quadrant III

10. (a) Quadrant IV   (b) Quadrant II

11. (a) Quadrant III   (b) Quadrant II

12. (a) Quadrant IV   (b) Quadrant II

13. (a)   (b)

14. (a)   (b)

15. (a)   (b)

16. (a)   (b)

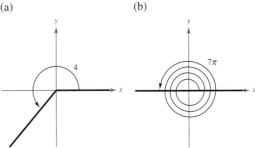

17. (a) $\dfrac{13\pi}{6}$, $-\dfrac{11\pi}{6}$ (b) $\dfrac{17\pi}{6}$, $-\dfrac{7\pi}{6}$

18. (a) $\dfrac{19\pi}{6}$, $-\dfrac{5\pi}{6}$ (b) $\dfrac{\pi}{6}$, $-\dfrac{23\pi}{6}$

19. (a) $\dfrac{8\pi}{3}$, $-\dfrac{4\pi}{3}$ (b) $\dfrac{25\pi}{12}$, $-\dfrac{23\pi}{12}$

20. (a) $\dfrac{7\pi}{4}$, $-\dfrac{\pi}{4}$ (b) $\dfrac{28\pi}{15}$, $-\dfrac{32\pi}{15}$

21. (a) Complement: $\dfrac{\pi}{6}$; Supplement: $\dfrac{2\pi}{3}$

(b) Complement: none; Supplement: $\dfrac{\pi}{4}$

22. (a) Complement: $\dfrac{5\pi}{12}$; Supplement: $\dfrac{11\pi}{12}$

(b) Complement: none; Supplement: $\dfrac{\pi}{12}$

23. (a) Complement: $\dfrac{\pi}{2} - 1 \approx 0.57$;

Supplement: $\pi - 1 \approx 2.14$

(b) Complement: none; Supplement: $\pi - 2 \approx 1.14$

24. (a) Complement: none; Supplement: $\pi - 3 \approx 0.14$

(b) Complement: $\dfrac{\pi}{2} - 1.5 \approx 0.07$;

Supplement: $\pi - 1.5 \approx 1.64$

25. $210°$ **26.** $120°$ **27.** $-60°$ **28.** $-330°$

29. $165°$ **30.** $10°$

31. (a) Quadrant II   (b) Quadrant IV

32. (a) Quadrant I   (b) Quadrant III

33. (a) Quadrant III   (b) Quadrant I

34. (a) Quadrant II   (b) Quadrant IV

35. (a)   (b)

36. (a) (b)

37. (a) (b)

38. (a) (b)

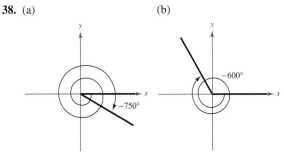

39. (a) $405°, -315°$ (b) $324°, -396°$
40. (a) $480°, -240°$ (b) $300°, -60°$
41. (a) $600°, -120°$ (b) $180°, -540°$
42. (a) $300°, -60°$ (b) $590°, -130°$
43. (a) Complement: $72°$; Supplement: $162°$
  (b) Complement: none; Supplement: $65°$
44. (a) Complement: $87°$; Supplement: $177°$
  (b) Complement: $26°$; Supplement: $116°$
45. (a) Complement: $11°$; Supplement: $101°$
  (b) Complement: none; Supplement: $30°$
46. (a) Complement: none; Supplement: $50°$
  (b) Complement: none; Supplement: $10°$
47. (a) $\frac{\pi}{6}$ (b) $\frac{5\pi}{6}$ **48.** (a) $\frac{7\pi}{4}$ (b) $\frac{2\pi}{3}$
49. (a) $-\frac{\pi}{9}$ (b) $-\frac{4\pi}{3}$ **50.** (a) $-\frac{3\pi}{2}$ (b) $\frac{4\pi}{5}$
51. (a) $270°$ (b) $210°$ **52.** (a) $-105°$ (b) $20°$
53. (a) $420°$ (b) $-66°$ **54.** (a) $330°$ (b) $408°$
55. 2.007 **56.** 1.525 **57.** -3.776 **58.** -0.842
59. 9.285 **60.** 6.021 **61.** -0.014 **62.** 0.009
63. $25.714°$ **64.** $81.818°$ **65.** $337.500°$

66. $1170.000°$ **67.** $-756.000°$ **68.** $864.000°$
69. $-114.592°$ **70.** $-32.659°$
71. (a) $54.75°$ (b) $-128.5°$
72. (a) $245.167°$ (b) $2.2°$
73. (a) $85.308°$ (b) $330.007°$
74. (a) $-135.01°$ (b) $-408.272°$
75. (a) $240°36'$ (b) $-145°48'$
76. (a) $-345°7'12''$ (b) $0°27'$
77. (a) $2°30'$ (b) $-3°34'48''$
78. (a) $-0°21'18''$ (b) $0°47'11.4''$
79. $\frac{6}{5}$ radians **80.** $\frac{29}{10}$ radians **81.** $\frac{32}{7}$ radians
82. $-\frac{4}{5}$ radian **83.** $\frac{2}{9}$ radian **84.** $\frac{4}{7}$ radian
85. $\frac{50}{29}$ radians **86.** 2 radians
87. 15π inches ≈ 47.12 inches **88.** 3π feet ≈ 9.42 feet
89. 3 meters **90.** 5π centimeters ≈ 15.71 centimeters
91. $\frac{8\pi}{3}$ square inches ≈ 8.38 square inches
92. 18π square millimeters ≈ 56.55 square millimeters
93. 12.27 square feet **94.** 5.64 square miles
95. 591.3 miles **96.** 686.2 miles
97. 0.071 radian $\approx 4.04°$ **98.** 0.063 radian $\approx 3.59°$
99. $\frac{5}{12}$ radian **100.** $275°$
101. (a) 728.3 revolutions per minute
  (b) 4576 radians per minute
102. (a) 3400π radians per minute; 1700π radians per minute
  (b) 850 revolutions per minute
103. (a) $10,400\pi$ radians per minute
    $\approx 32,672.56$ radians per minute
  (b) $9425\pi/3$ feet per minute ≈ 9869.84 feet per minute
104. (a) 8π radians per minute ≈ 25.13 radians per minute
  (b) 200π feet per minute ≈ 628.3 feet per minute
105. (a) $[400\pi, 1000\pi]$ radians per minute
  (b) $[2400\pi, 6000\pi]$ centimeters per minute
106. $A = 175\pi$ square inches ≈ 549.8 square inches
107.

$A = 476.39\pi$ square meters ≈ 1496.62 square meters
108. (a) $\frac{14\pi}{3}$ feet per second; ≈ 10 miles per hour

  (b) $d = \frac{7\pi}{7920}n$ (c) $d = \frac{7\pi}{7920}t$

  (d) The functions are both linear.
109. False. A measurement of 4π radians corresponds to two complete revolutions from the initial to the terminal side of an angle.
110. True. Let α and β represent coterminal angles, and let n represent an integer.
  $\alpha = \beta + n(360°)$
  $\alpha - \beta = n(360°)$

111. False. The terminal side of the angle lies on the x-axis.

112. (a) The vertex is at the origin and the initial side is on the positive x-axis.

(b) Clockwise rotation of the terminal side

(c) Two angles in standard position where the terminal sides coincide

(d) The magnitude of the angle is between $90°$ and $180°$.

113. Increases. The linear velocity is proportional to the radius.

114. Radian. 1 radian $\approx 57.3°$

115. The arc length is increasing. If θ is constant, the length of the arc is proportional to the radius $(s = r\theta)$.

116. Proofs will vary.

117. $\dfrac{\sqrt{2}}{2}$ **118.** $\dfrac{2\sqrt{3}}{3}$ **119.** $\sqrt{2}$ **120.** $\dfrac{5\sqrt{2}}{4}$

121. $2\sqrt{10}$ **122.** $6\sqrt{5}$ **123.** $12\sqrt{2}$ **124.** $4\sqrt{13}$

125.

126.

127.

128.

Section 1.2 *(page 147)*

Vocabulary Check *(page 147)*

1. unit circle **2.** periodic

3. period **4.** odd; even

---

1. $\sin\theta = \frac{15}{17}$ $\csc\theta = \frac{17}{15}$
  $\cos\theta = -\frac{8}{17}$ $\sec\theta = -\frac{17}{8}$
  $\tan\theta = -\frac{15}{8}$ $\cot\theta = -\frac{8}{15}$

2. $\sin\theta = \frac{5}{13}$ $\csc\theta = \frac{13}{5}$
  $\cos\theta = \frac{12}{13}$ $\sec\theta = \frac{13}{12}$
  $\tan\theta = \frac{5}{12}$ $\cot\theta = \frac{12}{5}$

3. $\sin\theta = -\frac{5}{13}$ $\csc\theta = -\frac{13}{5}$
  $\cos\theta = \frac{12}{13}$ $\sec\theta = \frac{13}{12}$
  $\tan\theta = -\frac{5}{12}$ $\cot\theta = -\frac{12}{5}$

4. $\sin\theta = -\frac{3}{5}$ $\csc\theta = -\frac{5}{3}$
  $\cos\theta = -\frac{4}{5}$ $\sec\theta = -\frac{5}{4}$
  $\tan\theta = \frac{3}{4}$ $\cot\theta = \frac{4}{3}$

5. $\left(\dfrac{\sqrt{2}}{2}, \dfrac{\sqrt{2}}{2}\right)$ **6.** $\left(\dfrac{1}{2}, \dfrac{\sqrt{3}}{2}\right)$ **7.** $\left(-\dfrac{\sqrt{3}}{2}, -\dfrac{1}{2}\right)$

8. $\left(-\dfrac{\sqrt{2}}{2}, -\dfrac{\sqrt{2}}{2}\right)$ **9.** $\left(-\dfrac{1}{2}, -\dfrac{\sqrt{3}}{2}\right)$ **10.** $\left(\dfrac{1}{2}, -\dfrac{\sqrt{3}}{2}\right)$

11. $(0, -1)$ **12.** $(-1, 0)$

13. $\sin\dfrac{\pi}{4} = \dfrac{\sqrt{2}}{2}$
  $\cos\dfrac{\pi}{4} = \dfrac{\sqrt{2}}{2}$
  $\tan\dfrac{\pi}{4} = 1$

14. $\sin\dfrac{\pi}{3} = \dfrac{\sqrt{3}}{2}$
  $\cos\dfrac{\pi}{3} = \dfrac{1}{2}$
  $\tan\dfrac{\pi}{3} = \sqrt{3}$

15. $\sin\left(-\dfrac{\pi}{6}\right) = -\dfrac{1}{2}$
  $\cos\left(-\dfrac{\pi}{6}\right) = \dfrac{\sqrt{3}}{2}$
  $\tan\left(-\dfrac{\pi}{6}\right) = -\dfrac{\sqrt{3}}{3}$

16. $\sin\left(-\dfrac{\pi}{4}\right) = -\dfrac{\sqrt{2}}{2}$
  $\cos\left(-\dfrac{\pi}{4}\right) = \dfrac{\sqrt{2}}{2}$
  $\tan\left(-\dfrac{\pi}{4}\right) = -1$

17. $\sin\left(-\dfrac{7\pi}{4}\right) = \dfrac{\sqrt{2}}{2}$
  $\cos\left(-\dfrac{7\pi}{4}\right) = \dfrac{\sqrt{2}}{2}$
  $\tan\left(-\dfrac{7\pi}{4}\right) = 1$

18. $\sin\left(-\dfrac{4\pi}{3}\right) = \dfrac{\sqrt{3}}{2}$
  $\cos\left(-\dfrac{4\pi}{3}\right) = -\dfrac{1}{2}$
  $\tan\left(-\dfrac{4\pi}{3}\right) = -\sqrt{3}$

19. $\sin\dfrac{11\pi}{6} = -\dfrac{1}{2}$
  $\cos\dfrac{11\pi}{6} = \dfrac{\sqrt{3}}{2}$
  $\tan\dfrac{11\pi}{6} = -\dfrac{\sqrt{3}}{3}$

20. $\sin\dfrac{5\pi}{3} = -\dfrac{\sqrt{3}}{2}$
  $\cos\dfrac{5\pi}{3} = \dfrac{1}{2}$
  $\tan\dfrac{5\pi}{3} = -\sqrt{3}$

21. $\sin\left(-\dfrac{3\pi}{2}\right) = 1$
  $\cos\left(-\dfrac{3\pi}{2}\right) = 0$
  $\tan\left(-\dfrac{3\pi}{2}\right)$ is undefined.

22. $\sin(-2\pi) = 0$
  $\cos(-2\pi) = 1$
  $\tan(-2\pi) = 0$

23. $\sin\dfrac{3\pi}{4} = \dfrac{\sqrt{2}}{2}$ $\csc\dfrac{3\pi}{4} = \sqrt{2}$
  $\cos\dfrac{3\pi}{4} = -\dfrac{\sqrt{2}}{2}$ $\sec\dfrac{3\pi}{4} = -\sqrt{2}$
  $\tan\dfrac{3\pi}{4} = -1$ $\cot\dfrac{3\pi}{4} = -1$

24. $\sin \dfrac{5\pi}{6} = \dfrac{1}{2}$ $\qquad\qquad$ $\csc \dfrac{5\pi}{6} = 2$

$\cos \dfrac{5\pi}{6} = -\dfrac{\sqrt{3}}{2}$ $\qquad$ $\sec \dfrac{5\pi}{6} = -\dfrac{2\sqrt{3}}{3}$

$\tan \dfrac{5\pi}{6} = -\dfrac{\sqrt{3}}{3}$ $\qquad$ $\cot \dfrac{5\pi}{6} = -\sqrt{3}$

25. $\sin\left(-\dfrac{\pi}{2}\right) = -1$ $\qquad$ $\csc\left(-\dfrac{\pi}{2}\right) = -1$

$\cos\left(-\dfrac{\pi}{2}\right) = 0$ $\qquad$ $\sec\left(-\dfrac{\pi}{2}\right)$ is undefined.

$\tan\left(-\dfrac{\pi}{2}\right)$ is undefined. $\qquad$ $\cot\left(-\dfrac{\pi}{2}\right) = 0$

26. $\sin \dfrac{3\pi}{2} = -1$ $\qquad\qquad$ $\csc \dfrac{3\pi}{2} = -1$

$\cos \dfrac{3\pi}{2} = 0$ $\qquad\qquad$ $\sec \dfrac{3\pi}{2}$ is undefined.

$\tan \dfrac{3\pi}{2}$ is undefined. $\qquad$ $\cot \dfrac{3\pi}{2} = 0$

27. $\sin \dfrac{4\pi}{3} = -\dfrac{\sqrt{3}}{2}$ $\qquad$ $\csc \dfrac{4\pi}{3} = -\dfrac{2\sqrt{3}}{3}$

$\cos \dfrac{4\pi}{3} = -\dfrac{1}{2}$ $\qquad$ $\sec \dfrac{4\pi}{3} = -2$

$\tan \dfrac{4\pi}{3} = \sqrt{3}$ $\qquad$ $\cot \dfrac{4\pi}{3} = \dfrac{\sqrt{3}}{3}$

28. $\sin \dfrac{7\pi}{4} = -\dfrac{\sqrt{2}}{2}$ $\qquad$ $\csc \dfrac{7\pi}{4} = -\sqrt{2}$

$\cos \dfrac{7\pi}{4} = \dfrac{\sqrt{2}}{2}$ $\qquad$ $\sec \dfrac{7\pi}{4} = \sqrt{2}$

$\tan \dfrac{7\pi}{4} = -1$ $\qquad$ $\cot \dfrac{7\pi}{4} = -1$

29. $\sin 5\pi = \sin \pi = 0$ $\qquad$ **30.** $\cos 5\pi = \cos \pi = -1$

31. $\cos \dfrac{8\pi}{3} = \cos \dfrac{2\pi}{3} = -\dfrac{1}{2}$

32. $\sin \dfrac{9\pi}{4} = \sin \dfrac{\pi}{4} = \dfrac{\sqrt{2}}{2}$

33. $\cos\left(-\dfrac{15\pi}{2}\right) = \cos \dfrac{\pi}{2} = 0$

34. $\sin \dfrac{19\pi}{6} = \sin \dfrac{7\pi}{6} = -\dfrac{1}{2}$

35. $\sin\left(-\dfrac{9\pi}{4}\right) = \sin \dfrac{7\pi}{4} = -\dfrac{\sqrt{2}}{2}$

36. $\cos\left(-\dfrac{8\pi}{3}\right) = \cos \dfrac{4\pi}{3} = -\dfrac{1}{2}$

37. (a) $-\dfrac{1}{3}$ (b) -3 $\quad$ **38.** (a) $-\dfrac{3}{8}$ (b) $-\dfrac{8}{3}$

39. (a) $-\dfrac{1}{5}$ (b) -5 $\quad$ **40.** (a) $-\dfrac{3}{4}$ (b) $-\dfrac{4}{3}$

41. (a) $\dfrac{4}{5}$ (b) $-\dfrac{4}{5}$ $\quad$ **42.** (a) $-\dfrac{4}{5}$ (b) $-\dfrac{4}{5}$

43. 0.7071 $\quad$ **44.** 1.7321 $\quad$ **45.** 1.0378 $\quad$ **46.** 0.6421

47. -0.1288 $\quad$ **48.** -0.8011 $\quad$ **49.** 1.3940

50. -4.4014 $\quad$ **51.** -1.4486 $\quad$ **52.** -0.7833

53. (a) -1 (b) -0.4 $\quad$ **54.** (a) 0.7 (b) -0.8

55. (a) 0.25 or 2.89 (b) 1.82 or 4.46

56. (a) 4.0 or 5.4 (b) 0.72 or 5.56

57. (a)

t	0	$\frac{1}{2}$	1	$\frac{3}{2}$	2
y	0	1.148	2.121	2.772	3

(b) $t \approx 2$ seconds $\quad$ (c) $t = 4$ seconds

58. (a) 0.25 foot (b) 0.02 foot (c) -0.25 foot

59. False. $\sin(-t) = -\sin t$ means that the function is odd, not that the sine of a negative angle is a negative number.

60. True. The tangent function has a period of π.

61. (a) y-axis symmetry (b) $\sin t_1 = \sin(\pi - t_1)$

(c) $\cos(\pi - t_1) = -\cos t_1$

62. Answers will vary. $\qquad$ **63.** $f^{-1}(x) = \frac{2}{3}(x + 1)$

64. $f^{-1}(x) = \sqrt[3]{4(x - 1)}$ $\qquad$ **65.** $f^{-1}(x) = \sqrt{x^2 + 4}, \quad x \geq 0$

66. $f^{-1}(x) = \dfrac{2(2x + 1)}{x - 1}$

Section 1.3 $\quad$ (page 156)

Vocabulary Check $\quad$ (page 156)

1. (a) v (b) iv (c) vi (d) iii (e) i (f) ii

2. opposite; adjacent; hypotenuse

3. elevation; depression

1. $\sin \theta = \frac{3}{5}$ $\qquad$ $\csc \theta = \frac{5}{3}$

$\cos \theta = \frac{4}{5}$ $\qquad$ $\sec \theta = \frac{5}{4}$

$\tan \theta = \frac{3}{4}$ $\qquad$ $\cot \theta = \frac{4}{3}$

2. $\sin \theta = \frac{5}{13}$ $\qquad$ $\csc \theta = \frac{13}{5}$

$\cos \theta = \frac{12}{13}$ $\qquad$ $\sec \theta = \frac{13}{12}$

$\tan \theta = \frac{5}{12}$ $\qquad$ $\cot \theta = \frac{12}{5}$

3. $\sin \theta = \frac{9}{41}$ $\qquad$ $\csc \theta = \frac{41}{9}$

$\cos \theta = \frac{40}{41}$ $\qquad$ $\sec \theta = \frac{41}{40}$

$\tan \theta = \frac{9}{40}$ $\qquad$ $\cot \theta = \frac{40}{9}$

4. $\sin \theta = \dfrac{\sqrt{2}}{2}$ $\qquad$ $\csc \theta = \sqrt{2}$

$\cos \theta = \dfrac{\sqrt{2}}{2}$ $\qquad$ $\sec \theta = \sqrt{2}$

$\tan \theta = 1$ $\qquad$ $\cot \theta = 1$

5. $\sin \theta = \dfrac{1}{3}$ $\qquad$ $\csc \theta = 3$

$\cos \theta = \dfrac{2\sqrt{2}}{3}$ $\qquad$ $\sec \theta = \dfrac{3\sqrt{2}}{4}$

$\tan \theta = \dfrac{\sqrt{2}}{4}$ $\qquad$ $\cot \theta = 2\sqrt{2}$

The triangles are similar, and corresponding sides are proportional.

6. $\sin \theta = \frac{8}{17}$ $\qquad$ $\csc \theta = \frac{17}{8}$

$\cos \theta = \frac{15}{17}$ $\qquad$ $\sec \theta = \frac{17}{15}$

$\tan \theta = \frac{8}{15}$ $\qquad$ $\cot \theta = \frac{15}{8}$

The triangles are similar, and corresponding sides are proportional.

CHAPTER 1

7. $\sin \theta = \frac{3}{5}$ $\csc \theta = \frac{5}{3}$
$\cos \theta = \frac{4}{5}$ $\sec \theta = \frac{5}{4}$
$\tan \theta = \frac{3}{4}$ $\cot \theta = \frac{4}{3}$

The triangles are similar, and corresponding sides are proportional.

8. $\sin \theta = \frac{\sqrt{5}}{5}$ $\csc \theta = \sqrt{5}$

$\cos \theta = \frac{2\sqrt{5}}{5}$ $\sec \theta = \frac{\sqrt{5}}{2}$

$\tan \theta = \frac{1}{2}$ $\cot \theta = 2$

The triangles are similar, and corresponding sides are proportional.

9.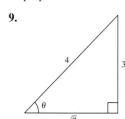
$\cos \theta = \frac{\sqrt{7}}{4}$ $\sec \theta = \frac{4\sqrt{7}}{7}$

$\tan \theta = \frac{3\sqrt{7}}{7}$ $\cot \theta = \frac{\sqrt{7}}{3}$

$\csc \theta = \frac{4}{3}$

10.
$\sin \theta = \frac{2\sqrt{6}}{7}$ $\sec \theta = \frac{7}{5}$

$\tan \theta = \frac{2\sqrt{6}}{5}$ $\cot \theta = \frac{5\sqrt{6}}{12}$

$\csc \theta = \frac{7\sqrt{6}}{12}$

11.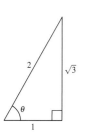
$\sin \theta = \frac{\sqrt{3}}{2}$ $\csc \theta = \frac{2\sqrt{3}}{3}$

$\cos \theta = \frac{1}{2}$ $\cot \theta = \frac{\sqrt{3}}{3}$

$\tan \theta = \sqrt{3}$

12.
$\sin \theta = \frac{\sqrt{26}}{26}$ $\csc \theta = \sqrt{26}$

$\cos \theta = \frac{5\sqrt{26}}{26}$ $\sec \theta = \frac{\sqrt{26}}{5}$

$\tan \theta = \frac{1}{5}$

13.
$\sin \theta = \frac{3\sqrt{10}}{10}$ $\sec \theta = \sqrt{10}$

$\cos \theta = \frac{\sqrt{10}}{10}$ $\cot \theta = \frac{1}{3}$

$\csc \theta = \frac{\sqrt{10}}{3}$

14.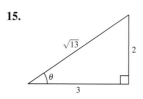
$\sin \theta = \frac{\sqrt{35}}{6}$ $\csc \theta = \frac{6\sqrt{35}}{35}$

$\cos \theta = \frac{1}{6}$ $\cot \theta = \frac{\sqrt{35}}{35}$

$\tan \theta = \sqrt{35}$

15.
$\sin \theta = \frac{2\sqrt{13}}{13}$ $\csc \theta = \frac{\sqrt{13}}{2}$

$\cos \theta = \frac{3\sqrt{13}}{13}$ $\sec \theta = \frac{\sqrt{13}}{3}$

$\tan \theta = \frac{2}{3}$

16.
$\sin \theta = \frac{4}{17}$ $\sec \theta = \frac{17\sqrt{273}}{273}$

$\cos \theta = \frac{\sqrt{273}}{17}$ $\cot \theta = \frac{\sqrt{273}}{4}$

$\tan \theta = \frac{4\sqrt{273}}{273}$

17. $\frac{\pi}{6}; \frac{1}{2}$ **18.** $\frac{\pi}{4}; \frac{\sqrt{2}}{2}$ **19.** $60°; \sqrt{3}$ **20.** $45°; \sqrt{2}$

21. $60°; \frac{\pi}{3}$ **22.** $45°; \frac{\pi}{4}$ **23.** $30°; \frac{\sqrt{3}}{2}$ **24.** $45°; \frac{\sqrt{2}}{2}$

25. $45°; \frac{\pi}{4}$ **26.** $30°; \frac{\pi}{6}$

27. (a) $\sqrt{3}$ (b) $\frac{1}{2}$ (c) $\frac{\sqrt{3}}{2}$ (d) $\frac{\sqrt{3}}{3}$

28. (a) 2 (b) $\frac{\sqrt{3}}{3}$ (c) $\frac{\sqrt{3}}{2}$ (d) $\sqrt{3}$

29. (a) $\frac{2\sqrt{13}}{13}$ (b) $\frac{3\sqrt{13}}{13}$ (c) $\frac{2}{3}$ (d) $\frac{\sqrt{13}}{2}$

30. (a) $\frac{1}{5}$ (b) $\frac{\sqrt{6}}{12}$ (c) $2\sqrt{6}$ (d) $\frac{2\sqrt{6}}{5}$

31. (a) 3 (b) $\frac{2\sqrt{2}}{3}$ (c) $\frac{\sqrt{2}}{4}$ (d) $\frac{1}{3}$

32. (a) $\frac{1}{5}$ (b) $\frac{\sqrt{26}}{26}$ (c) $\frac{1}{5}$ (d) $\frac{\sqrt{26}}{5}$

33–42. Answers will vary.
43. (a) 0.1736   (b) 0.1736
44. (a) 0.4348   (b) 0.4348
45. (a) 0.2815   (b) 3.5523
46. (a) 0.9598   (b) 0.9609
47. (a) 1.3499   (b) 1.3432
48. (a) 0.9964   (b) 1.0036
49. (a) 5.0273   (b) 0.1989

50. (a) 1.7946 (b) 0.5572
51. (a) 1.8527 (b) 0.9817
52. (a) 2.6695 (b) 0.0699
53. (a) $30° = \dfrac{\pi}{6}$ (b) $30° = \dfrac{\pi}{6}$
54. (a) $45° = \dfrac{\pi}{4}$ (b) $45° = \dfrac{\pi}{4}$
55. (a) $60° = \dfrac{\pi}{3}$ (b) $45° = \dfrac{\pi}{4}$
56. (a) $60° = \dfrac{\pi}{3}$ (b) $60° = \dfrac{\pi}{3}$
57. (a) $60° = \dfrac{\pi}{3}$ (b) $45° = \dfrac{\pi}{4}$
58. (a) $60° = \dfrac{\pi}{3}$ (b) $45° = \dfrac{\pi}{4}$
59. $30\sqrt{3}$ **60.** $9\sqrt{3}$ **61.** $\dfrac{32\sqrt{3}}{3}$
62. $20\sqrt{2}$ **63.** 443.2 meters; 323.3 meters
64. (a)

(b) $\tan\theta = \dfrac{6}{3} = \dfrac{h}{135}$

(c) 270 feet

Not drawn to scale

65. $30° = \dfrac{\pi}{6}$ **66.** 137.6 feet
67. (a) 371.1 feet (b) 341.6 feet
(c) Moving down line at 61.8 feet per second
   Dropping vertically at 24.2 feet per second
68. 1.3 miles **69.** $(x_1, y_1) = \left(28\sqrt{3},\ 28\right)$
   $(x_2, y_2) = \left(28,\ 28\sqrt{3}\right)$
70. 6.57 centimeters
71. (a)

(b) $\sin 85° = \dfrac{h}{20}$ (c) 19.9 meters
(d) The side of the triangle labeled h will become shorter.
(e)

Angle, θ	80°	70°	60°	50°
Height	19.7	18.8	17.3	15.3

Angle, θ	40°	30°	20°	10°
Height	12.9	10.0	6.8	3.5

(f) As $\theta \to 0°$, $h \to 0$.

72. $\sin 20° \approx 0.34$
   $\cos 20° \approx 0.94$
   $\tan 20° \approx 0.36$
   $\csc 20° \approx 2.92$
   $\sec 20° \approx 1.06$
   $\cot 20° \approx 2.75$
73. True, $\csc x = \dfrac{1}{\sin x}$. **74.** True, $\sec x = \csc(90° - x)$.
75. False, $\dfrac{\sqrt{2}}{2} + \dfrac{\sqrt{2}}{2} \ne 1$.
76. True, $\cot^2\theta - \csc^2\theta = -1$ for all θ.
77. False, $1.7321 \ne 0.0349$.
78. False, $\tan 25° \ne (\tan 5°)(\tan 5°)$.
79. Corresponding sides of similar triangles are proportional.
80. Yes, $\tan\theta$ is equal to opp/adj. You can find the value of the hypotenuse by the Pythagorean Theorem, then you can find $\sec\theta$, which is equal to hyp/adj.
81. (a)

θ	0.1	0.2	0.3	0.4	0.5
$\sin\theta$	0.0998	0.1987	0.2955	0.3894	0.4794

(b) θ (c) As θ approaches 0, $\sin\theta$ approaches 0.
82. (a)

θ	0°	18°	36°	54°	72°	90°
$\sin\theta$	0	0.3090	0.5878	0.8090	0.9511	1
$\cos\theta$	1	0.9511	0.8090	0.5878	0.3090	0

(b) Increasing function (c) Decreasing function
(d) As the angle increases, the length of the side opposite the angle increases relative to the length of the hypotenuse and the length of the side adjacent to the angle decreases relative to the length of the hypotenuse. Thus, the sine increases and the cosine decreases.
83. $\dfrac{x}{x - 2}$, $x \ne \pm 6$ **84.** $\dfrac{2t + 3}{4 - t}$, $t \ne \pm\dfrac{3}{2}$, -4
85. $\dfrac{2(x^2 - 5x - 10)}{(x - 2)(x + 2)^2}$ **86.** $\dfrac{1}{4}$, $x \ne 0, 12$

Section 1.4 *(page 166)*

1. (a) $\sin \theta = \frac{3}{5}$ (b) $\sin \theta = -\frac{15}{17}$

$\cos \theta = \frac{4}{5}$ $\cos \theta = \frac{8}{17}$

$\tan \theta = \frac{3}{4}$ $\tan \theta = -\frac{15}{8}$

$\csc \theta = \frac{5}{3}$ $\csc \theta = -\frac{17}{15}$

$\sec \theta = \frac{5}{4}$ $\sec \theta = \frac{17}{8}$

$\cot \theta = \frac{4}{3}$ $\cot \theta = -\frac{8}{15}$

2. (a) $\sin \theta = -\frac{5}{13}$ (b) $\sin \theta = \frac{\sqrt{2}}{2}$

$\cos \theta = -\frac{12}{13}$

$\tan \theta = \frac{5}{12}$ $\cos \theta = -\frac{\sqrt{2}}{2}$

$\csc \theta = -\frac{13}{5}$ $\tan \theta = -1$

$\sec \theta = -\frac{13}{12}$ $\csc \theta = \sqrt{2}$

$\cot \theta = \frac{12}{5}$ $\sec \theta = -\sqrt{2}$

                    $\cot \theta = -1$

3. (a) $\sin \theta = -\dfrac{1}{2}$ (b) $\sin \theta = \dfrac{\sqrt{17}}{17}$

$\cos \theta = -\dfrac{\sqrt{3}}{2}$ $\cos \theta = -\dfrac{4\sqrt{17}}{17}$

$\tan \theta = \dfrac{\sqrt{3}}{3}$ $\tan \theta = -\dfrac{1}{4}$

$\csc \theta = -2$ $\csc \theta = \sqrt{17}$

$\sec \theta = -\dfrac{2\sqrt{3}}{3}$ $\sec \theta = -\dfrac{\sqrt{17}}{4}$

$\cot \theta = \sqrt{3}$ $\cot \theta = -4$

4. (a) $\sin \theta = \dfrac{\sqrt{10}}{10}$ (b) $\sin \theta = -\dfrac{\sqrt{2}}{2}$

$\cos \theta = \dfrac{3\sqrt{10}}{10}$ $\cos \theta = \dfrac{\sqrt{2}}{2}$

$\tan \theta = \dfrac{1}{3}$ $\tan \theta = -1$

$\csc \theta = \sqrt{10}$ $\csc \theta = -\sqrt{2}$

$\sec \theta = \dfrac{\sqrt{10}}{3}$ $\sec \theta = \sqrt{2}$

$\cot \theta = 3$ $\cot \theta = -1$

5. $\sin \theta = \frac{24}{25}$ $\csc \theta = \frac{25}{24}$

$\cos \theta = \frac{7}{25}$ $\sec \theta = \frac{25}{7}$

$\tan \theta = \frac{24}{7}$ $\cot \theta = \frac{7}{24}$

6. $\sin \theta = \frac{15}{17}$ $\csc \theta = \frac{17}{15}$

$\cos \theta = \frac{8}{17}$ $\sec \theta = \frac{17}{8}$

$\tan \theta = \frac{15}{8}$ $\cot \theta = \frac{8}{15}$

7. $\sin \theta = \dfrac{5\sqrt{29}}{29}$ $\csc \theta = \dfrac{\sqrt{29}}{5}$

$\cos \theta = -\dfrac{2\sqrt{29}}{29}$ $\sec \theta = -\dfrac{\sqrt{29}}{2}$

$\tan \theta = -\dfrac{5}{2}$ $\cot \theta = -\dfrac{2}{5}$

8. $\sin \theta = -\dfrac{2\sqrt{29}}{29}$ $\csc \theta = -\dfrac{\sqrt{29}}{2}$

$\cos \theta = -\dfrac{5\sqrt{29}}{29}$ $\sec \theta = -\dfrac{\sqrt{29}}{5}$

$\tan \theta = \dfrac{2}{5}$ $\cot \theta = \dfrac{5}{2}$

9. $\sin \theta = \dfrac{68\sqrt{5849}}{5849}$ $\csc \theta = \dfrac{\sqrt{5849}}{68}$

$\cos \theta = -\dfrac{35\sqrt{5849}}{5849}$ $\sec \theta = -\dfrac{\sqrt{5849}}{35}$

$\tan \theta = -\dfrac{68}{35}$ $\cot \theta = -\dfrac{35}{68}$

10. $\sin \theta = -\dfrac{31\sqrt{1157}}{1157}$ $\csc \theta = -\dfrac{\sqrt{1157}}{31}$

$\cos \theta = \dfrac{14\sqrt{1157}}{1157}$ $\sec \theta = \dfrac{\sqrt{1157}}{14}$

$\tan \theta = -\dfrac{31}{14}$ $\cot \theta = -\dfrac{14}{31}$

11. Quadrant III   **12.** Quadrant I

13. Quadrant II   **14.** Quadrant IV

15. $\sin \theta = \frac{3}{5}$ $\csc \theta = \frac{5}{3}$

$\cos \theta = -\frac{4}{5}$ $\sec \theta = -\frac{5}{4}$

$\tan \theta = -\frac{3}{4}$ $\cot \theta = -\frac{4}{3}$

16. $\sin \theta = -\frac{3}{5}$ $\csc \theta = -\frac{5}{3}$

$\cos \theta = -\frac{4}{5}$ $\sec \theta = -\frac{5}{4}$

$\tan \theta = \frac{3}{4}$ $\cot \theta = \frac{4}{3}$

17. $\sin \theta = -\frac{15}{17}$ $\csc \theta = -\frac{17}{15}$

$\cos \theta = \frac{8}{17}$ $\sec \theta = \frac{17}{8}$

$\tan \theta = -\frac{15}{8}$ $\cot \theta = -\frac{8}{15}$

18. $\sin \theta = -\frac{15}{17}$ $\csc \theta = -\frac{17}{15}$

$\cos \theta = \frac{8}{17}$ $\sec \theta = \frac{17}{8}$

$\tan \theta = -\frac{15}{8}$ $\cot \theta = -\frac{8}{15}$

19. $\sin \theta = -\dfrac{\sqrt{10}}{10}$ $\csc \theta = -\sqrt{10}$

$\cos \theta = \dfrac{3\sqrt{10}}{10}$ $\sec \theta = \dfrac{\sqrt{10}}{3}$

$\tan \theta = -\dfrac{1}{3}$ $\cot \theta = -3$

20. $\sin \theta = \dfrac{1}{4}$ $\csc \theta = 4$

$\cos \theta = -\dfrac{\sqrt{15}}{4}$ $\sec \theta = -\dfrac{4\sqrt{15}}{15}$

$\tan \theta = -\dfrac{\sqrt{15}}{15}$ $\cot \theta = -\sqrt{15}$

21. $\sin \theta = \dfrac{\sqrt{3}}{2}$ $\qquad$ $\csc \theta = \dfrac{2\sqrt{3}}{3}$

$\cos \theta = -\dfrac{1}{2}$ $\qquad$ $\sec \theta = -2$

$\tan \theta = -\sqrt{3}$ $\qquad$ $\cot \theta = -\dfrac{\sqrt{3}}{3}$

22. $\sin \theta = 0$ $\qquad$ $\csc \theta$ is undefined.

$\cos \theta = -1$ $\qquad$ $\sec \theta = -1$

$\tan \theta = 0$ $\qquad$ $\cot \theta$ is undefined.

23. $\sin \theta = 0$ $\qquad$ $\csc \theta$ is undefined.

$\cos \theta = -1$ $\qquad$ $\sec \theta = -1$

$\tan \theta = 0$ $\qquad$ $\cot \theta$ is undefined.

24. $\sin \theta = -1$ $\qquad$ $\csc \theta = -1$

$\cos \theta = 0$ $\qquad$ $\sec \theta$ is undefined.

$\tan \theta$ is undefined. $\qquad$ $\cot \theta = 0$

25. $\sin \theta = \dfrac{\sqrt{2}}{2}$ $\qquad$ $\csc \theta = \sqrt{2}$

$\cos \theta = -\dfrac{\sqrt{2}}{2}$ $\qquad$ $\sec \theta = -\sqrt{2}$

$\tan \theta = -1$ $\qquad$ $\cot \theta = -1$

26. $\sin \theta = -\dfrac{\sqrt{10}}{10}$ $\qquad$ $\csc \theta = -\sqrt{10}$

$\cos \theta = -\dfrac{3\sqrt{10}}{10}$ $\qquad$ $\sec \theta = -\dfrac{\sqrt{10}}{3}$

$\tan \theta = \dfrac{1}{3}$ $\qquad$ $\cot \theta = 3$

27. $\sin \theta = -\dfrac{2\sqrt{5}}{5}$ $\qquad$ $\csc \theta = -\dfrac{\sqrt{5}}{2}$

$\cos \theta = -\dfrac{\sqrt{5}}{5}$ $\qquad$ $\sec \theta = -\sqrt{5}$

$\tan \theta = 2$ $\qquad$ $\cot \theta = \dfrac{1}{2}$

28. $\sin \theta = -\frac{4}{5}$ $\qquad$ $\csc \theta = -\frac{5}{4}$

$\cos \theta = \frac{3}{5}$ $\qquad$ $\sec \theta = \frac{5}{3}$

$\tan \theta = -\frac{4}{3}$ $\qquad$ $\cot \theta = -\frac{3}{4}$

29. 0 $\qquad$ **30.** -1 $\qquad$ **31.** Undefined $\qquad$ **32.** -1 $\qquad$ **33.** 1

34. Undefined $\qquad$ **35.** Undefined $\qquad$ **36.** 0

37. $\theta' = 23°$ $\qquad\qquad\qquad$ **38.** $\theta' = 51°$

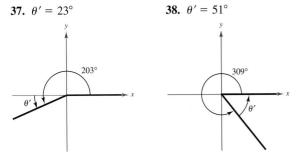

39. $\theta' = 65°$ $\qquad\qquad$ **40.** $\theta' = 35°$

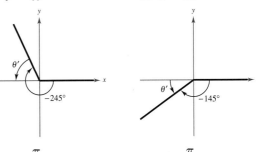

41. $\theta' = \dfrac{\pi}{3}$ $\qquad\qquad$ **42.** $\theta' = \dfrac{\pi}{4}$

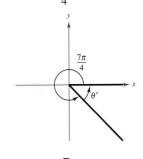

43. $\theta' = 3.5 - \pi$ $\qquad\qquad$ **44.** $\theta' = \dfrac{\pi}{3}$

45. $\sin 225° = -\dfrac{\sqrt{2}}{2}$ $\qquad$ **46.** $\sin 300° = -\dfrac{\sqrt{3}}{2}$

$\cos 225° = -\dfrac{\sqrt{2}}{2}$ $\qquad$ $\cos 300° = \dfrac{1}{2}$

$\tan 225° = 1$ $\qquad$ $\tan 300° = -\sqrt{3}$

47. $\sin 750° = \dfrac{1}{2}$ $\qquad$ **48.** $\sin(-405°) = -\dfrac{\sqrt{2}}{2}$

$\cos 750° = \dfrac{\sqrt{3}}{2}$ $\qquad$ $\cos(-405°) = \dfrac{\sqrt{2}}{2}$

$\tan 750° = \dfrac{\sqrt{3}}{3}$ $\qquad$ $\tan(-405°) = -1$

49. $\sin(-150°) = -\dfrac{1}{2}$ $\qquad$ **50.** $\sin(-840°) = -\dfrac{\sqrt{3}}{2}$

$\cos(-150°) = -\dfrac{\sqrt{3}}{2}$ $\qquad$ $\cos(-840°) = -\dfrac{1}{2}$

$\tan(-150°) = \dfrac{\sqrt{3}}{3}$ $\qquad$ $\tan(-840°) = \sqrt{3}$

CHAPTER 1

51. $\sin\dfrac{4\pi}{3} = -\dfrac{\sqrt{3}}{2}$

$\cos\dfrac{4\pi}{3} = -\dfrac{1}{2}$

$\tan\dfrac{4\pi}{3} = \sqrt{3}$

52. $\sin\dfrac{\pi}{4} = \dfrac{\sqrt{2}}{2}$

$\cos\dfrac{\pi}{4} = \dfrac{\sqrt{2}}{2}$

$\tan\dfrac{\pi}{4} = 1$

53. $\sin\left(-\dfrac{\pi}{6}\right) = -\dfrac{1}{2}$

$\cos\left(-\dfrac{\pi}{6}\right) = \dfrac{\sqrt{3}}{2}$

$\tan\left(-\dfrac{\pi}{6}\right) = -\dfrac{\sqrt{3}}{3}$

54. $\sin\left(-\dfrac{\pi}{2}\right) = -1$

$\cos\left(-\dfrac{\pi}{2}\right) = 0$

$\tan\left(-\dfrac{\pi}{2}\right)$ is undefined.

55. $\sin\dfrac{11\pi}{4} = \dfrac{\sqrt{2}}{2}$

$\cos\dfrac{11\pi}{4} = -\dfrac{\sqrt{2}}{2}$

$\tan\dfrac{11\pi}{4} = -1$

56. $\sin\dfrac{10\pi}{3} = -\dfrac{\sqrt{3}}{2}$

$\cos\dfrac{10\pi}{3} = -\dfrac{1}{2}$

$\tan\dfrac{10\pi}{3} = \sqrt{3}$

57. $\sin\left(-\dfrac{3\pi}{2}\right) = 1$

$\cos\left(-\dfrac{3\pi}{2}\right) = 0$

$\tan\left(-\dfrac{3\pi}{2}\right)$ is undefined.

58. $\sin\left(-\dfrac{25\pi}{4}\right) = -\dfrac{\sqrt{2}}{2}$

$\cos\left(-\dfrac{25\pi}{4}\right) = \dfrac{\sqrt{2}}{2}$

$\tan\left(-\dfrac{25\pi}{4}\right) = -1$

59. $\dfrac{4}{5}$ **60.** $\dfrac{\sqrt{10}}{10}$ **61.** $-\dfrac{\sqrt{13}}{2}$ **62.** $-\sqrt{3}$ **63.** $\dfrac{8}{5}$

64. $\dfrac{\sqrt{65}}{4}$ **65.** 0.1736   **66.** -1.4142 **67.** -0.3420

68. 2.0000   **69.** -1.4826 **70.** -28.6363

71. 3.2361   **72.** -0.1405 **73.** 4.6373   **74.** 0.2245

75. 0.3640   **76.** -0.3640 **77.** -0.6052

78. 1.0436   **79.** -0.4142 **80.** 4.4940

81. (a) $30° = \dfrac{\pi}{6}$, $150° = \dfrac{5\pi}{6}$ (b) $210° = \dfrac{7\pi}{6}$, $330° = \dfrac{11\pi}{6}$

82. (a) $45° = \dfrac{\pi}{4}$, $315° = \dfrac{7\pi}{4}$ (b) $135° = \dfrac{3\pi}{4}$, $225° = \dfrac{5\pi}{4}$

83. (a) $60° = \dfrac{\pi}{3}$, $120° = \dfrac{2\pi}{3}$ (b) $135° = \dfrac{3\pi}{4}$, $315° = \dfrac{7\pi}{4}$

84. (a) $60° = \dfrac{\pi}{3}$, $300° = \dfrac{5\pi}{3}$ (b) $120° = \dfrac{2\pi}{3}$, $240° = \dfrac{4\pi}{3}$

85. (a) $45° = \dfrac{\pi}{4}$, $225° = \dfrac{5\pi}{4}$ (b) $150° = \dfrac{5\pi}{6}$, $330° = \dfrac{11\pi}{6}$

86. (a) $60° = \dfrac{\pi}{3}$, $120° = \dfrac{2\pi}{3}$ (b) $240° = \dfrac{4\pi}{3}$, $300° = \dfrac{5\pi}{3}$

87. (a) $N = 22.099 \sin(0.522t - 2.219) + 55.008$
$F = 36.641 \sin(0.502t - 1.831) + 25.610$

(b) February: $N = 34.6°$, $F = -1.4°$
March: $N = 41.6°$, $F = 13.9°$
May: $N = 63.4°$, $F = 48.6°$
June: $N = 72.5°$, $F = 59.5°$
August: $N = 75.5°$, $F = 55.6°$
September: $N = 68.6°$, $F = 41.7°$
November: $N = 46.8°$, $F = 6.5°$

(c) Answers will vary.

88. (a) 26,134 units   (b) 31,438 units   (c) 21,452 units
(d) 26,756 units

89. (a) 270.63 feet   (b) 307.75 feet   (c) 270.63 feet

90. (a) 389.71 feet   (b) 289.25 feet   (c) 153.91 feet

91. (a) 2 centimeters   (b) 0.14 centimeter
(c) -1.98 centimeters

92. (a) 12 miles   (b) 6 miles   (c) 6.9 miles

93. False. In each of the four quadrants, the signs of the secant function and cosine function will be the same, because these functions are reciprocals of each other.

94. False. For θ in Quadrant II, $\theta' = 180° - \theta$. For θ in Quadrant III, $\theta' = \theta - 180°$. For θ in Quadrant IV, $\theta' = 360° - \theta$.

95. As θ increases from $0°$ to $90°$, x decreases from 12 cm to 0 cm and y increases from 0 cm to 12 cm. Therefore, $\sin\theta = y/12$ increases from 0 to 1 and $\cos\theta = x/12$ decreases from 1 to 0. Thus, $\tan\theta = y/x$ and increases without bound. When $\theta = 90°$, the tangent is undefined.

96. Determine the trigonometric function of the reference angle and prefix the appropriate sign.

97.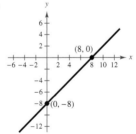

x-intercept: $(8, 0)$
y-intercept: $(0, -8)$
Domain: all real numbers x

98.

x-intercept: $\left(\frac{6}{7}, 0\right)$
y-intercept: $(0, 6)$
Domain: all real numbers x

99.

x-intercepts:
$(1, 0), (-4, 0)$
y-intercept: $(0, -4)$
Domain: all real numbers x

100.

Intercept: $(0, 0)$
x-intercept: $\left(\frac{5}{2}, 0\right)$
Domain: all real numbers x

101.

x-intercept: $(-2, 0)$
y-intercept: $(0, 8)$
Domain: all real numbers x

102.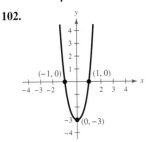

x-intercepts:
$(1, 0), (-1, 0)$
y-intercept: $(0, -3)$
Domain: all real numbers x

103.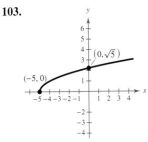

x-intercept: $(-5, 0)$
y-intercept: $\left(0, \sqrt{5}\right)$
Domain: all real numbers x
    such that $x \geq -5$

104.

x-intercept: $\left(\frac{1}{4}, 0\right)$
y-intercept: None
Domain: all real numbers x
    such that $x \geq \frac{1}{4}$

Section 1.5 *(page 176)*

1. Period: π **2.** Period: $\dfrac{2\pi}{3}$ **3.** Period: 4π

Amplitude: 3   Amplitude: 2   Amplitude: $\frac{5}{2}$

4. Period: 6π **5.** Period: 6   **6.** Period: 4

Amplitude: 3   Amplitude: $\frac{1}{2}$ Amplitude: $\frac{3}{2}$

7. Period: 2π **8.** Period: 3π **9.** Period: $\dfrac{\pi}{5}$

Amplitude: 2   Amplitude: 1   Amplitude: 3

10. Period: $\dfrac{\pi}{4}$ **11.** Period: 3π **12.** Period: 8π

Amplitude: $\dfrac{1}{3}$ Amplitude: $\dfrac{1}{2}$ Amplitude: $\dfrac{5}{2}$

13. Period: 1   **14.** Period: 20

Amplitude: $\frac{1}{4}$ Amplitude: $\frac{2}{3}$

15. g is a shift of f π units to the right.
16. g is a shift of f π units to the left.
17. g is a reflection of f in the x-axis.
18. g is a reflection of f in the x-axis.
19. The period of g is one-half the period of f.
20. The period of g is one-third the period of f.
21. g is a shift of f three units upward.
22. g is a shift of f two units downward.
23. The graph of g has twice the amplitude of the graph of f.
24. The period of g is $\frac{1}{3}$ the period of f.
25. The graph of g is a horizontal shift of the graph of f π units to the right.
26. g is a shift of f two units upward.

27.   **28.**

29.   **30.**

31.

32.

41.

42.

33.

34.

43.

44.

35.

36.

45.

46.

37.

38.

47.

48.

39.

40.

49.

50.

51.

52.

53.

54.

55.

56.

57.

58.

59.

60.

61.

62.

63. $a = 2, d = 1$　　**64.** $a = 2, d = -1$

65. $a = -4, d = 4$　　**66.** $a = -1, d = -3$

67. $a = -3, b = 2, c = 0$　　**68.** $a = 2, b = \frac{1}{2}, c = 0$

69. $a = 2, b = 1, c = -\dfrac{\pi}{4}$　　**70.** $a = 2, b = \dfrac{\pi}{2}, c = -\dfrac{\pi}{2}$

71.

72.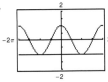

$x = -\dfrac{\pi}{6}, -\dfrac{5\pi}{6}, \dfrac{7\pi}{6}, \dfrac{11\pi}{6}$　　$x = \pi, -\pi$

73. (a) 6 seconds　(b) 10 cycles per minute

(c)

74. (a) 4 seconds　(b) 15 cycles per minute

(c)

75. (a) $C(t) = 56.55 + 26.95 \cos\left(\dfrac{\pi}{6}t - 3.67\right)$

(b) 　(c)

The model is a good fit.　　The model is a good fit.

(d) Tallahassee: 77.90°; Chicago: 56.55°

The constant term gives the annual average temperature.

(e) 12; yes; one full period is one year.

(f) Chicago; amplitude; the greater the amplitude, the greater the variability in temperature.

76. (a) $\frac{6}{5}$ seconds　(b) 50 heartbeats per minute

77. (a) $\frac{1}{440}$ second　(b) 440 cycles per second

CHAPTER 1

78. (a)–(c)

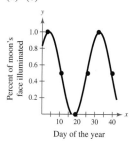

Day of the year

(b) $y = 0.5 + 0.5 \cos\left[\dfrac{2\pi}{29}(x - 3)\right]$

The model is a good fit.

(d) 29 days

(e) 0.22

79. (a) 365 days; Yes, the period is one year.

(b) 30.3 gallons; the constant term

(c)

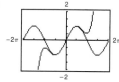

$124 < t < 252$;
this corresponds to
April–September

80. (a) 20 seconds; it takes 20 seconds to complete one revolution on the Ferris wheel.

(b) 50 feet; the diameter of the Ferris wheel is 100 feet.

(c)

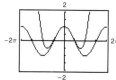

81. False. The graph of $f(x) = \sin(x + 2\pi)$ translates the graph of $f(x) = \sin x$ exactly one period to the left so that the two graphs look identical.

82. False. The function $y = \frac{1}{2} \cos 2x$ has an amplitude that is one-half that of $y = \cos x$. For $y = a \cos bx$, the amplitude is $|a|$.

83. True. Because $\cos x = \sin\left(x + \dfrac{\pi}{2}\right)$, $y = -\cos x$ is a reflection in the x-axis of $y = \sin\left(x + \dfrac{\pi}{2}\right)$.

84. Answers will vary.

85.

Conjecture:

$\sin x = \cos\left(x - \dfrac{\pi}{2}\right)$

86.

Conjecture:

$\sin x = -\cos\left(x + \dfrac{\pi}{2}\right)$

87. (a)

The graphs appear to coincide from $-\dfrac{\pi}{2}$ to $\dfrac{\pi}{2}$.

(b)

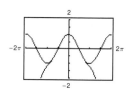

The graphs appear to coincide from $-\dfrac{\pi}{2}$ to $\dfrac{\pi}{2}$.

(c) $-\dfrac{x^7}{7!}, -\dfrac{x^6}{6!}$

The interval of accuracy increased.

88. (a) 0.4794, 0.4794 (b) 0.8417, 0.8415 (c) 0.5, 0.5

(d) 0.8776, 0.8776 (e) 0.5417, 0.5403

(f) 0.7074, 0.7071

The error increases as x moves farther away from 0.

89. Distributive Property

90. Commutative Property of Addition

91. Identity Property of Addition

92. Associative Property of Addition

93. $y = -\frac{6}{5}x + \frac{7}{5}$

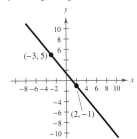

94. $y = -5x + 11$

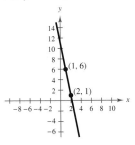

95. $y = \frac{3}{5}x + \frac{13}{5}$

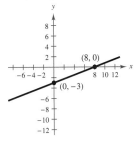

96. $y = \frac{3}{8}x - 3$

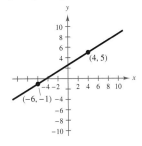

97. Answers will vary.

Section 1.6   *(page 187)*

Vocabulary Check   *(page 187)*

1. vertical   **2.** reciprocal   **3.** damping
4. π **5.** $x \neq n\pi$ **6.** $(-\infty, -1] \cup [1, \infty)$
7. 2π

1. e, π **2.** c, 2π **3.** a, 1   **4.** d, 2π
5. f, 4   **6.** b, 4

7.

8.

9.

10.

11.

12.

13.

14.

15.

16.

17.

18.

19.

20.

CHAPTER 1

21.

22.

23.

24.

25.

26.

27.

28.

29.

30.

31.

32.

33.

34.

35.

36.

37.

38.

39.

40.

41. $-\dfrac{7\pi}{4}, -\dfrac{3\pi}{4}, \dfrac{\pi}{4}, \dfrac{5\pi}{4}$ **42.** $-\dfrac{5\pi}{3}, -\dfrac{2\pi}{3}, \dfrac{\pi}{3}, \dfrac{4\pi}{3}$

43. $-\dfrac{4\pi}{3}, -\dfrac{\pi}{3}, \dfrac{2\pi}{3}, \dfrac{5\pi}{3}$ **44.** $-\dfrac{7\pi}{4}, -\dfrac{3\pi}{4}, \dfrac{\pi}{4}, \dfrac{5\pi}{4}$

45. $-\dfrac{4\pi}{3}, -\dfrac{2\pi}{3}, \dfrac{2\pi}{3}, \dfrac{4\pi}{3}$ **46.** $-\dfrac{5\pi}{3}, -\dfrac{\pi}{3}, \dfrac{\pi}{3}, \dfrac{5\pi}{3}$

47. $-\dfrac{7\pi}{4}, -\dfrac{5\pi}{4}, \dfrac{\pi}{4}, \dfrac{3\pi}{4}$ **48.** $-\dfrac{2\pi}{3}, -\dfrac{\pi}{3}, \dfrac{4\pi}{3}, \dfrac{5\pi}{3}$

49. Even   **50.** Odd

51. (a) 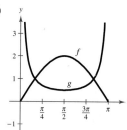   (b) $\dfrac{\pi}{6} < x < \dfrac{5\pi}{6}$

(c) f approaches 0 and g approaches $+\infty$ because the cosecant is the reciprocal of the sine.

52. (a)

(b) $\left(-1, \frac{1}{3}\right)$ (c) $\left(-1, \frac{1}{3}\right)$; The intervals are the same.

53.

The expressions are equivalent except that when $\sin x = 0$, y_1 is undefined.

54.

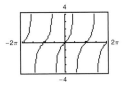

The expressions are equivalent.

55.

The expressions are equivalent.

56.

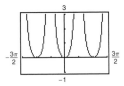

The expressions are equivalent.

57. d, $f \to 0$ as $x \to 0$.   **58.** a, $f \to 0$ as $x \to 0$.
59. b, $g \to 0$ as $x \to 0$.   **60.** c, $g \to 0$ as $x \to 0$.

61.

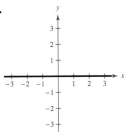

The functions are equal.

62.

The functions are equal.

63.

The functions are equal.

64.

The functions are equal.

65.

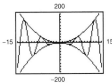

Damping factor: x
As $x \to \infty$, $g(x)$ oscillates and approaches $-\infty$ and ∞

66.

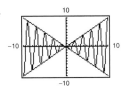

Damping factor: x^2
As $x \to \infty$, $f(x)$ oscillates and approaches $-\infty$ and ∞

67.

Damping factor: x^3
As $x \to \infty$, $g(x)$ oscillates and approaches $-\infty$ and ∞

68.

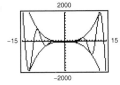

Damping factor: x^3
As $x \to \infty$, $h(x)$ oscillates and approaches $-\infty$ and ∞

69.

As $x \to 0$, $y \to \infty$

70.

As $x \to 0$, $y \to \infty$

71.

As $x \to 0$, $g(x) \to 1$

72.

As $x \to 0$, $y \to \infty$

73.

As $x \to 0$, $f(x)$ oscillates between 1 and -1.

74.

As $x \to 0$, $h(x)$ oscillates.

75. $d = 7 \cot x$

76. $d = 27 \sec x$

CHAPTER 1

77. (a)

(b) As the predator population increases, the number of prey decreases. When the number of prey is small, the number of predators decreases.

(c) C: 24 months; R: 24 months

78.

Month (1 ↔ January)

79. (a) H: 12 months; L: 12 months

(b) Summer; winter   (c) 1 month

80. (a)

(b) y approaches 0 as t increases.

81. True. For a given value of x, the y-coordinate of csc x is the reciprocal of the y-coordinate of sin x.

82. True. $y = \sec x$ is equal to $y = 1/\cos x$, and if the reciprocal of $y = \sin x$ is translated $\pi/2$ units to the left, then

$$\frac{1}{\sin\left(x + \dfrac{\pi}{2}\right)} = \frac{1}{\cos x} = \sec x.$$

83. As x approaches $\pi/2$ from the left, f approaches ∞. As x approaches $\pi/2$ from the right, f approaches $-\infty$.

84. As x approaches π from the left, f approaches ∞. As x approaches π from the right, f approaches $-\infty$.

85. (a)

0.7391

(b) 1, 0.5403, 0.8576, 0.6543, 0.7935, 0.7014, 0.7640, 0.7221, 0.7504, 0.7314, . . . ; 0.7391

86.

The graphs appear to coincide on the interval $-1.1 \le x \le 1.1$.

87.

The graphs appear to coincide on the interval $-1.1 \le x \le 1.1$.

88. (a)

(b) $y_3 = \dfrac{4}{\pi}\left[\sin(\pi x) + \dfrac{1}{3}\sin(3\pi x) + \dfrac{1}{5}\sin(5\pi x) + \dfrac{1}{7}\sin(7\pi x)\right]$

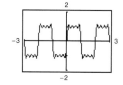

(c) $y_4 = \dfrac{4}{\pi}\left[\sin(\pi x) + \dfrac{1}{3}\sin(3\pi x) + \dfrac{1}{5}\sin(5\pi x) + \dfrac{1}{7}\sin(7\pi x) + \dfrac{1}{9}\sin(9\pi x)\right]$

89. ± 8 **90.** $5 \pm 2\sqrt{2}$ **91.** $\frac{3}{2}$ **92.** $-\frac{1}{3}, -1$
93. $3 \pm \sqrt{5}$ **94.** $3, -1$ **95.** $-\frac{10}{3}, 5$ **96.** $\frac{11}{8}$

Section 1.7 *(page 197)*

Vocabulary Check *(page 197)*

1. $y = \sin^{-1} x; -1 \le x \le 1$

2. $y = \arccos x; 0 \le y \le \pi$

3. $y = \tan^{-1} x; -\infty < x < \infty; -\dfrac{\pi}{2} < y < \dfrac{\pi}{2}$

1. $\dfrac{\pi}{6}$ **2.** 0   **3.** $\dfrac{\pi}{3}$ **4.** $\dfrac{\pi}{2}$ **5.** $\dfrac{\pi}{6}$ **6.** $-\dfrac{\pi}{4}$

7. $\dfrac{5\pi}{6}$ **8.** $-\dfrac{\pi}{4}$ **9.** $-\dfrac{\pi}{3}$ **10.** $\dfrac{\pi}{3}$ **11.** $\dfrac{2\pi}{3}$

12. $\dfrac{\pi}{4}$ **13.** $\dfrac{\pi}{3}$ **14.** $-\dfrac{\pi}{6}$ **15.** 0   **16.** 0

17.   **18.**

19. 1.29   **20.** 0.47   **21.** -0.85 **22.** 2.35
23. -1.25 **24.** 1.50   **25.** 0.32   **26.** 1.31
27. 1.99   **28.** -0.13 **29.** 0.74   **30.** 1.23
31. 0.85   **32.** 1.91   **33.** 1.29   **34.** -1.50

35. $-\dfrac{\pi}{3}, -\dfrac{\sqrt{3}}{3}, 1$ **36.** $\pi, \dfrac{2\pi}{3}, \dfrac{\sqrt{3}}{2}$

37. $\theta = \arctan \dfrac{x}{4}$ **38.** $\theta = \arccos \dfrac{4}{x}$

39. $\theta = \arcsin \dfrac{x+2}{5}$ **40.** $\theta = \arctan \dfrac{x+1}{10}$

41. $\theta = \arccos \dfrac{x+3}{2x}$ **42.** $\theta = \arctan \dfrac{1}{x+1}, \; x \neq 1$

43. 0.3   **44.** 25   **45.** -0.1 **46.** -0.2 **47.** 0

48. $\dfrac{\pi}{2}$ **49.** $\dfrac{3}{5}$ **50.** $\dfrac{5}{3}$ **51.** $\dfrac{\sqrt{5}}{5}$ **52.** $\dfrac{2\sqrt{5}}{5}$

53. $\dfrac{12}{13}$ **54.** $-\dfrac{13}{5}$ **55.** $\dfrac{\sqrt{34}}{5}$ **56.** $-\dfrac{3\sqrt{7}}{7}$

57. $\dfrac{\sqrt{5}}{3}$ **58.** $\dfrac{8}{5}$ **59.** $\dfrac{1}{x}$ **60.** $\dfrac{x}{\sqrt{x^2+1}}$

61. $\sqrt{1-4x^2}$ **62.** $\sqrt{9x^2+1}$ **63.** $\sqrt{1-x^2}$

64. $\dfrac{1}{\sqrt{2x-x^2}}$ **65.** $\dfrac{\sqrt{9-x^2}}{x}$ **66.** x

67. $\dfrac{\sqrt{x^2+2}}{x}$ **68.** $\dfrac{\sqrt{r^2-(x-h)^2}}{r}$

69.   **70.**

Asymptotes: $y = \pm 1$ Asymptote: $x = 0$

71. $\dfrac{9}{\sqrt{x^2+81}}, \; x > 0; \; \dfrac{-9}{\sqrt{x^2+81}}, \; x < 0$

72. $\dfrac{x}{6}$ **73.** $\dfrac{|x-1|}{\sqrt{x^2-2x+10}}$ **74.** $\dfrac{\sqrt{4x-x^2}}{x-2}$

75.   **76.**

The graph of g is a horizontal shift one unit to the right of f.

The graph of g is a horizontal stretch of the graph of f.

77.   **78.**

79.   **80.**

81.   **82.**

83.   **84.**

85.   **86.**

87.   **88.**

89. $3\sqrt{2} \sin\left(2t + \dfrac{\pi}{4}\right)$

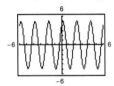

The graph implies that the identity is true.

90. $5 \sin\left(\pi t + \arctan \dfrac{4}{3}\right)$

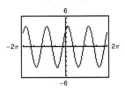

The graph implies that the identity is true.

91. (a) $\theta = \arcsin \dfrac{5}{s}$ (b) 0.13, 0.25

92. (a) $\theta = \arctan \dfrac{s}{750}$ (b) 21.8°, 58.0°

93. (a)

(b) 2 feet   (c) $\beta = 0$; As x increases, β approaches 0.

94. (a) $\theta = 32.9°$ (b) 12.94 feet

95. (a) $\theta \approx 26.0°$ (b) 24.4 feet

96. (a) $\theta = \arctan \dfrac{6}{x}$ (b) 40.6°, 80.5°

97. (a) $\theta = \arctan \dfrac{x}{20}$ (b) 14.0°, 31.0°

98. False. $\dfrac{5\pi}{6}$ is not in the range of the arcsine.

99. False. $\dfrac{5\pi}{4}$ is not in the range of the arctangent.

100. False. The graphs are not the same.

101. Domain: $(-\infty, \infty)$
Range: $(0, \pi)$

102. Domain: $(-\infty, -1] \cup [1, \infty)$
Range: $[0, \pi/2) \cup (\pi/2, \pi]$

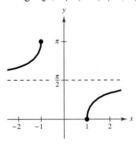

103. Domain: $(-\infty, -1] \cup [1, \infty)$
Range: $[-\pi/2, 0) \cup (0, \pi/2]$

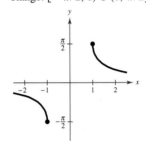

104. (a) $\dfrac{\pi}{4}$ (b) 0   (c) $\dfrac{5\pi}{6}$ (d) $\dfrac{\pi}{6}$

105. (a) $\dfrac{\pi}{4}$ (b) $\dfrac{\pi}{2}$ (c) 1.25   (d) 2.03

106.

As x increases to infinity, g approaches 3π, but f has no maximum.
$a \approx 87.54$

107. (a) $f \circ f^{-1}$ $f^{-1} \circ f$

(b) The domains and ranges of the functions are restricted. The graphs of $f \circ f^{-1}$ and $f^{-1} \circ f$ differ because of the domains and ranges of f and f^{-1}.

108. (a)–(e) Proofs will vary.

109.

$\cos \theta = \dfrac{\sqrt{7}}{4}$ $\sec \theta = \dfrac{4\sqrt{7}}{7}$

$\tan \theta = \dfrac{3\sqrt{7}}{7}$ $\cot \theta = \dfrac{\sqrt{7}}{3}$

$\csc \theta = \dfrac{4}{3}$

110.

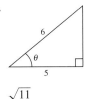

$\sin \theta = \dfrac{2\sqrt{5}}{5}$ $\sec \theta = \sqrt{5}$

$\cos \theta = \dfrac{\sqrt{5}}{5}$ $\cot \theta = \dfrac{1}{2}$

$\csc \theta = \dfrac{\sqrt{5}}{2}$

111.

$\sin \theta = \dfrac{\sqrt{11}}{6}$ $\sec \theta = \dfrac{6}{5}$

$\tan \theta = \dfrac{\sqrt{11}}{5}$ $\cot \theta = \dfrac{5\sqrt{11}}{11}$

$\csc \theta = \dfrac{6\sqrt{11}}{11}$

112.

$\sin \theta = \dfrac{2\sqrt{2}}{3}$ $\csc \theta = \dfrac{3\sqrt{2}}{4}$

$\cos \theta = \dfrac{1}{3}$ $\cot \theta = \dfrac{\sqrt{2}}{4}$

$\tan \theta = \dfrac{2\sqrt{2}}{1}$

113. Eight people   **114.** 3 miles per hour

Section 1.8   *(page 207)*

Vocabulary Check   *(page 207)*

1. elevation; depression   **2.** bearing

3. harmonic motion

1. $a \approx 3.64$ **2.** $a \approx 8.82$ **3.** $a \approx 8.26$
$c \approx 10.64$ $b \approx 12.14$ $c \approx 25.38$
$B = 70°$ $A = 36°$ $A = 19°$

4. $b \approx 274.27$ **5.** $c \approx 11.66$ **6.** $b \approx 24.49$
$c \approx 277.24$ $A \approx 30.96°$ $A \approx 45.58°$
$B = 81.6°$ $B \approx 59.04°$ $B \approx 44.42°$

7. $a \approx 49.48$ **8.** $a \approx 9.36$ **9.** $a \approx 91.34$
$A \approx 72.08°$ $A \approx 81.97°$ $b \approx 420.70$
$B \approx 17.92°$ $B \approx 8.03°$ $B = 77°45'$

10. $b \approx 30.73$ **11.** 2.56 inches   **12.** 1.62 meters
$c \approx 33.85$
$A = 24°48'$

13. 19.99 inches   **14.** 2.80 feet   **15.** 107.2 feet

16. 1648.5 feet   **17.** 19.7 feet   **18.** 81.2 feet

19. (a)

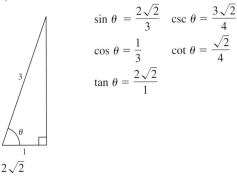

(b) $h = 50(\tan 47°40' - \tan 35°)$ (c) 19.9 feet

20. 123.5 feet   **21.** 2236.8 feet   **22.** 56.3°

23. (a)

(b) $\tan \theta = \dfrac{12\frac{1}{2}}{17\frac{1}{3}}$ (c) 35.8°

24. 71.34°   **25.** 2.06°

26. (a) 5099 feet   (b) 117.7 seconds   **27.** 0.73 mile

28. 6.8°; 2516.3 feet   **29.** 554 miles north; 709 miles east

30. (a) 429.26 miles north; 2434.44 miles east
(b) 280°

31. (a) 58.18 nautical miles west; 104.95 nautical miles south
(b) S 36.7° W; distance = 130.9 nautical miles

32. (a) 21.4 hours
(b) 5.86 nautical miles east; 239.93 nautical miles south
(c) 178.6°

33. (a) N 58° E   (b) 68.82 meters

34. $d \approx 5.46$ kilometers   **35.** N 56.31° W

36. 208°   **37.** 1933.3 feet   **38.** 11.8 kilometers

39. ≈ 3.23 miles or $\approx 17{,}054$ feet

40. ≈ 1.025 miles or ≈ 5412 feet

41. 78.7°   **42.** 52.1°   **43.** 35.3°   **44.** 54.7°

45. 29.4 inches   **46.** 25 inches   **47.** $y = \sqrt{3}\,r$

48. 9.06 centimeters   **49.** $a \approx 12.2, b \approx 7$

50. $a \approx 21.6$ feet, $b \approx 7.2$ feet, $c \approx 13$ feet

51. $d = 4 \sin(\pi t)$ **52.** $d = 3 \sin\left(\dfrac{\pi t}{3}\right)$

CHAPTER 1

53. $d = 3 \cos\left(\dfrac{4\pi t}{3}\right)$ **54.** $d = 2 \cos\left(\dfrac{\pi t}{5}\right)$

55. (a) 4   (b) 4   (c) 4   (d) $\frac{1}{16}$

56. (a) $\frac{1}{2}$ (b) 10   (c) $\frac{1}{2}$ (d) $\frac{1}{40}$

57. (a) $\frac{1}{16}$ (b) 60   (c) 0   (d) $\frac{1}{120}$

58. (a) $\frac{1}{64}$ (b) 396   (c) 0   (d) $\frac{1}{792}$

59. $\omega = 528\pi$ **60.** $d = \dfrac{7}{4} \cos \dfrac{\pi t}{5}$

61. (a)   (b) $\dfrac{\pi}{8}$ (c) $\dfrac{\pi}{32}$

62. (a)

θ	L_1	L_2	$L_1 + L_2$
0.1	$\dfrac{2}{\sin 0.1}$	$\dfrac{3}{\cos 0.1}$	23.0
0.2	$\dfrac{2}{\sin 0.2}$	$\dfrac{3}{\cos 0.2}$	13.1
0.3	$\dfrac{2}{\sin 0.3}$	$\dfrac{3}{\cos 0.3}$	9.9
0.4	$\dfrac{2}{\sin 0.4}$	$\dfrac{3}{\cos 0.4}$	8.4

(b)

θ	L_1	L_2	$L_1 + L_2$
0.5	$\dfrac{2}{\sin 0.5}$	$\dfrac{3}{\cos 0.5}$	7.6
0.6	$\dfrac{2}{\sin 0.6}$	$\dfrac{3}{\cos 0.6}$	7.2
0.7	$\dfrac{2}{\sin 0.7}$	$\dfrac{3}{\cos 0.7}$	7.0
0.8	$\dfrac{2}{\sin 0.8}$	$\dfrac{3}{\cos 0.8}$	7.1

7.0 (minimum length)

(c) $L = L_1 + L_2 = \dfrac{2}{\sin \theta} + \dfrac{3}{\cos \theta}$

(d)   7.0 (minimum length)

63. (a)

Base 1	Base 2	Altitude	Area
8	8 + 16 cos 30°	8 sin 30°	59.7
8	8 + 16 cos 40°	8 sin 40°	72.7
8	8 + 16 cos 50°	8 sin 50°	80.5
8	8 + 16 cos 60°	8 sin 60°	83.1
8	8 + 16 cos 70°	8 sin 70°	80.7
8	8 + 16 cos 80°	8 sin 80°	74.0
8	8 + 16 cos 90°	8 sin 90°	64.0

(b)

Base 1	Base 2	Altitude	Area
8	8 + 16 cos 56°	8 sin 56°	82.73
8	8 + 16 cos 58°	8 sin 58°	83.04
8	8 + 16 cos 59°	8 sin 59°	83.11
8	8 + 16 cos 60°	8 sin 60°	83.14
8	8 + 16 cos 61°	8 sin 61°	83.11
8	8 + 16 cos 62°	8 sin 62°	83.04

83.14 square feet

(c) $A = 64(1 + \cos \theta)(\sin \theta)$

(d)

≈ 83.1 square feet when $\theta = 60°$

The answers are the same.

64. (a)

(b) $S = 8 + 6.3 \cos\left(\dfrac{\pi}{6}t\right)$ or $S = 8 + 6.3 \sin\left(\dfrac{\pi}{6}t + \dfrac{\pi}{2}\right)$

The model is a good fit.

(c) 12. Yes, sales of outerwear are seasonal.

(d) Maximum displacement from average sales of $8 million

65. False. The tower is leaning, so it is not perfectly vertical and does not form a right angle with the ground.

66. False. One period is the time for one complete cycle of the motion.

67. No. N 24° E means 24 degrees east of north.

68. Air navigation is always measured clockwise from the north. Nautical navigation measures the acute angle a path makes with a fixed north-south line.

69. $y = 4x + 6$

70. $y = -\frac{1}{2}x + \frac{1}{6}$

71. $y = -\frac{4}{5}x + \frac{22}{5}$

72. $y = -\frac{4}{3}x - \frac{1}{3}$

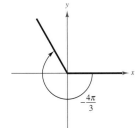

Review Exercises *(page 213)*

1. 0.5 radian   **2.** 4.5 radians

3. (a)

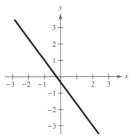

   (b) Quadrant II

   (c) $\dfrac{3\pi}{4}, -\dfrac{5\pi}{4}$

4. (a)

   (b) Quadrant I

   (c) $\dfrac{20\pi}{9}, -\dfrac{16\pi}{9}$

5. (a)

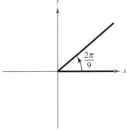

   (b) Quadrant II

   (c) $\dfrac{2\pi}{3}, -\dfrac{10\pi}{3}$

6. (a)

   (b) Quadrant I

   (c) $\dfrac{\pi}{3}, -\dfrac{17\pi}{3}$

7. (a)

   (b) Quadrant I

   (c) 430°, −290°

8. (a)

   (b) Quadrant IV

   (c) 640°, −80°

9. (a)

   (b) Quadrant III

   (c) 250°, −470°

10. (a)

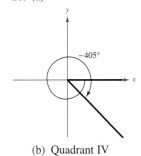

   (b) Quadrant IV

   (c) 315°, −45°

11. 8.378   **12.** −2.225   **13.** −0.589   **14.** 3.443

15. 128.571°   **16.** −330.000°   **17.** −200.535°

18. 326.586°   **19.** 48.17 inches   **20.** 11.52 meters

21. (a) $66\frac{2}{3}\pi$ radians per minute

   (b) 400π inches per minute

22. 212.1 inches per second

23. Area = 339.29 square inches

24. Area = 55.31 square millimeters   **25.** $\left(-\dfrac{1}{2}, \dfrac{\sqrt{3}}{2}\right)$

26. $\left(-\dfrac{\sqrt{2}}{2}, \dfrac{\sqrt{2}}{2}\right)$ **27.** $\left(-\dfrac{\sqrt{3}}{2}, \dfrac{1}{2}\right)$ **28.** $\left(-\dfrac{1}{2}, \dfrac{\sqrt{3}}{2}\right)$

29. $\sin\dfrac{7\pi}{6} = -\dfrac{1}{2}$ $\csc\dfrac{7\pi}{6} = -2$

$\cos\dfrac{7\pi}{6} = -\dfrac{\sqrt{3}}{2}$ $\sec\dfrac{7\pi}{6} = -\dfrac{2\sqrt{3}}{3}$

$\tan\dfrac{7\pi}{6} = \dfrac{\sqrt{3}}{3}$ $\cot\dfrac{7\pi}{6} = \sqrt{3}$

30. $\sin\dfrac{\pi}{4} = \dfrac{\sqrt{2}}{2}$ $\csc\dfrac{\pi}{4} = \sqrt{2}$

$\cos\dfrac{\pi}{4} = \dfrac{\sqrt{2}}{2}$ $\sec\dfrac{\pi}{4} = \sqrt{2}$

$\tan\dfrac{\pi}{4} = 1$ $\cot\dfrac{\pi}{4} = 1$

31. $\sin\left(-\dfrac{2\pi}{3}\right) = -\dfrac{\sqrt{3}}{2}$ $\qquad \csc\left(-\dfrac{2\pi}{3}\right) = -\dfrac{2\sqrt{3}}{3}$

$\cos\left(-\dfrac{2\pi}{3}\right) = -\dfrac{1}{2}$ $\qquad \sec\left(-\dfrac{2\pi}{3}\right) = -2$

$\tan\left(-\dfrac{2\pi}{3}\right) = \sqrt{3}$ $\qquad \cot\left(-\dfrac{2\pi}{3}\right) = \dfrac{\sqrt{3}}{3}$

32. $\sin 2\pi = 0$ $\qquad \csc 2\pi$ is undefined.

$\cos 2\pi = 1$ $\qquad \sec 2\pi = 1$

$\tan 2\pi = 0$ $\qquad \cot 2\pi$ is undefined.

33. $\sin\dfrac{11\pi}{4} = \sin\dfrac{3\pi}{4} = \dfrac{\sqrt{2}}{2}$

34. $\cos 4\pi = \cos 0 = 1$ $\qquad$ **35.** $\sin\left(-\dfrac{17\pi}{6}\right) = \sin\dfrac{7\pi}{6} = -\dfrac{1}{2}$

36. $\cos\left(-\dfrac{13\pi}{3}\right) = \cos\dfrac{5\pi}{3} = \dfrac{1}{2}$ $\qquad$ **37.** -75.3130

38. -1.1368 $\qquad$ **39.** 3.2361 $\qquad$ **40.** -0.3420

41. $\sin\theta = \dfrac{4\sqrt{41}}{41}$

$\cos\theta = \dfrac{5\sqrt{41}}{41}$

$\tan\theta = \dfrac{4}{5}$

$\csc\theta = \dfrac{\sqrt{41}}{4}$

$\sec\theta = \dfrac{\sqrt{41}}{5}$

$\cot\theta = \dfrac{5}{4}$

42. $\sin\theta = \dfrac{\sqrt{2}}{2}$

$\cos\theta = \dfrac{\sqrt{2}}{2}$

$\tan\theta = 1$

$\csc\theta = \sqrt{2}$

$\sec\theta = \sqrt{2}$

$\cot\theta = 1$

43. $\sin\theta = \dfrac{\sqrt{3}}{2}$

$\cos\theta = \dfrac{1}{2}$

$\tan\theta = \sqrt{3}$

$\csc\theta = \dfrac{2\sqrt{3}}{3}$

$\sec\theta = 2$

$\cot\theta = \dfrac{\sqrt{3}}{3}$

44. $\sin\theta = \dfrac{5}{9}$

$\cos\theta = \dfrac{2\sqrt{14}}{9}$

$\tan\theta = \dfrac{5\sqrt{14}}{28}$

$\csc\theta = \dfrac{9}{5}$

$\sec\theta = \dfrac{9\sqrt{14}}{28}$

$\cot\theta = \dfrac{2\sqrt{14}}{5}$

45. (a) 3 (b) $\dfrac{2\sqrt{2}}{3}$ (c) $\dfrac{3\sqrt{2}}{4}$ (d) $\dfrac{\sqrt{2}}{4}$

46. (a) $\dfrac{1}{4}$ (b) $\sqrt{17}$ (c) $\dfrac{\sqrt{17}}{17}$ (d) $\dfrac{\sqrt{17}}{4}$

47. (a) $\dfrac{1}{4}$ (b) $\dfrac{\sqrt{15}}{4}$ (c) $\dfrac{4\sqrt{15}}{15}$ (d) $\dfrac{\sqrt{15}}{15}$

48. (a) $\dfrac{1}{5}$ (b) $2\sqrt{6}$ (c) $\dfrac{\sqrt{6}}{12}$ (d) 5 $\qquad$ **49.** 0.6494

50. 5.2408 $\qquad$ **51.** 0.5621 $\qquad$ **52.** 5.3860 $\qquad$ **53.** 3.6722

54. 0.2045 $\qquad$ **55.** 71.3 meters $\qquad$ **56.** 19.5 feet

57. $\sin\theta = \dfrac{4}{5}$ $\qquad \csc\theta = \dfrac{5}{4}$

$\cos\theta = \dfrac{3}{5}$ $\qquad \sec\theta = \dfrac{5}{3}$

$\tan\theta = \dfrac{4}{3}$ $\qquad \cot\theta = \dfrac{3}{4}$

58. $\sin\theta = -\dfrac{4}{5}$ $\qquad \csc\theta = -\dfrac{5}{4}$

$\cos\theta = \dfrac{3}{5}$ $\qquad \sec\theta = \dfrac{5}{3}$

$\tan\theta = -\dfrac{4}{3}$ $\qquad \cot\theta = -\dfrac{3}{4}$

59. $\sin\theta = \dfrac{15\sqrt{241}}{241}$ $\qquad \csc\theta = \dfrac{\sqrt{241}}{15}$

$\cos\theta = \dfrac{4\sqrt{241}}{241}$ $\qquad \sec\theta = \dfrac{\sqrt{241}}{4}$

$\tan\theta = \dfrac{15}{4}$ $\qquad \cot\theta = \dfrac{4}{15}$

60. $\sin\theta = -\dfrac{\sqrt{26}}{26}$ $\qquad \csc\theta = -\sqrt{26}$

$\cos\theta = -\dfrac{5\sqrt{26}}{26}$ $\qquad \sec\theta = -\dfrac{\sqrt{26}}{5}$

$\tan\theta = \dfrac{1}{5}$ $\qquad \cot\theta = 5$

61. $\sin\theta = \dfrac{9\sqrt{82}}{82}$ $\qquad \csc\theta = \dfrac{\sqrt{82}}{9}$

$\cos\theta = \dfrac{-\sqrt{82}}{82}$ $\qquad \sec\theta = -\sqrt{82}$

$\tan\theta = -9$ $\qquad \cot\theta = -\dfrac{1}{9}$

62. $\sin\theta = \dfrac{4}{5}$ $\qquad \csc\theta = \dfrac{5}{4}$

$\cos\theta = \dfrac{3}{5}$ $\qquad \sec\theta = \dfrac{5}{3}$

$\tan\theta = \dfrac{4}{3}$ $\qquad \cot\theta = \dfrac{3}{4}$

63. $\sin\theta = \dfrac{4\sqrt{17}}{17}$ $\qquad \csc\theta = \dfrac{\sqrt{17}}{4}$

$\cos\theta = \dfrac{\sqrt{17}}{17}$ $\qquad \sec\theta = \sqrt{17}$

$\tan\theta = 4$ $\qquad \cot\theta = \dfrac{1}{4}$

64. $\sin\theta = -\dfrac{3\sqrt{13}}{13}$ $\qquad \csc\theta = -\dfrac{\sqrt{13}}{3}$

$\cos\theta = -\dfrac{2\sqrt{13}}{13}$ $\qquad \sec\theta = -\dfrac{\sqrt{13}}{2}$

$\tan\theta = \dfrac{3}{2}$ $\qquad \cot\theta = \dfrac{2}{3}$

65. $\sin\theta = -\dfrac{\sqrt{11}}{6}$

$\cos\theta = \dfrac{5}{6}$

$\tan\theta = -\dfrac{\sqrt{11}}{5}$

$\csc\theta = -\dfrac{6\sqrt{11}}{11}$

$\cot\theta = -\dfrac{5\sqrt{11}}{11}$

66. $\sin\theta = \dfrac{2}{3}$

$\cos\theta = -\dfrac{\sqrt{5}}{3}$

$\tan\theta = -\dfrac{2\sqrt{5}}{5}$

$\sec\theta = -\dfrac{3\sqrt{5}}{5}$

$\cot\theta = -\dfrac{\sqrt{5}}{2}$

67. $\cos \theta = -\dfrac{\sqrt{55}}{8}$

$\tan \theta = -\dfrac{3\sqrt{55}}{55}$

$\csc \theta = \dfrac{8}{3}$

$\sec \theta = -\dfrac{8\sqrt{55}}{55}$

$\cot \theta = -\dfrac{\sqrt{55}}{3}$

68. $\sin \theta = -\dfrac{5\sqrt{41}}{41}$

$\cos \theta = -\dfrac{4\sqrt{41}}{41}$

$\csc \theta = -\dfrac{\sqrt{41}}{5}$

$\sec \theta = -\dfrac{\sqrt{41}}{4}$

$\cot \theta = \dfrac{4}{5}$

69. $\sin \theta = \dfrac{\sqrt{21}}{5}$

$\tan \theta = -\dfrac{\sqrt{21}}{2}$

$\csc \theta = \dfrac{5\sqrt{21}}{21}$

$\sec \theta = -\dfrac{5}{2}$

$\cot \theta = -\dfrac{2\sqrt{21}}{21}$

70. $\cos \theta = \dfrac{\sqrt{3}}{2}$

$\tan \theta = -\dfrac{\sqrt{3}}{3}$

$\csc \theta = -2$

$\sec \theta = \dfrac{2\sqrt{3}}{3}$

$\cot \theta = -\sqrt{3}$

71. $\theta' = 84°$

72. $\theta' = 85°$

73. $\theta' = \dfrac{\pi}{5}$

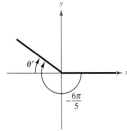

74. $\theta' = \dfrac{\pi}{3}$

75. $\sin \dfrac{\pi}{3} = \dfrac{\sqrt{3}}{2}$; $\cos \dfrac{\pi}{3} = \dfrac{1}{2}$; $\tan \dfrac{\pi}{3} = \sqrt{3}$

76. $\sin \dfrac{\pi}{4} = \dfrac{\sqrt{2}}{2}$; $\cos \dfrac{\pi}{4} = \dfrac{\sqrt{2}}{2}$; $\tan \dfrac{\pi}{4} = 1$

77. $\sin\left(-\dfrac{7\pi}{3}\right) = -\dfrac{\sqrt{3}}{2}$; $\cos\left(-\dfrac{7\pi}{3}\right) = \dfrac{1}{2}$;

$\tan\left(-\dfrac{7\pi}{3}\right) = -\sqrt{3}$

78. $\sin\left(-\dfrac{5\pi}{4}\right) = \dfrac{\sqrt{2}}{2}$; $\cos\left(-\dfrac{5\pi}{4}\right) = -\dfrac{\sqrt{2}}{2}$;

$\tan\left(-\dfrac{5\pi}{4}\right) = -1$

79. $\sin 495° = \dfrac{\sqrt{2}}{2}$; $\cos 495° = -\dfrac{\sqrt{2}}{2}$; $\tan 495° = -1$

80. $\sin(-150°) = -\dfrac{1}{2}$; $\cos(-150°) = -\dfrac{\sqrt{3}}{2}$;

$\tan(-150°) = \dfrac{\sqrt{3}}{3}$

81. $\sin(-240°) = \dfrac{\sqrt{3}}{2}$; $\cos(-240°) = -\dfrac{1}{2}$;

$\tan(-240°) = -\sqrt{3}$

82. $\sin 315° = -\dfrac{\sqrt{2}}{2}$; $\cos 315° = \dfrac{\sqrt{2}}{2}$; $\tan 315° = -1$

83. -0.7568 **84.** -0.1425 **85.** 0.0584

86. 0.0878 **87.** 3.2361 **88.** 4.3813

89.

90.

91.

92.

93.

94.

95.

96.

105.

106.

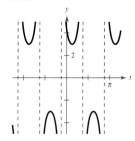

97. (a) $y = 2 \sin 528\pi x$ (b) 264 cycles per second

98. (a)

(b) 12. Yes. One period is one year.

(c) 1.41. 1.41 represents the maximum change in time from the average time $(d = 18.09)$ of sunset.

107.

108.

As $x \to +\infty, f(x) \to +\infty$ As $x \to +\infty, f(x) \to +\infty$

109. $-\dfrac{\pi}{6}$ **110.** $-\dfrac{\pi}{2}$ **111.** 0.41   **112.** 0.21

99.

100.

113. -0.46 **114.** 1.10   **115.** $\dfrac{\pi}{6}$ **116.** $\dfrac{\pi}{4}$

117. π **118.** $\dfrac{\pi}{6}$ **119.** 1.24   **120.** 2.66

121. -0.98 **122.** 1.45

123.

124.

101.

102.

125.

126.

127. $\dfrac{4}{5}$ **128.** $\dfrac{4}{3}$ **129.** $\dfrac{13}{5}$ **130.** $-\dfrac{5}{12}$

131. $\dfrac{\sqrt{4 - x^2}}{x}$ **132.** $\dfrac{1}{\sqrt{1 - (x - 1)^2}}$ **133.** $66.8°$

134. 9.6 feet   **135.** 1221 miles, $85.6°$

103.

104.

136. $d = 0.75 \cos\left(\dfrac{2\pi t}{3}\right)$

137. False. The sine or cosine function is often useful for modeling simple harmonic motion.

138. True. The inverse sine, $y = \arcsin x$, is defined by $y = \arcsin x$ if and only if $\sin y = x$ where $-1 \le x \le 1$ and $-\pi/2 \le y \le \pi/2$.

139. False. For each θ there corresponds exactly one value of y.

140. False. $3\pi/4$ is not in the range of the arctangent function.

141. d; The period is 2π and the amplitude is 3.

142. a; The period is 2π and, because $a < 0$, the graph is reflected in the *x*-axis.

143. b; The period is 2 and the amplitude is 2.

144. c; The period is 4π and the amplitude is 2.

145. The function is undefined because $\sec \theta = 1/\cos \theta$.

146. (a)

θ	0.1	0.4	0.7
$\tan\left(\theta - \dfrac{\pi}{2}\right)$	-9.9666	-2.3652	-1.1872
$-\cot \theta$	-9.9666	-2.3652	-1.1872

θ	1.0	1.3
$\tan\left(\theta - \dfrac{\pi}{2}\right)$	-0.6421	-0.2776
$-\cot \theta$	-0.6421	-0.2776

(b) $\tan\left(\theta - \dfrac{\pi}{2}\right) = -\cot \theta$

147. The ranges of the other four trigonometric functions are $(-\infty, \infty)$ or $(-\infty, -1] \cup [1, \infty)$.

148. (a) $A = 0.4r^2, r > 0;$ (b) $A = 50\theta, \theta > 0;$
$\quad\quad s = 0.8r, r > 0$ $\quad\quad s = 10\theta, \theta > 0$

The area function increases more rapidly.

149. Answers will vary.

Chapter Test   *(page 217)*

1. (a)   (b) $\dfrac{13\pi}{4}, -\dfrac{3\pi}{4}$

(c) $225°$

2. 3000 radians per minute   **3.** ≈ 709.04 square feet

4. $\sin \theta = \dfrac{3\sqrt{10}}{10}$ $\csc \theta = \dfrac{\sqrt{10}}{3}$

$\cos \theta = -\dfrac{\sqrt{10}}{10}$ $\sec \theta = -\sqrt{10}$

$\tan \theta = -3$ $\cot \theta = -\dfrac{1}{3}$

5. For $0 \le \theta < \dfrac{\pi}{2}$:   For $\pi \le \theta < \dfrac{3\pi}{2}$:

$\sin \theta = \dfrac{3\sqrt{13}}{13}$ $\sin \theta = -\dfrac{3\sqrt{13}}{13}$

$\cos \theta = \dfrac{2\sqrt{13}}{13}$ $\cos \theta = -\dfrac{2\sqrt{13}}{13}$

$\csc \theta = \dfrac{\sqrt{13}}{3}$ $\csc \theta = -\dfrac{\sqrt{13}}{3}$

$\sec \theta = \dfrac{\sqrt{13}}{2}$ $\sec \theta = -\dfrac{\sqrt{13}}{2}$

$\cot \theta = \dfrac{2}{3}$ $\cot \theta = \dfrac{2}{3}$

6. $\theta' = 70°$

7. Quadrant III   **8.** $150°, 210°$ **9.** 1.33, 1.81

10. $\sin \theta = -\dfrac{4}{5}$ **11.** $\sin \theta = \dfrac{15}{17}$

$\tan \theta = -\dfrac{4}{3}$ $\cos \theta = -\dfrac{8}{17}$

$\csc \theta = -\dfrac{5}{4}$ $\tan \theta = -\dfrac{15}{8}$

$\sec \theta = \dfrac{5}{3}$ $\csc \theta = \dfrac{17}{15}$

$\cot \theta = -\dfrac{3}{4}$ $\cot \theta = -\dfrac{8}{15}$

12.   **13.**

14.   **15.**

Period: 2   Not periodic

16. $a = -2, b = \dfrac{1}{2}, c = -\dfrac{\pi}{4}$ **17.** $\dfrac{\sqrt{5}}{2}$

18.

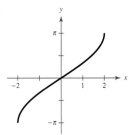

19. 310.1°   **20.** $d = -6 \cos \pi t$

Problem Solving   *(page 219)*

1. (a) $\dfrac{11\pi}{2}$ radians or 990°   (b) ≈ 816.42 feet

2. Gear 1: 270°, $\dfrac{3\pi}{2}$ radians

Gear 2: $\approx 332.3°$, $\dfrac{24\pi}{13}$ radians

Gear 3: $\approx 392.7°$, $\dfrac{24\pi}{11}$ radians

Gear 4: 450°, $\dfrac{5\pi}{2}$ radians

Gear 5: $\approx 454.7°$, $\dfrac{48\pi}{19}$ radians

3. (a) 4767 feet   (b) 3705 feet

(c) $w = 2183$ feet,

$$\tan 63° = \dfrac{w + 3705}{3000}$$

4. (a) Answers will vary.   (b) The ratios are equal.

(c) No; No

(d) Yes, because all six trigonometric functions are ratios of right triangles.

5. (a)   (b)

Even   Even

6. (a) $h(x)$ is even.   (b) $h(x)$ is even.

7. $h = 51 - 50 \sin\left(8\pi t + \dfrac{\pi}{2}\right)$

8. (a)

(b) Period $= \dfrac{3}{4}$ seconds; answers will vary.

(c) 20 millimeters; answers will vary.

(d) 80 beats per minute

(e) Period $= \dfrac{15}{16}$ seconds; $\dfrac{32\pi}{15}$

9. (a)

(b)

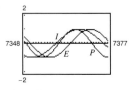

(c) $P(7369) = 0.631$

$E(7369) = 0.901$

$I(7369) = 0.945$

10. (a)

(b) Period of f: 2π; Period of g: π

(c) Yes, because the sine and cosine functions are periodic.

11. (a) 3.35, 7.35   (b) -0.65

(c) Yes. There is a difference of nine periods between the values.

12. (a) Equal; two-period shift

(b) Not equal; $f\left(t + \tfrac{1}{2}c\right)$ is a horizontal translation and $f\left(\tfrac{1}{2}t\right)$ is a period change.

(c) Not equal; For example, $\sin\left[\tfrac{1}{2}(\pi + 2\pi)\right] \neq \sin\left(\tfrac{1}{2}\pi\right)$.

13. (a) 40.5°   (b) $x \approx 1.71$ feet; $y \approx 3.46$ feet

(c) ≈ 1.75 feet

(d) As you move closer to the rock, d must get smaller and smaller. The angles θ_1 and θ_2 will decrease along with the distance y, so d will decrease.

14. (a)

The approximation is accurate over the interval $-1 \leq x \leq 1$.

(b) $\dfrac{x^9}{9}$

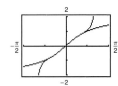

The accuracy improved.

Chapter 2
Section 2.1 (page 227)

1. $\sin x = \dfrac{\sqrt{3}}{2}$

$\cos x = -\dfrac{1}{2}$

$\tan x = -\sqrt{3}$

$\csc x = \dfrac{2\sqrt{3}}{3}$

$\sec x = -2$

$\cot x = -\dfrac{\sqrt{3}}{3}$

2. $\sin x = -\dfrac{1}{2}$

$\cos x = -\dfrac{\sqrt{3}}{2}$

$\tan x = \dfrac{\sqrt{3}}{3}$

$\csc x = -2$

$\sec x = -\dfrac{2\sqrt{3}}{3}$

$\cot x = \sqrt{3}$

3. $\sin \theta = -\dfrac{\sqrt{2}}{2}$

$\cos \theta = \dfrac{\sqrt{2}}{2}$

$\tan \theta = -1$

$\sec \theta = \sqrt{2}$

$\csc \theta = -\sqrt{2}$

$\cot \theta = -1$

4. $\sin \theta = \dfrac{3}{5}$

$\cos \theta = \dfrac{4}{5}$

$\tan \theta = \dfrac{3}{4}$

$\sec \theta = \dfrac{5}{4}$

$\csc \theta = \dfrac{5}{3}$

$\cot \theta = \dfrac{4}{3}$

5. $\sin x = -\dfrac{5}{13}$

$\cos x = -\dfrac{12}{13}$

$\tan x = \dfrac{5}{12}$

$\sec x = -\dfrac{13}{12}$

$\csc x = -\dfrac{13}{5}$

$\cot x = \dfrac{12}{5}$

6. $\sin \phi = \dfrac{\sqrt{10}}{10}$

$\cos \phi = -\dfrac{3\sqrt{10}}{10}$

$\tan \phi = -\dfrac{1}{3}$

$\cot \phi = -3$

$\csc \phi = \sqrt{10}$

$\sec \phi = -\dfrac{\sqrt{10}}{3}$

7. $\sin \phi = -\dfrac{\sqrt{5}}{3}$

$\cos \phi = \dfrac{2}{3}$

$\tan \phi = -\dfrac{\sqrt{5}}{2}$

$\sec \phi = \dfrac{3}{2}$

$\csc \phi = -\dfrac{3\sqrt{5}}{5}$

$\cot \phi = -\dfrac{2\sqrt{5}}{5}$

8. $\sin x = \dfrac{3}{5}$

$\cos x = \dfrac{4}{5}$

$\tan x = \dfrac{3}{4}$

$\csc x = \dfrac{5}{3}$

$\sec x = \dfrac{5}{4}$

$\cot x = \dfrac{4}{3}$

9. $\sin x = \dfrac{1}{3}$

$\cos x = -\dfrac{2\sqrt{2}}{3}$

$\tan x = -\dfrac{\sqrt{2}}{4}$

$\csc x = 3$

$\sec x = -\dfrac{3\sqrt{2}}{4}$

$\cot x = -2\sqrt{2}$

10. $\sin x = \dfrac{\sqrt{15}}{4}$

$\cos x = \dfrac{1}{4}$

$\tan x = \sqrt{15}$

$\sec x = 4$

$\csc x = \dfrac{4\sqrt{15}}{15}$

$\cot x = \dfrac{\sqrt{15}}{15}$

11. $\sin \theta = -\dfrac{2\sqrt{5}}{5}$

$\cos \theta = -\dfrac{\sqrt{5}}{5}$

$\tan \theta = 2$

$\csc \theta = -\dfrac{\sqrt{5}}{2}$

$\sec \theta = -\sqrt{5}$

$\cot \theta = \dfrac{1}{2}$

12. $\sin \theta = -\dfrac{1}{5}$

$\cos \theta = -\dfrac{2\sqrt{6}}{5}$

$\tan \theta = \dfrac{\sqrt{6}}{12}$

$\sec \theta = -\dfrac{5\sqrt{6}}{12}$

$\csc \theta = -5$

$\cot \theta = 2\sqrt{6}$

13. $\sin \theta = -1$
$\cos \theta = 0$
$\tan \theta$ is undefined.
$\cot \theta = 0$
$\csc \theta = -1$
$\sec \theta$ is undefined.

14. $\sin \theta = 1$
$\cos \theta = 0$
$\tan \theta$ is undefined.
$\csc \theta = 1$
$\sec \theta$ is undefined.
$\cot \theta = 0$

15. d **16.** a **17.** b **18.** f **19.** e **20.** c
21. b **22.** c **23.** f **24.** a **25.** e **26.** d
27. $\csc \theta$ **28.** $\sin \beta$ **29.** $\cos^2 \phi$ **30.** 1
31. $\cos x$ **32.** $\cot \theta$ **33.** $\sin^2 x$ **34.** $\cos^2 x$
35. 1 **36.** $\sin^2 \theta$ **37.** $\tan x$ **38.** $\sin x$
39. $1 + \sin y$ **40.** $\sec t$ **41.** $\sec \beta$ **42.** $2 \sec \phi$
43. $\cos u + \sin u$ **44.** $\csc \theta \sec \theta$ **45.** $\sin^2 x$
46. $\cos^2 x$ **47.** $\sin^2 x \tan^2 x$ **48.** 1 **49.** $\sec x + 1$
50. $\cos x + 2$ **51.** $\sec^4 x$ **52.** $\sin^4 x$
53. $\sin^2 x - \cos^2 x$ **54.** $\sec^2 x + \tan^2 x$

55. $\cot^2 x(\csc x - 1)$ **56.** $\tan^2 x(\sec x - 1)$
57. $1 + 2\sin x\cos x$ **58.** -1 **59.** $4\cot^2 x$
60. $9\cos^2 x$ **61.** $2\csc^2 x$ **62.** $-2\cot^2 x$ **63.** $2\sec x$
64. $-\cot x$ **65.** $1 + \cos y$ **66.** $5(\sec x - \tan x)$
67. $3(\sec x + \tan x)$ **68.** $\tan^4 x(\csc x - 1)$

69.

x	0.2	0.4	0.6	0.8	1.0
y_1	0.1987	0.3894	0.5646	0.7174	0.8415
y_2	0.1987	0.3894	0.5646	0.7174	0.8415

x	1.2	1.4
y_1	0.9320	0.9854
y_2	0.9320	0.9854

$y_1 = y_2$

70.

x	0.2	0.4	0.6	0.8	1.0
y_1	0.0403	0.1646	0.3863	0.7386	1.3105
y_2	0.0403	0.1646	0.3863	0.7386	1.3105

x	1.2	1.4
y_1	2.3973	5.7135
y_2	2.3973	5.7135

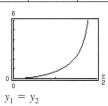

$y_1 = y_2$

71.

x	0.2	0.4	0.6	0.8	1.0
y_1	1.2230	1.5085	1.8958	2.4650	3.4082
y_2	1.2230	1.5085	1.8958	2.4650	3.4082

x	1.2	1.4
y_1	5.3319	11.6814
y_2	5.3319	11.6814

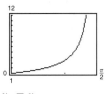

$y_1 = y_2$

72.

x	0.2	0.4	0.6	0.8	1.0
y_1	0.0428	0.2107	0.6871	2.1841	8.3087
y_2	0.0428	0.2107	0.6871	2.1841	8.3087

x	1.2	1.4
y_1	50.3869	1163.6143
y_2	50.3869	1163.6143

$y_1 = y_2$

73. $\csc x$ **74.** $\cot x$ **75.** $\tan x$ **76.** $\sec \theta$
77. $3\sin \theta$ **78.** $8\sin \theta$ **79.** $3\tan \theta$ **80.** $2\tan \theta$
81. $5\sec \theta$ **82.** $10\sec \theta$
83. $3\cos \theta = 3;\ \sin \theta = 0;\ \cos \theta = 1$
84. $6\cos \theta = 3;\ \sin \theta = \pm\dfrac{\sqrt{3}}{2};\ \cos \theta = \dfrac{1}{2}$
85. $4\sin \theta = 2\sqrt{2};\ \sin \theta = \dfrac{\sqrt{2}}{2};\ \cos \theta = \dfrac{\sqrt{2}}{2}$
86. $10\sin \theta = -5\sqrt{3};\ \sin \theta = -\dfrac{\sqrt{3}}{2};\ \cos \theta = \dfrac{1}{2}$
87. $0 \le \theta \le \pi$ **88.** $\dfrac{\pi}{2} \le \theta \le \dfrac{3\pi}{2}$

89. $0 \le \theta < \dfrac{\pi}{2},\ \dfrac{3\pi}{2} < \theta < 2\pi$ **90.** $0 < \theta < \pi$

91. (a) $\csc^2 132° - \cot^2 132° \approx 1.8107 - 0.8107 = 1$

    (b) $\csc^2 \dfrac{2\pi}{7} - \cot^2 \dfrac{2\pi}{7} \approx 1.6360 - 0.6360 = 1$

92. (a) $(\tan 346°)^2 + 1 = (\sec 346°)^2 \approx 1.0622$
    (b) $(\tan 3.1)^2 + 1 = (\sec 3.1)^2 \approx 1.0017$

93. (a) $\cos(90° - 80°) = \sin 80° \approx 0.9848$

    (b) $\cos\left(\dfrac{\pi}{2} - 0.8\right) = \sin 0.8 \approx 0.7174$

94. (a) $\sin(-250°) = -\sin 250° \approx 0.9397$
    (b) $\sin\left(-\tfrac{1}{2}\right) = -\sin\left(\tfrac{1}{2}\right) \approx -0.4794$
95. $\mu = \tan \theta$ **96.** Answers will vary.
97. True. For example, $\sin(-x) = -\sin x$.
98. False. A cofunction identity can be used to transform a tangent function so that it can be represented by a cotangent function.
99. $1, 1$ **100.** $1, 1$ **101.** $\infty, 0$ **102.** $0, -\infty$
103. Not an identity because $\cos \theta = \pm\sqrt{1 - \sin^2 \theta}$
104. Not an identity because $\cot \theta = \pm\sqrt{\csc^2 \theta - 1}$

105. Not an identity because $\dfrac{\sin k\theta}{\cos k\theta} = \tan k\theta$

106. Not an identity because $\dfrac{5}{\cos \theta} \neq \dfrac{1}{5 \cos \theta}$

107. An identity because $\sin \theta \cdot \dfrac{1}{\sin \theta} = 1$

108. Not an identity because $\dfrac{1}{\sin \theta} \neq 1$

109 and 110. Answers will vary.

111. $x - 25$ **112.** $4z + 12\sqrt{z} + 9$

113. $\dfrac{x^2 + 6x - 8}{(x + 5)(x - 8)}$ **114.** $\dfrac{3(2x + 1)}{x - 4}$

115. $\dfrac{-5x^2 + 8x + 28}{(x^2 - 4)(x + 4)}$ **116.** $\dfrac{x(x^2 + 5x + 1)}{x^2 - 25}$

117.   **118.**

119.   **120.**

Section 2.2   *(page 235)*

Vocabulary Check   *(page 235)*

1. identity   **2.** conditional equation   **3.** $\tan u$

4. $\cot u$ **5.** $\cos^2 u$ **6.** $\sin u$ **7.** $-\csc u$

8. $\sec u$

1–38. Answers will vary.

39. (a)   (b)

Identity

(c) Answers will vary.

40. (a)   (b)

Identity

(c) Answers will vary.

41. (a)   (b)

Not an identity

(c) Answers will vary.

42. (a)   (b)

Not an identity

(c) Answers will vary.

43. (a)   (b)

Identity

(c) Answers will vary.

44. (a)   (b)

Identity

(c) Answers will vary.

45. (a)   (b)

Not an identity

(c) Answers will vary.

CHAPTER 2

46. (a) (b)

Not an identity

(c) Answers will vary.

47–50. Answers will vary. **51.** 1

52. 1 **53.** 2 **54.** 2 **55.** Answers will vary.

56. (a) Answers will vary.

(b)

θ	10°	20°	30°	40°	50°
s	28.36	13.74	8.66	5.96	4.20

θ	60°	70°	80°	90°
s	2.89	1.82	0.88	0

(c) Greatest: 10°; Least: 90° (d) Noon

57. False. An identity is an equation that is true for all real values of θ.

58. True. An identity is an equation that is true for all real values in the domain of the variable.

59. The equation is not an identity because $\sin\theta = \pm\sqrt{1 - \cos^2\theta}$.

Possible answer: $\dfrac{7\pi}{4}$

60. The equation is not an identity because $\tan\theta = \pm\sqrt{\sec^2\theta - 1}$.

Possible answer: $\dfrac{3\pi}{4}$

61. $-3 \pm \sqrt{21}$ **62.** $\dfrac{-5 \pm \sqrt{53}}{2}$

63. $1 \pm \sqrt{5}$ **64.** $\frac{1}{4}\left(1 \pm \sqrt{7}\right)$

Section 2.3 (page 244)

Vocabulary Check (page 244)

1. general **2.** quadratic **3.** extraneous

1–6. Answers will vary. **7.** $\dfrac{2\pi}{3} + 2n\pi, \dfrac{4\pi}{3} + 2n\pi$

8. $\dfrac{7\pi}{6} + 2n\pi, \dfrac{11\pi}{6} + 2n\pi$ **9.** $\dfrac{\pi}{3} + 2n\pi, \dfrac{2\pi}{3} + 2n\pi$

10. $\dfrac{2\pi}{3} + n\pi$ **11.** $\dfrac{\pi}{6} + n\pi, \dfrac{5\pi}{6} + n\pi$

12. $\dfrac{\pi}{3} + n\pi, \dfrac{2\pi}{3} + n\pi$ **13.** $n\pi, \dfrac{3\pi}{2} + 2n\pi$

14. $\dfrac{\pi}{6} + n\pi, \dfrac{5\pi}{6} + n\pi, \dfrac{\pi}{3} + n\pi, \dfrac{2\pi}{3} + n\pi$

15. $\dfrac{\pi}{3} + n\pi, \dfrac{2\pi}{3} + n\pi$ **16.** $\dfrac{\pi}{3} + n\pi, \dfrac{2\pi}{3} + n\pi$

17. $\dfrac{\pi}{8} + \dfrac{n\pi}{2}, \dfrac{3\pi}{8} + \dfrac{n\pi}{2}$ **18.** $\dfrac{\pi}{9} + \dfrac{n\pi}{3}, \dfrac{2\pi}{9} + \dfrac{n\pi}{3}$

19. $\dfrac{n\pi}{3}, \dfrac{\pi}{4} + n\pi$ **20.** $\dfrac{\pi}{4} + \dfrac{n\pi}{2}, \dfrac{2\pi}{3} + 2n\pi, \dfrac{4\pi}{3} + 2n\pi$

21. $0, \dfrac{\pi}{2}, \pi, \dfrac{3\pi}{2}$ **22.** $0, \pi$ **23.** $0, \pi, \dfrac{\pi}{6}, \dfrac{5\pi}{6}, \dfrac{7\pi}{6}, \dfrac{11\pi}{6}$

24. $\dfrac{\pi}{2}, \dfrac{3\pi}{2}, \dfrac{2\pi}{3}, \dfrac{4\pi}{3}$ **25.** $\dfrac{\pi}{3}, \dfrac{5\pi}{3}, \pi$ **26.** $\dfrac{\pi}{3}, \dfrac{5\pi}{3}$

27. No solution **28.** 0 **29.** $\pi, \dfrac{\pi}{3}, \dfrac{5\pi}{3}$

30. $\dfrac{7\pi}{6}, \dfrac{3\pi}{2}, \dfrac{11\pi}{6}$ **31.** $\dfrac{\pi}{6}, \dfrac{5\pi}{6}, \dfrac{7\pi}{6}, \dfrac{11\pi}{6}$ **32.** $\dfrac{\pi}{3}, \dfrac{5\pi}{3}$

33. $\dfrac{\pi}{2}$ **34.** $\dfrac{\pi}{4}, \dfrac{5\pi}{4}$ **35.** $\dfrac{\pi}{6} + n\pi, \dfrac{5\pi}{6} + n\pi$

36. $\dfrac{2\pi}{3} + n\pi, \dfrac{5\pi}{6} + n\pi$ **37.** $\dfrac{\pi}{12} + \dfrac{n\pi}{3}$

38. $\dfrac{\pi}{12} + \dfrac{n\pi}{2}, \dfrac{5\pi}{12} + \dfrac{n\pi}{2}$ **39.** $\dfrac{\pi}{2} + 4n\pi, \dfrac{7\pi}{2} + 4n\pi$

40. $\dfrac{8\pi}{3} + 4n\pi, \dfrac{10\pi}{3} + 4n\pi$ **41.** $-1 + 4n$

42. $\frac{3}{4} + n$ **43.** $-2 + 6n, 2 + 6n$

44. $-2 + 8n, 2 + 8n$ **45.** 2.678, 5.820

46. 0.785, 2.356, 3.665, 3.927, 5.498, 5.760

47. 1.047, 5.236 **48.** 0.524, 2.618 **49.** 0.860, 3.426

50. 4.917 **51.** 0, 2.678, 3.142, 5.820

52. 0.515, 2.726, 3.657, 5.868

53. 0.983, 1.768, 4.124, 4.910

54. 0.524, 0.730, 2.412, 2.618

55. 0.3398, 0.8481, 2.2935, 2.8018

56. 0.5880, 2.0344, 3.7296, 5.1760

57. 1.9357, 2.7767, 5.0773, 5.9183 **58.** 1.7794, 4.5038

59. $\dfrac{\pi}{4}, \dfrac{5\pi}{4}$, arctan 5, arctan $5 + \pi$

60. arctan$(-2) + \pi$, arctan$(-2) + 2\pi, \dfrac{\pi}{4}, \dfrac{5\pi}{4}$

61. $\dfrac{\pi}{3}, \dfrac{5\pi}{3}$ **62.** $\dfrac{\pi}{6}, \dfrac{5\pi}{6}$

63. (a) (b) $\dfrac{\pi}{4} \approx 0.7854$

$\dfrac{5\pi}{4} \approx 3.9270$

Maximum: $(0.7854, 1.4142)$
Minimum: $(3.9270, -1.4142)$

64. (a)

Maximum: (0.5236, 1.5)
Maximum: (2.6180, 1.5)
Minimum: (1.5708, 1)
Minimum: (4.7124, −3)

(b) $\dfrac{\pi}{6} \approx 0.5236$

$\dfrac{\pi}{2} \approx 1.5708$

$\dfrac{5\pi}{6} \approx 2.618$

$\dfrac{3\pi}{2} \approx 4.7124$

65. 1   **66.** 0.739
67. (a) All real numbers x except $x = 0$
   (b) y-axis symmetry; Horizontal asymptote: $y = 1$
   (c) Oscillates   (d) Infinitely many solutions
   (e) Yes, 0.6366
68. (a) All real numbers x except $x = 0$
   (b) y-axis symmetry   (c) y approaches 1
   (d) Four solutions: $\pm\pi, \pm 2\pi$
69. 0.04 second, 0.43 second, 0.83 second
70.

1.91 second

71. February, March, and April
72. January, October, November, and December
73. 36.9° or 53.1°   **74.** 1.9°
75. (a) Between $t = 8$ seconds and $t = 24$ seconds
   (b) 5 times: $t = 16, 48, 80, 112,$ and 144 seconds
76. (a)

Year (0 ↔ 1990)

(b) Model 1
(c) The constant term, 5.45
(d) ≈ 13.37 years
(e) 2010

77. (a)

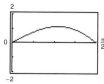

(b) $0.6 < x < 1.1$

$A \approx 1.12$
78. (a) $\dfrac{10}{3}$
   (b)

For $3.5 \le x \le 6$, the approximation appears to be good.
(c) 3.46, 8.81
   3.46 is close to the zero of f in the interval $[0, 6]$.
79. True. The first equation has a smaller period than the second equation, so it will have more solutions in the interval $[0, 2\pi)$.
80. False. There is no value of x for which $\sin x = 3.4$.
81. 1   **82.** 3
83. $C = 24°$ **84.** $C = 19°$
   $a \approx 54.8$ $a \approx 15.4$
   $b \approx 50.1$ $c \approx 5.0$
85. $\sin 390° = \dfrac{1}{2}$ **86.** $\sin 600° = -\dfrac{\sqrt{3}}{2}$

$\cos 390° = \dfrac{\sqrt{3}}{2}$ $\cos 600° = -\dfrac{1}{2}$

$\tan 390° = \dfrac{\sqrt{3}}{3}$ $\tan 600° = \sqrt{3}$

87. $\sin(-1845°) = -\dfrac{\sqrt{2}}{2}$ **88.** $\sin(-1410°) = \dfrac{1}{2}$

$\cos(-1845°) = \dfrac{\sqrt{2}}{2}$ $\cos(-1410°) = \dfrac{\sqrt{3}}{2}$

$\tan(-1845°) = -1$ $\tan(-1410°) = \dfrac{\sqrt{3}}{3}$

89. 1.36°   **90.** 30 feet
91. Answers will vary.

Section 2.4   *(page 252)*

Vocabulary Check   *(page 252)*

1. $\sin u \cos v - \cos u \sin v$

2. $\cos u \cos v - \sin u \sin v$ **3.** $\dfrac{\tan u + \tan v}{1 - \tan u \tan v}$

4. $\sin u \cos v + \cos u \sin v$

5. $\cos u \cos v + \sin u \sin v$ **6.** $\dfrac{\tan u - \tan v}{1 + \tan u \tan v}$

CHAPTER 2

1. (a) $\dfrac{-\sqrt{2} - \sqrt{6}}{4}$ (b) $\dfrac{-1 + \sqrt{2}}{2}$

2. (a) $\dfrac{\sqrt{6} + \sqrt{2}}{4}$ (b) $\dfrac{\sqrt{2} - \sqrt{3}}{2}$

3. (a) $\dfrac{\sqrt{2} - \sqrt{6}}{4}$ (b) $\dfrac{\sqrt{2} + 1}{2}$

4. (a) $-\dfrac{\sqrt{6} + \sqrt{2}}{4}$ (b) $\dfrac{\sqrt{2} + 1}{2}$

5. (a) $\dfrac{1}{2}$ (b) $\dfrac{-\sqrt{3} - 1}{2}$

6. (a) $\dfrac{-2 - \sqrt{6}}{4}$ (b) $\dfrac{-\sqrt{2} - \sqrt{3}}{2}$

7. $\sin 105^\circ = \dfrac{\sqrt{2}}{4}(\sqrt{3} + 1)$

$\cos 105^\circ = \dfrac{\sqrt{2}}{4}(1 - \sqrt{3})$

$\tan 105^\circ = -2 - \sqrt{3}$

8. $\sin 165^\circ = \dfrac{\sqrt{2}}{4}(\sqrt{3} - 1)$

$\cos 165^\circ = -\dfrac{\sqrt{2}}{4}(\sqrt{3} + 1)$

$\tan 165^\circ = -2 + \sqrt{3}$

9. $\sin 195^\circ = \dfrac{\sqrt{2}}{4}(1 - \sqrt{3})$

$\cos 195^\circ = -\dfrac{\sqrt{2}}{4}(\sqrt{3} + 1)$

$\tan 195^\circ = 2 - \sqrt{3}$

10. $\sin 255^\circ = -\dfrac{\sqrt{2}}{4}(\sqrt{3} + 1)$

$\cos 255^\circ = \dfrac{\sqrt{2}}{4}(1 - \sqrt{3})$

$\tan 255^\circ = 2 + \sqrt{3}$

11. $\sin \dfrac{11\pi}{12} = \dfrac{\sqrt{2}}{4}(\sqrt{3} - 1)$

$\cos \dfrac{11\pi}{12} = -\dfrac{\sqrt{2}}{4}(\sqrt{3} + 1)$

$\tan \dfrac{11\pi}{12} = -2 + \sqrt{3}$

12. $\sin \dfrac{7\pi}{12} = \dfrac{\sqrt{2}}{4}(\sqrt{3} + 1)$

$\cos \dfrac{7\pi}{12} = \dfrac{\sqrt{2}}{4}(1 - \sqrt{3})$

$\tan \dfrac{7\pi}{12} = -2 - \sqrt{3}$

13. $\sin \dfrac{17\pi}{12} = -\dfrac{\sqrt{2}}{4}(\sqrt{3} + 1)$

$\cos \dfrac{17\pi}{12} = \dfrac{\sqrt{2}}{4}(1 - \sqrt{3})$

$\tan \dfrac{17\pi}{12} = 2 + \sqrt{3}$

14. $\sin\left(-\dfrac{\pi}{12}\right) = \dfrac{\sqrt{2}}{4}(1 - \sqrt{3})$

$\cos\left(-\dfrac{\pi}{12}\right) = \dfrac{\sqrt{2}}{4}(\sqrt{3} + 1)$

$\tan\left(-\dfrac{\pi}{12}\right) = -2 + \sqrt{3}$

15. $\sin 285^\circ = -\dfrac{\sqrt{2}}{4}(\sqrt{3} + 1)$

$\cos 285^\circ = \dfrac{\sqrt{2}}{4}(\sqrt{3} - 1)$

$\tan 285^\circ = -(2 + \sqrt{3})$

16. $\sin(-105^\circ) = -\dfrac{\sqrt{2}}{4}(\sqrt{3} + 1)$

$\cos(-105^\circ) = \dfrac{\sqrt{2}}{4}(1 - \sqrt{3})$

$\tan(-105^\circ) = 2 + \sqrt{3}$

17. $\sin(-165^\circ) = -\dfrac{\sqrt{2}}{4}(\sqrt{3} - 1)$

$\cos(-165^\circ) = -\dfrac{\sqrt{2}}{4}(1 + \sqrt{3})$

$\tan(-165^\circ) = 2 - \sqrt{3}$

18. $\sin 15^\circ = \dfrac{\sqrt{2}}{4}(\sqrt{3} - 1)$

$\cos 15^\circ = \dfrac{\sqrt{2}}{4}(1 + \sqrt{3})$

$\tan 15^\circ = 2 - \sqrt{3}$

19. $\sin \dfrac{13\pi}{12} = \dfrac{\sqrt{2}}{4}(1 - \sqrt{3})$

$\cos \dfrac{13\pi}{12} = -\dfrac{\sqrt{2}}{4}(1 + \sqrt{3})$

$\tan \dfrac{13\pi}{12} = 2 - \sqrt{3}$

20. $\sin\left(-\dfrac{7\pi}{12}\right) = -\dfrac{\sqrt{2}}{4}(\sqrt{3} + 1)$

$\cos\left(-\dfrac{7\pi}{12}\right) = \dfrac{\sqrt{2}}{4}(1 - \sqrt{3})$

$\tan\left(-\dfrac{7\pi}{12}\right) = 2 + \sqrt{3}$

21. $\sin\left(-\dfrac{13\pi}{12}\right) = \dfrac{\sqrt{2}}{4}(\sqrt{3} - 1)$

$\cos\left(-\dfrac{13\pi}{12}\right) = -\dfrac{\sqrt{2}}{4}(\sqrt{3} + 1)$

$\tan\left(-\dfrac{13\pi}{12}\right) = -2 + \sqrt{3}$

22. $\sin \dfrac{5\pi}{12} = \dfrac{\sqrt{2}}{4}(1 + \sqrt{3})$

$\cos \dfrac{5\pi}{12} = \dfrac{\sqrt{2}}{4}(\sqrt{3} - 1)$

$\tan \dfrac{5\pi}{12} = \sqrt{3} + 2$

23. $\cos 40°$ **24.** $\sin 190°$ **25.** $\tan 239°$ **26.** $\tan 80°$

27. $\sin 1.8$ **28.** $\cos \dfrac{12\pi}{35}$ **29.** $\tan 3x$

30. $\cos(3x - 2y)$ **31.** $-\dfrac{\sqrt{3}}{2}$ **32.** $\dfrac{\sqrt{2}}{2}$ **33.** $\dfrac{\sqrt{3}}{2}$

34. $\dfrac{\sqrt{2}}{2}$ **35.** -1 **36.** $\dfrac{\sqrt{3}}{3}$ **37.** $-\dfrac{63}{65}$ **38.** $\dfrac{56}{65}$

39. $\dfrac{16}{65}$ **40.** $-\dfrac{33}{65}$ **41.** $-\dfrac{63}{16}$ **42.** $\dfrac{65}{33}$ **43.** $\dfrac{65}{56}$

44. $-\dfrac{16}{63}$ **45.** $\dfrac{3}{5}$ **46.** $\dfrac{4}{5}$ **47.** $-\dfrac{44}{117}$ **48.** $\dfrac{117}{44}$

49. $\dfrac{5}{3}$ **50.** $\dfrac{117}{125}$ **51.** 1 **52.** $\dfrac{2x^2 - \sqrt{1 - x^2}}{\sqrt{4x^2 + 1}}$

53. 0 **54.** $\dfrac{x + x\sqrt{1 - x^2}}{\sqrt{x^2 + 1}}$ **55–64.** Answers will vary.

65. $-\sin x$ **66.** $-\cos x$ **67.** $-\cos\theta$ **68.** $\tan\theta$

69. $\dfrac{\pi}{2}$ **70.** $\dfrac{\pi}{3}, \dfrac{5\pi}{3}$ **71.** $\dfrac{5\pi}{4}, \dfrac{7\pi}{4}$ **72.** $0, \dfrac{\pi}{3}, \pi, \dfrac{5\pi}{3}$

73. $\dfrac{\pi}{4}, \dfrac{7\pi}{4}$ **74.** $0, \pi$

75. (a) $y = \dfrac{5}{12} \sin(2t + 0.6435)$

(b) $\dfrac{5}{12}$ feet   (c) $\dfrac{1}{\pi}$ cycle per second

76. Answers will vary.
77. False. $\sin(u \pm v) = \sin u \cos v \pm \cos u \sin v$
78. False. $\cos(u \pm v) = \cos u \cos v \mp \sin u \sin v$
79. False.

$$\cos\left(x - \dfrac{\pi}{2}\right) = \cos x \cos\dfrac{\pi}{2} + \sin x \sin\dfrac{\pi}{2} = \sin x$$

80. True.

$$\sin\left(x - \dfrac{\pi}{2}\right) = \sin x \cos\dfrac{\pi}{2} - \cos x \sin\dfrac{\pi}{2} = -\cos x$$

81–84. Answers will vary.

85. (a) $\sqrt{2}\sin\left(\theta + \dfrac{\pi}{4}\right)$ (b) $\sqrt{2}\cos\left(\theta - \dfrac{\pi}{4}\right)$

86. (a) $5\sin(2\theta + 0.9273)$ (b) $5\cos(2\theta - 0.6435)$
87. (a) $13\sin(3\theta + 0.3948)$ (b) $13\cos(3\theta - 1.1760)$

88. (a) $\sqrt{2}\sin\left(2\theta - \dfrac{\pi}{4}\right)$ (b) $\sqrt{2}\cos\left(2\theta + \dfrac{\pi}{4}\right)$

89. $2\cos\theta$ **90.** $-\dfrac{5\sqrt{2}}{2}\sin\theta - \dfrac{5\sqrt{2}}{2}\cos\theta$

91. Proof
92. (a) All real numbers x except $x = 0$
(b)

h	0.01	0.02	0.05
$f(h)$	-0.504	-0.509	-0.521
$g(h)$	-0.504	-0.509	-0.521

h	0.1	0.2	0.5
$f(h)$	-0.542	-0.583	-0.691
$g(h)$	-0.542	-0.583	-0.691

(c)

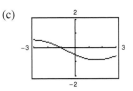

(d) As $h \to 0$, the function $f(h)$ approaches -0.5 and the function $g(h)$ approaches -0.5.

93. $15°$ **94.** $15°$
95.

$$\sin^2\left(\theta + \dfrac{\pi}{4}\right) + \sin^2\left(\theta - \dfrac{\pi}{4}\right) = 1$$

96. (a) and (b) Proofs will vary.

97. $f^{-1}(x) = \dfrac{x + 15}{5}$

98. $f^{-1}(x) = -8x + 7$
99. Because f is not one-to-one, f^{-1} does not exist.
100. $f^{-1}(x) = x^2 + 16$

Section 2.5   *(page 263)*

Vocabulary Check   *(page 263)*

1. $2\sin u \cos u$ **2.** $\cos^2 u$
3. $\cos^2 u - \sin^2 u = 2\cos^2 u - 1 = 1 - 2\sin^2 u$
4. $\tan^2 u$ **5.** $\pm\sqrt{\dfrac{1 - \cos u}{2}}$

6. $\dfrac{1 - \cos u}{\sin u} = \dfrac{\sin u}{1 + \cos u}$

7. $\frac{1}{2}[\cos(u - v) + \cos(u + v)]$
8. $\frac{1}{2}[\sin(u + v) + \sin(u - v)]$

9. $2\sin\left(\dfrac{u + v}{2}\right)\cos\left(\dfrac{u - v}{2}\right)$

10. $-2\sin\left(\dfrac{u + v}{2}\right)\sin\left(\dfrac{u - v}{2}\right)$

---

1. $\dfrac{\sqrt{17}}{17}$ **2.** $\dfrac{1}{4}$ **3.** $\dfrac{15}{17}$ **4.** $\dfrac{8}{17}$ **5.** $\dfrac{8}{15}$

6. $\dfrac{17}{15}$ **7.** $\dfrac{17}{8}$ **8.** $\dfrac{15}{8}$ **9.** $0, \dfrac{\pi}{3}, \pi, \dfrac{5\pi}{3}$

10. $\dfrac{\pi}{2}, \dfrac{7\pi}{6}, \dfrac{3\pi}{2}, \dfrac{11\pi}{6}$ **11.** $\dfrac{\pi}{12}, \dfrac{5\pi}{12}, \dfrac{13\pi}{12}, \dfrac{17\pi}{12}$

12. $\dfrac{\pi}{2}, \dfrac{3\pi}{2}, \dfrac{\pi}{4}, \dfrac{3\pi}{4}, \dfrac{5\pi}{4}, \dfrac{7\pi}{4}$ **13.** $0, \dfrac{2\pi}{3}, \dfrac{4\pi}{3}$

14. $\dfrac{\pi}{2}, \dfrac{7\pi}{6}, \dfrac{11\pi}{6}$ **15.** $\dfrac{\pi}{2}, \dfrac{\pi}{6}, \dfrac{5\pi}{6}, \dfrac{7\pi}{6}, \dfrac{3\pi}{2}, \dfrac{11\pi}{6}$

CHAPTER 2

16. $\dfrac{\pi}{6}, \dfrac{\pi}{2}, \dfrac{5\pi}{6}, \dfrac{3\pi}{2}$ **17.** $0, \dfrac{\pi}{2}, \pi, \dfrac{3\pi}{2}$

18. $0, \dfrac{\pi}{4}, \dfrac{\pi}{2}, \dfrac{3\pi}{4}, \pi, \dfrac{5\pi}{4}, \dfrac{3\pi}{2}, \dfrac{7\pi}{4}$ **19.** $3 \sin 2x$

20. $3 \cos 2x$ **21.** $4 \cos 2x$ **22.** $\cos 2x$

23. $\sin 2u = \dfrac{24}{25}$ **24.** $\sin 2u = -\dfrac{4\sqrt{5}}{9}$

$\cos 2u = -\dfrac{7}{25}$ $\cos 2u = -\dfrac{1}{9}$

$\tan 2u = -\dfrac{24}{7}$ $\tan 2u = 4\sqrt{5}$

25. $\sin 2u = \frac{24}{25}$ **26.** $\sin 2u = -\frac{8}{17}$

$\cos 2u = \frac{7}{25}$ $\cos 2u = \frac{15}{17}$

$\tan 2u = \frac{24}{7}$ $\tan 2u = -\frac{8}{15}$

27. $\sin 2u = -\dfrac{4\sqrt{21}}{25}$ **28.** $\sin 2u = -\dfrac{4\sqrt{2}}{3}$

$\cos 2u = -\dfrac{17}{25}$ $\cos 2u = \dfrac{7}{9}$

$\tan 2u = \dfrac{4\sqrt{21}}{17}$ $\tan 2u = -\dfrac{4\sqrt{2}}{7}$

29. $\frac{1}{8}(3 + 4 \cos 2x + \cos 4x)$

30. $\frac{1}{128}(35 - 48 \cos 2x + 28 \cos 4x - 16 \cos 2x \cos 4x + \cos 8x)$

31. $\frac{1}{8}(1 - \cos 4x)$ **32.** $\frac{1}{128}(3 - 4 \cos 4x + \cos 8x)$

33. $\frac{1}{16}(1 + \cos 2x - \cos 4x - \cos 2x \cos 4x)$

34. $\frac{1}{16}(1 - \cos 2x - \cos 4x + \cos 2x \cos 4x)$

35. $\dfrac{4\sqrt{17}}{17}$ **36.** $\dfrac{\sqrt{17}}{17}$ **37.** $\dfrac{1}{4}$ **38.** $\dfrac{\sqrt{17}}{4}$

39. $\sqrt{17}$ **40.** 4

41. $\sin 75° = \frac{1}{2}\sqrt{2 + \sqrt{3}}$ **42.** $\sin 165° = \frac{1}{2}\sqrt{2 - \sqrt{3}}$

$\cos 75° = \frac{1}{2}\sqrt{2 - \sqrt{3}}$ $\cos 165° = -\frac{1}{2}\sqrt{2 + \sqrt{3}}$

$\tan 75° = 2 + \sqrt{3}$ $\tan 165° = \sqrt{3} - 2$

43. $\sin 112° \, 30' = \frac{1}{2}\sqrt{2 + \sqrt{2}}$

$\cos 112° \, 30' = -\frac{1}{2}\sqrt{2 - \sqrt{2}}$

$\tan 112° \, 30' = -1 - \sqrt{2}$

44. $\sin 67° \, 30' = \dfrac{1}{2}\sqrt{2 + \sqrt{2}}$ **45.** $\sin \dfrac{\pi}{8} = \dfrac{1}{2}\sqrt{2 - \sqrt{2}}$

$\cos 67° \, 30' = \dfrac{1}{2}\sqrt{2 - \sqrt{2}}$ $\cos \dfrac{\pi}{8} = \dfrac{1}{2}\sqrt{2 + \sqrt{2}}$

$\tan 67° \, 30' = 1 + \sqrt{2}$ $\tan \dfrac{\pi}{8} = \sqrt{2} - 1$

46. $\sin \dfrac{\pi}{12} = \dfrac{1}{2}\sqrt{2 - \sqrt{3}}$ **47.** $\sin \dfrac{3\pi}{8} = \dfrac{1}{2}\sqrt{2 + \sqrt{2}}$

$\cos \dfrac{\pi}{12} = \dfrac{1}{2}\sqrt{2 + \sqrt{3}}$ $\cos \dfrac{3\pi}{8} = \dfrac{1}{2}\sqrt{2 - \sqrt{2}}$

$\tan \dfrac{\pi}{12} = 2 - \sqrt{3}$ $\tan \dfrac{3\pi}{8} = \sqrt{2} + 1$

48. $\sin \dfrac{7\pi}{12} = \dfrac{1}{2}\sqrt{2 + \sqrt{3}}$ **49.** $\sin \dfrac{u}{2} = \dfrac{5\sqrt{26}}{26}$

$\cos \dfrac{7\pi}{12} = -\dfrac{1}{2}\sqrt{2 - \sqrt{3}}$ $\cos \dfrac{u}{2} = \dfrac{\sqrt{26}}{26}$

$\tan \dfrac{7\pi}{12} = -2 - \sqrt{3}$ $\tan \dfrac{u}{2} = 5$

50. $\sin \dfrac{u}{2} = \dfrac{\sqrt{5}}{5}$ **51.** $\sin \dfrac{u}{2} = \sqrt{\dfrac{89 - 8\sqrt{89}}{178}}$

$\cos \dfrac{u}{2} = \dfrac{2\sqrt{5}}{5}$ $\cos \dfrac{u}{2} = -\sqrt{\dfrac{89 + 8\sqrt{89}}{178}}$

$\tan \dfrac{u}{2} = \dfrac{1}{2}$ $\tan \dfrac{u}{2} = \dfrac{8 - \sqrt{89}}{5}$

52. $\sin \dfrac{u}{2} = \dfrac{1}{2}\sqrt{\dfrac{10 + 3\sqrt{10}}{5}}$ **53.** $\sin \dfrac{u}{2} = \dfrac{3\sqrt{10}}{10}$

$\cos \dfrac{u}{2} = -\dfrac{1}{2}\sqrt{\dfrac{10 - 3\sqrt{10}}{5}}$ $\cos \dfrac{u}{2} = -\dfrac{\sqrt{10}}{10}$

$\tan \dfrac{u}{2} = -3 - \sqrt{10}$ $\tan \dfrac{u}{2} = -3$

54. $\sin \dfrac{u}{2} = \dfrac{3\sqrt{14}}{14}$ **55.** $|\sin 3x|$

$\cos \dfrac{u}{2} = \dfrac{\sqrt{70}}{14}$

$\tan \dfrac{u}{2} = \dfrac{3\sqrt{5}}{5}$

56. $|\cos 2x|$ **57.** $-|\tan 4x|$ **58.** $-\left|\sin\left(\dfrac{x-1}{2}\right)\right|$

59. π **60.** $0, \dfrac{\pi}{3}, \dfrac{5\pi}{3}$

61. $\dfrac{\pi}{3}, \pi, \dfrac{5\pi}{3}$ **62.** $0, \dfrac{\pi}{2}, \dfrac{3\pi}{2}$

63. $3\left(\sin \dfrac{\pi}{2} + \sin 0\right)$ **64.** $2\left(\sin \dfrac{7\pi}{6} + \sin \dfrac{\pi}{2}\right)$

65. $5(\cos 60° + \cos 90°)$ **66.** $3(\sin 60° + \sin 30°)$

67. $\frac{1}{2}(\sin 10\theta + \sin 2\theta)$ **68.** $\frac{3}{2}(\cos \alpha - \cos 5\alpha)$

69. $\frac{5}{2}(\cos 8\beta + \cos 2\beta)$ **70.** $\frac{1}{2}(\cos 2\theta + \cos 6\theta)$

71. $\frac{1}{2}(\cos 2y - \cos 2x)$ **72.** $\frac{1}{2}(\sin 2x + \sin 2y)$

73. $\frac{1}{2}(\sin 2\theta + \sin 2\pi)$

74. $\frac{1}{2}(\cos 2\pi - \cos 2\theta) = \frac{1}{2}(1 - \cos 2\theta)$

75. $2 \cos 4\theta \sin \theta$ **76.** $2 \sin 2\theta \cos \theta$

77. $2 \cos 4x \cos 2x$ **78.** $2 \sin 3x \cos 2x$

79. $2 \cos \alpha \sin \beta$

80. $2 \cos(\phi + \pi) \cos \pi = -2 \cos(\phi + \pi)$

81. $-2 \sin \theta \sin \dfrac{\pi}{2}$ **82.** $2 \sin x \cos \dfrac{\pi}{2} = 0$

83. $\dfrac{\sqrt{3} + 1}{2}$ **84.** $\dfrac{\sqrt{3} - 1}{2}$

85. $-\sqrt{2}$ **86.** $-\sqrt{2}$

87. $0, \dfrac{\pi}{4}, \dfrac{\pi}{2}, \dfrac{3\pi}{4}, \pi, \dfrac{5\pi}{4}, \dfrac{3\pi}{2}, \dfrac{7\pi}{4}$

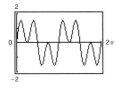

88. $0, \dfrac{\pi}{4}, \dfrac{\pi}{2}, \dfrac{3\pi}{4}, \pi, \dfrac{5\pi}{4}, \dfrac{3\pi}{2}, \dfrac{7\pi}{4}$

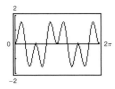

89. $\dfrac{\pi}{6}, \dfrac{5\pi}{6}$ **90.** $0, \dfrac{\pi}{2}, \pi, \dfrac{3\pi}{2}, \dfrac{\pi}{4}, \dfrac{3\pi}{4}, \dfrac{5\pi}{4}, \dfrac{7\pi}{4}$

 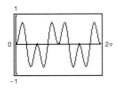

91. $\dfrac{25}{169}$ **92.** $\dfrac{144}{169}$ **93.** $\dfrac{4}{13}$ **94.** $\dfrac{36}{65}$

95–110. Answers will vary.

111.

112.

113.

114.

115.   **116.**

117. $2x\sqrt{1 - x^2}$ **118.** $2x^2 - 1$ **119.** $23.85°$

120. (a) $A = 100 \sin \dfrac{\theta}{2} \cos \dfrac{\theta}{2}$

(b) $A = 50 \sin \theta$

The area is maximum when $\theta = \pi/2$.

121. (a) π (b) 0.4482

(c) 760 miles per hour; 3420 miles per hour

(d) $\theta = 2 \sin^{-1}\left(\dfrac{1}{M}\right)$

122. $x = 2r(1 - \cos \theta)$

123. False. For $u < 0$,

$$\begin{aligned}\sin 2u &= -\sin(-2u) \\ &= -2 \sin(-u) \cos(-u) \\ &= -2(-\sin u) \cos u \\ &= 2 \sin u \cos u.\end{aligned}$$

124. False. $\sin \dfrac{u}{2} = \sqrt{\dfrac{1 - \cos u}{2}}$ when $\dfrac{u}{2}$ is in the second quadrant.

125. (a)   (b) π

Maximum: $(\pi, 3)$

126. (a)   (b) $\dfrac{\pi}{2}, \dfrac{7\pi}{6}, \dfrac{3\pi}{2}, \dfrac{11\pi}{6}$

Minima: $\left(\dfrac{\pi}{2}, -3\right), \left(\dfrac{3\pi}{2}, 1\right)$

Maxima: $\left(\dfrac{7\pi}{6}, \dfrac{3}{2}\right), \left(\dfrac{11\pi}{6}, \dfrac{3}{2}\right)$

127. (a) $\frac{1}{4}(3 + \cos 4x)$ (b) $2 \cos^4 x - 2 \cos^2 x + 1$

(c) $1 - 2 \sin^2 x \cos^2 x$ (d) $1 - \frac{1}{2} \sin^2 2x$

(e) No. There is often more than one way to rewrite a trigonometric expression.

CHAPTER 2

128. (a)

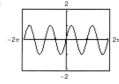

(b) $g(x) = \sin 2x$ (c) Answers will vary.

129. (a)

(b) Distance $= 2\sqrt{10}$ (c) Midpoint: $(2, 3)$

130. (a)

(b) Distance $= \sqrt{269}$ (c) Midpoint: $\left(1, \frac{7}{2}\right)$

131. (a)

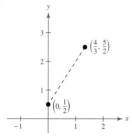

(b) Distance $= \frac{2}{3}\sqrt{13}$ (c) Midpoint: $\left(\frac{2}{3}, \frac{3}{2}\right)$

132. (a)

(b) Distance $= \frac{1}{6}\sqrt{233}$ (c) Midpoint: $\left(-\frac{1}{3}, -\frac{5}{12}\right)$

133. (a) Complement: $35°$; supplement: $125°$
   (b) No complement; supplement: $18°$

134. (a) No complement; supplement: $71°$
   (b) Complement: $12°$; supplement: $102°$

135. (a) Complement: $\dfrac{4\pi}{9}$; supplement: $\dfrac{17\pi}{18}$

   (b) Complement: $\dfrac{\pi}{20}$; supplement: $\dfrac{11\pi}{20}$

136. (a) Complement: 0.62; supplement: 2.19
   (b) No complement; supplement: 0.38

137. September: \$235,000; October: \$272,600

138. ≈ 15.7 gallons   **139.** ≈ 127 feet

Review Exercises   *(page 268)*

1. $\sec x$ **2.** $\csc x$ **3.** $\cos x$ **4.** $\cot x$

5. $\cot x$ **6.** $|\sec x|$

7. $\tan x = \dfrac{3}{4}$ **8.** $\sin\theta = \dfrac{2\sqrt{13}}{13}$

   $\csc x = \dfrac{5}{3}$ $\cos\theta = \dfrac{3\sqrt{13}}{13}$

   $\sec x = \dfrac{5}{4}$ $\csc\theta = \dfrac{\sqrt{13}}{2}$

   $\cot x = \dfrac{4}{3}$ $\cot\theta = \dfrac{3}{2}$

9. $\cos x = \dfrac{\sqrt{2}}{2}$ **10.** $\cos\theta = \dfrac{1}{9}$

   $\tan x = -1$ $\tan\theta = 4\sqrt{5}$

   $\csc x = -\sqrt{2}$ $\csc\theta = \dfrac{9\sqrt{5}}{20}$

   $\sec x = \sqrt{2}$

   $\cot x = -1$ $\sec\theta = 9$

              $\cot\theta = \dfrac{\sqrt{5}}{20}$

11. $\sin^2 x$ **12.** $\sec\theta\csc\theta$ **13.** 1 **14.** $\cos^2 x$

15. $\cot\theta$ **16.** $\tan u\sec u$ **17.** $\cot^2 x$ **18.** 1

19. $\sec x + 2\sin x$ **20.** $2\tan^2 x - 2\sec x\tan x + 1$

21. $-2\tan^2\theta$ **22.** $1 + \sin x$

23–32. Answers will vary.

33. $\dfrac{\pi}{3} + 2n\pi, \dfrac{2\pi}{3} + 2n\pi$ **34.** $\dfrac{\pi}{3} + 2n\pi, \dfrac{5\pi}{3} + 2n\pi$

35. $\dfrac{\pi}{6} + n\pi$ **36.** $\dfrac{\pi}{3} + 2n\pi, \dfrac{5\pi}{3} + 2n\pi$

37. $\dfrac{\pi}{3} + n\pi, \dfrac{2\pi}{3} + n\pi$ **38.** $\dfrac{\pi}{6} + n\pi, \dfrac{5\pi}{6} + n\pi$

39. $0, \dfrac{2\pi}{3}, \dfrac{4\pi}{3}$ **40.** $\dfrac{\pi}{6}, \dfrac{\pi}{2}, \dfrac{5\pi}{6}$ **41.** $0, \dfrac{\pi}{2}, \pi$ **42.** 0

43. $\dfrac{\pi}{8}, \dfrac{3\pi}{8}, \dfrac{9\pi}{8}, \dfrac{11\pi}{8}$ **44.** $0, \dfrac{\pi}{3}, \dfrac{2\pi}{3}, \pi, \dfrac{4\pi}{3}, \dfrac{5\pi}{3}$

45. $0, \dfrac{\pi}{8}, \dfrac{3\pi}{8}, \dfrac{5\pi}{8}, \dfrac{7\pi}{8}, \dfrac{9\pi}{8}, \dfrac{11\pi}{8}, \dfrac{13\pi}{8}, \dfrac{15\pi}{8}$

46. No solution   **47.** $0, \pi$ **48.** $\dfrac{\pi}{2}, \dfrac{3\pi}{2}$

49. $\arctan(-4) + \pi, \arctan(-4) + 2\pi, \arctan 3,$
   $\pi + \arctan 3$

50. $\dfrac{3\pi}{4}, \dfrac{7\pi}{4}, \arctan(-5) + \pi, \arctan(-5) + 2\pi$

51. $\sin 285° = -\dfrac{\sqrt{2}}{4}\left(\sqrt{3} + 1\right)$

$\cos 285° = \dfrac{\sqrt{2}}{4}\left(\sqrt{3} - 1\right)$

$\tan 285° = -2 - \sqrt{3}$

52. $\sin 345° = \dfrac{\sqrt{2}}{4}\left(1 - \sqrt{3}\right)$

$\cos 345° = \dfrac{\sqrt{2}}{4}\left(1 + \sqrt{3}\right)$

$\tan 345° = -2 + \sqrt{3}$

53. $\sin \dfrac{25\pi}{12} = \dfrac{\sqrt{2}}{4}\left(\sqrt{3} - 1\right)$

$\cos \dfrac{25\pi}{12} = \dfrac{\sqrt{2}}{4}\left(\sqrt{3} + 1\right)$

$\tan \dfrac{25\pi}{12} = 2 - \sqrt{3}$

54. $\sin \dfrac{19\pi}{12} = -\dfrac{\sqrt{2}}{4}\left(\sqrt{3} + 1\right)$

$\cos \dfrac{19\pi}{12} = \dfrac{\sqrt{2}}{4}\left(\sqrt{3} - 1\right)$

$\tan \dfrac{19\pi}{12} = -2 - \sqrt{3}$

55. $\sin 15°$ **56.** $\cos 165°$ **57.** $\tan 35°$

58. $\tan(-47°)$ **59.** $-\dfrac{3}{52}\left(5 + 4\sqrt{7}\right)$

60. $\dfrac{960 + 507\sqrt{7}}{1121}$ **61.** $\dfrac{1}{52}\left(5\sqrt{7} + 36\right)$

62. $\dfrac{12\sqrt{7} - 15}{52}$ **63.** $\dfrac{1}{52}\left(5\sqrt{7} - 36\right)$

64. $\dfrac{-960 + 507\sqrt{7}}{1121}$ **65–70.** Answers will vary.

71. $\dfrac{\pi}{4}, \dfrac{7\pi}{4}$ **72.** $\dfrac{3\pi}{2}$ **73.** $\dfrac{\pi}{6}, \dfrac{11\pi}{6}$ **74.** $0, \pi$

75. $\sin 2u = \dfrac{24}{25}$ **76.** $\sin 2u = -\dfrac{4}{5}$

$\cos 2u = -\dfrac{7}{25}$ $\cos 2u = \dfrac{3}{5}$

$\tan 2u = -\dfrac{24}{7}$ $\tan 2u = -\dfrac{4}{3}$

77.

78.

79. $\dfrac{1 - \cos 4x}{1 + \cos 4x}$ **80.** $\dfrac{1 + \cos 6x}{2}$

81. $\dfrac{3 - 4\cos 2x + \cos 4x}{4(1 + \cos 2x)}$ **82.** $\dfrac{1 - \cos 2x}{2}$

83. $\sin(-75°) = -\dfrac{1}{2}\sqrt{2 + \sqrt{3}}$

$\cos(-75°) = \dfrac{1}{2}\sqrt{2 - \sqrt{3}}$

$\tan(-75°) = -2 - \sqrt{3}$

84. $\sin 15° = \dfrac{1}{2}\sqrt{2 - \sqrt{3}}$

$\cos 15° = \dfrac{1}{2}\sqrt{2 + \sqrt{3}}$

$\tan 15° = 2 - \sqrt{3}$

85. $\sin \dfrac{19\pi}{12} = -\dfrac{1}{2}\sqrt{2 + \sqrt{3}}$

$\cos \dfrac{19\pi}{12} = \dfrac{1}{2}\sqrt{2 - \sqrt{3}}$

$\tan \dfrac{19\pi}{12} = -2 - \sqrt{3}$

86. $\sin\left(-\dfrac{17\pi}{12}\right) = \dfrac{1}{2}\sqrt{2 + \sqrt{3}}$ **87.** $\sin \dfrac{u}{2} = \dfrac{\sqrt{10}}{10}$

$\cos\left(-\dfrac{17\pi}{12}\right) = -\dfrac{1}{2}\sqrt{2 - \sqrt{3}}$ $\cos \dfrac{u}{2} = \dfrac{3\sqrt{10}}{10}$

$\tan\left(-\dfrac{17\pi}{12}\right) = -2 - \sqrt{3}$ $\tan \dfrac{u}{2} = \dfrac{1}{3}$

88. $\sin \dfrac{u}{2} = \sqrt{\dfrac{\sqrt{89} + 8}{2\sqrt{89}}} = \dfrac{\sqrt{178(8\sqrt{89} + 89)}}{178}$

$\cos \dfrac{u}{2} = -\sqrt{\dfrac{\sqrt{89} - 8}{2\sqrt{89}}} = \dfrac{\sqrt{-178(8\sqrt{89} - 89)}}{178}$

$\tan \dfrac{u}{2} = \dfrac{\sqrt{89} - 8}{5}$

89. $\sin \dfrac{u}{2} = \dfrac{3\sqrt{14}}{14}$

$\cos \dfrac{u}{2} = \dfrac{\sqrt{70}}{14}$

$\tan \dfrac{u}{2} = \dfrac{3\sqrt{5}}{5}$

90. $\sin \dfrac{u}{2} = \sqrt{\dfrac{7}{12}} = \dfrac{\sqrt{21}}{6}$

$\cos \dfrac{u}{2} = \sqrt{\dfrac{5}{12}} = \dfrac{\sqrt{15}}{6}$

$\tan \dfrac{u}{2} = \dfrac{7}{\sqrt{35}} = \dfrac{\sqrt{35}}{5}$

91. $-\left|\cos 5x\right|$ **92.** $\tan 3x$ **93.** $\dfrac{1}{2}\sin \dfrac{\pi}{3}$

94. $3\left[\cos(-30°) - \cos 60°\right]$ **95.** $\dfrac{1}{2}(\cos 2\theta + \cos 8\theta)$

96. $2(\sin 5\alpha + \sin \alpha)$ **97.** $2\cos 3\theta \sin \theta$

98. $2\cos \dfrac{5\theta}{2} \cos \dfrac{\theta}{2}$ **99.** $-2\sin x \sin \dfrac{\pi}{6}$

100. $2\cos x \sin \dfrac{\pi}{4}$ **101.** $\theta = 15°$ or $\dfrac{\pi}{12}$

102. (a) $V = \sin \dfrac{\theta}{2} \cos \dfrac{\theta}{2}$ cubic meters

(b) $V = \dfrac{1}{2}\sin \theta$ cubic meters

Volume is maximum when $\theta = \pi/2$.

103.

104. $y = \dfrac{1}{2}\sqrt{10} \sin\left(8t - \arctan \dfrac{1}{3}\right)$

105. $\dfrac{1}{2}\sqrt{10}$ feet

106. $4/\pi$ cycles per second

107. False. If $(\pi/2) < \theta < \pi$, then $\cos(\theta/2) > 0$. The sign of $\cos(\theta/2)$ depends on the quadrant in which $\theta/2$ lies.

108. False. Using the sum and difference formula, $\sin(x + y) = \sin x \cos y + \cos x \sin y$.

109. True. $4 \sin(-x) \cos(-x) = 4(-\sin x) \cos x$
$$= -4 \sin x \cos x$$
$$= -2(2 \sin x \cos x)$$
$$= -2 \sin 2x$$

110. True by the product-to-sum formula

111. Reciprocal identities:
$$\sin \theta = \frac{1}{\csc \theta}, \cos \theta = \frac{1}{\sec \theta}, \tan \theta = \frac{1}{\cot \theta},$$
$$\csc \theta = \frac{1}{\sin \theta}, \sec \theta = \frac{1}{\cos \theta}, \cot \theta = \frac{1}{\tan \theta}$$

Quotient identities: $\tan \theta = \dfrac{\sin \theta}{\cos \theta}, \cot \theta = \dfrac{\cos \theta}{\sin \theta}$

Pythagorean identities: $\sin^2 \theta + \cos^2 \theta = 1$,
$1 + \tan^2 \theta = \sec^2 \theta, 1 + \cot^2 \theta = \csc^2 \theta$

112. No. For an equation to be an identity, the equation must be true for all real numbers x. $\sin \theta = \frac{1}{2}$ has an infinite number of solutions but is not an identity.

113. $-1 \le \sin x \le 1$ for all x

114. (a) $49.91479°, 59.86118°$ (b) $54.735616°$

115. $y_1 = y_2 + 1$

116. $y_1 = 1 - y_2$

117. $-1.8431, 2.1758, 3.9903, 8.8935, 9.8820$

118. $-3.1395, -2.0000, -0.4378, 2.0000$

Chapter Test *(page 271)*

1. $\sin \theta = -\dfrac{3\sqrt{13}}{13}$ **2.** 1 **3.** 1 **4.** $\csc \theta \sec \theta$

$\cos \theta = -\dfrac{2\sqrt{13}}{13}$

$\csc \theta = -\dfrac{\sqrt{13}}{3}$

$\sec \theta = -\dfrac{\sqrt{13}}{2}$

$\cot \theta = \dfrac{2}{3}$

5. $\theta = 0, \dfrac{\pi}{2} < \theta \le \pi, \dfrac{3\pi}{2} < \theta < 2\pi$

6.

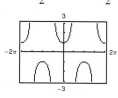

$y_1 = y_2$

7–12. Answers will vary.

13. $\dfrac{1}{16}\left(\dfrac{10 - 15 \cos 2x + 6 \cos 4x - \cos 6x}{1 + \cos 2x}\right)$

14. $\tan 2\theta$

15. $2(\sin 6\theta + \sin 2\theta)$ **16.** $-2 \cos \dfrac{7\theta}{2} \sin \dfrac{\theta}{2}$

17. $0, \dfrac{3\pi}{4}, \pi, \dfrac{7\pi}{4}$ **18.** $\dfrac{\pi}{6}, \dfrac{\pi}{2}, \dfrac{5\pi}{6}, \dfrac{3\pi}{2}$

19. $\dfrac{\pi}{6}, \dfrac{5\pi}{6}, \dfrac{7\pi}{6}, \dfrac{11\pi}{6}$ **20.** $\dfrac{\pi}{6}, \dfrac{5\pi}{6}, \dfrac{3\pi}{2}$

21. $-2.938, -2.663, 1.170$ **22.** $\dfrac{\sqrt{2} - \sqrt{6}}{4}$

23. $\sin 2u = \frac{4}{5}, \tan 2u = -\frac{4}{3}, \cos 2u = -\frac{3}{5}$

24. Day 123 to day 223

25. $t = 0.26$ minute
    0.58 minute
    0.89 minute
    1.20 minutes
    1.52 minutes
    1.83 minutes

Problem Solving *(page 275)*

1. (a) $\cos \theta = \pm\sqrt{1 - \sin^2 \theta}$

$\tan \theta = \pm\dfrac{\sin \theta}{\sqrt{1 - \sin^2 \theta}}$

$\cot \theta = \pm\dfrac{\sqrt{1 - \sin^2 \theta}}{\sin \theta}$

$\sec \theta = \pm\dfrac{1}{\sqrt{1 - \sin^2 \theta}}$

$\csc \theta = \dfrac{1}{\sin \theta}$

(b) $\sin \theta = \pm\sqrt{1 - \cos^2 \theta}$

$\tan \theta = \pm\dfrac{\sqrt{1 - \cos^2 \theta}}{\cos \theta}$

$\csc \theta = \pm\dfrac{1}{\sqrt{1 - \cos^2 \theta}}$

$\sec \theta = \dfrac{1}{\cos \theta}$

$\cot \theta = \pm\dfrac{\cos \theta}{\sqrt{1 - \cos^2 \theta}}$

2 and 3. Answers will vary.

4. (a) $p_1(t) = \sin(524\,\pi t)$

$p_2(t) = \frac{1}{2}\sin(1048\,\pi t)$

$p_3(t) = \frac{1}{3}\sin(1572\,\pi t)$

$p_5(t) = \frac{1}{5}\sin(2620\,\pi t)$

$p_6(t) = \frac{1}{6}\sin(3144\,\pi t)$

(b) $p_1: \frac{1}{262}$ $p_5: \frac{1}{1310}$

$p_2: \frac{1}{524}$ $p_6: \frac{1}{1572}$

$p_3: \frac{1}{786}$

p is periodic with period $\frac{1}{262}$.

(c) $(0, 0), (0.00096, 0), (0.00191, 0)$

$(0.00285, 0), (0.00382, 0)$

(d) Maximum: $(0.00024, 1.19524)$

Minimum: $(0.00357, -1.19525)$

5. $u + v = w$ **6.** $y = \frac{1}{64}v_0^2 \sin^2\theta$

7. $\sin\dfrac{\theta}{2} = \sqrt{\dfrac{1 - \cos\theta}{2}}$

$\cos\dfrac{\theta}{2} = \sqrt{\dfrac{1 + \cos\theta}{2}}$

$\tan\dfrac{\theta}{2} = \dfrac{\sin\theta}{1 + \cos\theta}$

8. (a) $F = \dfrac{0.6\,W\cos\theta}{\sin 12°}$

(b)

(c) Maximum: $\theta = 0°$

Minimum: $\theta = 90°$

9. (a)

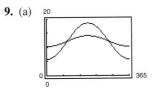

(b) $t = 91, t = 274$; Spring Equinox and Fall Equinox

(c) Seward; The amplitudes: 6.4 and 1.9

(d) 365.2 days

10. (a) High tides: 6:12 A.M., 6:36 P.M.

Low tides: 12:00 A.M., 12:24 P.M.

(b) The water depth never falls below 7 feet.

(c)

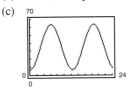

11. (a) $\dfrac{\pi}{6} \le x \le \dfrac{5\pi}{6}$

(b) $\dfrac{2\pi}{3} \le x \le \dfrac{4\pi}{3}$

(c) $\dfrac{\pi}{2} < x < \pi, \dfrac{3\pi}{2} < x < 2\pi$

(d) $0 \le x \le \dfrac{\pi}{4}, \dfrac{5\pi}{4} \le x \le 2\pi$

12. (a) $n = \dfrac{1}{2}\left(\cot\dfrac{\theta}{2} + \sqrt{3}\right)$

(b) $\theta \approx 76.5°$

13. (a) $\sin(u + v + w)$

$= \sin u \cos v \cos w - \sin u \sin v \sin w$

$+ \cos u \sin v \cos w + \cos u \cos v \sin w$

(b) $\tan(u + v + w)$

$= \dfrac{\tan u + \tan v + \tan w - \tan u \tan v \tan w}{1 - \tan u \tan v - \tan u \tan w - \tan v \tan w}$

14. (a) $\cos 3\theta = \cos\theta - 4\sin^2\theta\cos\theta$

(b) $\cos 4\theta = \cos^4\theta - 6\sin^2\theta\cos^2\theta + \sin^4\theta$

15. (a)

(b) 233.3 times per second

Chapter 3

Section 3.1 *(page 284)*

Vocabulary Check *(page 284)*

1. oblique   **2.** $\dfrac{b}{\sin B}$ **3.** $\dfrac{1}{2}ac\sin B$

---

1. $C = 105°, b \approx 28.28, c \approx 38.64$
2. $A = 35°, a \approx 11.88, b \approx 13.31$
3. $C = 120°, b \approx 4.75, c \approx 7.17$
4. $A = 35°, a \approx 36.50, b \approx 11.05$
5. $B \approx 21.55°, C \approx 122.45°, c \approx 11.49$
6. Two solutions:
  $B \approx 45.79°, C \approx 74.21°, b \approx 7.45$
  $B \approx 14.21°, C \approx 105.79°, b \approx 2.55$
7. $B = 60.9°, b \approx 19.32, c \approx 6.36$
8. $B = 101.1°, a \approx 1.35, b \approx 3.23$
9. $B = 42°4', a \approx 22.05, b \approx 14.88$
10. $C = 166°5', a \approx 3.30, c \approx 8.05$
11. $A \approx 10°11', C \approx 154°19', c \approx 11.03$
12. $A \approx 174°41', C \approx 2°34', a \approx 11.99$
13. $A \approx 25.57°, B \approx 9.43°, a \approx 10.53$
14. $B \approx 75.48°, C \approx 4.52°, b \approx 122.87$
15. $B \approx 18°13', C \approx 51°32', c \approx 40.06$
16. $A \approx 44°14', B \approx 50°26', b \approx 38.67$
17. $C = 83°, a \approx 0.62, b \approx 0.51$
18. $A = 48°, b \approx 2.29, c \approx 4.73$
19. $B \approx 48.74°, C \approx 21.26°, c \approx 48.23$
20. No solution   **21.** No solution
22. $B \approx 36.82°, C \approx 67.18°, c \approx 32.30$
23. Two solutions:
  $B \approx 72.21°, C \approx 49.79°, c \approx 10.27$
  $B \approx 107.79°, C \approx 14.21°, c \approx 3.30$
24. No solution

25. (a) $b \le 5,\ b = \dfrac{5}{\sin 36°}$ (b) $5 < b < \dfrac{5}{\sin 36°}$

  (c) $b > \dfrac{5}{\sin 36°}$

26. (a) $b \le 10, b = \dfrac{10}{\sin 60°}$ (b) $10 < b < \dfrac{10}{\sin 60°}$

  (c) $b > \dfrac{10}{\sin 60°}$

27. (a) $b \le 10.8, b = \dfrac{10.8}{\sin 10°}$ (b) $10.8 < b < \dfrac{10.8}{\sin 10°}$

  (c) $b > \dfrac{10.8}{\sin 10°}$

28. (a) $b \le 315.6, b = \dfrac{315.6}{\sin 88°}$ (b) $315.6 < b < \dfrac{315.6}{\sin 88°}$

  (c) $b > \dfrac{315.6}{\sin 88°}$

29. 10.4   **30.** 474.9   **31.** 1675.2   **32.** 4.5
33. 3204.5   **34.** 159.3   **35.** 15.3 meters
36. (a)

  (b) $\dfrac{16}{\sin 70°} = \dfrac{h}{\sin 32°}$ (c) 9 meters

37. 16.1°   **38.** 240°   **39.** 77 meters
40. (a)

  (b) 4385.71 feet
  (c) 3061.80 feet

41. (a)

  (b) 22.6 miles
  (c) 21.4 miles
  (d) 7.3 miles

42. From Pine Knob: 42.4 kilometers
  From Colt Station: 15.5 kilometers
43. 3.2 miles
44. (a) $\alpha \approx 5.36°$ (b) $\beta = \sin^{-1}\left(\dfrac{d\sin\theta}{58.36}\right)$

  (c) $d = \dfrac{58.36\sin(84.64 - \theta)}{\sin\theta}$

  (d)

θ	10°	20°	30°	40°	50°	60°
d	324.1	154.2	95.2	63.8	43.3	28.1

45. True. If an angle of a triangle is obtuse (greater than 90°), then the other two angles must be acute and therefore less than 90°. The triangle is oblique.
46. False. Two angles and one side determine a unique triangle, while two sides and one opposite angle do not necessarily determine a unique triangle.
47. (a) $\alpha = \arcsin(0.5\sin\beta)$
  (b)

Domain: $0 < \beta < \pi$

Range: $0 < \alpha < \dfrac{\pi}{6}$

(c) $c = \dfrac{18 \sin[\pi - \beta - \arcsin(0.5 \sin \beta)]}{\sin \beta}$

(d)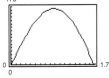

Domain: $0 < \beta < \pi$
Range: $9 < c < 27$

(e)

β	0.4	0.8	1.2	1.6
α	0.1960	0.3669	0.4848	0.5234
c	25.95	23.07	19.19	15.33

β	2.0	2.4	2.8
α	0.4720	0.3445	0.1683
c	12.29	10.31	9.27

As β increases from 0 to π, α increases and then decreases, and c decreases from 27 to 9.

48. (a) $A = 20\left(15 \sin \dfrac{3\theta}{2} - 4 \sin \dfrac{\theta}{2} - 6 \sin \theta\right)$

(b) ![graph]

(c) Domain: $0 \le \theta \le 1.6690$
The area would increase and the domain would increase in length.

49. $\cos x$ **50.** $\tan x$ **51.** $\sin^2 x$ **52.** $\sec^2 x$

Section 3.2 (page 291)

Vocabulary Check (page 291)

1. Cosines **2.** $b^2 = a^2 + c^2 - 2ac \cos B$
3. Heron's Area Formula

1. $A \approx 23.07°, B \approx 34.05°, C \approx 122.88°$
2. $A \approx 61.22°, B \approx 19.19°, C \approx 99.59°$
3. $B \approx 23.79°, C \approx 126.21°, a \approx 18.59$
4. $A \approx 53.73°, B \approx 21.27°, c \approx 11.98$
5. $A \approx 31.99°, B \approx 42.39°, C \approx 105.63°$
6. $A \approx 39.35°, B \approx 16.75°, C \approx 123.90°$
7. $A \approx 92.94°, B \approx 43.53°, C \approx 43.53°$
8. $A \approx 86.68°, B \approx 31.82°, C \approx 61.50°$
9. $B \approx 13.45°, C \approx 31.55°, a \approx 12.16$
10. $B \approx 16.53°, C \approx 108.47°, a \approx 8.64$

11. $A \approx 141°45', C \approx 27°40', b \approx 11.87$
12. $A \approx 37°6', C \approx 67°34', b \approx 9.94$
13. $A \approx 27°10', C \approx 27°10', b \approx 56.94$
14. $A \approx 157°2', B \approx 7°43', c \approx 4.21$
15. $A \approx 33.80°, B \approx 103.20°, c \approx 0.54$
16. $A \approx 23.65°, B \approx 53.35°, c \approx 0.91$

	a	b	c	d	θ	ϕ
17.	5	8	12.07	5.69	45°	135.1°
18.	25	35	52.20	31.22	60°	120°
19.	10	14	20	13.86	68.2°	111.8°
20.	40	60	63.30	80	104.5°	75.6°
21.	15	16.96	25	20	77.2°	102.8°
22.	35.18	25	50	35	68.8°	111.2°

23. 16.25 **24.** 54 **25.** 10.4 **26.** 1350.2
27. 52.11 **28.** 0.61
29.   N 37.1° E, S 63.1° E

30.   1357.8 miles, 56°

31. 373.3 meters **32.** 41.2°, 52.9° **33.** 72.3°
34. 127.2° **35.** 43.3 miles
36. 131.1 feet, 118.6 feet
37. (a) N 58.4° W (b) S 81.5° W
38. (a) N 59.7° E (b) N 72.8° E **39.** 63.7 feet
40. 103.9 feet **41.** 24.2 miles **42.** 3.8 miles
43. $\overline{PQ} \approx 9.4, \overline{QS} = 5, \overline{RS} \approx 12.8$
44. (a) $x^2 - 3x \cos \theta - 46.75 = 0$
(b) $x = \tfrac{1}{2}\left(3 \cos \theta + \sqrt{9 \cos^2 \theta + 187}\right)$
(c) ![graph] (d) 8.5 inches

CHAPTER 3

45.

d (inches)	9	10	12	13	14
θ (degrees)	60.9°	69.5°	88.0°	98.2°	109.6°
s (inches)	20.88	20.28	18.99	18.28	17.48

d (inches)	15	16
θ (degrees)	122.9°	139.8°
s (inches)	16.55	15.37

46. 3.95 feet   **47.** 46,837.5 square feet
48. 6577.8 square meters   **49.** $83,336.37
50. $62,340.71
51. False. For s to be the average of the lengths of the three sides of the triangle, s would be equal to $(a + b + c)/3$.
52. False. To solve an SSA triangle, the Law of Sines is needed.
53. False. The three side lengths do not form a triangle.
54. (a) and (b) Proofs will vary.
55. (a) 570.60   (b) 5910   (c) 177
56. 405.2 feet   **57 and 58.** Proofs will vary.
59. $-\dfrac{\pi}{2}$ **60.** $\dfrac{\pi}{2}$ **61.** $\dfrac{\pi}{3}$
62. $-\dfrac{\pi}{3}$ **63.** $-\dfrac{\pi}{3}$ **64.** $\dfrac{5\pi}{6}$ **65.** $\dfrac{1}{\sqrt{1-4x^2}}$
66. $\dfrac{\sqrt{1-9x^2}}{3x}$ **67.** $\dfrac{1}{x-2}$ **68.** $\dfrac{\sqrt{4-(x-1)^2}}{2}$
69. $\cos\theta = 1$
   $\sec\theta = 1$
   $\csc\theta$ is undefined.
70. $\sin\theta = -\dfrac{\sqrt{2}}{2}$
   $\sec\theta = \sqrt{2}$
   $\csc\theta = -\sqrt{2}$
71. $\tan\theta = -\dfrac{\sqrt{3}}{3}$
   $\sec\theta = \dfrac{2\sqrt{3}}{3}$
   $\csc\theta = -2$
72. $12 = 6\sec\theta$
   $\sec\theta = 2$
   $\csc\theta = \pm\dfrac{2\sqrt{3}}{3}$
73. $-2\sin\dfrac{7\pi}{12}\sin\dfrac{\pi}{4}$ **74.** $2\cos x\sin\left(-\dfrac{\pi}{2}\right)$

Section 3.3   *(page 304)*

Vocabulary Check   *(page 304)*

1. directed line segment   **2.** initial; terminal
3. magnitude   **4.** vector
5. standard position   **6.** unit vector
7. multiplication; addition   **8.** resultant
9. linear combination; horizontal; vertical

1. $\|\mathbf{u}\| = \|\mathbf{v}\| = \sqrt{17}$, slope$_\mathbf{u}$ = slope$_\mathbf{v}$ = $\frac{1}{4}$
   $\mathbf{u}$ and $\mathbf{v}$ have the same magnitude and direction, so they are equal.
2. $\|\mathbf{u}\| = \|\mathbf{v}\| = \sqrt{73}$, slope$_\mathbf{u}$ = slope$_\mathbf{v}$ = $\frac{8}{3}$
   $\mathbf{u}$ and $\mathbf{v}$ have the same magnitude and direction, so they are equal.
3. $\mathbf{v} = \langle 3, 2\rangle$; $\|\mathbf{v}\| = \sqrt{13}$
4. $\mathbf{v} = \langle -4, -2\rangle$; $\|\mathbf{v}\| = 2\sqrt{5}$ **5.** $\mathbf{v} = \langle -3, 2\rangle$; $\|\mathbf{v}\| = \sqrt{13}$
6. $\mathbf{v} = \langle 4, 6\rangle$; $\|\mathbf{v}\| = 2\sqrt{13}$ **7.** $\mathbf{v} = \langle 0, 5\rangle$; $\|\mathbf{v}\| = 5$
8. $\mathbf{v} = \langle 7, 0\rangle$; $\|\mathbf{v}\| = 7$ **9.** $\mathbf{v} = \langle 16, 7\rangle$; $\|\mathbf{v}\| = \sqrt{305}$
10. $\mathbf{v} = \langle 8, -8\rangle$; $\|\mathbf{v}\| = 8\sqrt{2}$ **11.** $\mathbf{v} = \langle 8, 6\rangle$; $\|\mathbf{v}\| = 10$
12. $\mathbf{v} = \langle 12, 29\rangle$; $\|\mathbf{v}\| = \sqrt{985}$
13. $\mathbf{v} = \langle -9, -12\rangle$; $\|\mathbf{v}\| = 15$
14. $\mathbf{v} = \langle 7, -24\rangle$; $\|\mathbf{v}\| = 25$

15.   **16.**

17.   **18.**

19.   **20.**

21. (a) $\langle 3, 4\rangle$ (b) $\langle 1, -2\rangle$

(c) $\langle 1, -7 \rangle$

22. (a) $\langle 6, 3 \rangle$

(b) $\langle -2, 3 \rangle$

(c) $\langle -8, 6 \rangle$

23. (a) $\langle -5, 3 \rangle$

(b) $\langle -5, 3 \rangle$

(c) $\langle -10, 6 \rangle$

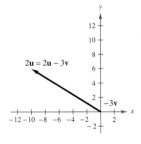

24. (a) $\langle 2, 1 \rangle$

(b) $\langle -2, -1 \rangle$

(c) $\langle -6, -3 \rangle$

25. (a) $3\mathbf{i} - 2\mathbf{j}$

(b) $-\mathbf{i} + 4\mathbf{j}$

(c) $-4\mathbf{i} + 11\mathbf{j}$

26. (a) $-3\mathbf{i} + 3\mathbf{j}$

(b) $-\mathbf{i} - \mathbf{j}$

CHAPTER 3

(c) $-\mathbf{i} - 4\mathbf{j}$

27. (a) $2\mathbf{i} + \mathbf{j}$ (b) $2\mathbf{i} - \mathbf{j}$

(c) $4\mathbf{i} - 3\mathbf{j}$

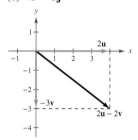

28. (a) $2\mathbf{i} + 3\mathbf{j}$ (b) $-2\mathbf{i} + 3\mathbf{j}$

(c) $-6\mathbf{i} + 6\mathbf{j}$

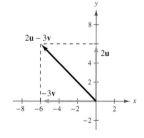

29. $\langle 1, 0 \rangle$ **30.** $\langle 0, -1 \rangle$ **31.** $\left\langle -\dfrac{\sqrt{2}}{2}, \dfrac{\sqrt{2}}{2} \right\rangle$

32. $\left\langle \dfrac{5}{13}, -\dfrac{12}{13} \right\rangle$ **33.** $\dfrac{3\sqrt{10}}{10}\mathbf{i} - \dfrac{\sqrt{10}}{10}\mathbf{j}$

34. $\dfrac{\sqrt{2}}{2}\mathbf{i} + \dfrac{\sqrt{2}}{2}\mathbf{j}$ **35.** $\mathbf{j}$ **36.** $-\mathbf{i}$

37. $\dfrac{\sqrt{5}}{5}\mathbf{i} - \dfrac{2\sqrt{5}}{5}\mathbf{j}$ **38.** $-\dfrac{3\sqrt{58}}{58}\mathbf{i} + \dfrac{7\sqrt{58}}{58}\mathbf{j}$

39. $\left\langle \dfrac{5\sqrt{2}}{2}, \dfrac{5\sqrt{2}}{2} \right\rangle$ **40.** $\langle -3\sqrt{2}, 3\sqrt{2} \rangle$

41. $\left\langle \dfrac{18\sqrt{29}}{29}, \dfrac{45\sqrt{29}}{29} \right\rangle$ **42.** $\langle -10, 0 \rangle$ **43.** $7\mathbf{i} + 4\mathbf{j}$

44. $3\mathbf{i} + 8\mathbf{j}$ **45.** $3\mathbf{i} + 8\mathbf{j}$ **46.** $6\mathbf{i} - 3\mathbf{j}$

47. $\mathbf{v} = \langle 3, -\frac{3}{2} \rangle$ **48.** $\mathbf{v} = \langle \frac{3}{4}, \frac{3}{2} \rangle$

49. $\mathbf{v} = \langle 4, 3 \rangle$ **50.** $\mathbf{v} = \langle -1, 3 \rangle$

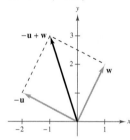

51. $\mathbf{v} = \langle \frac{7}{2}, -\frac{1}{2} \rangle$ **52.** $\mathbf{v} = \langle 0, -5 \rangle$

53. $\|\mathbf{v}\| = 3$; $\theta = 60°$ **54.** $\|\mathbf{v}\| = 8$; $\theta = 135°$
55. $\|\mathbf{v}\| = 6\sqrt{2}$; $\theta = 315°$ **56.** $\|\mathbf{v}\| = \sqrt{41}$; $\theta = 141.3°$

57. $\mathbf{v} = \langle 3, 0 \rangle$

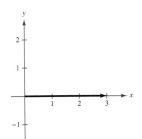

58. $\mathbf{v} = \left\langle \dfrac{\sqrt{2}}{2}, \dfrac{\sqrt{2}}{2} \right\rangle$

59. $\mathbf{v} = \left\langle -\dfrac{7\sqrt{3}}{4}, \dfrac{7}{4} \right\rangle$

60. $\mathbf{v} = \left\langle \dfrac{5\sqrt{2}}{4}, \dfrac{5\sqrt{2}}{4} \right\rangle$

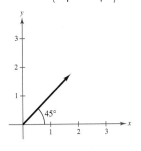

61. $\mathbf{v} = \left\langle -\dfrac{3\sqrt{6}}{2}, \dfrac{3\sqrt{2}}{2} \right\rangle$

62. $\mathbf{v} = \langle 0, 4\sqrt{3} \rangle$

63. $\mathbf{v} = \left\langle \dfrac{\sqrt{10}}{5}, \dfrac{3\sqrt{10}}{5} \right\rangle$

64. $\mathbf{v} = \left\langle \dfrac{9}{5}, \dfrac{12}{5} \right\rangle$

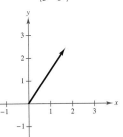

65. $\langle 5, 5 \rangle$ **66.** $\langle 2, 4 + 2\sqrt{3} \rangle$
67. $\langle 10\sqrt{2} - 50, 10\sqrt{2} \rangle$ **68.** $\langle 33.04, 53.19 \rangle$ **69.** $90°$
70. $90°$ **71.** $62.7°$ **72.** $47.4°$
73. $12.8°$; 398.32 newtons   **74.** $8.7°$; 2396.2 newtons
75. $71.3°$; 228.5 pounds   **76.** $37.5°$; 58.6 pounds

77. Vertical component: $70 \sin 35° \approx 40.15$ feet per second
Horizontal component: $70 \cos 35° \approx 57.34$ feet per second
78. Vertical component: $1200 \sin 6° \approx 125.4$ feet per second
Horizontal component:
$1200 \cos 6° \approx 1193.4$ feet per second
79. $T_{AC} \approx 1758.8$ pounds   **80.** $T_{AC} \approx 3611.1$ pounds
$T_{BC} \approx 1305.4$ pounds $T_{BC} \approx 2169.5$ pounds
81. 3154.4 pounds   **82.**

53.2 pounds

83. N $21.4°$ E; 138.7 kilometers per hour
84. (a)

(b) $60 \langle \cos 45°, \sin 45° \rangle$ (c) $580 \langle \cos 118°, \sin 118° \rangle$
(d) 600.3 miles per hour   (e) $337.5°$
85. 1928.4 foot-pounds   **86.** $\sqrt{2}$ pounds; 1 pound
87. True. See Example 1.
88. True, by the definition of a unit vector.
89. (a) $0°$ (b) $180°$
(c) No. The magnitude is at most equal to the sum when the angle between the vectors is $0°$.
90. (a) $5\sqrt{5 + 4 \cos \theta}$
(b)

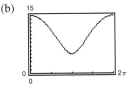

(c) Range: $[5, 15]$
Maximum is 15 when $\theta = 0$.
Minimum is 5 when $\theta = \pi$.
(d) The magnitudes of $\mathbf{F}_1$ and $\mathbf{F}_2$ are not the same.
91. Proofs will vary.   **92.** Answers will vary.
93. $\langle 1, 3 \rangle$ or $\langle -1, -3 \rangle$ **94.** $\langle 10, 50 \rangle$ or $\langle -10, -50 \rangle$
95. $8 \tan \theta$ **96.** $8 \cos \theta$ **97.** $6 \sec \theta$

98. $125 \tan^3 \theta$ **99.** $\dfrac{\pi}{2} + n\pi, \pi + 2n\pi$

CHAPTER 3

100. $n\pi, \dfrac{5\pi}{4} + 2n\pi, \dfrac{7\pi}{4} + 2n\pi$

101. $n\pi, \dfrac{\pi}{6} + 2n\pi, \dfrac{11\pi}{6} + 2n\pi$

102. $\dfrac{\pi}{2} + n\pi, \dfrac{5\pi}{4} + 2n\pi, \dfrac{7\pi}{4} + 2n\pi$

Section 3.4 *(page 315)*

Vocabulary Check *(page 315)*

1. dot product **2.** $\dfrac{\mathbf{u} \cdot \mathbf{v}}{\|\mathbf{u}\|\,\|\mathbf{v}\|}$ **3.** orthogonal

4. $\left(\dfrac{\mathbf{u} \cdot \mathbf{v}}{\|\mathbf{v}\|^2}\right)\mathbf{v}$ **5.** $\|\text{proj}_{\overrightarrow{PQ}}\mathbf{F}\|\,\|\overrightarrow{PQ}\|$; $\mathbf{F} \cdot \overrightarrow{PQ}$

---

1. -9 **2.** 9 **3.** -11 **4.** -8 **5.** 6
6. 13 **7.** -12 **8.** -4 **9.** 8; scalar
10. 6; scalar **11.** $\langle -6, 8 \rangle$; vector
12. $\langle 2, -4 \rangle$; vector **13.** $\langle -66, -66 \rangle$; vector
14. $\langle 4, -8 \rangle$; vector **15.** $\sqrt{5} - 1$; scalar
16. $2 - 2\sqrt{2}$; scalar **17.** 4; scalar
18. 13; scalar **19.** 13 **20.** $2\sqrt{5}$ **21.** $5\sqrt{41}$
22. 20 **23.** 6 **24.** 21 **25.** $90°$ **26.** $33.69°$
27. $143.13°$ **28.** $7.13°$ **29.** $60.26°$ **30.** $53.13°$
31. $90°$ **32.** $93.18°$ **33.** $\dfrac{5\pi}{12}$ **34.** $\dfrac{\pi}{4}$

35.   **36.**

$\approx 91.33°$

$\approx 108.43°$

37.   **38.**

$90°$

$\approx 67.15°$

39. $26.57°, 63.43°, 90°$ **40.** $41.41°, 64.14°, 74.45°$
41. $41.63°, 53.13°, 85.24°$ **42.** $21.80°, 41.63°, 116.57°$
43. -20 **44.** $12{,}500\sqrt{3}$ **45.** -229.1 **46.** 24

47. Parallel **48.** Neither **49.** Neither **50.** Neither
51. Orthogonal **52.** Orthogonal
53. $\frac{1}{37}\langle 84, 14 \rangle, \frac{1}{37}\langle -10, 60 \rangle$ **54.** $\langle 0, 0 \rangle, \langle 4, 2 \rangle$
55. $\frac{45}{229}\langle 2, 15 \rangle, \frac{6}{229}\langle -15, 2 \rangle$ **56.** $\frac{14}{17}\langle -4, -1 \rangle, \frac{5}{17}\langle 1, -4 \rangle$
57. 0 **58.** 0 **59.** $\langle -5, 3 \rangle, \langle 5, -3 \rangle$
60. $\langle 3, 8 \rangle, \langle -3, -8 \rangle$ **61.** $\frac{2}{3}\mathbf{i} + \frac{1}{2}\mathbf{j}, -\frac{2}{3}\mathbf{i} - \frac{1}{2}\mathbf{j}$
62. $3\mathbf{i} - \frac{5}{2}\mathbf{j}, -3\mathbf{i} + \frac{5}{2}\mathbf{j}$ **63.** 32 **64.** 14
65. (a) \$58,762.50

This value gives the total revenue that can be earned by selling all of the units.

(b) $1.05\mathbf{v}$

66. (a) \$8732.50; answers will vary. (b) $1.025\mathbf{v}$

67. (a) Force $= 30{,}000 \sin d$

(b)

d	$0°$	$1°$	$2°$	$3°$	$4°$	$5°$
Force	0	523.6	1047.0	1570.1	2092.7	2614.7

d	$6°$	$7°$	$8°$	$9°$	$10°$
Force	3135.9	3656.1	4175.2	4693.0	5209.4

(c) 29,885.8 pounds

68. 937.7 pounds; 5318.0 pounds
69. 735 newton-meters **70.** 12,000 foot-pounds
71. 779.4 foot-pounds **72.** 1,048,514.62 kilogram-meters
(10,282,651.78 newton-meters)
73. 21,650.64 foot-pounds **74.** 1174.62 foot-pounds
75. False. Work is represented by a scalar.
76. True. $\cos 90° = 0$

77. (a) $\theta = \dfrac{\pi}{2}$ (b) $0 \le \theta < \dfrac{\pi}{2}$ (c) $\dfrac{\pi}{2} < \theta \le \pi$

78. (a) $\mathbf{u}$ and $\mathbf{v}$ are parallel. (b) $\mathbf{u}$ and $\mathbf{v}$ are orthogonal.
79 and 80. Proofs will vary.

81. $0, \dfrac{\pi}{6}, \pi, \dfrac{11\pi}{6}$ **82.** $\dfrac{\pi}{2}, \dfrac{5\pi}{4}, \dfrac{3\pi}{2}, \dfrac{7\pi}{4}$ **83.** $0, \pi$

84. $\dfrac{7\pi}{6}, \dfrac{3\pi}{2}, \dfrac{11\pi}{6}$ **85.** $-\dfrac{253}{325}$ **86.** $-\dfrac{323}{325}$

87. $\frac{204}{325}$ **88.** $-\frac{253}{204}$

Review Exercises *(page 319)*

1. $C = 74°, b \approx 13.19, c \approx 13.41$
2. $C = 37°, b \approx 38.90, c \approx 27.31$
3. $A = 26°, a \approx 24.89, c \approx 56.23$
4. $A = 150°, a \approx 48.24, b \approx 16.75$
5. $C = 66°, a \approx 2.53, b \approx 9.11$
6. $C = 40°, a \approx 162.42, b \approx 115.29$
7. $B = 108°, a \approx 11.76, c \approx 21.49$
8. $A = 80°, b \approx 334.95, c \approx 219.04$
9. $A \approx 20.41°, C \approx 9.59°, a \approx 20.92$
10. No solution
11. $B \approx 39.48°, C \approx 65.52°, c \approx 48.24$

12. Two solutions:

$A \approx 40.92°, C \approx 114.08°, c \approx 8.64$

$A \approx 139.08°, C \approx 15.92°, c \approx 2.60$

13. 7.9 **14.** 15.8 **15.** 33.5 **16.** 44.1

17. 31.1 meters **18.** 4.8 **19.** 31.01 feet

20. 586.4 feet **21.** $A \approx 29.69°, B \approx 52.41°, C \approx 97.90°$

22. $A \approx 53.13°, B \approx 36.87°, C = 90°$

23. $A \approx 29.92°, B \approx 86.18°, C \approx 63.90°$

24. $A \approx 101.47°, B \approx 31.73°, C \approx 46.80°$

25. $A = 35°, C = 35°, b \approx 6.55$

26. $A \approx 9.90°, C \approx 20.10°, b \approx 29.09$

27. $A \approx 45.76°, B \approx 91.24°, c \approx 21.42$

28. $B \approx 35.20°, C \approx 82.80°, a \approx 17.37$

29. ≈ 4.3 feet, ≈ 12.6 feet

30. ≈ 11.3 meters, ≈ 33.5 meters

31. 615.1 meters

32.

1135.5 miles

33. 9.80 **34.** 36.98 **35.** 8.36 **36.** 242.63

37. $\|\mathbf{u}\| = \|\mathbf{v}\| = \sqrt{61}$, $\text{slope}_{\mathbf{u}} = \text{slope}_{\mathbf{v}} = \frac{5}{6}$

38. $\|\mathbf{u}\| = \|\mathbf{v}\| = 2\sqrt{10}$, $\text{slope}_{\mathbf{u}} = \text{slope}_{\mathbf{v}} = -3$

39. $\langle 7, -5 \rangle$ **40.** $\langle 6, \frac{5}{2} \rangle$ **41.** $\langle 7, -7 \rangle$ **42.** $\langle 14, 4 \rangle$

43. $\langle -4, 4\sqrt{3} \rangle$ **44.** $\left\langle -\frac{\sqrt{2}}{4}, -\frac{\sqrt{2}}{4} \right\rangle$

45. (a) $\langle -4, 3 \rangle$ (b) $\langle 2, -9 \rangle$ (c) $\langle -3, -9 \rangle$

(d) $\langle -11, -3 \rangle$

46. (a) $\langle 4, 4 \rangle$ (b) $\langle 4, 6 \rangle$ (c) $\langle 12, 15 \rangle$ (d) $\langle 20, 23 \rangle$

47. (a) $\langle -1, 6 \rangle$ (b) $\langle -9, -2 \rangle$ (c) $\langle -15, 6 \rangle$

(d) $\langle -17, 18 \rangle$

48. (a) $\langle 4, -10 \rangle$ (b) $\langle -2, -6 \rangle$ (c) $\langle 3, -24 \rangle$

(d) $\langle 11, -44 \rangle$

49. (a) $7\mathbf{i} + 2\mathbf{j}$ (b) $-3\mathbf{i} - 4\mathbf{j}$ (c) $6\mathbf{i} - 3\mathbf{j}$

(d) $20\mathbf{i} + \mathbf{j}$

50. (a) $-3\mathbf{i} - 4\mathbf{j}$ (b) $-11\mathbf{i} - 4\mathbf{j}$ (c) $-21\mathbf{i} - 9\mathbf{j}$

(d) $-27\mathbf{i} - 17\mathbf{j}$

51. (a) $3\mathbf{i} + 6\mathbf{j}$ (b) $5\mathbf{i} - 6\mathbf{j}$ (c) $12\mathbf{i}$ (d) $18\mathbf{i} + 12\mathbf{j}$

52. (a) $\mathbf{i} - 5\mathbf{j}$ (b) $\mathbf{i} - 7\mathbf{j}$ (c) $-18\mathbf{j}$ (d) $2\mathbf{i} - 28\mathbf{j}$

53. $\langle 22, -7 \rangle$

54. $\langle -26, -35 \rangle$

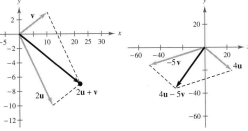

55. $\langle 30, 9 \rangle$

56. $\langle 5, \frac{3}{2} \rangle$

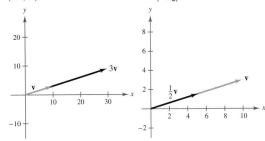

57. $-3\mathbf{i} + 4\mathbf{j}$ **58.** $-6\mathbf{i} - 8\mathbf{j}$ **59.** $6\mathbf{i} + 4\mathbf{j}$

60. $7\mathbf{i} - 16\mathbf{j}$ **61.** $10\sqrt{2}(\cos 135°\mathbf{i} + \sin 135°\mathbf{j})$

62. $\sqrt{17}(\cos 346°\mathbf{i} + \sin 346°\mathbf{j})$ **63.** $\|\mathbf{v}\| = 7; \theta = 60°$

64. $\|\mathbf{v}\| = 3; \theta = 150°$ **65.** $\|\mathbf{v}\| = \sqrt{41}; \theta = 38.7°$

66. $\|\mathbf{v}\| = \sqrt{65}; \theta = 119.7°$ **67.** $\|\mathbf{v}\| = 3\sqrt{2}; \theta = 225°$

68. $\|\mathbf{v}\| = \sqrt{65}; \theta = 352.9°$

69. The resultant force is 133.92 pounds and 5.6° from the 85-pound force.

70. 180 pounds each **71.** 422.30 miles per hour; 130.4°

72. 740.5 kilometers per hour; 32.1° **73.** 45

74. -140 **75.** -2 **76.** -136 **77.** 50; scalar

78. 5; scalar **79.** $\langle 6, -8 \rangle$; vector **80.** -6; scalar

81. $\dfrac{11\pi}{12}$ **82.** 105° **83.** 160.5° **84.** 22.4°

85. Orthogonal **86.** Parallel **87.** Neither

88. Orthogonal **89.** $-\frac{13}{17}\langle 4, 1 \rangle, \frac{16}{17}\langle -1, 4 \rangle$

90. $\langle 5, 0 \rangle, \langle 0, 6 \rangle$ **91.** $\frac{5}{2}\langle -1, 1 \rangle, \frac{9}{2}\langle 1, 1 \rangle$

92. $\frac{25}{29}\langle -5, 2 \rangle, \frac{19}{29}\langle 2, 5 \rangle$ **93.** 48 **94.** -132

95. 72,000 foot-pounds **96.** 281.9 foot-pounds

97. True. $\sin 90°$ is defined in the Law of Sines.

98. False. There may be no solution, one solution, or two solutions.

99. True. By definition, $\mathbf{u} = \dfrac{\mathbf{v}}{\|\mathbf{v}\|}$, so $\mathbf{v} = \|\mathbf{v}\|\mathbf{u}$.

100. False. If $\mathbf{v} = a\mathbf{i} + b\mathbf{j} = \mathbf{0}$, then $a = b = 0$.

101. $\dfrac{a}{\sin A} = \dfrac{b}{\sin B} = \dfrac{c}{\sin C}$

102. $a^2 = b^2 + c^2 - 2bc \cos A$, $b^2 = a^2 + c^2 - 2ac \cos B$, $c^2 = a^2 + b^2 - 2ab \cos C$

103. Direction and magnitude **104.** $\mathbf{A}$ and $\mathbf{C}$

105. a; The angle between the vectors is acute.

106. If $k > 0$, the direction is the same and the magnitude is k times as great.

If $k < 0$, the result is a vector in the opposite direction and the magnitude is $|k|$ times as great.

107. The diagonal of the parallelogram with **u** and **v** as its adjacent sides

Chapter Test *(page 322)*

1. $C = 88°, b \approx 27.81, c \approx 29.98$

2. $A = 43°, b \approx 25.75, c \approx 14.45$

3. Two solutions:
$B \approx 29.12°, C \approx 126.88°, c \approx 22.03$
$B \approx 150.88°, C \approx 5.12°, c \approx 2.46$

4. No solution   **5.** $A \approx 39.96°, C \approx 40.04°, c \approx 15.02$

6. $A \approx 23.43°, B \approx 33.57°, c \approx 86.46$

7. 2052.5 square meters   **8.** 606.3 miles; 29.1°

9. $\langle 14, -23 \rangle$ **10.** $\left\langle \dfrac{18\sqrt{34}}{17}, -\dfrac{30\sqrt{34}}{17} \right\rangle$

11. $\langle -4, 6 \rangle$ **12.** $\langle 10, 4 \rangle$

13. $\langle 36, 22 \rangle$ **14.** $\left\langle \dfrac{4}{5}, -\dfrac{3}{5} \right\rangle$

15. 14.9°; 250.15 pounds   **16.** 135°   **17.** No

18. $\dfrac{37}{26}\langle 5, 1 \rangle; \dfrac{29}{26}\langle -1, 5 \rangle$ **19.** ≈ 104 pounds

Cumulative Test for Chapters 1–3 *(page 323)*

1. (a)

(b) $240°$

(c) $-\dfrac{2\pi}{3}$

(d) $60°$

(e) $\sin(-120°) = -\dfrac{\sqrt{3}}{2}$ $\csc(-120°) = -\dfrac{2\sqrt{3}}{3}$

$\cos(-120°) = -\dfrac{1}{2}$ $\sec(-120°) = -2$

$\tan(-120°) = \sqrt{3}$ $\cot(-120°) = \dfrac{\sqrt{3}}{3}$

2. $134.6°$ **3.** $\dfrac{3}{5}$

4.

5.

6.
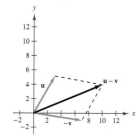

7. $a = -3, b = \pi, c = 0$

8.

9. 6.7   **10.** $\dfrac{3}{4}$

11. $\sqrt{1 - 4x^2}$ **12.** 1   **13.** $2 \tan \theta$

14–16. Answers will vary.   **17.** $\dfrac{\pi}{3}, \dfrac{\pi}{2}, \dfrac{3\pi}{2}, \dfrac{5\pi}{3}$

18. $\dfrac{\pi}{6}, \dfrac{5\pi}{6}, \dfrac{7\pi}{6}, \dfrac{11\pi}{6}$ **19.** $\dfrac{3\pi}{2}$ **20.** $\dfrac{16}{63}$ **21.** $\dfrac{4}{3}$

22. $\dfrac{\sqrt{5}}{5}, \dfrac{2\sqrt{5}}{5}$ **23.** $\dfrac{5}{2}\left(\sin \dfrac{5\pi}{2} - \sin \pi \right)$

24. $2 \cos 6x \cos 2x$

25. $B \approx 26.39°, C \approx 123.61°, c \approx 15.0$

26. $B \approx 52.48°, C \approx 97.52°, a \approx 5.04$

27. $B = 60°, a \approx 5.77, c \approx 11.55$

28. $A = 26.38°, B \approx 62.72°, C \approx 90.90°$

29. 36.4 square inches   **30.** 85.2 square inches

31. $3\mathbf{i} + 5\mathbf{j}$ **32.** $\left\langle \dfrac{\sqrt{2}}{2}, \dfrac{\sqrt{2}}{2} \right\rangle$ **33.** -5

34. $-\dfrac{1}{13}\langle 1, 5 \rangle; \dfrac{21}{13}\langle 5, -1 \rangle$

35. ≈ 395.8 radians per minute; ≈ 8312.6 inches per minute

36. Area $= 63.67$ square yards **37.** 5 feet **38.** $22.6°$

39. $d = 4 \cos \dfrac{\pi}{4} t$ **40.** $32.6°$; 543.9 kilometers per hour

41. 425 foot-pounds

Problem Solving *(page 329)*

1. 2.01 feet **2.** S $22.09°$ E; 1025.88 yards

3. (a)

(b) Station A: 27.45 miles; Station B: 53.03 miles

(c) 11.03 miles; S $21.7°$ E

4. (a)

(b) 50.5 feet

(c) 22 bags

5. (a) (i) $\sqrt{2}$ (ii) $\sqrt{5}$ (iii) 1
   (iv) 1 (v) 1 (vi) 1

(b) (i) 1 (ii) $3\sqrt{2}$ (iii) $\sqrt{13}$
   (iv) 1 (v) 1 (vi) 1

(c) (i) $\dfrac{\sqrt{5}}{2}$ (ii) $\sqrt{13}$ (iii) $\dfrac{\sqrt{85}}{2}$
   (iv) 1 (v) 1 (vi) 1

(d) (i) $2\sqrt{5}$ (ii) $5\sqrt{2}$ (iii) $5\sqrt{2}$
   (iv) 1 (v) 1 (vi) 1

6. (a) $\mathbf{u} = \langle 0, -120 \rangle$, $\mathbf{v} = \langle 40, 0 \rangle$

(b)

(c) 126.5 miles per hour; The magnitude gives the velocity of the skydiver's fall.

(d) $108.43°$

(e)

123.7 miles per hour

7. $\mathbf{w} = \frac{1}{2}(\mathbf{u} + \mathbf{v})$; $\mathbf{w} = \frac{1}{2}(\mathbf{v} - \mathbf{u})$

8. $\mathbf{u} \cdot \mathbf{v} = 0$ and $\mathbf{u} \cdot \mathbf{w} = 0$

$\mathbf{u} \cdot (c\mathbf{v} + d\mathbf{w})$
$= \mathbf{u} \cdot c\mathbf{v} + \mathbf{u} \cdot d\mathbf{w}$
$= c(\mathbf{u} \cdot \mathbf{v}) + d(\mathbf{u} \cdot \mathbf{w})$
$= 0$

9. (a) (b)
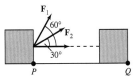

The amount of work done by $\mathbf{F}_1$ is equal to the amount of work done by $\mathbf{F}_2$.   The amount of work done by $\mathbf{F}_2$ is $\sqrt{3}$ times as great as the amount of work done by $\mathbf{F}_1$.

10. (a)

θ	$0.5°$	$1.0°$	$1.5°$
$\|\mathbf{v}\| \sin \theta$	0.873	1.745	2.618
$\|\mathbf{v}\| \cos \theta$	99.996	99.985	99.966

θ	$2.0°$	$2.5°$	$3.0°$
$\|\mathbf{v}\| \sin \theta$	3.490	4.362	5.234
$\|\mathbf{v}\| \cos \theta$	99.939	99.905	99.863

(b) No. Find the square root of the sum of the squares of the vertical and horizontal components.

(c) (i) 150 miles per hour
   (ii) 150 miles per hour

Chapter 4

Section 4.1 *(page 337)*

> **Vocabulary Check** *(page 337)*
>
> **1.** (a) iii (b) i (c) ii **2.** $\sqrt{-1}$; -1
>
> **3.** principal square **4.** complex conjugates

1. $a = -10$, $b = 6$ **2.** $a = 13$, $b = 4$

3. $a = 6$, $b = 5$ **4.** $a = 0$, $b = -\frac{5}{2}$ **5.** $4 + 3i$

6. $3 + 4i$ **7.** $2 - 3\sqrt{3}i$ **8.** $1 + 2\sqrt{2}i$

9. $5\sqrt{3}i$ **10.** $2i$ **11.** 8 **12.** 45

13. $-1 - 6i$ **14.** $4 + 2i$ **15.** $0.3i$ **16.** $0.02i$

17. $11 - i$ **18.** $8 + 4i$ **19.** 4 **20.** $-3 - 11i$

21. $3 - 3\sqrt{2}i$ **22.** 4 **23.** $-14 + 20i$

24. $17 + 18i$ **25.** $\frac{1}{6} + \frac{7}{6}i$ **26.** $-4.2 + 7.5i$

27. $5 + i$ **28.** $6 - 22i$ **29.** $12 + 30i$

30. $32 - 72i$ **31.** 24 **32.** 18 **33.** $-9 + 40i$

34. $-5 - 12i$ **35.** -10 **36.** $-8i$

37. $6 - 3i$, 45   **38.** $7 + 12i$, 193

39. $-1 + \sqrt{5}i$, 6   **40.** $-3 - \sqrt{2}i$, 11

41. $-2\sqrt{5}i$, 20   **42.** $-\sqrt{15}i$, 15

43. $\sqrt{8}$, 8   **44.** $1 + \sqrt{8}$, $9 + 4\sqrt{2}$ **45.** $-5i$

46. $7i$ **47.** $\frac{8}{41} + \frac{10}{41}i$ **48.** $\frac{5}{2} + \frac{5}{2}i$ **49.** $\frac{4}{5} + \frac{3}{5}i$

50. $4 + i$ **51.** $-5 - 6i$ **52.** $8 - 4i$

53. $-\frac{120}{1681} - \frac{27}{1681}i$ **54.** $\frac{60}{169} - \frac{25}{169}i$ **55.** $-\frac{1}{2} - \frac{5}{2}i$

56. $\frac{12}{5} + \frac{9}{5}i$ **57.** $\frac{62}{949} + \frac{297}{949}i$ **58.** $\frac{5}{17} - \frac{20}{17}i$

59. $-2\sqrt{3}$ **60.** $-5\sqrt{2}$ **61.** -10 **62.** -75

63. $\left(21 + 5\sqrt{2}\right) + \left(7\sqrt{5} - 3\sqrt{10}\right)i$ **64.** $-2 - 4\sqrt{6}i$

65. $1 \pm i$ **66.** $-3 \pm i$ **67.** $-2 \pm \frac{1}{2}i$

68. $\frac{1}{3} \pm 2i$ **69.** $-\frac{5}{2}, -\frac{3}{2}$ **70.** $\frac{1}{8} \pm \frac{\sqrt{11}}{8}i$

71. $2 \pm \sqrt{2}i$ **72.** $\frac{3}{7} \pm \frac{\sqrt{34}}{14}i$ **73.** $\frac{5}{7} \pm \frac{5\sqrt{15}}{7}$

74. $\frac{1}{3} \pm \frac{\sqrt{23}}{3}i$ **75.** $-1 + 6i$ **76.** $-4 + 2i$

77. $-5i$ **78.** $-i$ **79.** $-375\sqrt{3}i$ **80.** -8

81. i **82.** $\frac{1}{8}i$

83. (a) $z_1 = 9 + 16i, z_2 = 20 - 10i$

     (b) $z = \dfrac{11{,}240}{877} + \dfrac{4630}{877}i$

84. (a) 8   (b) 8   (c) 8

85. (a) 16   (b) 16   (c) 16   (d) 16

86. (a) 1   (b) i (c) -1 (d) $-i$

87. False. If the complex number is real, the number equals its conjugate.

88. True. $x^4 - x^2 + 14 = 56$

$$\left(-i\sqrt{6}\right)^4 - \left(-i\sqrt{6}\right)^2 + 14 \overset{?}{=} 56$$

$$36 + 6 + 14 \overset{?}{=} 56$$

$$56 = 56$$

89. False.

$$i^{44} + i^{150} - i^{74} - i^{109} + i^{61} = 1 - 1 + 1 - i + i = 1$$

90. $\sqrt{-6}\sqrt{-6} = \sqrt{6}i\sqrt{6}i = 6i^2 = -6$

91 and 92. Proofs will vary.   **93.** $-x^2 - 3x + 12$

94. $x^3 + x^2 + 2x - 6$

95. $3x^2 + \frac{23}{2}x - 2$ **96.** $4x^2 - 20x + 25$ **97.** -31

98. 14   **99.** $\dfrac{27}{2}$ **100.** $-\dfrac{4}{3}$ **101.** $a = \dfrac{\sqrt{3V\pi b}}{2\pi b}$

102. $r = \dfrac{\sqrt{\alpha m_1 m_2 F}}{F}$ **103.** 1 liter

Section 4.2   (page 344)

Vocabulary Check   (page 344)

1. Fundamental; Algebra   **2.** Linear Factorization

3. conjugates   **4.** discriminant

1. Three solutions   **2.** Six solutions

3. Four solutions   **4.** Five solutions

5. No real solutions   **6.** Two real solutions

7. Two real solutions   **8.** No real solutions

9. Two real solutions   **10.** Two real solutions

11. No real solutions   **12.** No real solutions

13. $\pm\sqrt{5}$ **14.** $\pm\dfrac{\sqrt{3}}{3}$ **15.** $-5 \pm \sqrt{6}$ **16.** $-3, 5$

17. 4   **18.** $-\frac{1}{2}$ **19.** $-1 \pm 2i$ **20.** $8 \pm \sqrt{118}$

21. $\frac{1}{2} \pm i$ **22.** $\frac{1}{2} \pm \sqrt{5}i$ **23.** $20 \pm 2\sqrt{215}$

24. $\dfrac{75 \pm 25\sqrt{7}}{2}$ **25.** $-3 \pm 2\sqrt{2}i$ **26.** $1 \pm \sqrt{6}$

27. (a)   (b) $4, \pm i$

   (c) The number of real zeros and the number of x-intercepts are the same.

28. (a)   (b) $\pm 2, 4$

   (c) The number of real zeros and the number of x-intercepts are the same.

29. (a)   (b) $\pm\sqrt{2}i$

   (c) The number of real zeros and the number of x-intercepts are the same.

30. (a)   (b) $\pm 2, \pm i$

   (c) The number of real zeros and the number of x-intercepts are the same.

31. $\pm 5i$; $(x + 5i)(x - 5i)$

32. $\dfrac{1 \pm \sqrt{223}i}{2}$; $\left(x - \dfrac{1 - \sqrt{223}i}{2}\right)\left(x - \dfrac{1 + \sqrt{223}i}{2}\right)$

33. $2 \pm \sqrt{3}$; $\left(x - 2 - \sqrt{3}\right)\left(x - 2 + \sqrt{3}\right)$

34. $-5 \pm \sqrt{2}$; $\left(x + 5 + \sqrt{2}\right)\left(x + 5 - \sqrt{2}\right)$

35. $\pm 3, \pm 3i$; $(x + 3)(x - 3)(x + 3i)(x - 3i)$

36. $\pm 5, \pm 5i$; $(y + 5)(y - 5)(y + 5i)(y - 5i)$

37. $1 \pm i$; $(z - 1 + i)(z - 1 - i)$

38. $3 \pm \sqrt{19}$; $(x - 3 + \sqrt{19})(x - 3 - \sqrt{19})$

39. $-3, \pm\sqrt{3}$; $(x + 3)(x + \sqrt{3})(x - \sqrt{3})$

40. $8, \pm 2\sqrt{3}$; $(x - 8)(x + 2\sqrt{3})(x - 2\sqrt{3})$

41. $4, \pm 4i$; $(x - 4)(x + 4i)(x - 4i)$

42. $-5, \pm\sqrt{2}i$; $(x + 5)(x + \sqrt{2}i)(x - \sqrt{2}i)$

43. $\frac{1}{2}, \pm 3\sqrt{2}i$; $(2x - 1)(x + 3\sqrt{2}i)(x - 3\sqrt{2}i)$

44. $-\frac{3}{4}, \pm 2\sqrt{6}i$; $(4x + 3)(x + 2\sqrt{6}i)(x - 2\sqrt{6}i)$

45. $0, 4, \pm 6i$; $x(x - 4)(x + 6i)(x - 6i)$

46. $0, -1, \pm 10i$; $x(x + 1)(x + 10i)(x - 10i)$

47. $\pm i, \pm 3i$; $(x + i)(x - i)(x + 3i)(x - 3i)$

48. $\pm 2i, \pm 5i$; $(x + 2i)(x - 2i)(x + 5i)(x - 5i)$

49. $-\frac{3}{2}, \pm 5i$ **50.** $-1, \pm 3i$ **51.** $\pm 2i, 1, -\frac{1}{2}$

52. $-3, 5 \pm 2i$ **53.** $-3 \pm i, \frac{1}{4}$ **54.** $-\frac{2}{3}, 1 \pm \sqrt{3}i$

55. $2, -3 \pm \sqrt{2}i, 1$ **56.** $-2, -1 \pm 3i$

57. $f(x) = x^3 - x^2 + 25x - 25$

58. $f(x) = x^3 - 4x^2 + 9x - 36$

59. $f(x) = x^3 + 4x^2 - 31x - 174$

60. $f(x) = x^3 - 10x^2 + 33x - 34$

61. $f(x) = 3x^4 - 17x^3 + 25x^2 + 23x - 22$

62. $f(x) = x^4 + 8x^3 + 9x^2 - 10x + 100$

63. $f(x) = -x^3 + x^2 - 4x + 4$

64. $f(x) = -x^3 + 2x^2 - x + 2$

65. $f(x) = -3x^3 + 9x^2 - 3x - 15$

66. $f(x) = \frac{1}{2}x^3 + \frac{1}{2}x + 5$

67. $f(x) = -2x^3 + 5x^2 - 10x + 4$

68. $f(x) = 4x^3 - 22x^2 + 48x - 36$

69. $f(x) = x^3 - 6x^2 + 4x + 40$

70. $f(x) = x^3 - 7x^2 + 16x - 10$

71. $f(x) = x^3 - 3x^2 + 6x + 10$

72. $f(x) = x^3 - 6x^2 + 17x - 18$

73. $f(x) = x^3 + x^2 - x - 33$

74. $f(x) = x^3 + 8x^2 + 23x + 22$

75. $f(x) = \frac{1}{2}x^4 + \frac{1}{2}x^3 + \frac{3}{2}x^2 + \frac{5}{2}x - 5$

76. $f(x) = -2x^4 + 2x^3 + 4x + 8$

77. (a)

t	0	0.5	1	1.5	2	2.5	3
h	0	20	32	36	32	20	0

(b) No

(c) When you set $h = 64$, the resulting equation yields imaginary roots. So, the projectile will not reach a height of 64 feet.

(d)

The graphs do not intersect, so the projectile does not reach 64 feet.

(e) The results all show that it is not possible for the projectile to reach a height of 64 feet.

78. (a)

t	0	1	2	3	4	5
h	5	68	99	98	65	0

(b) No

(c) When you set $h = 110$, the resulting equation yields imaginary roots. So the baseball does not reach a height of 110 feet.

(d)

The graphs do not intersect, so the baseball does not reach 110 feet.

(e) The results all show that it is not possible for the baseball to reach a height of 110 feet.

79. (a) $P = -0.0001x^2 + 60x - 150{,}000$

(b) $8,600,000   (c) $115

(d) It is not possible to have a profit of 10 million dollars.

80. (a) $S(t) = -0.116t^2 + 2.03t + 4.9$

(b)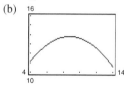

(c) The sales never reached $14 billion.

(d) When you set $S = 14$, the resulting equation yields imaginary roots. So, the sales never reached $14 billion.

81. False. The most complex zeros it can have is two, and the Linear Factorization Theorem guarantees that there are three linear factors, so one zero must be real.

82. False. f does not have real coefficients.

83. r_1, r_2, r_3 **84.** r_1, r_2, r_3

85. $5 + r_1, 5 + r_2, 5 + r_3$ **86.** $\frac{r_1}{2}, \frac{r_2}{2}, \frac{r_3}{2}$

87. The zeros cannot be determined.   **88.** $-r_1, -r_2, -r_3$

89. $x^2 + b$ **90.** $x^2 - 2ax + a^2 + b^2$ **91.** $-11 + 9i$

92. $12 + 11i$ **93.** $20 + 40i$ **94.** 106

95.   **96.**

97.

98.

99.

100.

101. 80.5° **102.** 102.5° **103.** 57.1° **104.** 45°
105. 57,000 foot-pounds **106.** 544 foot-pounds
107. Answers will vary.

Section 4.3 *(page 352)*

Vocabulary Check *(page 352)*

1. absolute value
2. trigonometric form; modulus; argument
3. $r_1 r_2[\cos(\theta_1 + \theta_2) + i \sin(\theta_1 + \theta_2)]$;
$\dfrac{r_1}{r_2}[\cos(\theta_1 - \theta_2) + i \sin(\theta_1 - \theta_2)]$

1.

7

2.

7

3.

$4\sqrt{2}$

4.

13

5.

$\sqrt{85}$

6.
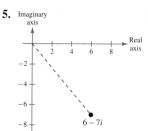

$\sqrt{73}$

7. $3\left(\cos\dfrac{\pi}{2} + i \sin\dfrac{\pi}{2}\right)$ **8.** $2(\cos \pi + i \sin \pi)$

9. $\sqrt{10}\,(\cos 5.96 + i \sin 5.96)$ **10.** $2\left(\cos\dfrac{2\pi}{3} + i \sin\dfrac{2\pi}{3}\right)$

11.

12.

$3\sqrt{2}\left(\cos\dfrac{7\pi}{4} + i \sin\dfrac{7\pi}{4}\right)$ $2\sqrt{2}\left(\cos\dfrac{\pi}{4} + i \sin\dfrac{\pi}{4}\right)$

13.

14.

$2\left(\cos\dfrac{\pi}{6} + i \sin\dfrac{\pi}{6}\right)$ $8\left(\cos\dfrac{5\pi}{3} + i \sin\dfrac{5\pi}{3}\right)$

15.

16.
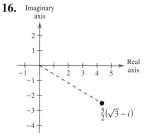

$4\left(\cos\dfrac{4\pi}{3} + i \sin\dfrac{4\pi}{3}\right)$ $5\left(\cos\dfrac{11\pi}{6} + i \sin\dfrac{11\pi}{6}\right)$

17.

$$5\left(\cos \frac{3\pi}{2} + i \sin \frac{3\pi}{2}\right)$$

18.

$$4\left(\cos \frac{\pi}{2} + i \sin \frac{\pi}{2}\right)$$

19.

$\sqrt{65}(\cos 2.62 + i \sin 2.62)$

20.

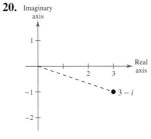

$\sqrt{10}(\cos 5.96 + i \sin 5.96)$

21.

$7(\cos 0 + i \sin 0)$

22.

$4(\cos 0 + i \sin 0)$

23.

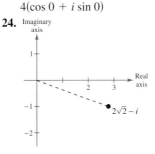

$$2\sqrt{3}\left(\cos \frac{\pi}{6} + i \sin \frac{\pi}{6}\right)$$

24.

$3(\cos 5.94 + i \sin 5.94)$

25.

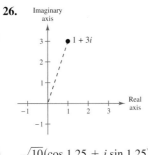

$\sqrt{10}(\cos 3.46 + i \sin 3.46)$

26.

$\sqrt{10}(\cos 1.25 + i \sin 1.25)$

27.

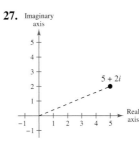

$5.39(\cos 0.38 + i \sin 0.38)$

28.

$8.54(\cos 0.36 + i \sin 0.36)$

29.

$11.79(\cos 3.97 + i \sin 3.97)$

30.

$11(\cos 3.75 + i \sin 3.75)$

31.

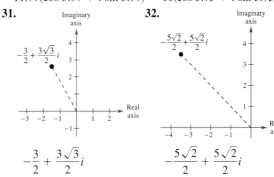

$$-\frac{3}{2} + \frac{3\sqrt{3}}{2}i$$

32.

$$-\frac{5\sqrt{2}}{2} + \frac{5\sqrt{2}}{2}i$$

33.

$$\frac{3}{4} - \frac{3\sqrt{3}}{4}i$$

34.

$$-\frac{\sqrt{2}}{8} - \frac{\sqrt{2}}{8}i$$

35.

$$-\frac{15\sqrt{2}}{8} + \frac{15\sqrt{2}}{8}i$$

36.

$1.5529 + 5.7956i$

37.

$8i$

38.

7

39.

$2.8408 + 0.9643i$

40.

$-3.8165 - 4.6297i$

41. $4.6985 + 1.7101i$ **42.** $3.0902 + 9.5106i$

43. $-2.9044 + 0.7511i$ **44.** $4.7693 + 7.6324i$

45.

46.

The absolute value of each is 1.

The absolute value of each is 1.

47. $12\left(\cos \dfrac{\pi}{3} + i \sin \dfrac{\pi}{3}\right)$ **48.** $3\left(\cos \dfrac{13\pi}{12} + i \sin \dfrac{13\pi}{12}\right)$

49. $\dfrac{10}{9}(\cos 200° + i \sin 200°)$ **50.** $0.4(\cos 40° + i \sin 40°)$

51. $0.27(\cos 150° + i \sin 150°)$ **52.** $\cos 25° + i \sin 25°$

53. $\cos 30° + i \sin 30°$ **54.** $\dfrac{1}{2}(\cos 80° + i \sin 80°)$

55. $\cos \dfrac{2\pi}{3} + i \sin \dfrac{2\pi}{3}$ **56.** $\dfrac{5}{4}(\cos 2.2 + i \sin 2.2)$

57. $4(\cos 302° + i \sin 302°)$ **58.** $\dfrac{6}{7}(\cos 300° + i \sin 300°)$

59. (a) $\left[2\sqrt{2}\left(\cos \dfrac{\pi}{4} + i \sin \dfrac{\pi}{4}\right)\right]\left[\sqrt{2}\left(\cos \dfrac{7\pi}{4} + i \sin \dfrac{7\pi}{4}\right)\right]$

(b) $4(\cos 0 + i \sin 0) = 4$ (c) 4

60. (a) $\left[2\left(\cos \dfrac{\pi}{6} + i \sin \dfrac{\pi}{6}\right)\right]\left[\sqrt{2}\left(\cos \dfrac{\pi}{4} + i \sin \dfrac{\pi}{4}\right)\right]$

(b) $2\sqrt{2}\left(\cos \dfrac{5\pi}{12} + i \sin \dfrac{5\pi}{12}\right) \approx 0.732 + 2.732i$

(c) $\left(\sqrt{3} - 1\right) + \left(\sqrt{3} + 1\right)i \approx 0.732 + 2.732i$

61. (a) $\left[2\left(\cos \dfrac{3\pi}{2} + i \sin \dfrac{3\pi}{2}\right)\right]\left[\sqrt{2}\left(\cos \dfrac{\pi}{4} + i \sin \dfrac{\pi}{4}\right)\right]$

(b) $2\sqrt{2}\left(\cos \dfrac{7\pi}{4} + i \sin \dfrac{7\pi}{4}\right) = 2 - 2i$

(c) $-2i - 2i^2 = -2i + 2 = 2 - 2i$

62. (a) $[4(\cos 0 + i \sin 0)]\left[2\left(\cos \dfrac{5\pi}{3} + i \sin \dfrac{5\pi}{3}\right)\right]$

(b) $8\left(\cos \dfrac{5\pi}{3} + i \sin \dfrac{5\pi}{3}\right) = 4 - 4\sqrt{3}\,i$

(c) $4 - 4\sqrt{3}\,i$

63. (a) $[5(\cos 0.93 + i \sin 0.93)] \div \left[2\left(\cos \dfrac{5\pi}{3} + i \sin \dfrac{5\pi}{3}\right)\right]$

(b) $\dfrac{5}{2}(\cos 1.97 + i \sin 1.97) \approx -0.982 + 2.299i$

(c) $\approx -0.982 + 2.299i$

64. (a) $\left[2\left(\cos \dfrac{\pi}{3} + i \sin \dfrac{\pi}{3}\right)\right] \div \left[3\sqrt{5}(\cos 5.82 + i \sin 5.82)\right]$

(b) $\dfrac{2\sqrt{5}}{15}(\cos 1.51 + i \sin 1.51) \approx 0.018 + 0.298i$

(c) $\approx 0.018 + 0.298i$

65. (a) $[5(\cos 0 + i \sin 0)] \div \left[\sqrt{13}(\cos 0.98 + i \sin 0.98)\right]$

(b) $\dfrac{5}{\sqrt{13}}(\cos 5.30 + i \sin 5.30) \approx 0.769 - 1.154i$

(c) $\dfrac{10}{13} - \dfrac{15}{13}i \approx 0.769 - 1.154i$

66. (a) $\left[4\left(\cos \dfrac{\pi}{2} + i \sin \dfrac{\pi}{2}\right)\right] \div \left[2\sqrt{5}(\cos 2.68 + i \sin 2.68)\right]$

(b) $\dfrac{2}{\sqrt{5}}(\cos 5.18 + i \sin 5.18) \approx 0.403 - 0.798i$

(c) $\dfrac{2}{5} - \dfrac{4}{5}i = 0.400 - 0.800i$

67.

68.

69.

70.

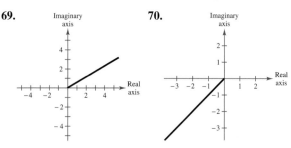

71. True, by the definition of the absolute value of a complex number.

72. True. $z_1z_2 = r_1r_2[\cos(\theta_1 + \theta_2) + i \sin(\theta_1 + \theta_2)] = 0$ if and only if $r_1 = 0$ and/or $r_2 = 0$.

73 and 74. Answers will vary.

75. (a) r^2 (b) $\cos 2\theta + i \sin 2\theta$

76. Answers will vary.

77. $B = 68°, b \approx 19.80, c \approx 21.36$

78. $A = 24°, b \approx 75.24, c \approx 82.36$

79. $B = 60°, a \approx 65.01, c \approx 130.02$

80. $A = 84°, a \approx 2009.43, c \approx 2020.50$

81. $B = 47°45', a \approx 7.53, b \approx 8.29$

82. $A = 8°30', a \approx 1.01, b \approx 6.73$

83. $16; 2$ **84.** $\frac{1}{8}; \frac{1}{24}$ **85.** $\frac{1}{16}; 0$ **86.** $\frac{1}{12}; 0$

Section 4.4   *(page 358)*

> **Vocabulary Check**   *(page 358)*
>
> **1.** DeMoivre's   **2.** nth root   **3.** unity

1. $-4 - 4i$ **2.** $-512i$ **3.** $-32i$

4. $-239 + 28,560i$ **5.** $-128\sqrt{3} - 128i$ **6.** -32

7. $\dfrac{125}{2} + \dfrac{125\sqrt{3}}{2}i$ **8.** $-\dfrac{81}{2} - \dfrac{81\sqrt{3}}{2}i$ **9.** -1

10. 256 **11.** $608.0204 + 144.6936i$ **12.** 1

13. $-597 - 122i$ **14.** $15,939 + 18,460i$

15. $-43\sqrt{5} + 4i$ **16.** $-47 - 8\sqrt{3}i$

17. $\dfrac{81}{2} + \dfrac{81\sqrt{3}}{2}i$ **18.** $44.4539 + 252.1108i$

19. $32.3524 - 120.7407i$

20. $44.4539 + 252.1108i$

21. $32i$ **22.** $-32\sqrt{2} + 32\sqrt{2}i$ **23.** 27

24. $62.8930 + 234.7200i$

25. (a) $\sqrt{5}(\cos 60° + i \sin 60°)$
$\sqrt{5}(\cos 240° + i \sin 240°)$

(b)

(c) $\dfrac{\sqrt{5}}{2} + \dfrac{\sqrt{15}}{2}i, \ -\dfrac{\sqrt{5}}{2} - \dfrac{\sqrt{15}}{2}i$

26. (a) $4(\cos 30° + i \sin 30°)$
$4(\cos 210° + i \sin 210°)$

(b)

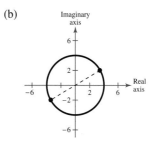

(c) $2\sqrt{3} + 2i, -2\sqrt{3} - 2i$

27. (a) $2\left(\cos \dfrac{2\pi}{9} + i \sin \dfrac{2\pi}{9}\right)$

$2\left(\cos \dfrac{8\pi}{9} + i \sin \dfrac{8\pi}{9}\right)$

$2\left(\cos \dfrac{14\pi}{9} + i \sin \dfrac{14\pi}{9}\right)$

(b)

(c) $1.5321 + 1.2856i, -1.8794 + 0.6840i,$
$0.3473 - 1.9696i$

28. (a) $4\left(\cos \dfrac{\pi}{9} + i \sin \dfrac{\pi}{9}\right)$

$4\left(\cos \dfrac{7\pi}{9} + i \sin \dfrac{7\pi}{9}\right)$

$4\left(\cos \dfrac{13\pi}{9} + i \sin \dfrac{13\pi}{9}\right)$

(b)

(c) $3.7588 + 1.3681i, -3.0642 + 2.5712i,$
$-0.6946 - 3.9392i$

29. (a) $3\left(\cos\dfrac{\pi}{30} + i\sin\dfrac{\pi}{30}\right)$

$3\left(\cos\dfrac{13\pi}{30} + i\sin\dfrac{13\pi}{30}\right)$

$3\left(\cos\dfrac{5\pi}{6} + i\sin\dfrac{5\pi}{6}\right)$

$3\left(\cos\dfrac{37\pi}{30} + i\sin\dfrac{37\pi}{30}\right)$

$3\left(\cos\dfrac{49\pi}{30} + i\sin\dfrac{49\pi}{30}\right)$

(b)

(c) $2.9836 + 0.3136i$, $0.6237 + 2.9344i$,
$-2.5981 + 1.5i$, $-2.2294 - 2.0074i$, $1.2202 - 2.7406i$

30. (a) $2\left(\cos\dfrac{\pi}{6} + i\sin\dfrac{\pi}{6}\right)$

$2\left(\cos\dfrac{17\pi}{30} + i\sin\dfrac{17\pi}{30}\right)$

$2\left(\cos\dfrac{29\pi}{30} + i\sin\dfrac{29\pi}{30}\right)$

$2\left(\cos\dfrac{41\pi}{30} + i\sin\dfrac{41\pi}{30}\right)$

$2\left(\cos\dfrac{53\pi}{30} + i\sin\dfrac{53\pi}{30}\right)$

(b)

(c) $\sqrt{3} + i$, $-0.4158 + 1.9563i$, $-1.9890 + 0.2091i$,
$-0.8135 - 1.8271i$, $1.4863 - 1.3383i$

31. (a) $5\left(\cos\dfrac{3\pi}{4} + i\sin\dfrac{3\pi}{4}\right)$

$5\left(\cos\dfrac{7\pi}{4} + i\sin\dfrac{7\pi}{4}\right)$

(b)

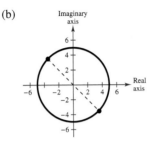

(c) $-\dfrac{5\sqrt{2}}{2} + \dfrac{5\sqrt{2}}{2}i$, $\dfrac{5\sqrt{2}}{2} - \dfrac{5\sqrt{2}}{2}i$

32. (a) $6\left(\cos\dfrac{3\pi}{4} + i\sin\dfrac{3\pi}{4}\right)$

$6\left(\cos\dfrac{7\pi}{4} + i\sin\dfrac{7\pi}{4}\right)$

(b)

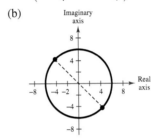

(c) $-3\sqrt{2} + 3\sqrt{2}i$, $3\sqrt{2} - 3\sqrt{2}i$

33. (a) $3\left(\cos\dfrac{\pi}{8} + i\sin\dfrac{\pi}{8}\right)$

$3\left(\cos\dfrac{5\pi}{8} + i\sin\dfrac{5\pi}{8}\right)$

$3\left(\cos\dfrac{9\pi}{8} + i\sin\dfrac{9\pi}{8}\right)$

$3\left(\cos\dfrac{13\pi}{8} + i\sin\dfrac{13\pi}{8}\right)$

(b)

(c) $2.7716 + 1.1481i$, $-1.1481 + 2.7716i$,
$-2.7716 - 1.1481i$, $1.1481 - 2.7716i$

34. (a) $5\left(\cos\dfrac{\pi}{8} + i\sin\dfrac{\pi}{8}\right)$

$5\left(\cos\dfrac{5\pi}{8} + i\sin\dfrac{5\pi}{8}\right)$

$5\left(\cos\dfrac{9\pi}{8} + i\sin\dfrac{9\pi}{8}\right)$

$5\left(\cos\dfrac{13\pi}{8} + i\sin\dfrac{13\pi}{8}\right)$

(b)

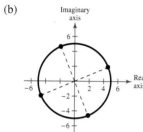

(c) $4.6194 + 1.9134i,\ -1.9134 + 4.6194i,$
$-4.6194 - 1.9134i,\ 1.9134 - 4.6194i$

35. (a) $5\left(\cos\dfrac{4\pi}{9} + i\sin\dfrac{4\pi}{9}\right)$

$5\left(\cos\dfrac{10\pi}{9} + i\sin\dfrac{10\pi}{9}\right)$

$5\left(\cos\dfrac{16\pi}{9} + i\sin\dfrac{16\pi}{9}\right)$

(b)

(c) $0.8682 + 4.9240i,\ -4.6985 - 1.7101i,$
$3.8302 - 3.2139i$

36. (a) $2\left(\cos\dfrac{\pi}{4} + i\sin\dfrac{\pi}{4}\right)$

$2\left(\cos\dfrac{11\pi}{12} + i\sin\dfrac{11\pi}{12}\right)$

$2\left(\cos\dfrac{19\pi}{12} + i\sin\dfrac{19\pi}{12}\right)$

(b)

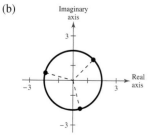

(c) $\sqrt{2} + \sqrt{2}i,\ -1.9319 + 0.5176i,\ 0.5176 - 1.9319i$

37. (a) $2(\cos 0 + i\sin 0)$

$2\left(\cos\dfrac{\pi}{2} + i\sin\dfrac{\pi}{2}\right)$

$2(\cos\pi + i\sin\pi)$

$2\left(\cos\dfrac{3\pi}{2} + i\sin\dfrac{3\pi}{2}\right)$

(b)

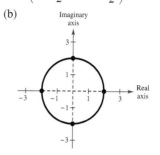

(c) $2,\ 2i,\ -2,\ -2i$

38. (a) $\cos\dfrac{\pi}{8} + i\sin\dfrac{\pi}{8}$

$\cos\dfrac{5\pi}{8} + i\sin\dfrac{5\pi}{8}$

$\cos\dfrac{9\pi}{8} + i\sin\dfrac{9\pi}{8}$

$\cos\dfrac{13\pi}{8} + i\sin\dfrac{13\pi}{8}$

(b)

(c) $0.9239 + 0.3827i,\ -0.3827 + 0.9239i,$
$-0.9239 - 0.3827i,\ 0.3827 - 0.9239i$

39. (a) $\cos 0 + i\sin 0$

$\cos\dfrac{2\pi}{5} + i\sin\dfrac{2\pi}{5}$

$\cos\dfrac{4\pi}{5} + i\sin\dfrac{4\pi}{5}$

$\cos\dfrac{6\pi}{5} + i\sin\dfrac{6\pi}{5}$

$\cos\dfrac{8\pi}{5} + i\sin\dfrac{8\pi}{5}$

(b)

(c) $1,\ 0.3090 + 0.9511i,\ -0.8090 + 0.5878i,$
$-0.8090 - 0.5878i,\ 0.3090 - 0.9511i$

CHAPTER 4

40. (a) $10(\cos 0 + i \sin 0)$

$$10\left(\cos \frac{2\pi}{3} + i \sin \frac{2\pi}{3}\right)$$

$$10\left(\cos \frac{4\pi}{3} + i \sin \frac{4\pi}{3}\right)$$

(b)

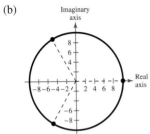

(c) $10, -5 + 5\sqrt{3}i, -5 - 5\sqrt{3}i$

41. (a) $5\left(\cos \frac{\pi}{3} + i \sin \frac{\pi}{3}\right)$

$$5(\cos \pi + i \sin \pi)$$

$$5\left(\cos \frac{5\pi}{3} + i \sin \frac{5\pi}{3}\right)$$

(b)

(c) $\dfrac{5}{2} + \dfrac{5\sqrt{3}}{2}i, -5, \dfrac{5}{2} - \dfrac{5\sqrt{3}}{2}i$

42. (a) $\sqrt{2}\left(\cos \frac{\pi}{4} + i \sin \frac{\pi}{4}\right)$

$$\sqrt{2}\left(\cos \frac{3\pi}{4} + i \sin \frac{3\pi}{4}\right)$$

$$\sqrt{2}\left(\cos \frac{5\pi}{4} + i \sin \frac{5\pi}{4}\right)$$

$$\sqrt{2}\left(\cos \frac{7\pi}{4} + i \sin \frac{7\pi}{4}\right)$$

(b)

(c) $1 + i, -1 + i, -1 - i, 1 - i$

43. (a) $2\sqrt{2}\left(\cos \frac{3\pi}{20} + i \sin \frac{3\pi}{20}\right)$

$$2\sqrt{2}\left(\cos \frac{11\pi}{20} + i \sin \frac{11\pi}{20}\right)$$

$$2\sqrt{2}\left(\cos \frac{19\pi}{20} + i \sin \frac{19\pi}{20}\right)$$

$$2\sqrt{2}\left(\cos \frac{27\pi}{20} + i \sin \frac{27\pi}{20}\right)$$

$$2\sqrt{2}\left(\cos \frac{7\pi}{4} + i \sin \frac{7\pi}{4}\right)$$

(b)

(c) $2.5201 + 1.2841i, -0.4425 + 2.7936i,$
$\quad -2.7936 + 0.4425i, -1.2841 - 2.5201i, 2 - 2i$

44. (a) $2\left(\cos \frac{\pi}{12} + i \sin \frac{\pi}{12}\right)$

$$2\left(\cos \frac{5\pi}{12} + i \sin \frac{5\pi}{12}\right)$$

$$2\left(\cos \frac{3\pi}{4} + i \sin \frac{3\pi}{4}\right)$$

$$2\left(\cos \frac{13\pi}{12} + i \sin \frac{13\pi}{12}\right)$$

$$2\left(\cos \frac{17\pi}{12} + i \sin \frac{17\pi}{12}\right)$$

$$2\left(\cos \frac{7\pi}{4} + i \sin \frac{7\pi}{4}\right)$$

(b)

(c) $1.9319 + 0.5176i, 0.5176 + 1.9319i,$
$\quad -\sqrt{2} + \sqrt{2}i, -1.9319 - 0.5176i,$
$\quad -0.5176 - 1.9319i, \sqrt{2} - \sqrt{2}i$

45. $\cos \dfrac{3\pi}{8} + i \sin \dfrac{3\pi}{8}$

$\cos \dfrac{7\pi}{8} + i \sin \dfrac{7\pi}{8}$

$\cos \dfrac{11\pi}{8} + i \sin \dfrac{11\pi}{8}$

$\cos \dfrac{15\pi}{8} + i \sin \dfrac{15\pi}{8}$

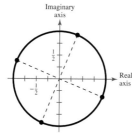

46. $\cos \dfrac{\pi}{6} + i \sin \dfrac{\pi}{6}$

$\cos \dfrac{5\pi}{6} + i \sin \dfrac{5\pi}{6}$

$\cos \dfrac{3\pi}{2} + i \sin \dfrac{3\pi}{2}$

47. $\cos \dfrac{\pi}{6} + i \sin \dfrac{\pi}{6}$

$\cos \dfrac{\pi}{2} + i \sin \dfrac{\pi}{2}$

$\cos \dfrac{5\pi}{6} + i \sin \dfrac{5\pi}{6}$

$\cos \dfrac{7\pi}{6} + i \sin \dfrac{7\pi}{6}$

$\cos \dfrac{3\pi}{2} + i \sin \dfrac{3\pi}{2}$

$\cos \dfrac{11\pi}{6} + i \sin \dfrac{11\pi}{6}$

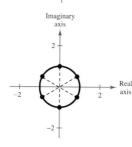

48. $\cos \dfrac{\pi}{3} + i \sin \dfrac{\pi}{3}$

$\cos \pi + i \sin \pi$

$\cos \dfrac{5\pi}{3} + i \sin \dfrac{5\pi}{3}$

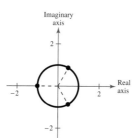

49. $3\left(\cos \dfrac{\pi}{5} + i \sin \dfrac{\pi}{5}\right)$

$3\left(\cos \dfrac{3\pi}{5} + i \sin \dfrac{3\pi}{5}\right)$

$3(\cos \pi + i \sin \pi)$

$3\left(\cos \dfrac{7\pi}{5} + i \sin \dfrac{7\pi}{5}\right)$

$3\left(\cos \dfrac{9\pi}{5} + i \sin \dfrac{9\pi}{5}\right)$

50. $5\left(\cos \dfrac{\pi}{3} + i \sin \dfrac{\pi}{3}\right)$

$5(\cos \pi + i \sin \pi)$

$5\left(\cos \dfrac{5\pi}{3} + i \sin \dfrac{5\pi}{3}\right)$

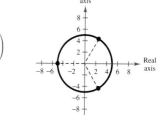

51. $2(\cos 0 + i \sin 0)$

$2\left(\cos \dfrac{2\pi}{5} + i \sin \dfrac{2\pi}{5}\right)$

$2\left(\cos \dfrac{4\pi}{5} + i \sin \dfrac{4\pi}{5}\right)$

$2\left(\cos \dfrac{6\pi}{5} + i \sin \dfrac{6\pi}{5}\right)$

$2\left(\cos \dfrac{8\pi}{5} + i \sin \dfrac{8\pi}{5}\right)$

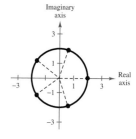

52. $3(\cos 0 + i \sin 0)$

$3\left(\cos \dfrac{2\pi}{3} + i \sin \dfrac{2\pi}{3}\right)$

$3\left(\cos \dfrac{4\pi}{3} + i \sin \dfrac{4\pi}{3}\right)$

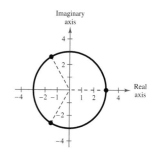

53. $2\left(\cos \dfrac{3\pi}{8} + i \sin \dfrac{3\pi}{8}\right)$

$2\left(\cos \dfrac{7\pi}{8} + i \sin \dfrac{7\pi}{8}\right)$

$2\left(\cos \dfrac{11\pi}{8} + i \sin \dfrac{11\pi}{8}\right)$

$2\left(\cos \dfrac{15\pi}{8} + i \sin \dfrac{15\pi}{8}\right)$

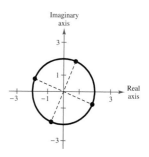

54. $3\left(\cos \dfrac{\pi}{2} + i \sin \dfrac{\pi}{2}\right)$

$3\left(\cos \dfrac{7\pi}{6} + i \sin \dfrac{7\pi}{6}\right)$

$3\left(\cos \dfrac{11\pi}{6} + i \sin \dfrac{11\pi}{6}\right)$

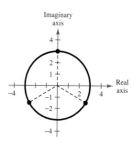

CHAPTER 4

55. $2\left(\cos\dfrac{\pi}{8} + i\sin\dfrac{\pi}{8}\right)$

$2\left(\cos\dfrac{5\pi}{8} + i\sin\dfrac{5\pi}{8}\right)$

$2\left(\cos\dfrac{9\pi}{8} + i\sin\dfrac{9\pi}{8}\right)$

$2\left(\cos\dfrac{13\pi}{8} + i\sin\dfrac{13\pi}{8}\right)$

56. $2\left(\cos\dfrac{\pi}{4} + i\sin\dfrac{\pi}{4}\right)$

$2\left(\cos\dfrac{7\pi}{12} + i\sin\dfrac{7\pi}{12}\right)$

$2\left(\cos\dfrac{11\pi}{12} + i\sin\dfrac{11\pi}{12}\right)$

$2\left(\cos\dfrac{5\pi}{4} + i\sin\dfrac{5\pi}{4}\right)$

$2\left(\cos\dfrac{19\pi}{12} + i\sin\dfrac{19\pi}{12}\right)$

$2\left(\cos\dfrac{23\pi}{12} + i\sin\dfrac{23\pi}{12}\right)$

57. $\sqrt[6]{2}\left(\cos\dfrac{7\pi}{12} + i\sin\dfrac{7\pi}{12}\right)$

$\sqrt[6]{2}\left(\cos\dfrac{5\pi}{4} + i\sin\dfrac{5\pi}{4}\right)$

$\sqrt[6]{2}\left(\cos\dfrac{23\pi}{12} + i\sin\dfrac{23\pi}{12}\right)$

58. $\sqrt[10]{2}\left(\cos\dfrac{7\pi}{20} + i\sin\dfrac{7\pi}{20}\right)$

$\sqrt[10]{2}\left(\cos\dfrac{3\pi}{4} + i\sin\dfrac{3\pi}{4}\right)$

$\sqrt[10]{2}\left(\cos\dfrac{23\pi}{20} + i\sin\dfrac{23\pi}{20}\right)$

$\sqrt[10]{2}\left(\cos\dfrac{31\pi}{20} + i\sin\dfrac{31\pi}{20}\right)$

$\sqrt[10]{2}\left(\cos\dfrac{39\pi}{20} + i\sin\dfrac{39\pi}{20}\right)$

59. $\sqrt[12]{2}\left(\cos\dfrac{5\pi}{24} + i\sin\dfrac{5\pi}{24}\right)$

$\sqrt[12]{2}\left(\cos\dfrac{13\pi}{24} + i\sin\dfrac{13\pi}{24}\right)$

$\sqrt[12]{2}\left(\cos\dfrac{7\pi}{8} + i\sin\dfrac{7\pi}{8}\right)$

$\sqrt[12]{2}\left(\cos\dfrac{29\pi}{24} + i\sin\dfrac{29\pi}{24}\right)$

$\sqrt[12]{2}\left(\cos\dfrac{37\pi}{24} + i\sin\dfrac{37\pi}{24}\right)$

$\sqrt[12]{2}\left(\cos\dfrac{15\pi}{8} + i\sin\dfrac{15\pi}{8}\right)$

60. $\sqrt[8]{2}\left(\cos\dfrac{5\pi}{16} + i\sin\dfrac{5\pi}{16}\right)$

$\sqrt[8]{2}\left(\cos\dfrac{13\pi}{16} + i\sin\dfrac{13\pi}{16}\right)$

$\sqrt[8]{2}\left(\cos\dfrac{21\pi}{16} + i\sin\dfrac{21\pi}{16}\right)$

$\sqrt[8]{2}\left(\cos\dfrac{29\pi}{16} + i\sin\dfrac{29\pi}{16}\right)$

61. False. They are equally spaced around the circle centered at the origin with radius $\sqrt[n]{r}$.

62. False. The complex number needs to be converted to trigonometric form before using DeMoivre's Theorem.

$\left(4 + \sqrt{6}i\right)^8 \approx \left[\sqrt{22}(\cos 0.55 + i\sin 0.55)\right]^8$

63 and 64. Answers will vary.

65. (a) $2(\cos 30° + i\sin 30°)$

$2(\cos 150° + i\sin 150°)$

$2(\cos 270° + i\sin 270°)$

(b) $8i$

66. (a) $3(\cos 45° + i\sin 45°)$

$3(\cos 135° + i\sin 135°)$

$3(\cos 225° + i\sin 225°)$

$3(\cos 315° + i\sin 315°)$

(b) -81

67. $m = \dfrac{1}{4},\ b = \dfrac{1}{4}$

68. $m = -\dfrac{7}{6},\ b = -\dfrac{4}{3}$

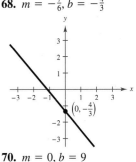

69. $m = 0,\ b = -5$

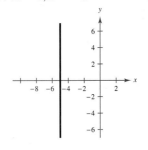

70. $m = 0,\ b = 9$

71. Yes. $f^{-1}(x) = \dfrac{x+1}{5}$ **72.** Yes. $g^{-1}(x) = \dfrac{2}{x}$

73. Yes. $h^{-1}(x) = \dfrac{x^2 - 3}{4}$, $x \geq 0$ **74.** No inverse

75. $\dfrac{\sqrt{34}}{136}(5\sqrt{7} - 9)$ **76.** $\dfrac{3\sqrt{34}}{136}(\sqrt{7} + 5)$

77. $\dfrac{3\sqrt{34}}{136}(\sqrt{7} - 5)$ **78.** $\dfrac{\sqrt{34}}{136}(5\sqrt{7} + 9)$

79. $\dfrac{120 + 51\sqrt{7}}{47}$ **80.** $\dfrac{51\sqrt{7} - 120}{47}$ **81.** $\dfrac{15}{8}$

82. $\dfrac{3\sqrt{7}}{8}$ **83.** $\dfrac{1}{2}\sqrt{\dfrac{4 - \sqrt{7}}{2}}$ **84.** $\pm\dfrac{1}{2}\sqrt{\dfrac{34 + 5\sqrt{34}}{17}}$

85. $(1, 0)$ **86.** $\left\langle -\dfrac{3\sqrt{58}}{58}, \dfrac{7\sqrt{58}}{58} \right\rangle$ **87.** $\dfrac{12}{13}\mathbf{i} - \dfrac{5}{13}\mathbf{j}$

88. $\mathbf{j}$ **89.** 5   **90.** $\sqrt{74}$ **91.** 41   **92.** 13

93. $\sqrt{493}$ **94.** $4\sqrt{17}$ **95.** $\sqrt{205}$ **96.** $8\sqrt{13}$

Review Exercises   *(page 361)*

1. $6 + 2i$ **2.** $3 - 5i$ **3.** $-1 + 3i$ **4.** $-1 - 5i$

5. $3 + 7i$ **6.** $-\sqrt{2}\,i$ **7.** $40 + 65i$

8. $17 + 28i$ **9.** $-4 - 46i$ **10.** $9 + 20i$

11. $\frac{23}{17} + \frac{10}{17}i$ **12.** $\frac{17}{26} + \frac{7}{26}i$ **13.** $\frac{21}{13} - \frac{1}{13}i$

14. $\dfrac{9}{85} + \dfrac{83}{85}i$ **15.** $\pm\dfrac{\sqrt{3}}{3}i$ **16.** $\pm\dfrac{1}{2}i$ **17.** $1 \pm 3i$

18. $-\dfrac{1}{4} \pm \dfrac{\sqrt{71}}{4}i$ **19.** Five solutions   **20.** Six solutions

21. Four solutions   **22.** Three solutions

23. Two real solutions   **24.** One repeated solution

25. No real solutions   **26.** One repeated solution

27. $0, 2$ **28.** $0, 6$ **29.** $\dfrac{3}{2} \pm \dfrac{\sqrt{11}i}{2}$ **30.** $2 \pm \sqrt{5}i$

31. $-4 \pm \sqrt{6}$ **32.** $2 \pm \sqrt{7}$ **33.** $-\dfrac{3}{4} \pm \dfrac{\sqrt{39}i}{4}$

34. $\dfrac{1}{8} \pm \dfrac{\sqrt{159}i}{8}$

35. Yes. A price of \$95.41 or \$119.59 per unit would yield a profit of 9 million dollars.

36. When you set $p = 800$, the resulting equation yields imaginary roots. So, the price will not drop to \$800.

37. $-\dfrac{1}{2} \pm \dfrac{\sqrt{5}}{2}i$; $\dfrac{1}{2}(2x + 1 - \sqrt{5}i)(2x + 1 + \sqrt{5}i)$

38. $\dfrac{-5 \pm \sqrt{7}i}{4}$; $\dfrac{1}{8}(4x + 5 + \sqrt{7}i)(4x + 5 - \sqrt{7}i)$

39. $\frac{3}{2}, \pm 5i$; $(2x - 3)(x + 5i)(x - 5i)$

40. $\frac{1}{4}, \pm 4\sqrt{2}i$; $(4x - 1)(x + 4\sqrt{2}i)(x - 4\sqrt{2}i)$

41. $\pm\dfrac{\sqrt{5}}{2}, \pm\sqrt{2}i$; $(2x + \sqrt{5})(2x - \sqrt{5})(x + \sqrt{2}i)(x - \sqrt{2}i)$

42. $\pm 5i, \pm\dfrac{\sqrt{5}}{5}i$; $\dfrac{1}{5}(x + 5i)(x - 5i)(5x + \sqrt{5}i)(5x - \sqrt{5}i)$

43. $-7, 2$; $(x + 7)(x - 2)^2$

44. $-2, -\frac{3}{5}, \frac{1}{2}$; $(x + 2)(5x + 3)(2x - 1)$

45. $-5, 1 \pm 2i$; $(x + 5)(x - 1 - 2i)(x - 1 + 2i)$

46. $4, 2 \pm 3i$; $(x - 4)(x - 2 - 3i)(x - 2 + 3i)$

47. $-\frac{1}{2}, 5 \pm 3i$; $(2x + 1)(x - 5 - 3i)(x - 5 + 3i)$

48. $\frac{4}{5}, \pm 2i$; $(5x - 4)(x - 2i)(x + 2i)$

49. $-2, 3, -3 \pm \sqrt{5}i$; $(x - 3)(x + 2)(x + 3 - \sqrt{5}i)(x + 3 + \sqrt{5}i)$

50. $2 \pm \sqrt{3}i, 1 \pm \sqrt{2}i$;

51. $f(x) = 12x^4 - 19x^3 + 9x - 2$

52. $f(x) = x^4 - 6x^3 + 5x^2 + 24x - 36$

53. $f(x) = x^3 - 7x^2 + 13x - 3$

54. $f(x) = x^3 - 7x^2 + 9x + 5$

55. $f(x) = 3x^4 - 14x^3 + 17x^2 - 42x + 24$

56. $f(x) = x^4 - x^3 - 3x^2 + 17x - 30$

57. $f(x) = x^4 + 27x^2 + 50$ **58.** $f(x) = x^4 + 20x^2 + 64$

59. $f(x) = 2x^3 - 14x^2 + 24x - 20$

60. $f(x) = 2x^3 - 20x^2 + 66x - 68$

61.

8

62.

63.
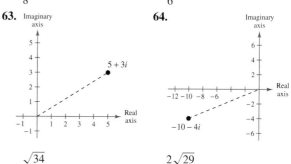
$\sqrt{34}$

64.
$2\sqrt{29}$

6

65. $5\sqrt{2}\left(\cos\dfrac{7\pi}{4} + i\sin\dfrac{7\pi}{4}\right)$

66. $13(\cos 1.176 + i\sin 1.176)$ **67.** $6\left(\cos\dfrac{5\pi}{6} + i\sin\dfrac{5\pi}{6}\right)$

68. $9(\cos\pi + i\sin\pi)$

69. (a) $z_1 = 4\left(\cos\dfrac{11\pi}{6} + i\sin\dfrac{11\pi}{6}\right)$

$\quad z_2 = 10\left(\cos\dfrac{3\pi}{2} + i\sin\dfrac{3\pi}{2}\right)$

(b) $z_1 z_2 = 40\left(\cos\dfrac{10\pi}{3} + i\sin\dfrac{10\pi}{3}\right)$

$\quad \dfrac{z_1}{z_2} = \dfrac{2}{5}\left(\cos\dfrac{\pi}{3} + i\sin\dfrac{\pi}{3}\right)$

70. (a) $z_1 = 3\sqrt{2}\left(\cos\dfrac{5\pi}{4} + i\sin\dfrac{5\pi}{4}\right)$

$z_2 = 4\left(\cos\dfrac{\pi}{6} + i\sin\dfrac{\pi}{6}\right)$

(b) $z_1 z_2 = 12\sqrt{2}\left(\cos\dfrac{17\pi}{12} + i\sin\dfrac{17\pi}{12}\right)$

$\dfrac{z_1}{z_2} = \dfrac{3\sqrt{2}}{4}\left(\cos\dfrac{13\pi}{12} + i\sin\dfrac{13\pi}{12}\right)$

71. $\dfrac{625}{2} + \dfrac{625\sqrt{3}}{2}i$ **72.** $-16 - 16\sqrt{3}i$

73. $2035 - 828i$ **74.** 16

75. (a) $3\left(\cos\dfrac{\pi}{4} + i\sin\dfrac{\pi}{4}\right)$

$3\left(\cos\dfrac{7\pi}{12} + i\sin\dfrac{7\pi}{12}\right)$

$3\left(\cos\dfrac{11\pi}{12} + i\sin\dfrac{11\pi}{12}\right)$

$3\left(\cos\dfrac{5\pi}{4} + i\sin\dfrac{5\pi}{4}\right)$

$3\left(\cos\dfrac{19\pi}{12} + i\sin\dfrac{19\pi}{12}\right)$

$3\left(\cos\dfrac{23\pi}{12} + i\sin\dfrac{23\pi}{12}\right)$

(b)

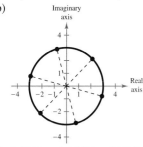

(c) $2.1213 + 2.1213i,\ -0.7765 + 2.8978i,$
$-2.8978 + 0.7765i,\ -2.1213 - 2.1213i,$
$0.7765 - 2.8978i,\ 2.8978 - 0.7765i$

76. (a) $\pm 4, \pm 4i$

(b)

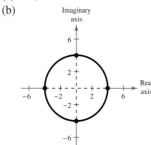

(c) $-4 + 0i, 4 + 0i, -4i, 4i$

77. $3\left(\cos\dfrac{\pi}{4} + i\sin\dfrac{\pi}{4}\right) = \dfrac{3\sqrt{2}}{2} + \dfrac{3\sqrt{2}}{2}i$

$3\left(\cos\dfrac{3\pi}{4} + i\sin\dfrac{3\pi}{4}\right) = -\dfrac{3\sqrt{2}}{2} + \dfrac{3\sqrt{2}}{2}i$

$3\left(\cos\dfrac{5\pi}{4} + i\sin\dfrac{5\pi}{4}\right) = -\dfrac{3\sqrt{2}}{2} - \dfrac{3\sqrt{2}}{2}i$

$3\left(\cos\dfrac{7\pi}{4} + i\sin\dfrac{7\pi}{4}\right) = \dfrac{3\sqrt{2}}{2} - \dfrac{3\sqrt{2}}{2}i$

78. $2(\cos 0 + i\sin 0) = 2$

$2\left(\cos\dfrac{2\pi}{5} + i\sin\dfrac{2\pi}{5}\right) \approx 0.6180 + 1.9021i$

$2\left(\cos\dfrac{4\pi}{5} + i\sin\dfrac{4\pi}{5}\right) \approx -1.6180 + 1.1756i$

$2\left(\cos\dfrac{6\pi}{5} + i\sin\dfrac{6\pi}{5}\right) \approx -1.6180 - 1.1756i$

$2\left(\cos\dfrac{8\pi}{5} + i\sin\dfrac{8\pi}{5}\right) \approx 0.6180 - 1.9021i$

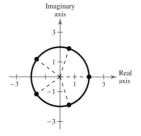

79. $2\left(\cos\dfrac{\pi}{2} + i\sin\dfrac{\pi}{2}\right) = 2i$

$2\left(\cos\dfrac{7\pi}{6} + i\sin\dfrac{7\pi}{6}\right) = -\sqrt{3} - i$

$2\left(\cos\dfrac{11\pi}{6} + i\sin\dfrac{11\pi}{6}\right) = \sqrt{3} - i$

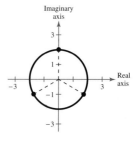

80. $\cos 0 + i \sin 0 = 1$

$\cos \dfrac{\pi}{2} + i \sin \dfrac{\pi}{2} = i$

$\cos \dfrac{2\pi}{3} + i \sin \dfrac{2\pi}{3} = -\dfrac{1}{2} + \dfrac{\sqrt{3}}{2} i$

$\cos \dfrac{4\pi}{3} + i \sin \dfrac{4\pi}{3} = -\dfrac{1}{2} - \dfrac{\sqrt{3}}{2} i$

$\cos \dfrac{3\pi}{2} + i \sin \dfrac{3\pi}{2} = -i$

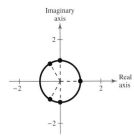

81. False.

$\sqrt{-18}\sqrt{-2} = 3\sqrt{2}i \sqrt{2}i$ and $\sqrt{(-18)(-2)} = \sqrt{36}$
$\qquad\qquad = 3\sqrt{4}i^2 \qquad\qquad\qquad\qquad = 6$
$\qquad\qquad = 6i^2$
$\qquad\qquad = -6$

82. False. The equation has no real solution. The solutions are:
$\dfrac{717}{650} \pm \dfrac{\sqrt{3311}\,i}{650}$.

83. False. A fourth-degree polynomial can have at most four zeros, and complex zeros occur in conjugate pairs.

84. Answers will vary.

85. (a) $4(\cos 60° + i \sin 60°)$ \qquad (b) -64
$4(\cos 180° + i \sin 180°)$
$4(\cos 300° + i \sin 300°)$

86. (a) $4(\cos 60° + i \sin 60°)$ \qquad (b) $-128 - 128\sqrt{3}i$
$4(\cos 150° + i \sin 150°)$
$4(\cos 240° + i \sin 240°)$
$4(\cos 330° + i \sin 330°)$

87. $z_1 z_2 = -4$, $\dfrac{z_1}{z_2} = -\cos 2\theta - i \sin 2\theta$

88. (a) Three roots
(b) On the circle 120°, 210°, and 300° from the positive real axis

Chapter Test   *(page 363)*

1. $-3 + 9i$ \qquad **2.** $-3 + 5i$ \qquad **3.** $-32 + 24i$ \qquad **4.** 7

5. $2 - i$ \qquad **6.** $\dfrac{1}{2} \pm \dfrac{\sqrt{5}}{2}i$ \qquad **7.** Five solutions

8. Four solutions \qquad **9.** $6, \pm\sqrt{5}i$ \qquad **10.** $\pm\sqrt{6}, \pm 2i$

11. $\pm 2, \pm\sqrt{2}i$; $(x + 2)(x - 2)(x + \sqrt{2}i)(x - \sqrt{2}i)$

12. $\dfrac{3}{2}, 2 \pm i$; $(2v - 3)(v - 2 - i)(v - 2 + i)$

13. $x^4 - 9x^3 + 28x^2 - 30x$

14. $x^4 - 8x^3 + 28x^2 - 60x + 63$

15. No. If $a + bi$ is a zero, its conjugate $a - bi$ is also a zero.

16. $5\sqrt{2}\left(\cos\dfrac{7\pi}{4} + i \sin\dfrac{7\pi}{4}\right)$ \qquad **17.** $-3 + 3\sqrt{3}i$

18. $-\dfrac{6561}{2} - \dfrac{6561\sqrt{3}}{2}i$ \qquad **19.** $5832i$

20. $4\sqrt[4]{2}\left(\cos\dfrac{\pi}{12} + i \sin\dfrac{\pi}{12}\right)$

$4\sqrt[4]{2}\left(\cos\dfrac{7\pi}{12} + i \sin\dfrac{7\pi}{12}\right)$

$4\sqrt[4]{2}\left(\cos\dfrac{13\pi}{12} + i \sin\dfrac{13\pi}{12}\right)$

$4\sqrt[4]{2}\left(\cos\dfrac{19\pi}{12} + i \sin\dfrac{19\pi}{12}\right)$

21. $3\left(\cos\dfrac{\pi}{6} + i \sin\dfrac{\pi}{6}\right)$

$3\left(\cos\dfrac{5\pi}{6} + i \sin\dfrac{5\pi}{6}\right)$

$3\left(\cos\dfrac{3\pi}{2} + i \sin\dfrac{3\pi}{2}\right)$

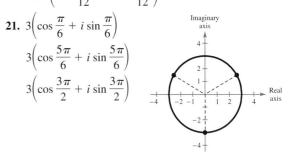

22. No. When you set $h = 125$, the resulting equation yields imaginary roots. So, the projectile will not reach a height of 125 feet.

Problem Solving   *(page 365)*

1. (a) $z^3 = 8$ for all three complex numbers.
(b) $z^3 = 27$ for all three complex numbers.
(c) The cube roots of a positive real number "a" are:
$\sqrt[3]{a}, \dfrac{-\sqrt[3]{a} + \sqrt[3]{a}\sqrt{3}i}{2}$, and $\dfrac{-\sqrt[3]{a} - \sqrt[3]{a}\sqrt{3}i}{2}$.

2. (a) $\dfrac{1}{2} - \dfrac{1}{2}i$ \quad (b) $\dfrac{3}{10} + \dfrac{1}{10}i$ \quad (c) $-\dfrac{1}{34} - \dfrac{2}{17}i$

3. $(a + bi)(a - bi) = a^2 + abi - abi - b^2 i^2$
$\qquad\qquad\qquad\qquad = a^2 + b^2$

4. Answers will vary. \qquad **5.** (a) $k > 1$ \quad (b) $k < 1$

6. (a) $0 < k < 4$ \quad (b) $k < 0$ \quad (c) $k > 4$

7. (a) No \quad (b) Yes

8. (a) $f(x) = -2x^3 + 3x^2 + 11x - 6$
(b)

(c) (Equations and graphs will vary.) There are infinitely many possible functions for f.

9. (a) Not correct because f has $(0, 0)$ as an intercept.
   (b) Not correct because the function must be at least a fourth-degree polynomial.
   (c) Correct function
   (d) Not correct because k has $(-1, 0)$ as an intercept.

10. (a) $-2, 1, 4$
   (b) The graph touches the x-axis at $x = 1$.
   (c) The least possible degree of the function is 4, because there are at least four real zeros (1 is repeated) and a function can have at most the number of real zeros equal to the degree of the function. The product of two odd powers and an even power will always be an even degree polynomial.
   (d) Positive. From the information in the table, it can be concluded that the graph will eventually rise to the left and rise to the right.
   (e) $f(x) = x^4 - 4x^3 - 3x^2 + 14x - 8$
   (f)

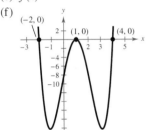

11. (a) Yes   (b) No   (c) Yes
12. (a)

Function	Zeros	Sum of zeros	Product of zeros
$f_1(x)$	$2, 3$	5	6
$f_2(x)$	$-3, 1, 2$	0	-6
$f_3(x)$	$-3, 1, \pm 2i$	-2	-12
$f_4(x)$	$0, 2, -3, 2 \pm \sqrt{3}$	3	0

   (b) The sum of the zeros is equal to the opposite of the coefficient of the $(n - 1)^{\text{th}}$ term.
   (c) The product of the zeros is equal to the constant term if the function is of an even degree and to the opposite of the constant term if the function is of an odd degree.

13. (a) $1 + i, 3 + i$ (b) $1 - i, 2 + 3i$
   (c) $1 + i, -\frac{7}{2} + 3i$ (d) $4 + 5i, -\frac{1}{3} - \frac{1}{3}i$

14. Answers will vary. Circle   **15.** Answers will vary.

Chapter 5

Section 5.1   *(page 376)*

Vocabulary Check   *(page 376)*

1. algebraic   **2.** transcendental

3. natural exponential; natural   **4.** $A = P\left(1 + \dfrac{r}{n}\right)^{nt}$

5. $A = Pe^{rt}$

1. 946.852   **2.** 3.488   **3.** 0.006   **4.** 0.544
5. 1767.767   **6.** 1.274×10^{25}
7. d   **8.** c   **9.** a   **10.** b
11.

x	-2	-1	0	1	2
$f(x)$	4	2	1	0.5	0.25

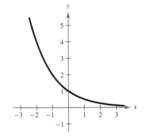

12.

x	-2	-1	0	1	2
$f(x)$	0.25	0.5	1	2	4

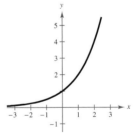

13.

x	-2	-1	0	1	2
$f(x)$	36	6	1	0.167	0.028

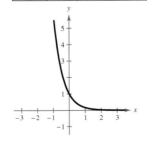

14.

x	-2	-1	0	1	2
$f(x)$	0.028	0.167	1	6	36

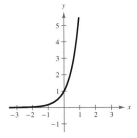

15.

x	-2	-1	0	1	2
$f(x)$	0.125	0.25	0.5	1	2

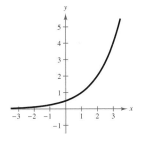

16.

x	-1	0	1	2	3
$f(x)$	3.004	3.016	3.063	3.25	4

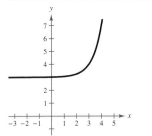

17. Shift the graph of f four units to the right.

18. Shift the graph of f one unit upward.

19. Shift the graph of f five units upward.

20. Reflect the graph of f in the y-axis and shift three units to the right.

21. Reflect the graph of f in the x-axis and y-axis and shift six units to the right.

22. Reflect f in the x-axis and shift five units upward.

23.   **24.**

25.   **26.**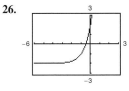

27. 0.472   **28.** 24.533   **29.** 3.857×10^{-22}

30. 1.956×10^{52} **31.** 7166.647   **32.** 679.570

33.

x	-2	-1	0	1	2
$f(x)$	0.135	0.368	1	2.718	7.389

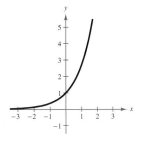

34.

x	-2	-1	0	1	2
$f(x)$	7.389	2.718	1	0.368	0.135

35.

x	-8	-7	-6	-5	-4
$f(x)$	0.055	0.149	0.406	1.104	3

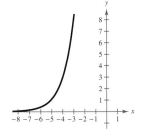

CHAPTER 5

36.

x	-2	-1	0	1	2
$f(x)$	5.437	3.297	2	1.213	0.736

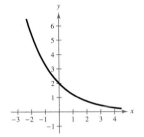

37.

x	-2	-1	0	1	2
$f(x)$	4.037	4.100	4.271	4.736	6

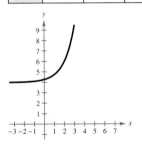

38.

x	0	2	4	5	6
$f(x)$	2.007	2.050	2.368	3	4.718

39. **40.**

41. **42.**

43. **44.**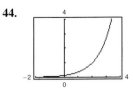

45. $x = 2$ **46.** $x = 7$ **47.** $x = -3$
48. $x = -4$ **49.** $x = \frac{1}{3}$ **50.** $x = \frac{5}{2}$
51. $x = 3, -1$ **52.** $x = 2, 3$

53.

n	1	2	4
A	\$3200.21	\$3205.09	\$3207.57

n	12	365	Continuous
A	\$3209.23	\$3210.06	\$3210.06

54.

n	1	2	4	12
A	\$1480.24	\$1485.95	\$1488.86	\$1490.83

n	365	Continuous
A	\$1491.79	\$1491.82

55.

n	1	2	4
A	\$4515.28	\$4535.05	\$4545.11

n	12	365	Continuous
A	\$4551.89	\$4555.18	\$4555.30

56.

n	1	2	4
A	\$10,285.72	\$10,640.89	\$10,828.46

n	12	365	Continuous
A	\$10,957.45	\$11,021.00	\$11,023.18

57.

t	10	20	30
A	\$17,901.90	\$26,706.49	\$39,841.40

t	40	50
A	\$59,436.39	\$88,668.67

58.

t	10	20	30
A	\$21,865.43	\$39,841.40	\$72,595.77

t	40	50
A	\$132,278.12	\$241,026.44

59.

t	10	20	30
A	$22,986.49	$44,031.56	$84,344.25

t	40	50
A	$161,564.86	$309,484.08

60.

t	10	20	30
A	$17,028.81	$24,165.03	$34,291.81

t	40	50
A	$48,662.40	$69,055.23

61. $222,822.57 **62.** $212,605.41 **63.** $35.45

64. (a)

(b) $421.12
(c) $350

65. (a) $V(1) = 10,000.298$
(b) $V(1.5) = 100,004.47$
(c) $V(2) = 1,000,059.6$

66. (a) Decreasing
(b) $P(8) = 147.58$ million
$P(10) = 146.44$ million
(c) $P(20) = 140.84$ million

67. (a) 25 grams (b) 16.21 grams
(c)

68. (a) 10 grams (b) 7.85 grams
(c)

69. (a)

(b)

x	0	25	50	75	100
Model	12.5	44.5	81.82	96.19	99.3
Actual	12	44	81	96	99

(c) 63.14% (d) 38 masses

70. (a) (b) 32,357 pascals

71. True. As $x \to -\infty$, $f(x) \to -2$ but never reaches -2.
72. False. e is an irrational number.
73. $f(x) = h(x)$ **74.** $g(x) = h(x)$
75. $f(x) = g(x) = h(x)$ **76.** None are equal.
77.

(a) $x < 0$ (b) $x > 0$

78. (a)

Decreasing: $(-\infty, 0)$, $(2, \infty)$
Increasing: $(0, 2)$
Relative maximum: $(2, 4e^{-2})$
Relative minimum: $(0, 0)$
(b)

Decreasing: $(1.44, \infty)$
Increasing: $(-\infty, 1.44)$
Relative maximum: $(1.44, 4.25)$

CHAPTER 5

79.

As $x \to \infty$, $f(x) \to g(x)$.
As $x \to -\infty$, $f(x) \to g(x)$.

80. c, d   **81.** $y = \pm\sqrt{25 - x^2}$

82. $y = x - 2$ and $y = -(x - 2)$, $x \geq 2$

83.   **84.**

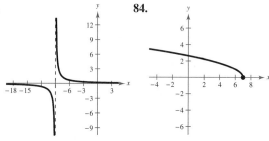

85. Answers will vary.

Section 5.2   *(page 386)*

Vocabulary Check   *(page 386)*

1. logarithmic   **2.** 10   **3.** natural; e
4. $a^{\log_a x} = x$ **5.** $x = y$

1. $4^3 = 64$ **2.** $3^4 = 81$ **3.** $7^{-2} = \frac{1}{49}$
4. $10^{-3} = \frac{1}{1000}$ **5.** $32^{2/5} = 4$ **6.** $16^{3/4} = 8$
7. $36^{1/2} = 6$ **8.** $8^{2/3} = 4$ **9.** $\log_5 125 = 3$
10. $\log_8 64 = 2$ **11.** $\log_{81} 3 = \frac{1}{4}$ **12.** $\log_9 27 = \frac{3}{2}$
13. $\log_6 \frac{1}{36} = -2$ **14.** $\log_4 \frac{1}{64} = -3$ **15.** $\log_7 1 = 0$
16. $\log 0.001 = -3$ **17.** 4   **18.** $\frac{1}{2}$ **19.** 0
20. 1   **21.** 2   **22.** -3 **23.** -0.097
24. -2.699 **25.** 1.097   **26.** 1.877   **27.** 4
28. 0   **29.** 1   **30.** 15

31.

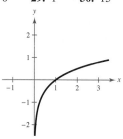

Domain: $(0, \infty)$
x-intercept: $(1, 0)$
Vertical asymptote: $x = 0$

32.

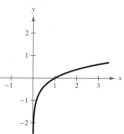

Domain: $(0, \infty)$
x-intercept: $(1, 0)$
Vertical asymptote: $x = 0$

33.

Domain: $(0, \infty)$
x-intercept: $(9, 0)$
Vertical asymptote: $x = 0$

34.

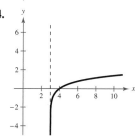

Domain: $(3, \infty)$
x-intercept: $(4, 0)$
Vertical asymptote: $x = 3$

35.

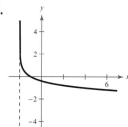

Domain: $(-2, \infty)$
x-intercept: $(-1, 0)$
Vertical asymptote: $x = -2$

36.

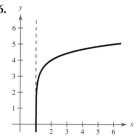

Domain: $(1, \infty)$
x-intercept: $\left(\frac{626}{625}, 0\right)$
Vertical asymptote: $x = 1$

37.

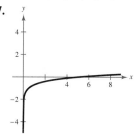

Domain: $(0, \infty)$
x-intercept: $(5, 0)$
Vertical asymptote: $x = 0$

38.

Domain: $(-\infty, 0)$
x-intercept: $(-1, 0)$
Vertical asymptote: $x = 0$

72.

Domain: $(-\infty, 3)$
x-intercept: $(2, 0)$
Vertical asymptote: $x = 3$

39. c   **40.** f   **41.** d
42. e   **43.** b   **44.** a
45. $e^{-0.693\ldots} = \frac{1}{2}$ **46.** $e^{-0.916\ldots} = \frac{2}{5}$
47. $e^{1.386\ldots} = 4$ **48.** $e^{2.302\ldots} = 10$
49. $e^{5.521\ldots} = 250$ **50.** $e^{6.520\ldots} = 679$
51. $e^0 = 1$ **52.** $e^1 = e$ **53.** $\ln 20.0855\ldots = 3$
54. $\ln 7.3890\ldots = 2$ **55.** $\ln 1.6487\ldots = \frac{1}{2}$
56. $\ln 1.3956\ldots = \frac{1}{3}$ **57.** $\ln 0.6065\ldots = -0.5$
58. $\ln 0.0165\ldots = -4.1$ **59.** $\ln 4 = x$
60. $\ln 3 = 2x$ **61.** 2.913
62. -3.418 **63.** -0.575
64. 0.693   **65.** 3   **66.** -2
67. $-\frac{2}{3}$ **68.** $-\frac{5}{2}$

73.   **74.**
75. (graph)   **76.**
77. (graph)   **78.**

79. $x = 3$ **80.** $x = 12$ **81.** $x = 7$ **82.** $x = \frac{9}{5}$
83. $x = 4$ **84.** $x = 6$ **85.** $x = -5, 5$
86. $x = -2, 3$
87. (a) 30 years; 20 years   (b) \$396,234; \$301,123.20
(c) \$246,234; \$151,123.20
(d) $x = 1000$; The monthly payment must be greater than \$1000.

69.

Domain: $(1, \infty)$
x-intercept: $(2, 0)$
Vertical asymptote: $x = 1$

70.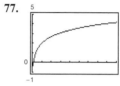

Domain: $(-1, \infty)$
x-intercept: $(0, 0)$
Vertical asymptote: $x = -1$

71.

Domain: $(-\infty, 0)$
x-intercept: $(-1, 0)$
Vertical asymptote: $x = 0$

88. (a)

K	1	2	4	6	8	10	12
t	0	7.3	14.6	18.9	21.9	24.2	26.2

The number of years required to multiply the original investment by K increases with K. However, the larger the value of K, the fewer the years required to increase the value of the investment by an additional multiple of the original investment.

(b) (graph)

89. (a)

(b) 80   (c) 68.1   (d) 62.3

90. (a) 120 decibels

(b) 100 decibels

(c) No, the difference results from the logarithmic relationship between intensity and number of decibels.

91. False. Reflecting $g(x)$ about the line $y = x$ will determine the graph of $f(x)$.

92. True. $\log_3 27 = 3 \Rightarrow 3^3 = 27$

93.   **94.**

The functions f and g are inverses.

The functions f and g are inverses.

95.   **96.**

The functions f and g are inverses.

The functions f and g are inverses.

97. (a)

$g(x)$; The natural log function grows at a slower rate than the square root function.

(b)

$g(x)$; The natural log function grows at a slower rate than the fourth root function.

98. (a)

x	1	5	10	10^2
$f(x)$	0	0.322	0.230	0.046

x	10^4	10^6
$f(x)$	0.00092	0.0000138

(b) 0

(c)

99. (a) False   (b) True   (c) True   (d) False

100. Answers will vary.

101. (a)   (b) Increasing: $(1, \infty)$

Decreasing: $(0, 1)$

(c) Relative minimum: $(1, 0)$

102. (a)   (b) Increasing: $(0, \infty)$

Decreasing: $(-\infty, 0)$

(c) Relative minimum: $(0, 0)$

103. 15   **104.** 1   **105.** 4300   **106.** -2

107. 1028   **108.** -344

Section 5.3   *(page 393)*

1. (a) $\dfrac{\log x}{\log 5}$ (b) $\dfrac{\ln x}{\ln 5}$ **2.** (a) $\dfrac{\log x}{\log 3}$ (b) $\dfrac{\ln x}{\ln 3}$

3. (a) $\dfrac{\log x}{\log \frac{1}{5}}$ (b) $\dfrac{\ln x}{\ln \frac{1}{5}}$ **4.** (a) $\dfrac{\log x}{\log \frac{1}{3}}$ (b) $\dfrac{\ln x}{\ln \frac{1}{3}}$

5. (a) $\dfrac{\log \frac{3}{10}}{\log x}$ (b) $\dfrac{\ln \frac{3}{10}}{\ln x}$ **6.** (a) $\dfrac{\log \frac{3}{4}}{\log x}$ (b) $\dfrac{\ln \frac{3}{4}}{\ln x}$

7. (a) $\dfrac{\log x}{\log 2.6}$ (b) $\dfrac{\ln x}{\ln 2.6}$ **8.** (a) $\dfrac{\log x}{\log 7.1}$ (b) $\dfrac{\ln x}{\ln 7.1}$

9. 1.771   **10.** 0.712   **11.** -2.000 **12.** -1.161

13. -0.417 **14.** -0.694 **15.** 2.633   **16.** -3.823

17. $\frac{3}{2}$ **18.** $4 + 4\log_2 3$ **19.** $-3 - \log_5 2$

20. $\log 3 - 2$ **21.** $6 + \ln 5$ **22.** $\ln 6 - 2$ **23.** 2

24. -3 **25.** $\frac{3}{4}$ **26.** $\frac{1}{3}$ **27.** 2.4   **28.** -0.8

29. -9 is not in the domain of $\log_3 x$.

30. -16 is not in the domain of $\log_2 x$.

31. 4.5   **32.** 12   **33.** $-\frac{1}{2}$ **34.** $\frac{3}{4}$ **35.** 7

36. 7   **37.** 2   **38.** 3   **39.** $\log_4 5 + \log_4 x$

40. $\log_3 10 + \log_3 z$ **41.** $4\log_8 x$ **42.** $\log y - \log 2$

43. $1 - \log_5 x$ **44.** $-3\log_6 z$ **45.** $\frac{1}{2}\ln z$

46. $\frac{1}{3}\ln t$ **47.** $\ln x + \ln y + 2\ln z$

48. $\log 4 + 2\log x + \log y$ **49.** $\ln z + 2\ln(z - 1)$

50. $\ln(x + 1) + \ln(x - 1) - 3\ln x$

51. $\frac{1}{2}\log_2(a - 1) - 2\log_2 3$ **52.** $\ln 6 - \frac{1}{2}\ln(x^2 + 1)$

53. $\frac{1}{3}\ln x - \frac{1}{3}\ln y$ **54.** $\ln x - \frac{3}{2}\ln y$

55. $4\ln x + \frac{1}{2}\ln y - 5\ln z$

56. $\frac{1}{2}\log_2 x + 4\log_2 y - 4\log_2 z$

57. $2\log_5 x - 2\log_5 y - 3\log_5 z$

58. $\log x + 4\log y - 5\log z$

59. $\frac{3}{4}\ln x + \frac{1}{4}\ln(x^2 + 3)$ **60.** $\ln x + \frac{1}{2}\ln(x + 2)$

61. $\ln 3x$ **62.** $\ln yt$ **63.** $\log_4 \dfrac{z}{y}$ **64.** $\log_5 \dfrac{8}{t}$

65. $\log_2(x + 4)^2$ **66.** $\log_7(z - 2)^{2/3}$

67. $\log_3 \sqrt[4]{5x}$ **68.** $\log_6 \dfrac{1}{16x^4}$ **69.** $\ln \dfrac{x}{(x + 1)^3}$

70. $\ln 64(z - 4)^5$ **71.** $\log \dfrac{xz^3}{y^2}$ **72.** $\log_3 \dfrac{x^3 y^4}{z^4}$

73. $\ln \dfrac{x}{(x^2 - 4)^4}$ **74.** $\ln \dfrac{z^4(z + 5)^4}{(z - 5)^2}$

75. $\ln \sqrt[3]{\dfrac{x(x + 3)^2}{x^2 - 1}}$ **76.** $\ln\left(\dfrac{x^3}{x^2 - 1}\right)^2$

77. $\log_8 \dfrac{\sqrt[3]{y(y + 4)^2}}{y - 1}$ **78.** $\log_4\left[x^6(x - 1)\sqrt{x + 1}\right]$

79. $\log_2 \frac{32}{4} = \log_2 32 - \log_2 4$; Property 2

80. $\log_7 \sqrt{70} = \frac{1}{2}(\log_7 7 + \log_7 10)$

$\phantom{\log_7 \sqrt{70}} = \frac{1}{2} + \log_7 \sqrt{10}$; Properties 1 and 3

81. $\beta = 10(\log I + 12)$; 60 dB   **82.** 24 dB difference

83. ≈ 3

84. (a) $90 - \log(t + 1)^{15}$ (b) 90

(c) 79.5   (d) 73.3

(e)

(f) 9 months   (g) $90 - 15\log 10 = 75$

85. $y = 256.24 - 20.8\ln x$

86. (a) and (b)

(c)

$T = 21 + e^{-0.037t + 3.997}$

The results are similar.

(d)

$T = 21 + \dfrac{1}{0.001t + 0.016}$

(e) Answers will vary.

87. False. $\ln 1 = 0$ **88.** True; Property 1

89. False. $\ln(x - 2) \neq \ln x - \ln 2$

90. False. $f(\sqrt{x}) = \frac{1}{2}f(x)$ **91.** False. $u = v^2$

92. True   **93 and 94.** Proofs will vary.

95. $f(x) = \dfrac{\log x}{\log 2} = \dfrac{\ln x}{\ln 2}$ **96.** $f(x) = \dfrac{\log x}{\log 4} = \dfrac{\ln x}{\ln 4}$

97. $f(x) = \dfrac{\log x}{\log \frac{1}{2}} = \dfrac{\ln x}{\ln \frac{1}{2}}$ **98.** $f(x) = \dfrac{\log x}{\log \frac{1}{4}} = \dfrac{\ln x}{\ln \frac{1}{4}}$

99. $f(x) = \dfrac{\log x}{\log 11.8} = \dfrac{\ln x}{\ln 11.8}$

100. $f(x) = \dfrac{\log x}{\log 12.4} = \dfrac{\ln x}{\ln 12.4}$

101. $f(x) = h(x)$; Property 2

102.
$\ln 2 \approx 0.6931$	$\ln 10 \approx 2.3025$
$\ln 3 \approx 1.0986$	$\ln 12 \approx 2.4848$
$\ln 4 \approx 1.3862$	$\ln 15 \approx 2.7080$
$\ln 5 \approx 1.6094$	$\ln 16 \approx 2.7724$
$\ln 6 \approx 1.7917$	$\ln 18 \approx 2.8903$
$\ln 8 \approx 2.0793$	$\ln 20 \approx 2.9956$
$\ln 9 \approx 2.1972$	

103. $\dfrac{3x^4}{2y^3}$, $x \neq 0$ **104.** $\dfrac{27y^3}{8x^6}$ **105.** $1, x \neq 0, y \neq 0$

106. $\dfrac{(xy)^2}{x + y}$ **107.** $-1, \dfrac{1}{3}$ **108.** $1, \dfrac{1}{4}$

109. $\dfrac{-1 \pm \sqrt{97}}{6}$ **110.** $\dfrac{1 \pm \sqrt{31}}{2}$

Section 5.4 (page 403)

1. (a) Yes (b) No **2.** (a) No (b) No
3. (a) No (b) Yes (c) Yes, approximate
4. (a) Yes (b) No (c) Yes, approximate
5. (a) Yes, approximate (b) No (c) Yes
6. (a) Yes (b) No (c) No
7. (a) No (b) Yes (c) Yes, approximate
8. (a) Yes (b) Yes, approximate (c) No
9. 2 **10.** 5 **11.** -5 **12.** -3 **13.** 2
14. 5 **15.** $\ln 2 \approx 0.693$ **16.** $\ln 4 \approx 1.386$
17. $e^{-1} \approx 0.368$ **18.** $e^{-7} \approx 0.000912$ **19.** 64
20. 0.008 **21.** $(3, 8)$ **22.** $\left(\frac{2}{3}, 9\right)$ **23.** $(9, 2)$
24. $(5, 0)$ **25.** $2, -1$ **26.** $-2, 4$

27. $\approx 1.618, \approx -0.618$ **28.** 0, 1 **29.** $\dfrac{\ln 5}{\ln 3} \approx 1.465$

30. $\dfrac{\ln 16}{\ln 5} \approx 1.723$ **31.** $\ln 5 \approx 1.609$

32. $\ln \dfrac{91}{4} \approx 3.125$ **33.** $\ln 28 \approx 3.332$

34. $\dfrac{\ln 37}{\ln 6} \approx 2.015$ **35.** $\dfrac{\ln 80}{2 \ln 3} \approx 1.994$

36. $\dfrac{\ln 3000}{5 \ln 6} \approx 0.894$ **37.** 2 **38.** $-\dfrac{\ln(0.10)}{3 \ln 4} \approx 0.554$

39. 4 **40.** 8 **41.** $3 - \dfrac{\ln 565}{\ln 2} \approx -6.142$

42. $\dfrac{-\ln 64 - \ln 431}{\ln 8} \approx -4.917$ **43.** $\dfrac{1}{3} \log\left(\dfrac{3}{2}\right) \approx 0.059$

44. $6 + \log \dfrac{7}{5} \approx 6.146$ **45.** $1 + \dfrac{\ln 7}{\ln 5} \approx 2.209$

46. $6 - \dfrac{\ln 5}{\ln 3} \approx 4.535$ **47.** $\dfrac{\ln 12}{3} \approx 0.828$

48. $\dfrac{\ln 50}{2} \approx 1.956$ **49.** $-\ln \dfrac{3}{5} \approx 0.511$

50. $-\dfrac{1}{4} \ln \dfrac{3}{40} \approx 0.648$ **51.** 0 **52.** $\ln \dfrac{25}{3} \approx 2.120$

53. $\dfrac{\ln \frac{8}{3}}{3 \ln 2} + \dfrac{1}{3} \approx 0.805$ **54.** $3 - \dfrac{\ln \frac{7}{2}}{2 \ln 4} \approx 2.548$

55. $\ln 5 \approx 1.609$ **56.** $\ln 2 \approx 0.693$; $\ln 3 \approx 1.099$
57. $\ln 4 \approx 1.386$ **58.** No solution
59. $2 \ln 75 \approx 8.635$ **60.** $\ln 7 \approx 1.946$

61. $\dfrac{1}{2} \ln 1498 \approx 3.656$ **62.** $\dfrac{\ln 31}{6} \approx 0.572$

63. $\dfrac{\ln 4}{365 \ln\left(1 + \frac{0.065}{365}\right)} \approx 21.330$ **64.** $\dfrac{\ln 21}{9 \ln 3.938225} \approx 0.247$

65. $\dfrac{\ln 2}{12 \ln\left(1 + \frac{0.10}{12}\right)} \approx 6.960$ **66.** $\dfrac{\ln 30}{3 \ln\left(16 - \frac{0.878}{26}\right)} \approx 0.409$

67.

-0.427

68.

-2.322

69.

3.847

70.

-0.478

71.

12.207

72.

1.081

73.
16.636

74.
1.236

75. $e^{-3} \approx 0.050$ **76.** $e^2 \approx 7.389$ **77.** $\dfrac{e^{2.4}}{2} \approx 5.512$

78. $\dfrac{e}{4} \approx 0.680$ **79.** $1,000,000$ **80.** $\dfrac{100}{3} \approx 33.333$

81. $\dfrac{e^{10/3}}{5} \approx 5.606$ **82.** $e^{7/2} \approx 33.115$

83. $e^2 - 2 \approx 5.389$ **84.** $e^{10} + 8 = 22,034.466$

85. $e^{-2/3} \approx 0.513$ **86.** $e^{-4/3} \approx 0.264$

87. $2(3^{11/6}) \approx 14.988$ **88.** $10^{11/5} + 2 \approx 160.489$

89. No solution   **90.** $\dfrac{-1 + \sqrt{1 + 4e}}{2} \approx 1.223$

91. $1 + \sqrt{1 + e} \approx 2.928$ **92.** $\dfrac{-3 + \sqrt{9 + 4e}}{2} \approx 0.729$

93. No solution   **94.** $\dfrac{3 + \sqrt{13}}{2} \approx 3.303$ **95.** 7

96. No solution   **97.** $\dfrac{-1 + \sqrt{17}}{2} \approx 1.562$ **98.** 2

99. 2   **100.** 9   **101.** $\dfrac{725 + 125\sqrt{33}}{8} \approx 180.384$

102. $\dfrac{1225 + 125\sqrt{73}}{2} \approx 1146.500$

103.
2.807

104.
2.197

105.
20.086

106.
14.182

107. (a) 8.2 years   (b) 12.9 years
108. (a) 5.8 years   (b) 9.2 years
109. (a) 1426 units   (b) 1498 units
110. (a) 303 units   (b) 528 units
111. (a)

(b) $V = 6.7$; The yield will approach 6.7 million cubic feet per acre.
(c) 29.3 years

112. 12.76 inches   **113.** 2001   **114.** 2001
115. (a) $y = 100$ and $y = 0$; The range falls between 0% and 100%.
(b) Males: 69.71 inches   Females: 64.51 inches
116. (a)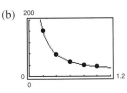

(b) Horizontal asymptotes: $P = 0$, $P = 0.83$
The proportion of correct responses will approach 0.83 as the number of trials increases.
(c) ≈ 5 trials

117. (a)

x	0.2	0.4	0.6	0.8	1.0
y	162.6	78.5	52.5	40.5	33.9

(b)

The model appears to fit the data well.
(c) 1.2 meters
(d) No. According to the model, when the number of g's is less than 23, x is between 2.276 meters and 4.404 meters, which isn't realistic in most vehicles.

118. (a) $T = 20$; Room temperature   (b) ≈ 0.81 hour
119. $\log_b uv = \log_b u + \log_b v$
True by Property 1 in Section 5.3.
120. $\log_b(u + v) = (\log_b u)(\log_b v)$
False.
$2.04 \approx \log(10 + 100) \neq (\log 10)(\log 100) = 2$
121. $\log_b(u - v) = \log_b u - \log_b v$
False.
$1.95 \approx \log(100 - 10) \neq \log 100 - \log 10 = 1$
122. $\log_b \dfrac{u}{v} = \log_b u - \log_b v$

True by Property 2 in Section 5.3.
123. Yes. See Exercise 93.
124. For $rt < \ln 2$ years, double the amount you invest. For $rt > \ln 2$ years, double your interest rate or double the number of years, because either of these will double the exponent in the exponential function.
125. Yes. Time to double: $t = \dfrac{\ln 2}{r}$;

Time to quadruple: $t = \dfrac{\ln 4}{r} = 2\left(\dfrac{\ln 2}{r}\right)$

126. Answers will vary.   **127.** $4|x|y^2\sqrt{3y}$

128. $4\sqrt{2} - 10$ **129.** $5\sqrt[3]{3}$ **130.** $\frac{1}{2}\sqrt{10} + 1$

131. **132.**

133. **134.**

135. 1.226 **136.** 1.262 **137.** -5.595 **138.** 1.486

Section 5.5 *(page 414)*

Vocabulary Check *(page 414)*

1. $y = ae^{bx}$; $y = ae^{-bx}$
2. $y = a + b \ln x$; $y = a + b \log x$
3. normally distributed **4.** bell; average value
5. sigmoidal

1. c **2.** e **3.** b **4.** a **5.** d **6.** f

	Initial Investment	Annual % Rate	Time to Double	Amount After 10 years
7.	$1000	3.5%	19.8 yr	$1419.07
8.	$750	10.5%	6.60 yr	$2143.24
9.	$750	8.9438%	7.75 yr	$1834.33
10.	$10,000	5.7762%	12 yr	$17,817.93
11.	$500	11.0%	6.3 yr	$1505.00
12.	$600	34.66%	2 yr	$19,205.00
13.	$6376.28	4.5%	15.4 yr	$10,000.00
14.	$1637.46	2%	34.7 yr	$2000.00

15. $112,087.09 **16.** $4214.16
17. (a) 6.642 years (b) 6.330 years
(c) 6.302 years (d) 6.301 years
18. (a) 6.94 years (b) 6.63 years
(c) 6.602 years (d) 6.601 years
19.

r	2%	4%	6%	8%	10%	12%
t	54.93	27.47	18.31	13.73	10.99	9.16

20.

Use PwrReg: $t = 1.099r^{-1}$

21.

r	2%	4%	6%	8%	10%	12%
t	55.48	28.01	18.85	14.27	11.53	9.69

22.

Use PwrReg: $t = 1.222r^{-1}$

23.

Continuous compounding

24.

Daily compounding

	Half-life (years)	Initial Quantity	Amount After 1000 Years
25.	1599	10 g	6.48 g
26.	1599	2.31 g	1.5 g
27.	5715	2.26 g	2 g
28.	5715	3 g	2.66 g
29.	24,100	2.16 g	2.1 g
30.	24,100	0.41 g	0.4 g

31. $y = e^{0.7675x}$ **32.** $y = \frac{1}{2}e^{0.5756x}$
33. $y = 5e^{-0.4024x}$ **34.** $y = e^{-0.4621x}$
35. (a) Decreasing due to the negative exponent.
(b) 2000: population of 2430 thousand
2003: population of 2408.95 thousand
(c) 2018
36. (a) Bulgaria: $y = 7.8e^{-0.00940t}$; 5.9 million
Canada: $y = 31.3e^{0.00915t}$; 41.2 million
China: $y = 1268.9e^{0.00602t}$; 1520.1 million
United Kingdom: $y = 59.5e^{0.00282t}$; 64.8 million
United States: $y = 282.3e^{0.00910t}$; 370.9 million
(b) b; The greater the rate of growth, the greater the value of b.
(c) b determines whether the population is increasing ($b > 0$) or decreasing ($b < 0$).
37. $k = 0.2988$; $\approx 5,309,734$ hits
38. $k = 0.1337$; $144.98 million **39.** 3.15 hours

40. 61.16 hours

41. (a) ≈ 12,180 years old   (b) ≈ 4797 years old

42. 15,642 years

43. (a) $V = -6394t + 30,788$ (b) $V = 30,788e^{-0.268t}$

(c)

The exponential model depreciates faster.

(d)

t	1	3
$V = -6394t + 30,788$	24,394	11,606
$V = 30,788e^{-0.268t}$	23,550	13,779

(e) Answers will vary.

44. (a) $V = -300t + 1150$ (b) $V = 1150e^{-0.368799t}$

(c)

The exponential model depreciates faster.

(d)

t	1 year	3 years
$V = -300t + 1150$	850	250
$V = 1150e^{-0.368799t}$	795	380

(e) Answers will vary.

45. (a) $S(t) = 100(1 - e^{-0.1625t})$

(b)   (c) 55,625

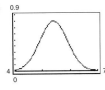

46. (a) $N = 30(1 - e^{-0.050t})$ (b) 36 days

47. (a)   (b) 100

48. (a)   (b) 5.4 hours per week

49. (a) 203 animals   (b) 13 years

(c)

Horizontal asymptotes: $y = 0$, $y = 1000$. The population size will approach 1000 as time increases.

50. (a) $S = \dfrac{500,000}{1 + 0.6e^{0.026t}}$ (b) 287,567 units sold

51. (a) $10^{7.9} \approx 79,432,823$ (b) $10^{8.3} \approx 199,526,231$

(c) $10^{4.2} \approx 15,849$

52. (a) 7.91   (b) 7.68   (c) 5.40

53. (a) 20 decibels   (b) 70 decibels

(c) 40 decibels   (d) 120 decibels

54. (a) 10 decibels   (b) 140 decibels

(c) 80 decibels   (d) 100 decibels

55. 95%   **56.** 97%   **57.** 4.64   **58.** 4.95

59. 1.58×10^{-6} moles per liter

60. $10^{-3.2} \approx 6.3 \times 10^{-4}$ moles per liter   **61.** $10^{5.1}$

62. 10   **63.** 3:00 A.M.

64. (a)   (b) Interest; $t \approx 26$ years

(c)

Interest; $t \approx 11$ years; The interest is still the majority of the monthly payment in the early years, but now the principal and interest are nearly equal when $t \approx 11$ years.

65. (a) 150,000

(b) ≈ 21 years; Yes

66. (a) $t_3 = 0.2729s - 6.0143$

$t_4 = 1.5385e^{0.02913s}$ or $t_4 = 1.5385(1.0296)^s$

(b)

(c)

s	30	40	50	60	70	80	90
t_1	3.6	4.6	6.7	9.4	12.5	15.9	19.6
t_2	3.3	4.9	7.0	9.5	12.5	15.9	19.9
t_3	2.2	4.9	7.6	10.4	13.1	15.8	18.5
t_4	3.7	4.9	6.6	8.8	11.8	15.8	21.2

(d) Model t_1: Sum = 2.0
    Model t_2: Sum = 1.1
    Model t_3: Sum = 5.6
    Model t_4: Sum = 2.7
    The quadratic (model t_2) fits best.

67. False. The domain can be the set of real numbers for a logistic growth function.

68. False. A logistic growth function never has an x-intercept.

69. False. The graph of $f(x)$ is the graph of $g(x)$ shifted upward five units.

70. True. The graph of a Gaussian model will never have an x-intercept.

71. (a) Logarithmic   (b) Logistic   (c) Exponential
    (d) Linear   (e) None of the above   (f) Exponential

72. Answers will vary.

73. (a)
    (b) $\sqrt{10}$
    (c) $\left(-\frac{1}{2}, \frac{7}{2}\right)$
    (d) 3

74. (a)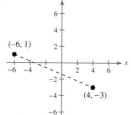
    (b) $2\sqrt{29}$
    (c) $(-1, -1)$
    (d) $-\frac{2}{5}$

75. (a)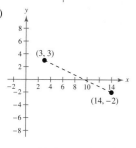
    (b) $\sqrt{146}$
    (c) $\left(\frac{17}{2}, \frac{1}{2}\right)$
    (d) $-\frac{5}{11}$

76. (a)
    (b) 5
    (c) $\left(\frac{17}{2}, 2\right)$
    (d) $\frac{4}{3}$

77. (a)
    (b) $\sqrt{\frac{1}{8}}$
    (c) $\left(\frac{5}{8}, -\frac{1}{8}\right)$
    (d) 1

78. (a)
    (b) $\sqrt{9.25}$
    (c) $\left(\frac{5}{6}, -\frac{1}{12}\right)$
    (d) $\frac{1}{6}$

79.

80.

81.

82.

83.

84.

85.

86.

87.

88.

89.

90.

91.

92.

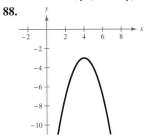

93. Answers will vary.

Review Exercises   *(page 421)*

1. 76.699   **2.** 361.784   **3.** 0.337
4. 4.181   **5.** 1456.529   **6.** −3.863
7. c   **8.** d   **9.** a   **10.** b
11. Shift the graph of *f* one unit to the right.
12. Shift the graph of *f* three units downward.

13. Reflect *f* in the *x*-axis and shift two units to the left.
14. Reflect *f* in the *x*-axis and shift eight units upward.

15.

x	-1	0	1	2	3
$f(x)$	8	5	4.25	4.063	4.016

16.

x	-2	-1	0	1	2
$f(x)$	-3.063	-3.25	-4	-7	-19

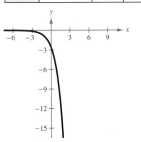

17.

x	-2	-1	0	1	2
$f(x)$	-0.377	-1	-2.65	-7.023	-18.61

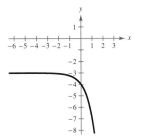

18.

x	-3	-1	0	1	3
$f(x)$	0.020	0.142	0.377	1	7.023

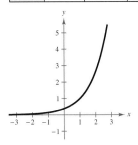

CHAPTER 5

19.

x	-1	0	1	2	3
$f(x)$	4.008	4.04	4.2	5	9

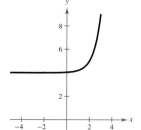

20.

x	0	5	6	7	8	9
$f(x)$	-4.984	-4.5	-4	-3	-1	3

21.

x	-2	-1	0	1	2
$f(x)$	3.25	3.5	4	5	7

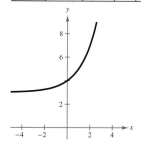

22.

x	-3	-2	-1	0	2
$f(x)$	3	-4	-4.875	-4.984	-5

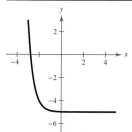

31.

x	-2	-1	0	1	2
$h(x)$	2.72	1.65	1	0.61	0.37

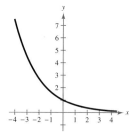

32.

x	-2	-1	0	1	2
$h(x)$	-0.72	0.35	1	1.39	1.63

33.

x	-3	-2	-1	0	1
$f(x)$	0.37	1	2.72	7.39	20.09

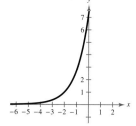

34.

t	$\frac{1}{2}$	1	2	3	4
$s(t)$	0.07	0.54	1.47	2.05	2.43

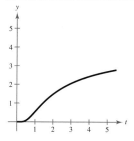

23. $x = -4$ **24.** $x = -2$ **25.** $x = \frac{22}{5}$ **26.** $x = \frac{11}{2}$

27. 2980.958   **28.** 1.868   **29.** 0.183   **30.** 1.320

35.

n	1	2	4	12
A	\$6569.98	\$6635.43	\$6669.46	\$6692.64

n	365	Continuous
A	\$6704.00	\$6704.39

36.

n	1	2	4	12
A	\$8643.88	\$8799.58	\$8880.43	\$8935.49

n	365	Continuous
A	\$8962.46	\$8963.38

37. (a) 0.154   (b) 0.487   (c) 0.811
38. (a)

   (b) \$7875

   (c) At the beginning. Yes
39. (a) \$1,069,047.14   (b) 7.9 years
40. (a) 100 grams   (b) 61.8 grams
   (c)

41. $\log_4 64 = 3$ **42.** $\log_{25} 125 = \frac{3}{2}$
43. $\ln 2.2255\ldots = 0.8$ **44.** $\ln 1 = 0$
45. 3   **46.** $\frac{1}{2}$ **47.** -3 **48.** -1 **49.** $x = 7$
50. $x = 5$ **51.** $x = -5$ **52.** $x = 6$
53. Domain: $(0, \infty)$
   x-intercept: $(1, 0)$
   Vertical asymptote: $x = 0$

54. Domain: $(0, \infty)$
   x-intercept: $(1, 0)$
   Vertical asymptote: $x = 0$

55. Domain: $(0, \infty)$
   x-intercept: $(3, 0)$
   Vertical asymptote: $x = 0$

56. Domain: $(0, \infty)$
   x-intercept: $(10^{-6}, 0)$
   Vertical asymptote: $x = 0$

57. Domain: $(-5, \infty)$
   x-intercept: $(9995, 0)$
   Vertical asymptote: $x = -5$

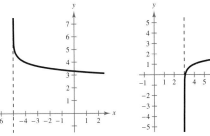

58. Domain: $(3, \infty)$
   x-intercept: $(3.1, 0)$
   Vertical asymptote: $x = 3$

59. 3.118   **60.** -0.020 **61.** -12 **62.** 7
63. 2.034   **64.** -1.530
65. Domain: $(0, \infty)$
   x-intercept: $(e^{-3}, 0)$
   Vertical asymptote: $x = 0$

66. Domain: $(3, \infty)$
   x-intercept: $(4, 0)$
   Vertical asymptote: $x = 3$

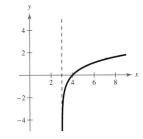

67. Domain: $(-\infty, 0), (0, \infty)$
   x-intercept: $(\pm 1, 0)$
   Vertical asymptote: $x = 0$

68. Domain: $(0, \infty)$
   x-intercept: $(1, 0)$
   Vertical asymptote: $x = 0$

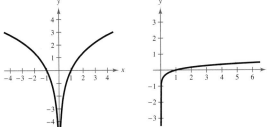

69. 53.4 inches   **70.** 27.16 miles   **71.** 1.585
72. 2.132   **73.** -2.322 **74.** -1.159

CHAPTER 5

75. $\log 2 + 2 \log 3 \approx 1.255$

76. $\dfrac{\log 1 - \log 3 - 2 \log 2}{\log 2} \approx -3.585$

77. $2 \ln 2 + \ln 5 \approx 2.996$ **78.** $\ln(3) - 4 \approx -2.901$

79. $1 + 2 \log_5 x$ **80.** $\log 7 + 4 \log x$

81. $1 + \log_3 2 - \frac{1}{3} \log_3 x$ **82.** $\frac{1}{2} \log_7 x - \log_7 4$

83. $2 \ln x + 2 \ln y + \ln z$ **84.** $\ln 3 + \ln x + 2 \ln y$

85. $\ln(x + 3) - \ln x - \ln y$ **86.** $2 \ln(y - 1) - \ln 16$

87. $\log_2 5x$ **88.** $\log_6 \dfrac{y}{z^2}$ **89.** $\ln \dfrac{x}{\sqrt[4]{y}}$

90. $\ln x^3(x + 1)^2$ **91.** $\log_8 y^7 \sqrt[3]{x + 4}$

92. $\log \dfrac{1}{x^2(x + 6)^5}$ **93.** $\ln \dfrac{\sqrt{2x - 1}}{(x + 1)^2}$

94. $\ln \dfrac{(x - 2)^5}{x^3(x + 2)}$

95. (a) $0 \le h < 18,000$

(b)

Vertical asymptote: $h = 18,000$

(c) The plane is climbing at a slower rate, so the time required increases.

(d) 5.46 minutes

96. $s = 84.66 - 11 \ln t$ **97.** 3   **98.** -3

99. $\ln 3 \approx 1.099$ **100.** $\ln 6$ **101.** 16

102. $\frac{1}{6}$ **103.** $e^4 \approx 54.598$ **104.** e^{-3}

105. $\ln 12 \approx 2.485$ **106.** $\dfrac{\ln 25}{3} \approx 1.073$

107. $x = 1, 3$ **108.** $\frac{1}{3}(\ln 40 - 2) \approx 0.563$

109. $\dfrac{\ln 22}{\ln 2} \approx 4.459$ **110.** $\dfrac{\ln 20}{\ln 6} \approx 1.672$

111. $\dfrac{\ln 17}{\ln 5} \approx 1.760$ **112.** $\dfrac{\ln 95}{\ln 12} \approx 1.833$

113. $\ln 2 \approx 0.693$, $\ln 5 \approx 1.609$

114. $\ln 2 \approx 0.693$, $\ln 4 \approx 1.386$

115.

7.480; 0.392

116.

-1.527; -7.038

117.

2.447

118.

0.676

119. $\frac{1}{3}e^{8.2} \approx 1213.650$ **120.** $\frac{1}{5}e^{7.2} \approx 267.886$

121. $\frac{1}{4}e^{7.5} \approx 452.011$ **122.** $\frac{1}{3}e^{15/4} \approx 14.174$

123. $3e^2 \approx 22.167$ **124.** $e^6 - 8 \approx 395.429$

125. $e^4 - 1 \approx 53.598$ **126.** $5e^4 \approx 272.991$

127. No solution   **128.** $-2 + \sqrt{6} \approx 0.449$

129. 0.900   **130.** -104

131.

1.643

132.

0, 0.416, 13.627

133.

No solution

134.

-3.990, 1.477

135. 15.2 years   **136.** $d \approx 221$ miles   **137.** e

138. b   **139.** f   **140.** d   **141.** a   **142.** c

143. $y = 2e^{0.1014x}$ **144.** $y = \frac{1}{2}e^{0.4605x}$

145. 2008   **146.** 98.6%

147. (a) 13.8629%   (b) \$11,486.98   **148.** 1243

149. (a)

(b) 71

150. (a) 7.7 weeks   (b) 13.3 weeks

151. $10^{-3.5}$ watt per square centimeter

152. (a) 251,188,643   (b) 7,079,458   (c) 1,258,925,412

153. True by the inverse properties

154. False. $\ln x + \ln y = \ln(xy) \ne \ln(x + y)$

155. b and d are negative.

a and c are positive.

Answers will vary.

Chapter Test   *(page 425)*

1. 1123.690   **2.** 687.291   **3.** 0.497   **4.** 22.198

5.

x	-1	$-\frac{1}{2}$	0	$\frac{1}{2}$	1
$f(x)$	10	3.162	1	0.316	0.1

6.

x	-1	0	1	2	3
$f(x)$	-0.005	-0.028	-0.167	-1	-6

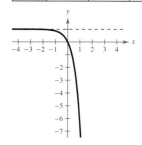

7.

x	-1	$-\frac{1}{2}$	0	$\frac{1}{2}$	1
$f(x)$	0.865	0.632	0	-1.718	-6.389

8. (a) -0.89 (b) 9.2

9.

x	$\frac{1}{2}$	1	$\frac{3}{2}$	2	4
$f(x)$	-5.699	-6	-6.176	-6.301	-6.602

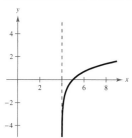

Vertical asymptote: $x = 0$

10.

x	5	7	9	11	13
$f(x)$	0	1.099	1.609	1.946	2.197

Vertical asymptote: $x = 4$

11.

x	-5	-3	-1	0	1
$f(x)$	1	2.099	2.609	2.792	2.946

Vertical asymptote: $x = -6$

12. 1.945 **13.** 0.115 **14.** 1.328

15. $\log_2 3 + 4\log_2 |a|$ **16.** $\ln 5 + \frac{1}{2}\ln x - \ln 6$

17. $(\log 7 + 2\log x) - (\log y + 3\log z)$

18. $\log_3 13y$ **19.** $\ln \dfrac{x^4}{y^4}$ **20.** $\ln\dfrac{x^2(x-5)}{y^3}$

21. $x = -2$ **22.** $x = \dfrac{\ln 44}{-5} \approx -0.757$

23. $\dfrac{\ln 197}{4} \approx 1.321$ **24.** $e^{1/2} \approx 1.649$

25. $e^{-11/4} \approx 0.0639$ **26.** $\frac{800}{501} \approx 1.597$

27. $y = 2745e^{0.1570x}$ **28.** 55%

29. (a)

x	$\frac{1}{4}$	1	2	4	5	6
H	58.720	75.332	86.828	103.43	110.59	117.38

(b) 103 centimeters; 103.43 centimeters

Problem Solving *(page 427)*

1.

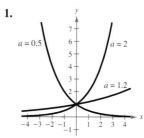

$y = 0.5^x$ and $y = 1.2^x$

$0 \le a \le 1.44$

2.

y_3 increases at the fastest rate.

3. As $x \to \infty$, the graph of e^x increases at a greater rate than the graph of x^n.

4–6. Answers will vary.

7. (a)

(b)

(c)

8. $y_4 = 1 + \dfrac{x}{1!} + \dfrac{x^2}{2!} + \dfrac{x^3}{3!} + \dfrac{x^4}{4!}$

The pattern implies that

$e^x = 1 + \dfrac{x}{1!} + \dfrac{x^2}{2!} + \dfrac{x^3}{3!} + \cdots$

9.

$f^{-1}(x) = \ln\!\left(\dfrac{x + \sqrt{x^2 + 4}}{2}\right)$

10. $f^{-1}(x) = \dfrac{\ln\!\left(\dfrac{x + 1}{x - 1}\right)}{\ln a}$ **11.** c

12. (a) The upper graph represents the investment compounded annually and the lower graph represents simple interest.

(b) $y_1 = 500(1.07)^t$
$y_2 = 35t + 500$

13. $t = \dfrac{\ln c_1 - \ln c_2}{\left(\dfrac{1}{k_2} - \dfrac{1}{k_1}\right)\ln \dfrac{1}{2}}$

14. $B = 500\!\left(\frac{2}{5}\right)^{t/2}$

15. (a) $y_1 = 252{,}606(1.0310)^t$

(b) $y_2 = 400.88t^2 - 1464.6t + 291{,}782$

(c)

(d) The exponential model is a better fit. No, because the model is rapidly approaching infinity.

16. Answers will vary.   **17.** $1, e^2$

18. (a)

(b)

(c)

19. $y_4 = (x - 1) - \frac{1}{2}(x - 1)^2 + \frac{1}{3}(x - 1)^3 - \frac{1}{4}(x - 1)^4$

The pattern implies that
$\ln x = (x - 1) - \frac{1}{2}(x - 1)^2 + \frac{1}{3}(x - 1)^3 - \cdots$

20. Slope $= \ln b$; y-intercept: $(0, \ln a)$
Slope $= b$; y-intercept: $(0, \ln a)$

21.

17.7 cubic feet per minute

22. (a) 15 cubic feet per minute
   (b) 382.0 cubic feet
   (c) 382.0 square feet

23. (a)

(b)–(e) Answers will vary.

24. (a)

(b)–(e) Answers will vary.

25. (a)

(b)–(e) Answers will vary.

26. (a)

(b)–(e) Answers will vary.

Chapter 6
Section 6.1   *(page 434)*

Vocabulary Check   *(page 434)*

1. inclination   **2.** $\tan \theta$

3. $\left| \dfrac{m_2 - m_1}{1 + m_1 m_2} \right|$ **4.** $\dfrac{|Ax_1 + By_1 + C|}{\sqrt{A^2 + B^2}}$

---

1. $\dfrac{\sqrt{3}}{3}$ **2.** 1   **3.** -1 **4.** $-\sqrt{3}$ **5.** $\sqrt{3}$

6. $-\dfrac{\sqrt{3}}{3}$ **7.** 3.2236   **8.** -0.2677

9. $\dfrac{3\pi}{4}$ radians, $135°$ **10.** 2.0344 radians, $116.6°$

11. $\dfrac{\pi}{4}$ radian, $45°$ **12.** 1.1071 radians, $63.4°$

13. 0.6435 radian, $36.9°$ **14.** 1.9513 radians, $111.8°$
15. 1.0517 radians, $60.3°$ **16.** 0.6023 radian, $34.5°$
17. 2.1112 radians, $121.0°$ **18.** 2.0344 radians, $116.6°$
19. 1.2490 radians, $71.6°$ **20.** 2.4669 radians, $141.3°$

21. 2.1112 radians, $121.0°$ **22.** $\dfrac{\pi}{4}$ radian, $45°$

23. 1.1071 radians, $63.4°$ **24.** $\dfrac{\pi}{4}$ radian, $45°$

25. 0.1974 radian, $11.3°$ **26.** 1.1071 radians, $63.4°$
27. 1.4289 radians, $81.9°$ **28.** 1.4109 radians, $80.8°$
29. 0.9273 radian, $53.1°$ **30.** 1.0808 radians, $61.9°$
31. 0.8187 radian, $46.9°$ **32.** 1.0240 radians, $58.7°$
33. $(2, 1) \leftrightarrow (4, 4)$: slope $= \frac{3}{2}$
   $(4, 4) \leftrightarrow (6, 2)$: slope $= -1$
   $(6, 2) \leftrightarrow (2, 1)$: slope $= \frac{1}{4}$
   $(2, 1)$: $42.3°$; $(4, 4)$: $78.7°$; $(6, 2)$: $59.0°$
34. $(-3, 2) \leftrightarrow (1, 3)$: slope $= \frac{1}{4}$
   $(1, 3) \leftrightarrow (2, 0)$: slope $= -3$
   $(2, 0) \leftrightarrow (-3, 2)$: slope $= -\frac{2}{5}$
   $(-3, 2)$: $35.8°$; $(1, 3)$: $94.4°$; $(2, 0)$: $49.8°$
35. $(-4, -1) \leftrightarrow (3, 2)$: slope $= \frac{3}{7}$
   $(3, 2) \leftrightarrow (1, 0)$: slope $= 1$
   $(1, 0) \leftrightarrow (-4, -1)$: slope $= \frac{1}{5}$
   $(-4, -1)$: $11.9°$; $(3, 2)$: $21.8°$; $(1, 0)$: $146.3°$
36. $(-2, 2) \leftrightarrow (-3, 4)$: slope $= -2$
   $(-3, 4) \leftrightarrow (2, 1)$: slope $= -\frac{3}{5}$
   $(2, 1) \leftrightarrow (-2, 2)$: slope $= -\frac{1}{4}$
   $(-3, 4)$: $32.5°$; $(2, 1)$: $16.9°$; $(-2, 2)$: $130.6°$

37. 0   **38.** $\dfrac{4\sqrt{5}}{5} \approx 1.7889$ **39.** $\dfrac{7}{5}$

40. $\dfrac{5\sqrt{2}}{2} \approx 3.5355$ **41.** 7   **42.** 4

43. $\dfrac{8\sqrt{37}}{37} \approx 1.3152$ **44.** $9\sqrt{2} \approx 12.7279$

45. (a)   (b) 4   (c) 8

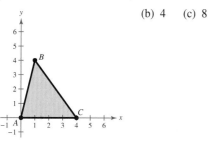

46. (a)   (b) $\dfrac{33\sqrt{29}}{29}$ (c) $\dfrac{33}{2}$

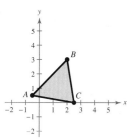

47. (a)   (b) $\dfrac{35\sqrt{37}}{74}$ (c) $\dfrac{35}{8}$

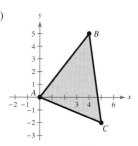

48. (a)   (b) $\dfrac{31\sqrt{389}}{389}$ (c) $\dfrac{31}{2}$

49. $2\sqrt{2}$ **50.** $\dfrac{9}{5}$ **51.** 0.1003, 1054 feet
52. 0.2027, 1049 feet   **53.** 31.0°
54. (a)

(b) 18.4°
(c) 15.8 meters
55. $\alpha \approx 33.69°$; $\beta \approx 56.31°$
56. (a) 0.6167 radian, 35.3°   (b) 518.5 feet
(c) $y = 0.709x$

(d)

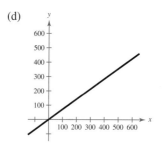

57. True. The inclination of a line is related to its slope by $m = \tan\theta$. If the angle is greater than $\pi/2$ but less than π, then the angle is in the second quadrant, where the tangent function is negative.
58. False. Substitute $\tan\theta_1$ and $\tan\theta_2$ for m_1 and m_2 in the formula for the angle between two lines.
59. (a) $d = \dfrac{4}{\sqrt{m^2 + 1}}$

(b)

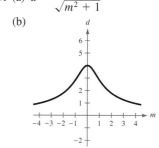

(c) $m = 0$
(d) The graph has a horizontal asymptote of $d = 0$. As the slope becomes larger, the distance between the origin and the line, $y = mx + 4$, becomes smaller and approaches 0.
60. (a) $d = \dfrac{3|m + 1|}{\sqrt{m^2 + 1}}$

(b)

(c) $m = 1$ (d) Yes. $m = -1$
(e) $d = 3$. As the line approaches the vertical, the distance approaches 3.
61. x-intercept: $(7, 0)$ **62.** x-intercept: $(-9, 0)$
y-intercept: $(0, 49)$ y-intercept: $(0, 81)$
63. x-intercepts: $(5 \pm \sqrt{5}, 0)$ **64.** No x-intercepts
y-intercept: $(0, 20)$ y-intercept: $(0, 133)$
65. x-intercepts: $\left(\dfrac{7 \pm \sqrt{53}}{2}, 0\right)$
y-intercept: $(0, -1)$

66. x-intercepts: $(-11, 0), (2, 0)$
y-intercept: $(0, -22)$

67. $f(x) = 3\left(x + \frac{1}{3}\right)^2 - \frac{49}{3}$
Vertex: $\left(-\frac{1}{3}, -\frac{49}{3}\right)$

68. $f(x) = 2\left(x - \frac{1}{4}\right)^2 - \frac{169}{8}$
Vertex: $\left(\frac{1}{4}, -\frac{169}{8}\right)$

69. $f(x) = 5\left(x + \frac{17}{5}\right)^2 - \frac{324}{5}$
Vertex: $\left(-\frac{17}{5}, -\frac{324}{5}\right)$

70. $f(x) = -(x + 4)^2 + 1$
Vertex: $(-4, 1)$

71. $f(x) = 6\left(x - \frac{1}{12}\right)^2 - \frac{289}{24}$
Vertex: $\left(\frac{1}{12}, -\frac{289}{24}\right)$

72. $f(x) = -8\left(x + \frac{17}{8}\right)^2 + \frac{121}{8}$
Vertex: $\left(-\frac{17}{8}, \frac{121}{8}\right)$

73.

74.

75.

76.

Section 6.2 *(page 442)*

Vocabulary Check *(page 442)*

1. conic   **2.** locus   **3.** parabola; directrix; focus
4. axis   **5.** vertex   **6.** focal chord   **7.** tangent

1. A circle is formed when a plane intersects the top or bottom half of a double-napped cone and is perpendicular to the axis of the cone.

2. An ellipse is formed when a plane intersects only the top or bottom half of a double-napped cone but is not parallel or perpendicular to the axis of the cone, is not parallel to the side of the cone, and does not intersect the vertex.

3. A parabola is formed when a plane intersects the top or bottom half of a double-napped cone, is parallel to the side of the cone, and does not intersect the vertex.

4. A hyperbola is formed when a plane intersects both halves of a double-napped cone, is parallel to the axis of the cone, and does not intersect the vertex.

5. e   **6.** b   **7.** d   **8.** f   **9.** a   **10.** c

11. Vertex: $(0, 0)$
Focus: $\left(0, \frac{1}{2}\right)$
Directrix: $y = -\frac{1}{2}$

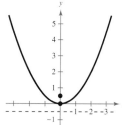

12. Vertex: $(0, 0)$
Focus: $\left(0, -\frac{1}{8}\right)$
Directrix: $y = \frac{1}{8}$

13. Vertex: $(0, 0)$
Focus: $\left(-\frac{3}{2}, 0\right)$
Directrix: $x = \frac{3}{2}$

14. Vertex: $(0, 0)$
Focus: $\left(\frac{3}{4}, 0\right)$
Directrix: $x = -\frac{3}{4}$

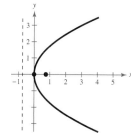

15. Vertex: $(0, 0)$
Focus: $\left(0, -\frac{3}{2}\right)$
Directrix: $y = \frac{3}{2}$

16. Vertex: $(0, 0)$
Focus: $\left(-\frac{1}{4}, 0\right)$
Directrix: $x = \frac{1}{4}$

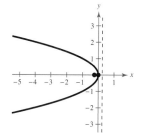

17. Vertex: $(1, -2)$
Focus: $(1, -4)$
Directrix: $y = 0$

18. Vertex: $(-5, 1)$
Focus: $\left(-\frac{21}{4}, 1\right)$
Directrix: $x = -\frac{19}{4}$

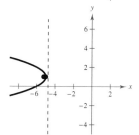

19. Vertex: $\left(-\frac{3}{2}, 2\right)$
Focus: $\left(-\frac{3}{2}, 3\right)$
Directrix: $y = 1$

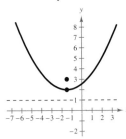

20. Vertex: $\left(-\frac{1}{2}, 1\right)$
Focus: $\left(-\frac{1}{2}, 2\right)$
Directrix: $y = 0$

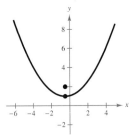

21. Vertex: $(1, 1)$
Focus: $(1, 2)$
Directrix: $y = 0$

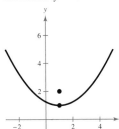

22. Vertex: $(8, -1)$
Focus: $(9, -1)$
Directrix: $x = 7$

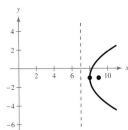

23. Vertex: $(-2, -3)$
Focus: $(-4, -3)$
Directrix: $x = 0$

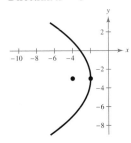

24. Vertex: $(-1, 2)$
Focus: $(0, 2)$
Directrix: $x = -2$

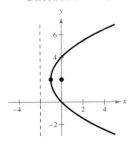

25. Vertex: $(-2, 1)$
Focus: $\left(-2, -\frac{1}{2}\right)$
Directrix: $y = \frac{5}{2}$

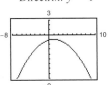

26. Vertex: $(1, -1)$
Focus: $(1, -3)$
Directrix: $y = 1$

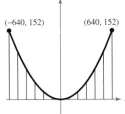

27. Vertex: $\left(\frac{1}{4}, -\frac{1}{2}\right)$
Focus: $\left(0, -\frac{1}{2}\right)$
Directrix: $x = \frac{1}{2}$

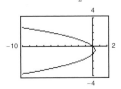

28. Vertex: $(-1, 0)$
Focus: $(0, 0)$
Directrix: $x = -2$

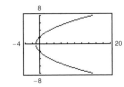

29. $x^2 = \frac{3}{2}y$ **30.** $y^2 = -18x$ **31.** $x^2 = -6y$

32. $y^2 = 10x$ **33.** $y^2 = -8x$ **34.** $x^2 = -8y$

35. $x^2 = 4y$ **36.** $x^2 = -12y$ **37.** $y^2 = -8x$

38. $y^2 = 12x$ **39.** $y^2 = 9x$ **40.** $x^2 = -3y$

41. $(x - 3)^2 = -(y - 1)$ **42.** $(y - 3)^2 = -2(x - 5)$

43. $y^2 = 4(x + 4)$ **44.** $(x - 3)^2 = 3(y + 3)$

45. $(y - 2)^2 = -8(x - 5)$ **46.** $(x + 1)^2 = -8(y - 2)$

47. $x^2 = 8(y - 4)$ **48.** $(y - 1)^2 = -12(x + 2)$

49. $(y - 2)^2 = 8x$ **50.** $x^2 = -16(y - 4)$

51. $y = \sqrt{6(x + 1)} + 3$ **52.** $y = -\sqrt{2(x - 4)} - 1$

53.

$(2, 4)$

54.

$(6, -3)$

55. $4x - y - 8 = 0;\ (2, 0)$ **56.** $6x + 2y + 9 = 0;\ \left(-\frac{3}{2}, 0\right)$

57. $4x - y + 2 = 0;\ \left(-\frac{1}{2}, 0\right)$ **58.** $8x + y - 8 = 0;\ (1, 0)$

59.

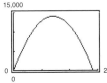

$x = 106$ units

60.

$x = 135$ units

61. $y = \frac{1}{18}x^2$

62. (a)

(−640, 152)   (640, 152)

(b) $y = \dfrac{19x^2}{51,200}$

(c)

Distance, x	0	250	400	500	1000
Height, y	0	23.19	59.38	92.77	371.09

63. (a) $y = -\frac{1}{640}x^2$ (b) 8 feet   **64.** $y^2 = 640x$

65. (a) $17,500\sqrt{2}$ miles per hour $\approx 24,750$ miles per hour

(b) $x^2 = -16,400(y - 4100)$

66. (a)

(b) Highest point: $(6.25, 7.125)$

Range: 15.69 feet

67. (a) $x^2 = -64(y - 75)$ (b) 69.3 feet

68. 34,294.6 feet

69. False. If the graph crossed the directrix, there would exist points closer to the directrix than the focus.

70. True. If the axis (line connecting the vertex and focus) is horizontal, then the direction must be vertical.

71. (a)

As p increases, the graph becomes wider.

(b) $(0, 1)$, $(0, 2)$, $(0, 3)$, $(0, 4)$

(c) $4, 8, 12, 16$; $4|p|$

(d) Easy way to determine two additional points on the graph

72. (a) $\dfrac{64\sqrt{2}}{3} \approx 30.17$

(b) As p approaches zero, the parabola becomes narrower and narrower, thus the area becomes smaller and smaller.

73. $m = \dfrac{x_1}{2p}$ **74.** Answers will vary.

75. $f(x) = x^3 - 7x^2 + 17x - 15$ **76.** $\frac{3}{2}, \pm 5i$

77. $\frac{1}{2}, -\frac{5}{3}, \pm 2$

78.

$\frac{1}{2}, -1, \pm 3$

79. $B \approx 23.67°$, $C \approx 121.33°$, $c \approx 14.89$

80. $A \approx 96.37°$, $C \approx 29.63°$, $a \approx 22.11$

81. $C = 89°$, $a \approx 1.93$, $b \approx 2.33$

82. $A = 50°$, $b \approx 10.87$, $c \approx 24.07$

83. $A \approx 16.39°$, $B \approx 23.77°$, $C \approx 139.84°$

84. $A \approx 43.53°$, $B \approx 19.42°$, $C \approx 117.05°$

85. $B \approx 24.62°$, $C \approx 90.38°$, $a \approx 10.88$

86. $A \approx 41.85°$, $C \approx 67.15°$, $b \approx 29.76$

Section 6.3 *(page 452)*

1. b **2.** c **3.** d **4.** f **5.** a **6.** e

7. Ellipse

Center: $(0, 0)$

Vertices: $(\pm 5, 0)$

Foci: $(\pm 3, 0)$

Eccentricity: $\dfrac{3}{5}$

8. Ellipse

Center: $(0, 0)$

Vertices: $(0, \pm 12)$

Foci: $(0, \pm 3\sqrt{7})$

Eccentricity: $\dfrac{\sqrt{7}}{4}$

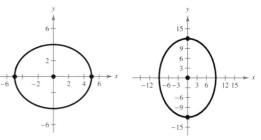

9. Circle

Center: $(0, 0)$

Radius: 5

10. Circle

Center: $(0, 0)$

Radius: 3

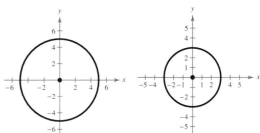

11. Ellipse

Center: $(0, 0)$

Vertices: $(0, \pm 3)$

Foci: $(0, \pm 2)$

Eccentricity: $\frac{2}{3}$

12. Ellipse

Center: $(0, 0)$

Vertices: $(\pm 8, 0)$

Foci: $(\pm 6, 0)$

Eccentricity: $\frac{3}{4}$

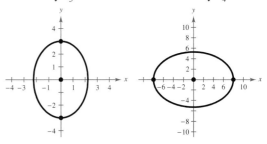

CHAPTER 6

13. Ellipse

Center: $(-3, 5)$

Vertices: $(-3, 10), (-3, 0)$

Foci: $(-3, 8), (-3, 2)$

Eccentricity: $\frac{3}{5}$

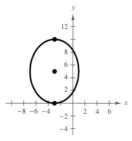

19. Ellipse

Center: $(-2, 3)$

Vertices: $(-2, 6), (-2, 0)$

Foci: $\left(-2, 3 \pm \sqrt{5}\right)$

Eccentricity: $\dfrac{\sqrt{5}}{3}$

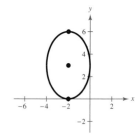

14. Ellipse

Center: $(4, -3)$

Vertices: $(4, 1), (4, -7)$

Foci: $(4, -1), (4, -5)$

Eccentricity: $\frac{1}{2}$

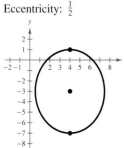

15. Circle

Center: $(0, -1)$

Radius: $\frac{2}{3}$

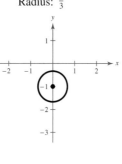

20. Ellipse

Center: $(3, -5)$

Vertices: $(3, 1), (3, -11)$

Foci: $\left(3, -5 \pm 2\sqrt{5}\right)$

Eccentricity: $\dfrac{\sqrt{5}}{3}$

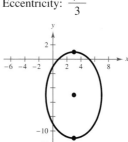

21. Circle

Center: $(1, -2)$

Radius: 6

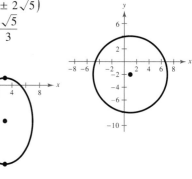

16. Ellipse

Center: $(-5, 1)$

Vertices: $\left(-\dfrac{7}{2}, 1\right), \left(-\dfrac{13}{2}, 1\right)$

Foci: $\left(-5 \pm \dfrac{\sqrt{5}}{2}, 1\right)$

Eccentricity: $\dfrac{\sqrt{5}}{3}$

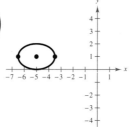

22. Ellipse

Center: $(4, 3)$

Vertices: $(14, 3), (-6, 3)$

Foci: $\left(4 \pm 4\sqrt{5}, 3\right)$

Eccentricity: $\dfrac{2\sqrt{5}}{5}$

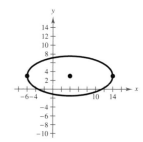

17. Ellipse

Center: $(-2, -4)$

Vertices: $(-3, -4), (-1, -4)$

Foci: $\left(\dfrac{-4 \pm \sqrt{3}}{2}, -4\right)$

Eccentricity: $\dfrac{\sqrt{3}}{2}$

23. Ellipse

Center: $(-3, 1)$

Vertices: $(-3, 7), (-3, -5)$

Foci: $\left(-3, 1 \pm 2\sqrt{6}\right)$

Eccentricity: $\dfrac{\sqrt{6}}{3}$

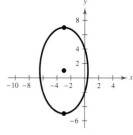

18. Circle

Center: $(3, 1)$

Radius: $\frac{5}{2}$

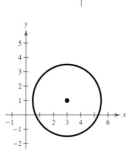

24. Ellipse

Center: $\left(-\dfrac{3}{2}, \dfrac{5}{2}\right)$

Vertices: $\left(-\dfrac{3}{2}, \dfrac{5}{2} \pm 2\sqrt{3}\right)$

Foci: $\left(-\dfrac{3}{2}, \dfrac{5}{2} \pm 2\sqrt{2}\right)$

Eccentricity: $\dfrac{\sqrt{6}}{3}$

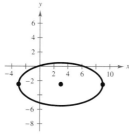

25. Ellipse

Center: $\left(3, -\dfrac{5}{2}\right)$

Vertices: $\left(9, -\dfrac{5}{2}\right), \left(-3, -\dfrac{5}{2}\right)$

Foci: $\left(3 \pm 3\sqrt{3}, -\dfrac{5}{2}\right)$

Eccentricity: $\dfrac{\sqrt{3}}{2}$

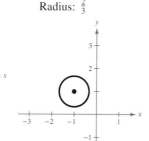

26. Circle

Center: $(2, -3)$

Radius: 4

27. Circle

Center: $(-1, 1)$

Radius: $\dfrac{2}{3}$

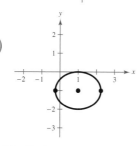

28. Ellipse

Center: $(1, -1)$

Vertices: $\left(\dfrac{9}{4}, -1\right), \left(-\dfrac{1}{4}, -1\right)$

Foci: $\left(\dfrac{7}{4}, -1\right), \left(\dfrac{1}{4}, -1\right)$

Eccentricity: $\dfrac{3}{5}$

29. Ellipse

Center: $(2, 1)$

Vertices: $\left(\dfrac{7}{3}, 1\right), \left(\dfrac{5}{3}, 1\right)$

Foci: $\left(\dfrac{34}{15}, 1\right), \left(\dfrac{26}{15}, 1\right)$

Eccentricity: $\dfrac{4}{5}$

30. Circle

Center: $(2, -1)$

Radius: $\dfrac{5}{4}$

31.

Center: $(0, 0)$

Vertices: $\left(0, \pm\sqrt{5}\right)$

Foci: $\left(0, \pm\sqrt{2}\right)$

32.

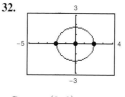

Center: $(0, 0)$

Vertices: $(\pm 2, 0)$

Foci: $(\pm 1, 0)$

Center: $\left(\dfrac{1}{2}, -1\right)$

Vertices: $\left(\dfrac{1}{2} \pm \sqrt{5}, -1\right)$

Foci: $\left(\dfrac{1}{2} \pm \sqrt{2}, -1\right)$

33.

34.

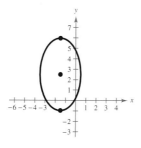

Center: $\left(-\dfrac{2}{3}, 2\right)$

Vertices: $\left(-\dfrac{2}{3}, 2 \pm \dfrac{2\sqrt{31}}{3}\right)$

Foci: $\left(-\dfrac{2}{3}, 2 \pm \dfrac{\sqrt{93}}{3}\right)$

35. $\dfrac{x^2}{4} + \dfrac{y^2}{16} = 1$ **36.** $\dfrac{x^2}{4} + \dfrac{4y^2}{9} = 1$ **37.** $\dfrac{x^2}{36} + \dfrac{y^2}{32} = 1$

38. $\dfrac{x^2}{48} + \dfrac{y^2}{64} = 1$ **39.** $\dfrac{x^2}{36} + \dfrac{y^2}{11} = 1$

40. $\dfrac{x^2}{16} + \dfrac{y^2}{12} = 1$ **41.** $\dfrac{21x^2}{400} + \dfrac{y^2}{25} = 1$

42. $\dfrac{x^2}{4} + \dfrac{y^2}{16} = 1$ **43.** $\dfrac{(x-2)^2}{1} + \dfrac{(y-3)^2}{9} = 1$

44. $\dfrac{(x-4)^2}{9} + \dfrac{y^2}{16} = 1$ **45.** $\dfrac{(x+2)^2}{16} + \dfrac{(y-3)^2}{9} = 1$

46. $\dfrac{(x-2)^2}{4} + \dfrac{(y+1)^2}{1} = 1$

47. $\dfrac{(x-2)^2}{4} + \dfrac{(y-4)^2}{1} = 1$ **48.** $\dfrac{(x-2)^2}{16} + \dfrac{y^2}{12} = 1$

49. $\dfrac{x^2}{48} + \dfrac{(y-4)^2}{64} = 1$ **50.** $\dfrac{(x-2)^2}{1} + \dfrac{4(y+1)^2}{9} = 1$

51. $\dfrac{x^2}{16} + \dfrac{(y-4)^2}{12} = 1$ **52.** $\dfrac{(x-3)^2}{36} + \dfrac{(y-2)^2}{32} = 1$

53. $\dfrac{(x-2)^2}{4} + \dfrac{(y-2)^2}{1} = 1$

54. $\dfrac{(x-5)^2}{16} + \dfrac{(y-6)^2}{36} = 1$ **55.** $\dfrac{x^2}{25} + \dfrac{y^2}{16} = 1$

56. $\dfrac{x^2}{48} + \dfrac{y^2}{64} = 1$

57. (a)

(−25, 0)   (0, 10)   (25, 0)

(b) $\dfrac{x^2}{625} + \dfrac{y^2}{100} = 1$

(c) Yes

58. Positions: $\left(\pm\sqrt{5}, 0\right)$;
Length of string: 6 feet

59. (a) $\dfrac{x^2}{321.84} + \dfrac{y^2}{20.89} = 1$

(b)

(c) Aphelion:
35.29 astronomical units
Perihelion:
0.59 astronomical unit

60. $e \approx 0.052$

61. (a) $\dfrac{x^2}{0.04} + \dfrac{y^2}{2.56} = 1$

(b)

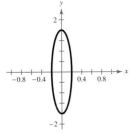

(c) The bottom half

62. Answers will vary.

63.

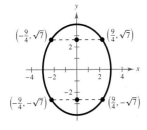

$\left(-\frac{9}{4}, \sqrt{7}\right)$ $\left(\frac{9}{4}, \sqrt{7}\right)$
$\left(-\frac{9}{4}, -\sqrt{7}\right)$ $\left(\frac{9}{4}, -\sqrt{7}\right)$

64.

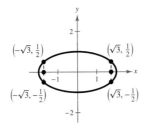

$\left(-\sqrt{3}, \frac{1}{2}\right)$ $\left(\sqrt{3}, \frac{1}{2}\right)$
$\left(-\sqrt{3}, -\frac{1}{2}\right)$ $\left(\sqrt{3}, -\frac{1}{2}\right)$

65.

$\left(-\frac{3\sqrt{5}}{5}, \sqrt{2}\right)$ $\left(\frac{3\sqrt{5}}{5}, \sqrt{2}\right)$
$\left(-\frac{3\sqrt{5}}{5}, -\sqrt{2}\right)$ $\left(\frac{3\sqrt{5}}{5}, -\sqrt{2}\right)$

66.

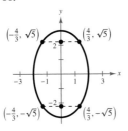

$\left(-\frac{4}{3}, \sqrt{5}\right)$ $\left(\frac{4}{3}, \sqrt{5}\right)$
$\left(-\frac{4}{3}, -\sqrt{5}\right)$ $\left(\frac{4}{3}, -\sqrt{5}\right)$

67. False. The graph of $x^2/4 + y^4 = 1$ is not an ellipse. The degree of y is 4, not 2.

68. True. If e is close to 1, the ellipse is elongated and the foci are close to the vertices.

69. (a) $A = \pi a(20 - a)$ (b) $\dfrac{x^2}{196} + \dfrac{y^2}{36} = 1$

(c)

a	8	9	10	11	12	13
A	301.6	311.0	314.2	311.0	301.6	285.9

$a = 10$, circle

(d)

The shape of an ellipse with a maximum area is a circle.

70. (a) $2a$

(b) The sum of the distances from the two fixed points is constant.

71. 1.414   **72.** 1.701   **73** −4.459
74. −4.419   **75.** $3 + \log_3 5$ **76.** $-\log 150$
77. $\ln 9 + 4$ **78.** $\ln 4 - 1.5$

Section 6.4   (page 462)

1. b   **2.** c   **3.** a   **4.** d
5. Center: $(0, 0)$
Vertices: $(\pm 1, 0)$
Foci: $\left(\pm\sqrt{2}, 0\right)$
Asymptotes: $y = \pm x$

6. Center: $(0, 0)$
Vertices: $(\pm 3, 0)$
Foci: $\left(\pm\sqrt{34}, 0\right)$
Asymptotes: $y = \pm\frac{5}{3}x$

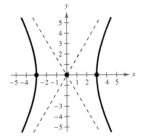

7. Center: $(0, 0)$
Vertices: $(0, \pm 5)$
Foci: $\left(0, \pm \sqrt{106}\right)$
Asymptotes: $y = \pm \frac{5}{9}x$

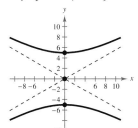

8. Center: $(0, 0)$
Vertices: $(\pm 6, 0)$
Foci: $\left(\pm 2\sqrt{10}, 0\right)$
Asymptotes: $y = \pm \frac{1}{3}x$

13. Center: $(2, -3)$
Vertices: $(3, -3), (1, -3)$
Foci: $\left(2 \pm \sqrt{10}, -3\right)$
Asymptotes:
$y = -3 \pm 3(x - 2)$

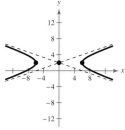

14. Center: $(0, 2)$
Vertices: $(\pm 6, 2)$
Foci: $\left(\pm 2\sqrt{10}, 2\right)$
Asymptotes: $y = 2 \pm \frac{1}{3}x$

9. Center: $(1, -2)$
Vertices: $(3, -2), (-1, -2)$
Foci: $\left(1 \pm \sqrt{5}, -2\right)$
Asymptotes:
$y = -2 \pm \frac{1}{2}(x - 1)$

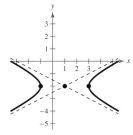

15. The graph of this equation is two lines intersecting at $(-1, -3)$.

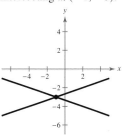

16. The graph of this equation is two lines intersecting at $(1, -2)$.

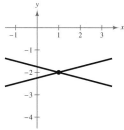

10. Center: $(-3, 2)$
Vertices: $(9, 2), (-15, 2)$
Foci: $(10, 2), (-16, 2)$
Asymptotes:
$y = 2 \pm \frac{5}{12}(x + 3)$

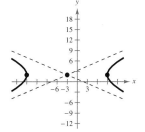

17. Center: $(0, 0)$
Vertices: $\left(\pm \sqrt{3}, 0\right)$
Foci: $\left(\pm \sqrt{5}, 0\right)$
Asymptotes: $y = \pm \frac{\sqrt{6}}{3}x$

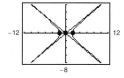

18. Center: $(0, 0)$
Vertices: $\left(0, \pm \sqrt{3}\right)$
Foci: $(0, \pm 3)$
Asymptotes: $y = \pm \frac{\sqrt{2}}{2}x$

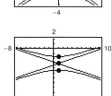

11. Center: $(2, -6)$
Vertices:
$\left(2, -\frac{17}{3}\right), \left(2, -\frac{19}{3}\right)$
Foci: $\left(2, -6 \pm \frac{\sqrt{13}}{6}\right)$
Asymptotes:
$y = -6 \pm \frac{2}{3}(x - 2)$

19. Center: $(1, -3)$
Vertices: $\left(1, -3 \pm \sqrt{2}\right)$
Foci: $\left(1, -3 \pm 2\sqrt{5}\right)$
Asymptotes:
$y = -3 \pm \frac{1}{3}(x - 1)$

12. Center: $(-3, 1)$
Vertices: $\left(-3, \frac{3}{2}\right), \left(-3, \frac{1}{2}\right)$
Foci: $\left(-3, 1 \pm \frac{\sqrt{5}}{4}\right)$
Asymptotes:
$y = 1 \pm 2(x + 3)$

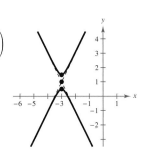

CHAPTER 6

20. Center: $(-3, 5)$

Vertices: $\left(-\dfrac{10}{3}, 5\right), \left(-\dfrac{8}{3}, 5\right)$

Foci: $\left(-3 \pm \dfrac{\sqrt{10}}{3}, 5\right)$

Asymptotes: $y = 5 \pm 3(x + 3)$

21. $\dfrac{y^2}{4} - \dfrac{x^2}{12} = 1$ **22.** $\dfrac{x^2}{16} - \dfrac{y^2}{20} = 1$

23. $\dfrac{x^2}{1} - \dfrac{y^2}{25} = 1$ **24.** $\dfrac{y^2}{9} - \dfrac{x^2}{1} = 1$

25. $\dfrac{17y^2}{1024} - \dfrac{17x^2}{64} = 1$ **26.** $\dfrac{x^2}{64} - \dfrac{y^2}{36} = 1$

27. $\dfrac{(x - 4)^2}{4} - \dfrac{y^2}{12} = 1$ **28.** $\dfrac{y^2}{9} - \dfrac{(x - 2)^2}{27} = 1$

29. $\dfrac{(y - 5)^2}{16} - \dfrac{(x - 4)^2}{9} = 1$ **30.** $\dfrac{x^2}{4} - \dfrac{(y - 1)^2}{5} = 1$

31. $\dfrac{y^2}{9} - \dfrac{4(x - 2)^2}{9} = 1$ **32.** $\dfrac{x^2}{4} - \dfrac{7(y - 1)^2}{12} = 1$

33. $\dfrac{(y - 2)^2}{4} - \dfrac{x^2}{4} = 1$ **34.** $\dfrac{y^2}{4} - \dfrac{(x - 1)^2}{4} = 1$

35. $\dfrac{(x - 2)^2}{1} - \dfrac{(y - 2)^2}{1} = 1$

36. $\dfrac{(y - 3)^2}{9} - \dfrac{(x - 3)^2}{9} = 1$

37. $\dfrac{(x - 3)^2}{9} - \dfrac{(y - 2)^2}{4} = 1$

38. $\dfrac{(y - 2)^2}{4} - \dfrac{(x - 3)^2}{9} = 1$

39. (a) $\dfrac{x^2}{1} - \dfrac{y^2}{169/3} = 1$ (b) ≈ 2.403 feet

40. $\dfrac{x^2}{98,010,000} - \dfrac{y^2}{13,503,600} = 1$ **41.** $(3300, -2750)$

42. (a) $x \approx 110.3$ miles   (b) 57.0 miles

    (c) 0.00129 second

    (d) The ship is at the position $(144.2, 60)$.

43. $\left(12\left(\sqrt{5} - 1\right), 0\right) \approx (14.83, 0)$

44. (a) Circle

    (b) $\dfrac{(x - 100)^2}{62,500} + \dfrac{y^2}{62,500} = 1$

    (c) Approximately 180.28 meters

45. Circle   **46.** Ellipse   **47.** Hyperbola

48. Parabola   **49.** Hyperbola   **50.** Circle

51. Parabola   **52.** Ellipse   **53.** Ellipse

54. Hyperbola   **55.** Parabola   **56.** Circle

57. Ellipse   **58.** Parabola   **59.** Circle   **60.** Hyperbola

61. True. For a hyperbola, $c^2 = a^2 + b^2$. The larger the ratio of b to a, the larger the eccentricity of the hyperbola, $e = c/a$.

62. False. For the trivial solution of two intersecting lines to occur, the standard form of the equation of the hyperbola would be equal to zero,

$$\dfrac{(x - h)^2}{a^2} - \dfrac{(y - k)^2}{b^2} = 0 \text{ or } \dfrac{(y - k)^2}{a^2} - \dfrac{(x - h)^2}{b^2} = 0.$$

63. Answers will vary.

64. The extended diagonals of the central rectangle are asymptotes of the hyperbola.

65. $y = 1 - 3\sqrt{\dfrac{(x - 3)^2}{4} - 1}$

66.

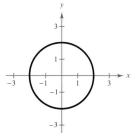

The equation $y = x^2 + C$ is a parabola that could intersect the circle in zero, one, two, three, or four places depending on its location on the y-axis.

    (a) $C > 2$ and $C < -\dfrac{17}{4}$ (b) $C = 2$

    (c) $-2 < C < 2$, $C = -\dfrac{17}{4}$ (d) $C = -2$

    (e) $-\dfrac{17}{4} < C < -2$

67. $x(x + 4)(x - 4)$ **68.** $(x + 7)^2$ **69.** $2x(x - 6)^2$

70. $x(3x + 2)(2x - 5)$ **71.** $2(2x + 3)(4x^2 - 6x + 9)$

72. $-(x^2 + 1)(x - 4)$

73.   **74.**

75.   **76.**

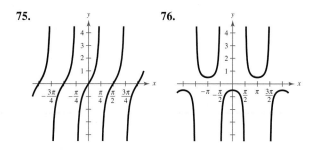

Section 6.5 *(page 471)*

Vocabulary Check *(page 471)*

1. rotation of axes

2. $A'(x')^2 + C'(y')^2 + D'x' + E'y' + F' = 0$

3. invariant under rotation   **4.** discriminant

1. $(3, 0)$ **2.** $\left(3\sqrt{2}, 0\right)$ **3.** $\left(\dfrac{3 + \sqrt{3}}{2}, \dfrac{3\sqrt{3} - 1}{2}\right)$

4. $\left(\dfrac{3 + \sqrt{3}}{2}, \dfrac{1 - 3\sqrt{3}}{2}\right)$ **5.** $\left(\dfrac{3\sqrt{2}}{2}, -\dfrac{\sqrt{2}}{2}\right)$

6. $\left(\sqrt{3} + 2, 2\sqrt{3} - 1\right)$

7. $\dfrac{(y')^2}{2} - \dfrac{(x')^2}{2} = 1$ **8.** $\dfrac{(x')^2}{4} - \dfrac{(y')^2}{4} = 1$

   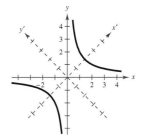

9. $y' = \pm \dfrac{\sqrt{2}}{2}$

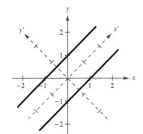

10. $\dfrac{\left[y' + \left(3\sqrt{2}/2\right)\right]^2}{10} - \dfrac{\left[x' - \left(\sqrt{2}/2\right)\right]^2}{10} = 1$

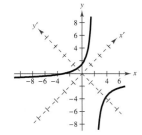

11. $\dfrac{\left(x' - 3\sqrt{2}\right)^2}{16} - \dfrac{\left(y' - \sqrt{2}\right)^2}{16} = 1$

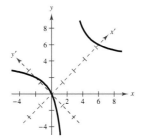

12. $\dfrac{(x')^2}{4} - \dfrac{(y')^2}{4} = 1$ **13.** $\dfrac{(x')^2}{6} + \dfrac{(y')^2}{\frac{3}{2}} = 1$

   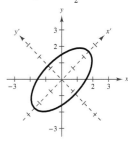

14. $\dfrac{(x')^2}{1} + \dfrac{(y')^2}{4} = 1$ **15.** $(y')^2 = -x'$

16. $(y')^2 = 4(x' - 1)$ **17.** $(x' - 1)^2 = 6\left(y' + \frac{1}{6}\right)$

18. $y' = \frac{1}{4}(x')^2$

19.

$\theta = 45°$

(c)

35. (a) Ellipse

(b) $y = \dfrac{6x \pm \sqrt{36x^2 - 28(12x^2 - 45)}}{14}$

(c)

20.

$\theta \approx 37.98°$

21.

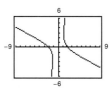

$\theta \approx 26.57°$

36. (a) Ellipse

(b) $y = \dfrac{-(4x - 4) \pm \sqrt{(4x - 4)^2 - 20(2x^2 + 3x - 20)}}{10}$

(c)

22.

$\theta \approx 33.69°$

23.

$\theta \approx 31.72°$

37. (a) Hyperbola

(b) $y = \dfrac{6x \pm \sqrt{36x^2 + 20(x^2 + 4x - 22)}}{-10}$

(c)

24.

$\theta \approx 28.16°$

25.

$\theta \approx 33.69°$

38. (a) Parabola

(b) $y = \dfrac{60x - 9 \pm \sqrt{(-60x + 9)^2 - 3600x^2}}{50}$

(c)

26.

$\theta = 31.72°$

27. e **28.** f **29.** b **30.** a **31.** d **32.** c

33. (a) Parabola

(b) $y = \dfrac{(8x - 5) \pm \sqrt{(8x - 5)^2 - 4(16x^2 - 10x)}}{2}$

(c)

34. (a) Hyperbola

(b) $y = \dfrac{4x \pm \sqrt{16x^2 + 8(x^2 - 6)}}{-4}$

39. (a) Parabola

(b) $y = \dfrac{-(4x - 1) \pm \sqrt{(4x - 1)^2 - 16(x^2 - 5x - 3)}}{8}$

(c)

40. (a) Ellipse

(b) $y = \dfrac{-(x+1) \pm \sqrt{(x+1)^2 - 16(x^2 + x - 4)}}{8}$

(c)

41.   **42.**

43.   **44.**

45. $(2, 2), (2, 4)$ **46.** $(-7, 0), (-1, 0)$ **47.** $(-8, 12)$

48. $(14, -8), (6, -8)$ **49.** $(0, 8), (12, 8)$ **50.** $(1, 0)$

51. $(0, 4)$ **52.** $(-2, 8), \left(5, 8 \pm 4\sqrt{21}\right)$

53. $\left(1, \sqrt{3}\right), \left(1, -\sqrt{3}\right)$ **54.** $(\pm 3, 2)$ **55.** No solution

56. $(1, 1), (3, 1)$ **57.** $\left(0, \frac{3}{2}\right), (-3, 0)$

58. $\left(\frac{1}{6}\left(3 + \sqrt{30}\right), \frac{1}{6}\left(3 - \sqrt{30}\right)\right), \left(\frac{1}{6}\left(3 - \sqrt{30}\right), \frac{1}{6}\left(3 + \sqrt{30}\right)\right)$

59. True. The graph of the equation can be classified by finding the discriminant. For a graph to be a hyperbola, the discriminant must be greater than zero. If $k \geq \frac{1}{4}$, then the discriminant would be less than or equal to zero.

60. False. The coefficients of the new equation after it has been rotated are obtained by making the substitutions $x = x' \cos \theta - y' \sin \theta$ and $y = x' \sin \theta + y' \cos \theta$.

61. Answers will vary.   **62.** Major axis: 4; Minor axis: 2

63.   **64.**

65.   **66.**

67.   **68.**

69.   **70.**

71. Area = 45.11 square units

72. Area = 187.94 square units

73. Area = 48.60 square units

74. Area = 310.39 square units

Section 6.6   *(page 478)*

Vocabulary Check   *(page 478)*

1. plane curve; parametric; parameter

2. orientation   **3.** eliminating the parameter

1. (a)

t	0	1	2	3	4
x	0	1	$\sqrt{2}$	$\sqrt{3}$	2
y	3	2	1	0	-1

(b)

(c) $y = 3 - x^2$

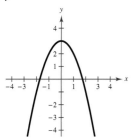

The graph of the rectangular equation shows the entire parabola rather than just the right half.

2. (a)

θ	$-\dfrac{\pi}{2}$	$-\dfrac{\pi}{4}$	0	$\dfrac{\pi}{4}$	$\dfrac{\pi}{2}$
x	0	2	4	2	0
y	-2	$-\sqrt{2}$	0	$\sqrt{2}$	2

(b)

(c) $x = -y^2 + 4$

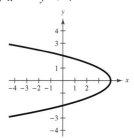

The graph of the rectangular equation continues the graph into the second and third quadrants.

3. (a)

(b) $y = \frac{2}{3}x + 3$

4. (a)

(b) $y = -\frac{3}{2}x + \frac{13}{2}$

5. (a)

(b) $y = 16x^2$

6. (a)

(b) $y = x^3$

7. (a)

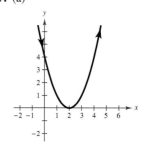

(b) $y = x^2 - 4x + 4$

8. (a)

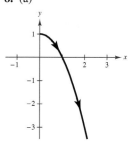

(b) $y = 1 - x^2, \quad x \geq 0$

9. (a)

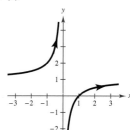

(b) $y = \dfrac{(x - 1)}{x}$

10. (a)

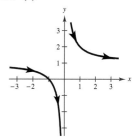

(b) $y = \dfrac{x + 1}{x}$

11. (a)

(b) $y = \left| \dfrac{x}{2} - 3 \right|$

12. (a)

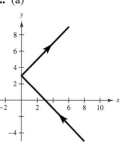

(b) $y = -x + 3, \ x > 0$
and $y = x + 3, \ x \geq 0$

13. (a)

(b) $\dfrac{y^2}{9} + \dfrac{x^2}{9} = 1$

14. (a)

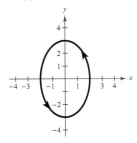

(b) $\dfrac{x^2}{4} + \dfrac{y^2}{9} = 1$

19. (a)

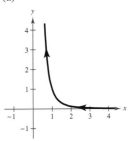

(b) $y = \dfrac{1}{x^3}, \quad x > 0$

20. (a)

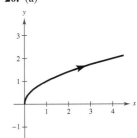

(b) $y = \sqrt{x}$

15. (a)

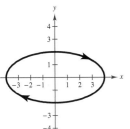

(b) $\dfrac{x^2}{16} + \dfrac{y^2}{4} = 1$

16. (a)

(b) $y = \pm 4x\sqrt{1 - x^2}$

21. (a)

(b) $y = \ln x$

22. (a)

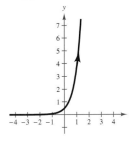

(b) $y = \dfrac{e^{2x}}{2}$

17. (a)

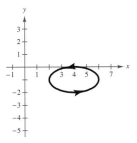

(b) $\dfrac{(x - 4)^2}{4} + (y + 1)^2 = 1$

18. (a)

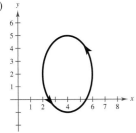

(b) $\dfrac{(x - 4)^2}{4} + \dfrac{(y - 2)^2}{9} = 1$

23. Each curve represents a portion of the line $y = 2x + 1$.

	Domain	Orientation
(a)	$(-\infty, \infty)$	Left to right
(b)	$[-1, 1]$	Depends on θ
(c)	$(0, \infty)$	Right to left
(d)	$(0, \infty)$	Left to right

24. Each curve represents a portion of the parabola $y = x^2 - 1$.

	Domain	Orientation
(a)	$(-\infty, \infty)$	Left to right
(b)	$[0, \infty)$	Depends on t
(c)	$[-1, 1]$	Depends on t
(d)	$(0, \infty)$	Left to right

25. $y - y_1 = m(x - x_1)$ **26.** $(x - h)^2 + (y - k)^2 = r^2$

27. $\dfrac{(x - h)^2}{a^2} + \dfrac{(y - k)^2}{b^2} = 1$

28. $\dfrac{(x - h)^2}{a^2} - \dfrac{(y - k)^2}{b^2} = 1$

29. $x = 6t$
$y = -3t$

30. $x = 2 + 4t$
$y = 3 - 6t$

31. $x = 3 + 4 \cos \theta$
$y = 2 + 4 \sin \theta$

32. $x = -3 + 5 \cos \theta$
$y = 2 + 5 \sin \theta$

33. $x = 4 \cos \theta$
$y = \sqrt{7} \sin \theta$

34. $x = 4 + 5 \cos \theta$
$y = 2 + 4 \sin \theta$

35. $x = 4 \sec \theta$
$y = 3 \tan \theta$

36. $x = 2 \sec \theta$
$y = 2\sqrt{3} \tan \theta$

37. (a) $x = t, \ y = 3t - 2$ (b) $x = -t + 2, \ y = -3t + 4$

CHAPTER 6

38. (a) $x = t, y = \frac{1}{3}(t + 2)$ (b) $x = -t + 2, y = -\frac{1}{3}(t - 4)$
39. (a) $x = t, y = t^2$ (b) $x = -t + 2, y = t^2 - 4t + 4$
40. (a) $x = t, y = t^3$ (b) $x = -t + 2, y = (-t + 2)^3$
41. (a) $x = t, y = t^2 + 1$ (b) $x = -t + 2, y = t^2 - 4t + 5$
42. (a) $x = t, y = 2 - t$ (b) $x = -t + 2, y = t$

43. (a) $x = t, y = \dfrac{1}{t}$ (b) $x = -t + 2, y = -\dfrac{1}{t - 2}$

44. (a) $x = t, y = \dfrac{1}{2t}$ (b) $x = -t + 2, y = \dfrac{1}{-2t + 4}$

45.

46.

47.

48.

49.

50.

51.

52.

53. b
Domain: $[-2, 2]$
Range: $[-1, 1]$

54. c
Domain: $[-4, 4]$
Range: $[-6, 6]$

55. d
Domain: $(-\infty, \infty)$
Range: $(-\infty, \infty)$

56. a
Domain: $(-\infty, \infty)$
Range: $[-2, 2]$
Maximum height: 90.7 feet
Range: 209.6 feet

57. (a)

(b)
Maximum height: 204.2 feet
Range: 471.6 feet

(c)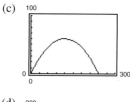
Maximum height: 60.5 feet
Range: 242.0 feet

(d)
Maximum height: 136.1 feet
Range: 544.5 feet

58. (a)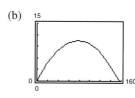
Maximum height: 3.8 feet
Range: 56.3 feet

(b)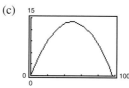
Maximum height: 10.5 feet
Range: 156.3 feet

(c)
Maximum height: 14.1 feet
Range: 97.4 feet

(d)
Maximum height: 39.1 feet
Range: 270.6 feet

59. (a) $x = (146.67 \cos \theta)t$
$y = 3 + (146.67 \sin \theta)t - 16t^2$

(b) No

(c) 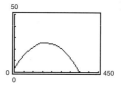 Yes

(d) $19.3°$

60. (a) $x = (240 \cos 10°)t$
$y = 5 + (240 \sin 10°)t - 16t^2$

(b) 643 feet

(c) 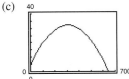 32.1 feet

(d) 2.72 seconds

61. Answers will vary.

62. (a) $h = 7$, $v_0 = 40$, $\theta = 45°$
$x = (40 \cos 45°)t$
$y = 7 + (40 \sin 45°)t - 16t^2$

(b)

(c) Maximum height: 19.5 feet
Range: 56.2 feet

63. $x = a\theta - b \sin \theta$
$y = a - b \cos \theta$

64. $x = 3 \cos \theta - \cos 3\theta$
$y = 3 \sin \theta - \sin 3\theta$

65. True
$x = t$
$y = t^2 + 1 \Longrightarrow y = x^2 + 1$
$x = 3t$
$y = 9t^2 + 1 \Longrightarrow y = x^2 + 1$

66. False. $y = x$ for $x \geq 0$

67. Parametric equations are useful when graphing two functions simultaneously on the same coordinate system. For example, they are useful when tracking the path of an object so that the position and the time associated with that position can be determined.

68. Sketching a plane curve starts by choosing a numeric value for the parameter. Then, the coordinates can be determined from the value chosen for the parameter. Finally, plotting the resulting points in the order of increasing parameter values shows the direction, or orientation, of the curve.

69. $\theta' = 75°$

70. $\theta' = 50°$

71. $\theta' = \dfrac{\pi}{3}$

72. $\theta' = \dfrac{\pi}{6}$

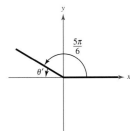

Section 6.7 *(page 485)*

Vocabulary Check *(page 485)*

1. pole   **2.** directed distance; directed angle

3. polar   **4.** $x = r \cos \theta$ $\tan \theta = \dfrac{y}{x}$

$y = r \sin \theta$ $r^2 = x^2 + y^2$

1.

$\left(4, \dfrac{5\pi}{3}\right), \left(-4, -\dfrac{4\pi}{3}\right)$

2.

$\left(-1, \dfrac{5\pi}{4}\right), \left(1, \dfrac{\pi}{4}\right)$

3.

$\left(0, \dfrac{5\pi}{6}\right), \left(0, -\dfrac{13\pi}{6}\right)$

4.

$\left(16, \dfrac{\pi}{2}\right), \left(-16, \dfrac{3\pi}{2}\right)$

5.

6.

$\left(\sqrt{2}, 8.64\right), \left(-\sqrt{2}, -0.78\right)$ $(-3, 4.7132), (3, 1.5716)$

7.

8.

$\left(2\sqrt{2}, 10.99\right), \left(-2\sqrt{2}, 7.85\right)$ $(-5, 3.9232), (5, 0.7816)$

9. $(0, 3)$ **10.** $(0, -3)$ **11.** $\left(\dfrac{\sqrt{2}}{2}, \dfrac{\sqrt{2}}{2}\right)$ **12.** $(0, 0)$

13. $\left(-\sqrt{2}, \sqrt{2}\right)$ **14.** $\left(\sqrt{3}, 1\right)$

15. $(-1.1340, -2.2280)$ **16.** $(-7.7258, -2.8940)$

17. $\left(\sqrt{2}, \dfrac{\pi}{4}\right)$ **18.** $\left(3\sqrt{2}, \dfrac{5\pi}{4}\right)$ **19.** $(6, \pi)$

20. $\left(5, \dfrac{3\pi}{2}\right)$ **21.** $(5, 2.2143)$ **22.** $\left(\sqrt{10}, 5.9614\right)$

23. $\left(\sqrt{6}, \dfrac{5\pi}{4}\right)$ **24.** $\left(2, \dfrac{11\pi}{6}\right)$ **25.** $\left(3\sqrt{13}, 0.9828\right)$

26. $(13, 1.1760)$ **27.** $\left(\sqrt{13}, 5.6952\right)$

28. $\left(\sqrt{29}, 2.7611\right)$ **29.** $\left(\sqrt{7}, 0.8571\right)$ **30.** $\left(6, \dfrac{\pi}{4}\right)$

31. $\left(\frac{17}{6}, 0.4900\right)$ **32.** $(2.3049, 0.7086)$ **33.** $r = 3$

34. $r = 4$ **35.** $r = 4 \csc \theta$ **36.** $\theta = \dfrac{\pi}{4}$

37. $r = 10 \sec \theta$ **38.** $r = 4a \sec \theta$

39. $r = \dfrac{-2}{3 \cos \theta - \sin \theta}$ **40.** $r = \dfrac{2}{3 \cos \theta + 5 \sin \theta}$

41. $r^2 = 16 \sec \theta \csc \theta = 32 \csc 2\theta$

42. $r^2 = \frac{1}{2} \sec \theta \csc \theta = \csc 2\theta$

43. $r = \dfrac{4}{1 - \cos \theta}$ or $-\dfrac{4}{1 + \cos \theta}$ **44.** $r^2 = 9 \cos 2\theta$

45. $r = a$ **46.** $r = 3a$ **47.** $r = 2a \cos \theta$

48. $r = 2a \sin \theta$ **49.** $x^2 + y^2 - 4y = 0$

50. $x^2 + y^2 - 2x = 0$ **51.** $\sqrt{3}x + y = 0$

52. $\sqrt{3}x + y = 0$ **53.** $x^2 + y^2 = 16$

54. $x^2 + y^2 = 100$ **55.** $y = 4$ **56.** $x = -3$

57. $x^2 + y^2 - x^{2/3} = 0$ **58.** $(x^2 + y^2)^2 = 2xy$

59. $(x^2 + y^2)^2 = 6x^2y - 2y^3$ **60.** $(x^2 + y^2)^3 = 9(x^2 - y^2)^2$

61. $x^2 + 4y - 4 = 0$ **62.** $y^2 = 2x + 1$

63. $4x^2 - 5y^2 - 36y - 36 = 0$ **64.** $2x - 3y = 6$

65. The graph of the polar equation consists of all points that are six units from the pole.
$x^2 + y^2 = 36$

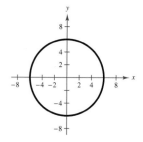

66. The graph of the polar equation consists of all points that are eight units from the pole.
$x^2 + y^2 = 64$

67. The graph of the polar equation consists of all points on the line that make an angle of $\pi/6$ with the positive polar axis.
$-\sqrt{3}x + 3y = 0$

68. The graph of the polar equation consists of all points on the line that make an angle of $3\pi/4$ with the positive polar axis.
$x + y = 0$

69. The graph of the polar equation is not evident by simple inspection, so convert to rectangular form.
$x - 3 = 0$

70. The graph of the polar equation is not evident by simple inspection, so convert to rectangular form.
$y - 2 = 0$

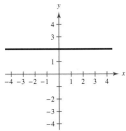

71. True. Because r is a directed distance, the point (r, θ) can be represented as $(r, \theta \pm 2\pi n)$.

72. False. If $r_1 = -r_2$, then (r_1, θ) and (r_2, θ) are different points.

73. $(x - h)^2 + (y - k)^2 = h^2 + k^2$
Radius: $\sqrt{h^2 + k^2}$
Center: (h, k)

74. $\left(x - \frac{1}{2}\right)^2 + \left(y - \frac{3}{2}\right)^2 = \frac{5}{2}$; circle

75. (a) Answers will vary.
(b) $(r_1, \theta_1), (r_2, \theta_2)$ and the pole are collinear.
$d = \sqrt{r_1^2 + r_2^2 - 2r_1r_2} = |r_1 - r_2|$
This represents the distance between two points on the line $\theta = \theta_1 = \theta_2$.
(c) $d = \sqrt{r_1^2 + r_2^2}$
This is the result of the Pythagorean Theorem.
(d) Answers will vary. For example:
Points: $(3, \pi/6), (4, \pi/3)$
Distance: 2.053
Points: $(-3, 7\pi/6), (-4, 4\pi/3)$
Distance: 2.053

76. (a) Horizontal: x-coordinate changes
Vertical: y-coordinate changes
(b) Horizontal: r and θ both change
Vertical: r and θ both change
(c) Unlike r and θ, x and y measure horizontal and vertical changes, respectively.

77. $2 \log_6 x + \log_6 z - \log_6 3 - \log_6 y$

78. $\frac{1}{4} + \frac{1}{2} \log_4 x - \log_4 y$ **79.** $\ln x + 2 \ln(x + 4)$

80. $\ln 5 + 2 \ln x + \ln(x^2 + 1)$ **81.** $\log_7 \dfrac{x}{3y}$

82. $\log_5 a(x + 1)^8$ **83.** $\ln \sqrt{x}(x - 2)$ **84.** $\ln \dfrac{6y}{x - 3}$

Section 6.8   *(page 493)*

Vocabulary Check   *(page 493)*

1. $\theta = \dfrac{\pi}{2}$ **2.** polar axis   **3.** convex limaçon

4. circle   **5.** lemniscate   **6.** cardioid

1. Rose curve with 4 petals   **2.** Cardioid
3. Limaçon with inner loop   **4.** Lemniscate
5. Rose curve with 4 petals   **6.** Circle

7. Polar axis   **8.** Polar axis   **9.** $\theta = \dfrac{\pi}{2}$

10. Polar axis   **11.** $\theta = \dfrac{\pi}{2}$, polar axis, pole   **12.** Pole

13. Maximum: $|r| = 20$ when $\theta = \dfrac{3\pi}{2}$

Zero: $r = 0$ when $\theta = \dfrac{\pi}{2}$

14. Maximum: $|r| = 18$ when $\theta = 0$

Zeros: $r = 0$ when $\theta = \dfrac{2\pi}{3}, \dfrac{4\pi}{3}$

15. Maximum: $|r| = 4$ when $\theta = 0, \dfrac{\pi}{3}, \dfrac{2\pi}{3}$

Zero: $r = 0$ when $\theta = \dfrac{\pi}{6}, \dfrac{\pi}{2}, \dfrac{5\pi}{6}$

16. Maximum: $|r| = 3$ when $\theta = \dfrac{\pi}{4}, \dfrac{3\pi}{4}, \dfrac{5\pi}{4}, \dfrac{7\pi}{4}$

Zeros: $r = 0$ when $\theta = 0, \dfrac{\pi}{2}, \pi, \dfrac{3\pi}{2}$

17.

18.

19.

20.

21.

22.

31.

32.

23.

24.

33.

34.

25.

26.

35.

36.

27.

28.

37.

38.

29.

30.

39.

40.

41.

42.

43.

44.

45.

46.

47.

48.

$0 \le \theta < 2\pi$ $0 \le \theta < 2\pi$

49.

50.

$0 \le \theta < 4\pi$ $0 \le \theta < 4\pi$

51.

52.

$0 \le \theta < \pi$ $0 \le \theta < \infty$

53.

54.

55.   **56.**

57. True. For a graph to have polar axis symmetry, replace (r, θ) by $(r, -\theta)$ or $(-r, \pi - \theta)$.

58. False. For a graph symmetric with respect to the pole, one portion of the graph coincides with the other portion when rotated π radians about the pole.

59. (a)   (b)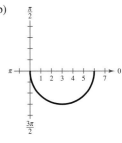

     Upper half of circle   Lower half of circle

(c)   (d)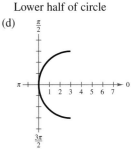

     Full circle   Left half of circle

60. (a)   (b)

(c)

The angle ϕ controls rotation of the axis of symmetry. $r = 6(1 + \sin \theta)$

61 and 62. Answers will vary.

63. (a) $r = 2 - \dfrac{\sqrt{2}}{2}(\sin \theta - \cos \theta)$ (b) $r = 2 + \cos \theta$

     (c) $r = 2 + \sin \theta$ (d) $r = 2 - \cos \theta$

64. (a) $r = 4 \sin\left(\theta - \dfrac{\pi}{6}\right) \cos\left(\theta - \dfrac{\pi}{6}\right)$

     (b) $r = -4 \sin \theta \cos \theta$

     (c) $r = 4 \sin\left(\theta - \dfrac{2\pi}{3}\right) \cos\left(\theta - \dfrac{2\pi}{3}\right)$

     (d) $r = 4 \sin \theta \cos \theta$

65. (a)   (b)

66. (a)

(b)

(c)

(d)

67.

$k = 0$, circle
$k = 1$, convex limaçon
$k = 2$, cardioid
$k = 3$, limaçon with inner loop

68. (a)

(b)

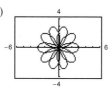

$0 \leq \theta < 4\pi$ $0 \leq \theta < 4\pi$

(c) Yes. Explanations will vary.

69. $\dfrac{(x + 1)^2}{9} + \dfrac{(y - 2)^2}{4} = 1$ **70.** $\dfrac{(x - 3)^2}{7} + \dfrac{(y + 1)^2}{16} = 1$

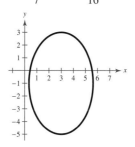

Section 6.9   *(page 499)*

Vocabulary Check   *(page 499)*

1. conic   **2.** eccentricity; e **3.** vertical; right
4. (a) iii   (b) i   (c) ii

1. $e = 1$: $r = \dfrac{4}{1 + \cos \theta}$, parabola

$e = 0.5$: $r = \dfrac{2}{1 + 0.5 \cos \theta}$, ellipse

$e = 1.5$: $r = \dfrac{6}{1 + 1.5 \cos \theta}$, hyperbola

2. $e = 1$: $r = \dfrac{4}{1 - \cos \theta}$, parabola

$e = 0.5$: $r = \dfrac{2}{1 - 0.5 \cos \theta}$, ellipse

$e = 1.5$: $r = \dfrac{6}{1 - 1.5 \cos \theta}$, hyperbola

3. $e = 1$: $r = \dfrac{4}{1 - \sin \theta}$, parabola

$e = 0.5$: $r = \dfrac{2}{1 - 0.5 \sin \theta}$, ellipse

$e = 1.5$: $r = \dfrac{6}{1 - 1.5 \sin \theta}$, hyperbola

4. $e = 1$: $r = \dfrac{4}{1 + \sin \theta}$, parabola

$e = 0.5$: $r = \dfrac{2}{1 + 0.5 \sin \theta}$, ellipse

$e = 1.5$: $r = \dfrac{6}{1 + 1.5 \sin \theta}$, hyperbola

5. f   **6.** c   **7.** d   **8.** e   **9.** a   **10.** b

11. Parabola

12. Parabola

13. Parabola

14. Parabola

15. Ellipse

16. Ellipse

17. Ellipse

18. Ellipse

19. Hyperbola

20. Hyperbola

21. Hyperbola

22. Hyperbola

23. Ellipse

24. Hyperbola

25.

Parabola

26.

Hyperbola

27.

Ellipse

28.

Hyperbola

29.

30.

31.

32.

33. $r = \dfrac{1}{1 - \cos\theta}$ **34.** $r = \dfrac{2}{1 - \sin\theta}$

35. $r = \dfrac{1}{2 + \sin\theta}$ **36.** $r = \dfrac{9}{4 - 3\sin\theta}$

37. $r = \dfrac{2}{1 + 2\cos\theta}$ **38.** $r = \dfrac{3}{2 - 3\cos\theta}$

39. $r = \dfrac{2}{1 - \sin\theta}$ **40.** $r = \dfrac{12}{1 + \cos\theta}$

41. $r = \dfrac{10}{1 - \cos\theta}$ **42.** $r = \dfrac{20}{1 + \sin\theta}$

43. $r = \dfrac{10}{3 + 2\cos\theta}$ **44.** $r = \dfrac{8}{3 + \sin\theta}$

45. $r = \dfrac{20}{3 - 2\cos\theta}$ **46.** $r = \dfrac{16}{3 + 5\cos\theta}$

47. $r = \dfrac{9}{4 - 5\sin\theta}$ **48.** $r = \dfrac{8}{3 + 5\sin\theta}$

49 and 50. Answers will vary.

51. $r = \dfrac{9.5929 \times 10^7}{1 - 0.0167\cos\theta}$

Perihelion: 9.4354×10^7 miles
Aphelion: 9.7558×10^7 miles

52. $r = \dfrac{1.4228 \times 10^9}{1 - 0.0542\cos\theta}$

Perihelion: 1.3497×10^9 kilometers
Aphelion: 1.5043×10^9 kilometers

53. $r = \dfrac{1.0820 \times 10^8}{1 - 0.0068\cos\theta}$

Perihelion: 1.0747×10^8 kilometers
Aphelion: 1.0894×10^8 kilometers

54. $r = \dfrac{3.4459 \times 10^7}{1 - 0.2056\cos\theta}$

Perihelion: 2.8583×10^7 miles
Aphelion: 4.3377×10^7 miles

55. $r = \dfrac{1.4039 \times 10^8}{1 - 0.0934\cos\theta}$

Perihelion: 1.2840×10^8 miles
Aphelion: 1.5486×10^8 miles

56. $r = \dfrac{7.7659 \times 10^8}{1 - 0.0484\cos\theta}$

Perihelion: 7.4073×10^8 kilometers
Aphelion: 8.1609×10^8 kilometers

57. $r = \dfrac{0.624}{1 + 0.847\sin\pi/2}$; $r = 0.338$ astronomical unit

58. (a) $r = \dfrac{8200}{1 + \sin\theta}$

(b)

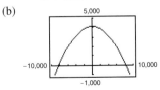

(c) 1467 miles (d) 394 miles

59. True. The graphs represent the same hyperbola.
60. False. The graph has a horizontal directrix below the pole.

61. True. The conic is an ellipse because the eccentricity is less than 1.
62–64. Answers will vary.

65. $r^2 = \dfrac{24{,}336}{169 - 25\cos^2\theta}$ **66.** $r^2 = \dfrac{400}{25 - 9\cos^2\theta}$

67. $r^2 = \dfrac{144}{25\cos^2\theta - 9}$ **68.** $r^2 = \dfrac{36}{10\cos^2\theta - 9}$

69. $r^2 = \dfrac{144}{25\sin^2\theta - 16}$ **70.** $r^2 = \dfrac{225}{25 - 16\cos^2\theta}$

71. (a) Ellipse
(b) The given polar equation, r, has a vertical directrix to the left of the pole. The equation, r_1, has a vertical directrix to the right of the pole, and the equation, r_2, has a horizontal directrix below the pole.

(c)

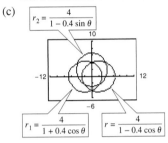

72. If e remains fixed and p changes, then the lengths of both the major axis and the minor axis change. For example, graph
$$r = \dfrac{5}{1 - \frac{2}{3}\sin\theta}, \text{ with } e = \tfrac{2}{3} \text{ and } p = \tfrac{15}{2}, \text{ and graph}$$
$$r = \dfrac{6}{1 - \frac{2}{3}\sin\theta}, \text{ with } e = \tfrac{2}{3} \text{ and } p = 9, \text{ on the same set of}$$
coordinate axes.

73. $\dfrac{\pi}{6} + n\pi$ **74.** $\dfrac{\pi}{3} + 2n\pi, \dfrac{5\pi}{3} + 2n\pi$

75. $\dfrac{\pi}{3} + n\pi, \dfrac{2\pi}{3} + n\pi$ **76.** $\dfrac{\pi}{3} + n\pi, \dfrac{2\pi}{3} + n\pi$

77. $\dfrac{\pi}{2} + n\pi$ **78.** $\dfrac{\pi}{3} + 2n\pi, \dfrac{5\pi}{3} + 2n\pi$ **79.** $\dfrac{\sqrt{2}}{10}$

80. $-\dfrac{7\sqrt{2}}{10}$ **81.** $\dfrac{7\sqrt{2}}{10}$ **82.** $\dfrac{\sqrt{2}}{10}$

83. $\sin 2u = -\dfrac{24}{25}$ **84.** $\sin 2u = -\dfrac{\sqrt{3}}{2}$

$\cos 2u = -\dfrac{7}{25}$ $\cos 2u = -\dfrac{1}{2}$

$\tan 2u = \dfrac{24}{7}$ $\tan 2u = \sqrt{3}$

Review Exercises *(page 503)*

1. $\dfrac{\pi}{4}$ radian, $45°$ **2.** 2.6012 radians, $149.04°$

3. 1.1071 radians, $63.43°$ **4.** 0.7086 radian, $40.60°$
5. 0.4424 radian, $25.35°$ **6.** 0.4424 radian, $25.35°$

7. 0.6588 radian, 37.75°   **8.** 1.4309 radians, 81.98°

9. $2\sqrt{2}$ **10.** $\dfrac{6\sqrt{5}}{5}$ **11.** Hyperbola   **12.** Parabola

13. $y^2 = 16x$

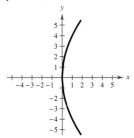

14. $y^2 = -8(x - 2)$

15. $(y - 2)^2 = 12x$

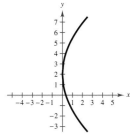

16. $(x - 2)^2 = 8(y - 2)$

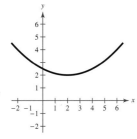

17. $y = -2x + 2$; $(1, 0)$ **18.** $y = 4x + 8$; $(-2, 0)$

19. $8\sqrt{6}$ meters   **20.** $y^2 = 6x$

21. $\dfrac{(x - 2)^2}{25} + \dfrac{y^2}{21} = 1$ **22.** $\dfrac{(x - 2)^2}{3} + \dfrac{(y - 2)^2}{4} = 1$

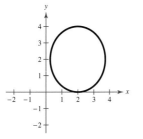

23. $\dfrac{(x - 2)^2}{4} + (y - 1)^2 = 1$ **24.** $\dfrac{(x + 4)^2}{4} + \dfrac{(y - 5)^2}{36} = 1$

25. The foci occur 3 feet from the center of the arch on a line connecting the tops of the pillars.

26. Longest distance: 36 feet
Shortest distance: 28 feet
Distance between foci: $16\sqrt{2}$ feet

27. Center: $(-2, 1)$
Vertices: $(-2, 11), (-2, -9)$
Foci: $\left(-2, 1 \pm \sqrt{19}\right)$
Eccentricity: $\dfrac{\sqrt{19}}{10}$

28. Center: $(5, -3)$
Vertices: $(5, 3), (5, -9)$
Foci: $\left(5, -3 \pm \sqrt{35}\right)$
Eccentricity: $\dfrac{\sqrt{35}}{6}$

29. Center: $(1, -4)$
Vertices: $(1, 0), (1, -8)$
Foci: $\left(1, -4 \pm \sqrt{7}\right)$
Eccentricity: $\dfrac{\sqrt{7}}{4}$

30. Center: $(-2, 3)$
Vertices: $(3, 3), (-7, 3)$
Foci: $\left(-2 \pm \sqrt{21}, 3\right)$
Eccentricity: $\dfrac{\sqrt{21}}{5}$

31. $y^2 - \dfrac{x^2}{8} = 1$ **32.** $\dfrac{x^2}{4} - \dfrac{(y - 2)^2}{12} = 1$

33. $\dfrac{5(x - 4)^2}{16} - \dfrac{5y^2}{64} = 1$ **34.** $\dfrac{5y^2}{16} - \dfrac{5(x - 3)^2}{4} = 1$

35. Center: $(3, -5)$
Vertices: $(7, -5), (-1, -5)$
Foci: $\left(3 \pm 2\sqrt{5}, -5\right)$
Asymptotes:
$y = -5 \pm \frac{1}{2}(x - 3)$

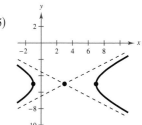

36. Center: $(0, 1)$
Vertices: $(0, 3), (0, -1)$
Foci: $\left(0, 1 \pm \sqrt{5}\right)$
Asymptotes: $y = 1 \pm 2x$

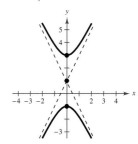

37. Center: $(1, -1)$
Vertices: $(5, -1), (-3, -1)$
Foci: $(6, -1), (-4, -1)$
Asymptotes:
$y = -1 \pm \frac{3}{4}(x - 1)$

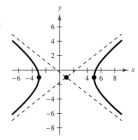

CHAPTER 6

38. Center: $(-1, -3)$
Vertices: $(-1, -1), (-1, -5)$
Foci: $\left(-1, -3 \pm \sqrt{29}\right)$
Asymptotes:
$y = -3 \pm \frac{2}{5}(x + 1)$

39. 72 miles

40. $\dfrac{576x^2}{25} - \dfrac{576y^2}{227} = 1$

$\dfrac{64(x - 1)^2}{25} - \dfrac{64y^2}{39} = 1$

41. Hyperbola   **42.** Parabola   **43.** Ellipse

44. Circle

45. $\dfrac{(x')^2}{8} - \dfrac{(y')^2}{8} = 1$ **46.** $\dfrac{(x')^2}{1/4} - \dfrac{(y')^2}{1/6} = 1$

    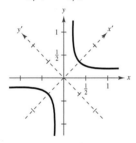

47. $\dfrac{(x')^2}{3} + \dfrac{(y')^2}{2} = 1$ **48.** $y' = -4(x')^2 - 8x'$

49. (a) Parabola
(b) $y = \dfrac{24x + 40 \pm \sqrt{(24x + 40)^2 - 36(16x^2 - 30x)}}{18}$
(c)

50. (a) Ellipse
(b) $y = \dfrac{8x \pm \sqrt{64x^2 - 28(13x^2 - 45)}}{14}$

(c)

51. (a) Parabola
(b)
$$y = \dfrac{-\left(2x - 2\sqrt{2}\right) \pm \sqrt{\left(2x - 2\sqrt{2}\right)^2 - 4\left(x^2 + 2\sqrt{2}x + 2\right)}}{2}$$

(c)

52. (a) Hyperbola   (b) $y = \dfrac{10x \pm \sqrt{100x^2 - 4(x^2 + 1)}}{2}$

(c)

53.

t	-3	-2	-1	0	1	2	3
x	-11	-8	-5	-2	1	4	7
y	19	15	11	7	3	-1	-5

54.

t	-1	0	2	3	4	5
x	$-\frac{1}{5}$	0	$\frac{2}{5}$	$\frac{3}{5}$	$\frac{4}{5}$	1
y	-2	-4	4	2	$\frac{4}{3}$	1

55. (a)

(b) $y = 2x$

56. (a)

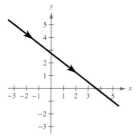

(b) $y = -\frac{3}{4}x + \frac{11}{4}$

57. (a)

(b) $y = \sqrt[4]{x}$

58. (a)

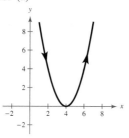

(b) $y = (x - 4)^2$

59. (a)

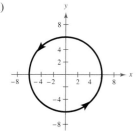

(b) $x^2 + y^2 = 36$

60. (a)

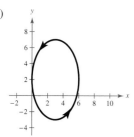

(b) $\dfrac{(x - 3)^2}{9} + \dfrac{(y - 2)^2}{25} = 1$

61. $x = 5 + 6 \cos \theta$
$y = 4 + 6 \sin \theta$

62. $x = -3 + 4 \cos \theta$
$y = 4 + 3 \sin \theta$

63. $x = 3 \tan \theta$
$y = 4 \sec \theta$

64. Answers will vary.

65.

$\left(2, \dfrac{9\pi}{4}\right), \left(-2, \dfrac{5\pi}{4}\right)$

66.

$\left(-5, \dfrac{5\pi}{3}\right), \left(5, \dfrac{2\pi}{3}\right)$

67.

$(7, 1.05), (-7, 10.47)$

68.

$\left(\sqrt{3}, 8.90\right), \left(-\sqrt{3}, 5.76\right)$

69. $\left(-\dfrac{1}{2}, -\dfrac{\sqrt{3}}{2}\right)$ **70.** $\left(-\sqrt{2}, -\sqrt{2}\right)$

71. $\left(-\dfrac{3\sqrt{2}}{2}, \dfrac{3\sqrt{2}}{2}\right)$ **72.** $(0, 0)$ **73.** $\left(2, \dfrac{\pi}{2}\right)$

74. $\left(\sqrt{10}, \dfrac{3\pi}{4}\right)$ **75.** $\left(2\sqrt{13}, 0.9828\right)$ **76.** $(5, 5.3559)$

77. $r = 7$ **78.** $r = 2\sqrt{5}$ **79.** $r = 6 \sin \theta$

80. $r = 4 \cos \theta$ **81.** $r^2 = 10 \csc 2\theta$

82. $r^2 = -4 \csc 2\theta$ **83.** $x^2 + y^2 = 25$

84. $x^2 + y^2 = 144$ **85.** $x^2 + y^2 = 3x$

86. $x^2 + y^2 = 8y$ **87.** $x^2 + y^2 = y^{2/3}$

88. $(x^2 + y^2)^2 = x^2 - y^2$

89. Symmetry: $\theta = \dfrac{\pi}{2}$, polar axis, pole

Maximum value of $|r|$: $|r| = 4$ for all values of θ
No zeros of r

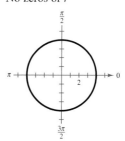

CHAPTER 6

90. Symmetry: $\theta = \dfrac{\pi}{2}$, polar axis, pole

Maximum value of $|r|$: $|r| = 11$ for all values of θ
No zeros of r

91. Symmetry: $\theta = \dfrac{\pi}{2}$, polar axis, pole

Maximum value of $|r|$: $|r| = 4$ when $\theta = \dfrac{\pi}{4}, \dfrac{3\pi}{4}, \dfrac{5\pi}{4}, \dfrac{7\pi}{4}$

Zeros of r: $r = 0$ when $\theta = 0, \dfrac{\pi}{2}, \pi, \dfrac{3\pi}{2}$

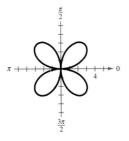

92. Symmetry: polar axis
Maximum value of $|r|$:

$|r| = 1$ when $\theta = 0, \dfrac{2\pi}{5}, \dfrac{4\pi}{5}, \dfrac{6\pi}{5}, \dfrac{8\pi}{5}$

Zeros of r: $r = 0$ when $\theta = \dfrac{\pi}{10}, \dfrac{3\pi}{10}, \dfrac{5\pi}{10}, \dfrac{7\pi}{10}, \dfrac{9\pi}{10}$

93. Symmetry: polar axis
Maximum value of $|r|$: $|r| = 4$ when $\theta = 0$
Zeros of r: $r = 0$ when $\theta = \pi$

94. Symmetry: polar axis
Maximum value of $|r|$: $|r| = 7$ when $\theta = \pi$
Zeros of r: $r = 0$ when $\theta = \arccos \frac{3}{4}, 2\pi - \arccos \frac{3}{4}$

95. Symmetry: $\theta = \dfrac{\pi}{2}$

Maximum value of $|r|$: $|r| = 8$ when $\theta = \dfrac{\pi}{2}$

Zeros of r: $r = 0$ when $\theta = 3.4814, 5.9433$

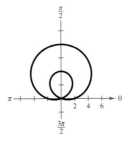

96. Symmetry: polar axis
Maximum value of $|r|$: $|r| = 10$ when $\theta = \pi$
Zero of r: $r = 0$ when $\theta = 0$

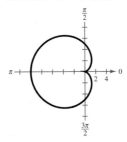

97. Symmetry: $\theta = \dfrac{\pi}{2}$, polar axis, pole

Maximum value of $|r|$: $|r| = 3$ when $\theta = 0, \dfrac{\pi}{2}, \pi, \dfrac{3\pi}{2}$

Zeros of r: $r = 0$ when $\theta = \dfrac{\pi}{4}, \dfrac{3\pi}{4}, \dfrac{5\pi}{4}, \dfrac{7\pi}{4}$

98. Symmetry: $\theta = \dfrac{\pi}{2}$, polar axis

Maximum value of $|r|$: $|r| = 1$ when $\theta = 0, \dfrac{\pi}{2}, \pi, \dfrac{3\pi}{2}$

Zeros of r: $r = 0$ when $\theta = \dfrac{\pi}{4}, \dfrac{3\pi}{4}, \dfrac{5\pi}{4}, \dfrac{7\pi}{4}$

99. Limaçon

100. Limaçon

101. Rose curve

102. Lemniscate

103. Hyperbola

104. Parabola

105. Ellipse

106. Hyperbola

107. $r = \dfrac{4}{1 - \cos \theta}$

108. $r = \dfrac{4}{1 + \sin \theta}$

109. $r = \dfrac{5}{3 - 2\cos \theta}$

110. $r = \dfrac{7}{3 + 4\cos \theta}$

111. $r = \dfrac{7978.81}{1 - 0.937 \cos \theta}$; 11,011.87 miles

112. $r = \dfrac{12,000,000}{1 + \sin \theta}$; 89,600,000 miles

113. False. When classifying an equation of the form $Ax^2 + Bxy + Cy^2 + Dx + Ey + F = 0$, its graph can be determined by its discriminant. For a graph to be a parabola, its discriminant, $B^2 - 4AC$, must equal zero. So, if $B = 0$, then A or C equals 0.

114. False. The equation of a hyperbola is a second-degree equation.

115. False. The following are two sets of parametric equations for the line.

$x = t, \ y = 3 - 2t$

$x = 3t, \ y = 3 - 6t$

116. False. $(2, \pi/4), (-2, 5\pi/4)$, and $(2, 9\pi/4)$ all represent the same point.

117. 5. The ellipse becomes more circular and approaches a circle of radius 5.

118. The orientation would be reversed.

119. (a) The speed would double.

(b) The elliptical orbit would be flatter; the length of the major axis would be greater.

120. (a) Symmetric to the pole

(b) Symmetric to the polar axis

(c) Symmetric to $\theta = \pi/2$

121. (a) The graphs are the same.
(b) The graphs are the same.

122. 40

Chapter Test   *(page 507)*

1. 0.2783 radian, 15.9°   **2.** 0.8330 radian, 47.7°

3. $\dfrac{7\sqrt{2}}{2}$

4. Parabola: $y^2 = 4(x - 1)$
Vertex: $(1, 0)$
Focus: $(2, 0)$

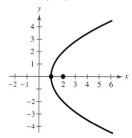

5. Hyperbola: $\dfrac{(x - 2)^2}{4} - y^2 = 1$

Center: $(2, 0)$
Vertices: $(0, 0), (4, 0)$
Foci: $\left(2 \pm \sqrt{5}, 0\right)$
Asymptotes: $y = \pm\frac{1}{2}(x - 2)$

6. Ellipse: $\dfrac{(x + 3)^2}{16} + \dfrac{(y - 1)^2}{9} = 1$

Center: $(-3, 1)$
Vertices: $(1, 1), (-7, 1)$
Foci: $\left(-3 \pm \sqrt{7}, 1\right)$

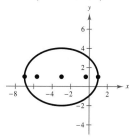

7. Circle: $(x - 2)^2 + (y - 1)^2 = \frac{1}{2}$
Center: $(2, 1)$

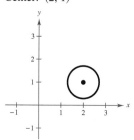

8. $(x - 3)^2 = \dfrac{3}{2}(y + 2)$ **9.** $\dfrac{5(y - 2)^2}{4} - \dfrac{5x^2}{16} = 1$

10. (a) 45°
(b)

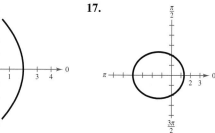

11.

12. $x = 6 + 4t$
$y = 4 + 7t$

$\dfrac{(x - 2)^2}{9} + \dfrac{y^2}{4} = 1$

13. $\left(\sqrt{3}, -1\right)$

14. $\left(2\sqrt{2}, \dfrac{7\pi}{4}\right), \left(-2\sqrt{2}, \dfrac{3\pi}{4}\right), \left(2\sqrt{2}, -\dfrac{\pi}{4}\right)$

15. $r = 4 \sin \theta$

16.   **17.**

Parabola   Ellipse

18.

Limaçon with inner loop

19.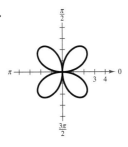

Rose curve

20. Answers will vary. For example: $r = \dfrac{1}{1 + 0.25 \sin \theta}$

21. Slope: 0.1511; Change in elevation: 789 feet

22. No; Yes

Cumulative Test for Chapters 4–6 *(page 508)*

1. $3 - 5i$ **2.** $-2 - 3i$ **3.** $5 - 12i$ **4.** 4

5. $\frac{8}{5} + \frac{4}{5}i$ **6.** $-2, \pm 2i$ **7.** $-7, 0, 3$

8. $x^4 + 3x^3 - 11x^2 + 9x + 70$

9. $2\sqrt{2}\left(\cos \dfrac{3\pi}{4} + i \sin \dfrac{3\pi}{4} \right)$ **10.** $-12\sqrt{3} + 12i$

11. $-8 + 8\sqrt{3}\,i$ **12.** -64

13. $\cos 0 + i \sin 0$

$\cos \dfrac{2\pi}{3} + i \sin \dfrac{2\pi}{3}$

$\cos \dfrac{4\pi}{3} + i \sin \dfrac{4\pi}{3}$

14. $3\left(\cos \dfrac{\pi}{8} + i \sin \dfrac{\pi}{8} \right)$

$3\left(\cos \dfrac{5\pi}{8} + i \sin \dfrac{5\pi}{8} \right)$

$3\left(\cos \dfrac{9\pi}{8} + i \sin \dfrac{9\pi}{8} \right)$

$3\left(\cos \dfrac{13\pi}{8} + i \sin \dfrac{13\pi}{8} \right)$

15. Reflect f in the x-axis and y-axis, and shift three units to the right.

16. Reflect f in the x-axis, and shift four units upward.

17. 1.991 **18.** -0.067 **19.** 1.717 **20.** 0.281

21. 0.302 **22.** -1.733 **23.** -4.087

24. $\ln(x + 4) + \ln(x - 4) - 4 \ln x,\ x > 4$

25. $\ln \dfrac{x^2}{\sqrt{x + 5}},\ x > 0$ **26.** $\dfrac{\ln 12}{2} \approx 1.242$

27. $\dfrac{\ln 9}{\ln 4} + 5 \approx 6.585$ **28.** $\dfrac{64}{5} = 12.8$

29. $\frac{1}{2}e^8 \approx 1490.479$

30.

Horizontal asymptotes: $y = 0,\ y = 1000$

31. 6.3 hours **32.** 2006 **33.** $81.87°$ **34.** $\dfrac{11\sqrt{5}}{5}$

35. Ellipse; $\dfrac{(x - 2)^2}{4} + \dfrac{(y + 1)^2}{9} = 1$

Center: $(2, -1)$

Vertices: $(2, 2),\ (2, -4)$

Foci: $\left(2, -1 \pm \sqrt{5} \right)$

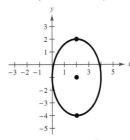

36. Hyperbola; $x^2 - \dfrac{y^2}{4} = 1$

Center: $(0, 0)$

Vertices: $(1, 0),\ (-1, 0)$

Foci: $\left(\sqrt{5}, 0 \right),\ \left(-\sqrt{5}, 0 \right)$

Asymptotes: $y = \pm 2x$

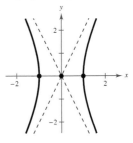

37. Circle; $(x + 1)^2 + (y - 3)^2 = 22$

Center: $(-1, 3)$

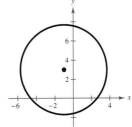

38. Parabola; $y^2 = -2(x + 1)$

Vertex: $(-1, 0)$

Focus: $\left(-\dfrac{3}{2}, 0 \right)$

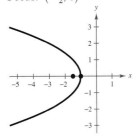

CHAPTER 6

39. $x^2 - 4x + y^2 + 8y - 48 = 0$

40. $5y^2 - 4x^2 - 30y + 25 = 0$

41. (a) $45°$

(b)

42.

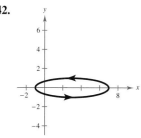

$$\frac{(x - 3)^2}{16} + y^2 = 1$$

43. $x = 3 - 6t$

$y = -2 + 6t$

44.

$\left(-2, \frac{5\pi}{4}\right), \left(2, -\frac{7\pi}{4}\right),$ and $\left(2, \frac{\pi}{4}\right)$

45. $r = 6 \sin \theta$ **46.** $9x^2 - 16y^2 + 20x + 4 = 0$

47.   **48.**

49. (a) iii   (b) i   (c) ii

Problem Solving   *(page 513)*

1. (a) 1.2016 radians   (b) 2420 feet, 5971 feet

2. (a) $\dfrac{x^2}{2352.25} + \dfrac{y^2}{529} = 1$ (b) ≈ 85.4 feet

(c) 1115.5π square feet ≈ 3504.45 square feet

3. $y^2 = 4p(x + p)$ **4.** $A = \dfrac{4a^2 b^2}{a^2 + b^2}$

5. (a) Since $d_1 + d_z \leq 20$, by definition, the outer bound that the boat can travel is an ellipse. The islands are the foci.

(b) Island 1: $(-6, 0)$;
   Island 2: $(6, 0)$

(c) 20 miles; Vertex: $(10, 0)$

(d) $\dfrac{x^2}{100} + \dfrac{y^2}{64} = 1$

6. $\dfrac{(x - 6)^2}{9} - \dfrac{(y - 2)^2}{7} = 1$

7. Answers will vary.

8. (a) The first set of parametric equations models projectile motion along a straight line. The second set of parametric equations models projectile motion of an object launched at a height of h units above the ground that will eventually fall back to the ground.

(b) $y = (\tan \theta)x;\ y = h + x \tan \theta - \dfrac{16x^2 \sec^2 \theta}{v_0^2}$

(c) In the first case, the path of the moving object is not affected by changing the velocity because eliminating the parameter removes v_0.

9. Answers will vary. For example:

$x = \cos(-t)$

$y = 2 \sin(-t)$

10. (a)   (b)

The graph is a line between -2 and 2 on the x-axis.

The graph is a three-sided figure with counterclockwise orientation.

(c)   (d)

The graph is a four-sided figure with counterclockwise orientation.

The graph is a 10-sided figure with counterclockwise orientation.

(e)

(f)

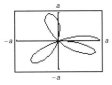

The graph is a three-sided figure with clockwise orientation.

The graph is a four-sided figure with clockwise orientation.

11. (a) $y^2 = x^2\left(\dfrac{1 - x}{1 + x}\right)$ (b) $r = \cos 2\theta \sec \theta$

(c)

12.

$r = a \cos(3.5\theta)$ $r = a \sin(2.63\theta)$

The graphs are rose curves where the petals never retrace, so there are infinitely many petals.

13. Circle

14. (a) No. Because of the exponential, the graph will continue to trace the butterfly curve at larger values of r.

(b) $r \approx 4.1$. This value will increase if θ is increased.

15.

For $n \geq 1$, a bell is produced.

For $n \leq -1$, a heart is produced.

For $n = 0$, a rose curve is produced.

16. (a) $r_{\text{Neptune}} = \dfrac{4.4997 \times 10^9}{1 - 0.0086 \cos \theta}$

$r_{\text{Pluto}} = \dfrac{5.07 \times 10^{10}}{1 - 0.2488 \cos \theta}$

(b) Neptune: Aphelion $= 4.539 \times 10^9$ km
  Perihelion $= 4.461 \times 10^9$ km

Pluto: Aphelion $= 6.752 \times 10^9$ km
  Perihelion $= 4.061 \times 10^9$ km

(c)

(d) If the orbits were in the same plane, then they would intersect. Furthermore, since the orbital periods differ (Neptune $= 164.79$ years, Pluto $= 247.68$ years), then the two planets would ultimately collide if the orbits intersect.

The orbital inclination of Pluto is significantly larger than that of Neptune ($17.16°$ vs. $1.769°$), so further analysis is required to determine if the orbits intersect.

(e) Yes, at times Pluto can be closer to the sun than Neptune. Pluto is called the ninth planet because it has the longest orbit around the sun and therefore also reaches the furthest distance away from the sun.

Explorations

Chapter P
(page 5)

a. 6   **b.** 1   **c.** 3   **d.** 3

Absolute value expressions are never negative.

(page 40)

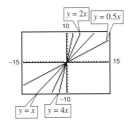

The line $y = 4x$ rises most quickly.

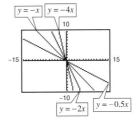

The line $y = -4x$ falls most quickly.

As $|m|$ increases, the line rises or falls faster.

(page 45)

$d_1 = \sqrt{1 + m_1^2}, d_2 = \sqrt{1 + m_2^2}; m_1 = -\dfrac{1}{m_2}$

(page 74)

$\dfrac{s(9) - s(0)}{9 - 0} = \dfrac{540}{9} = 60$ feet per second

As the time traveled increases, the distance increases rapidly, causing the average speed to increase with each time increment. From $t = 0$ to $t = 4$, the average speed is less than from $t = 4$ to $t = 9$, therefore the overall average from $t = 0$ to $t = 9$ falls below the average found in part (b).

EXPLORATIONS

(page 75)

Even

Neither

Odd

Even

Neither

Odd

Equations of odd functions contain only odd powers of x. Equations of even functions contain only even powers of x. Odd functions have all variables raised to odd powers and even functions have all variables raised to even powers. A function that has variables raised to even and odd powers is neither odd nor even.

(page 90)

a.

g is a right shift of four units. h is a right shift of four units and an upward shift of three units.

b.

g is a left shift of one unit. h is a left shift of one unit and a downward shift of two units.

c.

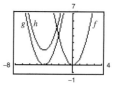

g is a left shift of four units. h is a left shift of four units and an upward shift of two units.

(page 91)

No. $g(x) = -x^4 - 2$. Yes. $h(x) = -(x - 3)^4$.

(page 109)

x	-10	0	7	45
$f(f^{-1}(x))$	-10	0	7	45
$f^{-1}(f(x))$	-10	0	7	45

The functions are inverses of each other.

(page 112)

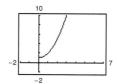

Yes. There is an inverse function $f^{-1}(x) = \sqrt{x - 1}$ since the domain of f is equal to the range of f^{-1} and the range of f is equal to the domain of f^{-1}.

Chapter 1

(page 145)

1.

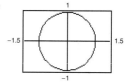

The graph is a circle.

2. The t-values represent the central angle in radians. The x- and y-values represent the location in the coordinate plane.

3. $-1 \le x \le 1, -1 \le y \le 1$

(page 171)

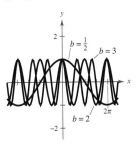

b affects the period of the graph.

$b = \frac{1}{2}$: $\frac{1}{2}$ cycle;

$b = 2$: 2 cycles;

$b = 3$: 3 cycles

(page 172)

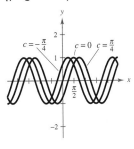

c shifts the graph horizontally.

Chapter 2

(page 239)

$\cos x = \pm\sqrt{2}$

No solution is obtained because $\pm\sqrt{2}$ are outside the range of the cosine function. No, first try to collect all the terms on one side and then try to separate the functions by factoring.

(page 241)

Yes. Preferences will vary.

(page 248)

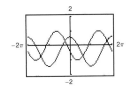

The graphs are different. No, y_1 does not equal y_2.

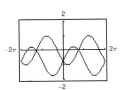

The graphs are different. No, y_1 does not equal y_2.

Chapter 3

(page 288)

Pythagorean Theorem.

The Pythagorean Theorem is just a special case of the more general Law of Cosines.

Chapter 4

(page 334)

$i, -1, -i, 1, i, -1, -i, 1$;

The pattern repeats the first four results. Divide the exponent by 4.

If the remainder is 1, the result is i.

If the remainder is 2, the result is -1.

If the remainder is 3, the result is $-i$.

If the remainder is 0, the result is 1.

(page 355)

The given equation can be written as

$x^4 = -16 = 16(\cos \pi + i \sin \pi)$

which means that you can solve the equation by finding the four fourth roots of -16. Each of these roots has the form

$$\sqrt[4]{16}\left(\cos\frac{\pi + 2\pi k}{4} + i \sin\frac{\pi + 2\pi k}{4}\right).$$

Finally, using $k = 0, 1, 2,$ and 3, you obtain the roots

$$2\left(\cos\frac{\pi}{4} + i \sin\frac{\pi}{4}\right) = 2\left(\frac{\sqrt{2}}{2} + \frac{\sqrt{2}}{2}i\right) = \sqrt{2} + \sqrt{2}i$$

$$2\left(\cos\frac{3\pi}{4} + i \sin\frac{3\pi}{4}\right) = 2\left(-\frac{\sqrt{2}}{2} + \frac{\sqrt{2}}{2}i\right) = -\sqrt{2} + \sqrt{2}i$$

$$2\left(\cos\frac{5\pi}{4} + i \sin\frac{5\pi}{4}\right) = 2\left(-\frac{\sqrt{2}}{2} - \frac{\sqrt{2}}{2}i\right) = -\sqrt{2} - \sqrt{2}i$$

$$2\left(\cos\frac{7\pi}{4} + i \sin\frac{7\pi}{4}\right) = 2\left(\frac{\sqrt{2}}{2} - \frac{\sqrt{2}}{2}i\right) = \sqrt{2} - \sqrt{2}i.$$

Chapter 5

(page 369)

a.

b.

In both viewing windows, the exponential function (a constant raised to a variable power) increases more rapidly than the power function (a variable raised to a constant power).

EXPLORATIONS

(page 372)

As the *x*-value in y_1 increases, y_1 approaches the value of *e*.

(page 373)

(a) $A = \$5466.09$ (b) $A = \$5466.35$
(c) $A = \$5466.36$ (d) $A = \$5466.38$
No. Explanations will vary.

(page 380)

x	-2	-1	0	1	2
$f(x) = 10^x$	$\frac{1}{100}$	$\frac{1}{10}$	1	10	100

x	$\frac{1}{100}$	$\frac{1}{10}$	1	10	100
$f(x) = \log x$	-2	-1	0	1	2

The domain of $f(x) = 10^x$ is equal to the range of $f(x) = \log x$ and vice versa. $f(x) = 10^x$ and $f(x) = \log x$ are inverses of each other.

(page 391)

No; the domain of y_1 is $(3, \infty)$ and the domain of y_2 is $(-\infty, 0) \cup (3, \infty)$. Explanations will vary.

(page 401)

a. 7%   **b.** 7.251%   **c.** 7.186%   **d.** 7.45%

Savings plan (d) will have the greatest effective yield. Savings plan (d) will have the highest balance after 5 years.

Chapter 6

(page 475)

t should be greater than -1. The upper bound of *t* varies, but one possibility is $t = 20$.

(page 482)

a. Yes. $\theta \approx 3.927$, $x \approx -2.121$, $y \approx -2.121$
b. Yes. Answers and explanations will vary.

(page 490)

8 petals; 3 petals. For $r = 2 \cos n\theta$ and $r = 2 \sin n\theta$, there are *n* petals if *n* is odd, 2*n* petals if *n* is even.

Technology

Chapter P

(page 27)

(page 45)

The lines appear perpendicular with the square setting.

(page 59)

Domain: $[-2, 2]$ Domain: $(-\infty, -2] \cup [2, \infty)$
                Yes, for -2 and 2.

Chapter 1

(page 170)

No graph is visible. Try $-\pi \le x \le \pi$ and $-0.5 \le y \le 0.5$ as a viewing window.

Chapter 5

(page 408)

$D = 30.92(1.124)^t$. The models are very similar.

Chapter 6

(page 477)

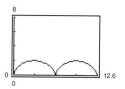

Writing About Mathematics

Chapter P

(page 62)

Answers will vary.

(page 103)

a.

x	1	2	3	4	5	6
$g(x)$	0	1	2	0	1	-1

Answers will vary.

b.

x	1	2	3	4	5	6
$h(x)$	2	1	-3	-1	0	5

Answers will vary.

(page 113)

a. The function does not have an inverse, because the retail price cannot be determined exactly because the sales tax is rounded.

b. The function does have an inverse because $f(f^{-1}(x)) = x$ and $f^{-1}(f(x)) = x$.

Chapter 1

(page 186)

$h(x) = x + \sin x$

(a) $f(x) = x$ and $g(x) = \sin x$

(b) Tables will vary. One example is:

x	-3	-2	-1
$f(x) = x$	-3	-2	-1
$g(x) = \sin x$	-0.14112	-0.90930	-0.84147
$h(x) = f(x) + g(x)$	-3.14112	-2.90930	-1.84147

x	0	1	2
$f(x) = x$	0	1	2
$g(x) = \sin x$	0	0.84147	0.90930
$h(x) = f(x) + g(x)$	0	1.84147	2.90930

x	3
$f(x) = x$	3
$g(x) = \sin x$	0.14112
$h(x) = f(x) + g(x)$	3.14112

(c) A graph of h can be formed by adding the graphs of f and g point by point.

$h(x) = \cos x - \sin 3x$

(a) $f(x) = \cos x$ and $g(x) = \sin 3x$

(b) Tables will vary. One example is:

x	-3	-2	-1
$f(x) = \cos x$	-0.98999	-0.41615	0.54030
$g(x) = \sin 3x$	-0.41212	0.27942	-0.14112
$h(x) = f(x) - g(x)$	-0.57787	-0.69557	0.68142

x	0	1	2
$f(x) = \cos x$	1	0.54030	-0.41615
$g(x) = \sin 3x$	0	0.14112	-0.27942
$h(x) = f(x) - g(x)$	1	0.39918	-0.13673

x	3
$f(x) = \cos x$	-0.98999
$g(x) = \sin 3x$	0.41212
$h(x) = f(x) - g(x)$	-1.40211

(c) A graph of h can be formed by subtracting the graph of g from the graph of f point by point.

Examples will vary. One example is $f(x) = \sin x$ and

$$g(x) = \cos\left(x + \frac{\pi}{2}\right).$$

Chapter 2

(page 234)

This particular set of range settings actually shows the statement's left side and right side as coinciding, when an algebraic verification shows that they are not equal (using a different set of range settings will confirm this). A trigonometric identity should always be verified algebraically first, and then graphically. Showing that the graphs of the left and right sides of a statement appear to coincide is not strong enough proof to conclude that the statement is an identity.

(page 243)

Equations a and b do not have solutions. Either $-2 \le \sqrt{b^2 - 4c} + b \le 2$ or $-2 \le \sqrt{b^2 - 4c} - b \le 2$ must be true.

(page 262)

Let c be the length of the two equal sides of an isosceles triangle, and let θ be the angle between the two equal sides.

$$\text{Area} = \left(\frac{1}{2}\text{ base}\right)(\text{height})$$

$$= \left(c \sin \frac{\theta}{2}\right)\left(c \cos \frac{\theta}{2}\right)$$

$$= c^2 \sqrt{\frac{1 - \cos \theta}{2}} \sqrt{\frac{1 + \cos \theta}{2}}$$

$$= c^2 \sqrt{\frac{1 - \cos^2 \theta}{4}}$$

$$= \frac{1}{2}c^2\sqrt{\sin^2 \theta}$$

$$= \frac{1}{2}c^2 \sin \theta$$

Examples will vary.

Chapter 3

(page 283)

Yes

a. $A = 40°$, $a \approx 12.86$, $b \approx 15.32$

b. $A = 40°$, $b \approx 11.92$, $c \approx 15.56$

It is probably easier to use the right triangle definitions of sine, cosine, and tangent to solve the triangle.

(page 290)

a. Area $= \frac{1}{2}(2)(4) \sin 50° \approx 3.064$ square feet

b. Area $= \sqrt{\frac{9}{2}\left(\frac{9}{2} - 2\right)\left(\frac{9}{2} - 3\right)\left(\frac{9}{2} - 4\right)} \approx 2.905$ square feet

c. Area $= \frac{1}{2}(2)(4) = 4$ square feet

d. Area $= \sqrt{6(6 - 3)(6 - 4)(6 - 5)} = 6$ square feet

Chapter 4

(page 343)

Answers will vary.

(page 351)

To approximate the product of two complex numbers graphically, add the two angles.

(a) $\left(\cos \frac{\pi}{3} + i \sin \frac{\pi}{3}\right)\left(\cos \frac{2\pi}{3} + i \sin \frac{2\pi}{3}\right) = -1$

(b) $\left(\cos \frac{\pi}{6} + i \sin \frac{\pi}{6}\right)\left(\cos \frac{\pi}{3} + i \sin \frac{\pi}{3}\right) = i$

(page 357)

$e^{a+bi} = e^a(\cos b + i \sin b)$

Let $a = 0$ and $b = \pi$.

$e^{0+\pi i} = e^0(\cos \pi + i \sin \pi)$

$e^{\pi i} = -1$

$e^{\pi i} + 1 = 0$

Chapter 5

(page 375)

$g(x)$ can be generated by f_5.

$h(x)$ can be generated by f_2 or f_3.

All six functions have $(0, 8)$ as their y-intercept.

The functions f_2 and f_3 are the same because

$$f_2(x) = 8\left(\frac{1}{2}\right)^x = 2^3(2)^{-x} = 2^{(3-x)} = \left(\frac{1}{2}\right)^{-(3-x)}$$

$$= \left(\frac{1}{2}\right)^{(x-3)} = f_3(x).$$

The functions f_1 and f_6 are the same because

$$f_1(x) = 2^{(x+3)} = 2^x 2^3 = (8)2^x = f_6(x).$$

Examples of exponential functions with y-intercepts of $(0, -3)$ will vary. Possibilities include $f(x) = -3(2^x)$ and $f(x) = \left(\frac{1}{3}\right)^x - 4$.

(page 385)

Approximately 11 months. Explanations will vary.

(page 402)

a.

Linear model:

$y = 0.358x + 9.30$

2007

b.

ln x	1.6094	2.0794	2.3979	2.7081
ln y	2.3749	2.4849	2.6355	2.7081

ln x	2.9444	3.0445	3.0910
ln y	2.7568	2.7726	2.8820

Logarithmic model: $\ln y = 0.31179 \ln x + 1.8631$

2012

c.

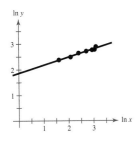

$y = 6.4437x^{0.31179}$ seems to fit the data better and will probably better predict future rates.

(page 413)

The quadratic model is better.

Chapter 6

(page 433)

The angle of inclination is the positive angle made with the x-axis measured counterclockwise. Therefore, there are two cases: (1) a line with positive slope, implying $\theta < 90°$, and (2) a line with negative slope, implying $90° < \theta < 180°$. The angle between two lines is, by definition, the smaller of the two angles formed by the intersecting lines. The two angles formed are supplementary; therefore, one angle must be acute—i.e., less than 90°—or, if the lines are perpendicular, both angles equal 90°.

    False. The inclination is the *positive angle measured counterclockwise* from the x-axis, not necessarily the angle between the line and the x-axis.

(page 441)

When a television signal hits the antenna dish, the parabolic shape of the dish reflects the signal inward to the focus of the parabola. The signal is then transferred to the receiving equipment.

(page 451)

a. Answers will vary.

b.

As e approaches 0, the shape of the ellipse approaches the shape of a circle.

c. When $e = 0$, the graph of the equation is circular. The equation of the ellipse is

$$\frac{(x - h)^2}{a^2} + \frac{(y - k)^2}{a^2} = 1,$$

otherwise known as the equation of a circle.

(page 461)

a.

b.

c.

d.

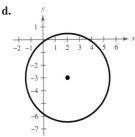

Discussions will vary.

(page 470)

$f(x) = 1/x$

$y = 1/x$

$xy = 1$

$xy - 1 = 0 \Rightarrow B = 1$ and $F = -1$.

So, $B^2 - 4AC = (1)^2 - 4(0)(0) = 1 > 0$.

Therefore, the graph of $f(x) = 1/x$ is a hyperbola.

Index

Definition of the Six Trigonometric Functions

Right triangle definitions, where $0 < \theta < \pi/2$

$$\sin \theta = \frac{\text{opp.}}{\text{hyp.}} \qquad \csc \theta = \frac{\text{hyp.}}{\text{opp.}}$$

$$\cos \theta = \frac{\text{adj.}}{\text{hyp.}} \qquad \sec \theta = \frac{\text{hyp.}}{\text{adj.}}$$

$$\tan \theta = \frac{\text{opp.}}{\text{adj.}} \qquad \cot \theta = \frac{\text{adj.}}{\text{opp.}}$$

Circular function definitions, where θ *is any angle*

$$\sin \theta = \frac{y}{r} \qquad \csc \theta = \frac{r}{y}$$

$$\cos \theta = \frac{x}{r} \qquad \sec \theta = \frac{r}{x}$$

$$\tan \theta = \frac{y}{x} \qquad \cot \theta = \frac{x}{y}$$

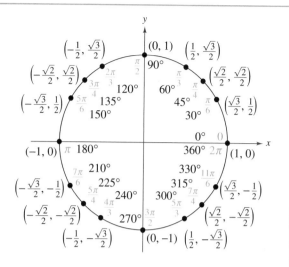

Reciprocal Identities

$$\sin u = \frac{1}{\csc u} \qquad \cos u = \frac{1}{\sec u} \qquad \tan u = \frac{1}{\cot u}$$

$$\csc u = \frac{1}{\sin u} \qquad \sec u = \frac{1}{\cos u} \qquad \cot u = \frac{1}{\tan u}$$

Quotient Identities

$$\tan u = \frac{\sin u}{\cos u} \qquad \cot u = \frac{\cos u}{\sin u}$$

Pythagorean Identities

$$\sin^2 u + \cos^2 u = 1$$

$$1 + \tan^2 u = \sec^2 u \qquad 1 + \cot^2 u = \csc^2 u$$

Cofunction Identities

$$\sin\left(\frac{\pi}{2} - u\right) = \cos u \qquad \cot\left(\frac{\pi}{2} - u\right) = \tan u$$

$$\cos\left(\frac{\pi}{2} - u\right) = \sin u \qquad \sec\left(\frac{\pi}{2} - u\right) = \csc u$$

$$\tan\left(\frac{\pi}{2} - u\right) = \cot u \qquad \csc\left(\frac{\pi}{2} - u\right) = \sec u$$

Even/Odd Identities

$$\sin(-u) = -\sin u \qquad \cot(-u) = -\cot u$$

$$\cos(-u) = \cos u \qquad \sec(-u) = \sec u$$

$$\tan(-u) = -\tan u \qquad \csc(-u) = -\csc u$$

Sum and Difference Formulas

$$\sin(u \pm v) = \sin u \cos v \pm \cos u \sin v$$

$$\cos(u \pm v) = \cos u \cos v \mp \sin u \sin v$$

$$\tan(u \pm v) = \frac{\tan u \pm \tan v}{1 \mp \tan u \tan v}$$

Double-Angle Formulas

$$\sin 2u = 2 \sin u \cos u$$

$$\cos 2u = \cos^2 u - \sin^2 u = 2 \cos^2 u - 1 = 1 - 2 \sin^2 u$$

$$\tan 2u = \frac{2 \tan u}{1 - \tan^2 u}$$

Power-Reducing Formulas

$$\sin^2 u = \frac{1 - \cos 2u}{2}$$

$$\cos^2 u = \frac{1 + \cos 2u}{2}$$

$$\tan^2 u = \frac{1 - \cos 2u}{1 + \cos 2u}$$

Sum-to-Product Formulas

$$\sin u + \sin v = 2 \sin\left(\frac{u + v}{2}\right) \cos\left(\frac{u - v}{2}\right)$$

$$\sin u - \sin v = 2 \cos\left(\frac{u + v}{2}\right) \sin\left(\frac{u - v}{2}\right)$$

$$\cos u + \cos v = 2 \cos\left(\frac{u + v}{2}\right) \cos\left(\frac{u - v}{2}\right)$$

$$\cos u - \cos v = -2 \sin\left(\frac{u + v}{2}\right) \sin\left(\frac{u - v}{2}\right)$$

Product-to-Sum Formulas

$$\sin u \sin v = \frac{1}{2}[\cos(u - v) - \cos(u + v)]$$

$$\cos u \cos v = \frac{1}{2}[\cos(u - v) + \cos(u + v)]$$

$$\sin u \cos v = \frac{1}{2}[\sin(u + v) + \sin(u - v)]$$

$$\cos u \sin v = \frac{1}{2}[\sin(u + v) - \sin(u - v)]$$

FORMULAS FROM GEOMETRY

Triangle:

$h = a \sin \theta$

$\text{Area} = \dfrac{1}{2}bh$

$c^2 = a^2 + b^2 - 2ab \cos \theta$ (Law of Cosines)

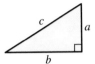

Right Triangle:

Pythagorean Theorem
$c^2 = a^2 + b^2$

Equilateral Triangle:

$h = \dfrac{\sqrt{3}s}{2}$

$\text{Area} = \dfrac{\sqrt{3}s^2}{4}$

Parallelogram:

$\text{Area} = bh$

Trapezoid:

$\text{Area} = \dfrac{h}{2}(a + b)$

Circle:

$\text{Area} = \pi r^2$

$\text{Circumference} = 2\pi r$

Sector of Circle:

$\text{Area} = \dfrac{\theta r^2}{2}$

$s = r\theta$

θ in radians

Circular Ring:

$\text{Area} = \pi(R^2 - r^2)$

$\quad\quad\quad = 2\pi pw$

$p = \text{average radius,}$

$w = \text{width of ring}$

Sector of Circular Ring:

$\text{Area} = \theta pw$

$p = \text{average radius,}$

$w = \text{width of ring,}$

θ in radians

Ellipse:

$\text{Area} = \pi ab$

$\text{Circumference} \approx 2\pi \sqrt{\dfrac{a^2 + b^2}{2}}$

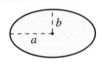

Cone:

$\text{Volume} = \dfrac{Ah}{3}$

$A = \text{area of base}$

Right Circular Cone:

$\text{Volume} = \dfrac{\pi r^2 h}{3}$

$\text{Lateral Surface Area} = \pi r \sqrt{r^2 + h^2}$

Frustum of Right Circular Cone:

$\text{Volume} = \dfrac{\pi(r^2 + rR + R^2)h}{3}$

$\text{Lateral Surface Area} = \pi s(R + r)$

Right Circular Cylinder:

$\text{Volume} = \pi r^2 h$

$\text{Lateral Surface Area} = 2\pi rh$

Sphere:

$\text{Volume} = \dfrac{4}{3}\pi r^3$

$\text{Surface Area} = 4\pi r^2$

Wedge:

$A = B \sec \theta$

$A = \text{area of upper face,}$

$B = \text{area of base}$

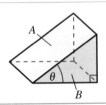